MEDICAL RADIOLOGY

Diagnostic Imaging and Radiation Oncology

Springer
Berlin
Heidelberg
New York
Barcelona
Budapest
Hong Kong
London
Milan
Paris
Santa Clara
Singapore
Tokyo

Biliary Tract Radiology

With Contributions by

A. Adam · A.L. Baert · H. Balon · M. Bazzocchi · M. Bezzi · H.-J. Brambs
P.M. Bret · W.R. Castañeda-Zuñiga · H. Coons · G. Costamagna · M. Cremer
R. Di Nardo · K.E. Fellows · D. Fink-Bennett · R. Galeotti · G. Gandini
L. Grenacher · M.C. Han · J. Lammer · J.M. Llerena · D. Lomanto · N.J. Lygidakis
F. Maccioni · P. Mannella · Y. Nimura · A. Pissas · P. Rabischong · D.C.B. Redd
J.W.A.J. Reeders · C. Reinhold · G.M. Richter · M. Rossi · P. Rossi · V. Speranza
L. Tipaldi · L. van Hoe · R.L. Vogelzang · A.F. Watkinson · H. Yoshimura
A.L. Zerbey · C.L. Zollikofer

With the Collaboration of

F.M. Arata · G. Bonomo · L. Broglia · M.C. Cassinis · B.I. Choi · S.L. Dawson
J. Deviere · F. Fiocca · P. Fonio, J.K. Han · J. Maass · R. Merlino · P.R. Mueller
M. Natrella · J.H. Park · E. Quaia · E.A.J. Rauws · A. Rieber · A. Rigamonti
D. Righi, H. Sakaguchi · F.M. Salvatori · N.J. Smits · A. Tortora · H. Uchida
D. Vanbeckevoort · O.J.M. van Delden · J.-L. van Laethem · C. Zulani

Edited by P. Rossi
Co-edited by M. Bezzi

Foreword by
Albert L. Baert

With 362 Figures in 682 Separate Illustrations, Some in Color

Springer

Plinio Rossi, MD
Professor and Chairman

Mario Bezzi, MD
Assistant Professor

Department of Radiology
University of Rome "La Sapienza"
Policlinico Umberto I
00161 Rome
Italy

MEDICAL RADIOLOGY · Diagnostic Imaging and Radiation Oncology

Continuation of
Handbuch der medizinischen Radiologie
Encyclopedia of Medical Radiology

ISSN 0942-5373
ISBN 978-3-642-64363-7 Springer-Verlag Berlin Heidelberg New York

Library of Congress Cataloging-in-Publication Data. Biliary tract radiology/with contributions by A. Adam; edited by P. Rossi; foreword by Albert L. Baert. p. cm. – (Medical radiology) Includes bibliographical references and index. ISBN 978-3-642-64363-7 (alk. paper) 1. Biliary tract – Imaging. 2. Biliary tract – Diseases – Diagnosis. 3. Interventional radiology. I. Adam, Andy. II. Rossi, Plinio. III. Series. [DNLM: 1. Gallbladder Diseases. 2. Bile Duct Diseases. 3. Cholecystography. 4. Cholangiography. 5. Radiography, Interventional. WI 750 B5956 1997] RC847.5.I42B54 1997 616.3'60757 – dc20 DNLM/DLC for Library of Congress 97-29096

ISBN-13: 978-3-642-64363-7 e-ISBN-13: 978-3-642-60343-3
DOI: 10.1007/978-3-642-60343-3

Softcover reprint of the hardcover 1st edition 1997

Cover design: Design & Production GmbH, Heidelberg

Typesetting: Best-set Typesetter Ltd., Hong Kong

SPIN: 10471431 21/3135/SPS – 5 4 3 2 1 0 – Printed on acid-free paper

Foreword

This book, edited by Prof. Plinio Rossi, an internationally well-known expert in biliary tract radiology, provides in 31 chapters a very comprehensive update on both diagnostic and therapeutic radiology of the biliary tract.

First, a very complete and interesting overview of the normal morphology of the anatomical area is presented by an anatomist, a radiologist, and a surgeon. This is followed by a discussion not only of the well-known and accepted imaging modalities, but also of the newer ones such as, magnetic resonance imaging and endoluminal and laparoscopic ultrasound. The book thus illustrates very nicely the diversity of diagnostic radiological techniques that are currently available.

Considerable technical progress and remarkable new insights in biliary tract radiology have been achieved in the field of the radiological interventional approach during recent years. Therefore, much emphasis has been placed in this book on the therapeutic possibilities of the radiologist using the percutaneous minimally invasive approach for the treatment of the various pathological conditions of the biliary tract.

It is most impressive to note the substantial progress in terms of patient comfort achieved by using the percutaneous as compared to the surgical methods. These new methods also result in shorter hospitalization times and thereby help to reduce health care costs. This book provides the interested reader with a fascinating and very complete overview of these new procedures.

Professor P. Rossi has been very actively engaged in all aspects of biliary tract radiology for many years as documented by his outstanding and numerous contributions to the radiological and surgical literature in this field. This exceptional personal expertise is reflected in the chapter that he has written on the radiological treatment of benign biliary strictures. Moreover he has been able to bring together an outstanding group of international experts to contribute from their personal experience to this book.

I am convinced that this new volume in our *Medical Radiology* series will be highly useful for radiologists, clinicians, and surgeons who deal with patients suffering from biliary tract disease.

Leuven

Albert L. Baert

Preface

It is now more than half a century since a percutaneous diagnostic approach to the biliary ducts was first described, yet biliary intervention began only in the early 1970s and has evolved rapidly over the past 25 years. Many procedures once not even considered feasible are now performed as a matter of routine in the majority of hospitals, thanks in part to the new materials which have been produced and continuously improved in the interim. The use of balloons for dilatation and of metallic stents in the treatment of benign and malignant diseases has completely changed the results and the outlook for the patients.

Improvements in surgical techniques and operators' abilities, on the other hand, have also expanded the indications for surgery, necessitating the acquisition of more detailed information on the anatomy and the extension of any given pathologic lesion.

In approaching the diagnosis of biliary pathology, MR cholangiography has now become the technique of choice for the morphologic representation and diagnosis of the biliary tree. The information obtained is so accurate, in regard to both anatomy and pathology, that this technique can replace invasive diagnostic procedures and limits the role of ERCP to therapeutic applications only.

This book assembles a number of contributors from all over the world who address specific topics relating to their personal experience in their own countries. The anatomy of the biliary tree is extensively described to indicate all the important information before any surgical or radiological procedure.

Therapeutic endoscopy and interventional radiology play an important role in the treatment of benign and malignant lesions; the results of both are described, analyzed, and compared.

Percutaneous treatment of benign lesions by balloon dilatation and stenting is a successful therapeutic procedure that compares favorably with surgery, as shown by a 10-year follow-up reported in a chapter dedicated to this topic.

Intraluminal radiotherapy combined with intra-arterial chemotherapy and external radiation is fully described as an additional therapeutic possibility, in order to increase survival in patients with malignant strictures.

The importance of cholangioscopy is stressed, since it has become a routine study in the evaluation of benign or malignant lesions, in determining tumor extension, and in guiding biopsy. Moreover, in patients with benign lesions, cholangioscopy can be the guiding instrument for electrohydraulic lithotripsy of intrahepatic stones, since it allows excellent results to be obtained without extensive and complicated surgery.

The book attempts to present all the newest technical and therapeutic modalities for the biliary tree as they are seen by a group of acknowledged experts in the field. The editor hopes that readers will find it useful not only in order to add to their knowledge of the treatment and imaging of the biliary tree, but also as a reference point for daily work.

As a last word, the editor wishes to extend his warmest thanks to all the collaborators, secretaries, and photographers who have contributed their labors to this book.

Rome

PLINIO ROSSI

Contents

Anatomy

Diagnostic Imaging

Anatomy

1 Anatomy

P. Rabischong and A. Pissas

CONTENTS

1.1 Introduction

Knowledge of the anatomical details of a particular system is a real need if one is to be able to make correct diagnoses and institute appropriate therapy. Anatomy as a medical discipline has made substantial progress with the aid of modern imaging techniques. Nevertheless, clinical anatomy must cover not only a common description of structures but also an analysis of normal variations. In addition, only by identifying the technical problems related to the performance of a function can a comprehensive approach be made to the subject, allowing anatomical features to be understood before they are memorized. The biliary tract has to be seen as integrated in a whole system, each component of which is interrelated in a logical scheme of construction that guarantees reliability and durability, two major characteristics of human living organs.

1.2 Functional Technical Problems

The biliary tract is like a tree, conveying the secretion of the very large hepatic gland to the poststomach duodenal tube. Because of the difference between the permanent production of bile and its intermittent metabolic use in the digestive tract, two additional systems are needed: a secondary way of storing bile in a cistern able to allow active draining, and a regulation flow control device to inject the bile product at the appropriate time. Pancreatic and hepatic secretions are needed at the same time for metabolic transformation of nutrients, which dictates the joining of the two excretion canals just before the digestive tract. This explains the presence of the contractile gallbladder and the sphincter of Oddi.

Then again, there are three necessary components of the hepatic circulation of the blood. The first system carries the blood from the intestine to the hepatic lobules for metabolic purposes and represents a functional afferent blood flow operating in a portal system, requiring a second venous system going to the right heart. The trophic arterial blood supply is the third system, volumetrically of much less importance than the two functional ones. All these components have to enter the hepatic parenchyma by the same entrance, the hilum.

Finally, in accordance with the fact that human beings can perform very complex tasks without knowing biology or physiology, the whole system has to be automatically regulated using the vegetative nervous system and, in case of dysregulation, an

P. Rabischong, Professor, Laboratoire d'Anatomie, Faculté de Médecine, 34059 Montpellier, France
A. Pissas, MD, Laboratoire d'Anatomie, Faculté de Médecine, 34059 Montpellier, France

alarm system that conveys painful, uncomfortable sensations to the skin, the only conscious interface for the totally unaware pilot of the machine.

1.3 Description of the Biliary Tree

1.3.1 The Common Bile Duct

Following the direction of flow, a very large intrahepatic biliary network converges into a common hepatic duct as far as the duodenal junction.

1.3.1.1 The Intrahepatic Biliary Network

The distribution of the intrahepatic biliary network is exactly the same as that of the portal venous network (Fig. 1.1), corresponding to functional sectors and segments that COUINAUD described in 1957. It should be stressed that the number of ducts seems invariable and there are no anastomoses, like in the portal system, interconnecting the biliary ducts. Segments II, III, and IV correspond roughly to the left lobe, V, VI, VII, and VIII to the right lobe, and I to the caudate lobe. All the branches of the intrahepatic bile network can be identified. Some are of particular importance for cholangiographic identification, particularly the one in segment III, which can very often be seen making a large curve behind the right paramedian pedicle (HJORTSJÖ 1951). Therefore the general appearance of these ducts is different on the right side, where they are superimposed on each other because the paramedian sector is anterior to the lateral, and on the left, where the network is more transversely disposed. Anyway, all these ducts are contained together with the portal and arterial branches in a special fibrous tissue, GLISSON's capsule, which is in continuity with the hilar plate and the vesicular plate. Bile ducts, arteries, and nerves are included within the fibrous tissue, but veins are separated from them by a connective tissue layer, which makes them easier to identify on sections. Some lymph vessels are also present in this space. It is interesting to note that three-dimensional reconstruction of these elements using an automatic segmentation procedure will in future make it possible to choose nonconventional segmentation following the individual variations of the distribution.

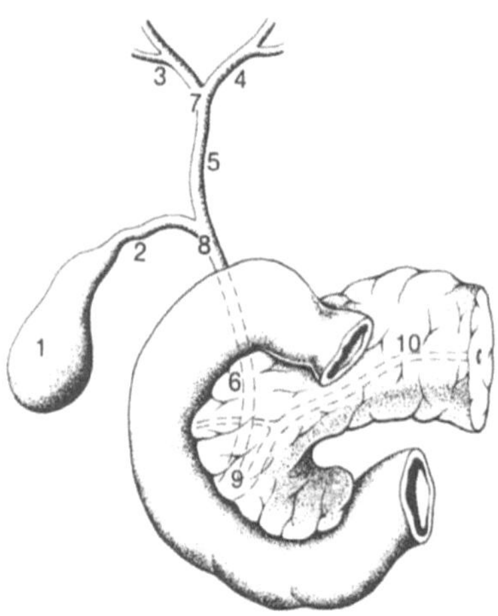

Fig. 1.1. General view of the biliary tract. *1*, Gallbladder; *2*, cystic duct; *3*, right hepatic duct; *4*, left hepatic duct; *5*, common hepatic duct; *6*, retropancreatic portion of the choledochous duct; *7*, superior biliary confluence; *8*, junction with accessory storage circuit; *9*, hepatopancreatic canal with the sphincter of Oddi; *10*, pancreatic canal. (Adapted from DEIXONNE and LOPEZ 1988)

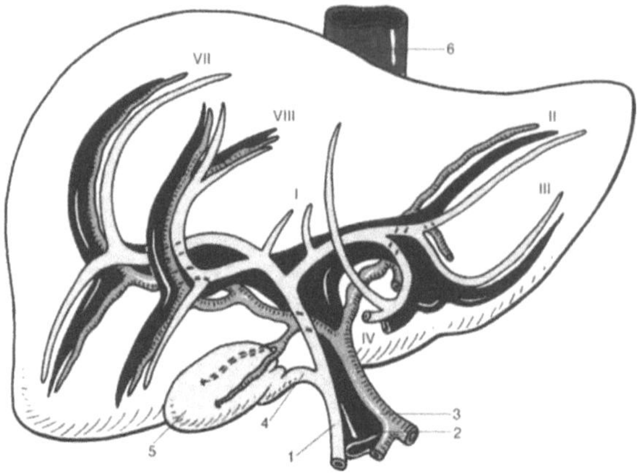

Fig. 1.2. Anterior aspect of the liver and intrahepatic biliary network. *1*, Choledochous duct; *2*, portal vein; *3*, hepatic artery; *4*, cystic duct; *5*, gallbladder; *6*, vena cava. The roman numbers refer to the hepatic segments as identified by COUINAUD. (Adapted from CHAMPETIER 1994)

1.3.1.2 The Hepatic Ducts

All the intrahepatic bile ducts finally converge into the two main right and left collectors (Fig. 1.2). The right hepatic duct is normally relatively short and vertically aligned with the common bile duct. It is made up of the anterior ramus, draining segments V and VIII, and the posterior ramus, for segments VI and VII. The left hepatic duct is more horizontal, in front of the left portal branch. Via a common trunk it receives ducts from segments II and III, in which segments I and IV are connected. It is important to realize that in 40% of subjects the right hepatic duct is missing, and the right ducts may be connected with the left duct or the common duct. In 3% of subjects, all the segmental ducts converge at the origin of the choledochous duct.

1.3.1.3 The Common Bile Duct

The common bile duct has two parts: the common hepatic duct, going from the superior biliary confluence to the junction with the cystic canal, and the ductus choledochus or choledochous duct, which ends in the second portion of the duodenum. Its total length is normally between 40 and 60 mm, the length of each part varying greatly according to the level of the cystic canal connection. The general orientation is in a large curve directed anteriorly and to the right. The inferior part crosses behind the duodenopancreas, making a groove in the head of the pancreatic gland, which can in some cases be a real intraparenchymal tunnel. Then the bile duct penetrates the duodenal wall obliquely, 3–4 cm from the pylorus, in the posterior and median zone of the second descending portion of the duodenum, in 60% of cases. A special window through the muscular layer of the duodenal wall allows the choledochous duct to join the major pancreatic duct to constitute a common canal. This hepatopancreatic canal has at its end a small dilatation which was described by VATER and is therefore known as the ampulla of VATER. It is of very variable size, and sometimes the junction of the two canals occurs within the pancreas, or else they arrive separately at the level of the papilla. A special sphincter described by ODDI, made of smooth muscle fibers situated around the end of the two canals as well as the ampulla, is a special feature, embryologically and histologically different from the muscular layer of the duodenum. The size of the sphincter and the disposition of the fibers – whether circular or spiral – are variable. The mucosa within the junction has a different shape than in the common duct. Small folds, corresponding to special glandular systems, may be visible. The opening itself in the duodenal lumen is variable. In most cases it looks like a papilla surrounded superiorly by a mucosal fold and inferiorly by a vertical fold, called the papilla string. However, in some cases, no protrusion of the papilla is visible and only the flow of bile makes possible the identification of the mucosal orifice.

1.3.2 The Bile Reservoir

An accessory circuit allows storage of bile. The reservoir itself, the gallbladder (vesica fellea or vesica biliaris) is normally 8–10 cm long and 3–4 cm wide. It is located in the cystic fossa, a groove in the inferior aspect of the liver. Three different parts can be identified. The *fundus* is emerges from the anterior border of the liver, which means it can be manually palpated from outside. The peritoneum completely covers the fundus, and in certain cases it is long enough to create a mesocyst, explaining why pathological torsion is possible (CARTER et al. 1963). The *middle part* or *body* (corpus vesicae) is more strongly fixed by thicker peritoneum, representing the vesicular plate in continuity with the hilar plate. The posterior part is constricted, forming the *neck*, which is very close to the right part of the hilum, making an angulation with the body. The cystic artery, a branch of the hepatic artery, reaches the neck with a variable course. The gallbladder is connected with the common bile duct via the cystic duct, 3–5 cm in length.

It is important to note the main differences in the histological structure of the gallbladder, in which three different layers can be identified – mucosa, submucosa, and muscular wall with some circular reinforcements – whereas the common bile duct has only a mucosa with glands inside and an external wall with connective tissue and few muscle fibers. Within the cystic duct, HEISTER described a spiral arrangement of the mucosa, playing the role of a valve (HEISTER's valve or the spiral fold), and LÜTKENS spoke of a special reinforcement of muscle fibers considered as a sphincter (LÜTKENS' sphincter). All these systems are modified by age, which is why one sees passive dilatation of all ducts and the gallbladder in elderly subjects.

1.4 Anatomical Topography

Some important anatomical relationships must be known before exploration of biliary tract. Classically, three different portions of this tract can be isolated. The first, the *hilar portion*, consists of the superior biliary confluence, anterior to the division of the portal artery, closer to the right branch, and superior to the division of the hepatic artery. If segment IV is retracted, the biliary convergence appears very clearly, without interposition of any vessels. The second part is the *pedicular portion* – really the surgical part – contained within the hepatoduodenal liga-

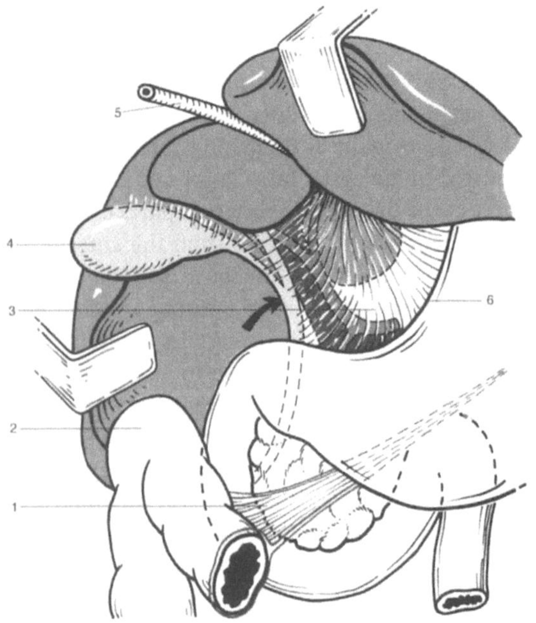

Fig. 1.3. Relations between the inferior aspect of the liver and stomach, duodenum, and colon. *1*, Transverse mesocolon; *2*, right colic flexure; *3*, small omentum (hepatogastric ligament); *4*, gallbladder; *5*, ligamentum teres; *6*, right border of the stomach. (Adapted from CHAMPETIER 1994)

ment, a peritoneal fascia going from the transverse hepatic sulcus to the posterior part of the superior portion of the duodenum (Fig. 1.3). Inside, schematically, the portal vein is posterior, with on the left side the hepatic artery and on the right the common biliary duct and the cystic duct. The right branch of the hepatic artery normally crosses behind the common hepatic duct, but in 13% of cases it crosses in front (TSUCHIYA et al. 1984). The right border of the hepatoduodenal fascia consists of an orifice, the epiploic foramen or Winslow's hiatus, which is the lateral opening of the bursa omentalis, a serous space allowing the stomach to move freely. Just behind the hepatoduodenal fascia is the inferior vena cava, covered by the parietal peritoneum. The third part is the *duodenopancreatic portion*, in which the choledochous duct crosses the duodenum. Among these three portions, attention should be focused on the following four particular elements.

1.4.1 The Triangle of Cholecystectomy

First described by CALOT in 1891 as having three sides – the cystic artery, cystic duct, and common hepatic duct – this "triangle" is now considered (ROCKO et al. 1981) to be limited superiorly by the inferior aspect of the liver. By retracting the cystic duct laterally, it is possible to enlarge the triangle and identify the cystic artery as well as the lymph node described by MASCAGNI.

1.4.2 The Interportocholedochal Triangle

Located behind the duodenum, this triangle is formed by the divergence of the choledochous duct and the portal vein. During the surgical maneuver of Kocher, separating the duodenum and the pancreas, two vascular anatomical landmarks have to be known: anteriorly, between choledochous duct and the duodenum, the arterial arch made by the superior and posterior pancreaticoduodenal artery, and, posteriorly, the venous arch made by the homologous vein. In 20% of cases, the gastroduodenal artery itself crosses the choledochous duct.

1.4.3 The Parietal Relationships of the Biliary Tract

Obviously, many variations related to patient morphotype, in the spectrum between the two extremes of leptosomatic and pyknic, can change the parietal projection of the different parts of the biliary tract (Fig. 1.4). Normally, the biliary axis is projected on the lateral border of the spine from T11 to L3, in the front of the lateral processes. The pancreatic portion of the choledochous duct is at the level of L2. The gallbladder can be palpated in a triangle represented by the inferior border of the right chondrocostal junction and the lateral border of the right rectus abdominis muscle in its fibrous sheath. If there is biliary dysfunction, the skin in this area may be painful when pinched.

1.4.4 The Visceral Relationships of the Biliary Tract

As we have seen before, the biliary tract has roughly two parts, one obliquely located under the visceral aspect of the liver and one contained vertically in the hepatoduodenal ligament. Therefore the first part, represented by the gallbladder, is in close relation with the following peritoneal spaces: the right subphrenic recess and the right subhepatic recess. The right part of the transverse colon and the superior angle of the duodenum are in contact, explaining why biliary fistulas sometimes are found in those organs. The second part has visceral contact by the way of the omentum. On the left, the gastrohepatic ligament represents the junction between stomach and liver, named pars condensa and flaccida of the small omentum, being in relation also with the right border of the stomach. Behind is the bursa omentalis, a large serous space going from the epiploic foramen to the posterior aspect of the stomach and pancreas. This space is in communication on the right with the right subhepatic recess, and in relation anteriorly with the right lobe of the liver and posteriorly with the right suprarenal gland on the superior pole of the right kidney. This particular hepatorenal recess, called MORISON's pouch, is the lowest part of the peritoneal cavity in the prone position.

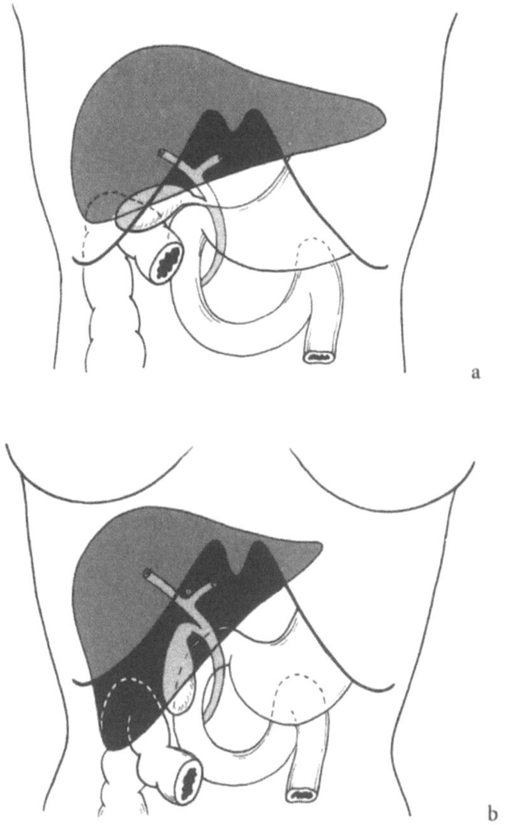

Fig. 1.4a,b. Variations of liver morphology in the two extreme morphotypes: **a** pyknic and **b** leptosomatic. (From CHAMPETIER 1994)

1.5 Vascularization

1.5.1 Arteries

The arterial blood supply of the biliary tract is very rich, combining a superficial vascular plexus with many anastomoses and two deep plexuses (PARKE et al. 1963) (Fig. 1.5). Sometime small independent arteries forming one or two parallel arcades replace the plexus. It is fed by the posterior and superior pancreaticoduodenal artery, a branch of the gas-

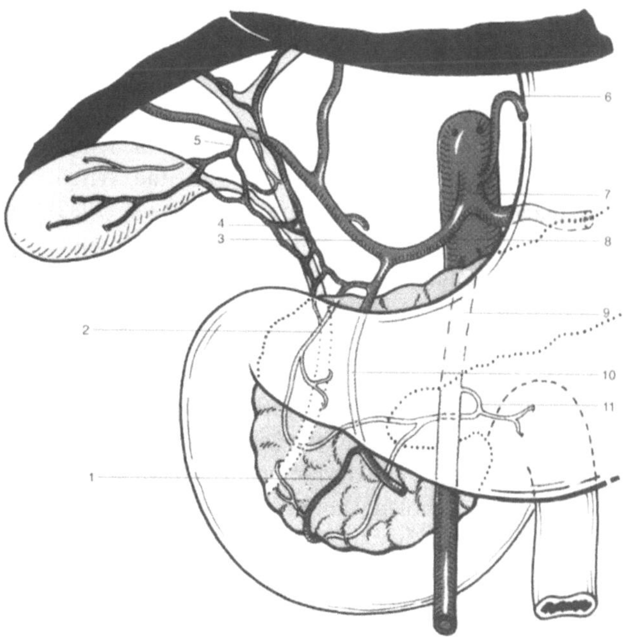

Fig. 1.5. Biliary tract arteries. *1*, Anterior superior pancreaticoduodenal artery; *2*, posterior superior pancreaticoduodenal artery; *3*, hepatic artery; *4*, arterial plexuses of the biliary tract; *5*, cystic artery; *6*, left gastric artery; *7*, celiac trunk; *8*, common hepatic artery; *9*, superior mesenteric artery; *10*, gastroduodenal artery; *11*, inferior pancreaticoduodenal artery. (Adapted from CHAMPETIER 1994)

troduodenal artery, as well as by branches of the hepatic artery. Via a special pedicle the cystic artery reaches the gallbladder at the level of the neck, dividing into right and left branches. When doing surgery, it is important to recognize the two most frequent versions of the cystic artery: short, coming from the right branch of the hepatic artery, or long, coming from the trunk of the hepatic artery or, sometimes, from the posterior and superior pancreaticoduodenal artery. All these branches make a large anastomotic network penetrating with the biliary ducts into the parenchyma and making vascular connections between the liver and the pancreas. It is at all events important to remember that the main source of the arterial blood supply of the biliary tract is located at the inferior part of the pedicular portion, which demands great attention during biliary surgery (Rath et al. 1993).

1.5.2 Veins

A parabiliary venous arcade (Couinaud 1988) runs along the biliary tract anteriorly on the left side, collecting venous blood from pancreaticoduodenal and gastric veins into the portal system. From the gallbladder some veins in the middle part of the cystic fossa go directly to the intrahepatic portal drainage, representing accessory portal veins. In cases of portal obstruction, marked dilatation of the biliary veins can be observed (Couinaud 1988).

1.5.3 Lymph Nodes and Vessels

A rich mucosal plexus in the gallbladder and common bile duct is in close relation with the hepatic lymph vessels. Some of the cystic lymph vessels are directly connected with the lymph drainage of segments IV and V. Caplan (1982) described four different flows, two of them going into the pedicular portion, passing through the gallbladder neck node and then the hiatus node, close to the right border of the hepatoduodenal ligament. The proximity of the choledochous duct explains why it may be compressed by the lymph node. Drainage is then to the celiac nodes following the hepatic artery, and the lumboaortic lymph nodes, following the choledochous duct behind the pancreatic head. In this area there is a major convergence of nodes, draining the stomach, pancreas, and biliary tract, just before the thoracic canal (ductus thoracicus).

1.6 Innervation

The double antagonist function of the biliary tract – contraction of the gallbladder with relaxation of the sphincter of Oddi and vice versa – dictates double innervation by sympathetic and parasympathetic fibers. Sympathetic branches from the right splanchnic nerve mixed with nerves from the celiac plexus reach the different parts of the tract and are connected to spinal centers located in T6–10. The anterior trunk of the vagal nerve goes along the right border of the stomach, sending branches into the pedicle. All these nerves follow vessels until they reach the hepatic parenchyma. Careful dissection of nerves can allow them to be preserved before arterial section. The possible projection of pain to the skin in the right scapular region in gallbladder dysfunction was earlier attributed to the right phrenic nerve (Couinaud 1963). However, it seems more complex than this, and needs to be integrated in the general problem of visceral projections on skin, which could possibly be used therapeutically by manipulation of skin for reversible functional disorders.

1.7 Main Variations

Embryological data can explain some normal variations. The biliary tract is formed between the fifth and seventh weeks of embryonic life. Two different conceptions have been presented in an attempt to explain its organogenetic development. According to Streeter (1951), Patten (1953), and Severn (1972), transformation of some hepatocytes in the embryological area of the gallbladder and cystic anlage in canalicular cells creates a large interconnected network. In this initial wide-mesh net, descending from liver to intestine, the fusion of certain elements creates a duct connected with the primitive gallbladder by a long cystic canal. Lassau and Hureau (1967), however, described a cholecystic axis, sprouts from which go up to colonize the hepatic parenchyma. Anyway, the initial plexiform structure of the biliary duct can explain most of the observable variations. It seems better to analyze them part by part.

1.7.1 Superior Biliary Confluence

In almost 60% of cases, a right and left hepatic duct can be identified. The right hepatic duct may be shifted to the left or to the common hepatic duct (Fig.

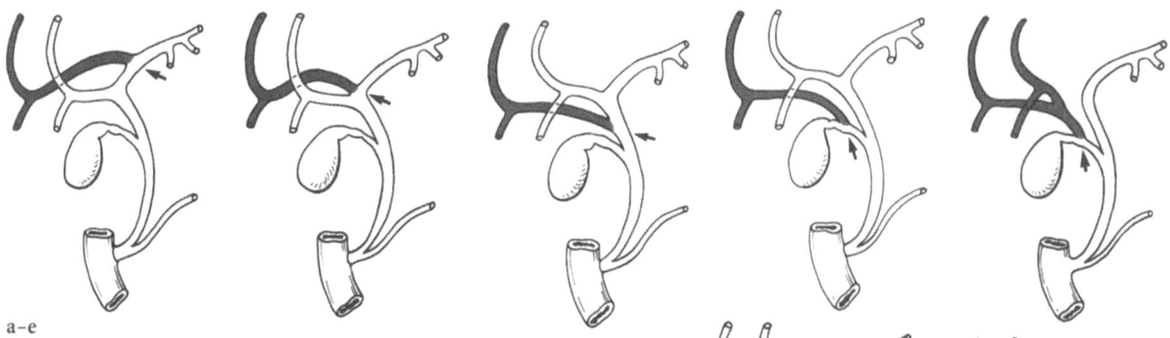

a–e

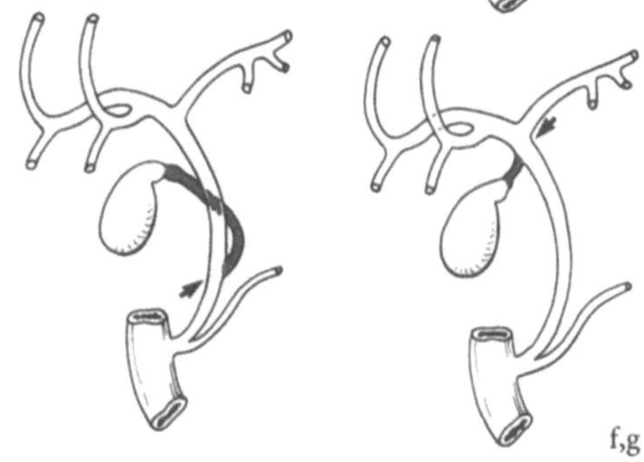

f,g

Fig. 1.6a–g. Main variations of the intrahepatic biliary network and cystic duct junction: **a–d** variable junction of the lateral hepatic duct, **d,e** hepatocystic ducts, **f,g** variations of the cystic duct junction. (From CHAMPETIER 1994)

1.6). Sometimes the lateral intrahepatic right duct may be inserted into the left duct or the hepatic common duct or the choledochous duct (CHAMPETIER et al. 1989). The confluence can also be made up of three or four branches.

1.7.2 Gallbladder

1.7.2.1 Topographical Variations

Topographical variations are not very frequent; the most usual are an intrahepatic location with the neck opening outside the liver, and a position on the left, medial to the ligamentum teres. Exceptionally, the gallbladder may be included in the falciform ligament, behind the parietal peritoneum, or within the abdominal wall.

1.7.2.2 Numerical Variations

Complete agenesis of the gallbladder can be associated with duodenal atresia in the new-born. In adults, an abnormal topography or unreported cholecystectomy has to be ruled out before agenesis is diagnosed. However, duplication, with two separate cystic ducts or a common one, or a supernumerary gallbladder implanted on the left side of the biliary tract may be observed.

1.7.2.3 Morphological Variations

An interior septum can divide the gallbladder into two different parts communicating at the level of the neck. A bilobed vesicle seen from outside, or mutilobed internal cavities, have also been described. Diverticula of pathological origin may also exist.

1.7.3 Cystic Duct

In addition to having two extreme types of course, short or long, the cystic duct may be absent, with the direct insertion of the neck in the right hepatic duct. Duplication is very rare.

1.7.4 Choledochous Duct

1.7.4.1 Duodenal Junction

Depending on the possible variations of the position of the biliary anlage on the intestine, the junction can be on the first portion of the duodenum, close to the pylorus – explaining intestinal reflex – on the descending protion, or on the horizontal portion. ENGELSKIRCHEN and KUHLS (1982) have described an exceptional case of choledochous duct junction within the stomach. The pancreatic and biliary ducts

Fig. 1.7a–d. Variations of the choledochopancreatic junction. **a** Normal type with ampulla, **b** short junction, **c** long junction, **d** no choledochopancreatic junction, each duct having a separate junction with the duodenum. (Adapted from CHAMPETIER 1994)

may be completely separate, and many variations may occur in regard to the respective length of the two sphincters (Fig. 1.7).

1.7.4.2 Morphological Variations

Five different types of common bile duct atresia were described by LADD (1928). In almost 50% of cases, atresia of the whole system leads to irreversible cirrhotic degeneration. In the other direction, cystic dilatations of the biliary tract of congenital origin have been classified in five types by TODANI (1988). In 86% of cases, the dilatation is visible all along the common biliary duct, in 5% only at the terminal part. The disease described by CAROLI is related to cystic dilatation of the intrahepatic ducts.

1.8 Conclusion

The introduction of celiosurgery in biliary surgery makes precise preoperative identification of all the key strategic points of the biliary tract even more crucial, including the vascular system, which has many variations. Modern imaging techniques allow us to approach the anatomical data as closely as possible; they then have to be integrated in 3D. In the future computer reconstruction will provide new facilities for teaching clinical anatomy and preparing, in the safest conditions, minimally invasive surgical procedures.

References

Alonso-Lej F, Rever WB, Pessagno D (1959) Collective review: congenital choledochal cyst with report of two analysis of 94 cases. Int Abstr Surg 108:1–30 (in: Surg Gynecol Obstet)

Barraya L, Pujol-Soler R, Yvergneaux JP (1971) La région odienne. Anatomie millimétrique. Presse Med 79:2527–2534

Caplan I (1982) Drainage lymphatique intra- et extra-hépathique de la vésicule biliaire. Bull Mem Acad Belg 137:324–334

Carter R Jr, Thompson RJ, Brennan LP, Hinshaw DB (1963) Volvulus of the gallbladder. Surg Gynecol Obstet 116:105–108

Champetier J (1994) Les voies biliaires. In: Chevrel JP (ed) Anatomie clinique: le tronc. Springer, Paris, pp 407–420

Champetier J, Davin JL, Letoublon C, Laborde Y, Yver R, Cousset F (1982) Aberrant biliary ducts (vasa aberrantia): surgical implications. Anat Clin 4:137–145

Champetier J, Letoublon C, Arvieux C, Gerard P, Labrosse PA (1989) Les variations de division des voies biliares extrahépatiques: signification et origine, conséquences chirurgicales. J Chir 126:147–154

Champetier J, Letoublon C, Alnaasan I, Charvin B (1991) Cystohepatic ducts: surgical implications. Surg Radiol Anat 3:203–212

Chevallier JM, Hannoun L (1991) Anatomic bases for liver transplantation. Surg Radiol Anat 13:7–16

Couinaud C (1957) Le foie. Etudes anatomiques et chirurgicales. Masson, Paris

Couinaud C (1988) The parabiliary venous system. Surg Radiol Anat 10:311–316

Deixonne B, Lopez F-M (eds) (1988) Operative ultrasonography. Springer, Berlin Heidelberg New York

Engelskirchen R, Kuhls HJ (1982) Rarissimum: in den Magen fehlmündender Ductus hepato-entericus. Chirurg 53:520–534

Flanigan DP (1975) Biliary cysts. Ann Surg 182:635–643

Goor DA, Ebert PA (1972) Anomalies of the biliary tree. Report of a repair of an accessory bile duct and review of the literature. Arch Surg 104:302–309

Hjortsjö CH (1951) The topography of the intrahepatic duct systems. Acta Anat 2:599–615

Ito M, Mishima Y, Sato T (1991) An anatomical study of the lymphatic drainage of the gallbladder. Surg Radiol Anat 13:89–104

Keddie NC, Taylor AW, Sykes PA (1974) The termination of the common bile duct. Br J Surg 61:623–625

Ladd WE (1928) Congenital atresia and stenosis of the bile ducts. J Am Med Assoc 91:1082–1085

Lane RJ, Coupland GA (1982) Ultrasonic indications to explore the common bile duct. Surgery 91:268–274

Lassau JP, Hureau J (1967) Remarques sur l'organogénèse des voies biliaires de l'homme. Bull Assoc Anat 138:750–754

Lindner HH, Pena VA, Ruggeri RA (1976) A clinical and anatomical study of anomalous terminations of the common bile duct into the duodenum. Ann Surg 184:626–632

Mentzer SH (1929) Anomalous bile ducts in man. Based on a study of comparative anatomy. J Am Med Assoc 93:1273–1277

Moosman DA (1975) Where and how to find the cystic artery during cholecystectomy. Surg Gynecol Obstet 141:769–772

Northover JMA, Terblanche J (1979) A new look at the arterial supply of the bile duct in man and its surgical implications. Br J Surg 66:379–384

Parke WW, Michels NA, Ghosh GM (1963) Blood supply of the common bile duct. Surg Gynecol Obstet 117:47–55
Puente SG, Bannura GC (1983) Radiological anatomy of the biliary tract: variations and congenital abnormalities. World J Surg 7:271–276
Rocko JM, Swan KG, Di Gioia JM (1981) Calot's triangle revisited. Surg Gynecol Obstet 153:410–414
Santos Ferreira AD, Caria Mendes J (1969) Le triangle de Budde-Calot et l'artère cystique. Bull Assoc Anat 143:1487–1494
Severn CB (1972) A morphological study of the development of the human liver. II. Establishment of liver parenchyma, extrahepatic ducts and associated venous channels. Am J Anat 133:85–108
Sigel B, Machi J, Beitler C, Donahue PE, Bombeck CT, Baker RJ, Duarte B (1983) Comparative accuracy of operative ultrasonography and cholangiography in detecting common duct calculi. Surgery 94:715–720
Streeter GL (1951) Developmental horizons in human embryos. Age groups XI to XXIII. Contributions to embriology. Carnegie Institution, Washington, pp 177–181
Todani T, Watanabe Y, Toki A, Urushihara N, Sato Y (1988) Reoperation for congenital choledochal cyst. Ann Surg 207:142–147
Tsuchiya R, Eto T, Harada N, Yamamoto K, Matsumoto T, Tsunoda T, Yamaguchi T, Noda T, Izawa K (1984) Compression of the common hepatic duct by the right hepatic artery in intrahepatic gallstones. World J Surg 8:321–326
Wong KC, Lister J (1981) Human fetal development of the hepato-pancreatic duct junction. A possible explanation of congenital dilatation of the biliary tract. J Pediatr Surg 16:139–145

2 Radiological Anatomy

P. MANNELLA and R. GALEOTTI

CONTENTS

2.1 Introduction

Knowledge of the anatomy of the biliary system is essential not only for the proper performance of surgical treatment (COUINAUD 1994), but also for the diagnostic and morphological assessment which the radiologist has to provide for the planning of surgery, and for the detection and control of postsurgical complications (ADAMS 1993).

The increasing number of ever more complex procedures becoming widely used in biliary interventional radiology require a deep understanding of the biliary segmentary anatomy and of the possible and the more common anatomic variations. (GAZELLE et al. 1994).

In the last 20 years diagnostic imaging has become progressively sharper, and radiological documentation of the biliary tree nowadays consists not only of traditional 2D, contrast-enhanced imaging, but also of multiplanar imaging [computed tomography (CT), magnetic resonance imaging (MRI), ultrasonography (US)], with new software for 3D reconstructions (Fig. 2.1), giving us new anatomic information, and a greater amount of it. One result of this has been that the new technologies confirm the impossibility of any classification and summarization of all the anatomic variations.

It has been shown (COUINAUD 1993) that the variations of the right hepatic ducts occur with different frequencies in different studies, so a complete classification is impossible; the most that can be given is a description or a list of the most commonly encountered variants.

P. MANNELLA, Professor, Istituto di Radiologia, Università di Ferrara, Corso Giovecca n. 203, 44100 Ferrara, Italy
R. GALEOTTI, MD, Istituto di Radiologia, Università di Ferrara, Corso Giovecca n. 203, 44100 Ferrara, Italy

2.2 Intrahepatic Biliary Ducts

Intrahepatic biliary ducts are radiologically detectable only using contrast enhancement. In traditional cholangiograms they appear like thin little canals flowing into one another, progressively increasing their diameter, with smooth walls and arcuate pathways (Fig. 2.2). Because of their small dimensions these ducts are not detectable on normal US, CT, and MR scans. Sometimes they are visible in the hepatic hilum, near the confluence (WEINSTEIN et al. 1986), where they can reach a diameter of 3–5 mm (Fig. 2.3). Their intrahepatic course is anterior to the portal vessels, which are their anatomic landmark (Fig. 2.4).

Normally, US can show the intrahepatic biliary ducts at the hilum as small, hypoechoic canals anterior to the portal vessels.

CT can show the biliary ducts as hypodense canals, better detectable if they are slightly dilated (Fig. 2.5), when they are axial to the scan slice, and when intravenous contrast medium is used for enhancement of the liver parenchyma and portal vessels.

With MRI they are better detected if they are slightly dilated and the scan is axial, as for CT. In MRI the ducts are of low signal intensity on T1-weighted images and high signal intensity on T2-weighted images.

Obviously, it is possible to visualize the narrower biliary ducts also by using contrast media with specific biliary trophism during CT, or by using specific MR sequences (turbo spin echo). They appear like a thin tree without definite walls and in good contrast to surrounding parenchyma (Figs. 2.6, 2.7).

The radiological anatomy of the intrahepatic biliary ducts is of particular importance not only in regard the morphology of individual small canals, but for the evaluation of the course and overall pat-

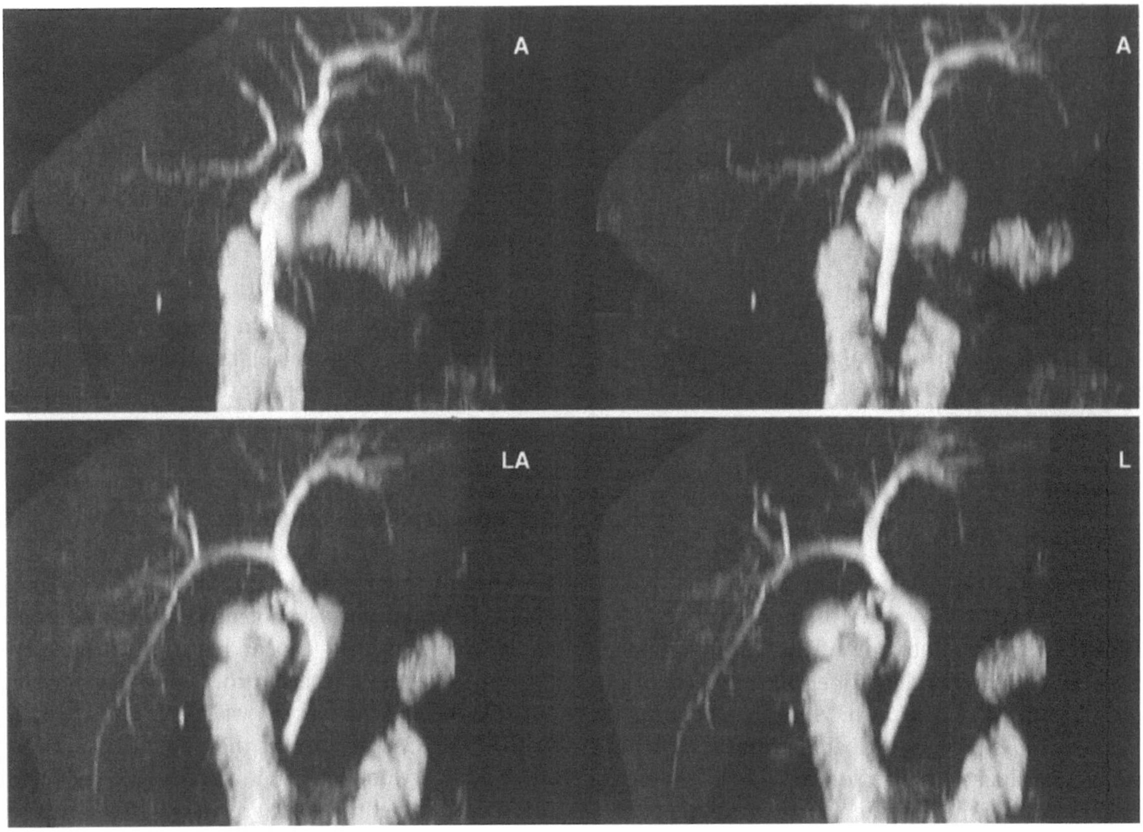

Fig. 2.1. 3D maximum-intensity-projection reconstructions of CT cholangiographic images. It is possible to reconstruct the anatomy of the biliary system in various projections

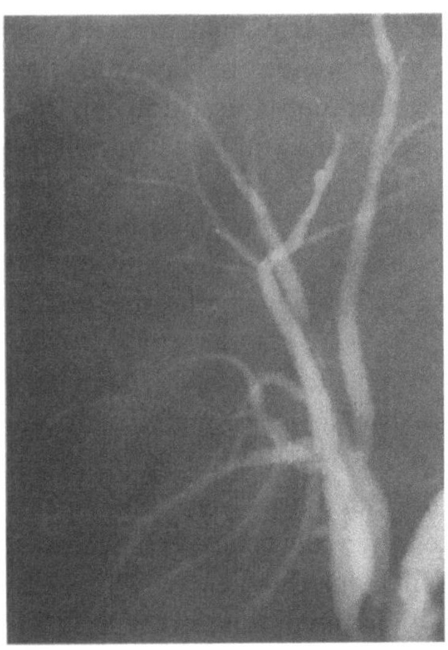

Fig. 2.2. Direct cholangiography: intrahepatic biliary ducts. Thin canals flow into one another, becoming progressively larger

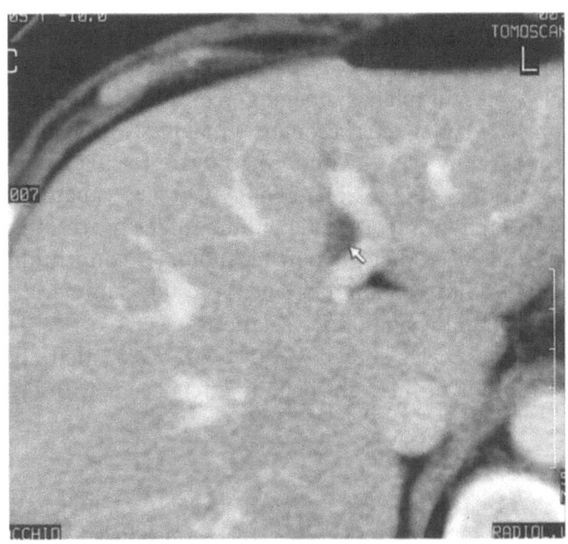

Fig. 2.3. CT. The left main duct (*arrow*) is detectable near the hilum

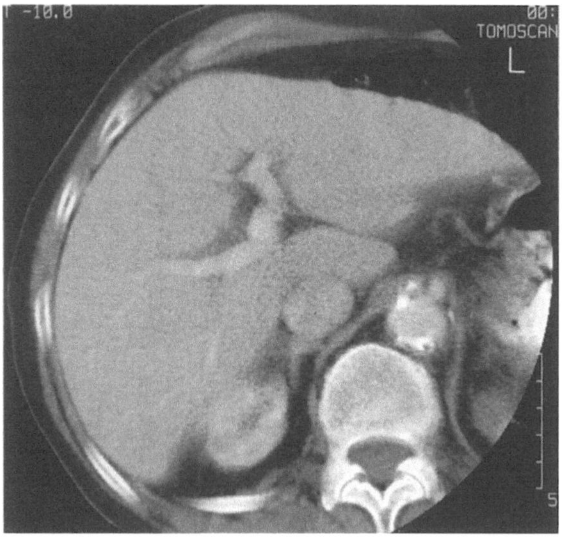

Fig. 2.4. CT. The confluence of the intrahepatic biliary ducts is anterior to the confluence of the portal vessels

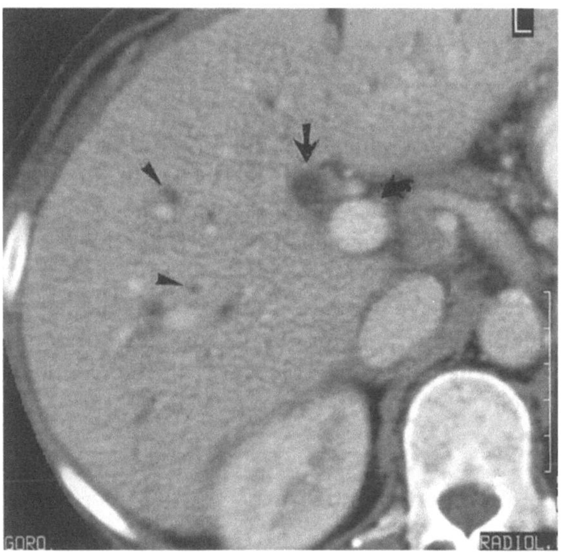

Fig. 2.5. CT. Dilated intrahepatic biliary ducts appear as hypodense circular spots (*arrowheads*). The common hepatic duct (*straight arrow*) is detectable at the hilum, anterolateral to the portal vein (*curved arrow*)

tern of the intrahepatic ducts, which usually have many anatomic variations.

According to the anatomofunctional classification into segments (Couinaud 1957), the intrahepatic biliary ducts drain into two main ducts, left and right, which at the hilum drain in the common hepatic duct. The ducts of the posterior right segments (VI inferior and VII superior) drain in the posterior segmental duct, and the ducts of the anterior right segments (V inferior and VIII superior) into an anterior segmental duct: these two segmental ducts drain into the right main duct just before this meets the left lobe main duct, which drains anterior (III lateral and IV medial) and posterior (II) segments (Fig. 2.8). The

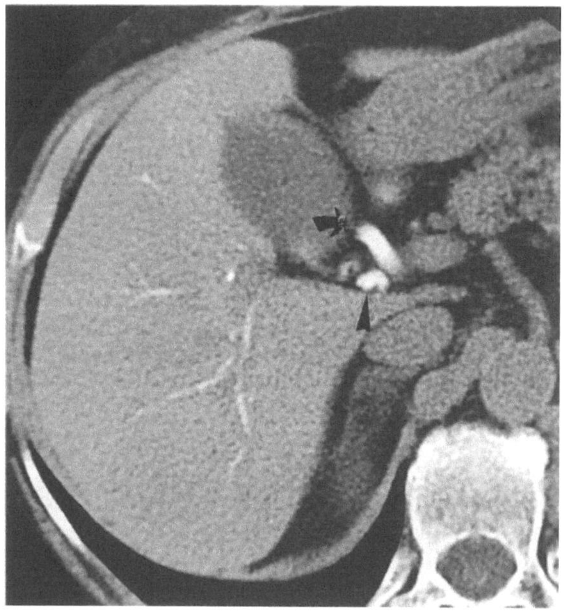

Fig. 2.6. CT cholangiographic appearance of intrahepatic biliary ducts of normal diameter. The cystic duct (*arrowhead*) and common hepatic duct (*curved arrow*) are recognizable

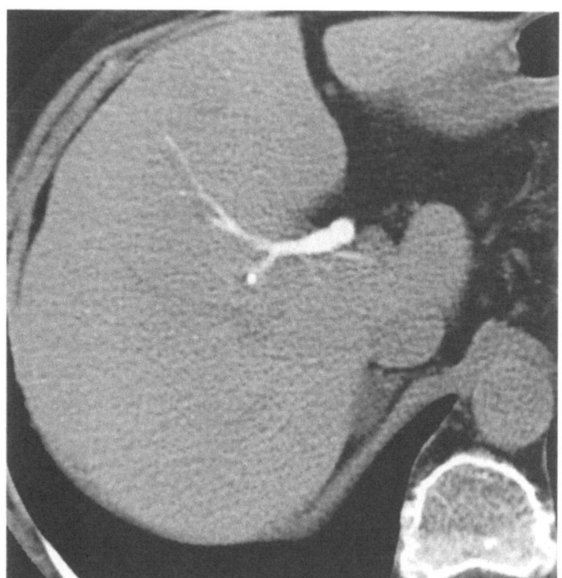

Fig. 2.7. CT cholangiography. Normal appearance of the confluence of the right intrahepatic biliary ducts

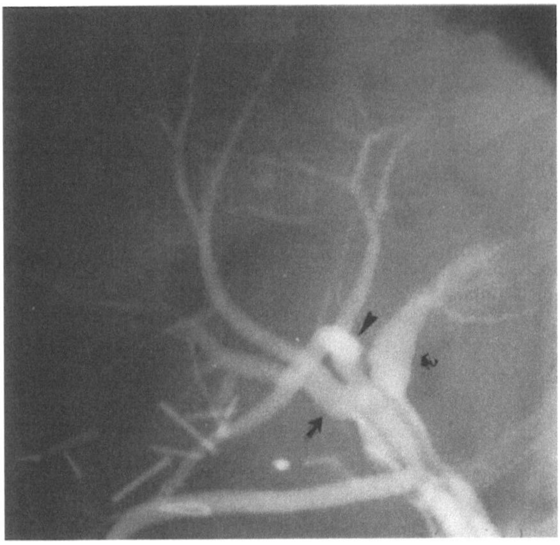

Fig. 2.8. Cholangiography. Confluence of the intrahepatic ducts. The right posterior segmental duct (*straight arrow*) and right anterior segmental duct (*arrowhead*) converge and drain into the left main duct (*curved arrow*)

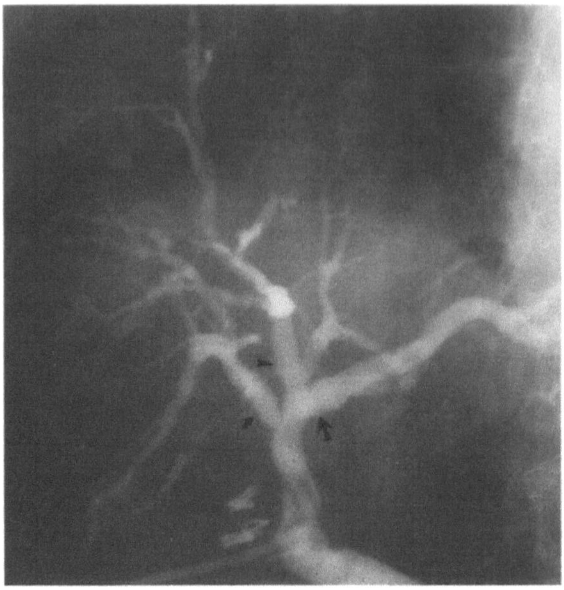

Fig. 2.10. Anomalous confluence of the right anterior segmental duct (*arrowhead*), which drains into the left main duct (*arrow*). The *curved arrow* indicates the right posterior segmental duct

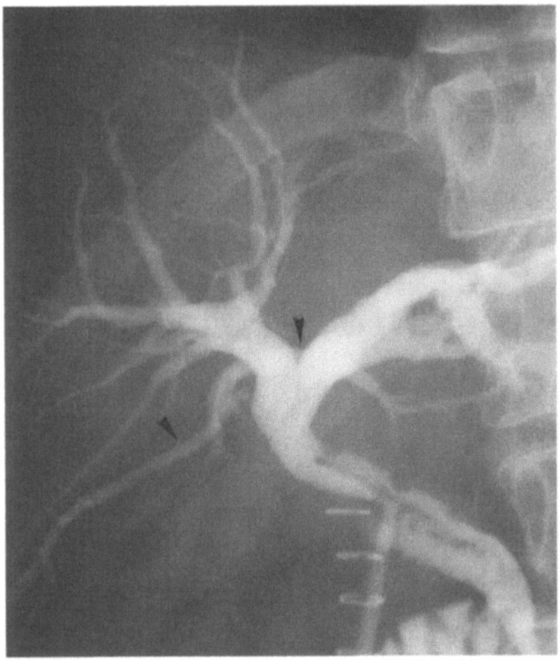

Fig. 2.9. Anomalous outlet of the VI segmental duct (*arrowheads*), draining into the left main duct

ducts of the caudal lobe (I segment) drain into the right and left main ducts.

This pattern, which is the most common, has many anatomic variations (HELOURG et al. 1985) as to the confluence at the hilum (Fig. 2.9). The most frequent are confluence of the posterior or anterior right segmental duct with the left main duct (Fig. 2.10), or confluence of the segmental ducts directly with the common hepatic duct (i.e., three hilar ducts).

A knowledge of these complex relations is particularly important for the performance of transhepatic percutaneous procedures in interventional radiology (NICHOLSON and ADAM 1994).

2.3 Common Hepatic Duct

The common hepatic duct originates from the junction of the intrahepatic ducts at the hilum. The conventional cholangiographic appearance is of a canal 4–5 mm in diameter coursing anteroposteriorly and cranio-caudally until it reaches the cystic duct junction. The walls are smooth, sometimes with thin mucous folds.

On US the common hepatic duct appears as a tubular structure anterolateral to the portal vein, 3 mm in diameter (Fig. 2.11), which can be confused with the hepatic artery (BEHAN and KAZAN 1978).

On CT scans it is detectable at the hilum, anterolateral to the portal vein, when scanned in a plane axial to its course, and thus is well detectable (Fig. 2.5) (WEINSTEIN et al. 1986). It then crosses the

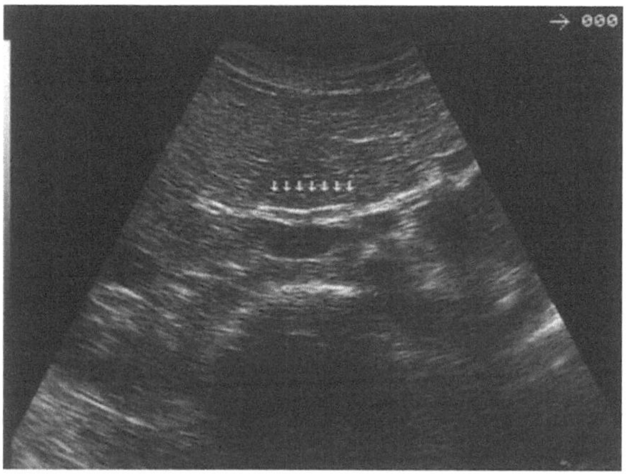

Fig. 2.11. US appearance of the common hepatic duct (*arrows*)

hepatoduodenal ligament and arches caudally (Fig. 2.6), sometimes disappearing in the surrounding fatty tissue (FOLEY et al. 1980).

On MRI the common hepatic duct, like the intrahepatic ducts, has a low signal on T1-weighted images and a high signal in T2-weighted images, owing to the long T1 relaxation time of bile.

2.4 Gallbladder and Cystic Duct

On conventional cholangiography the gallbladder appears as an ovoid of variable dimension, depending on its functional status and on fasting by the patient. It is 5–10 cm long, coursing anteroposteriorly, mediolaterally, and more or less obliquely in a craniocaudal plane. It is pendulous in long-limbed individuals and horizontal in short persons. The internal wall is irregular because of permanent mucous folds.

The gallbladder is regarded as consisting of three parts: the fundus, the body, and the neck.

US shows an ovoidal or elliptical shape on sagittal scans and a circular one on axial scans, of variable hypoechogenicity depending the density of the bile (Fig. 2.12). The wall is detectable as a hyperechoic line, and in normal patients is 3 mm thick (DODDS et al. 1985). The neck region may sometimes go into the region of the hilum and may be in contact with the portal vein and its right branch, while the fundus is often in contact with the colon.

In CT scans the gallbladder appears as an ovoid structure in the subhepatic region, on the medial side of the right lobe, containing liquid-density bile (0–20 HU). Wall detection depends on the fatty tissue surrounding it. The morphology of the organ depends on the angle at which its principal axes cross the scan plane (Fig. 2.13).

Its appearance on MRI, which in axial scans is morphologically similar to that on CT, depends on the bile juice concentration: after a long fast, when the bile is very concentrated, the signal intensity on T1-weighted acquired images is high (HRICAK et al. 1983).

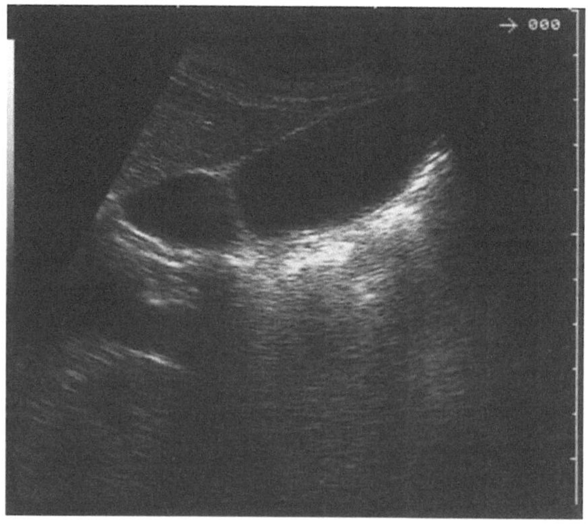

Fig. 2.12. Gallbladder US. Presence of an inside septation

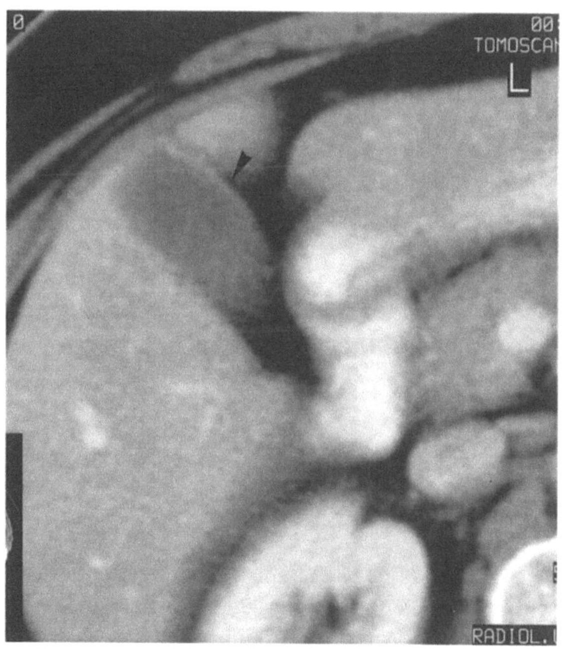

Fig. 2.13. CT: ovoid shape and hypodense appearance of the gallbladder. The walls are easily recognized when surrounded by fat tissue (*arrowhead*)

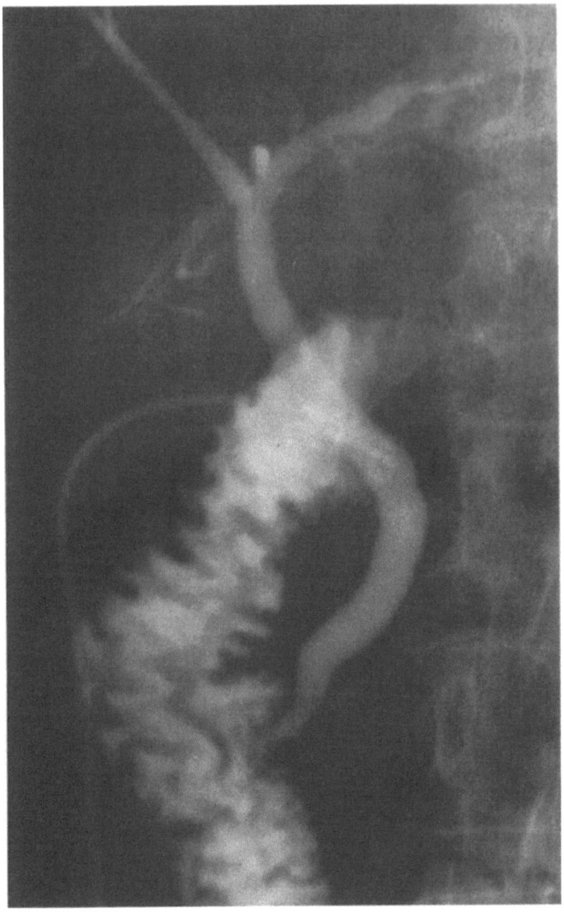

Fig. 2.14. Normal cholangiographic appearance of the relationship between the common bile duct and the duodenum

There is great variation in the morphology and course of the cystic duct in relation to the common bile duct, and variations in its junction with the latter, with the gallbladder, and with the intrahepatic biliary ducts (SHAW et al. 1993).

When the mucous folds are accentuated, the gallbladder assumes a spiral ("cork-screw") appearance on conventional cholangiography.

It is difficult to detect the cystic duct with US, CT and MR.

2.5 Common Bile Duct

The common bile duct appears on cholangiography with a diameter slightly larger than that of the hepatic duct (5–6 mm), coursing craniocaudally and anteroposteriorly, and mediolaterally in the distal tract. In this pathway (about 4 cm), it crosses the pancreatic head axially and then penetrates the posteromedial wall of the second part of the duodenum, in the ampulla of Vater (Figs. 2.14, 2.15). The outlet anatomy is variable, especially in relation to the pancreatic duct.

On US the common bile duct shows a diameter of 4–5 mm (up to 6 mm in cholecystectomized patients), and has a variable relationship with the hepatic artery.

On CT the common bile duct is easily detectable inside the pancreatic head when this is opacified with

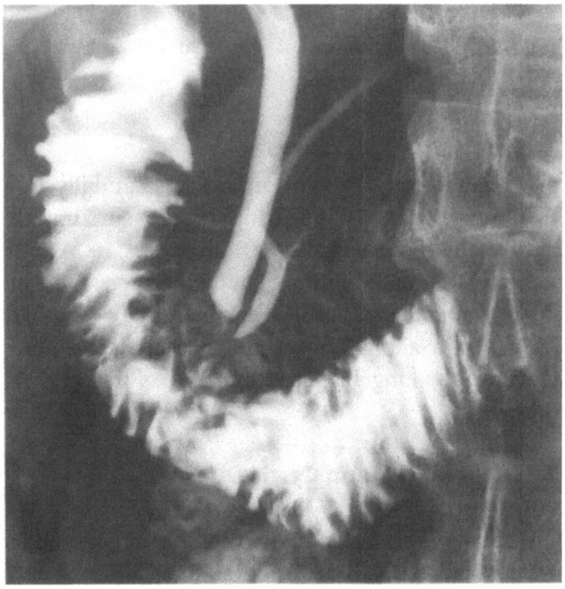

Fig. 2.15. Normal cholangiographic appearance of the ampulla of Vater: common outlet of the common hepatic duct and the pancreatic duct

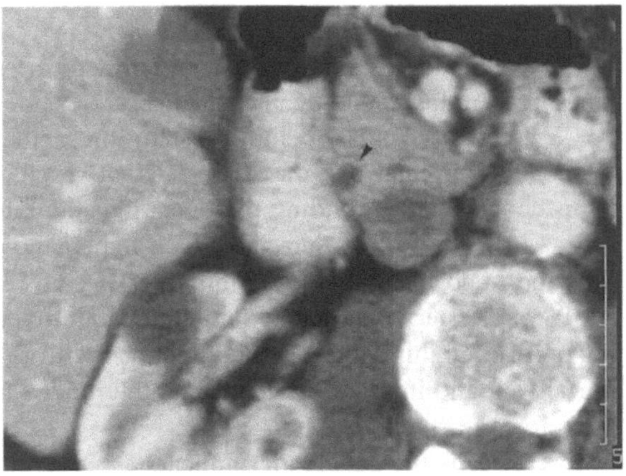

Fig. 2.16. CT: the intrapancreatic common bile duct, near the outlet of the papilla, appears as a small hypodense area (*arrowhead*)

intravenous contrast medium, appearing as a small (2–3 mm), circular area of hypodensity (Fig. 2.16).

Differentiating between the common bile duct and the pancreas is easy on T2-weighted MRI as well, because the hyperintense signal from the bile can be differentiated from blood vessels and pancreas (Outwater and Gordon 1994).

Obviously, all pathological conditions that cause dilatation of the common bile duct make it easier to detect the duct on multiplanar studies.

References

Adams DB (1993) The importance of extrahepatic biliary anatomy in preventing complications. Surg Clin North Am 73:861–871

Behan M, Kazan E (1978) Sonography of the common bile duct: value of the right anterior oblique view. AJR Am J Roentgenol 130:701–709

Couinaud C (1957) Le foie. Etudes anatomiques et chirurgicales. Masson, Paris

Couinaud C (1993) Les variations des canaux biliaires droits. De la vanité des classifications anatomiques complètes. Chirurgie 119 (6–7):354–356

Couinaud C (1994) Anatomie intra-hépatique. Application à la transplantation du foie. Ann Radiol (Paris) 37:323–333

Dodds JW, Gorh WJ, Darwesh RM (1985) Sonographic measurement of gallbladder volume. AJR Am J Roentgenol 145:1009–1013

Foley WD, Wilson CR, Quiroz FA (1980) Demonstration of the normal extrahepatic biliary tract with computed tomography. J Comput Assist Tomogr 4:48–52

Gazelle GS, Lee MJ, Mueller PR (1994) Cholangiographic segmental anatomy of the liver. Radiographics 14(5):1005–1013

Helourg Y, Leborgue J, Roges JM, et al. (1985) Radiological anatomy of the bile ducts based on intraoperative investigation in 250 cases. Anat Clin 7:93–102

Hricak H, Filly RA, Margulis AR, Moon KL, Crooks LE, Kaufman L (1983) Work in progress: nuclear magnetic resonance imaging of the gallbladder. Radiology 147:481–484

Ishizaki I, Wakayama T, Okada Y, Kobayashi T (1993) Magnetic resonance cholangiography for evaluation of obstructive jaundice. Am J Gastroenterol 88:2072–2077

Kubota Y, Yamaguchi T, Tani K, Takaoka M, Fujimura K, Ogura M, Yamamoto S (1993) Anatomical variation of pancreatobiliary ducts in biliary stone disease. Abdom Imaging 18(2):145–149

Nicholson DA, Adam A (1994) A practical guide to percutaneous transhepatic cholangiography and biliary drainage. J Intervent Radiol 9:3–14

Outwater EK, Gordon SJ (1994) Imaging the pancreatic and biliary ducts with MR. Radiology 192:19–21

Schumann-Giampieri G, Schmitt-Willich H, Rudiger W, Negishi C, Weinmann HJ, Speck U (1992) Preclinical evaluation of Gd-EOB-DTPA as a contrast agent in MR imaging of the hepatobiliary system. Radiology 183:59–64

Shaw MJ, Dorsher PJ, Vennes JA (1993) Cystic duct anatomy: an endoscopic perspective. Am J Gastroenterol 88:2102–2106

Yamamoto M, Stiegmann GV, Durham J, Berguer R, Oba Y, Fujiyama Y (1993) Laparoscopy-guided intracorporeal ultrasound accurately delineates hepatobiliary anatomy. Surg Endosc 7:325–330

Weinstein JB, Heiken JP, Lee JKT (1986) High resolution of the porta hepatis and hepatoduodenal ligament. Radiographics 6:55–74

3 Surgical Anatomy of the Biliary Ducts

Y. NIMURA

CONTENTS

3.1 Introduction

Diagnostic and therapeutic procedures for biliary tract diseases have been improving; many surgeons have used new equipment in endoscopic biliary surgery and employed aggressive surgical approaches to difficult biliary cancer such as hilar cholangiocarcinoma, with varying degrees of success (TIO et al. 1991; VERBEEK et al. 1992; NIMURA et al. 1988, 1989; NIMURA 1993). Laparoscopic cholecystectomy has brought not only various advantages but also, unfortunately, complications in the form of proximal biliary stricture (CARROLL et al. 1994; BRANUM et al. 1993). Detailed anatomical studies of the hepatic ducts including the hepatic hilus have contributed new ideas and approaches for further technical developments in hepatobiliary surgery to treat difficult proximal biliary complications and hilar cholangiocarcinoma (HEALEY and SCHROY 1953; HEALEY 1954; COUINAUD 1954, 1957, 1989; BISMUTH 1982; BLUMGART et al. 1984; MIZUMOTO et al. 1986; SAKAGUCHI and NAKAMURA 1986; NIMURA et al. 1990). This chapter describes recent anatomical studies which are relevant to the surgical treatment of difficult biliary lesions at the hepatic hilus.

Y. NIMURA, MD, Professor and Chairman, The First Department of Surgery, Nagoya University School of Medicine, 65 Tsurumai-cho, Showa-ku, Nagoya 466, Japan

3.2 Gross Anatomy of the Hepatic Hilus

The hepatic hilus consists of the hilar plate above the hepatic confluence, which lies on the portal bifurcation. The umbilical portion of the left portal vein (UP) is located at the left extremity of the hilus and the gallbladder covers the right extremity with the serosa. The left lateral segmental ducts (B2, lateral posterior branch; B3, lateral anterior branch) usually join just behind the UP. The left medial segmental duct (B4) enters this common trunk on the right of the UP and forms the left hepatic duct in front of the transverse portion of the left portal vein.

The right anterior hepatic duct lies on the right anterior portal vein and comes down to join with the right posterior hepatic duct, which emerges from the cranial aspect of the bifurcation of the right anterior and posterior portal vein. The left hepatic artery (LHA) usually enters the liver through the left of the UP and the middle hepatic artery (MHA) through the right of the UP. The right hepatic artery (RHA) lies on the main trunk of the portal vein behind the common hepatic duct and bifurcates into anterior and posterior branches. The latter turns caudally and enters the liver through the right extremity of the hepatic hilus, while the anterior branch runs between the right anterior segmental bile duct and the portal vein (Fig. 3.1).

3.3 Subsegmental Anatomy of the Intrahepatic Bile Duct

I established the subsegmental anatomy of the intrahepatic bile duct on the basis of clinical cholangiograms, this cholangiographic anatomy in turn being based on the radiological study of the intrahepatic portal vein branches carried out by TAKAYASU et al. (1985). Hepatobiliary radiologists, endoscopists, and surgeons should have a thorough knowledge of the segmental and subsegmental branches of the intrahepatic bile duct for preoperative diagnosis of proximal tumor extension, for an understanding of the anatomical variations of the intrahepatic biliary tree, and for decision making about therapeutic approaches to difficult hilar biliary lesions. Further, knowledge of the surgical anatomy allows intraoperative identification of the segments and subsegments of the bile duct during liver dissection and segmental bile duct resection in the dissection plane.

The right anterior segmental branches are superimposed over the right posterior segmental branches, and are thus hard to distinguish from each other on clinical cholangiograms taken with the patient in a supine position. They are clearly divided by the right hepatic vein on cholangiograms taken with the patient in the right lateral position, the former in the left cranial and the latter in the right caudal area. Pre- and intraoperative identification of the right anterior subsegmental ducts is particularly important for selecting a duct during percutaneous transhepatic biliary drainage (PTBD) and for dividing the proximal subsegmental ducts with free margins during left hepatectomy, which is the usual surgical approach to hilar cholangiocarcinoma (Fig. 3.2).

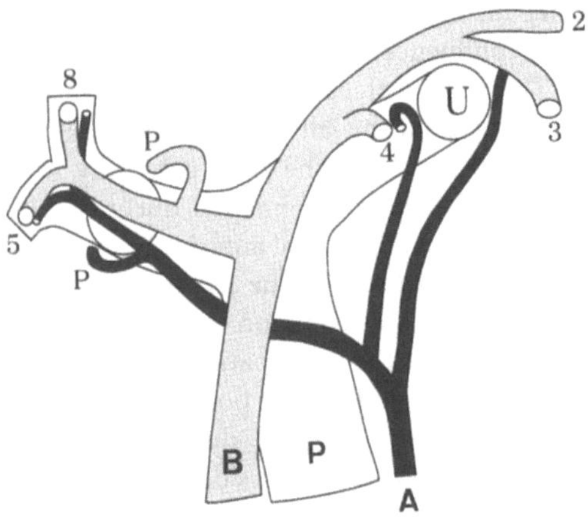

Fig. 3.1. Surgical anatomy of the hepatic hilus. Numerals refer to COUINAUD's segments. *A*, Hepatic artery; *B*, bile duct; *P* (*bottom*), portal vein; *U*, umbilical portion of the left portal vein; *P* (*top left*), posterior branch. (From NIMURA et al. 1995a)

3.4 Surgical Anatomy of the Caudate Lobe

The caudate lobe is the "key lobe" for an understanding of the surgical anatomy of the hepatic hilus (KAMIYA et al. 1994). Although COUINAUD (1994) recently described the dorsal liver to consist of the caudate lobe (segment I) and the paracaval segment (segment IX), the portal ramification and the biliary drainage of these segments are very complex and they have a common trunk contributing to both segments. The arrangement of these branches is very variable and they join with the left and right hepatic ducts and their confluence. For this reason, the present author classifies these biliary branches of segment I into three groups. Most of the branches from the cranial (superior) part of the left caudate

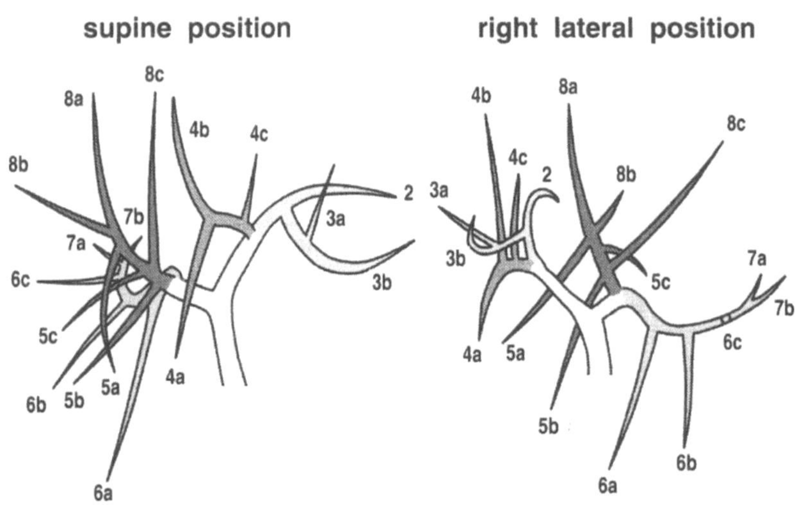

Fig. 3.2. Cholangiographic anatomy of the intrahepatic segmental bile duct according to the patient's position. Numerals refer to COUINAUD's segments. *3a*, Superior branch; *3b*, inferior branch; *4a*, inferior branch; *4b*, superior branch; *4c*, dorsal branch; *5a*, ventral branch; *5b*, dorsal branch; *5c*, lateral branch; *6a*, ventral branch; *6b*, dorsal branch; *6c*, lateral branch; *7a*, ventral branch; *7b*, dorsal branch; *8a*, ventral branch; *8b*, lateral branch; *8c*, dorsal branch. (From NIMURA et al. 1995a)

lobe (B1ls) enter the left hepatic duct (Fig. 3.3). About 40% of the bile ducts of the inferior portion of the left caudate lobe (B1li) join with the right posterior segmental duct. Biliary branches of the right caudate lobe (Couinaud's segment IX) (B1r) enter both the right posterior segmental duct and the hepatic duct, and a tiny branch of the caudate process (B1c) usually joins with the right posterior segmental duct (NIMURA et al. 1990, 1995a). These caudate lobe branches are easily involved by hilar cholangiocarcinoma and cannot be demonstrated in cholangiography. Therefore if the caudate lobe branches are not visualized in cholangiograms, caudate lobe resection should be mandatory in the surgical treatment of hilar cholangiocarcinoma (NIMURA et al. 1990) (Fig. 3.4).

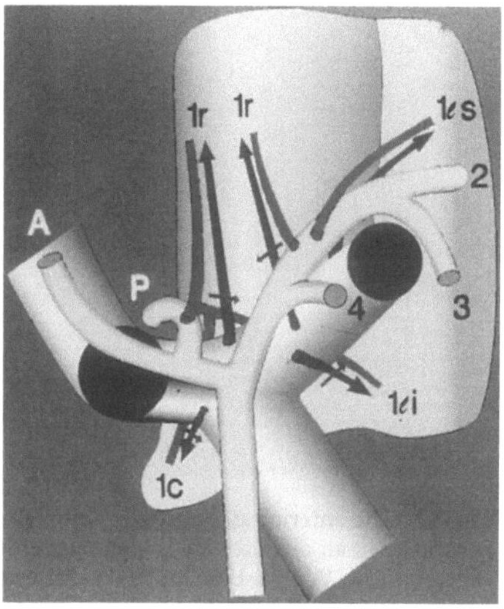

Fig. 3.3. Surgical anatomy of the hepatic hilus, including biliary and portal branches of the caudate lobe. Numerals refer to COUINAUD's segments. *P*, Right posterior branch; *A*, right anterior branch; *1li*, left inferior branch; *1ls*, left superior branch; *1r*, right branch; *1c*, caudate process branch. (From NIMURA et al. 1995a)

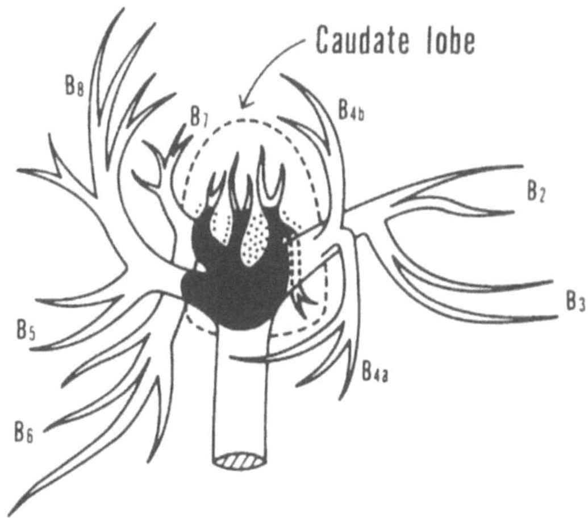

Fig. 3.4. Hilar cholangiocarcinoma involving the branches of the caudate lobe. Numerals refer to COUINAUD's segments. *B4a*, Medial inferior branch; *B4b*, medial superior branch. (From NIMURA et al. 1990)

3.5 Clinical Demonstration of Selective Cholangiography

In the case of advanced hilar cholangiocarcinoma, intrahepatic bile ducts are interrupted and separated into multiple units due to the spread of cancer along

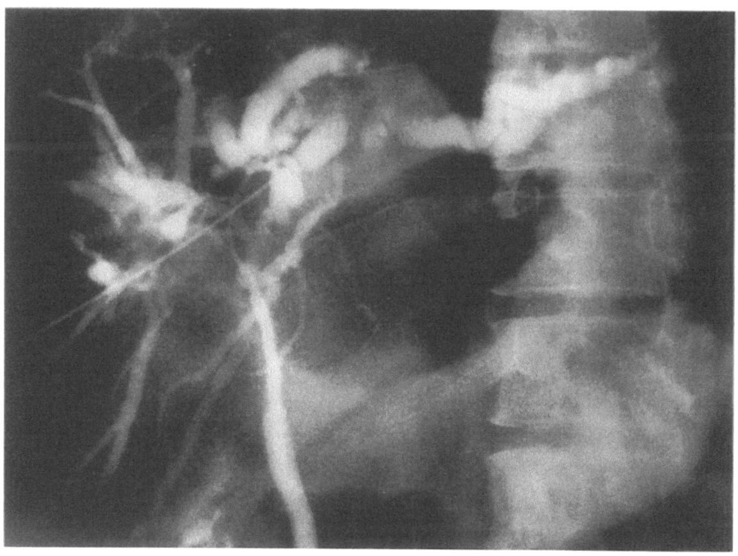

Fig. 3.5. Conventional percutaneous transhepatic cholangiography (PTC) shows diffuse hilar stenosis, but the extent of the lesion cannot be defined

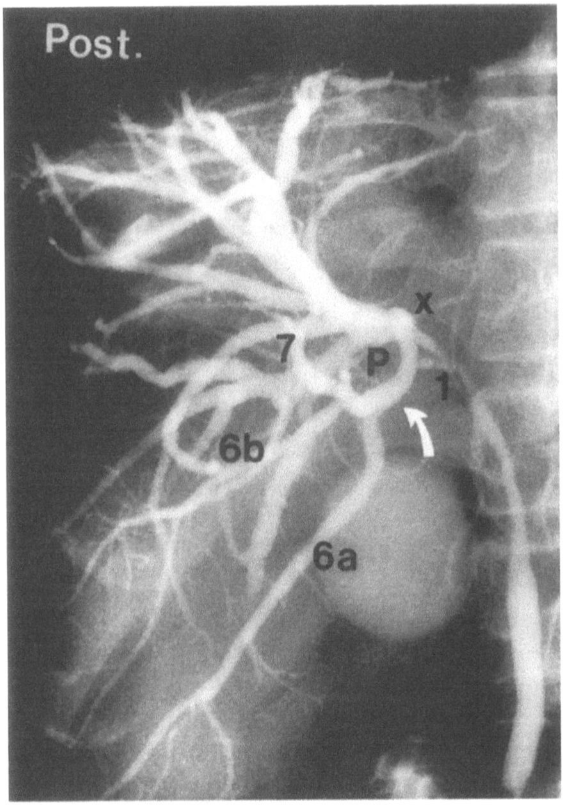

Fig. 3.6. The anterior and posterior segmental ducts are superimposed on each other in a tube cholangiogram taken with the patient supine. Numerals refer to COUINAUD's segments. *P*, Right posterior branch; *6a*, ventral branch; *6b*, dorsal branch. *X* indicates the confluence of the right anterior and posterior segmental ducts; *arrow* indicates proximal extension of the tumor along the right posterior branch

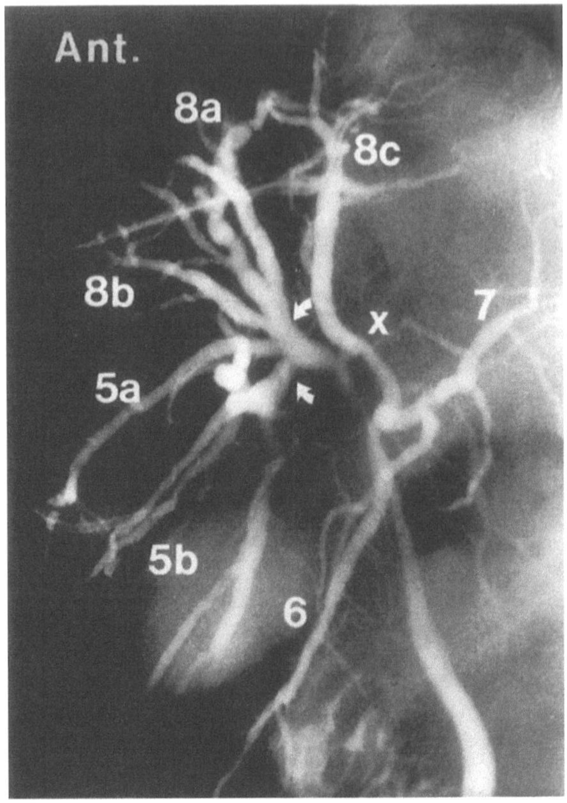

Fig. 3.7. The anterior and posterior segmental ducts are seen separately in an oblique view. Numerals refer to COUINAUD's segments. *5a*, Ventral branch; *5b*, dorsal branch; *8a*, ventral branch; *8b*, lateral branch; *8c*, dorsal branch. *Arrows* show tumor extension into the right anterior segmental ducts. *X* indicates the confluence of the right anterior and posterior segmental ducts

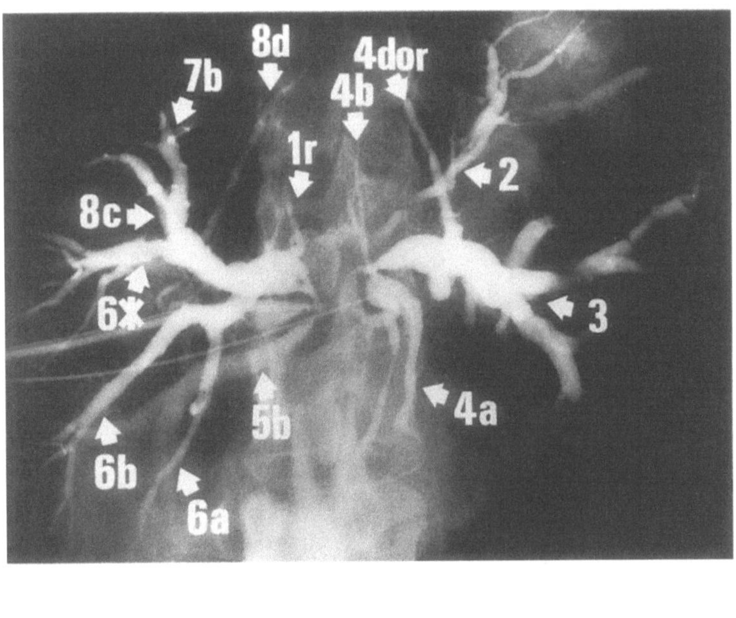

Fig. 3.8. Combined tube cholangiography and PTC demonstrate that bilaterally segmental branches are separated into multiple units. Numerals refer to COUINAUD's segments. *1r*, Right branch; *4a*, inferior branch; *4b*, superior branch; *4 dor*, dorsal branch; *5b*, dorsal branch; *6a*, ventral branch; *6b*, dorsal branch; *6**, lateral branch; *7b*, dorsal branch; *8c*, dorsal branch; *8d*, paracaval branch

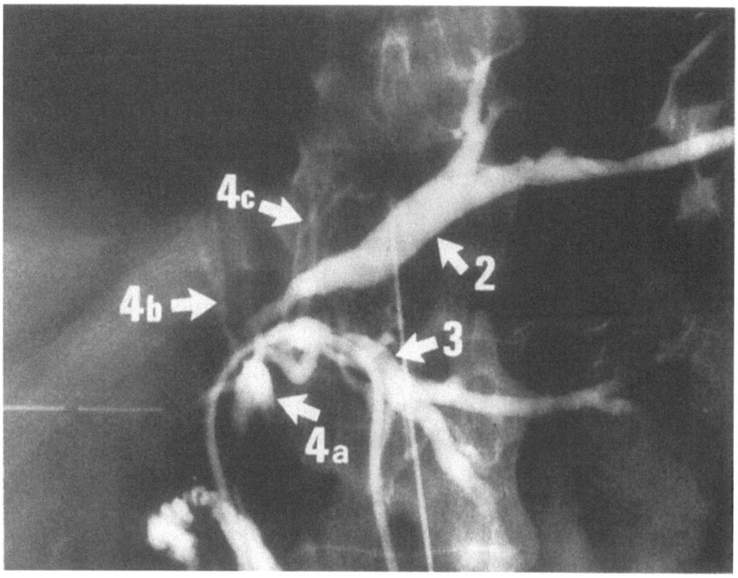

Fig. 3.9. Combined tube cholangiography and PTC show diffuse cancer extension into each segmental branch of the left liver. Numerals refer to COUINAUD's segments. *4a*, Inferior branch; *4b*, superior branch; *4c*, dorsal branch

the bile duct. Therefore it is usually impossible not only to diagnose the main lesion and cancer extension but also to decide about the possibility of curative resection using conventional percutaneous transhepatic cholangiography (PTC) (Fig. 3.5). Preoperative selective cholangiography through multiple PTBD demonstrates precisely the anatomical location, extent, and nature of the obstructing lesion in each separate part of the segmental ducts (NIMURA et al. 1995b). The right anterior and posterior segmental ducts are hardly to be distinguished from each other in a tube cholangiogram taken in a supine position (Fig. 3.6), but they are clearly isolated from each other in the right lateral or the right anterior oblique position. An anatomical variation may also be detected with some selective cholangiograms taken in several positions (Fig. 3.7).

3.5.1 Left Hepatectomy

Combined right tube cholangiography and left PTC demonstrate a diffuse carcinoma at the hepatic hilus which has separated intrahepatic segmental ducts bilaterally into multiple units (Fig. 3.8). The precise proximal extension of the tumor cannot be defined from this picture. Combined tube cholangiography through B3 and PTC of B2 (the left lateral posterior branch) show extensive cancer development into B2, B3 (the left lateral anterior branch), and B4 (the

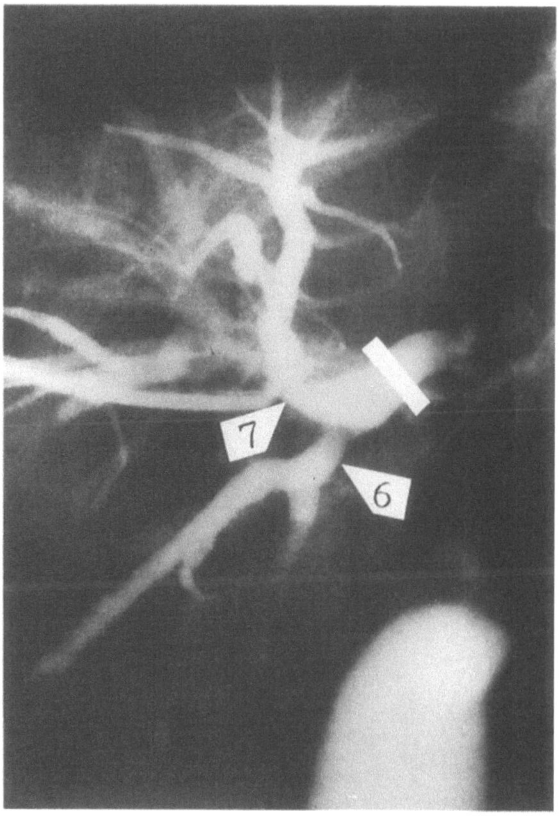

Fig. 3.10. Selective right posterior tube cholangiography reveals limited cancer invasion in the main trunk of the right posterior hepatic duct. The *white bar* indicates the intended line of resection of the duct. Numerals refer to COUINAUD's segments

left medial branch; Fig. 3.9). Selective right tube cholangiography reveals limited cancer invasion in the main trunk of the right posterior segmental duct and a resection line can be designed distal to the confluence of B6 (posterior inferior branch) and B7 (posterior superior; Fig. 3.10). Another selective tube cholangiogram of the right liver demonstrates diffuse cancer progress toward the interrupted right anterior segmental branches: B8c (dorsal branch of the anterior superior segment), B8a+b (ventral and lateral branch), and B5a (ventral branch of the anterior inferior segment), and a proposed line of resection for each subsegmental bile duct can be defined (Fig. 3.11). Combined selective cholangiography through the right anterior and posterior PTBD catheters taken in the right lateral position show tumor extension up to the anomalous confluence of B8c and the right posterior segmental duct (Fig. 3.12). In this particular case, another quite rare variation of the subsegmental duct B5b (dorsal branch of the anterior inferior segment) was found joining independently with the main right hepatic duct more distally, and a resection line for this isolated subsegmental duct was also designed (Fig. 3.13). From these selective cholangiograms the extent of the cancer was defined as predominantly in the left hepatic duct; rare variations of the B8c confluence with the right posterior segmental duct and the B5b with the main right hepatic duct were diagnosed preoperatively, and an extended left hepatectomy with total caudate lobectomy was carried out. The six interrupted intrahepatic segmental and subsegmental bile ducts were resected with free margins, and resected specimen cholangiography confirmed the resected margins of the right intrahepatic subsegmental branches (Fig. 3.14). Postoperative histological examination of the resected specimen revealed diffuse cancer invasion predominantly in the left hepatic duct and no invasion in the six resected margins of the isolated segmental ducts of the right liver (Fig. 3.15).

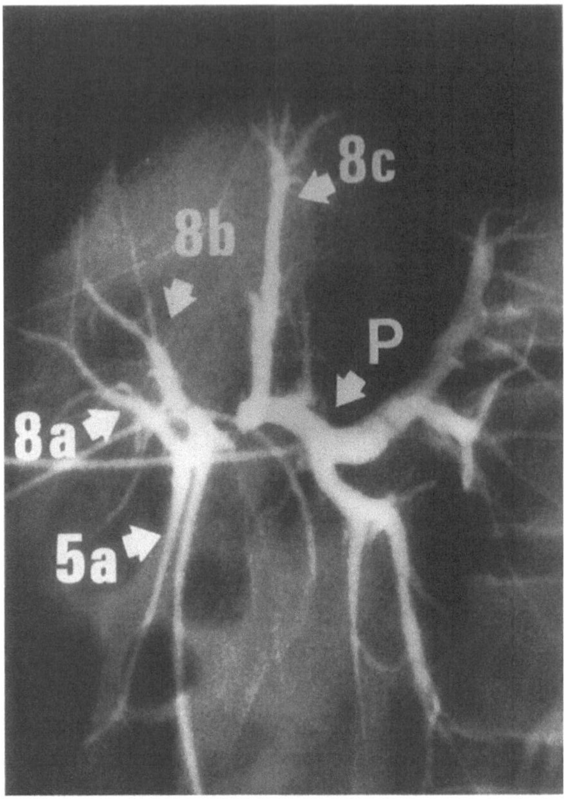

Fig. 3.12. Combined right anterior and posterior cholangiography taken with the patient in the right lateral position shows a rare variation of the confluence of B8c with the posterior segmental duct (*P*). Numerals refer to COUINAUD's segments. *5a*, Ventral branch; *8a*, ventral branch; *8b*, lateral branch; *8c*, dorsal branch

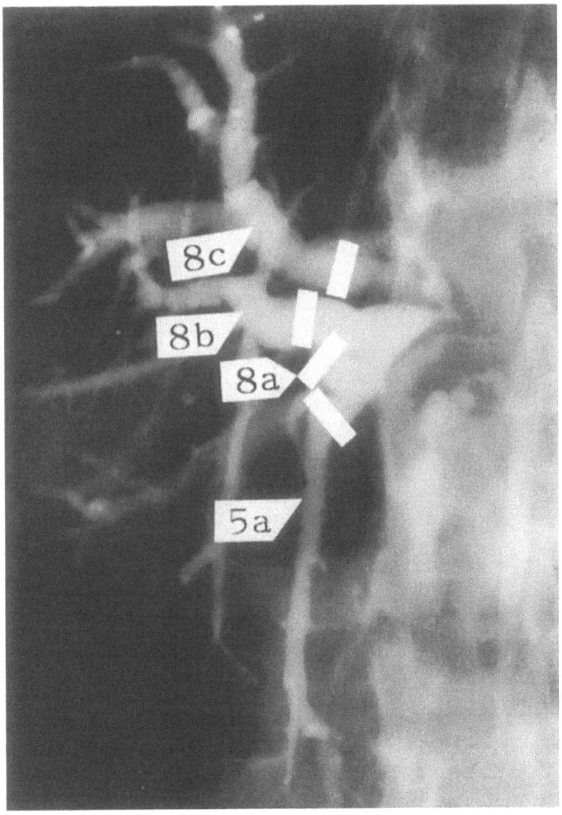

Fig. 3.11. Selective right anterior tube cholangiography demonstrates diffuse cancer progress toward the separated anterior segmental branches. Numerals refer to COUINAUD's segments. *5a*, Ventral branch; *8a*, ventral branch; *8b*, lateral branch; *8c*, dorsal branch. *White bars* indicate intended resection lines of the subsegmental branches

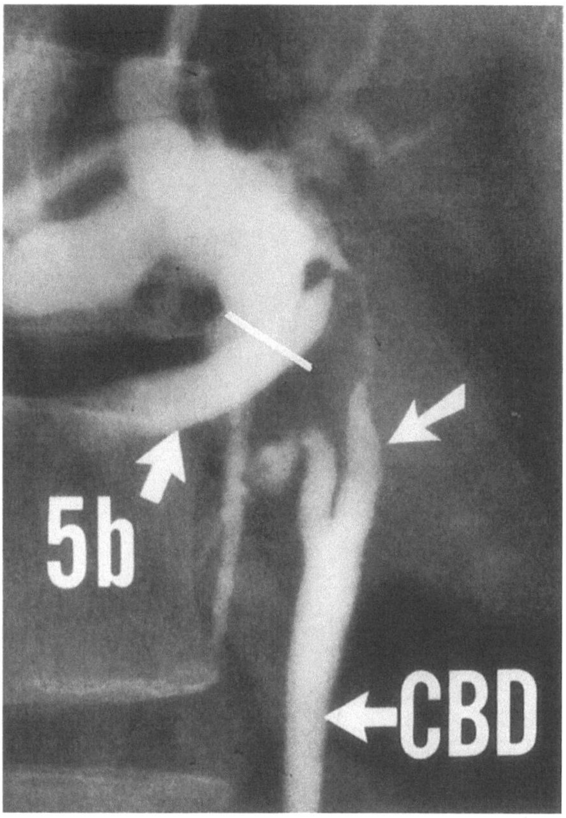

Fig. 3.13. A quite rare variation of a dorsal branch of the right anterior inferior segment (B5b) joining with the involved main right hepatic duct. *CBD*, Common bile duct. The *white bar* indicates the intended resection line of the dorsal branch of the anterior inferior segmental duct (*5b*). An *arrow* shows the distal extension of the tumor in the common hepatic duct

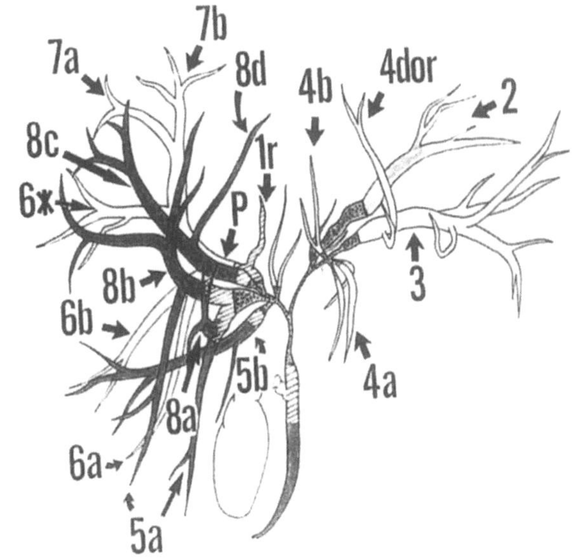

Fig. 3.15. Representation of tumor involvement in the case shown in Figs. 3.8–3.14. Histologically tumor invasion (▦) was determined to be predominantly in the left hepatic duct. Resected margins (▨) were free from cancer invasion. Numerals refer to COUINAUD's segments. *4a*, Anterior branch; *4b*, posterior branch; *4 dor*, d2orsal branch; *5a*, ventral branch; *5b*, dorsal branch; *6a*, ventral branch; *6b*, dorsal branch; 6*, lateral branch; *7a*, ventral branch; *7b*, dorsal branch; *8a*, ventral branch; *8b*, lateral branch; *8c*, dorsal branch; *8d*, paracaval branch; *P*, right posterior segmental duct

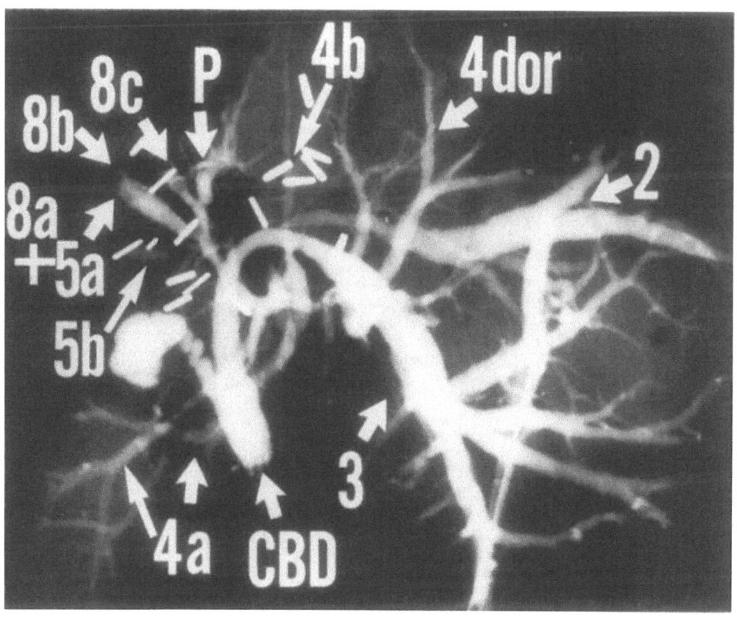

Fig. 3.14. Resected specimen cholangiography shows the resected margins of the right anterior subsegmental branches and the posterior segmental duct (P). Numerals refer to COUINAUD's segments. *5a*, Ventral branch; *5b*, dorsal branch; *8a*, ventral branch; *8b*, lateral branch; *8c*, dorsal branch; *4a*, inferior branch; *4b*, superior branch; *4 dor*, dorsal branch; *CBD*, common bile duct

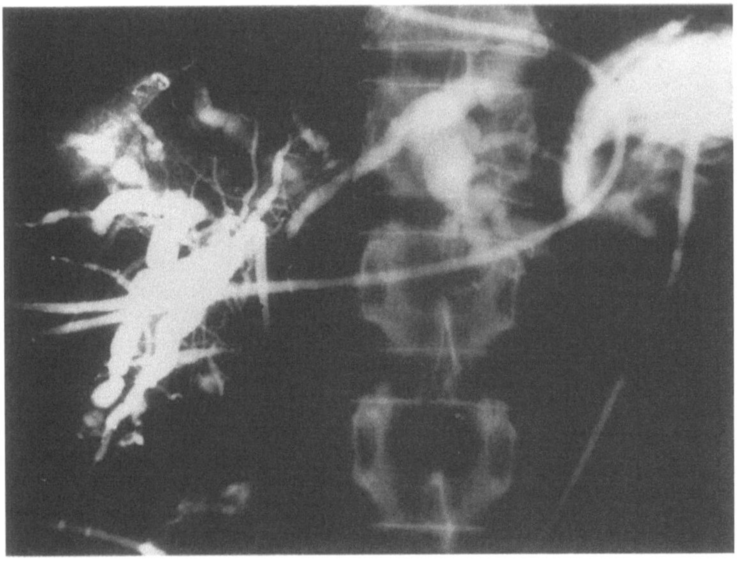

Fig. 3.16. Tube cholangiography through a right posterior PTBD catheter shows hilar biliary obstruction. The left hepatic duct is faintly seen

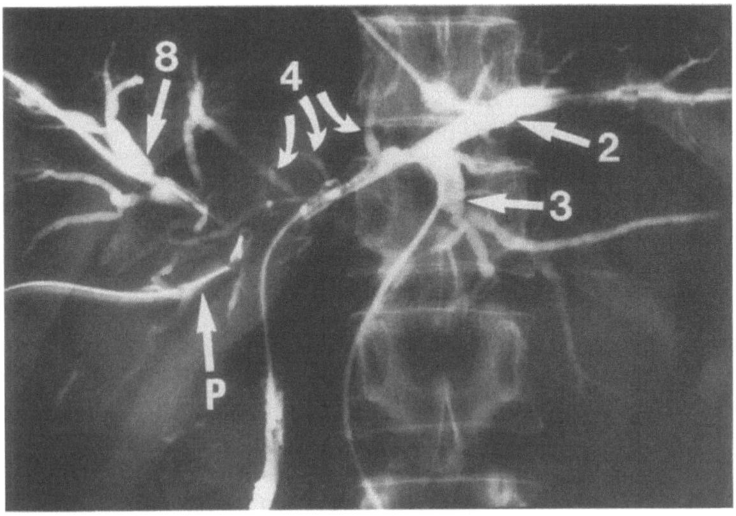

Fig. 3.17. Selective cholangiography through right anterior and left lateral PTBD catheters demonstrates marked cancer progress toward the right anterior and left medial segmental ducts. Numerals refer to COUINAUD's segments *P*, A PTBD catheter introduced into the right posterior segmental duct

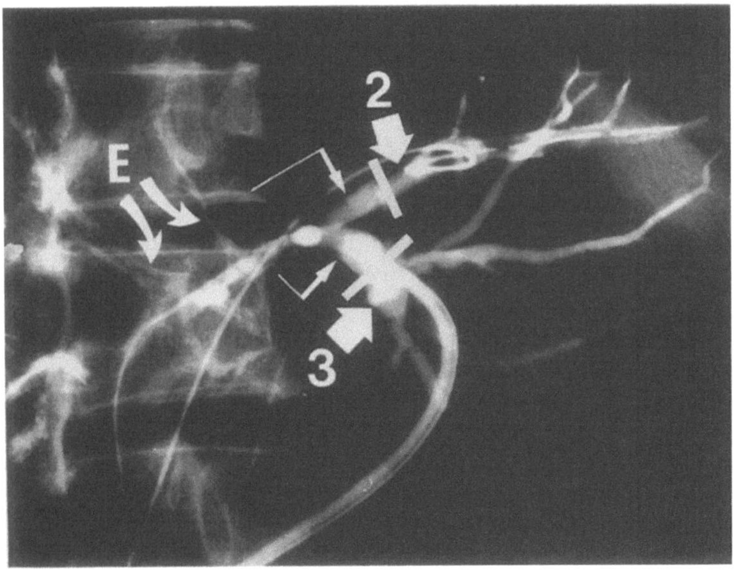

Fig. 3.18. Selective tube cholangiography through the left lateral PTBD catheters reveals cancer extension above the confluence of B2 and B3 (*angled arrows*). *White bars* indicate the intended resection line of the left lateral segmental ducts. *E* indicates embolic materials (fibrin glue mixed with lipiodol) in the left medial branches of the portal vein

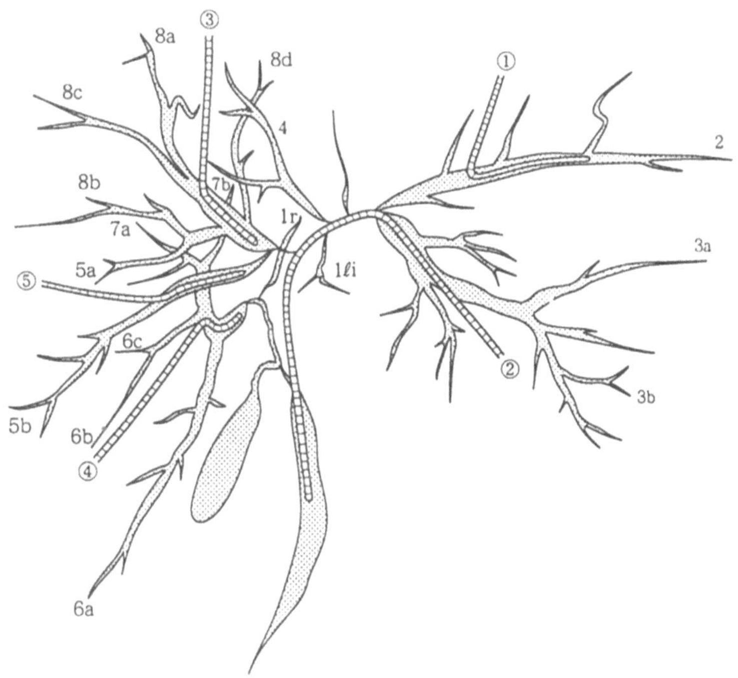

Fig. 3.19. Five PTBD catheters were placed through B2 (①), B3(②), B8(③), B6(④), and B5b(⑤). Numerals refer to COUINAUD's segments. *1r*, Right branch; *1li*, left inferior branch; *3a*, superior branch; *3b*, inferior branch; *5a*, ventral branch; *5b*, dorsal branch; *6a*, ventral branch; *6b*, dorsal branch; *6c*, lateral branch; *7a*, ventral branch; *7b*, dorsal branch; *8a*, ventral branch; *8b*, lateral branch; *8c*, dorsal branch; *8d*, paracaval branch

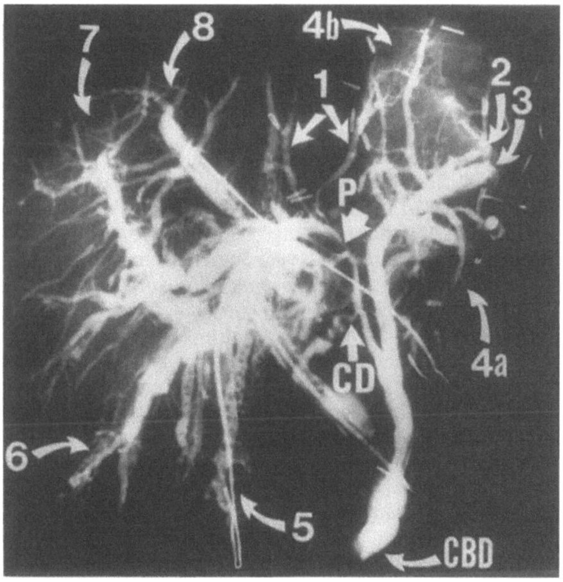

Fig. 3.20. Resected specimen cholangiography demonstrates sufficient length of the resected margins of the left lateral segmental ducts. Numerals refer to COUINAUD's segments. *4a*, Inferior branch; *4b*, superior branch; *CD*, cystic duct; *CBD*, common bile duct; *P*, right posterior segmental duct

3.5.2 Right Hepatectomy

Selective cholangiography through a right PTBD catheter shows isolated right posterior segmental ducts, and the left hepatic duct is faintly seen (Fig. 3.16). Multiple PTBD is mandatory for preoperative management and staging of the disease (NAGINO et al. 1992, 1995a). Selective tube cholangiography through the right anterior and left lateral PTBD catheters demonstrates marked cancer progress toward B8 and B4 (Fig. 3.17). Another selective cholangiogram taken through PTBD catheters of B2 and B3 in the right anterior oblique position disclosed cancer extension up to B2 and B3 proximal to their confluence (Fig. 3.18). In this case, five PTBD catheters were finally placed through B2, B3, B8, B6, and B5b, and cancer progress was observed predominantly in the right liver (Fig. 3.19), with involvement of the right hepatic artery and the right anterior portal vein. Therefore right hepatic trisegmentectomy with caudate lobectomy (NIMURA 1994) and portal vein resection and reconstruction were performed after preoperative right trisegment portal vein embolization (NAGINO et al. 1993, 1995b) and the left lateral segmental ducts, B2 and B3, were resected with free margins. The resected specimen cholangiogram clearly demonstrated sufficient length of the resected left lateral segmental ducts, B2 and B3 (Fig. 3.20).

3.6 Conclusion

As shown in the clinical cases of hilar cholangiocarcinoma above, intrahepatic segmental and subsegmental bile ducts must be anatomically identified in selective cholangiograms and the proximal extension of the cancer must be precisely defined after placement of multiple biliary drains. Then anatomical liver and segmental bile duct resection can be designed according to a definitive preoperative diagnosis of the progress of the cancer toward the segmental and subsegmental bile ducts, thus avoiding palliative interventional treatment and prolonging and increasing the quality of patients' survival. This is why hepatobiliary radiologists, endoscopists, and surgeons need a thorough knowledge of the segmental anatomy of the intrahepatic bile ducts in order to overcome the difficulties posed by disease of the proximal bile duct.

References

Bismuth H (1982) Surgical anatomy and anatomical surgery of the liver. World J Surg 6:3–9

Blumgart LH, Hadjis NS, Benjamin IS, Beazley RM (1984) Surgical approaches to cholangiocarcinoma at the confluence of the hepatic ducts. Lancet I:66–69

Branum G, Schmitt C, Baillie J, Suhocki P, Baker M, Davidoff A, Branch S, Chari R, Cucchiaro G, Murray E (1993) Management of major biliary complications after laparoscopic cholecystectomy. Ann Surg 217:532–540

Carroll BJ, Fallas MJ, Phillips EH (1994) Laparoscopic transcystic choledochoscopy. Surg Endosc 8:310–314

Couinaud C (1954) Lobes et segments hépatiques. Notes sur l'architecture anatomique et chirurgicale du foie. Presse Med 62:709–712

Couinaud C (1957) Le foie. Etudes anatomiques et chirurgicales. Masson, Paris

Couinaud C (1989) Surgical anatomy of the liver revisited. Couinaud, Paris

Couinaud C (1994) The paracaval segment of the liver. J Hep Bil Pancr Surg 2:145–151

Healey JE Jr (1954) Clinical anatomic aspects of radical hepatic surgery. J Int Coll Surg 22:542–550

Healey JE Jr, Schroy PC (1953) Anatomy of the biliary ducts within the human liver. Arch Surg 66:599–616

Kamiya J, Nimura Y, Hayakawa N, Kondo S, Nagino M, Kanai M (1994) Preoperative cholangiography of the caudate lobe: surgical anatomy and staging for biliary carcinoma. J Hep Bil Pancr Surg 1:385–389

Mizumoto R, Kawarada Y, Suzuki H (1986) Surgical treatment of hilar carcinoma of the bile duct. Surg Gynecol Obstet 162:153–158

Nagino M, Hayakawa N, Nimura Y, Dohke M, Kitagawa S (1992) Percutaneous transhepatic biliary drainage in patients with malignant biliary obstruction of the hepatic confluence. Hepatogastroenterology 39:296–300

Nagino M, Nimura Y, Hayakawa N (1993) Percutaneous transhepatic portal vein embolization using newly devised catheters: preliminary report. World J Surg 17:520–524

Nagino M, Nimura Y, Kamiya J, Kondo S, Kanai M, Miyachi M, Yamamoto H, Hayakawa N (1995a) Preoperative management of hilar cholangiocarcinoma. J Hep Bil Pancr Surg 2:239–248

Nagino M, Nimura Y, Kamiya J, Kondo S, Uesaka K, Kin Y, Kutusna Y, Hayakawa N, Yamamoto H (1995b) Right or left trisegment portal vein embolization before hepatic trisegmentectomy for hilar bile duct carcinoma. Surgery 117:677–681

Nimura Y (1993) Staging of biliary carcinoma: cholangiography and cholangioscopy. Endoscopy 25:76–80

Nimura Y (1994) Hepatectomy for proximal bile duct cancer. In: Braasch JW, Tompkins RK (ed) Surgical disease of the biliary tract and pancreas. Mosby-Year Book, St Louis, pp 251–261

Nimura Y, Shionoya S, Hayakawa N, Kamiya J, Kondo S, Yasui A (1988) Value of percutaneous transhepatic cholangioscopy (PTCS). Surg Endosc 2:213–219

Nimura Y, Kamiya J, Hayakawa N, Shionoya S (1989) Cholangioscopic differentiation of biliary strictures and polyps. Endoscopy 21:315–356

Nimura Y, Hayakawa N, Kamiya J, Kondo S, Shionoya S (1990) Hepatic segmentectomy with caudate lobe resection for bile duct carcinoma of the hepatic hilus. World J Surg 14:535–544

Nimura Y, Hayakawa N, Kamiya J, Kondo S, Nagino M, Kanai M (1995a) Hilar cholangiocarcinoma – surgical anatomy and curative resection. J Hep Bil Pancr Surg 2:239–248

Nimura Y, Kamiya J, Kondo S, Nagino M, Kanai M (1995b) Technique of inserting multiple biliary drains and management. Hepatogastroenterology 42:323–331

Sakaguchi S, Nakamura S (1986) Surgery of the portal vein in resection of cancer of the hepatic hilus. Surgery 90:344–349

Takayasu K, Moriyama N, Muramatsu Y, Shima Y, Goto H, Yamada T (1985) Intrahepatic portal vein branches studied by percutaneous transhepatic portography. Radiology 154:31–36

Tio TL, Cheng J, Wijers OB, Sars PRA, Tytgat GNJ (1991) Endosonographic TNM staging of extrahepatic bile duct cancer: comparison with pathological staging. Gastroenterology 100:1351–1361

Verbeek PCM, Leeuwen JV, Wit LT, Reeders JWAJ, Smits NJ, Bosma A, Huibregtse K, van der Heyde MN (1992) Benign fibrosing disease at the hepatic confluence mimicking Klatskin tumors. Surgery 112:866–871

Diagnostic Imaging

4 Oral Cholecystography

M. Bazzocchi, E. Quaia, C. Zuiani, and A. Rigamonti

CONTENTS

4.1 Introduction

Oral cholecystography (OCG) has undergone a rejuvenation with the development of nonsurgical therapeutic techniques: extracorporeal shock wave lithotripsy (ESWL), oral solvents, contact solvents (MTBE), and minimally invasive (laparoscopic) surgery (Brakel et al. 1992; Maglinte et al. 1991; Marzio et al. 1992). The reason for this "rediscovery" of OCG is chiefly the information it can provide about the function of the gallbladder.

OCG contrast agents are absorbed by passive diffusion across the intestinal mucosa. Since the intestinal blood barrier is lipoidal, the lipid-soluble OCG agent is readily absorbed in to the blood stream, binds to albumin, and is transported to the liver. In the liver, OCG contrast agent is conjugated with glucuronic acid, is rendered water-soluble, and is excreted into the bile canaliculi, and ultimately the majority of it enters the gallbladder through a patent cystic duct. Reabsorption of water by the gallbladder mucosa leads to contrast agent and gallbladder opacification. A concentration of 0.25%–1.0% iodine is required in the gallbladder for radiographic visualization (Edholm and Jacobson 1972; Jaffe and Wachoski 1942). Failure to visualize the gallbladder in the absence of cystic duct obstruction or extrabiliary causes reflects failure of the gallbladder to reabsorb water and concentrate the contrast material (Graham et al. 1924).

The possibility that reabsorption of the contrast material itself by the gallbladder could result in nonvisualization was recognized long ago (Ivy 1934), and later it was shown that the OCG contrast agent is indeed readily absorbed **from** the inflamed gallbladder (Berk and Lasser 1964), and that this mechanism is a cause of nonvisualization.

When OCG contrast agent at high dosage is employed, failure of the gallbladder to opacify, in the absence of extrabiliary causes of nonvisualization, is highly suggestive of intrinsic gallbladder disease (obstruction of cystic duct or nonobstructive cholecystitis).

4.2 Poor or Absent Opacification of the Gallbladder

A variety of possibilities (extrinsic causes) must be considered if the gallbladder does not opacify, or does so only faintly (Hamilton 1989; Mujahed 1976; Nelson 1972; Ochsner 1964; Sickles 1977; Ochsner and Buchtel 1967; Perrin and Boehnke 1974; Teplick and Adelman 1955; Koehler et al. 1960): (1) Failure of the patient to take the contrast agent; (2) obstructive upper gastrointestinal disease; (3) failure of intestinal absorption or rapid transit time; (4) hepatocellular diseases and biliary obstruction; (5) prior cholecystoenterostomy or cholecystectomy. If all these can be ruled out, nonvisualization of the gallbladder indicates intrinsic gallbladder disease. Nonopacification of the gallbladder with opacification of the common bile duct is evidence of cystic duct obstruction.

M. Bazzocchi, MD, Director, Cattedra di Radiologia, Policlinico Universitario, Piazzale S. Maria della Misericordia, 33100 Udine, Italy
E. Quaia, MD, Cattedra di Radiologia, Policlinico Universitario, Piazzale S. Maria della Misericordia, 33100 Udine, Italy
C. Zuiani, MD, Cattedra di Radiologia, Policlinico Universitario, Piazzale S. Maria della Misericordia, 33100 Udine, Italy
A. Rigamonti, MD, Cattedra di Radiologia, Policlinico Universitario, Piazzale S. Maria della Misericordia, 33100 Udine, Italy

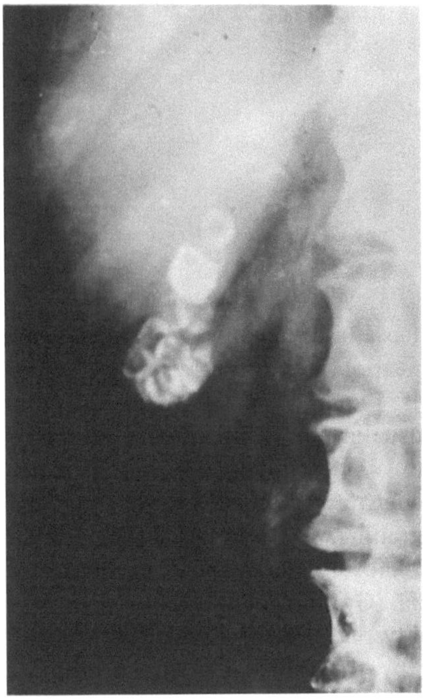

Fig. 4.1. Preliminary abdominal radiograph. Multiple opaque stones

Nonopacification of both the gallbladder and the common bile duct, following identification of absorbed and excreted contrast agent in the bowel, suggests impaired water absorption by the gallbladder or diffusion of contrast material through the inflamed wall. Nonopacification of both the gallbladder and the bile duct without the presence of opaque stones and without absorbed contrast agent in the bowel suggests causes of nonopacification other than gallbladder disease.

4.3 Alterations of Gallbladder Content

The alterations of the gallbladder content seen in OCG are due to (1) gallstones, (2) adenomyomatosis, or (3) cholesterolosis. Very rarely it is possible to observe a defect of filling due to a neoplasm, the gallbladder being not visualized in these cases.

4.3.1 Gallstones

A preliminary abdominal radiograph taken on the day before OCG obviates the latter in only 9% of patients who have opaque stones (Fig. 4.1); in less than 2% of patients, gallstones will be obscured by contrast agent (Harned and Le Ven 1978). Visible calcifications, seen in 10%–15% of patients (Donner and Weiner 1964) and 27% (Marzio et al. 1992) are believed to be due to calcium carbonate and calcium phosphate; this information is necessary for many of the nonsurgical alternatives to elective cholecystectomy. Calcifications can be mimicked by iopanoic acid, which can cause "staining" of radiolucent gallstones after repeated doses (Salzman and Warden 1958). The stones appear as defects of filling (radiolucency) except for the radiopaque component (Fig. 4.2) and are usually movable in the different positions except for big stones in the gallbladder fundus (Fig. 4.3), where complete opacification is prevented. If the stone obstructs the infundibulum, the gallbladder is not visualized.

Distinguishing between cholesterol stones and pigment stones has now become critically important, since the newer alternatives to elective cholecystectomy are effective only on cholesterol stones. Eighty-six percent of radiolucent stones are composed of cholesterol, while 14% are pigment gallstones (Trotman et al. 1975). On the other hand, of calcified gallstones visible on plain film, two-thirds are composed of pigment (central calcification more frequent) and one-third predominantly of choles-

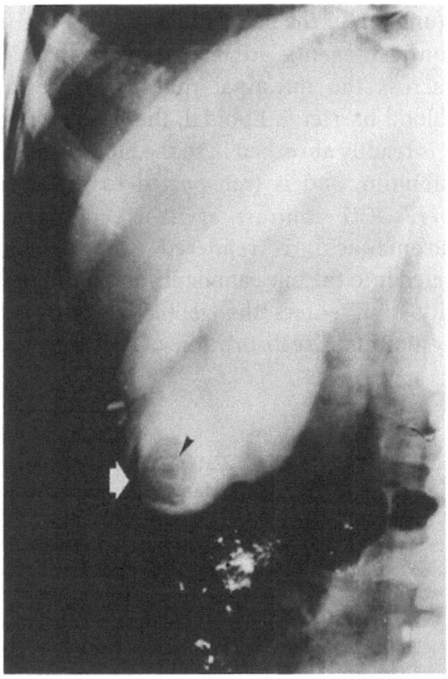

Fig. 4.2. Oral cholecystogram (OCG). A single radiolucent gallstone (*arrow*) with a central opaque component (*arrowhead*)

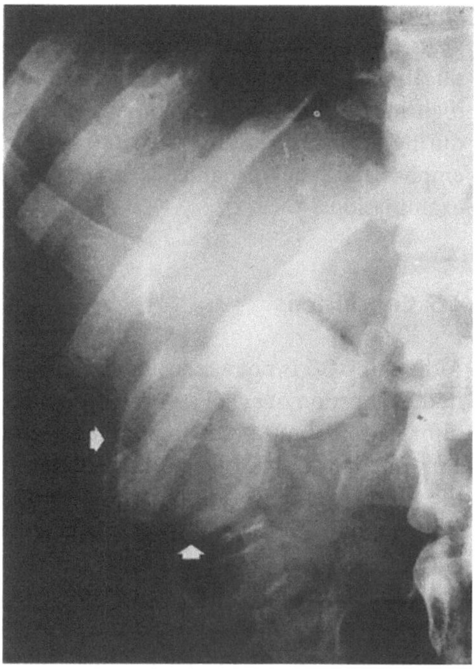

Fig. 4.3. OCG. Large stone (*arrows*) in the gallbladder fundus which prevents complete opacification

terol (rim calcification more frequent) (ZEMAN and BURREL 1987). OCG aids in the determination of stone composition (BELL et al. 1975), suggesting a cholesterol composition in the case of multiple stones that are radiolucent without central calcification and are floating (Fig. 4.4). The cholesterol stones may have a calcified rim. The floating or buoyancy is the only reliable sign of cholesterol stones in OCG, and is seen in only 35% of patients with such stones (DOLGIN et al. 1981).

Buoyancy is demonstrated more frequently in OCG images than on ultrasonography, presumably because the contrast agent increases the specific gravity of the bile produced, allowing the stones to float more freely (LEBENSART et al. 1984).

A specific but rare sign of cholesterol stones is the "Mercedes-Benz" sign, produced by gas-containing fissures within gallstones (MEYERS and O'DONOHUE 1973). The fissures result from shrinkage of the cholesterol crystals composing the gallstones and are filled with nitrogen.

The accuracy of OCG in detecting gallstones is around 85%–90% (BURHENNE 1975). A technically good OCG is theoretically superior in sensitivity and in specificity to ultrasound in the diagnosis of gallbladder disease (KROOK et al. 1980); however, the OCG images must be of excellent quality and must be interpreted properly. Moreover, the rate of nonvisualization of the gallbladder in OCG is 30% (Rome Group for the Epidemiology and Prevention of Cholelithiasis 1987), and this nonvisualization cannot be assumed to be an indirect sign of gallstones (MUJAHED et al. 1974). Therefore, in the diagnosis of gallstones OCG is used predominantly to complement ultrasonagraphy when the latter has failed to demonstrate evidence of gallbladder disease in the presence of strong clinical symptoms, or when ultrasonography has shown nonspecific abnormalities (AMBERG and LEOPOLD 1988). OCG is also indicated for counting and sizing of stones and fragments after ESWL, and for assessing cystic duct patency, because its resolution is better than that of ultrasonography (TORRES and STEINBERG 1989; MATHIESON et al. 1990; SIMEONE et al. 1989).

4.3.2 Adenomyomatosis and Cholesterolosis

Adenomyomatois and cholesterolosis are both forms of hyperplastic cholecystosis and appear as radiolucent images in the lumen of the gallbladder (Fig. 4.5) or as stenosis of various portions of the gallbladder (Fig. 4.6). In many cases hypertrophy of the Rokitansky-Aschoff sinuses (pseudodiverticular pattern) is seen, characteristic of adenomyomatosis (Fig. 4.7). Radiographs obtained after a fatty meal will also help to show these findings.

In recent reports OCG has been considered more sensitive and accurate than ultrasonography in detecting benign disorders of the gallbladder, especially adenomyomatosis (GELFAND et al. 1988; MAGLINTE et al. 1991).

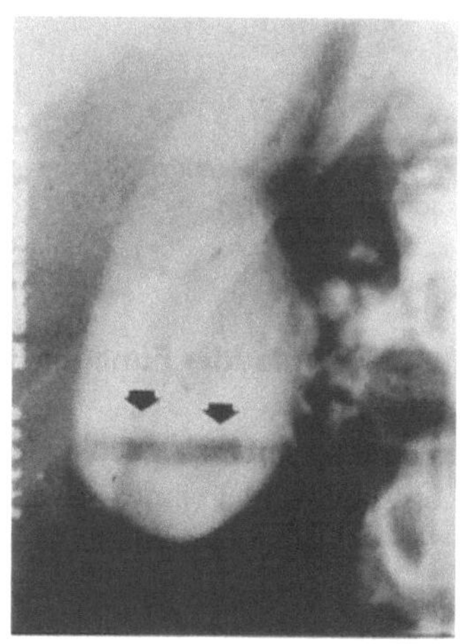

Fig. 4.4. OCG. Multiple cholesterol stones (*arrows*) floating in the gallbladder because they are lighter than opacified bile

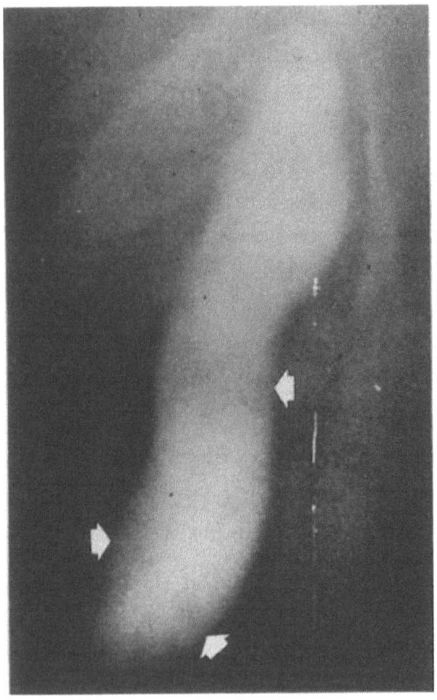

Fig. 4.5. OCG. Multiple filling defects of the gallbladder (*arrows*) that do not move with changing decubitus positions of the patient, due to diffuse cholesterolosis

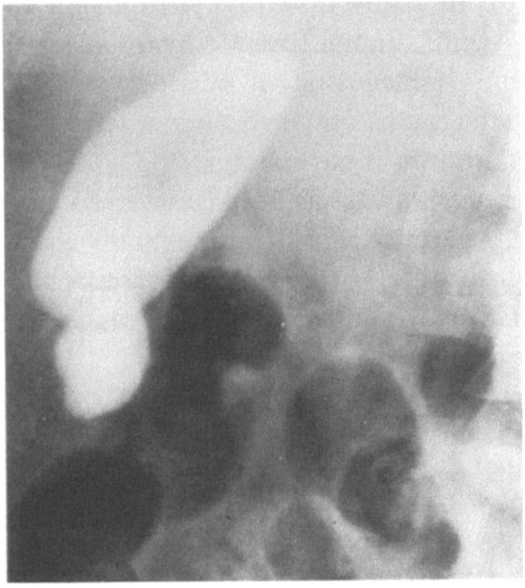

Fig. 4.6. OCG. Stenosis of the gallbladder fundus due to adenomyomatosis

4.4. Gallbladder Function

The assessment of gallbladder function on the basis of visualization vs nonvisualization has been shown to be a predictor of the risk of developing future complications of gallstones (WENCKERT and ROBERTSON 1966). Complications of gallstones have been shown to occur nearly twice as frequently in patients with nonvisualized gallbladder, and patients above 60 years of age in whom the gallbladder is not visualized are four times more likely to develop complications than those below this age. Considering the high incidence of incidental discovery of gallstones during ultrasonography, MAGLINTE et al. (1991) emphasized the prognostic value of the OCG examination.

4.5 Conclusions

Although OCG is not as sensitive as ultrasonagraphy in the detection of small stones, it provides adequate information about cystic duct patency and gallbladder function and allows accurate determination of the number and size of stones. In some instances, stone composition can also be predicted accurately by OCG and plain radiography of the abdomen. Although, of the treatment methods for gallstones presently undergoing clinical evaluation that do not require removal of the gallbladder, many may be rendered obsolete by the recent introduction of laparoscopic cholecystectomy, the approval in 1988 by the United States Food and Drug Administration of ursodeoxycholic acid for oral dissolution of cholesterol gallstones will ensure the continued use of OCG (GELFAND et al. 1988; MAGLINTE et al. 1991). Unfortunately other authors disagree and belive the replacement of the OCG with ultrasonography in the

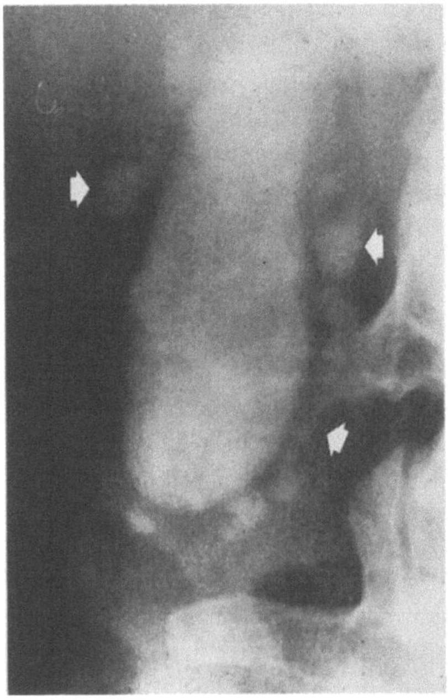

Fig. 4.7. Multiple diverticula due to hypertrophy of the Rokitansky-Aschoff sinuses (*arrows*)

majority of the circumstances (Gelfand et al. 1988; Mathieson et al. 1989).

References

Amberg JR, Leopold GR (1988) Is oral cholecystography still useful? AJR 151:73–74

Bell GD, Dowling RH, Whitney B, Sutoz DJ (1975) The value of radiology in predicting gallstone type when selecting patients for medical treatment. Gut 16:359–364

Berk RN, Lasser EC (1964) Altered concepts of the mechanism of nonvisualization of the gallbladder. Radiology 82:296–302

Brakel K, Laméris JS, Nijs GT, Ginai AZ, Terpstra T (1992) Accuracy of ultrasound and oral cholecystography in assessing the number and size of gallstones: implications for non-surgical therapy. Br J Radiol 65:779–783

Burhenne HJ (1975) Problem areas in the biliary tract. Curr Probl Radiol 5

Dolgin SM, Schwartz JS, Kressel HY (1986) Identification of patients with cholesterol or pigment gallstones by discriminant analysis of radiographic features. N Engl J Med 304:808–811

Donner MW, Weiner S (1964) Diagnostic evaluation of abdominal calcifications in acute abdominal disorders. Radiol Clin North Am 2:145–147

Edholm P, Jacobson B (1972) Quantitative determination of iodine in vivo. Acta Radiol 52:337–346

Gelfand DW, Wolfman NT, Ott DJ, Watson NE Jr, Chen YM, Dale WJ (1988) Oral cholecystography vs gallbladder sonography: a prospective, blinded reappraisal. AJR 151:69–72

Graham EA, Cole WH, Copher GH (1924) Visualization of the gallbladder by the sodium salt of the tetrabromophenolphthalein. JAMA 82:177–178

Hamilton S (1989) Gastroduodenal retention of Telepaque: a useful sign in oral cholecystography. Irish Med J 82:28–29

Harned RK, Le Ven RS (1978) Preliminary abdominal films in oral cholecystography: are they necessary? AJR 130:477–479

Ivy AC (1934) The physiology of the gallbladder. Physiol Rev 14:1–102

Jaffe H, Wachoski TJ (1942) Relation of density of cholecystographic shadows on the gallbladder to the iodine content. Radiology 28:43–46

Koehler R, Fabrikant JI, Dana ER (1960) Gastric retention during oral cholecystography due to underlying lesions of the stomach and duodenum. Surg Gynecol Obstet 110:409–412

Krook PM, Allen FH, Bush WH, Malmer G, McLean MD (1980) Comparison of real-time cholecystosonography and oral cholecystography. Radiology 135:145–148

Lebensart P, Bloom RA, Meretyk S, Landau E, Shiloni E (1984) Oral cholecystosonography: a method for facilitating the diagnosis of cholesterol gallstones. Radiology 153:255–256

Maglinte DT, Torres WE, Laufer I (1991) Oral cholecystography in contemporary gallstone imaging: a review. Radiology 178:49–58

Marzio L, Innocenti P, Genovesi N, Di Felice F, Napolitano AM, Costantini R, Di Giandomenico E (1992) Role of oral cholecystography, real-time ultrasound, and CT in evaluation of gallstones and gallbladder function. Gastrointest Radiol 17:257–261

Mathieson JR, So BC, Malone DE, Becker CD, Burhenne HJ (1989) Accuracy of sonography for determining the number and size of gallbladder stones before and after lithotripsy. AJR 153:977–980

Mathieson JR, So CB, Malone DE, Becker CD, Burhenne HJ (1990) Accuracy of ultrasound imaging: implications for biliary extracorporeal shock wave lithotripsy. In: Burhenne JH, Paumgartner G, Ferrucci JT (eds) Biliary lithotripsy II. Year Book Medical, Chicago, pp 45–49

Meyers MA, O'Donohue N (1973) The Mercedes-Benz sign: an insight into the dynamics of formation and disappearance of gallstones. AJR 119:63–66

Mujahed Z (1976) Factors interfering with the opacification of a normal gallbladder. Gastrointest Radiol 1:183–185

Mujahed Z, Evans JA, Whalen JP (1974) The nonopacified gallbladder on oral cholecystography. Radiology 112:1–7

Nelson SW (1972) A crescent-shaped collection of residual cholecystography contrast material: a new sign of benign gastric ulcer? AJR 116:293–303

Ochsner SF (1964) Esophageal diverticulum as cause for unsuccessful cholecystography. AJR 91:866–868

Ochsner SF, Buchtel BC (1967) Nonvisualization of gallbladder cuased by hiatal hernia. AJR 101:589–591

Perrin RL, Boehnke M (1974) Collection of cholecystographic contrast material in a large ulcerating leiomyoma of the stomach. Gastroenterology 66:601–603

Rome Group for the Epidemiology and Prevention of Cholelithiasis (GREPCO) (1987) Radiologic appearance of gallstones and its relationship with biliary symptoms and awareness of having gallstones: observations during epidemiological studies. Dig Dis Sci 32:349–353

Salzman E, Warden MR (1958) Telepaque opacification of radiolucent biliary calculi. Radiology 71:85–88

Sickles EA (1977) Cholecystographic diagnosis of duodenal ulcer: the incomplete ring sign. Radiology 124:27–30

Simeone JF, Mueller PR, Ferrucci JT (1989) Non-surgical therapy of gallstones: implications for imaging. AJR 152: 11–17

Teplick JG, Adelman BP (1955) Retention of the opaque medium during cholecystography. AJR 74:256–261

Torres WE, Steinberg HS (1989) The use of gallbladder ultrasonography and oral cholecystography in the evaluation of potential candidates for biliary lithotripsy. Presented at the Annual Meeting of the American Roentgen Ray Society, New Orleans, May 8–11

Trotman BW, Petrella EJ, Soloway RD, Sanchez HM, Morris TA, Miller WT (1975) Evaluation of radiographic lucency or opaqueness of gallstones as a means of identifying cholesterol or pigment stones. Gastroenterology 68:1563–1566

Wenckert A, Robertson B (1966) The natural course of gallstone disease: 11 year review of 781 non operated cases. Gastroenterology 50:376–381

Zeman RK, Burrel MI (1987) Gallbladder and bile duct imaging: a clinical and radiologic approach, Churchill Livingstone, New York

5 Ultrasonography of the Gallbladder

M. Bazzocchi, C. Zuiani, A. Rigamonti, and E. Quaia

CONTENTS

5.1 Introduction

The value of ultrasonography compared with oral cholecystography in the diagnosis of gallstones has been addressed in many papers (Goldberg 1974; Lawson 1977; Anderson and Harned 1977; Arnon and Rosenquist 1976; Cammisa and Armillotta 1981; Crade et al. 1976; Croce et al. 1981; Crow et al. 1976; Hessler et al. 1981; Marano et al. 1981). At the present date, it is universally agreed that ultrasonography should be the first examination for the diagnosis of gallstones. Moreover, ultrasound is useful in the diagnosis of acute and chronic inflammatory disease (Handler 1979; Kane 1980; Marchal 1979, 1980; Findberg and Birnholz 1979; Gonzales and Johnson 1978; Dalla Palma et al. 1979; Engel et al. 1980; Raghavendra et al. 1981a; Ruhe et al. 1979), benign noninflammatory disease, and malignant tumors of the gallbladder (Sanders 1980; Yum and Fink 1980, Carter et al. 1978; Dalla Palma et al. 1980; Ruhe et al. 1979).

Recently the introduction of new methods, surgical and nonsurgical, for treatment of gallstones has caused the exact role of ultrasonography to be reassessed. Furthermore, technological developments - Doppler, color, and power Doppler - have brought changes to the contribution that can be made by ultrasonography.

5.2 Gallstones

The typical sonographic appearance of a gallbladder calculus is an intraluminal echo casting an acoustic shadow, which is important for a high degree of confidence in the diagnosis of cholelithiasis. Small calculi may not produce acoustic shadowing, especially with poor focusing of the ultrasonic beam, as occurs with low-frequency transducers (Fig. 5.1). The presence of the shadow does not depend on the composition of the stones, but rather on their combined ability to both reflect and absorb the majority of the sound energy (Filly et al. 1979; Taylor et al. 1979) (Fig. 5.2). Even if the correct focusing and frequency are employed, many small calculi do not throw an acoustic shadow. However, if they can be shown to move when the patient changes position, they can confidently be diagnosed as calculi (the "rolling stone" sign; Fig. 5.1).

On the other hand, in many cases the gallstones may be fixed in the gallbladder infudibulum or in the cystic duct; this pattern is associated with a hydropic gallbladder and pain in the right upper quadrant of the abdomen (Fig. 5.3). Most gallstones lie in the dependent portion of the gallbladder; very rarely, the stones will have a lower specific gravity than the bile and will float (buoyancy sign). This sign is more likely to occur if the patient has received oral cholecystographic contrast agents, which increase the specific gravity of the bile (Fig. 5.4).

A focal area of echoes with acoustic shadowing in the region of the gallbladder bed increases the likeli-

M. Bazzocchi, MD, Director, Cattedra di Radiologia, Policlinico Universitario, Piazzale S. Maria della Misericordia 33100 Udine, Italy
C. Zuiani, MD, Cattedra di Radiologia, Policlinico Universitario, Piazzale S. Maria della Misericordia, 33100 Udine, Italy
A. Rigamonti, MD, Cattedra di Radiologia, Policlinico Universitario, Piazzale S. Maria della Misericordia, 33100 Udine, Italy
E. Quaia, MD, Cattedra di Radiologia, Policlinico Universitario, Piazzale S. Maria della Misericordia, 33100 Udine, Italy

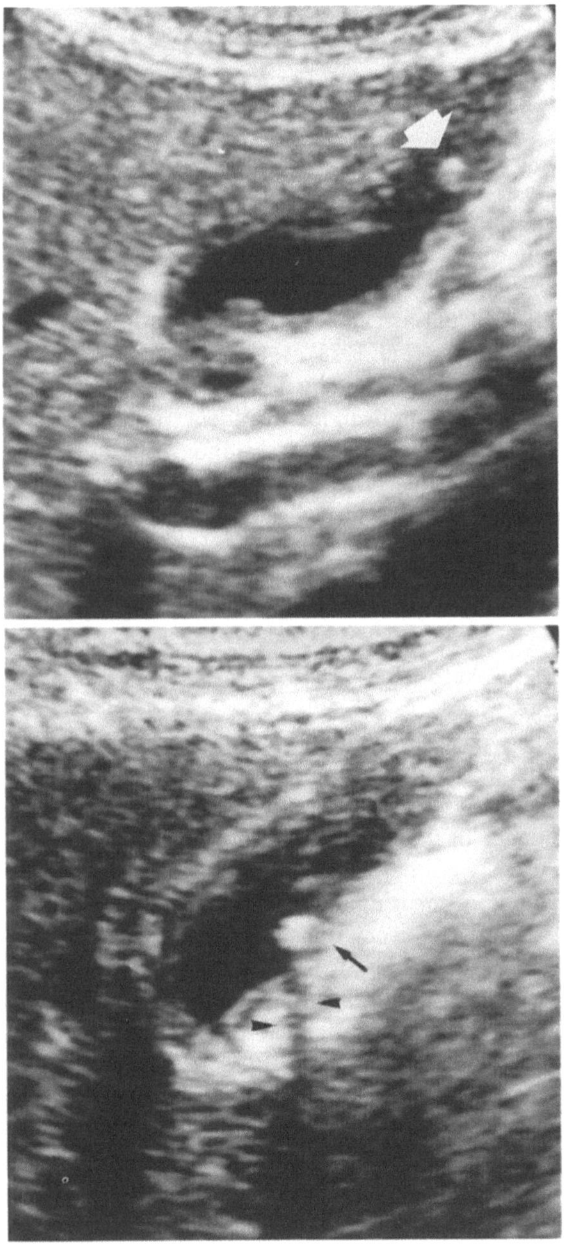

a

b

Fig. 5.1a,b. Gallstones. **a** A Small gallstone (*arrow*) in the gallbladder fundus produces no acoustic shadow. **b** When the patient changes position, the stone (*arrow*) rolls into the body of the gallbladder and now produces acoustic shadow (*arrowheads*) with better focusing

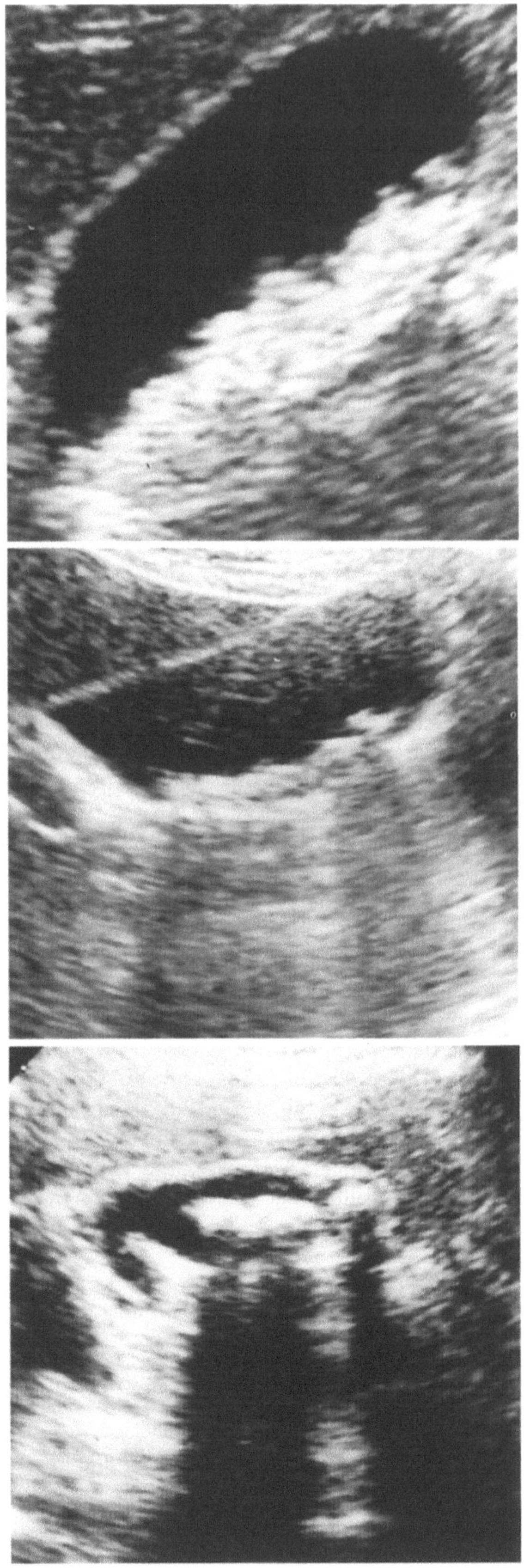

a

b

c

Fig. 5.2a–c. Gallstones. **a** Multiple stones without acoustic shadow. **b** Multiple stones with a weak acoustic shadow. **c** Three stones larger than those in **a** and **b** with a clear acoustic shadow

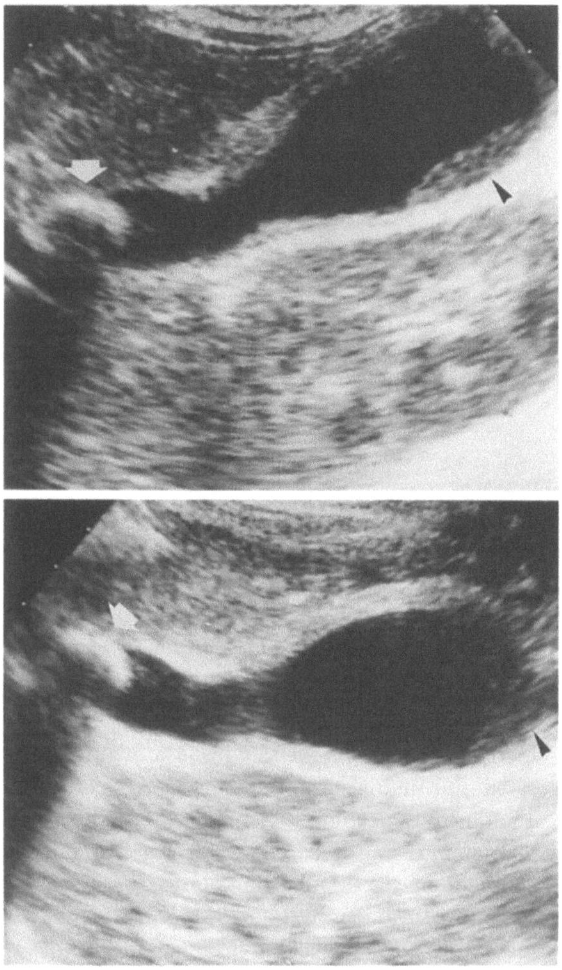

a

b

Fig. 5.3a,b. Gallstones. **a** With the patient in the supine position a large stone (*arrow*) is seen in the gallbladder infundibulum; in the fundus there is echogenic material (debris, *arrowhead*). **b** When the patient is standing erect, the stone remains fixed in the infundibulum, whereas the debris moves (*arrowhead*)

hood that there is a contracted gallbladder containing calculi. However, gas in the duodenal bulb or hepatic flexure of the colon may simulate this appearance. A less common cause of this appearance is air within the gallbladder lumen, associated with either emphysematous cholecystitis or gallstone ileus. If the gallbladder wall is calcified ("porcelain gallbladder") the appearance is similar too (KANE et al. 1984). The most specific sign of a contracted, stone-filled gallbladder is hypoechoic wall superficial to a curvilinear echo from the stones and an acoustic shadow: the WES triad (Wall, Echo, Shadow) or the "double arc-shadow sign" (MACDONALD et al. 1981; RAPTOPOULOS et al. 1982).

Ultrasonography has been shown to have an accuracy of 96% in the diagnosis of gallstones (COOPERBERG and GIBNEY 1987). Ultrasonography is widely accepted to be 15%–20% more sensitive than oral cholecystography (COOPERBERG and BURHENNE 1980) and has a lower false-negative rate (LEOPOLD et al. 1976; CRADE et al. 1976), especially in the detection of stones measuring in the range of a few millimeters. However, some authors had shown that measurement of gallstones size on ultrasound scans is accurate when the stones are smaller than 2 cm, but is less so if the stones are larger than 2 cm (SIMEONE et al. 1989a). Furthermore, on ultrasound scans gallstones will often be contiguous or overlap, making their size and number difficult to assess (BRINK et al. 1989, SIMEONE et al. 1989a). Other authors maintain that ultrasonography can be used to count and measure gallstones and postlithotripsy fragments as accurately as oral cholecystography and CT (MATHIESON et al. 1989; MARZIO et al. 1992). The most difficult problem encountered with ultrasound was in calculating the size of those gallstones not surrounded by bile, but the error was small and relatively unimportant (MARZIO et al. 1992).

5.3 Acute Cholecystitis

In cases of acute right upper quadrant pain the diagnosis of cholecystitis is important since most surgeons now favor early cholecystectomy (VAN DER LINDEN and SUNZEL 1970; SALTZSTEIN et al. 1983). Ultrasonography is complementary to the clinical and laboratory data, and it is particularly helpful when the clinical pattern is ambiguous. The difficul-

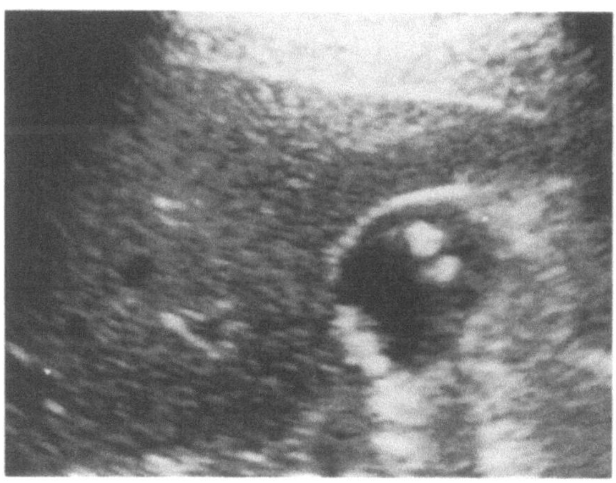

Fig. 5.4. Gallstones. Some stones are in the most sloping portion of the gallbladder, while two stones are floating because they are of a lower specific gravity than the bile

ties in the differential diagnosis between acute cholecystitis, acute appendicitis, acute pancreatitis, and perforated gastroduodenal ulcer in some cases are known and have been reported in a large series (STANILAND et al. 1972). In addition, more than half of the patients with clinical features suggestive of acute cholecystitis do not in fact have acute inflammation of the gallbladder (FREEMAN et al. 1982; LAING et al. 1981; SAMUELS et al. 1983; SHUMAN et al. 1982; SHUMAN et al. 1984). Further diagnostic difficulties consist in differentiating acute calculous from acute acalculous cholecystitis and in identifying the evolution of acute cholecystitis into empyema, ischemic necrosis, or perforation.

The sonographic signs of acute cholecystitis are (BAZZOCCHI et al. 1984a):

- *Gallstones* in 85%–90% of cases, which may be impacted in the neck or the cystic duct (Fig. 5.5). Stones in this location can frequently be missed since they are not surrounded by bile and may be situated as far as 1–2 cm from the bile-filled gallbladder (LAING and JEFFREY 1983).

  Acalculous cholecystitis most commonly occurs in critically ill patients, especially following surgery, trauma, or burns, or in association with hyperalimentation. AAC is said to account for 47% of cases of postoperative cholecystitis (GLENN and BECKER 1982; MacGAHAN and WALTER 1985). In these settings the complicated forms are more common.

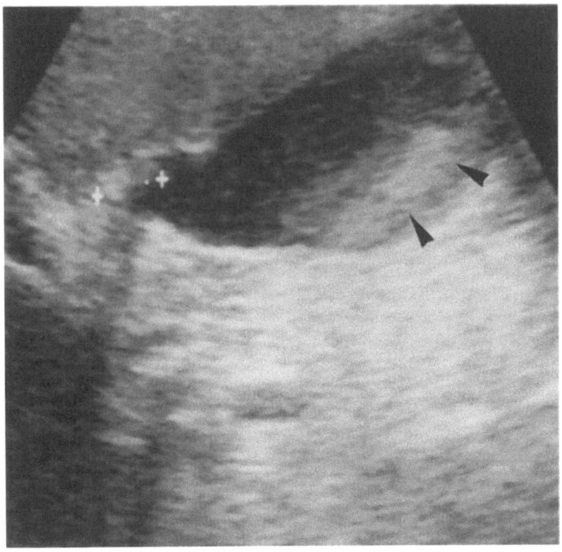

Fig. 5.5. Acute cholecystitis. The gallbladder is increasing in size because a stone has become wedged in the neck. Sludge can be seen in the gallbladder body (*arrowheads*)

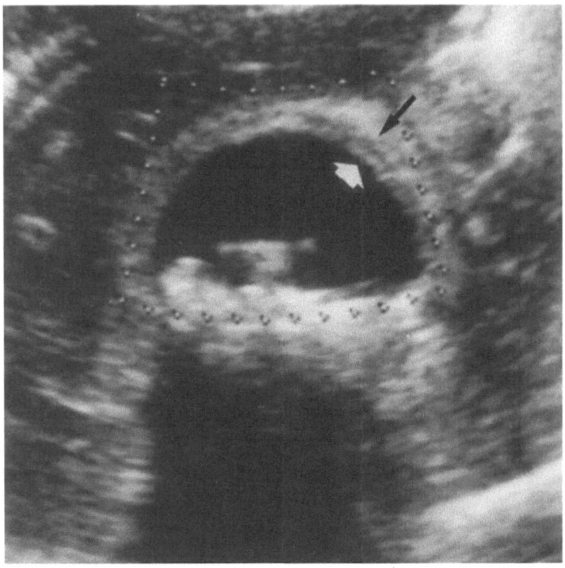

Fig. 5.6. Acute cholecystitis. The gallbladder shows uniform wall thickening (*arrows*) containing a hypoechoic band. In the lower portion of the gallbladder there are multiple stones

- *Increase in gallbladder size* in 58% of uncomplicated and in 76% of complicated cases. The gallbladder may become distended with a short-axis diameter greater than 4 cm; the majority of normal gallbladders are smaller than 4 cm across the short axis (RAGHAVENDRA et al. 1981b). Unfortunately this finding lacks specificity.
- *Wall thickening*, uniform in most uncomplicated cases (Fig. 5.6), not uniform in most complicated cases (64%). In the fasting patient the gallbladder wall usually measures less than 3 mm thick, but it may appear thicker when contracted (RAGHAVENDRA et al. 1981b). In acute cholecystitis, but also in many other conditions, the gallbladder wall may be thickened (BRIZZI et al. 1982) and contains a hypoechoic band (Fig. 5.6). In acute cholecystitis the sign is related to the edema and cellular infiltration of subserosa and submucosa (MARCHAL 1979). In complicated forms of acute cholecystitis (empyema, gangrenous cholecystitis) the thickened gallbladder wall shows discontinuous, irregularly margined hypoechoic and echogenic intramural bands and foci due to hemorrhage and intramural abscesses (BAZZOCCHI et al. 1984a) (Fig. 5.7). In cases of perforation a fluid collection may be seen in the gallbladder bed or in the intraperitoneal cavity (pericholecystic abscesses) (TAKADA et al. 1989) (Fig. 5.8). Distinguishing between the two is important to determine the proper treatment: elec-

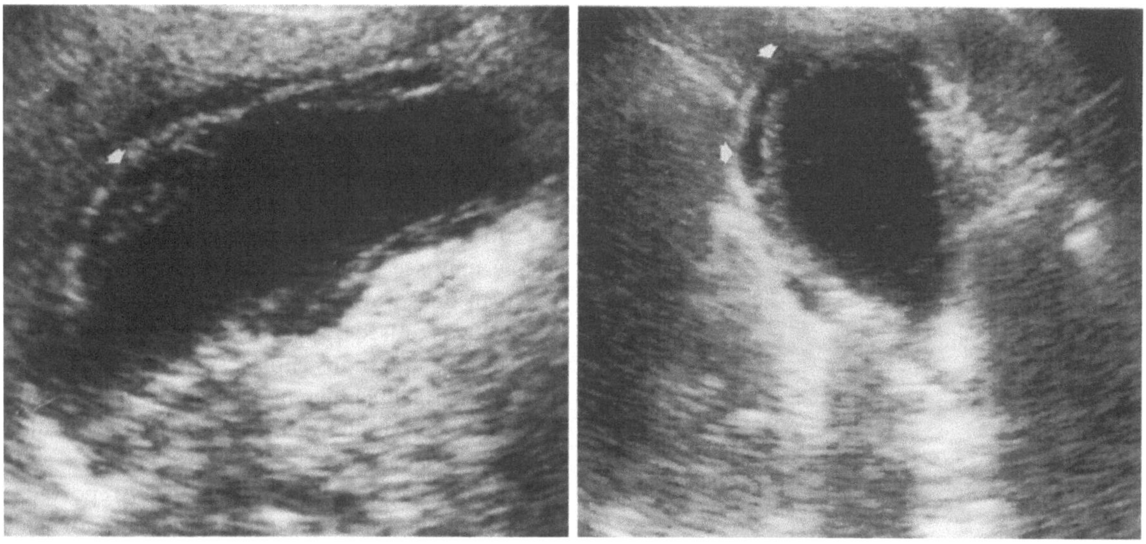

Fig. 5.7. Acute cholecystitis, complicated form. The gallbladder wall is irregularly thickened because of multiple intramural bands, hemorrhagic foci, and intramural abscesses (*arrows*)

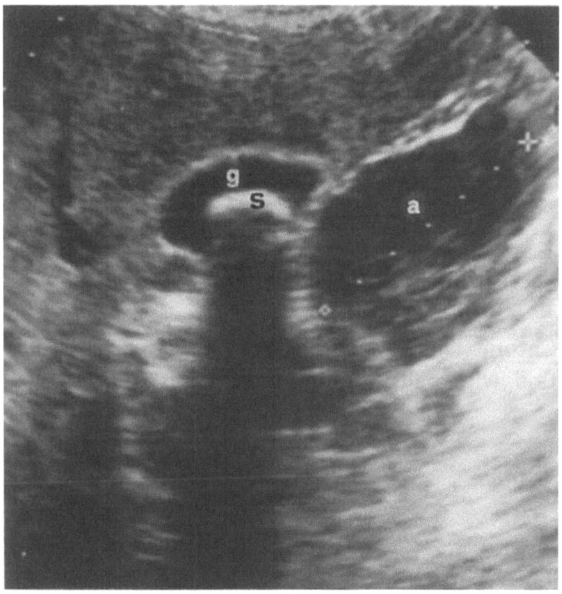

Fig. 5.8. Acute cholecystitis. Pericholecystic abscess (*a*) due to perforation of the gallbladder (*g*). *S*, Stone

tive surgery in cases of abscess of gallbladder bed, and emergency surgery or ultrasound-guided drainage followed by surgery in cases of abscess of intraperitoneal cavity (TAKADA et al. 1989).

The presence of gas in the gallbladder wall and/or lumen may be seen in emphysematous cholecystitis (effervescent gallbladder) (NEMCEK et al. 1988).

Thickening of the wall of the gallbladder is particularly evident in AIDS, varying from 4 to 15 mm (ROMANO et al. 1988) (Fig. 5.9). This finding in AIDS may indicate opportunistic infection of the biliary tract (ROMANO et al. 1988).

Color and power Doppler, recent introductions, may demonstrate hypervascularization of the gallbladder wall in acute cholecystitis (Fig. 5.10).

Many conditions unrelated to gallbladder disease may thicken the wall of the gallbladder, especially hepatitis, hypoalbuminemia, ascites (LEWANDOWSKI and WINSBERG 1981; SHALER et al. 1981) and dilatation of the cystic veins in portal hypertension. These conditions are associated

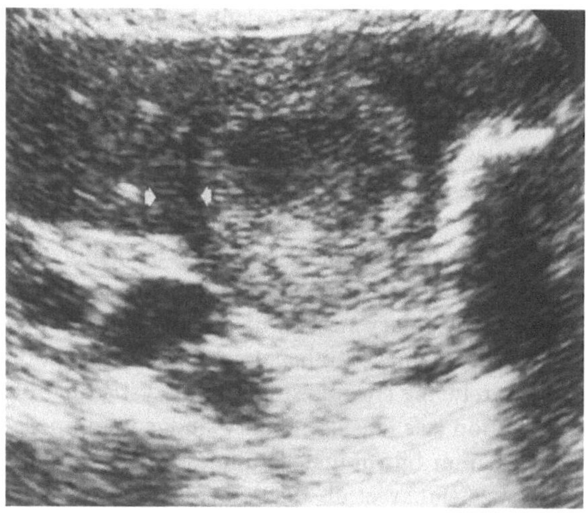

Fig. 5.9. Acute cholecystitis. Conspicuous thickening (*arrows*) of the gallbladder wall in an AIDS patient, due to opportunistic infection

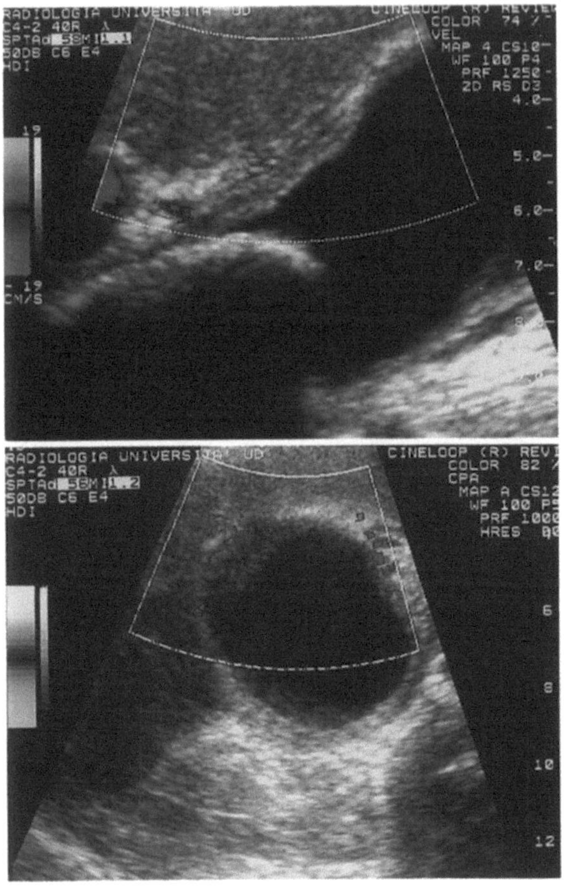

a

b

Fig. 5.10a,b. Acute cholecystitis. **a** Color Doppler flow mapping shows hypervascularization of the gallbladder wall. **b** Using the power Doppler technique, this pattern is particularly conspicuous

with a single hypoechoic zone between two relatively echogenic layers (Cohan et al. 1986). Nonetheless, the combined finding of gallstones with thickening of the gallbladder wall in a patient with the appropriate clinical presentation has a high positive predictive value, in excess of 90%, for the diagnosis of acute cholecystitis (Ralls et al. 1985).

- *Sludge and debris* in 25% of uncomplicated cases and 85% of complicated cases. The gallbladder sludge is nonshadowing echogenic material caused by biliary stasis, which may be due to prolonged fasting, extrahepatic biliary obstruction, or, rarely, the chronic situation of cystic duct obstruction (see Fig. 5.5). Biliary sludge is composed of calcium bilirubinate granules and cholesterol crystals (Filly et al. 1980). Typically, sludge appears as homogeneous low-amplitude-dependent echoes which tend to form a fluid-fluid level that moves very slowly when the patient changes position (Cover et al. 1985) (Fig. 5.11). If the echogenic material completely fills the gallbladder lumen, the gallbladder becomes similar in appearance to the adjacent liver ("hepatization of the bile"; Fig. 5.12) or even more echogenic than the liver. Occasionally the sludge forms round, echogenic masses ("sludge balls"; Fig. 5.13).

  In the complicated forms of acute cholecystitis one may see echogenic material without a level, which does not move or moves reluctantly. This finding is caused by pus, debris, and fibrinous exudate. In gangrenous cholecystitis one may see the sloughed gallbladder mucosa as a thin membranous echo parallel to the wall of the gallbladder (Jeffrey et al. 1983).
- *Sonographic Murphy's sign* has been reported as having a high positive predictive value for acute cholecystitis, greater than 90% (Ralls et al. 1985). The sonographic Murphy's sign is defined as the presence of maximal tenderness elicited by direct pressure of the transducer over a sonographically located gallbladder. This sign is present in only 33% of cases of gangrenous cholecystitis; for that reason Simeone et al. (1989b) suggest that the absence of Murphy's sign increases the possibility of gangrenous cholecystitis in patients with abdominal pain and sonographic findings of cholecystitis.

5.4 Chronic Cholecystitis

Chronic cholecystitis has been observed in association with gallstones in most cases (Kane 1980).

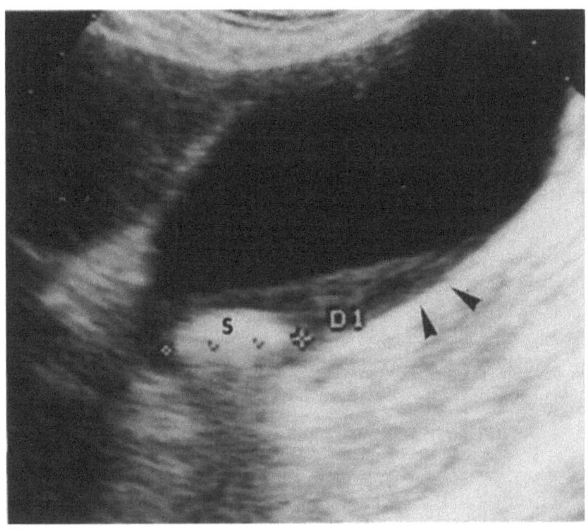

Fig. 5.11. Acute cholecystitis. In the lower portion of the gallbladder there are homogeneous low-amplitude echoes forming a fluid-fluid level (sludge, *arrows*) S, Stone

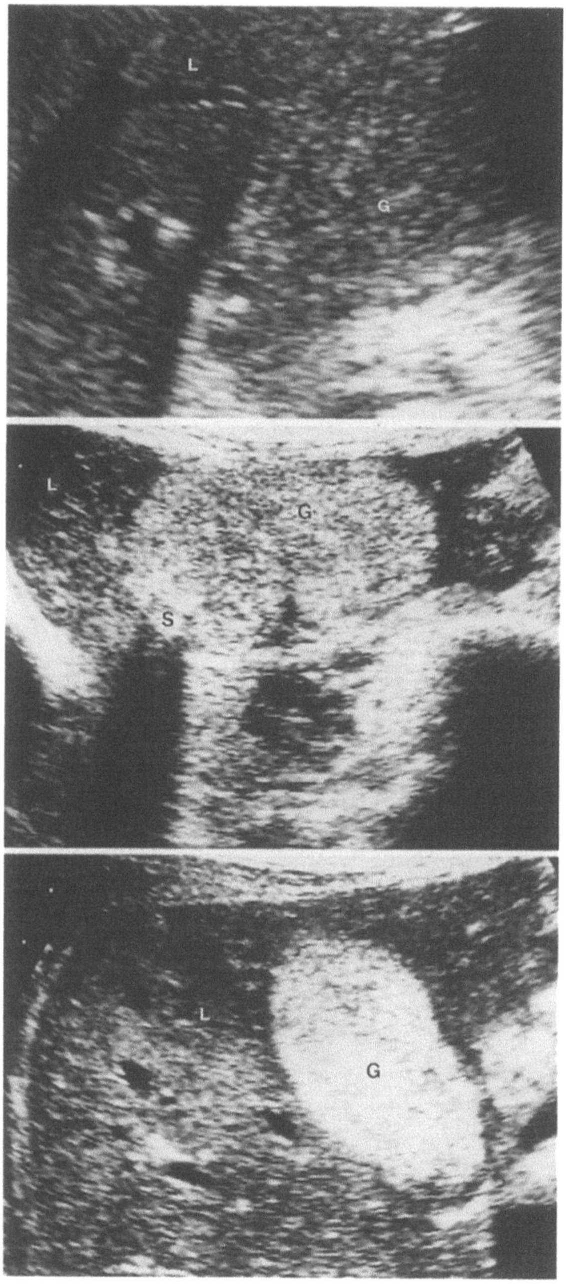

a

b

c

Fig. 5.12a–c. Acute cholecystitis. The sludge completely fills the gallbladder lumen. Its echogenicity may vary: in **a**, it is similar to that of the adjacent liver ("hepatization" of the bile), while in **b** and **c**, it is more echogenic than liver. *L*, liver; *G*, gallbladder; *S*, stone

Intraluminal nonshadowing echogenic material, corresponding to biliary sludge, is seen in 26% of cases (FINDBERG and BIRNHOLZ 1979). In 40% of cases of chronic cholecystitis and in 69% of cases of reacutizing chronic cholecystitis we have observed thickening of the wall of the gallbladder (>4 mm). In these cases one may see a single hypoechoic zone in the wall. This finding is related in chronic cholecystitis to glassy and myxoid degeneration in a perimuscular ring, or to fatty infiltration (BAZZOCCHI et al. 1984b) (Fig. 5.14c). In reacutizing chronic cholecystitis, the hypoechoic zone in the wall of the gallbladder is due to edema, phlogistic infiltration, or hemorrhage (BAZZOCCHI et al. 1984b) (Fig. 5.14a,b).

5.5 Cholecystoses

Cholecystoses are a group of noninflammatory conditions of the gallbladder wall. Figures reported for the prevalence of the disease vary from 2% to 40% in autopsy studies (LUBERA et al. 1967; RAPACCINI et al. 1993).

Cholecystoses can be divided into hyperplastic and "thesaurismotic" or accumulating forms (JUTRAS et al. 1958, 1960). The hyperplastic forms are characterized by normal growth of the wall components, while in the accumulating forms deposits of organic and inorganic materials are seen in the gallbladder wall.

The most common forms of cholecystosis are: adenomyomatosis, cholesterolosis, and porcelain gallbladder.

Adenomyomatosis is the most frequently seen of the cholecystoses. It is characterized by segmental or diffuse thickening of mucosa and muscular wall (BERK et al. 1983) (Fig. 5.15). In the thickened wall, more specifically, small cystic spaces representing the dilated Rokitansky-Aschoff sinuses may be identified (KIDNEY et al. 1986) (Fig. 5.15b).

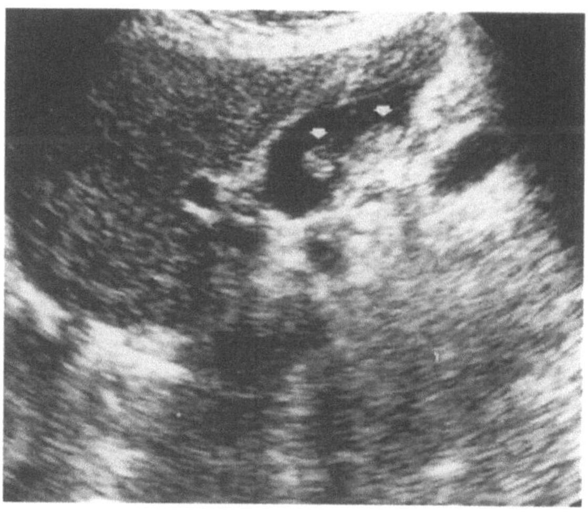

Fig. 5.13. The sludge may form round, echogenic masses – sludge balls (*arrows*)

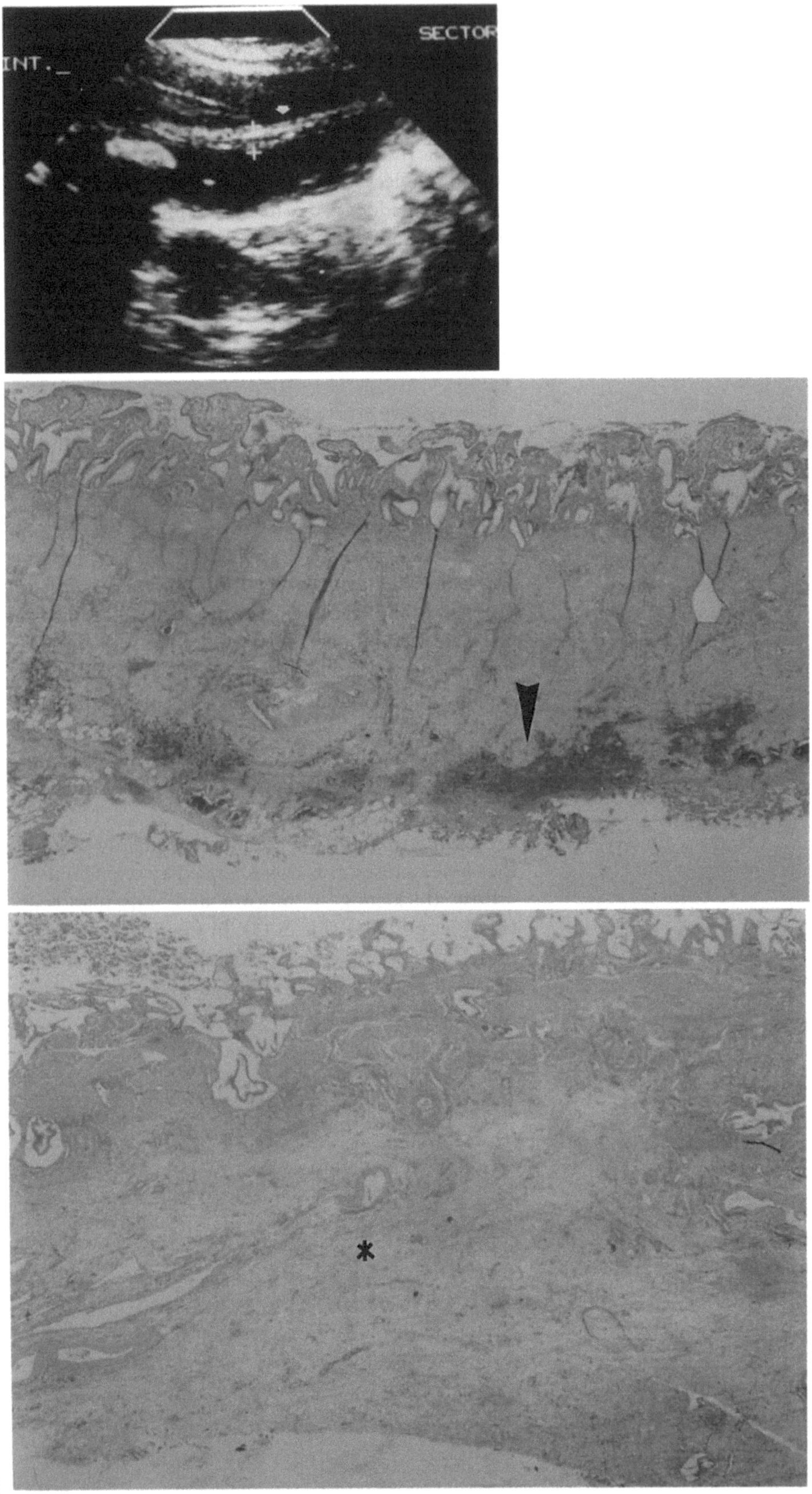

a

b

c

Fig. 5.14a–c. Chronic cholecystitis. **a** In reacutizing forms the gallbladder wall may be thick (*arrow*) with a hypoechoic band like acute cholecystitis. **b** Histologic specimen of the gallbladder wall (same case as **a**). This appearance is due to edema, phlogistic infiltration, or hemorrhage (*arrowhead*). **c** Histologic specimen from nonreacutizing chronic cholecystitis. The same appearance as in **b** may be caused, in nonreacutizing forms, by glassy and myxoid or fatty infiltration of the perimuscular ring (*asterisk*)

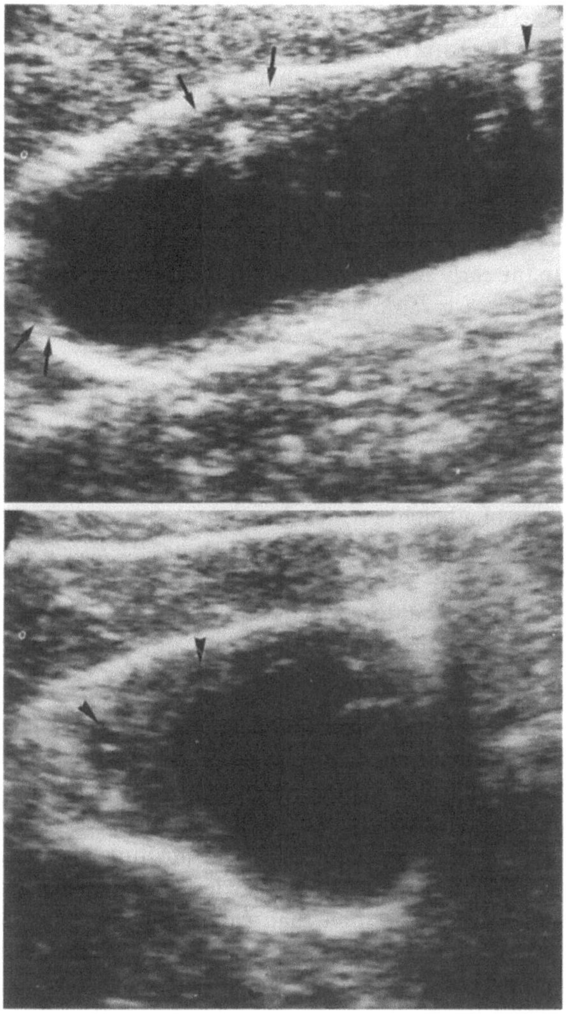

a

b

Fig. 5.15a,b. Cholecystoses: adenomyomatosis. The gallbladder shows: **a** thickening of the front wall (*arrows*) and the fundus (*arrows*) and **b** of the right portion. In the wall there are some cystic zones and echogenic spots with "comet tail artifacts," which correspond respectively to dilated Rokitansky-Aschoff sinuses and to wall stones (*arrowheads*)

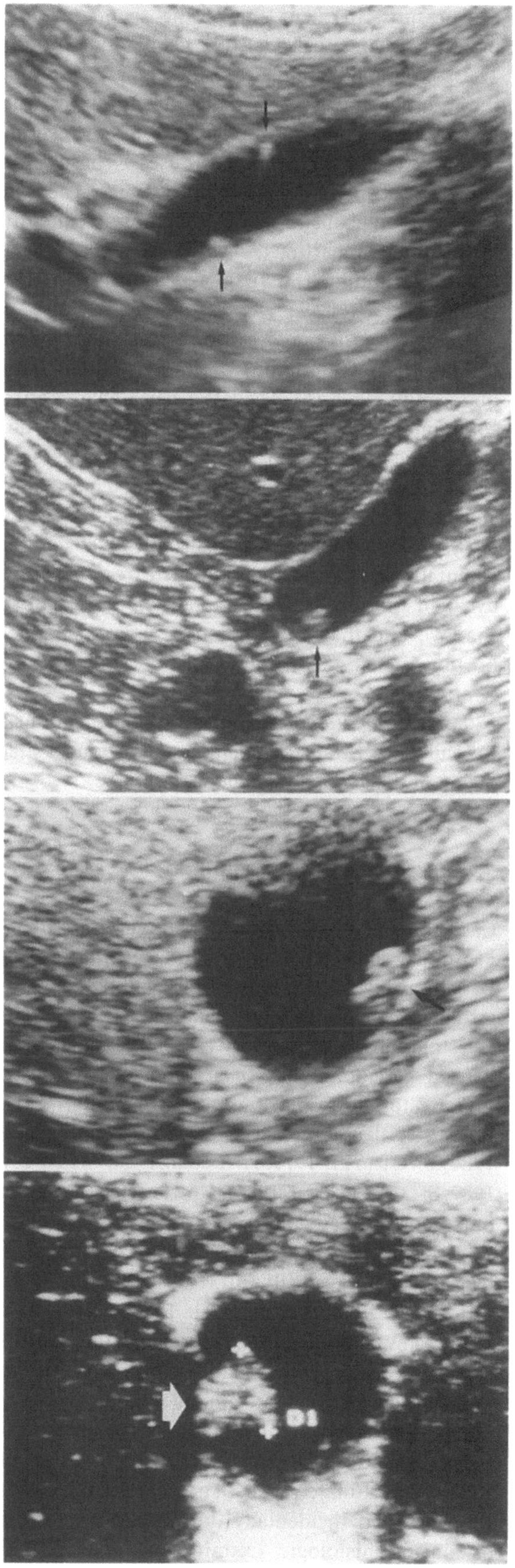

a

b

c

d

Fig. 5.16a–d. Cholecystoses: cholesterolosis. **a** Diffuse form: the internal surface of the wall is irregular. There are also two little polyps 1–2 mm in diameter (*arrows*). **b–d** Solitary cholesterol pseudopolyps (*arrows*) appearing as echogenic foci attached to a nondependent portion of the gallbladder. In **d**, the largest one caused some problems of differential diagnosis from carcinoma

The most frequent site of the segmental form of adenomyomatosis is the gallbladder fundus. The sonographic signs are an irregular contour of the internal mucosa and wall thickening, in which may be seen hypoechoic cyst-like zones corresponding to Rokitansky-Aschoff sinuses. These cystic zones may contain debris or little stones. The latter may have an acoustic shadow or, more frequently , the "comet tail artifact," which we believe to be a specific finding (Fowler and Reid 1988) (Fig. 5.15b).

Cholesterolosis may be diffuse and micronodular (nodules less than 1 mm) or focal (nodules larger than 3 mm; Fig. 5.16). The latter has a characteristic sonographic appearance and is called the polypoid form, which accounts for 90% of all gallbladders polyps. Rarely, the cholesterol pseudopolyp has a diameter larger than 1 cm (De Albertis et al. 1995; Fig. 5.16). These pseudopolyps are usually discovered incidentally and can be distinguished from gallstones by their fixed position, the absence of acoustic shadow, and their attachment to a nondependent portion of the gallbladder wall (Cooperberg and Gibney 1987).

Porcelain gallbladder consists of calcifications of the gallbladder wall; the calcified wall produces shadowing of the acoustic signal (see Sect. 5.2).

5.6 Carcinoma of the Gallbladder

Carcinoma of the gallbladder is not very rare (0.55% of autopsy subjects) (Piehler and Crichlow 1978). Usually, while in these cases the gallbladder is not visualized or is poorly visualized on oral cholecystography, a satisfactory rate of diagnosis of advanced carcinoma has been reported by various investigators who have used ultrasound (Dalla Palma et al. 1980; Ruiz et al. 1980; Weiner et al. 1984) or CT (Yeh 1979; Itai et al. 1980; Muto et al. 1987; Smathers et al. 1984). However, the prognosis of patients with advanced carcinoma of the gallbladder is still poor even if an aggressive surgical approach is employed (Bergdahl 1980; Piehler and Crichlow 1978; Araki et al. 1988; Nevin et al. 1976). Nevin et al. (1976) correlated the depth of cancer invasion with postoperative survival and found a high survival rate among patients whose tumor was confined to the mucosa or muscularis mucosae (American Joint Committee on Cancer 1988). Unfortunately, most such early-stage lesions were detected incidentally during cholecystectomy or during histologic examination of specimens resected for what had been diagnosed as benign disease (Bergdahl 1980; Solan and Jackson 1971).

The sonographic findings varying depending on whether the disease is at an early stage or advanced.

5.6.1 Early Stage

In the early stage there are four types of macroscopic forms of carcinoma of the gallbladder (Tsuchiya 1991):

1. The *pedunculated type*, often associated with cholesterosis. Tumor size ranges from 2–3 mm to 30 mm maximum diameter.
2. The *sessile type*. In this case the protruding lesion is of a height greater than 5 mm.
3. The *superficial raised type*. This has been seen as a smoothly elevated lesion rising 1.5–3 mm from the surrounding mucosa.
4. The *flat type*. These lesions are almost flush with the normal mucosa.

The most frequent appearance of early-stage cancer on ultrasonography is as a fungating mass consisting of a hyperechoic lesion protruding into the gallbladder lumen with various levels and shapes of echo (Fig. 5.17a). Unfortunately they resemble benign polypoid lesions, which are much more common than carcinoma (Jørgensen and Jensen 1990; Heyder et al. 1990). The most imporant criterion for the differential diagnosis is the volume of the polypoid mass. Although in some cases of early-stage cancer of the gallbladder the lesion is of a diameter smaller than 1 cm (Tsuchiya 1991), it seems practical to consider polypoid lesions of a diameter greater than 1 cm to be rapidly growing and they should be strongly suspected of malignancy (Deitch and Engel 1980; Koga et al. 1988). A helpful procedure is color Doppler flow mapping, which should show pathological findings in the gallbladder wall if tumoral vessels are present (Fig. 5.17b). Unfortunately, less than 10% of gallbladder cancers are diagnosed at the T1 stage (Kuo et al. 1990).

5.6.2 Advanced Stage

In some cases of advanced gallbladder cancer an inhomogeneous echogenic mass is seen, seldom with acoustic shadowing, allowing a normal shape of the gallbladder and normal adjacent liver parenchyma to be recognized (Yeh 1979) (Fig. 5.17). In other cases

a c b d

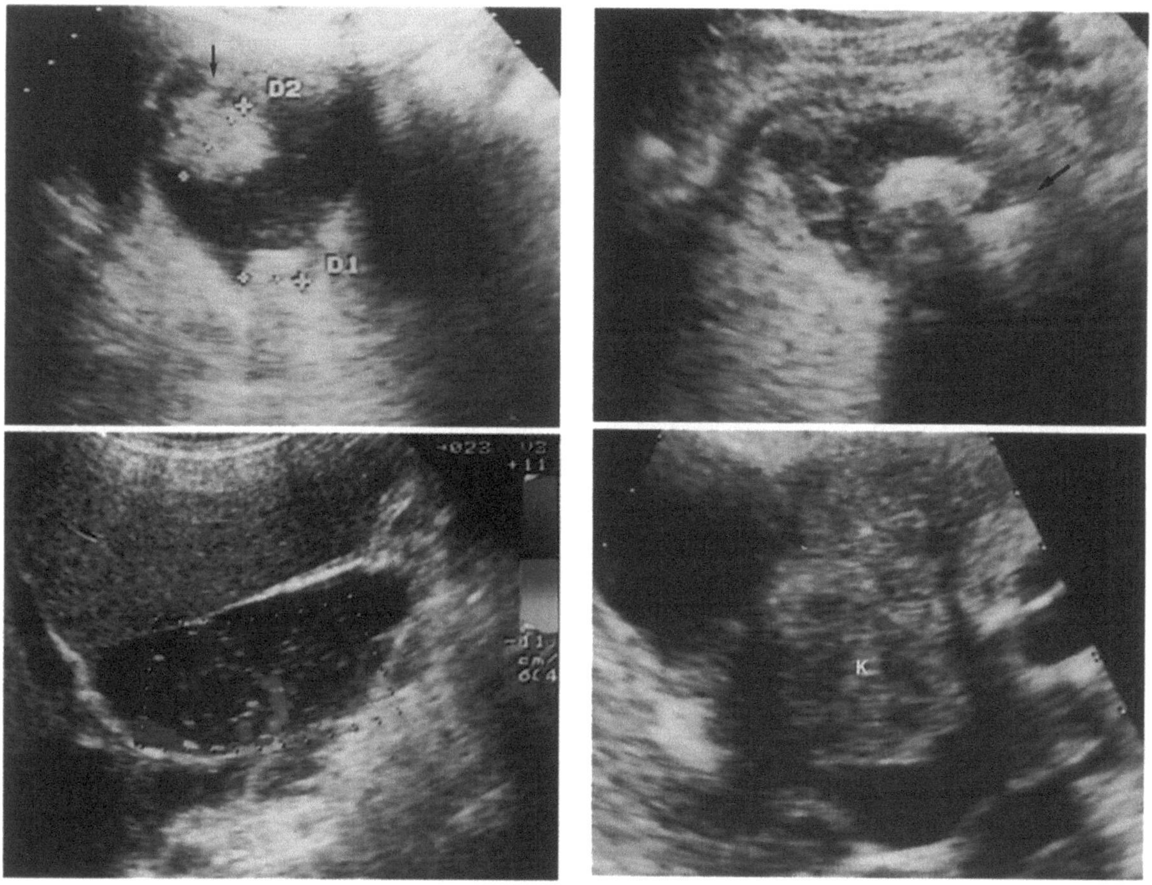

Fig. 5.17a–d. Carcinoma of the gallbladder. **a** Early-stage cancer: a fungating mass (*arrow*). Diagnostically, this is difficult to differentiate from cholesterolosis polyps. **b** Advanced-stage cancer. Color Doppler shows rich vascularization in the mass in the visceral lumen. **c** Advanced-stage cancer. The tumor occupies the visceral wall, extending to the liver bed (*arrows*). **d** Advanced-stage cancer. A large mass (*K*) extends from the gallbladder lumen to the adjacent liver parenchyma

the tumor may be so large as to occupy the entire visceral lumen, and may infiltrate the adjacent liver parenchyma. In these cases the tumor appears as a subhepatic mass and the shape of the gallbladder is rarely recognizable.

Liver and nodal metastases are often present at this stage.

Gallstones are present in 73%–88% of cases of carcinoma of the gallbladder, and a porcelain gallbladder in up to 25% of cases (Kane et al. 1984; Weiner et al. 1984).

The accuracy of ultrasonography in detecting large tumors of the gallbladder has varied between reports from 50% to 89% (Yeh 1979; Dalla Palma et al. 1980; Weiner et al. 1984). On the other hand, the incidence of sonographic findings that simulate gallbladder carcinoma is not known with any accuracy. Most reports mention the difficulties of differential diagnosis between inflammatory changes of the gallbladder, carcinoma of the pancreas, and carcinoma of the gallbladder (Dalla Palma et al. 1980; Soiva et al. 1987).

Unfortunately, Soiva et al. (1987) found a mean survival time after the diagnosis of carcinoma of the gallbladder, in patients with the largest tumors, of 172 days.

5.7 Conclusions

Ultrasonography remains today the principal tool in the diagnosis of pathological conditions of the gallbladder. In certain cases oral cholecystography (see Chap. 4) or CT and/or MRI may be used in the staging of malignancy.

References

American Joint Committee on Cancer (1988) Manual for staging of cancer, 3rd edn. Lippincott, Philadelphia

Anderson JC, Harned RK (1977) Gray scale ultrasonography of gallbladder: an evaluation of accuracy and report of additional ultrasonographic signs. AJR 129:975–977

Araki T, Hihora T, Karikomi M, Kachi K, Uchiyama G (1988) Intraluminal papillary carcinoma of the gallbladder: prognostic value of computed tomography and sonography. Gastrointest Radiol 13:261–265

Arnon JC, Rosenquist CJ (1976) Gray scale ultrasonography: an evaluation of accuracy. AJR 127:817–818

Bazzocchi M, Pozzi-Mucelli RS, Brizzi F, Ricci C, Dalla Palma L (1984a) Aspects échographiques des Maladies inflammatories de la vesicule biliaire. J Fr Echo 1:23–28

Bazzocchi M, Ricci C, Puhali N, Marconi R, Tuttobene P (1984b) Aspetti ecografici della colecistite cronica: correlazioni anatomo-istologiche. Ecomodena 2:31–36

Bergdahl L (1980) Gallbladder carcinoma first diagnosed at microscopic examination of gallbladder removed for presumed benign disease. Ann Surg 191:19–22

Berk RN, Van Der Vegt JH, Lichtenstein JE (1983) The hyperplastic cholecystoses: cholesterolosis and adenomyomatosis: Radiology 146:593–601

Brink JA, Simeone JF, Mueller PR, Saini S, Tung G, Spell NO, Ferrucci JT (1989) Routine sonographic techniques fail to quantify gallstone size and number: a retrospective study of 111 surgically proved cases. AJR 153:503–506

Brizzi F, Pozzi-Mucelli RS, Rizzato G, Maffesanti M, Bazzocchi M, Dalla Palma L (1982) Ultrasonographic aspects of inflammatory and neoplastic diseases of the gallbladder. Eur J Radiol 2:214–220

Cammisa M, Armillotta M (1981) Strategia di studio della calcolosi colecistica: ruolo dell' ecotomografia. Radiol Med 65:289–293

Carter SJ, Rutledge J, Hirsh JH, Vracko R, Chickos PM (1978) Papillary adenoma of the gallbladder ultrasonic demonstration: case report. JCU 6:377–382

Cohan RH, Mahony BS, Illescas FF, Baker ME, Bowie JD (1986) Sonographic evaluation of gallbladder wall thickening: analysis of specific patterns. Presented at the 72nd Scientific Assembly and Annual Meeting of the Radiological Society of North America, Chicago, Nov 30 to Dec 5

Cooperberg PL, Burhenne HJ (1980) Real-time ultrasonography. Diagnostic technique of choice in calculous gallbladder disease. N Engl J Med 302:1277–1279

Cooperberg PL, Gibney RG (1987) Imaging of the gallbladder, 1987. Radiology 163:605–613

Cover KL, Slasky BS, Skolnick ML (1985) Sonography of cholesterol in the biliary system. J Ultrasound Med 4:647–653

Crade M, Taylor KJU, Rosenfield AJ, De Graaf CS, Minihan P (1976) Surgical and pathologic correlation of cholecystography and cholecystosonography. AJR 131: 227–229

Croce F, Montali G, Colbiati L, Marinoni G (1981) Ultrasonography in acute cholecystitis. Br J Radiol 54:927–931

Crow HC, Bartrum RJ, Foote SR (1976) Expanded criteria for the ultrasonic diagnosis of gallstones. JCU 4:289–292

Dalla Palma L, Rizzato C, Bazzocchi M (1979) L' ecografia nella valutazione delle emergenze addominali non traumatiche. Radiol Med 65:693–696

Dalla Palma L, Rizzato G, Pozzi-Mucelli RS, Bazzocchi M (1980) Gray scale ultrasonography in the evaluation of carcinoma of the gallbladder. Br J Radiol 53:662–667

De Albertis P, Quadri D, Serafini G, Cajano A (1995) Le lesioni benigne e maligne della parete della colecisti. In: Gnocchi G (ed) Colecisti. Liviana Medicina, Naples, pp 57–72

Deitch EA, Engel JM (1980) Ultrasound in elective biliary tract surgery. Am J Surg 140:277–283

Engel JM, Deitch EA, Sikkema W (1980) Gallbladder wall thickness: sonographic accuracy and relation to disease. AJR 134:907–909

Filly RA, Moss AA, Way LW (1979) in vitro investigation of gallstone shadowing with ultrasound tomography. J Clin Ultrasound 7:255–262

Filly RA, Allen B, Minton MJ (1980) In vitro investigation of the origin of echoes within biliary sludge. J Clin Ultrasound 8:193–200

Findberg MG, Birnholz UC (1979) Ultrasound evaluation of gallbladder wall. Radiology 133:693–698

Fowler RC, Reid WA (1988) Ultrasound diagnosis of adenomyomatosis of the gallbladder: ultrasonic and pathological correlation. Clin Radiol 39:402–406

Freeman LM, Weissman HS, Rosenblatt R (1982) Ultrasound versus radionuclide imaging in the evaluation of patients with acute right upper quadrant pain (letter). Radiology 143:280–282

Glenn F, Becker CG (1982) Acute acalculous cholecystitis: an increasing entity. Ann Surg 195:131–136

Goldberg BB (1974) Ultrasonic and radiologic cholecystography. Radiology 111:405–409

Gonzales AC, Johnson JA (1978) Ultrasonic examination of the gallbladder: a review. Cl in Radiol 29:171–176

Handler SJ (1979) Ultrasound of gallbladder wall thickening and its relation to cholecystitis. AJR 132:581–585

Hessler PC, Mill S, Detorie F, Bosco AF (1981) High accuracy sonographic recognition of gallstones. AJR 136:517–520

Heyder N, Gunter E, Giedl J, Obenauf A, Hahn EG (1990) Polypoide Läsionen der Gallenblase. Dtsch Med Wochenschr 115:243–247

Itai Y, Araki T, Yoshikawa K, Furui S, Yoshiro N, Tasaka A (1980) Computed tomography of gallbladder carcinoma. Radiology 137:713–718

Jeffrey RB Jr, Laing FC, Wong W, Callen PW (1983) Gangrenous cholecystitis: diagnosis by ultrasound. Radiology 148:219–222

Jørgensen T, Jensen KH (1990) Polyps in the gallbladder: a prevalence study. Scand J Gastroenterol 25:281–286

Jutras AJ, Longtin M, Levesque HF (1958) La colesterolose et manifestations radiologiques. Ann Radiol 1:179–184

Jutras AJ, Longtin M, Levesque HF (1960) Hyperplastic cholecystoses. AJR 83:795–800

Kane RA (1980) Ultrasonographic diagnosis of gangrenous cholecystitis and empyema of gallbladder. Radiology 134:191–194

Kane RA, Jacobs R, Katz J, Costello P (1984) Porcelain gallbladder: ultrasound and CT apperance. Radiology 152:137–141

Kidney MR, Goiney R, Cooperberg PL (1986) Adenomyomatosis of the gallbladder: a pictorial exhibit. J Ultrasound Med 5:331–333

Koga A, Watanabe K, Fukuyama T, Takiguchi S, Nakayama F (1988) Diagnosis and operative indications for polypoid lesions of the gallbladder. Arch Surg 123:26–29

Kuo YC, Liu JY, Sheen IS (1990) Ultrasonographic difficulties and pitfalls in diagnosing primary carcinoma of the gallbladder. JCU 18:639–644

Laing FC, Jeffrey RB Jr (1983) Choledocholithiasis and cystic duct obstruction: difficult ultrasonographic diagnosis. Radiology 146:475–479

Laing FC, Federle MP, Jeffrey RB Jr, Brown TW (1981) Ultrasonic evaluation of patients with right upper quadrant pain. Radiology 140:449–455

Lewandowski BJ, Winsberg F (1981) Gallbladder wall thickness distortion by ascites. AJR 137:519–521

Lawson JC (1977) Gray scale cholecystosonography: diagnostic criteria and accuracy. Radiology 122:247–251
Leopold GR, Amberg J, Gosink BB, Mittelstaedt C (1976) Gray scale ultrasonic cholecystography: a comparison with conventional radiographic techniques. Radiology 121:445–448
Lubera RJ, Clihie ARW, Kling JE (1967) Cholecystitis and the hyperplastic cholesteroloses: a clinical, radiologic and pathology study. Am J Dig Dis 12:696–500
MacDonald FR, Cooperberg PL, Cohen MM (1981) The WES triad: a specific sonographic sign of gallstones in contracted gallbladder. Gastrointest Radiol 6:39–41
MacGahan JP, Walter JP (1985) Diagnostic percutaneous aspiration of the gallbladder. Radiology 155:619–622
Marano P, Digiancomenico G, Renda F, Trapani AR (1981) Studio controllato colecistografico ed ecografico della calcolosi della colecisti. Radiol Med 65:23–27
Marchal G (1979) Gallbladder wall sonolucency in acute cholecystitis. Radiology 133:429–433
Marchal G (1980) Gallbladder wall thickening: a new sign of gallbladder disease visualized by gray scale cholecystosonography. JCU 6:177–179
Marzio L, Innocenti P, Genovesi N, Di Felice F, Napolitano AM, Costantini R, Di Giandomenico E (1992) Role of oral cholecystography, real-time ultrasound, and CT in evaluation of gallstones and gallbladder function. Gastrointest Radiol 17:257–261
Mathieson JR, So BC, Malone DE, Becker CD, Burhenne HJ (1989) Accuracy of sonography for determining the number and size of gallbladder stones before and after lithotripsy. AJR 153:977–980
Muto Y, Yamada M, Uchimura M, Okamoto K (1987) Polypoid lesions of the gallbladder. Ital J Surg Sci 17:171–178
Nemcek A, Gore RM, Vogelzang RL, Grant M (1988) The effervescent gallbladder: a sonographic sign of emphysematous cholecystitis. AJR 150:575–577
Nevin JE, Moran TJ, Kay S, King R (1976) Carcinoma of the gallbladder: staging, treatment, and prognosis. Cancer 37:141–148
Piehler JM, Crichlow RW (1978) Primary carcinoma of the gallbladder. Surg Gynecol Obstet 147:929–942
Raghavendra BN, Feiner HD, Subramanyan BR, Ranson JHC, Toder SP, Horii SC, Madamba MR (1981a) Acute cholecystitis: sonographic-pathologic analysis. AJR 137:327–332
Raghavendra BN, Feiner HD, Subramanyan BR (1981b) Acute cholecystitis: sonographic pathologic analysis. AJR 137: 327–332
Ralls PW, Glolletti PW, Lapin SA (1985) Real-time sonography in suspected acute cholecystitis. Radiology 155:767–771
Rapaccini GL, Pompili M, Caturelli E (1993) Colecisti e vie biliari. In: Poletto (ed) Trattato italiano di ecografia, vol I. Societa' Italiana di Ultrasonologia in Medicina e Biologia, Milan, pp 271–295
Raptopoulos VD, D'Orsi CJ, Smith EH, Renter K, Moss L, Kleinman P (1982) Dynamic cholecystosonography of the contracted gallbladder: the double arc shadow sign. AJR 138:275–278
Romano AJ, VanSonnenberg E, Casola G, Gosink BB, Withers CE, MacCutchan JA, Leopold GR (1988) Gallbladder and bile duct abnormalities in AIDS: sonographic findings in eight patients. AJR 150:123–127
Ruhe AH, Zachman JP, Hudler BD, Rime AE (1979) Cholesterol polyps of the gallbladder: ultrasonic demonstration. JCU 7:386–388
Ruiz R, Teyssou H, Fernandez N (1980) Ultrasonic diagnosis of primary carcinoma of the gallbladder: a review of 16 cases. JCU 8:489–495
Saltzstein EC, Peacock JB, Mercer LC (1983) Early operation for acute biliary tract stone disease. Surgery 94:704–707
Samuels BI, Freitas JE, Bree RL, Schwab RE, Heller ST (1983) A comparison of radionuclide hepatobiliary imaging and real time ultrasound for the detection of acute cholecystitis. Radiology 147:207–210
Sanders RC (1980) The significance of sonographic gallbladder wall thickening. JCU 8:143–146
Shaler WJ, Leopold GR, Scheible FW (1981) Sonography of the thickened gallbladder wall: a nonspecific finding. AJR 136:337–339
Shuman WP, Mack LA, Rudd TG, Rogers JV, Gibbs P (1982) Evaluation of acute right upper quadrant pain: sonography and Tc99m-PIPIDA cholescintigraphy. AJR 139:61–64
Shuman WP, Rudd TG, Rogers JV, Stevenson JK, Larson EB (1984) Radionuclide hepatobiliary imaging and real time ultrasound in the detection of acute cholecystitis (letter). Radiology 152:238–239
Simeone JF, Mueller PR, Ferrucci JT (1989a) Non-surgical therapy of gallstones: implications for imaging. AJR 152:11–17
Simeone JF, Brink JA, Mueller PR, Compton C, Hahn PF, Saini S, Silverman SG. Tung G, Ferrucci JT (1989b) The sonographic diagnosis of acute cholecystitis: importance of the Murphy sign. AJR 152:289–290
Smathers RL, Lee JK, Heiken JP (1984) Differentiation of complicated cholecystitis from gallbladder carcinoma by computed tomography. AJR 143:255–259
Soiva M, Aro K, Pamilo M, Paivansalo M, Suramo I, Taavitsainen (1987) Ultrasonography in carcinoma of the gallbladder. Acta Radiol 28:711–714
Solan MJ, Jackson BT (1971) Carcinoma of the gallbladder. Br J Surg 58:593–597
Staniland JR, Dichburn F, De Domball T (1972) Clinical presentation of acute abdomen: study of 600 patients. Br Med 3:393–398
Takada T, Yasuda H, Uchiyama K, Hasegawa H, Asagoe T, Shikata J (1989) Pericholecystic abscess: classification of US findings to determine the proper therapy. Radiology 172:693–697
Taylor KJW, Jacobsen P, Jaffe CC (1979) Lack of an acoustic shadow on scans of gallstones: a possible artifact. Radiology 131:463–464
Tsuchiya Y (1991) Early carcinoma of the gallbladder: macroscopic features and US findings. Radiology 179:171–175
Van der Linden W, Sunzel H (1970) Early versus delayed operation for acute cholecystitis: a controlled clinical trial. Am J Surg 120:7–13
Weiner SN, Koenigsberg M, Morehous H, Hoffman J (1984) Sonography and computed tomography in the diagnosis of carcinoma of the gallbladder. AJR 142:735–739
Yeh HC (1979) Ultrasonography and computed tomography of carcinoma of the gallkbladder. Radiology 133:167–173
Yum MY, Fink DM (1980) Sonographic findings in primary carcinoma of gallbladder. Radiology 134:693–695

6 CT of the Gallbladder

L. van Hoe, D. Vanbeckevoort, and A.L. Baert

CONTENTS

6.1 Normal Anatomy and Variants

The gallbladder is an oval or elliptical structure of near-water density on computed tomography (CT) that is usually located at the inferior surface of the liver, between the right and left lobes. The wall of the gallbladder is normally very thin and is particularly well seen after intravenous administration of contrast material.

Anomalies in the position of the gallbladder are relatively rare. The most commonly encountered ectopic positions are: beneath the left liver lobe, intrahepatic, retrodisplaced, and suprahepatic (Morse et al. 1985). Another very rare entity is duplication of the gallbladder, which corresponds to complete division of the fundic portion of the organ with fusion of the two lobes at the neck (Martinoli et al. 1993).

6.2 Pathology

6.2.1 Gallstones

It is well known that CT is inferior to ultrasound (US) for detection of (noncalcified) gallstones. The appearance of gallstones on CT images is variable and depends on their composition (cholesterol versus calcium bilirubinate). In a study of 39 patients, Bova et al. (1992) found that the information obtained from CT can be used to predict the cholesterol content of gallstones and consequently their solubility. They identified six different CT patterns as follows: pattern 1: negative defect within the bile (2.5%); pattern 2: gallstone invisible (56.5%); pattern 3: faint homogeneous central calcification (10%); pattern 4: thin rim of calcification (21%); pattern 5: thick rim of calcification (7%); pattern 6: dense homogeneous central calcification (2.5%).

6.2.2 Adenomyomatosis of the Gallbladder

Adenomyomatosis is characterized by a proliferation of epithelial and smooth muscle elements, with outpouchings of the mucosa into a thickened muscular layer. There are three types of adenomyomatosis: diffuse, localized, and segmental. Although adenomyomatosis is not commonly diagnosed with imaging studies, the condition is not uncommon: it has been found in 2%–5% of cholecystectomy specimens. It carries no malignant potential. CT shows a soft-density mass in the gallbladder fossa. The diagnosis may be suggested when intramural diverticula or calcified stones are seen within the mass. Also typical and important in the differential diagnosis with gallbladder carcinoma is the presence of a distinct tumor-free space between the mass and the liver (Miyake et al. 1992; Gerard et al. 1990).

6.2.3 Acute Cholecystitis

The diagnosis of acute cholecystitis is usually made by clinical examination or US. The major disadvantage of CT is that the presence or absence of Murphy's sign (pain during compression of the gallbladder) cannot be assessed. Typical signs of acute cholecystitis on CT are focal or diffuse thickening of

L. van Hoe, MD, Department of Radiology, University Hospitals K.U. Leuven, Herestraat 49, 3000 Leuven, Belgium
D. Vanbeckevoort, MD, Department of Radiology, University Hospitals K.U. Leuven, Herestraat 49, 3000 Leuven, Belgium
A.L. Baert, MD, Department of Radiology, University Hospitals K.U. Leuven, Herestraat 49, 3000 Leuven, Belgium

the wall of the gallbladder, fluid in the gallbladder fossa, enlargement of the gallbladder, and infiltration of the surrounding fat (Fig. 6.1a). When helical (spiral) CT is performed, hyperdense areas may be seen in the liver parenchyma adjacent to the gallbladder on images obtained early after contrast injection (arterial phase) (GRYSPEERDT et al. 1995) (Fig. 6.1b). This phenomenon is related to hyperemia secondary to cholecystitis. It is important to differentiate fluid in the gallbladder fossa from marked hypodense thickening of the gallbladder wall (Fig. 6.2). The latter condition may be secondary to other disease processes such as hepatitis, malignancy, and hypoalbuminemia and thus unrelated to inflammation of the gallbladder. Typical CT findings in cases of concentric gallbladder wall thickening are: (1) two concentric enhancing rims with intervening low-density material, (2) small enhancing punctuate structures within the apparent fluid, corresponding to vessels in the gallbladder wall, and (3) diffusely thickened gallbladder wall (GOLDSTEIN et al. 1986).

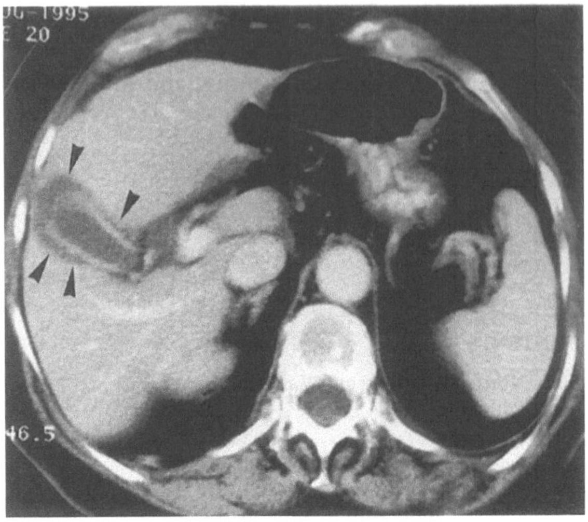

Fig. 6.2. Concentric thickening of gallbladder wall simulating pericholecystic fluid on CT. The CT image shows a thin enhancing inner layer and a nonenhancing outer layer (*arrowheads*) of the gallbladder wall in this patient with acute hepatitis. Ultrasound revealed no free fluid in the gallbladder fossa

a

b

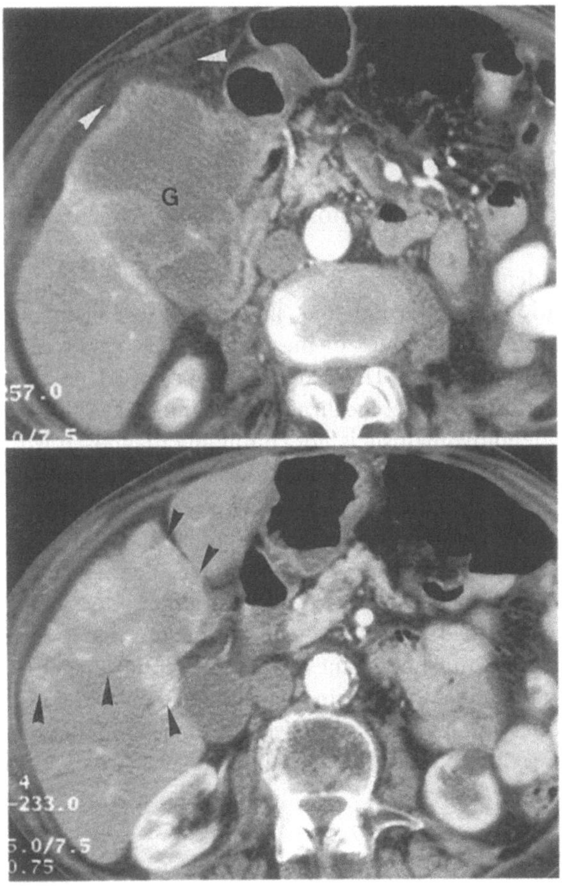

Fig. 6.1a,b. CT findings in acute cholecystitis. **a** Arterial-phase image obtained at the level of the gallbladder shows an enlarged gallbladder (*G*) with thickened wall. Also visible is infiltration of the pericholecystic fat (*arrowheads*). **b** Arterial-phase image obtained further craniad shows heterogeneous enhancement of the hepatic parenchyma (*arrowheads*), which probably represents hyperemia secondary to the inflammation of the gallbladder

While the diagnosis of acute cholecystitis can usually be made by clinical examination or US, diagnosis of *acalculous cholecystitis* by these methods is usually very difficult. In such cases, CT may be useful by showing pericholecystic fluid collections, perforation of the gallbladder wall, or inflammation of pericholecystic fat (BLANKENBERG et al. 1991).

Emphysematous cholecystitis is a complication of acute cholecystitis. In this condition gas is present in the gallbladder lumen or wall or in the pericholecystic space. There is an association with diabetes, as in emphysematous pyelonephritis. The diagnosis can be made by CT in the clinical setting of acute cholecystitis (Fig. 6.3). The differential diagnosis includes biliary-antral anastomosis from previous surgery, fistula secondary to gallstone erosion into bowel, and reflux of air through an incompetent sphincter of Oddi or following endoscopic retrograde cholangiopancreatography (MCMILLIN 1985).

Gallbladder perforation is a common complication of acute cholecystitis. Common signs are a pericholecystic fluid collection, streaky omentum or mesentery, and a gallbladder wall defect on CT (KIM et al. 1994) (Fig. 6.4). While US is well suited to the

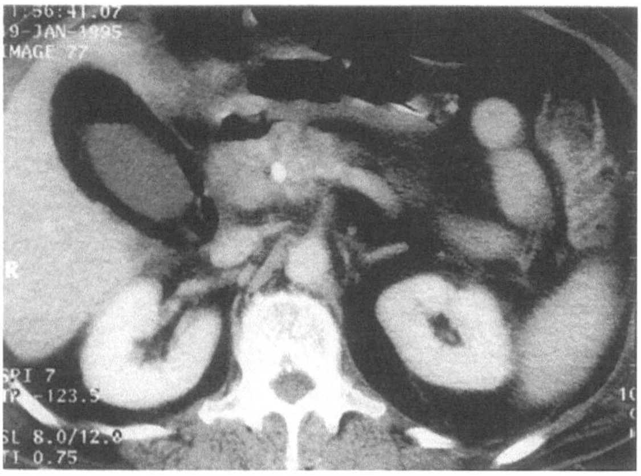

Fig. 6.3. Emphysematous cholecystitis. The CT image shows multiple air bubbles in the wall of the gallbladder

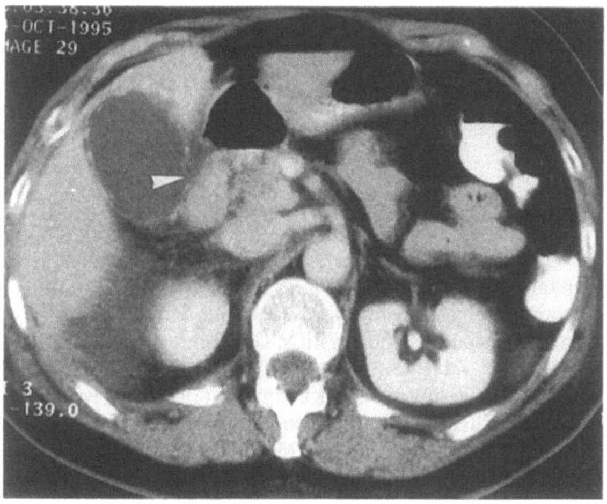

Fig. 6.4. Gallbladder wall perforation in acute cholecystitis. The CT image shows enlargement of the gallbladder and infiltration of the pararenal fat. Note also the presence of a small fluid collection medial to the gallbladder and a defect in the gallbladder wall (*arrowhead*), findings which point to the diagnosis of perforation

diagnosis of acute cholecystitis, CT is superior to US for diagnosis of gallbladder perforation (Kim et al. 1994).

6.2.4 Chronic and Xanthogranulomatous Cholecystitis

Xanthogranulomatous cholecystitis is an inflammatory disease of the gallbladder characterized histologically by the infiltration of round cells, lipid-laden histiocytes, and multinucleated giant cells, and by the proliferation of fibroblasts in the muscle layer. The pathogenesis of this disease is thought to be similar to that of xanthogranulomatous pyelonephritis: chronic infection and calculi, associated with bile stasis. These elements cause degeneration and necrosis of the wall and formation of microabscesses. The intramural microabscesses are replaced eventually by xanthogranulomata. The reported prevalence of this disease ranges from 0.7% to 10.6% (Hanada et al. 1987). Its importance resides in the difficult differential diagnosis with gallbladder carcinoma; marked thickening of the gallbladder wall is common finding in both conditions (Fig. 6.5).

Very low-grade chronic inflammation has also been postulated to be a cause of mural calcifications in *porcelain gallbladder*. CT is well suited to making the specific diagnosis of this condition (Gale and Robbins 1985).

6.2.5 Trauma to the Gallbladder

Gallbladder injuries following blunt trauma are uncommon. Approximately 2% of the patients who undergo exploratory laparotomy for abdominal injuries are found to have gallbladder injury. Isolated injury of the gallbladder is extremely rare. Traumatic lesions to the gallbladder can be classified as contusion, laceration or perforation, and avulsion. Hemorrhage into the gallbladder lumen is also commonly seen. The CT findings depend on the type of gallbladder injury. Common findings are pericholecystic fluid, an ill-defined contour of the gallbladder wall,

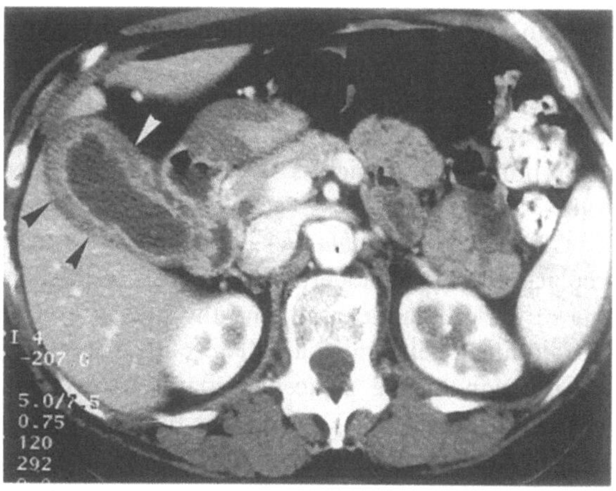

Fig. 6.5. Xanthogranulomatous cholecystitis. The CT image shows marked thickening of the gallbladder wall (*arrowheads*); note the absence of pericholecystic fluid and normal aspect of surrounding fat. Xanthogranulomatous cholecystitis was proven histologically

and high-density material in the gallbladder lumen. The differential diagnosis of the latter finding includes vicarious excretion of intravenous contrast material in patients with suboptimal renal function, milk of calcium bile, and cholelithiasis. When pericholecystic fluid is the only CT finding, differential diagnosis with periportal lymph edema, ascites, and extravasated bile may be difficult (ERB et al. 1994).

6.2.6 Gallbladder Wall Varices

Varices in the gallbladder wall are typically seen in patients with thrombosis of the portal vein. The varices in the gallbladder wall may act as a bypass around a focally thrombosed extrahepatic segment of the portal vein. Retrograde flow from the cystic vein to the gallbladder varices may give rise to flow across the gallbladder bed directly into the hepatic parenchyma and ultimately into the right portal vein. Varices of the gallbladder wall are easily recognised on US and are sometimes seen as dense tumor structures on contrast-enhanced CT scans (WEST et al. 1991).

6.2.7 Gallbladder Carcinoma

Gallbladder carcinoma accounts for 1%–2% of all gastrointestinal malignancies. Its incidence is higher in patients of advanced age. The clinical presentation of gallbladder carcinoma may be atypical. Unfortunately the prognosis for primary carcinoma of the gallbladder remains uniformly grim. In most cases the cancer is far advanced at operation (OHTANI et al. 1993). An important predisposing factor for gallbladder carcinoma is *porcelain gallbladder*; the incidence of gallbladder carcinoma in patients with this condition may be as high as 22% (GALE and ROBBINS 1985).

CT findings of *gallbladder carcinoma* are a mass replacing the gallbladder, focal or diffuse thickening of the gallbladder wall, and the presence of an intraluminal polypoid mass (Fig. 6.6). There may be invasion of the adjacent hepatic parenchyma (FRANQUET et al. 1991). CT may be used for staging of gallbladder carcinoma and thus for treatment planning. It has been demonstrated that CT has a high positive predictive value in the diagnosis of lymph node involvement. Lymph nodes commonly involved in gallbladder carcinoma are cystic nodes, pericholedochal nodes, pancreaticoduodenal nodes, retroportal nodes, nodes along the hepatic artery, and interaortocaval nodes (OHTANI et al. 1993).

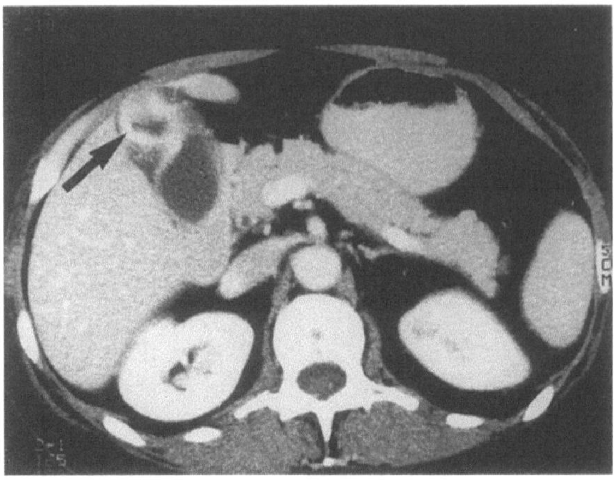

Fig. 6.6. Gallbladder carcinoma. The CT image shows a markedly enhancing "mass" in the gallbladder. This mass proved to be a gallbladder carcinoma

References

Blankenberg F, Wirth R, Brooke Jeffrey R (1991) Computed tomography as an adjunct to ultrasound in the diagnosis of acute acalculous cholecystitis. Gastrointest Radiol 16:149–153

Bova JG, Schwesinger WH, Kurtin WE (1992) In vivo analysis of gallstone composition by computed tomography. Gastrointest Radiol 17:253–256

Erb RE, Mirvis SE, Shanmuganathan K (1994) Gallbladder injury secondary to blunt trauma: CT findings. J Comput Assist Tomogr 18:778–784

Franquet T, Montes M, Ruiz de Azua Y, Jimenez FJ, Cozcolluela R (1991) Primary gallbladder carcinoma: imaging findings in 50 patients with pathologic correlation. Gastrointest Radiol 16:143–148

Gale ME, Robbins AH (1985) Computed tomography of the gallbladder: unusual diseases. J Comput Assist Tomogr 9:439–443

Gerard PS, Berman D, Zafaranloo S (1990) CT and ultrasound of gallbladder adenomyomatosis mimicking carcinoma. J Comput Assist Tomogr 14:490–491

Goldstein RB, Wing VW, Laing FC, Brooke Jeffrey R (1986) Computed tomography of thick-walled gallbladder mimicking pericholecystic fluid. J Comput Assist Tomogr 10:55–56

Gryspeerdt S, Van Hoe L, Marchal G, Baert AL (1995) Imaging of hepatic perfusion disorders with double phase spiral CT. Radiology 193(P):511

Hanada K, Nakata H, Nakayama T, Tsukamoto Y, Terashima H, Kuroda Y, Okuma R (1987) AJR 14:727–730

Kim PN, Lee KS, Bae WK, Lee BH (1994) Gallbladder perforation: comparison of US findings with CT. Abdom Imaging 19:239–242

Martinoli C, Derchi LE, Pastorino C, Cittadini G (1993) Case report: imaging of a bilobed gallbladder. Br J Radiol 66:734–736

McMillin K (1985) Computed tomography of emphysematous cholecystitis. J Comput Assist Tomogr 9:330–332
Miyake H, Aikawa H, Hori Y et al (1992) Adenomyomatosis of the gallbladder with subserosal fatty proliferation: T findings in two cases. Gastrointest Radiol 17:21–23
Morse JMD, Lakshman S, Thomas E (1985) Gallbladder ectopia simulating pancreatic mass on CT. Gastrointest Radiol 10:111–113
Ohtani T, Shirai Y, Tsukada K, Hatakeyama K, Muto T (1993) Carcinoma of the gallbladder: CT evaluation of lymphatic spread. Radiology 189:875–880
West MS, Garra BS, Horii SC et al. (1991) Gallbladder varices: imaging findings in patients with portal hypertension. Radiology 179:179–182

7 MRI of the Gallbladder

P.M. Bret and C. Reinhold

CONTENTS

7.1 Introduction

Cholelithiasis accounts for over 90% of gallbadder pathology, and is diagnosed with a high degree of accuracy using ultrasound. Other imaging techniques play a secondary role in the diagnosis of gallbladder disease, and are indicated either in a specific clinical context or as problem-solving modalities. To date, the role of magnetic resonance imaging (MRI) has been limited in the diagnosis of diseases of the biliary system, although the availability of fast imaging techniques has considerably improved the image quality of abdominal magnetic resonance (MR) examinations. In addition, the advent of MR cholangiopancreatography, which provides a three-dimensional representation of the biliary tree and pancreatic duct, without the need for contrast medium administration, has shown promising preliminary results.

7.2 Technique

The usual contraindications to MR imaging also apply to examinations of the gallbladder. Although the presence of surgical clips in the right upper abdominal quadrant is not a contraindication to MR imaging, signal void artifacts may result and can be considerable if the clips are numerous.

MR examinations of the gallbladder should be performed after the patient has fasted for at least 8–12h in order to promote gallbladder filling and concentration of the bile. An intravenous or intramuscular injection of an antispasmodic agent is usually administered during the examination to decrease motion artifacts from bowel peristaltism. MR examinations of the gallbladder should be tailored to the clinical question at hand. A combination of T2-weighted and T1-weighted sequences are generally used. T2-weighted sequences can be obtained using conventional spin-echo (CSE) or fast spin-echo (FSE) sequences. Advantages of the FSE sequence include (1) an increase in the bile-to-liver signal intensity ratio and (2) improved conspicuity of the gallbladder contours, due to the shorter acquisition times and, consequently, decreased motion artifact. The section thickness should be maintained at 5mm with an intersection spacing of 1–2mm, in order to avoid significant volume averaging. Other imaging parameters are similar to those used for examinations of the liver or pancreas.

A recent development of MR imaging in the biliary tract, MR cholangiography uses heavily T2-weighted sequences that produce images similar to those of direct cholangiography. With this technique, solid organs and moving fluids have a low signal intensity while stationary fluids such as gallbladder bile have a resultant high signal intensity. The technique of MR cholangiography is described in more detail in Chap. 10.

T1-weighted sequences can be obtained using CSE or spoiled gradient-echo sequences. As the contents of water in bile play a significant role in the resulting signal intensity of the bile on T1-weighted sequences, it is important that patients be examined after a period of fasting that allows bile concentration by the gallbladder. It is also important that the repetition time (TR) be maintained as low as possible to avoid

P.M. Bret, MD, Department of Diagnostic Radiology, Montreal General Hospital, McGill University, 1650 Cedar Avenue, Montreal, Quebec, H3G 1A4 Canada
C. Reinhold MD, Department of Diagnostic Radiology, Montreal General Hospital, McGill University, 1650 Cedar Avenue, Montreal, Quebec, H3G 1A4 Canada

artificially increasing the signal of bile by an added T2 component to the image. Paramagnetic agents such as gadolinium diethylenetriamine pentaacetate (Gd-DTPA) can be used to study enhancement of the gallbladder wall when needed, most often to better delineate the extent of a tumor of the gallbladder or one that has developed in the vicinity of the gallbladder. After intravenous injection of contrast agents, early scanning is usually performed using breath-hold T1-weighted fast gradient-echo sequences. Finally, new hepatobiliary contrast agents such as manganese (II) *N,N'*-dipyridoxylethylenediamine-*N,N'*-diacetate 5,5'-bis(phosphate) (Mn-DPDP) or gadopentetate dimeglumine, which are excreted by the biliary system, produce significant enhancement of the biliary contents after intravenous injection. These agents, however, are not yet widely available and their clinical role remains to be determined.

7.3 Results

7.3.1 Normal Anatomy

7.3.1.1 Gallbladder

On T2-weighted sequences, the gallbladder wall is not visualized (Fig. 7.1). The gallbladder contents are uniformly hyperintense in comparison with the liver parenchyma. On T1-weighted sequences, the gallbladder contents exhibit various signal intensities, depending on the composition of the bile (Figs. 7.2, 7.3) (Demas et al. 1985).

7.3.1.2 Cystic Duct

Because it is a thin, relatively short duct, the cystic duct is best depicted on the heavily T2-weighted sequences used for MR cholangiography. On serial sections, especially when using multiplanar reformatting software, the cystic duct can be followed from the gallbladder infundibulum to its insertion in the common hepatic duct (Fig. 7.4). Variations in the anatomy of the cystic duct are a well-established risk factor in cholecystectomy, especially in laparoscopic cholecystectomy. The following variants of the cystic duct anatomy increase the risk of bile duct injury at the time of surgery: low cystic duct insertion, medial insertion of the cystic duct (anterior or, most commonly, posterior spiral cystic duct), parallel course of the cystic duct and the common hepatic duct, and short cystic duct. In a

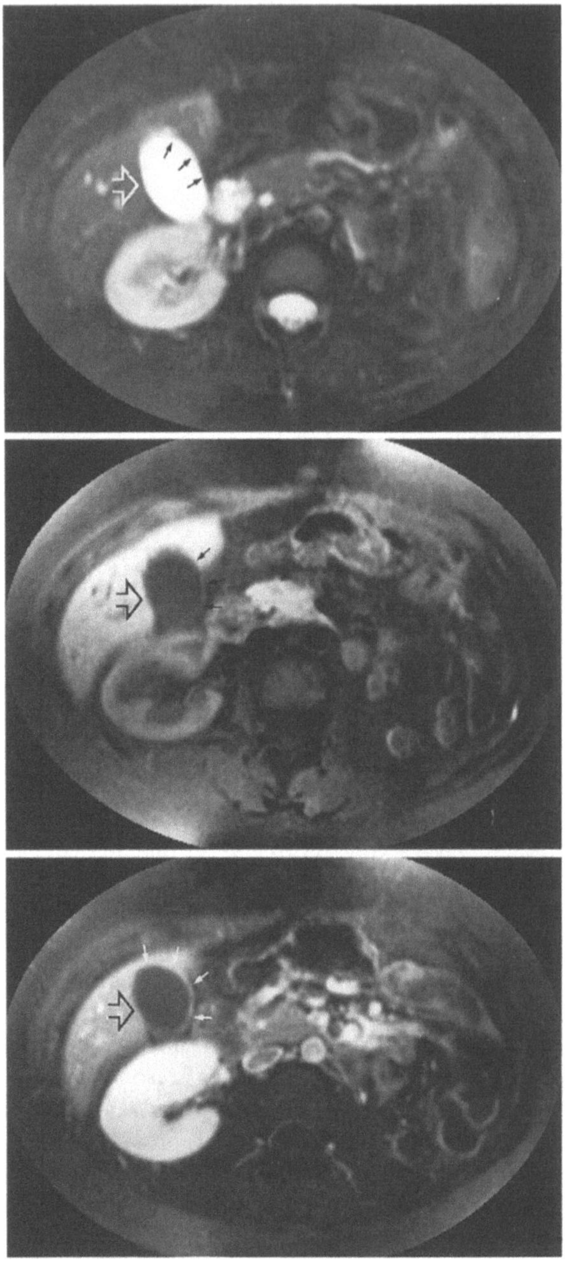

a

b

c

Fig. 7.1a–c. Normal gallbladder. **a** T2-weighted conventional spin-echo (CSE) sequence with fat saturation. The gallbladder contents (*open arrow*) are of high signal intensity. No distinct gallbladder wall is seen (*arrows*). **b** T1-weighted CSE sequence with fat saturation. The gallbladder contents (*open arrow*) are of low signal intensity and a gallbladder wall (*arrows*) of intermediate signal intensity is seen. **c** T1-weighted CSE sequence with fat saturation after intravenous injection of Gd-DTPA. The gallbladder contents (*open arrow*) do not enhance, but the gallbladder wall (*arrows*) is enhanced and more clearly visualized

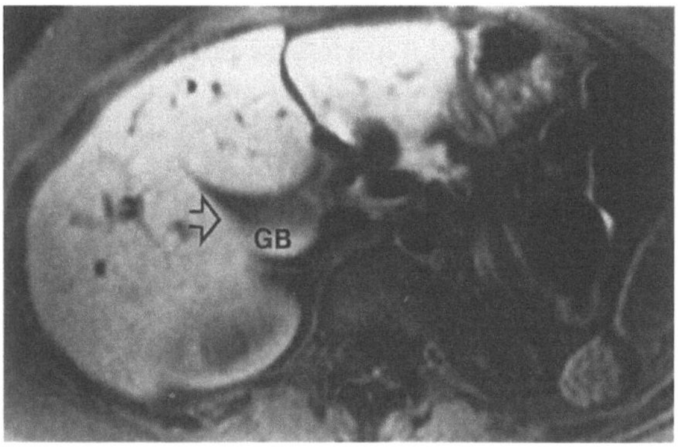

Fig. 7.2. Normal gallbladder. T1-weighted CSE sequence with fat saturation. The dependent portion of the gallbladder contents is filled with bile of high signal intensity (*arrow*). *GB*, Gallbladder

ultrasonography. With the development of nonsurgical gallstone therapies, it was hoped that MRI would be useful to predict gallstone composition, and several studies have evaluated the in vitro appearance of gallstones, using different field strengths. Results have been contradictory: With low field strength, the majority of stones studied had a low signal intensity, and less than 20% of stones showed a faint increased signal centrally (Moon et al. 1983; Moriyasu et al. 1987). In a more recent study performed at 1.5 T, Baron et al. (1989) found foci of hypersignal intensity in 46 of 63, or 73% of stones. Correlation with spectroscopy and measures of T1 relaxation times as well as chemical analysis of the

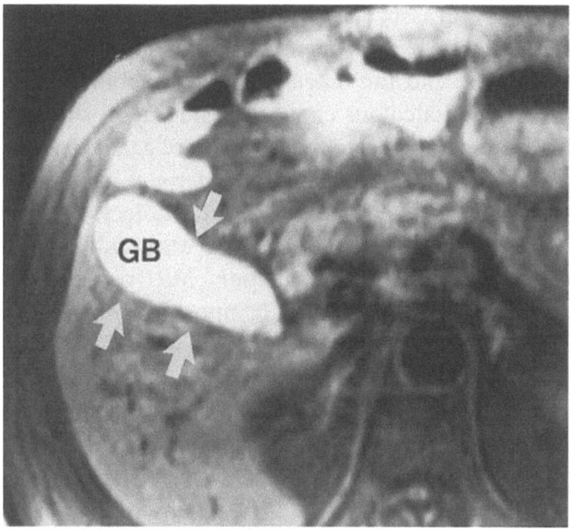

Fig. 7.3. Normal gallbladder. T1-weighted CSE sequence with fat saturation. The gallbladder contents are of very high signal intensity (*arrows*). *GB*, Gallbladder

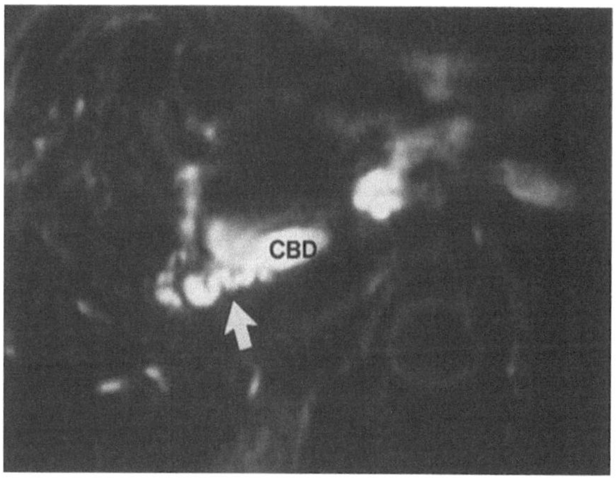

Fig. 7.4. Normal cystic duct anatomy. Heavily T2-weighted fast spin-echo (FSE) sequence with fat saturation. The cystic duct (*arrow*) is seen at the level of the cysticohepatic junction. *CBD*, Common bile duct

recent study on the role of MR cholangiography in the detection of normal variants of the cystic duct anatomy, the cystic duct was visualized in 126 of 171 patients (74%) studied with MRI (Taourel et al. 1996), and when available, correlation with endoscopic retrograde cholangiopancreatography was excellent.

7.3.2 Gallstones (Figs. 7.5–7.12)

The accuracy of MRI in the detection of cholelithiasis is not known, the reason being that MRI plays no role in the diagnosis of gallstones. Cholelithiasis is usually diagnosed with a high degree of accuracy by

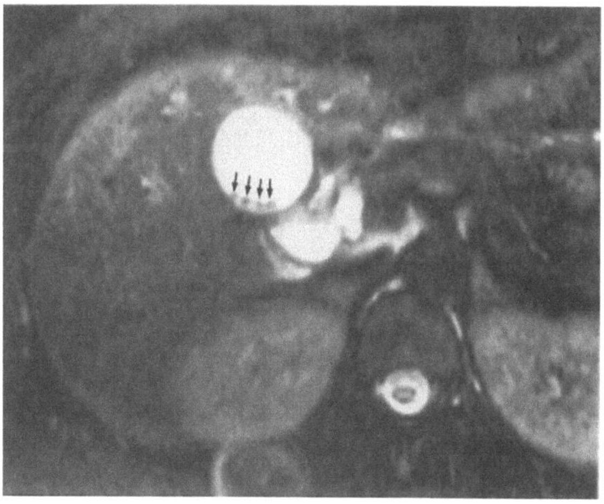

Fig. 7.5. Gallstones. Heavily T2-weighted FSE sequence with fat saturation. Four tiny gallstones (*arrows*) are demonstrated in the dependent portion of the gallbladder

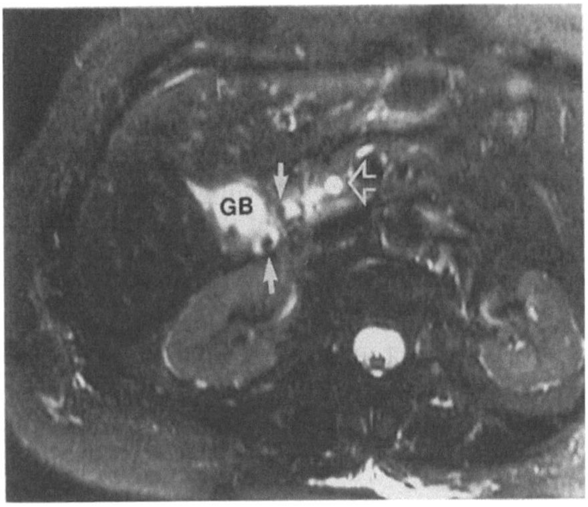

Fig. 7.6. Cystic duct stones. Heavily T2-weighted FSE sequence with fat saturation. Several foci of low signal intensity (*arrows*) are seen within the lumen of the cystic duct which is situated to the right of the common bile duct (*open arrow*). *GB*, Gallbladder

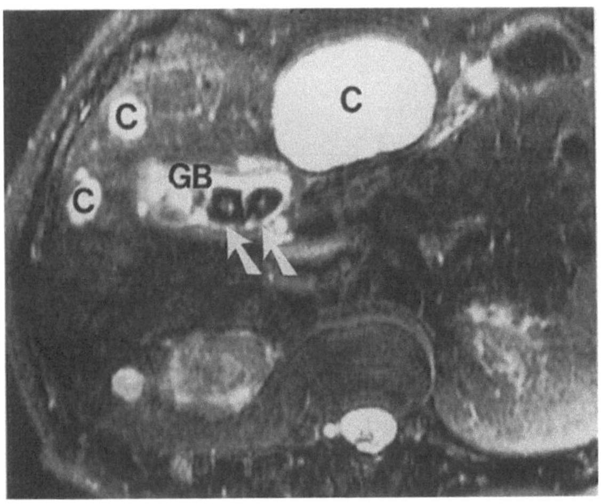

Fig. 7.7. Gallstones and liver cysts. Heavily T2-weighted FSE sequence with fat saturation. Several gallstones (*arrows*) are visible in the lumen of the gallbladder. Note the foci of high signal intensity in the center of the gallstones. Multiple cysts are also visible in the liver. *GB*, Gallbladder; *C*, liver cysts

stones showed that the foci of increased signal did not correspond with areas of high lipid content.

Since the advent of laparoscopic cholecystectomy, the need to characterize gallstone composition with imaging has substantially decreased, since the therapeutic indications are no longer based on the composition of the stones present. In our in vivo experience with a high field-strength magnet, the majority of gallstones appear as areas of signal void in both T1- and T2-weighted spin-echo images. In the heavily T2-weighted sequences used to obtain MR cholangiography, even small gallstones are depicted in the background of hyperintense bile (Figs. 7.5, 7.6, 7.10). Occasionally areas of high signal are present in the center of the stones on T1- and T2-weighted sequences, and less commonly, the stones are entirely hyperintense on T1-weighted sequences (Fig. 7.9) (Moeser et al. 1988). The ability of MRI to image the entire biliary tree allows evaluation of the gallbladder and cystic duct as well as the intra- and extrahepatic bile ducts for the presence of lithiasis (Figs. 7.6, 7.11, 7.12).

7.3.3 Acute Cholecystitis (Figs. 7.13–7.15)

Morphologic features of acute cholecystitis include the presence of gallstones and thickening of the gallbladder wall associated with an ultrasound Murphy's sign. However, these signs lack both sensitivity and specificity, and the diagnosis of acute cholecystitis, especially acalculous cholecystitis, remains difficult. MRI can demonstrate the presence of gallstones and gallbladder wall thickening sometimes associated with intramural abscesses or pericholecystic fluid (Weissleder et al. 1988). However, in addition to this morphological information, MRI can provide data on (1) the chemical composition of the bile and (2) the vascularization of the gallbladder wall. Pathophysiologically, in acute cholecystitis the water content of the gallbladder bile is higher than in either chronic cholecystitis or normal states (Svanvik et al. 1981). This results in a decrease in signal intensity of the gallbladder bile on T1-weighted sequences. Although an initial in vitro study performed on bile

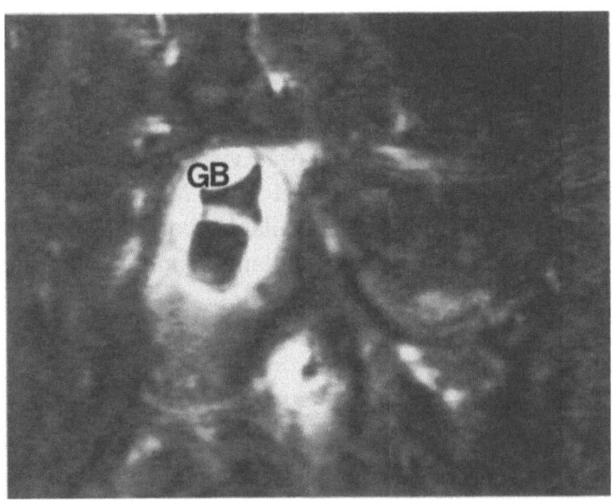

Fig. 7.8. Gallstones. Heavily T2-weighted FSE sequence with fat saturation. A quadrangular and a triangular gallstone are demonstrated. *GB*, Gallbladder

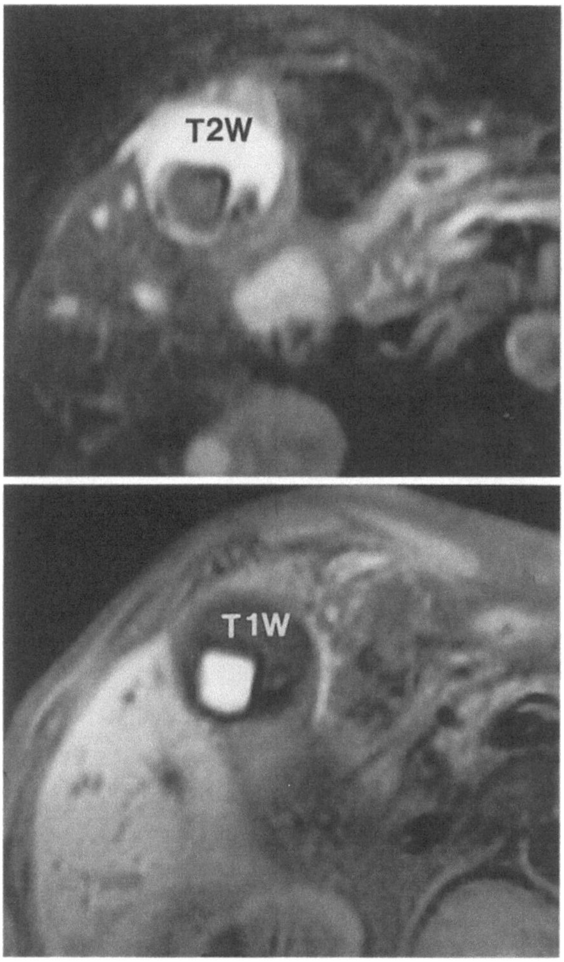

Fig. 7.9a,b. Gallstone. **a** T2-weighted CSE sequence with fat saturation. A quadrangular gallstone of intermediate signal intensity is demonstrated. **b** T1-weighted CSE sequence with fat saturation in the same patient. The stone is of very high signal intensity relative to the bile contents and the liver parenchyma. *T2W*, T2-weighted sequence; *T1W*, T1-weighted sequence

obtained from surgical specimens did not demonstrate a clear correlation between signal intensity of bile and presence of acute cholecystitis (Loflin et al. 1985), two clinical series have shown promising results: McCarthy et al. (1986), comparing the signal intensity of the gallbladder bile with that of the liver, and using hyperintensity of the gallbladder contents relative to the liver as a diagnostic criterion for normality, reported a 75% sensitivity and a 100% specificity in the diagnosis of acute cholecystitis using a 0.35-T magnet and a TR/TE of 500/56 ms. More recently, Pu et al. (1994) found a significantly higher liver/gallbladder signal intensity ratio in acute cholecystitis than in chronic cholecystitis or normal gallbladder: The study included 72 patients, 11 with proven acute cholecystitis and 13 with chronic cholecystitis, and was performed with a 0.5-T superconducting system using two different T1-weighted sequences. Using a ROC curve, the cutoff value of the liver/gallbladder signal intensity ratio that provided a 100% sensitivity in the diagnosis of acute cholecystitis was 2.13 with a TR/TE = 500/20 ms, and was associated with a 100% specificity. With a TR/TE = 620/25 ms, the liver/gallbladder signal intensity ratio that provided a 100% sensitivity was 1.38 and was associated with a 98% specificity. It is to be noted that these numbers need to be validated at different field strengths, since T1 is proportional to the field strength of the system used. Also, since it appears that the concentration of water in the gallbladder plays a pivotal role in these results, it is crucial to examine only patients who have fasted for 8–12 h, in order to allow the normal gallbladder to concentrate bile.

A further advantage of MRI over other imaging techniques is the superior contrast resolution that allows it to detect and quantify contrast enhancement. In their study, Loud et al. (1996) compared ten cases of acute cholecystitis, two cases of chronic cholecystitis, and two cases of gallbladder carcinoma with ten control subjects, using fat-suppressed T1-weighted and breath-hold spoiled gradient-echo sequences before and after bolus injection of 0.1 mmol/kg gadopenetate dimeglumine. They found a significantly higher percentage of con-

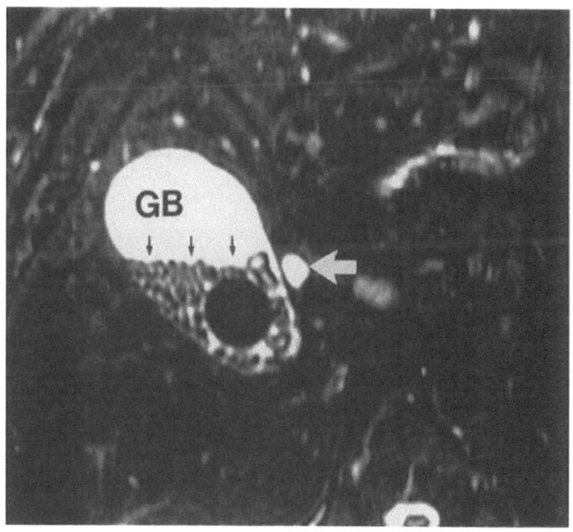

Fig. 7.10. Microlithiasis. Heavily T2-weighted FSE sequence with fat saturation. Innumerable tiny stones are seen in the dependent portion of the gallbladder (*small arrows*). A larger stone is also present. The common bile duct (*arrow*) is demonstrated on the left side of the gallbladder. *GB*, Gallbladder

a b c d

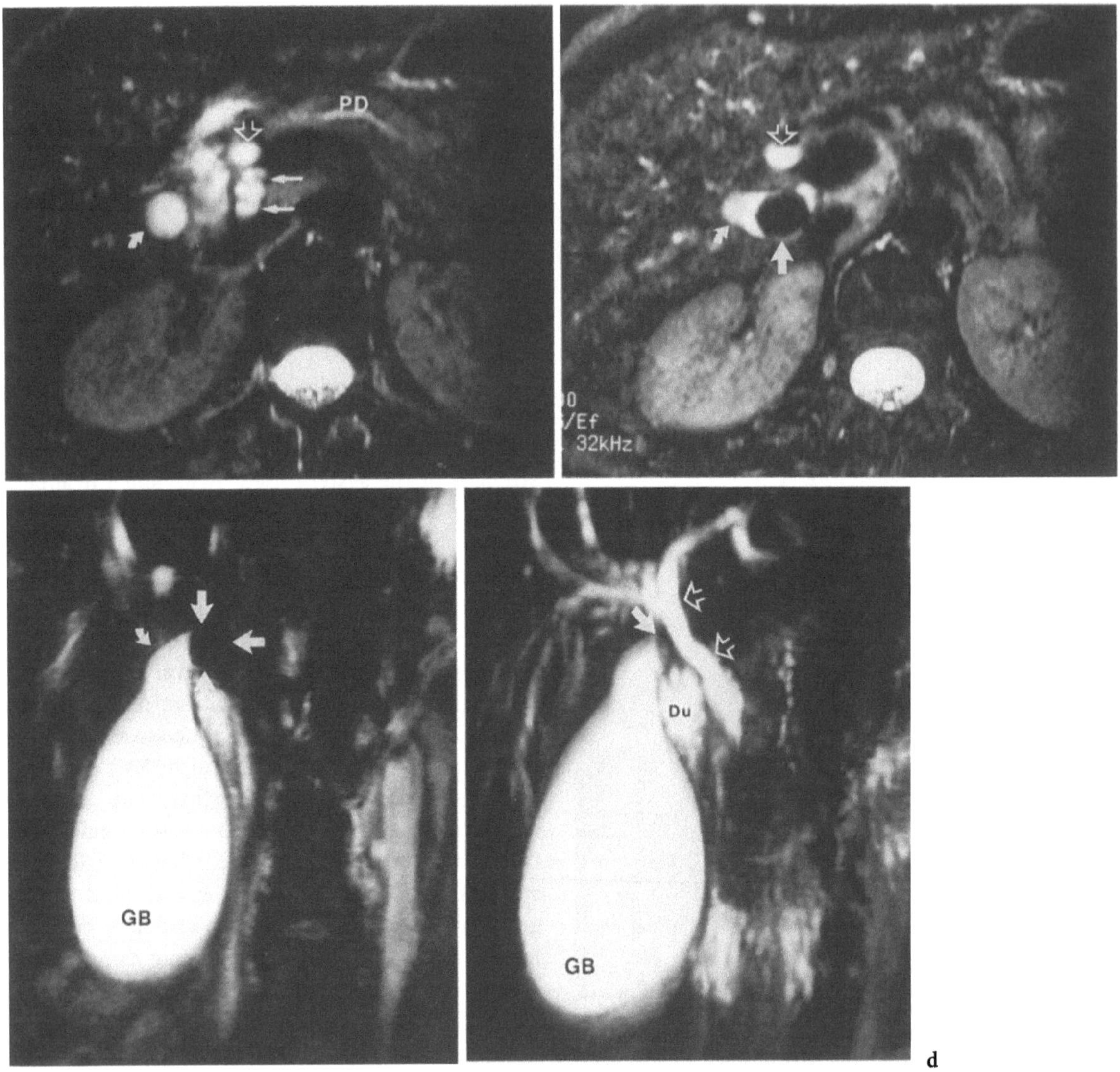

Fig. 7.11a–d. Impacted stone in the infundibulum of the gallbladder. Heavily T2-weighted FSE sequence with fat saturation. **a,b** Axial source images demonstrate a round stone (*large arrow*) in the infundibulum of the gallbladder (*curved arrow*). The cystic duct (*arrows*) is shown posterior to the common hepatic duct (*open arrow*). **c,d** Targeted coronal maximum intensity projection (MIP) reconstructions. **c** The gallstone (*arrows*) obstructs the gallbladder infundibulum (*curved arrow*). **d** On the more anterior data set, the common bile duct (*open arrows*) masks the gallstone. *GB*, Gallbladder; *Du*, duodenum; *PD*, pancreatic duct

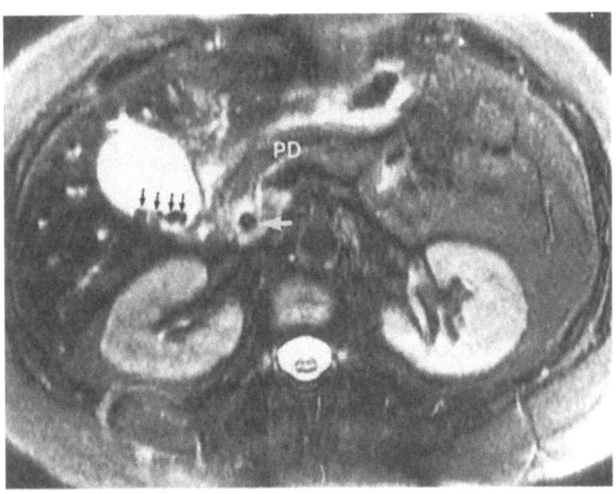

Fig. 7.12. Gallstones and choledocholithiasis. Heavily T2-weighted FSE sequence with fat saturation. Gallstones (*small arrows*) are demonstrated in the dependent portion of the gallbladder. In addition, a round stone is visible in the distal common bile duct (*arrow*) at the level of the junction with the pancreatic duct. *PD*, Pancreatic duct

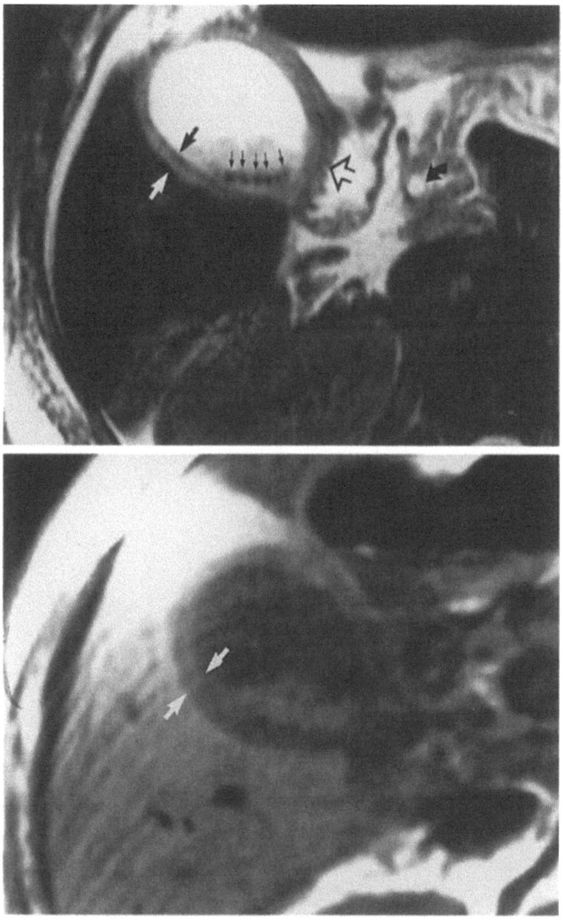

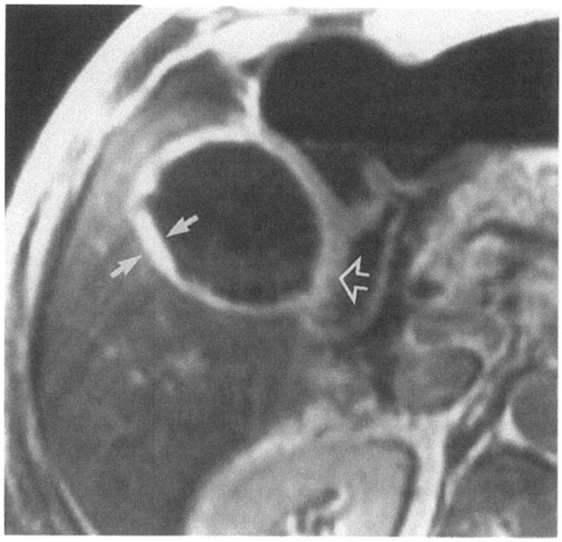

Fig. 7.13a–c. Acute cholecystitis. **a** T2-weighted CSE sequence. There is marked thickening of the gallbladder wall (*arrows*). Sludge and microlithiasis (*small arrows*) are seen in the dependent portion of the gallbladder. Note the close proximity between the medial aspect of the gallbladder wall and the duodenum which shows focal wall thickening (*open arrow*). The common bile duct (*curved arrow*) is not dilated. **b** T1-weighted CSE sequence. The thickened wall remains visible (*arrows*) while the gallstones are less conspicuous. **c** T1-weighted CSE sequence after intravenous injection of Gd-DTPA. There is marked enhancement of the gallbladder wall (*arrows*). Note the extension of the inflammatory process to the duodenum (*open arrow*)

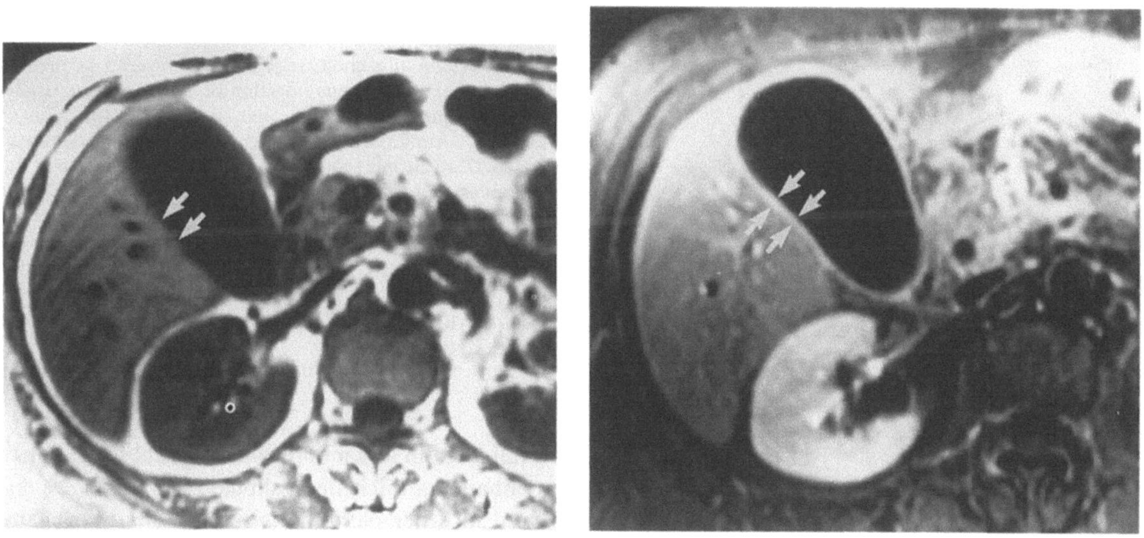

Fig. 7.14a,b. Acute cholecystitis. **a** T1-weighted CSE sequence. The gallbladder wall (*arrows*) cannot be reliably assessed. **b** T1-weighted CSE sequence after intravenous injection of Gd-DTPA. There is marked enhancement of the gallbladder wall, which is mildly thickened (*arrows*)

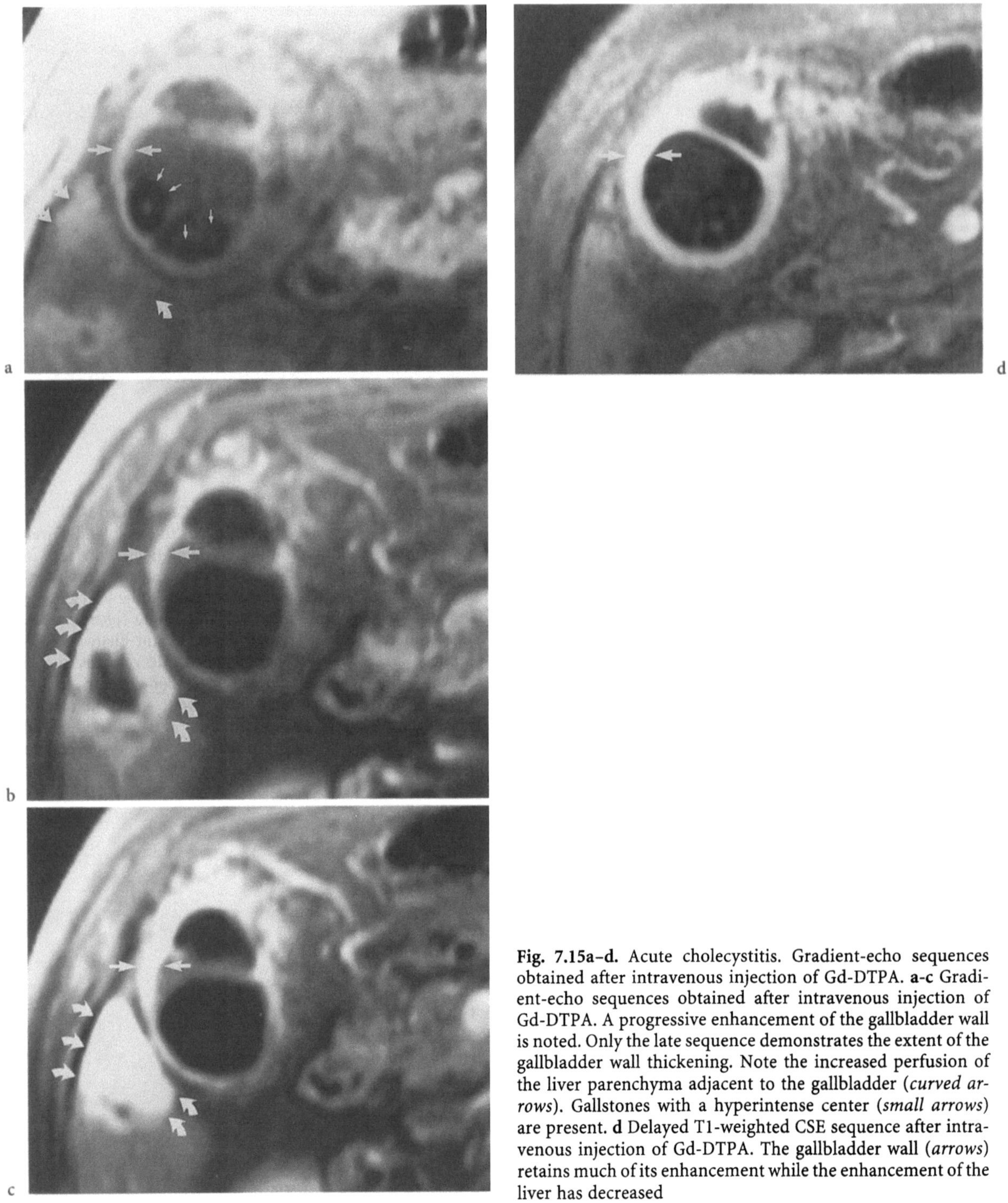

Fig. 7.15a–d. Acute cholecystitis. Gradient-echo sequences obtained after intravenous injection of Gd-DTPA. **a-c** Gradient-echo sequences obtained after intravenous injection of Gd-DTPA. A progressive enhancement of the gallbladder wall is noted. Only the late sequence demonstrates the extent of the gallbladder wall thickening. Note the increased perfusion of the liver parenchyma adjacent to the gallbladder (*curved arrows*). Gallstones with a hyperintense center (*small arrows*) are present. **d** Delayed T1-weighted CSE sequence after intravenous injection of Gd-DTPA. The gallbladder wall (*arrows*) retains much of its enhancement while the enhancement of the liver has decreased

trast enhancement (over 80%) in the patients with acute cholecystitis than in those with chronic cholecystitis or those in the control group. In addition, transient pericholecystic enhancement of hepatic parenchyma was found in seven of the ten patients with acute cholecystitis. These findings must be validated in larger controlled clinical series, since an enhancing thickened gallbladder wall can be observed in other pathological conditions than acute cholecystitis (Fig. 7.16).

In summary, although ultrasound has the distinct advantage over all other imaging techniques in that

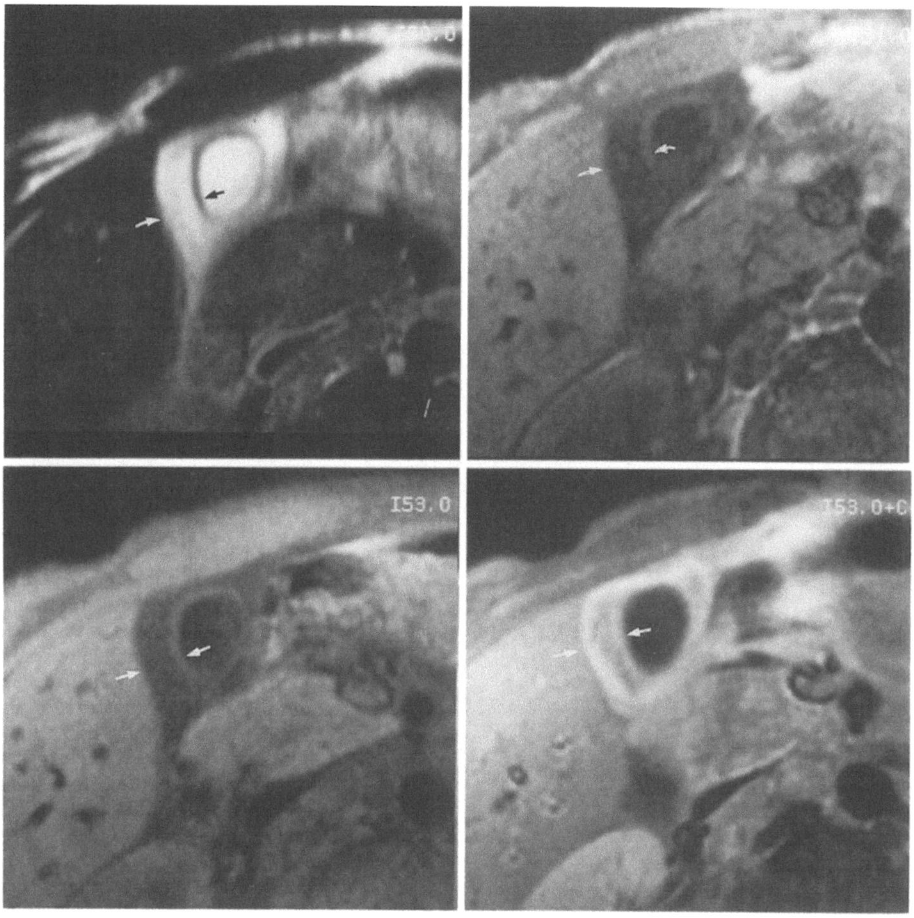

Fig. 7.16a–d. Gallbladder wall thickening in acute hepatitis. Patient presenting with lymphoma and jaundice of recent onset. **a** T2-weighted, **b** T1-weighted CSE sequences, **c** T1-weighted gradient-echo sequence, and **d** T1-weighted CSE sequence after intravenous injection of Gd-DTPA. There is marked thickening of the gallbladder wall (*arrows*), which enhances inhomogeneously after contrast injection. Gallbladder wall had returned to normal thickness on a follow-up ultrasound examination performed 2 months after this study

it can accurately localize the patient's pain with the ultrasound transducer (ultrasound Murphy's sign), and is the most sensitive technique in gallstone detection, MRI can also demonstrate classic morphological features of acute cholecystitis. In addition, the relative liver/gallbladder bile signal intensity ratio may be significantly decreased on T1-weighted sequences in patients with acute cholecystitis, compared with that of patients with chronic cholecystitis or normal gallbladders, provided that the patients have been fasting for at least 8h prior to the examination. The percentage of contrast enhancement of the gallbladder wall is also higher in patients with acute cholecystitis. If validated in large controlled clinical series, these findings could be of great use, especially in the diagnosis of acalculous acute cholecystitis.

7.3.4 Gallbladder Carcinoma (Fig. 7.17)

Cross-sectional images through the gallbladder and liver obtained with MRI show similar findings to those of computed tomography (CT) in cases of gallbladder carcinoma (Rooholamini et al. 1994). In the majority of cases, the lesion is advanced and presents as a large mass that is centered around gallbladder stones and invades the liver. The ability of MRI to detect gallstones trapped within the tumor mass may provide a clue with respect to the nature of a subhepatic mass of unknown origin. On T2-weighted sequences, the tumor is usually of increased signal intensity relative to the liver with poorly delineated contours, and is either iso- or hypointense to the liver on T1-weighted sequences (Sagoh et al. 1990). MRI is indicated in tumor stag-

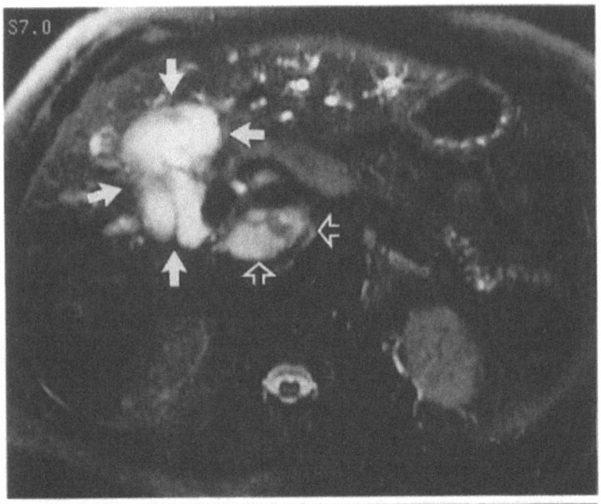

a

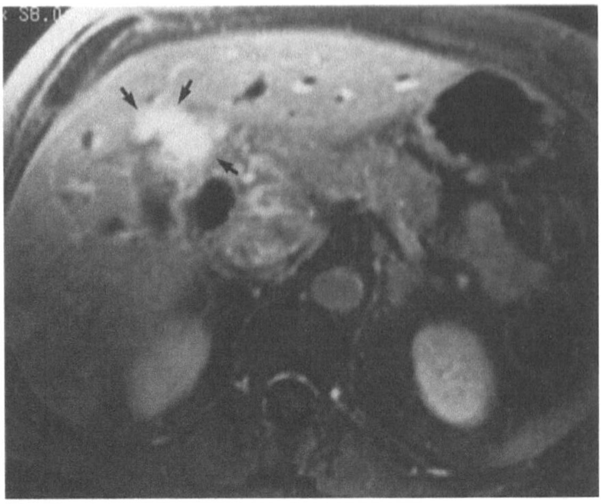

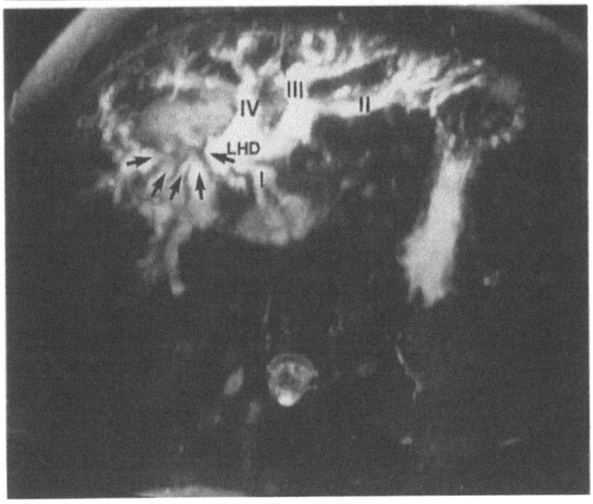

b

Fig. 7.17a–c. Gallbladder carcinoma. **a** Heavily T2-weighted FSE sequence with fat saturation. Targeted axial MIP reconstruction. A lobulated mass is seen in the gallbladder fossa (*arrows*). Lymph nodes are also seen posteriorly (*open arrows*). **b** Heavily T2-weighted FSE sequence with fat saturation. Axial MIP reconstruction using a thicker slab. The 3D representation demonstrates the relationship of the tumor with the intrahepatic bile ducts of the various segments of the liver (*arrows*). On the *left* side, the tumor only extends to the left hepatic duct, while on the *right* side, the tumor extends to the four segmental branches. **c** T1-weighted CSE after intravenous injection of Gd-DTPA. The mass is enhancing (*arrows*) and invasion of the liver parenchyma is clearly demonstrated. *I, II, III, IV*, Segments I, II, III, IV of the liver; *LHD*, left hepatic duct

ing rather than for diagnosis. Tumoral extension to the liver, the duodenum, the intrahepatic bile ducts, the presence of nodal metastases to the hepatoduodenal ligament or liver metastases usually preclude surgical resection. In a study evaluating the value of MRI in 19 patients with gallbladder carcinomas, SAGOH et al. (1990) found that T1-weighted and T2-weighted sequences were equally accurate in the diagnosis of tumor extension to the adjacent liver. Hepatic invasion, either in the vicinity of the primary tumor or distant liver metastases, had a similar signal intensity to the gallbladder tumor. Extension to the duodenum was better evaluated on T1-weighted images, which provide good signal contrast between fatty tissue, the tumor, and the duodenum. However, obliteration of the fat plane between the tumor and the duodenum was seen without duodenal invasion because of motion artifact, partial volume effect, or paucity of fat. In the study by SAGOH et al., tumoral extension to the hepatoduodenal ligament was better demonstrated with MRI than with CT, especially on T1-weighted sequences. The use of fast imaging sequences, as well as contrast medium injection, may improve the accuracy of MRI in staging gallbladder carcinomas in the future. The MR cholangiographic sequences may also prove useful in demonstrating extension of the tumor to the intrahepatic bile ducts.

In summary, MRI has no role to play in the routine diagnosis of gallbladder diseases, especially gallstones. Its findings in predicting the composition of gallstones have been disappointing and are of little use in the era of laparoscopic cholecystectomy. The normal anatomy of the cystic duct and its variants are well demonstrated on heavily T2-weighted sequences. In acute cholecystitis, MRI may be useful as a problem-solving modality by demonstrating a low signal intensity of bile relative to liver, and an increased contrast enhancement of the gallbladder wall. In gallbladder carcinomas, MRI can be useful in evaluating tumor extension to the liver, the intrahepatic bile ducts, and the hepatoduodenal ligament.

References

Baron RL, Shuman WP, Lee SP, Rohrmann CA Jr, Golden RN, Richards TL, Richardson ML, Nelson JA (1989) MR appearance of gallstones in vitro at 1.5 T: correlation with chemical composition. Am J Roentgenol 153:497–502

Demas BE, Hricak H, Moseley M, Wall SD, Moon K, Goldberg HI, Margulis AR (1985) Gallbladder bile: an experimental study in dogs using MR imaging and proton MR spectroscopy. Radiology 157:453–455

Loflin TG, Simeone JF, Mueller PR, Saini S, Stark DD, Butch RJ, Brady TJ, Ferrucci JT Jr (1985) Gallbladder bile in cholecystitis: in vitro MR evaluation. Radiology 157:457–459

Loud PA, Semelka RC, Kettritz U, Brown JJ, Reinhold C (1996) MRI of acute cholecystitis: comparison with the normal gallbladder and other entities. Magn Reson Imaging 14:349–355

McCarthy S, Hricak H, Cohen M, Fisher MR, Winkler ML, Filly RA, Margulis AR (1986) Cholecystitis: detection with MR imaging. Radiology 158:333–336

Moeser PM Julian, Karstaedt N, Sterchi M (1988) Unusual presentation of cholelithiasis on T1-weighted MR imaging. J Comput Assist Tomogr 12:150–152

Moon KL Jr, Hricak H, Margulis AR, et al (1983) Nuclear magnetic resonance imaging characteristics of gallstones in vitro. Radiology 148:753–756

Moriyasu F, Ban N, Nishida O, et al (1987) Central signals of gallstones in magnetic resonance imaging. Am J Gastroenterol 82:139–142

Pu Y, Yamamoto F, Igimi H, Shilpakar SK, Kojima T, Yamamoto S, Luo D (1994) A comparative study of the usefulness of magnetic resonance imaging in the diagnosis of acute cholecystitis. J Gastroenterol 29:192–198

Rooholamini SA, Tehrani NS, Razavi MK, Au AH, Hansen GC, Ostrzega N, Verma RC (1994) Imaging of gallbladder carcinoma. Radiographics 14:291–306

Sagoh T, Itoh K, Togashi K, Shibata T, Minami S, Noma S, Yamashita K, Nishimura K, Asato R, Mori K, Nishikawa T, Kakano Y, Konishi J (1990) Gallbladder carcinoma: evaluation with MR imaging. Radiology 174:131–136

Svanik J, Thornell E, Zettergren L (1981) Gallbladder function in experimental cholecystitis. Surgery 89:500–506

Taourel P, Bret PM, Reinhold C, Barkun AN, Atri M (1986) Anatomic variants of the biliary tree: Diagnosis with MR cholangiopancreatography. Radiology 199:521–527

Weissleder R, Stark DD, Compton CC, Simeone JF, Ferrucci JT (1988) Cholecystitis: diagnosis by MR imaging. Magn Reson Imaging 6:345–348

8 Ultrasonography of the Bile Ducts

R. Di Nardo, L. Broglia, M. Bezzi, and A. Tortora

CONTENTS

8.1 Introduction

Ultrasonography (US) is a primary tool in imaging of the hepatobiliary system. It is a reliable technique, whether in assessing the size of the bile ducts or in evaluating the liver, structures at the porta hepatis, gallbladder, or pancreas. It can be performed in any patient; there are no contraindications since no contrast material or radiation is required. Moreover, it can be performed rapidly and it may be repeated as often as necessary.

This chapter will describe Ultrasonographic technique, normal bile duct anatomy, pitfalls of diagnosis, and the more common pathologies of the bile duct.

8.2 Examination Technique

Both 3.5-MHz and 5-MHz sector and curved-array transducers allow good assessment of the biliary tree through a subcostal or intercostal acoustic access. A 5-MHz transducer generally provides sufficiently good resolution to reveal small biliary stones and their acoustic shadows. In large or obese patients, however the 3.5-MHz frequency is usually more appropriate.

For good visualization of the biliary system, patients should have followed an appropriate diet for a few days and fasted for a minimum of 6–10 h before the examination. This is to reduce gas content in the stomach, duodenum, and transverse colon, which can obscure the pancreas and the common bile duct, and to allow distention of the gallbladder and bile ducts by bile. In nonfasted patients, the gallbladder is normally contracted and therefore presents thick walls. This condition can usually be distinguished from pathological wall thickening, but usually it is preferable to reexamine the patient in a fasting condition before concluding that the viscus is definitely normal.

A comprehensive study of the biliary system should always include visualization of the intra- and extrahepatic bile ducts and demonstration of the liver and pancreas. The examination generally starts with the patient in a supine position and continues with the patient in a supine left posterior oblique position.

R. Di Nardo, MD, Senior Staff Radiologist, Department of Radiology, University of Rome "La Sapienza", 00161 Rome, Italy
L. Broglia, MD, Resident in Diagnostic Radiology, Department of Radiology, University of Rome "La Sapienza", 00161 Rome, Italy
M. Bezzi, MD, Assistant Professor, Department of Radiology, University of Rome "La Sapienza", 00161 Rome, Italy
A. Tortora, MD, Resident in Diagnostic Radiology, Department of Radiology, University of Rome "La Sapienza", 00161 Rome Italy

To visualize the intrahepatic bile ducts, axial and longitudinal scans obtained during suspended deep inspiration are performed through the epigastrium and the right upper quadrant, starting from the subxiphoid area. Sometimes intercostal scans are necessary for better visualization of the right lobe and the porta hepatis, which may be obscured by overlying gas in other positions.

The extrahepatic bile duct is demonstrated in its long axis by scanning along a plane perpendicular to the right costal margin and angled medially (Figs. 8.1, 8.2). This is best achieved by turning the patient into the left posterior oblique position. This maneuver allows the liver to drop below the costal margin and provides a larger acoustic window (Behan and Kazam 1978). On the other hand, this position may sometimes impede study of the most distal portion of the extrahepatic duct since it may bring an excessive amount of gas into the duodenum. To reduce gas content in the gastric antrum and duodenum, the patient can be placed in a semierect position or in a right posterior oblique position (Laing et al. 1984). If this maneuver is not sufficient, the patient should drink three or four cups of water (300–500 ml), assume a right lateral position for a few minutes, and then be reexamined in the semierect position described above.

Longitudinal views are integrated with transverse scans of the pancreatic head and uncinate process; these views are often more accurate for determining whether the distal duct is dilated and for assessing endoluminal contents.

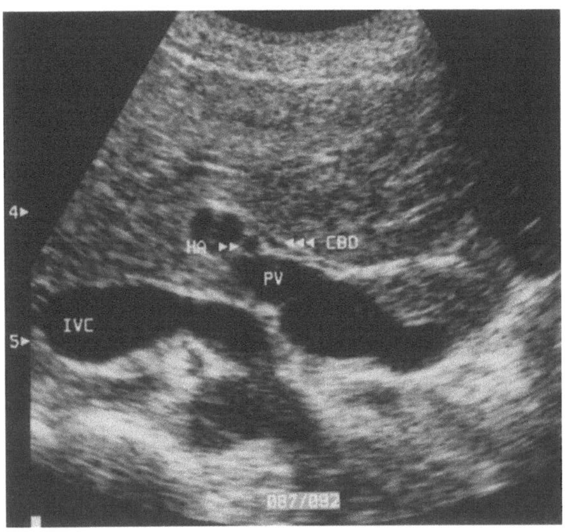

Fig. 8.1. Ultrasonography (US) demonstrates a normal common bile duct (*CBD*) passing anterior to the hepatic artery (*HA*) and portal vein (*PV*). *IVC*, Inferior vena cava

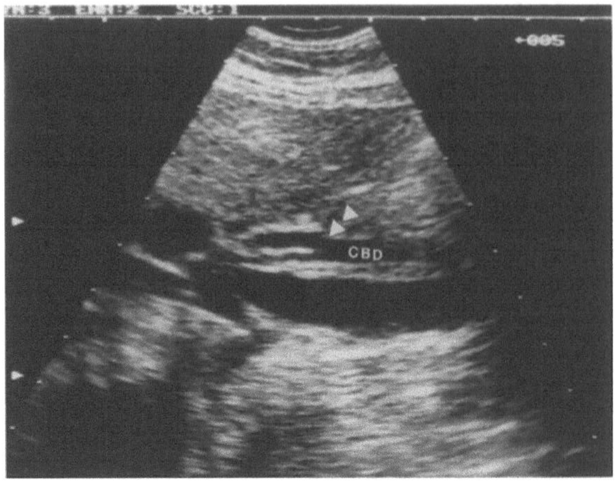

Fig. 8.2. Confluence of the hepatic duct and the cystic duct (*arrowheads*) to form the common bile duct (*CBD*)

8.3 Ultrasonographic Appearance

8.3.1 Normal Anatomy

The intrahepatic bile ducts run in the portal triad close to the portal vein and branches of the hepatic artery and drain into the right and left hepatic ducts.

Ultrasonographically the portal triad is hyperechoic. This hyperechogenicity is due to the fibrous liver capsule (hepatobiliary capsule of Glisson) and by the wall of the tubular structures contained within the triad (artery, portal vein branch, bile duct, and lymphatic); of all these structures, however, the portal vein is usually the only one which is distinguishable. Nondilated intrahepatic bile ducts are generally not identified because they are smaller than 1 mm in size. More common is visualization of the main right and left hepatic ducts, which appear as thin tubular structures, approximately 1 mm in diameter, running anterior to the right and left portal vein branches and converging toward the hepatic duct, anterior to the common portal trunk.

The hepatic duct is generally 3–4 cm in length and lies anterior to the right margin of the main portal vein (Fig. 8.1). As the duct leaves the porta hepatis it joins the cystic duct and forms the common bile duct (Fig. 8.2). However, the exact level of the confluence of the two ducts is often not seen on US, and therefore the generally accepted term of "common duct" comprises both the common hepatic duct and the common bile duct (CBD) (Figs. 8.2, 8.3). The ability of US to distinguish between a normal and a dilated CBD makes US the initial screening modality in patients with jaundice or a suspected abnormality of the biliary tree.

For assessment of size, the CBD is usually measured at the level of the porta hepatis where the duct crosses the right hepatic artery; the hyperreflective walls are not included in the measurement. The normal size of the CBD when measured at this level in adults with the gallbladder in place is 4–5 mm (COOPERBERG et al. 1980). However, several conditions may affect this "normal" value. In patients older than 60 years, due to a loss of elasticity of the ductal wall, the upper limit of normal is 7–8 mm (WU et al. 1984). In patients who have undergone cholecystectomy the CBD is generally of normal caliber (i.e., less than 5–6 mm); in a certain percentage, however, the duct is dilated, ranging in size from 7 to 11 mm, without evidence of obstruction. It may be said that ducts that were normal before surgery remain normal, while ducts that were either obstructed or chronically inflamed tend to loose elasticity and assume some of the reservoir functions of the gallbladder (GLAZER et al. 1981; MULLER et al. 1982). The proportion of these patients with "dilated" but normal ducts ranges from 10% to 58% (BRUNETON et al. 1981; NIEDERAU et al. 1983; GRAHAM et al. 1980).

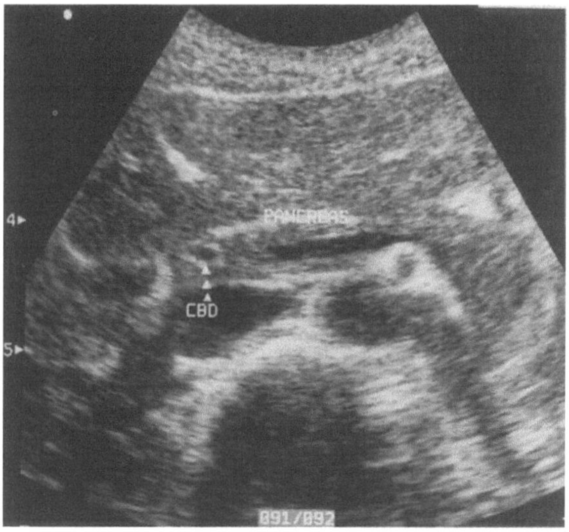

Fig. 8.3. Transverse scan showing the common bile duct (*CBD*) within the head of the pancreas

Frequently a discrepancy is seen between the measurements taken at US and those obtained during cholangiography, either percutaneous, intravenous, or retrograde endoscopic. This should not be a cause for surprise and may be due to radiological magnification, a choleretic effect of the intravenous contrast medium, or to distension produced by direct contrast medium injection into the bile duct (BEHAN and KAZAM 1978; SAUERBREI et al. 1980; GRAHAM et al. 1980).

8.3.2 Anatomical Variants

One of the variants that can be encountered is a variation in the relative position of the CBD and right hepatic artery. While the latter usually lies posterior to the duct, in 30% of patients it is found anterior to it (SAUERBREI et al. 1992). If it is necessary to distinguish between the two, one can use pulsed Doppler to check for blood flow.

Anatomical variants of the bile ducts are common. The point at which the cystic duct joins the

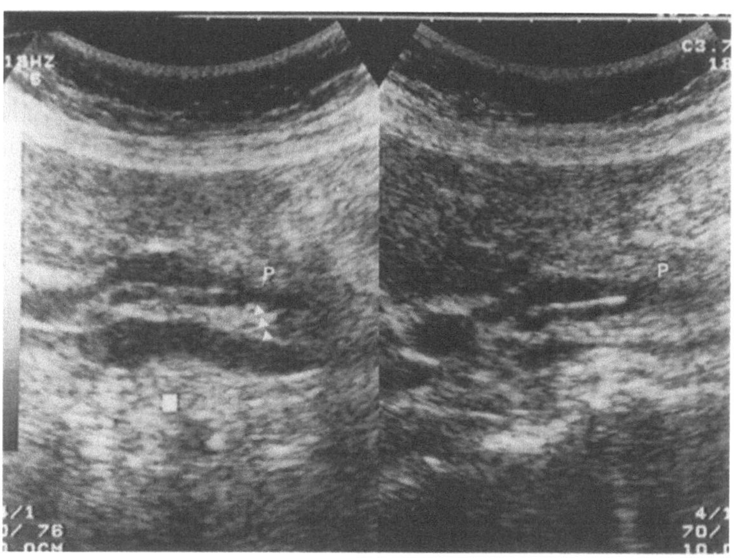

Fig. 8.4. Anomalous confluence of the cystic duct and common bile duct (*arrowheads*) at the head of the pancreas (*P*)

common hepatic duct may vary from the level of the porta hepatis to the head of the pancreas (Fig. 8.4). An accessory or aberrant right hepatic duct may also exist. In this case the aberrant duct generally enters either the cystic duct or the common duct above or below the cystic duct confluence (IRVING and BATES 1993). Occasionally two accessory ducts may be present, and one of them may insert directly into the gallbladder. These anatomical variants place the patient at a higher risk for bile duct injuries during laparoscopic cholecystectomy; however, in preoperative US they are difficult to recognize if the biliary system is not dilated.

8.4 Bile Duct Obstruction

In patients with jaundice any decision making is based on the presence or absence of biliary obstruction. Furthermore, in any case of obstructive jaundice, knowledge of the level and cause of obstruction may determine the strategy of any further diagnostic test or therapeutic procedures.

Ultrasound, due to its safety, low cost, accuracy and wide availability is universally regarded as the initial imaging test in evaluating patients with jaundice.

8.4.1 Presence of Obstruction

Obstructive jaundice is diagnosed when bile duct dilation is seen. The specificity and sensitivity of US in determining whether obstructive jaundice is present are 99% and 87% respectively (COOPERBERG et al. 1980).

A typical ultrasonographic finding of intrahepatic bile duct dilation is an appearance of "parallel channels" inside the liver, when the dilated bile duct becomes appreciable and may have a caliber similar to the adjacent portal vein in scans performed parallel to the longitudinal axis of the duct (Fig. 8.5). In scans perpendicular to the longitudinal axis of the duct the typical appearance is that of a "double-barrelled shot-gun" (MULLER et al. 1982; COOPERBERG et al. 1980).

Another common appearance of the dilated intrahepatic ducts is the immediate impression that there are too many tubular structures within the liver which do not correspond to vessels, either portal or arterial. The branching pattern of these tubular structures gives a characteristic "spider" or "stellate" appearance (TAYLOR et al. 1979) (Fig. 8.6).

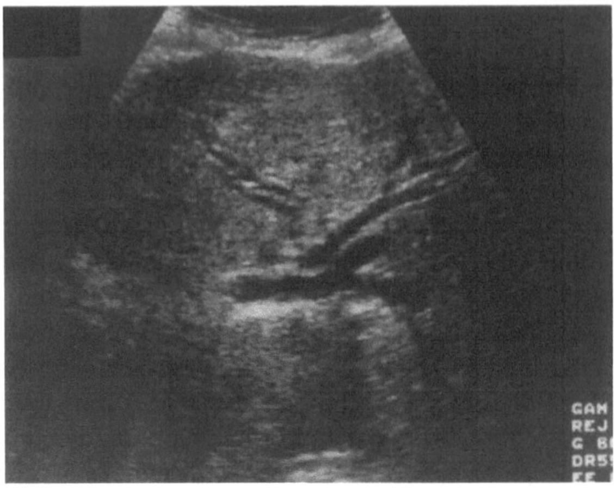

Fig. 8.5. "Parallel channels": dilated intrahepatic bile ducts have a caliber similar to that of the adjacent branches of the portal vein

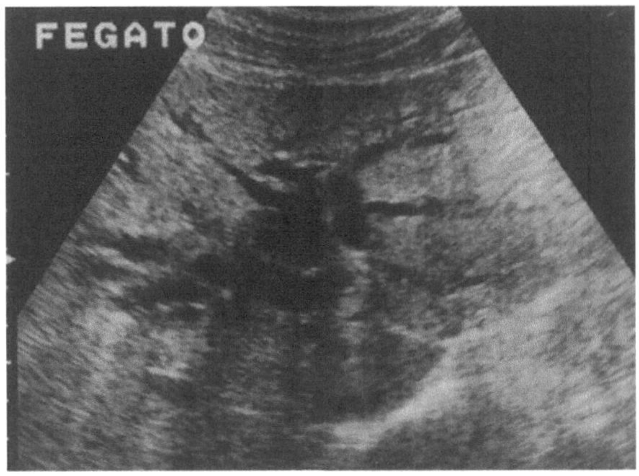

Fig. 8.6. "Spider" or "stellate" branching pattern: dilated bile ducts are evident within the liver

The extrahepatic bile duct is usually considered dilated when it measures more than 6 mm in internal diameter.

The normally larger size of the CBD in older patients and in patients who have undergone cholecystectomy should obviously be taken into account when examining such patients.

In some patients, bile duct obstruction may occur without dilation. In early obstruction (within 24–48 h of acute onset) or in intermittent obstruction, the CBD may be normal. Choledocholithiasis and ampullary stenosis, either inflammatory or neoplastic, may be associated with incomplete and/or intermittent obstruction and with nondilated CBD. Of 131 patients referred because of right upper quadrant symptoms or abnormal liver function tests, 8% had obstructing lesions but a normal-sized bile duct (SIMEONE et al. 1985).

When obstruction is suspected, but difficult to demonstrate because of normal-sized ducts, a fatty meal test may be performed. The patient is given a fatty meal, usually consisting of three egg yolks. As a consequence, cholecystokinin is released from the duodenal mucosa, producing gallbladder contraction, relaxation of Oddi's sphincter, and an increase in the flow of bile from the liver. In normal patients, after about 45–60 min, the common duct diameter is unchanged or reduced in size. If it increases in size by more than 2 mm there is a high probability of biliary obstruction. This technique has a sensitivity of 74% and specificity of 100% for biliary obstruction (DARWEESH et al. 1988). It is also indicated to rule out obstruction in patients without evidence of bile duct obstruction, with normal laboratory values, but with a mildly increased CBD caliber (SIMEONE et al. 1985).

Other causes of normal-sized ducts in obstructive jaundice are sclerosing cholangitis and cirrhosis. In both cases the scarring and fibrosis around the ducts may prevent or delay an appreciable dilation.

8.4.2 Level and Cause of Obstruction

The level of obstruction is determined by examining the whole biliary tree from the periphery to the papilla. The possible locations of obstruction can be schematically divided into the intrahepatic tree, the porta hepatis, the suprapancreatic CBD, and the intrapancreatic common duct. In intrahepatic obstruction a segmentary or lobar dilation of the duct may occur (Fig. 8.7). In such instances, the level of obstruction is determined by carefully following the dilated branches and identifying the point where they converge. If the obstruction involves the hilar confluence or the proximal common hepatic duct, all the intrahepatic ducts will be dilated, but the gallbladder and the choledochus will be normal.

When the obstruction involves the lower two thirds of the CBD, it will be easy to identify the dilated duct within the hepatoduodenal ligament. When the obstruction is below the join between the cystic duct and the CBD, the cystic duct can be seen to be dilated as well. The gallbladder, according to the COURVOISIER's law, will also be overly distended if not affected by chronic cholangitis. The gallbladder finding, however, should always be regarded as supportive one, since there are well-known exceptions to the COURVOISIER's law.

US is fairly accurate in determining the level of obstruction. The most recent studies on the subject

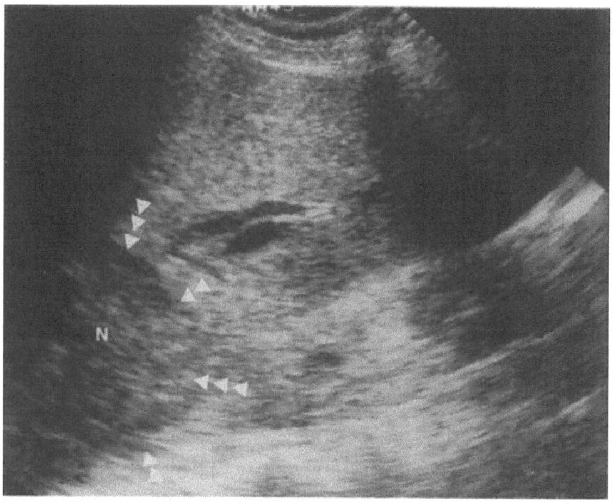

Fig. 8.7. Intrahepatic duct obstruction: segmentary dilated intrahepatic ducts, due to a hepatic neoplasm (*arrowheads*), have a "parallel channel" appearance

Table 8.1. Most frequent causes of biliary obstruction

Intrahepatic obstruction
Lithiasis
Neoplasms:
Hepato cellular carcinoma
Cholangiocarcinoma
Metastases
Inflammation:
Sclerosing cholangitis
AIDS cholangitis
Congenital anomalies:
Atresia
Caroli's disease
Extrahepatic obstruction
Lithiasis
Neoplasms:
Cholangiocarcinoma
Gallbladder carcinoma
Pancreatic carainoma
Ampullary carcinoma
Lymph node metastasis
Inflammation:
Pancreatitis
Ampullary stenosis
Lymph node inflammation

(LAING et al. 1986; GIBSON et al. 1986) were published in late 1980s. LAING et al. reported 92% accuracy in indicating the level of obstruction and 71% accuracy in assessing the cause. GIBSON et al. reported 95% and 88% accuracy respectively for the level and the cause. OKUDA and OSHIBUCHI (1989) reported a very high accuracy in establishing the location of obstruction (99.3%), with an 81% accuracy in determining the cause.

The most frequent causes of biliary obstruction are reported in Table 8.1, classified by the level of the obstruction.

8.5 Lithiasis

8.5.1 Intrahepatic Lithiasis and Pneumobilia

Intrahepatic stones are rare in Western countries, but are seen more frequently in the Far East, where they may be the consequence of parasitic infestation (NAKAYAMA 1982). Choledocholithiasis, congenital abnormalities of the biliary ducts, and biliary strictures are predisposing conditions for intrahepatic calculi formation.

Ultrasonographically, intrahepatic stones appear as highly reflective echoes, with or without a prominent posterior acoustic shadow. They are often associated with segmentary dilation of the intrahepatic ducts, and can therefore be differentiated from calcific granuloma, which is generally seen far from the intrahepatic portal branches (Fig 8.8). Sometimes these stones do not shadow, and this makes their identification difficult.

In our experience intrahepatic stones are seen more frequently secondary to CBD stone or to either an iatrogenic bile duct stricture or a narrowed biliary-enteric anastomosis.

The US examination is very difficult in patients with suspected intrahepatic lithiasis and cholangitis, who have previously undergone to sphincterotomy or biliary-enteric anastomosis. This is because gas bubbles in the bile ducts produce echogenic foci, which may show acoustic shadow or a posterior "comet-tail" artifact, due to reverberation (Fig. 8.9). This may lead to a false-positive diagnosis of intrahepatic stones. However, when the decubitus of

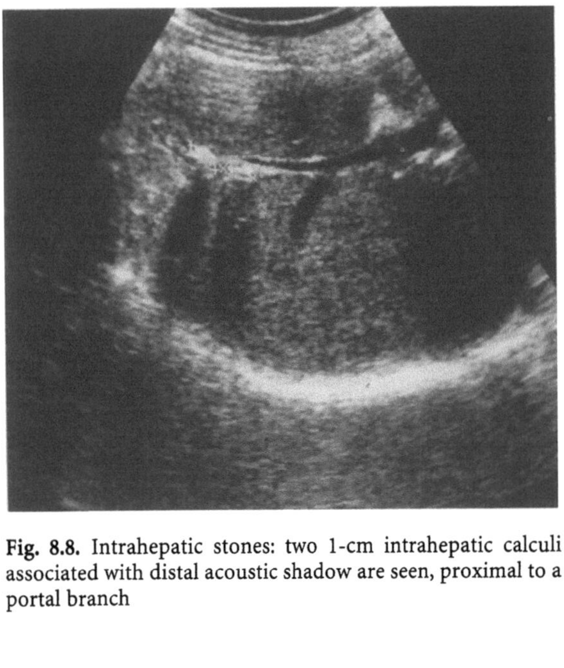

Fig. 8.8. Intrahepatic stones: two 1-cm intrahepatic calculi associated with distal acoustic shadow are seen, proximal to a portal branch

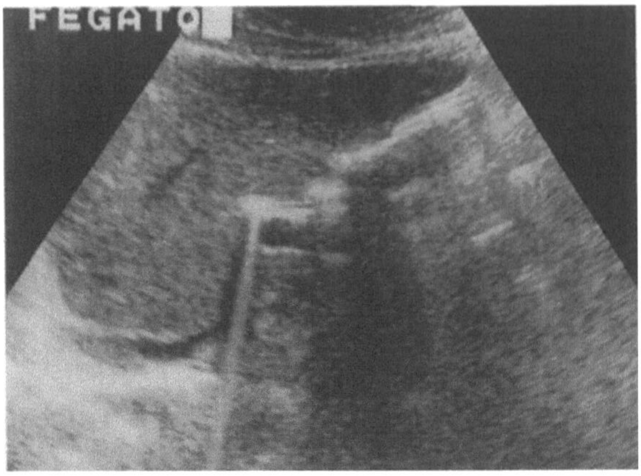

Fig. 8.9. A strong "comet tail" artifact is visible posterior to biliary gas bubbles, which appears as hyperechoic linear foci

the patient is changed, gas bubbles tend to move within the ducts, and this can be observed by performing a scan parallel to the longitudinal axis of the duct. On the other hand, extensive pneumobilia may mask the presence of intrahepatic calculi, leading to a false-negative diagnosis (IRVING 1993).

Air within the bile ducts must be distinguished from air in the portal system, which is seen in several pathologic conditions. In both conditions multiple intrahepatic hyperechoic foci are the main ultrasonographic findings. However they differ in their distribution: gas in the venous system has a peripheral distribution, close to the hepatic capsule, whereas pneumobilia has a more central distribution.

When air is seen in the biliary ducts, it is mandatory to check the patient history for previous sphincterotomy or surgery. If there is nothing of relevance in the history, pneumobilia may be suggestive of a spontaneous biliary-enteric fistula: this is the consequence of cholelithiasis with chronic cholecystitis, where inflammation produces migration of the stone into the bowel lumen through erosion of the gallbladder wall (SAUERBREI et al. 1992).

8.5.2 Cystic Duct Stones

At US, diagnosis of impacted stones in the cystic duct is extremely difficult and indeed is often unsuccessful (LAING and JEFFREY 1983). This is because the cystic duct, when impacted by stones, does not contain enough bile to produce a sufficient acoustic contrast for their visualization.

Echogenic fat or calcific masses in the porta hepatis and duodenal gas may mimic the appearance of a cystic duct stone. However, a dilation of the cystic duct above the stone is usually seen, often associated with gallbladder distension and a hyperechoic density with posterior shadow outside the gallbladder and close to the CBD (Figs. 8.10, 8.11).

8.5.3 Choledocholithiasis

Choledocholithiasis occurs in approximately 15% of patients with gallstones (Way and Sleisenger 1983) and in 4% of postcholecystectomy patients (Glenn 1972). It is the second most common cause of biliary obstruction after malignant neoplasms (Baron et al. 1982; Salem and Vas 1981).

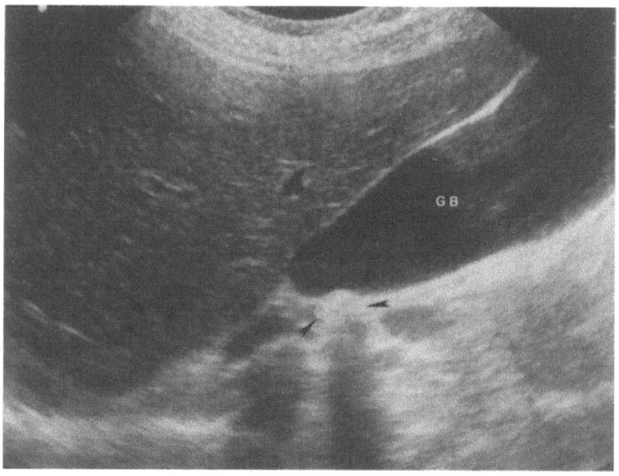

Fig. 8.10. Cystic duct stone: a hyperechoic density with posterior shadowing (*arrowheads*) is seen outside the gallbladder (*GB*), which appears distended

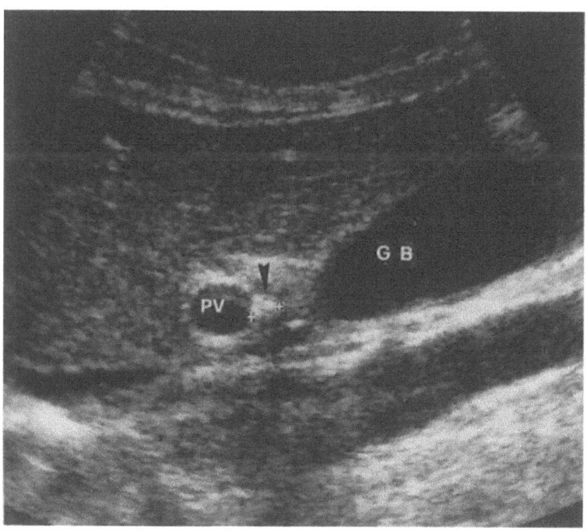

Fig. 8.11. A small hyperechoic stone (*arrowheads*) is seen in the distal portion of a dilated cystic duct close to the portal vein (*PV*). *GB*, Gallbladder

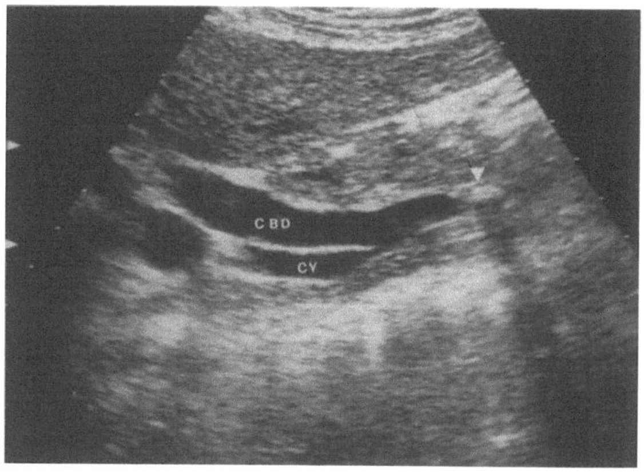

Fig. 8.12. US demonstrates a dilated common bile duct (*CBD*) with an obstructing hyperechoic stone (*arrowhead*) in its distal portion. The stone produces a posterior acoustic shadow. The cystic duct is also dilated (*CY*)

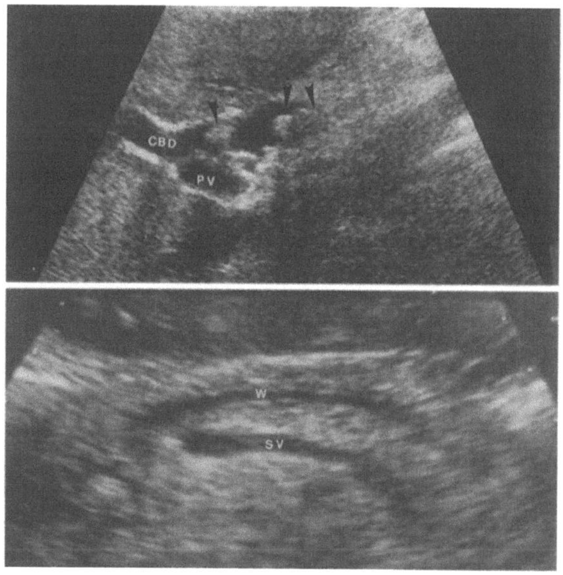

Fig. 8.13. Longitudinal scan showing a dilated common bile duct (*CBD*) which contains slightly echogenic stones (*arrowheads*), with poor posterior shadowing. In the same patient a dilated Wirsung's duct (*W*) is also seen. *PV*, Portal vein; *SV*, splenic vein

CBD stones are again identified by the presence of echogenic material within the duct. Stones may be a few millimeters in size or may reach a size of 2–3 cm. They may be obstructive, nonobstructive, or cause only intermittent obstruction. They can be located anywhere in the CBD, but are most often found in the intrapancreatic duct, where they may be associated with dilation of Wirsung's duct (Laing et al. 1986) (Figs. 8.12, 8.13).

The role of US in the diagnosis of choledocholithiasis is not well established. Early studies reported a low sensitivity of US in detecting CBD stones, ranging from 11% to 25% (CRONAN et al. 1983; GROSS et al. 1983). Improvements in equipment and transducer technology, together with optimization of scanning techniques (LAING et al. 1984), have increased sensitivity to 55%–80% (LAING et al. 1984, 1986; CRONAN 1986). In most series the lowest values are reported for detecting stones in the intrapancreatic bile duct and stones in nondilated ducts. The following conditions increase difficulties in CBD stone detections:

- Air in the duodenal lumen. This may impede ultrasonographic demonstration of stone in the more distal portion of the common duct (LAING et al. 1984, 1986). Air can be displaced by a water load as described in sect. 8.2 above.
- Air in the CBD after sphincterotomy or biliary-enteric anastomosis. This obstructs visualization of the CBD or mimics the presence of stones. This artifact is eliminated by changing the decubitus of the patient.
- Lack of bile surrounding the stone within the duct. There is a natural contrast between echo-free bile and hyperechoic stone in the intrapancreatic duct. If bile is lacking, the stone is more difficult to perceive, because it is surrounded by glandular parenchyma and tissue-stone contrast is less than bile-stone contrast (Fig. 8.14). Sometimes the common duct, if filled by a large number of stones, may produce a highly reflective image with a distal acoustic shadow that can be misdiagnosed as gas in the duodenum.
- Absence of distal acoustic shadowing. This occurs in about 10% of cases, especially when stones are in the distal portion of the common duct (EINSTEIN et al. 1984; KANE 1988; DEWBURY and SMITH 1983) (Fig. 8.14). It may be due to the stone content (most CBD stones are cholesterol and pigment stones, with little or no calcium) or to technical factors such as transducer frequency gain setting and reflection/refraction of the ultrasound by the duct walls (IRVING 1993). It is extremely difficult to distinguish a stone without a distal acoustic shadow from blood clots, neoplasms, or sludge.
- Calcifications in the pancreatic head. Calcifications due to chronic pancreatitis or calcified arterial vessels, if located along the course of the CBD, may simulate stones.
- Undilated common duct containing stones. This is reported to be present in about 25% of cases of acute biliary obstruction (CRONAN et al. 1983; LAING and JEFFREY 1983), and is considered to be due to the so-called "ball-valve effect," where the intermittent nature of the obstruction means that the bile duct does not become dilated. Another reason why ductal dilation may be absent in choledocholithiasis is that the examination is performed early after the acute onset of symptoms, in the "temporal lag" between obstruction and dilation (FRIED et al. 1981).

8.5.4 Mirizzi's Syndrome

Mirizzi's syndrome is an uncommon condition, occurring especially in elderly people, in which longstanding impaction of a stone in the ampulla of the gallbladder or in the cystic duct is associated with chronic inflammation. Two different types of

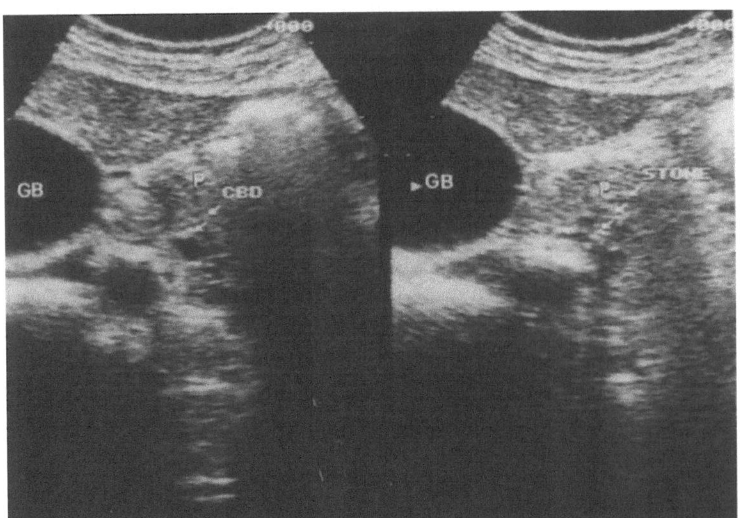

Fig. 8.14. Transverse scan at the head of the pancreas (*P*) shows mild dilation of the common bile duct (*CBD*) above an obstructing small stone (*arrowheads*) in the distal portion of the duct. *GB*, Gallbladder

Mirrizzi's syndrome are known (Mc Sherry et al. 1982). In the first type the impacted stone causes extrinsic compression of the common duct. Ultrasonographic findings are intrahepatic bile duct dilation, stones within the cystic duct or the gallbladder, and a normal-caliber common duct below the stone (Becker et al. 1984). In the second type, the stone penetrates the common hepatic duct by inflammation and erosion, causing a cholecystobiliary fistula. In such instances it is extremely difficult to make out whether the stone is in the gallbladder or in the CBD.

The ultrasonographic findings of bile duct dilation associated with a contracted gallbladder and stone in the CBD may indicate the presence of the second type of Mirizzi's syndrome.

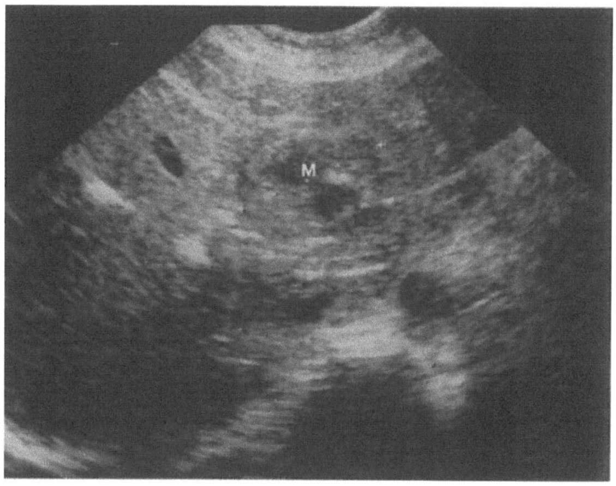

Fig. 8.15. Intrahepatic cholangiocarcinoma: a transverse scan shows an inhomogeneous solid mass (*M*) in the left hepatic lobe, close to the hepatic hilum

8.6 Bile Duct Neoplasms

8.6.1 Benign Neoplasms

Benign bile duct neoplasms, such as adenoma, papilloma, and cystoadenoma, are extremely rare (Kane 1988). Papilloma and solid adenomas appear as solid intraluminal masses with no acoustic shadow. Cystoadenoma appears as a multiloculated cystic mass; it may have malignant transformation.

8.6.2 Cholangiocarcinoma

Although rare, cholangiocarcinoma represents one of the most common neoplastic causes of biliary obstruction. Jaundice generally occurs when the tumor is relatively small and therefore it needs to be differentiated from lithiasis. Cholangiocarcinoma may be associated with congenital abnormalities of the biliary system such as Caroli's disease, or with inflammatory conditions such as sclerosing cholangitis (Ritchie et al. 1974).

Cholangiocarcinoma originates from epithelial cells of the biliary duct and may develop at any level within the biliary system, either intra- or extrahepatically. When it arises at the hepatic duct bifurcation it is called Klatskin's tumor (Klatskin 1965).

8.6.2.1 Intrahepatic Cholangiocarcinoma

There are no ultrasonographic findings that are specific for intrahepatic cholangiocarcinoma, which should be suspected when dilated bile ducts are seen in one or both hepatic lobes and there is no evidence of any other cause of obstruction. Typical features include marked biliary duct dilation in the presence of normal extrahepatic bile ducts. In some cases a hepatic mass is identified usually ill-defined inhomogeneous, and with involvement of dilated intrahepatic bile ducts (Fig. 8.15). Diagnosis, however, requires cytological or histological confirmation obtained under ultrasonographic guidance.

8.6.2.2 Klatskin's Tumor

With Klatskin's tumor, a focal mass is often not directly appreciated at the hepatic hilum. Diagnosis will be based on the ultrasonographic findings of no union of dilated intrahepatic ducts at the hepatic hilum with a normal extrahepatic bile duct (Meyer and Weinstein 1983) (Fig. 8.16).

8.6.2.3 Cholangiocarcinoma of the CBD

Cholangiocarcinoma of the CBD, also called ductal carcinoma (Braasch 1973; Dalla Palma et al. 1980), is not always detectable on US, especially when it is of small size or does not cause bile duct obstruction.

If the tumor has a polypoidal growth inside the duct, US will show increased echogenicity within the dilated common duct, difficult to differentiate from choledocholithiasis or mud. However, findings typical with lithiasis, including posterior acoustic shadowing and dilation of the duct distal to the

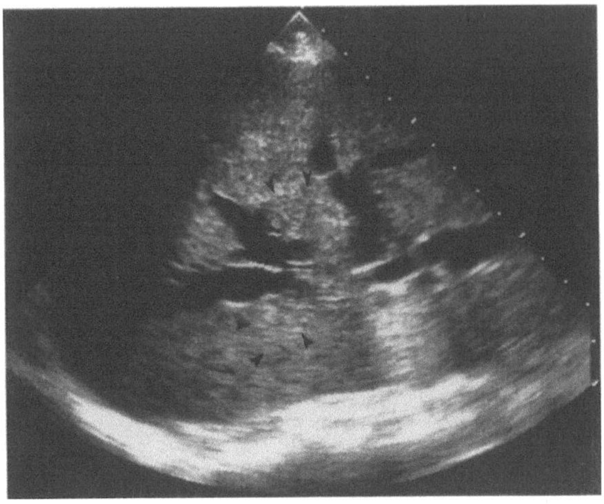

Fig. 8.16. Klatskin's tumor: dilated right and left hepatic ducts at the hilum are surrounded by inhomogeneous hepatic tissue (*arrowheads*)

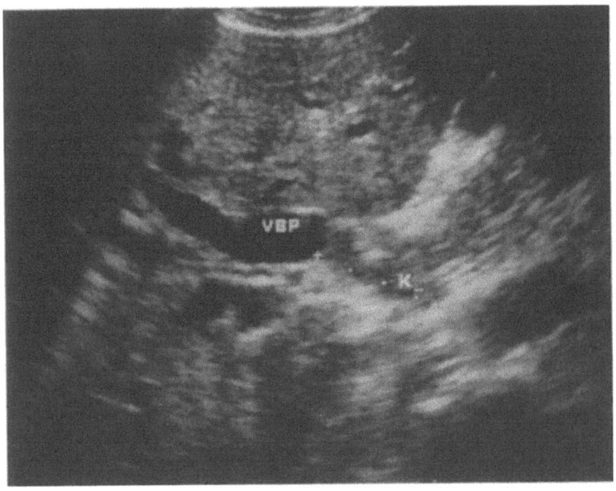

Fig. 8.17. Cholangiocarcinoma of the common bile duct: US demonstrates a solid mass (*K*) inside the lumen of the common bile duct (*VBP*), growing along the long axis of the duct

obstruction, may be used as differential criteria (Okuda and Oshibuchi 1989).

Sometimes the neoplasm has submucosal growth along the long axis of the duct (Fig. 8.17). However, with the more infiltrating type, a solid, ill-defined mass involving the CBD may be seen on US (Fig. 8.18). In such instances, differentiating tumor from lymphadenopathy or spread to the porta hepatis from other primary neoplasms, such as stomach or breast cancer or pancreatic or gallbladder tumors, is often impossible (Mittelstaedt 1987)

In the diagnosis of biliary and pancreatic neoplasms US has shown a sensitivity of 84%, a specificity of 95%, and an accuracy of 88% (Lindsell 1990) (Fig. 8.19). In addition to being diagnostic, US has a relevant role in assessing the operability of neoplasms, expecially in assessing tumor spread and invasion of other structures such as the portal vein, hepatic artery, liver parenchyma, and regional lymph nodes.

In a comparison of US, computed tomography, percutaneous transhepatic cholangiography, and endoscopic retrograde cholangiopancreatography in assessing the operability of biliary neoplasms, their respective accuracies were shown to be 71%, 42%, 58%, and 25% (Gibson et al. 1986).

8.6.3 Ampullary Carcinoma

This neoplasm is extremely difficult to demonstrate on US; however, it should be suspected whenever a dilated common duct and Wirsung's duct are seen. Since these ultrasonographic findings may also be

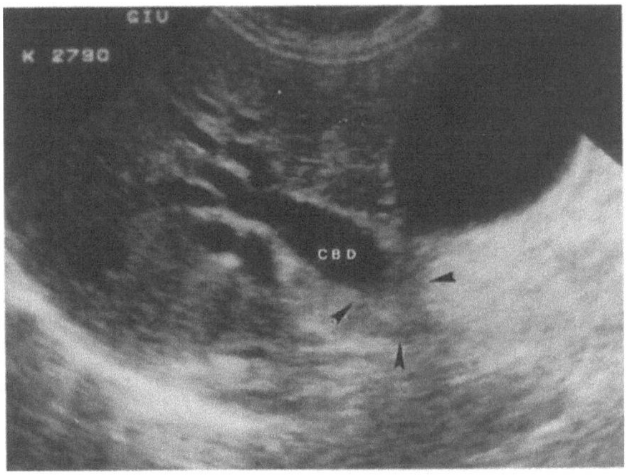

Fig. 8.18. Cholangiocarcinoma of the common bile duct. A long axis view of the dilated ocmmon bile duct (*CBD*) shows a solid mass involving the common bile duct (*arrowheads*)

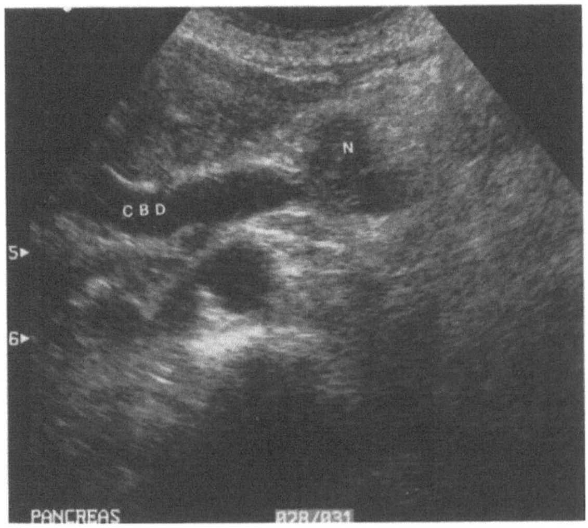

Fig. 8.19. Neoplasm (*N*) of the head of the pancreas involving the distal common bile duct (*CBD*)

found in the presence of inflammation of the ampulla, small tumor, or pancreatitis in the head of the pancreas, diagnosis is only to be achieved by endoscopy.

8.7 Inflammation

8.7.1 Recurrent Cholangitis

Recurrent cholangitis is an inflammatory condition of the biliary system caused by recurrent obstruction of the bile ducts. Symptoms are biliary colic, fever, chills, and jaundice. The condition is usually related to previous surgery on the biliary system, or it may represent a complication of choledocholithiasis, congenital abnormalities, or neoplasm.

Bile duct dilation is the result of repeated inflammatory episodes, with progressive destruction and loss of elasticity of the duct. (LIM et al. 1990; CHAN et al. 1989). It may be seen independently in the presence of a mechanical obstruction, and it may involve the biliary system either above or below a stone. Extrahepatic bile duct dilation occurs in 85%–100% of patients, and intrahepatic duct dilation may be found in up to 66% of cases (LIM 1991; LIM et al. 1990).

Ultrasonographic findings include some particular features, as follows:

- Thickened and more echoic portal triads are frequently seen, due to periportal fibrosis resulting from repeated infections of the biliary system (Fig. 8.20).

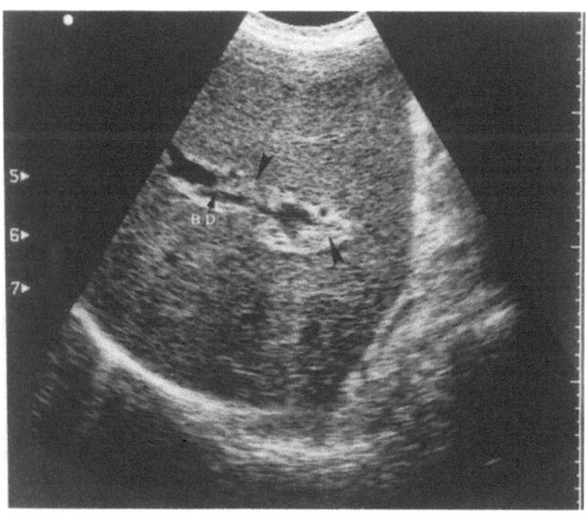

Fig. 8.20. Recurrent cholangitis: periportal fibrosis (*arrowheads*) is associated with segmentary dilation of intrahepatic bile ducts (*BD*)

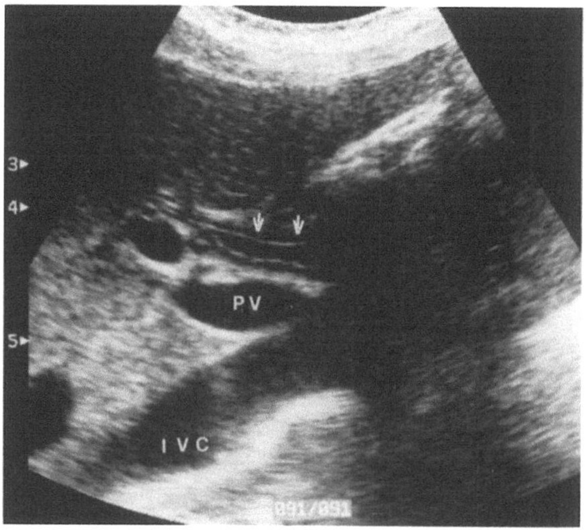

Fig. 8.21. Recurrent cholangitis: US demonstrates a common hepatic duct with thickened wall (*arrowheads*). *PV*, Portal vein; *IVC*, inferior vena cava

- Thickening of the bile duct is more easily appreciated at the level of the CBD than at the level of the intrahepatic biliary radicles (Fig. 8.21).
- Biliary mud, due to biliary stasis, appears as a hypoechoic collection of material with no posterior acoustic shadowing. Usually it layers in the dependent portion of the ducts and moves slowly when the patient's position is changed (BEHAN and KAZAM 1978). Echoes produced by biliary mud, especially by bilirubinate granules, may be confused with reverberation or noise related to the equipment. Some authors suggest that the presence of mud in screening US should be considered a warning sign of development of suppurative cholangitis and should be followed by biliary system decompression (ISHIDA et al. 1987).
- Calculi inside the bilated ducts are often difficult to visualize, because of their small size and soft and sludge-like nature, and they may be confused with gas bubbles.
- Gallstones may be seen in about 72% of patients with recurrent cholangitis (LIM 1991).

8.7.2 Primary Sclerosing Cholangitis

The features of primary sclerosing cholangitis are inflammation, sclerosis, and obstruction extending progressively from the intra- to the extrahepatic ducts. The condition may occur at any age in life Features of this disease in children are discussed elsewhere in this book.

In the early stage of the disease, it is difficult to show any typical ultrasonographic findings. Later,

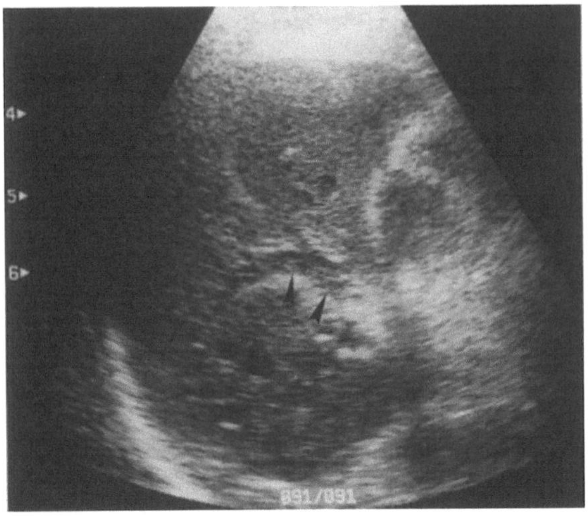

Fig. 8.22. Sclerosing cholangitis: US shows a dilated intrahepatic duct containing biliary mud (*arrowheads*)

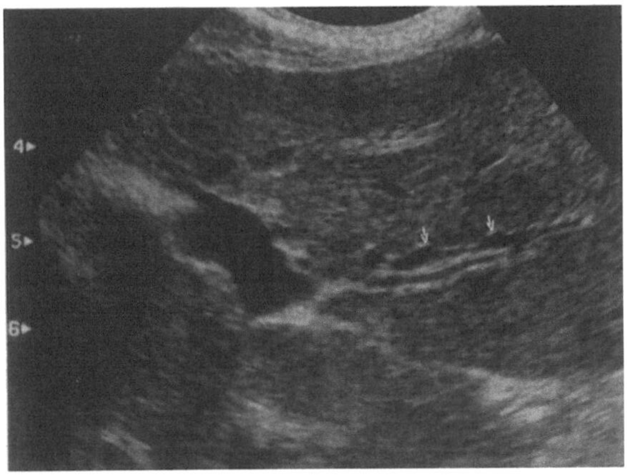

Fig. 8.23. Sclerosing cholangitis: multisegmental intrahepatic biliary dilation, with typical strictures and beaded appearing cholangectases, is seen (*arrows*)

when jaundice occurs, irregular multisegmental intra- and extrahepatic biliary dilation may be seen, associated with thickening of the common duct walls and tight stenosis of the lumen. The common duct generally shows an echogenic intraluminal content consisting of pus, biliary sludge, or desquamated bile duct epithelium (Fig. 8.22). However, because of difficulties in demonstrating the typical strictures and "beaded-type" cholangiectases on US (Fig. 8.23), the diagnosis should be made by endoscopic retrograde cholangiopancreatography.

Similar biliary changes on US are reported in patients affected by AIDS. Some opportunistic organisms, such as *Cryptosporidium*, *Cytomegalovirus*, and *Candida albicans* (Dolmatch et al. 1987; Schneiderman 1988), produce cholangitis with elevated cholestasis parameters and bile duct dilation even in the absence of mechanical obstruction. Thickening of the wall is another common finding. However, in AIDS patients, bile duct dilation may also be caused by enlarged lymph nodes at the hepatic hilum, secondary to lymphoma or Kaposi's sarcoma.

8.7.3 Parasitic Infestations

Ascariasis is widespread in the Far East, Latin America, and Africa, and is due to the wide distribution of a worm with a propensity to enter the biliary system. However, the large-scale emigration of people from these parts of the world to Europe and the United States of America has increased the geographical diffusion of this infection. Other parasites probably related to bile duct infection and endemic in South Est Asia (Wastie and Cunningham 1973) are *Clonorchis sinensis*, *Opisthorchis viverrini*, *Entamoeba coli*, and *Fasciola hepatica* (Federle et al. 1982).

The adult worms are 10 cm long and 3–6 cm thick. They are easily detected on US and appear as nonshadowing strips or mobile, echoic, tubular structures with an anechoic central tube when scanned along their longitudinal axis. They can be seen inside the common duct or the gallbladder (Khuroo et al. 1992; Schulman et al. 1982; Cerri et al. 1983). Lack of mobility of the worm is considered a criterion of efficaceous medical therapy.

8.8 Postsurgical Complications

On ultrasonographic examination, various complications may be found after surgery, occurring early or late as shown in Table 8.2. The role of US is to assess the presence of intra- or extrahepatic fluid collection and bile duct dilation. During the first week after surgery, the most frequent

Table 8.2. Complications after surgery on the biliary tract

Early complications
Biloma
CBD ligation
CBD transection
Late complications
Stenosis
Lithiasis
Inflammation

complication is biloma, often associated with jaundice. Delayed complications usually include stenosis of the bilioenteric anastomosis or choledochocholedochostomy.

8.8.1 Biloma

Bilomas are collections of bile within the hepatic parenchyma or peritoneal cavity, generally occurring at a variable delayed interval after surgery. They may also occur after blunt or penetrating abdominal trauma or interventional procedures.

On US a biloma usually appears as a sharply defined anechoic mass, with or without internal loculations, with acoustic enhancement (Fig. 8.24). The presence of bile generates an inflammatory reaction, producing loculations and a pseudocapsule all around the fluid collection.

Bilomas generally show a continuity with the liver or the biliary structure and they must be differentiated from hematoma and abscesses, which have a more echoic content.

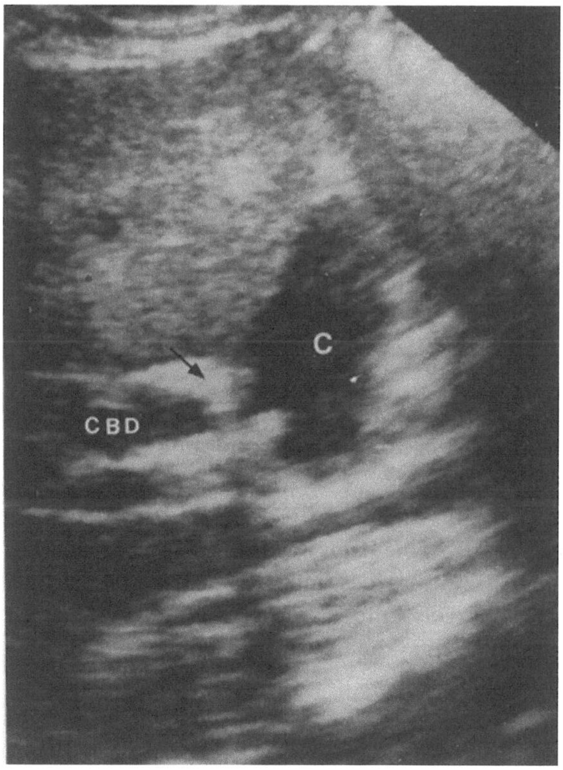

Fig. 8.24. Bile duct obstruction occurring 18 days after cholecystectomy due to a surgical clip which shows as a small echogenic focus with posterior shadowing (*arrow*). An anechoic bile collection with posterior acoustic enhancement (*C*) is seen

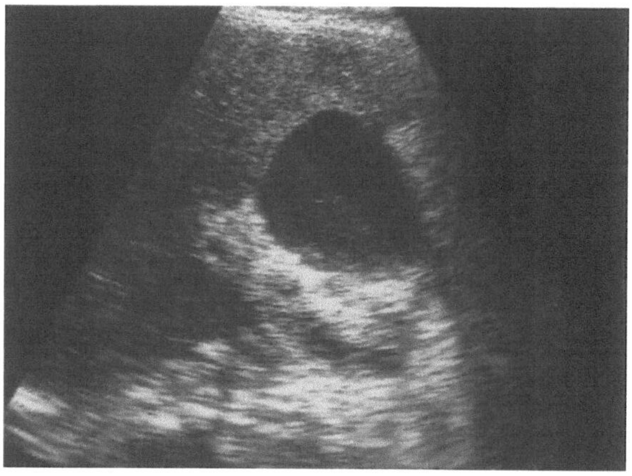

Fig. 8.25. Biloma. US shows a fluid collection with posterior acoustic enhancement and echogenic content

Bilomas may be complicated by inflammation. In this case the fluid collection may contain gas bubbles and echogenic material (Fig. 8.25).

8.8.2 Stenosis of Surgical Biliary Anastomosis

In patients who have undergone surgical bilioenteric anastomosis, ultrasonographic examination is limited, since the presence of air may obscure the bile duct or may mimic the presence of stones in the intrahepatic bile ducts. However, almost always these artifacts may be eliminated by changing the patient's position.

Some authors suggest that US allows an assessment of the patency of the anastomosis, in that the presence of gas in the bile ducts implies patency, whereas bile duct dilation without evidence of air will suggest a partial or tight stenosis of the anastomosis (Sauerbrei et al. 1980). We have used US during follow-up of such patients to evaluate the size of the biliary tree and to assess the presence of luminal narrowing or obstruction at the site of the anastomosis (Fig. 8.26). Bile obstruction may be caused by tumor recurrence (Fig. 8.27) if surgery has been performed for a neoplasm.

After liver transplantation, biliary strictures may occur due to stenosis of the biliary anastomosis. Other causes include hepatic artery occlusion, pretransplantation primary sclerosing cholangitis, or the use of particular organ preservation solutions (Campbell et al. 1994) (Fig. 8.28).

The incidence of biliary strictures is higher in young people, after choledochojejunostomy, or in patients who have received allografts for sclerosing cholangitis.

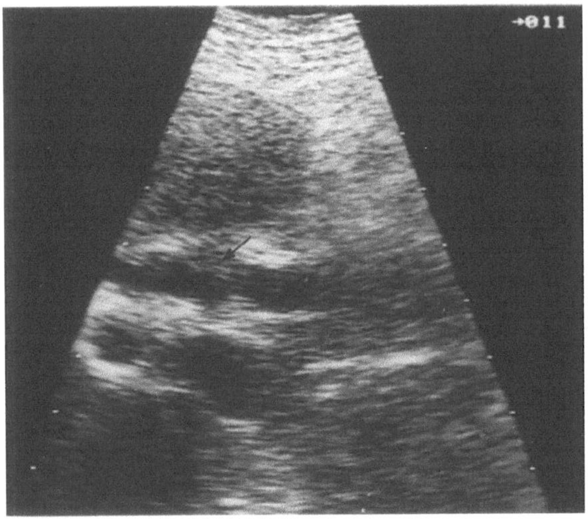

Fig. 8.26. Choledocho-choledochostomy: granulation tissue produces a narrowing of the anastomosis (*arrow*)

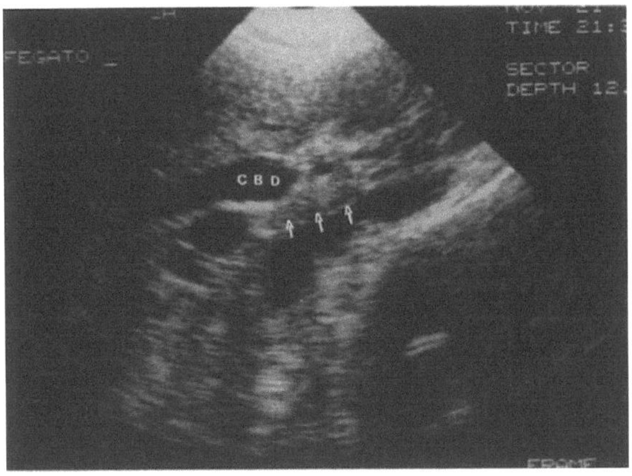

Fig. 8.27. One month after Whipple's procedure for carcinoma of the head of the pancreas, US demonstrates common bile duct (*CBD*) dilation due to tumor recurrence (*arrows*)

Intrahepatic strictures may be associated with extrahepatic strictures and are seen in 11% of patients with nonanastomotic extrahepatic strictures and in 17% of patients with anastomotic strictures (Sheng et al. 1993; Campbell et al. 1994).

8.9 Follow-Up of Biliary Interventional Procedures

US is very useful in patient follow-up after percutaneous or endoscopic biliary drainage. It allows assessment of the caliber and content of the biliary tree, and the presence of complication such as fluid collection. In cases of the latter, US may also be used to guide percutaneous drainage.

US can clearly visualize any endoprostheses whether plastic or metallic; it will appear as two par-

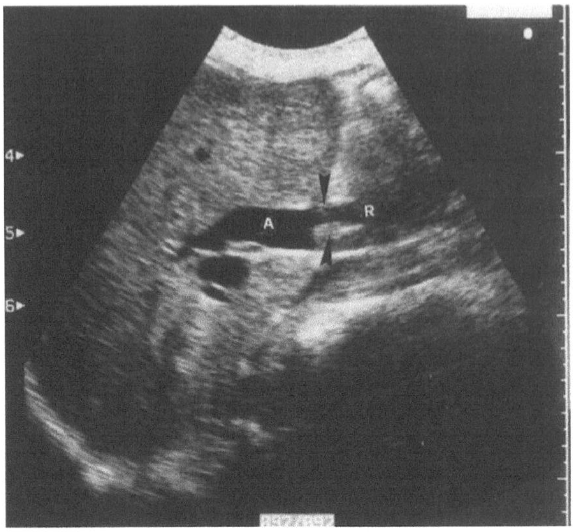

Fig. 8.28. Transplanted liver: narrowing of the anastomosis (*arrowheads*) between allograft choledochus (*A*) and recipient's choledochus (*R*)

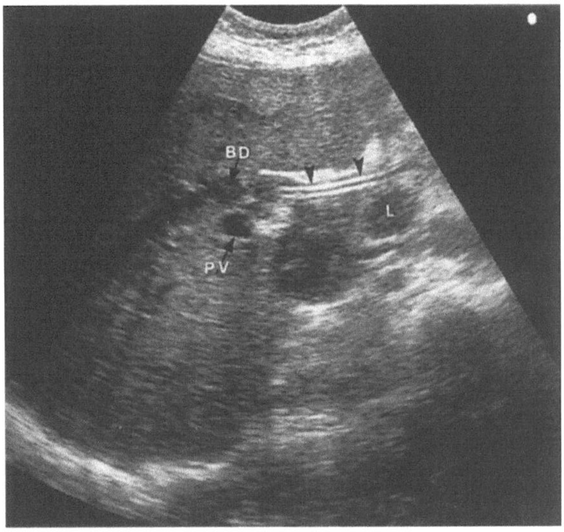

Fig. 8.29. US demonstrates a biliary stent (*arrowheads*) inside the common duct in a patient with gastric carcinoma and enlarged lymph nodes (*L*) at the hepatic hilum. A dilated bile duct (*BD*) with echogenic content is also seen. *PV*, Portal vein

allel hyperechoic strips inside the common duct (Fig. 8.29). However, it is impossible to assess stent patency by US, although progressive dilation of the bile ducts may be considered an indirect sign of malfunctioning.

References

Baron RL, Stanley RJ, Lee JKT, Koehler RE, Melson GL, Balfe DM, Weyman PJ (1982) A prospective comparison of the evaluation of biliary obstruction using computed tomography and ultrasonography. Radiology 145:91–98

Becker CD, Hassler H, Terrier F (1984) Preoperative diagnosis of the Mirizzi syndrome: limitations of sonography and computed tomography. AJR 143:591–596

Behan M, Kazam E (1978) Sonography of the common bile duct: value of the right anterior oblique view. AJR 130:701–709

Braasch JW (1973) Carcinoma of the bile duct. Surg Clin North Am 53:1217

Bruneton JN, Roux P, Fenart D, Caramella E, Occelli JP (1981) Ultrasound evaluation of common bile duct size in normal adult patients and following cholecystectomy. A report of 750 cases. Eur J Radiol 1:171–172

Campbell WL, Sheng R, Zajko AB, Abu-Elmagd K, Demetris AJ, (1994) Intrahepatic biliary strictures after liver transplantation. Radiology 191:735–740

Cerri GG, Leite GJ, Simoes JB, Correia Da Rocha DJ, Albuquerque FP, Machado MCC, Magalhaes A (1983) Ultrasonic evaluation of *Ascaris* in the biliary tract. Radiology 146:753–754

Chan FL, Man SW, Leong LLY, Fan ST (1989) Evaluation of recurrent pyogenic cholangitis with CT: analysis of 50 patients. Radiology 170:165–169

Cooperberg PL, Li D, Wong P, Cohen MM, Burhenne HJ (1980) Accuracy of common duct size in the evaluation of extrahepatic biliary obstruction. Radiology 135:141–144

Cronan JJ (1986) US diagnosis of choledocholithiasis: a reappraisal. Radiology 161:133–134

Cronan JJ, Mueller PR, Simeone JF, O'Connell RS, van Sonnenberg E, Wittenberg J, Ferrucci JT Jr (1983) Prospective diagnosis of choledocholithiasis. Radiology 146:467–469

Dalla Palma L, Rizzatto G, Pozzi-Mucelli RS, Bazzocchi M (1980) Gray-scale ultrasonography in the evaluation of carcinoma of the gallbladder. Br J Radiol 53:662

Darweesh RMA, Dodds WJ, Hogan WJ, et al (1988) Fatty-meal sonography for evaluating patients with suspected partial common duct obstruction. AJR 151:63–68

Dewbury KC, Smith CL (1983) The misdiagnosis of common bile duct stones with ultrasound. Br J Radiol 56:625–630

Dolmatch BL, Liang FC, Federle MP, Jeffrey RB, Cello J (1987) AIDS-related cholangitis: radiographic findings in nine patients. Radiology 163:313–316

Einstein DM, Lapin SA, Ralls PW, Halls JM (1984) The insensitivity of sonography in the detection of choledocholithiasis. AJR 142:725–728

Federle MP, Cello JP, Laing FC (1982) Recurrent pyogenic cholangitis in Asian immigrants: use of ultrasonography, computed tomography, and cholangiography. Radiology 143:151–156

Fried AM, Bell RM, Bivins BA (1981) Biliary obstruction in a canine model: sequential study of the sonographic threshold. Invest Radiol 16:317–319

Gibson RN, Yeung E, Thompson JN, et al (1986) Bile duct obstruction; radiologic evaluation of level, cause and tumor resectability. Radiology 160:43–47

Glazer GM, Filly RA, Laing FC, (1981) Rapid change in caliber of the non-obstructed common duct. Radiology 140:161–162

Glenn F(1972) Postcholecystectomy choledocholithiasis. Surg Gynecol Obstet 134:249–252

Graham MF, Cooperberg PL, Cohen MM, Burhenne HJ (1980) The size of the normal common hepatic duct following cholecystectomy: an ultrasonographic study. Radiology 135;137–139

Gross BH, Harter LP, Gore RM, Callen PW, Filly RA, Shapiro HA, Goldberg HI (1983) Ultrasonic evaluation of common bile duct stones: prospective comparison with endoscopic retrograde cholangiography. Radiology 146:471–474

Irving HC (1993) Gallbladder and biliary tree. In: Cosgrave D, Meire H, Dewbury K (eds) Abdominal and general ultrasound, vol 1. Churchill Livingstone, Edinburgh, pp 209–226

Irving HC, Bates J (1993) Bile duct pathology. In: Cosgrave D, Meire H, Dewbury K (eds) Abdominal and general ultrasound, vol 1. Churchill Livingstone, Edinburgh, pp 172–184

Ishida H, Yagisawa H, Nasu H, Arakawa H, Masamune O (1987) Ultrasonography of acute obstructive suppurative cholangitis: serial observation by ultrasound. J Clin ultrasound 15:51

Kane RA (1988) The biliary system. In: Kurtz AB, Goldberg BB (eds) Gastrointestinal ultrasonography. Churchill Livingstone, Edinburgh, p 75–137 Clinics in diagnostic ultrasound

Khuroo MS, Zargar SA, Yattoo GN, et al (1992) Sonographic findings in gallbladder ascariasis. J Clin Ultrasound 20:587–591

Klatskin G (1965) Adenocarcinoma of the hepatic duct at its bifurcation within the porta hepatis. Am J Med 38:241–256

Laing FC, Jeffrey RB (1983) Choledocholithiasis and cystic duct obstruction: difficult ultrasonographic diagnosis. Radiology 146:475–479

Laing FC, Jeffrey RB, Wing VW (1984) Improved visualisation of choledocholithiasis by sonography. AJR 143:949–952

Laing FC, Jeffrey RB, Wing VW, Nyberg DA (1986) Biliary dilatation: defining the level and cause by real-time US. Radiology 160:39–42

Lim JH (1991) Oriental cholangiohepatitis: pathologic, clinical, and radiologic features. AJR 157:1–8

Lim JH, Ko YT, Lee DH, Hong KS (1990) Oriental cholangiohepatitis: sonographic findings in 48 cases. AJR 155:511–514

Lindsell DRM (1990) Ultrasound imaging of pancreas and biliary tract. Lancet 335:390

McSherry CK, Ferstumberg H, Virshup M (1982) Mirzzi's syndrome: suggesated classification and surgical therapy. Surg Gastroenterol 1:219–225

Meyer DG, Weinstein BJ (1983) Klatskin tumors of the bile ducts: sonographic appearance. Radiology 148;803–804

Mittelstaedt CA (ed) (1987) Abdominal ultrasound. Churchill Livingstone, New York, p 81–162

Muller PR, Ferrucci JT, Simeone JF, van Sonnenberg E, Hall PDA, Wittenberg J (1982) Observations on the distensibility of the common bile duct. Radiology 142:467–472

Nakayama F (1982) Intrahepatic calculi. A special problem in East Asia. World J Surg 6:802

Niederau C, Muller J, Sonnenberg A, et al (1983) Extrahepatic bile ducts in healthy subjects, in patients with cholelithiasis, and in post cholecystectomy patients: a prospective ultrasonic study. J Clin ultrasound 11:23–27

Okuda K, Oshibuchi M (1989) Imaging diagnosis of obstructive jaundice. Hepatogastroenterology 36:398

Ritchie JK, Allan RN, MaCartney J, Thompson H, Hawley PR, Cooke WT (1974) Biliary tract carcinoma associated with ulcerative colitis. Q J Med 43:263

Salem S, Vas W (1981) Ultrasonography in evaluation of the jaundiced patient. J Can Assoc Radiol 31:30–34

Sauerbrei EE, Cooperberg PL, Gordon P, Li D, Cohen MM, Burhenne HJ (1980) The discrepancy between radiographic and sonographic bile duct measurements. Radiology 137:751–755

Sauerbrei EE, Nguyen KT, Nolan RL (eds) (1992) Abdominal sonography. Raven, New York, pp 51–72

Schneiderman DJ (1988) Hepatobiliary abnormalities of AIDS Gastroenterol Clin North Am 17:615–630

Schulman A, Loxton AJ, Heydenrych JJ, Abdurahman KE (1982) Sonographic diagnosis of biliary ascariasis. AJR 139:485–489

Sheng R, Zajko AB, Campbell WL, Abu-Elmagd K (1993) Biliary strictures in hepatic transplants: prevalence and types in patients with primary sclerosing cholangitis vs those with other liver diseases. AJR 161:297–300

Simeone JF, Butch RJ, Mueller PR, et al (1985) The bile ducts after a fatty meal: further sonographic observations. Radiology 154:763–768

Taylor KJW, Rosenfield AT, De Graaff CS (1979) Anatomy and pathology of the biliary tree as demonstrated by ultrasound. In: Taylor KJW (ed) Diagnostic ultrasound in gastrointestinal disease. Churchill Livingstone, Edinburgh, pp 103–121. (Chinics in diagnostic ultrasound)

Way LW, Sleisenger MH (1983) Biliary obstruction, cholangitis, and choledocholithiasis. In: Sleisenger SH, Fordtran JS (eds) Gastrointestinal disease. Saunders, Philadelphia, pp 1389–1403

Wastie ML, Cunningham IGE (1973) Roentgenologic findings in recurrent pyogenic cholangitis. AJR 119:71–77

Wu CC, Ho YH, Chen CY (1984) Effect of aging on common bile duct diameter: a real time sonographic study. J Clin ultrasound 12:473

9 CT of the Biliary Tree

G.M. Richter and L. Grenacher

CONTENTS

9.1 Introduction

Anatomically, the biliary tree really originates at the level of the hepatocyte draining its bile into the intercellular microtubular structures. These microtubuli join at the lobular level to form small ducts, and from then on towards the papilla of Vater are lined with cholangiocellular epithelium. This cholangiocytic cell type accompanies the entire ductular arborization and is also found in the segmental ducts and the common bile duct near the papilla of Vater, with the same microscopic and biochemical properties. For practical reasons, therefore, this chapter is organized according to the disease types of this epithelio-ductular substrate.

During the last ten years computed tomography (CT) has developed into one of the primary techniques in the diagnostic approach to biliary tract disease. It is the method of choice for detection and staging of malignant tumors. CT also plays a major role in the differential diagnosis of biliary tract lesions by identifying and showing the nature of disseminated or focal duct dilatation, parenchymal lesions, and intra- or extraductal calcifications. In a variety of diagnostic problems, the combination of CT with other techniques such as retrograde or transhepatic cholangiography or external or intraductal ultrasonography helps to establish a final diagnosis.

The basic prerequisite for obtaining reliable diagnostic information is individual planning and adaptation of CT technique to the diagnostic problem at hand, out of a host of different technical approaches, the right one needs to be chosen. Fourth-generation scanners allow consecutive arterial and venous phase contrast-enhanced studies within a single examination. Volume scanning of the liver after biliary tract opacification by intravenously administered bile duct contrast agent gives simultaneous 3D imaging of both the liver parenchyma and central and peripheral ducts, resulting in what is known as CT cholangiography, with an image quality almost comparable to that of direct cholangiography. In general, the right technical approach includes nonenhanced scanning of the entire liver from the diaphragm distal to the ampulla, including the caudal parts of the right liver lobe. Contrast-enhanced scanning is adjusted to the individual diagnostic problem. Up-to-date scanning methodology is based on spiral scanning modes with imaging parameters that allow both high spatial and high contrast resolution.

G.M. Richter, MD, Abteilung Radiodiagnostik, Radiologische Universitätsklinik Heidelberg, Im Neuenheimer Feld 110, 69120 Heidelberg, Germany
L. Grenacher, MD, Abteilung Radiodiagnostik, Radiologische Universitätsklinik Heidelberg, Im Neuenheimer Feld 110, 69120 Heidelberg, Germany

Therefore, the smallest increment possible in reference to the time frame imposed by the heat capacity of the tube and the length of the scanning region should be applied for the contrast-enhanced study. In selected situations, appropriate phasing of the contrast bolus profile should be provided by bolus tracking to obtain maximum lesion-to-liver contrast.

3D CT cholangiography is a newly developed technique which is presently under intensive evaluation to define its practical value. More technical details will be given individually in reference to the lesions discussed within this chapter. The chapter is subdivided into five main parts dealing with the following topics:

- Biliary tract tumors
- Inflammatory disease of the bile ducts
- Traumatic lesions of the bile ducts
- Anomalies of the biliary tree
- Future developments

The discussion of these five main topics has been restricted to those lesions or conditions that are reasonably amenable to diagnosis by CT.

9.2 Biliary Tract Tumors

9.2.1 Benign Tumors

Benign tumors of the biliary tree are very rare. Only 0.1% of biliary tract surgery is performed for such lesions (Burhans and Myers 1971). Polypoid or adenomatous hyperproliferations are usually found only in the larger ducts. The adenoma of the papilla of Vater should be included in this class as it originates in the cholangiocellular epithelium.

9.2.1.1 Papillomas

General Aspects

Two out of three of all benign tumors of the biliary tree are polyps, papillomas, or adenomas, which develop as a benign proliferation of ductular cholangiocytes. The papilloma is the most frequent subtype. Transformation from benign to malignant is uncommon. Its most frequent site is the papilla of Vater or the ampullar (47%), followed by the common bile duct (27%). No gender predilection of the tumor has been described.

Clinical Findings

The vast majority of patients (90%) with these tumors present with obstructive jaundice as the initial clinical symptom (McIntyre and Cheng 1968). Commonly patients complain about right upper quadrant pain. Tumor makers are uncharacteristic. The liver type of phosphatase and bilirubin levels may be elevated, depending on the extent of biliary obstruction.

CT Findings

As stated before, two out of three of these tumors develop within the pancreatic head and hence pose the problem of the differential diagnosis of a pancreatic head lesion. The most typical finding is a circumscribed hypodense mass in the projection of the distal course of the biliary tree. Centrally it may even appear cystic as a result of mucin production. This is a very specific feature of ampullary tumors (Fig. 9.1). Small polyps, however, usually are not directly depicted by CT. When a mass is present near the ampulla, differentiation from malignant disease, particularly from ampullary carcinoma, is important. For a long time the general opinion was that such differentiation is not possible by CT. However, by employing a combination of thin-slice spiral scanning, water distension of the duodenum, and pharmacological paralysis by intravenous administration of *N*-buthylscopolaminebromide ("hydro-CT"), it does seem possible to delineate the tumor and characterize its nature (Richter et al. 1996). Recently, CT cholangiography has been advocated as an adjunct to regular CT for these tumors.

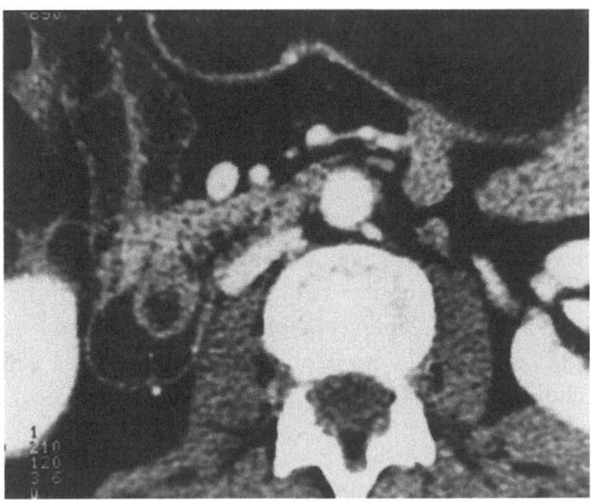

Fig. 9.1. Papillary adenoma projecting into the lumen of the duodenum, demonstrated using the hydro-CT technique with a fully expanded duodenal wall and maximum mucosal contrast in the wall. The center of the lesion is markedly hypodense because of mucinous contents (confirmed at surgery)

9.2.1.2 Cystadenomas

General Aspects

Cystadenomas of the biliary tree are very rare and are predominantly found in women. Morphologically a cystadenoma presents as an intrahepatic and focally well circumscribed lobulated cystic mass, sometimes multilocular. Sizes may vary. In larger tumors multiple septi may be found, underlining the polycystic character, which is the result of the proliferating and mucin-producing cholangiocytic epithelium that forms the cellular basis of the lesion. The intercystic tissue is composed of a well-vascularized interstitium lined by cholangiocytes, sometimes in multiple layers. However, within one lesion transition to malignancy may be found. Practically, the lesion should then be called cystadenocarcinoma. From both the morphologic and the biologic point of view biliary cystadenomas are very similar to pancreatic cystadenomas. Usually, the tumors develop at a very peripheral level of the biliary tree. In some reports they have also been considered as bile duct hamartomas.

Clinical Findings

Usually, if the lesion is truly benign, there is no specific clinical sign. In the vast majority of cases the lesions are discovered accidentally as a result of imaging studies for unrelated reasons. Usually, tumor markers are absent.

CT Findings

The cystic nature of the lesions determines its CT morphology. In general it presents with a cystic mass in which several or multiple septi and solid tumor nodules may be discovered. In nonenhanced studies the intralesional density is significantly higher than in normal liver cysts. The shape may be polygonal. In contrast-enhanced studies the relatively high level of interstitial vascularization is reflected by significant contrast uptake in septal and capsular tissue. Presently, no CT technique is known that would be specific for the differential diagnosis of benign cystadenoma versus cystadenocarcinoma.

9.2.2 Malignant Tumors

The incidence of malignant bile duct tumors shows wide geographic and demographic variations. In countries with an established tumor and autopsy registry malignant bile duct tumors are recorded as making up between 0.1% and 2% of all tumors. SAKO et al. (1957) described an 0.12% overall incidence of malignant extrahepatic bile duct tumors. In an old review KIRSCHBAUM (1941) reported a widely varying incidence of 0.01%–0.2% of bile duct tumors depending on the ethnic background and structural organization of the reviewed studies. KUWAYTI et al., in a more recent study, however, found an incidence of 2% (KUWAYTI et al. 1957).

A classification of malignant bile duct tumors according to their level of origin within the biliary tree has been widely accepted and reflects not merely anatomic considerations but also the different pathophysiologic properties and the biologic behavior of the tumors. The following five subgroups have been identified:

1. Parenchymal or intrahepatic bile duct carcinoma (= cholangiocellular carcinoma)
2. Hilar bile duct carcinoma (Klatskin's tumor)
3. Infrahilar bile duct carcinoma (ductal cholangiocarcinoma, common bile duct carcinoma)
4. Ampullary carcinoma
5. Secondary tumors (bile duct metastasis)

9.2.2.1 Cholangiocellular Carcinoma

General Aspects

The typical microscopic finding in cholangiocellular carcinoma (CCC) is a mucin-producing adenocarcinoma (e.g., in 98% of cases, ALBORES-SAAVEDRA and HENSON 1986). It may induce desmoplastic reactions to a widely varying extent that can result in a scirrhous tumor type with abundant fibroplastic tissue. Therefore, the degree of vascularization also varies to the extent that the tumor may be characterized as hypo- or hypervascular. Hypervascularity is seen in tumors with high cellular density Usually CCC develops as a monotopic tumor. A multilocular appearance is regarded as intraductular metastatic spread. By contrast to hepatocellular carcinoma (HCC), cholangiocarcinoma is not associated with viral infection of the liver or liver cirrhosis. However, ulcerative colitis was early identified as a predisposing factor (ROBERTS-THOMPSON et al. 1973). In the thorotrastosis complex CCC plays also an important role: it appears to be the most frequent hepatic tumor resulting from intravenous injection of thorium dioxide. Tumor markers are not specific. A male preponderance is reported with varying male-to-female ratios, e.g., 4:3, 3:2. The TNM classification of CCC is as follows:

- T1: Solitary tumor smaller than 2 cm in diameter, no major vessel infiltration
- T2: Solitary tumor smaller than 2 cm in diameter with vessel infiltration

or: Multiple tumors smaller than 2 cm in diameter within one lobe
T3: Solitary tumor larger than 2 cm in diameter with vessel infiltration
or: Multiple tumors smaller than 2 cm in diameter within one lobe with vessel infiltration
T4: Multiple tumors in both liver lobes
or: Tumors infiltrating major vessel structures (portal, hepatic vein, hepatic artery)
N1: Regional lymph node infiltration
M1: Distant metastasis

This classification is parallel to the staging system of HCC.

Clinical Findings

The typical patient is older than 60 years and has *no specific symptoms.* In rare cases upper quadrant pain or a palpable mass is found. Usually these findings are associated with advanced disease and result from infiltration of adjacent organs or tissue. The same is true of elevated enzyme or bilirubin levels, which are seen when central bile ducts and/or vessel structures are involved. In advanced stages weight loss is a regular finding.

CT Findings

The variable CT morphology of CCC reflects not only the varying cellular organization of the tumor as mentioned above, but also the different possible levels of involvement of the biliary tree. If the tumor is of segmental origin close to the bile duct bifurcation, peripheral duct dilatation accompanies the mass effect of the tumor. With a more peripheral origin,

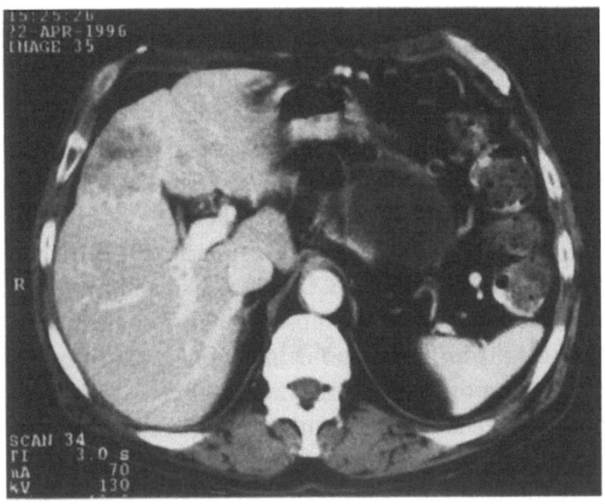

Fig. 9.2. Typical cholangiocellular carcinoma in segment IVa/b (stage T3) with intrahepatic metastasis after intravenous contrast bolus injection: ill-defined tumor margins, central hypoattenuation, focal areas with hyperattenuation

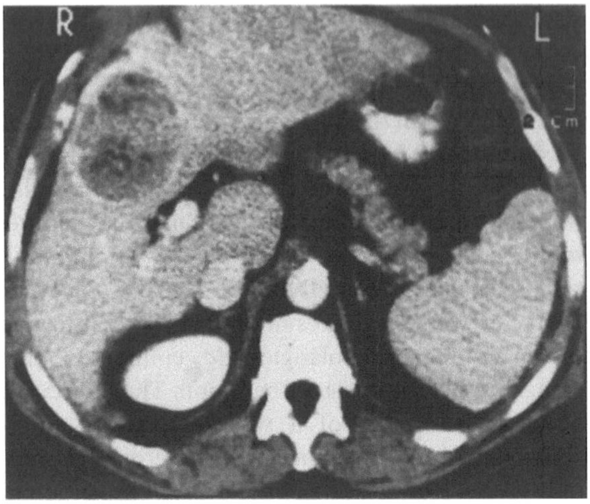

Fig. 9.3. Encapsulated cholangiocellular carcinoma, also in segment IVb (stage T3), after intravenous contrast bolus injection: well-defined tumor margin and hypervascular tumor pseudocapsule

only an intrahepatic mass is present (Figs. 9.2, 9.3). On nonenhanced CT, CCC usually is hypodense with morphologically a wide range of shape (Ros et al. 1988), and thus differentiation from HCC is difficult. Furthermore, sometimes encapsulated tumor nodes may be present (Fig. 9.3), which are also considered as a feature of well-differentiated HCC. As already mentioned, the degree of vascularization varies. In the majority of cases contrast uptake of the tumor on CT is poor because of a prevailing malignant (tumor-induced) fibroplastic response (Fig. 9.3). In some cases, however, delayed CT might depict a late enhancement pattern as a possible result of this microscopic structure (Adam 1991). About one-third of CCCs and hypervascular because they have a more adenomatous structure and greater cell density (Fig. 9.2). In almost all tumors necrotic changes are present, reflected by areas of low attenuation in nonenhanced scanning and focal lack of enhancement after contrast injection. A very rare finding is a well-circumscribed tumor with a clear, distinct ductal origin and spread. Calcification, intratumoral hemorrhage, or tumor rupture with extrahepatic bleeding are occasionally reported. In advanced stages lymph node infiltration of the hepatoduodenal ligament or, more distantly, of the peripancreatic region is a frequent finding on CT.

9.2.2.2 Hilar Duct Carcinoma (Klatskin's Tumor)

General Aspects

This tumor type also originates in the cholangiocytic epithelium. In the majority of the cases it is characterized by scattered tumor cell noduli embedded in abundant malignant fibroplastic tissue which results from the induction of a desmoplastic reaction to tumor cell invasion. It is still unclear whether this is a product of the tumor cell itself or an induced reaction of interstitial cells mediated through tissue factor production by the tumor. Complete ductal obstruction at microscopy is usually associated with lengthy spread along and within the ductal wall. Longitudinal growth and expansion, therefore, is a highly pertinent feature of this tumor type. A rare subtype is formed by a tumor cell population secreting mucin to a much higher degree than the usual type. Accordingly, the term "mucinous ductal carcinoma" is applied to this subtype.

An anatomical approach to the classification of ductal hilar carcinoma was made in 1975 by BISMUTH, who subdivided hilar duct carcinoma into the following subtypes: trunk type (originating in the main hepatic duct), bifurcation type (originating within the bifurcation), and radicle type (originating in either the left or right hepatic duct (BISMUTH and CORLETTE 1975)). This classification has a strong influence both on therapeutic options in reference to resection planes and on the morphologic characteristics depicted by CT.

Another essential feature of hilar duct carcinoma is its early lymphogenic spread along the hepatoduodenal ligament. Interestingly, when spreading via lymphatic vessels, tumor cells may spare nodular structures close to the tumor origin and settle in nodes distant to the tumor. This explains the sometimes conflicting finding of hilar duct carcinoma associated with lymphadenopathy in the wall of the common bile duct or close to the ampulla. As a consequence of this, a reliable TNM classification of Klatskin's tumor is still lacking.

As to causative factors, primary sclerosing cholangitis and secondary sclerosing cholangitis mediated through ulcerative colitis or other inflammatory disease of the biliary tree have been discussed (THOMPSON et al. 1972). This may also explain the potential association with biliary stone disease (BLUMGART et al. 1984). CHAPMAN reported an incidence of ductal carcinoma of 20%–30% in patients with sclerosing cholangitis (DORUDI et al. 1991).

Clinical Findings

Klatskin's tumor is typically found in Causasian men older than 60 years. At primary admission it is associated with obstructive jaundice in 74%, weight loss in 81%, and bilirubin levels frequently higher than 10 mg/dl (THORSEN et al. 1984). Tumor markers are unspecific. Increased serum levels of liver enzymes, e.g., liver type phosphatase, usually correspond to the extent of ductal obstruction and bile congestion. In some cases ascending bacterial infection of congested bile fluid may be the initial clinical finding. Pain is a very unreliable sign. It is induced by splanchnic nerve fiber infiltration in the hepatoduodenal ligament or liver hilum and represents a far advanced stage of disease.

CT Findings

In general, CT diagnosis of hilar duct carcinoma is difficult; or, in other words, CT does not depict early stages. This is explained, as stated above, particular by the specific biologic behavior of the tumor. The CT morphology is essentially dependent on the size and the primary tumor site: small or medium size tumors are not seen in either nonenhanced or contrast-enhanced studies. In a radicle type tumor originating in either of the hepatic ducts, an "early" finding may be a combination of segmental or lobar atrophy and ductal dilatation of the involved part of the liver. Primary trunk type tumors usually induce bilobar ductal congestion (Fig. 9.4). The clinical sign of obstructive jaundice in dicates that disease is already at an advanced stage, although in the trunk type this sign may appear somewhat earlier than in the other types. In far advanced disease stages, then, a hypodense tumor mass at the liver hilum is present on nonenhanced CT. In routine parenchymal phase contrast-enhanced CT studies the tumor is seen to be hypovascular, as a result of its abundant fibroplastic tissue. However, in arterial phase spiral mode, or in CT arteriography with contrast injection into the hepatic artery during scanning, a faintly stronger and earlier arterial enhancement pattern than in normal liver parenchyma may be depicted (Fig. 9.4) In later disease stages involvement of portal vein branches is seen on CT. Then, as a result of intensive longitudinal spread, tumor masses may be present along the ductal bifurcation (Fig. 9.5). Interestingly, even in such advanced stages large nodal infiltration is a rare finding on CT, although micronodular involvement is often found microscopically.

9.2.2.3 Common Bile Duct Carcinoma

General Aspects

Like hilar duct carcinoma, common bile duct carcinoma originates in the cholangiocytic epithelium of the bile ducts. Its microscopic features and growth pattern and its biologic behavior in general

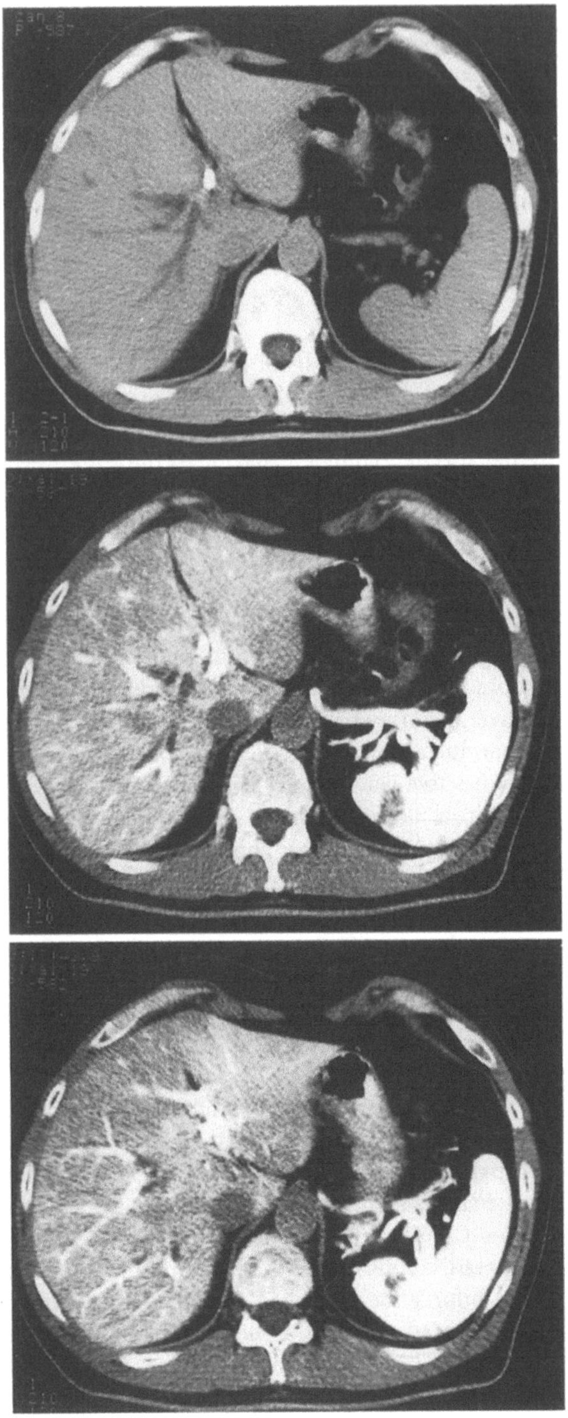

a

b

c

Fig. 9.4a–c. Advanced-stage Klatskin's tumor. **a** Nonenhanced scan in the hilar plane. The top of a plastic stent is visible projecting into the left main duct. In the right lobe the segmental ducts are dilated. The "empty" space between the dilated ducts and the stent may considered an indirect sign of tumor. **b** CT arteriography with contrast injection via a catheter inserted in the celiac trunk at the same level as **a**. The delay is 12 s, providing arterial and, already, portal phase enhancement. The tumor is shown by its faint hyperattenuation. **c** Scan 2 cm below **b**

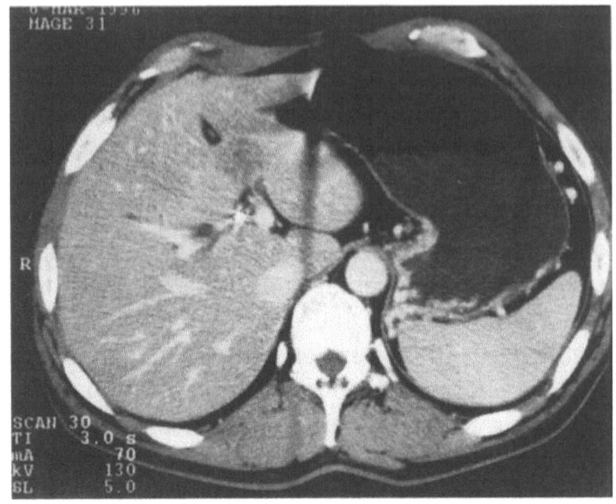

Fig. 9.5. Klatskin's tumor with an unusually large intrahepatic hypoattenuating tumor mass (type III according to BISMUTH's classification)

are parallel to those of hilar duct carcinoma, including desmoplastic reaction to tumor cell invasion, longitudinal growth and expansion along the ductal wall, skip-type lymphogenic spread, and rare differentiation as a mucinous subtype. Because of the common bile duct's anatomic relationship to the cystic duct and the pancreatic head, involvement of these structures and/or infiltration of them by common bile duct carcinoma determines clinical symptoms, therapeutic options, and, obviously, imaging characteristics on CT. As in the Klatskin type, a large tumor mass represents a far advanced stage of disease irrespective of the primary tumor site.

By contrast to hilar duct carcinoma a TNM classification has been formulated for common bile duct carcinoma which is as follows:

- Tis: Carcinoma in situ
- T1: Tumor restricted to the wall
  - T1a: Mucosa infiltrated
  - T1b: Muscle layer infiltrated
- T2: Tumor invades perimuscular connective tissue
- T3: Tumor invades adjacent structures (liver, duodenum, pancreas, gallbladder, colon, or stomach)
- N1: Regional lymph node invasion
- M1: Distant metastasis

There are no clear ideas about causative or predisposing factors for the development of common bile duct cancer. The relationship with inflammatory biliary tract disease is less distinct than in hilar duct carcinoma.

Clinical Findings

Demographically and epidemiologically this tumor type is quite similar to hilar duct carcinoma, discussed above. As with all tumors of the biliary tree, there are no early clinical signs. Tumors at an early stage are found accidentally as a result of routine gall-bladder resection or Whipple's operation for chronic pancreatitis. In cases of more advanced tumor (≥ stage T2) the primary tumor site determines the clinical symptoms: distal common bile duct carcinoma may present with symptoms and signs like those of pancreatic head carcinoma: painless obstructive jaundice and hydropic gallbladder ("Courvoisier's sign"). However, the biological situation is different: pancreatic head carcinoma infiltrating the distal common bile duct and producing obstructive jaundice is staged as a T2 carcinoma, while distal common bile duct carcinoma infiltrating the pancreatic head is classified as a T3 tumor. Juxtapancreatic duct involvement causes symptoms similar to those of the trunk type of hilar duct carcinoma. Usually, however, a hydropic gallbladder that can be easily palpated without pain is found in addition. Tumor markers are unspecific. Liver enzyme and bilirubin level increases correspond with the degree of ductal obstruction and bile congestion.

CT Findings

As already stated above, early-stage tumors usually are not depicted on CT. However, it should be pointed out that, because clinically no symptoms are caused until stage T2 is reached, small tumors do not enter the typical diagnostic cascade early enough to challenge this dispiriting statement. Theoretically, thin-slice spiral CT may have the potential to visualize small tumors. Routine CT, however, technically performed with medium slice thicknesses (5 – 8 mm) in the search for a reason for obstructive jaundice, fails to identify a tumor before it reaches a significant size. In such a setting, distal bile duct tumors need to produce a mass effect larger than 2 cm to be visualized (Fig. 9.6). In other words, decision making on the basis of CT morphology in the differential diagnosis of obstructive jaundice needs to take into account the size and filling density of the gallbladder, the congestion pattern of the extrahepatic and intrahepatic bile ducts, the diameter and appearance of the pancreatic duct, and the condition of the parenchyma in the pancreatic head. To summarize, diagnosis of common bile duct carcinoma by routine CT can only be indirect, by putting together the individual CT signs mentioned (Levine et al. 1979). In an attempt to achieve a higher level of accuracy or

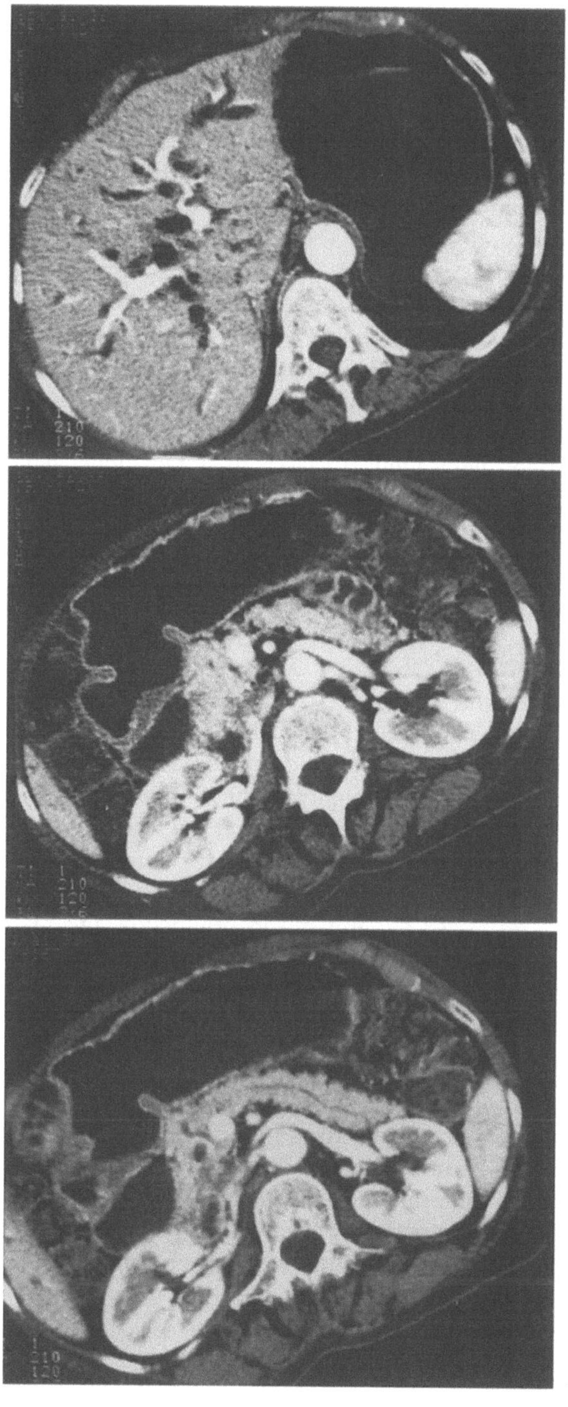

a

b

c

Fig. 9.6a–c. Stage T3 tumor of the distal common bile duct, hydro-CT technique. **a** CT scan higher than the level of the liver hilum, showing massive bilobar intrahepatic bile duct dilatation. **b** CT scan just above the venous splenomesenteric junction, demonstrating dilatation of the pancreatic and common bile duct. The common bile duct wall shows hyperattenuation. **c** Juxtapapillary tumor mass almost isodense to the pancreatic head parenchyma. The duodenal wall is obviously infiltrated improve tumor depiction at earlier stages, our group has shown that the method of hydro-CT discussed above (Sect. 9.2.1.1), applying water as a negative intraluminal contrast agent for the duodenum in combination with pharmacologic intestinal paralysis and thin-slice spiral CT, might have the potential to discriminate distal bile duct carcinoma from pancreatic head carcinoma (Fig. 9.6) at resectable stages (Richter et al. 1996). Other promising approaches using recent CT developments include 3D CT cholangiography as an adjunct to thin-slice CT in order to identify simultaneously the level and the source of obstruction, and double-phase spiral CT with arterial and venous phase imaging for better tumor delineation. However, these techniques have not yet been investigated in large clinical studies.

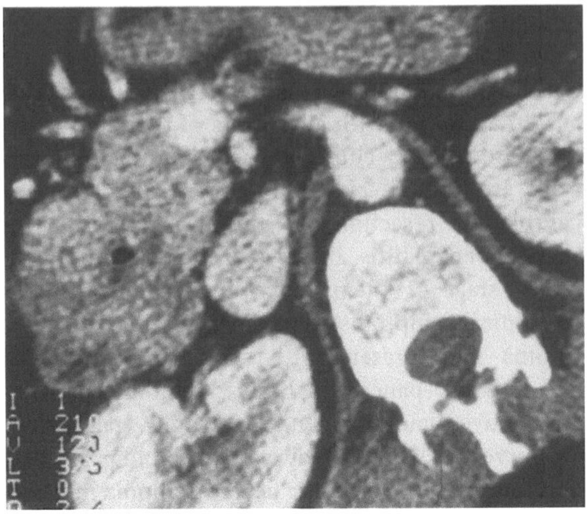

Fig. 9.7. Stage T3 ampullary carcinoma, hydro-CT technique. The tumor infiltrates almost the entire circumference of the duodenal wall with complete luminal obstruction. There is marked contrast enhancement

9.2.2.4 Ampullary Carcinoma

General Aspects

The term "ampullary carcinoma" is restricted to adenocarcinoma of the papilla of Vater, although sometimes a lack of understanding of the nature and origin of this highly specific tumor is revealed when it is confused with duodenal carcinoma or juxtapapillary pancreatic head carcinoma. The cellular origin, again, is the cholangiocytic epithelium characterized by faint mucin production and high mucin output in the rare mucinous subtype. In comparison to the other carcinoma types described so far, the desmoplastic reaction in ampullary carcinoma is less significant. The following TNM classification system has been formulated:

Tis: Carcinoma in situ
T1: Tumor restricted to the papilla of Vater
T2: Tumor infiltrates the duodenal wall
T3: Tumor invades 2 cm or less into the pancreas
T4: Tumor invades more than 2 cm into the pancreas or other organs
N1: Regional lymph node invasion
M1: Distant metastasis

Clinical Findings

Demographically and epidemiologically this tumor type is also quite similar to the other bile duct carcinomas. Clinical symptoms such as obstructive jaundice and enlarged and congested gallbladder without episodes of abdominal pain are features of advanced stages. In contrast to the common bile duct tumors mentioned so far, however, pancreatitis and/or duodenal obstruction are an essential part of the clinical presentation, although usually seen only in advanced stages of disease. In its induction of pancreatitis the tumor mimick juxtapapillary pancreatic head carcinoma.

CT Findings

The CT morphology of ampullary carcinoma reflects the primary site of the tumor at the entrance point of the distal common duct into the duodenum. Therefore, characteristic CT findings include a mass within the duodenal wall associated with an enlarged distal common bile duct and intrahepatic bile congestion (Fig. 9.7). As already stated, however, this represents the morphology of advanced-stage disease (≥T3). Without identification of duodenal wall involvement, which is already stage T2, CT diagnosis of ampullary carcinoma is not possible. Another CT feature is dilatation of the pancreatic duct, with or without pancreatitis (Fig. 9.7). Sometimes atrophy of the pancreatic tail and body are seen. Larger tumors may appear either cystic, owing to intratumoral mucin production, or more solid, having a higher cell density (Fig. 9.7). The latter is also associated with a higher postcontrast increase in density (Fig. 9.7). Again the hydro-CT technique has proved helpful for differential diagnosis and identification of ampullary carcinoma (Richter et al. 1996). However, in general the role of CT in the diagnostic cascade is confined to the staging and description of local spread in advanced tumors. Endoscopy is the diagnostic method of choice, giving early tumor depiction and biopsy proof of malignancy.

9.2.2.5 Metastasis

General Aspects
In metastatic disease of the biliary tree, direct wall involvement should be distinguished from lymphadenopathy along the hepatoduodenal ligament and at the liver hilum. While the latter is commonly found in a variety of intestinal neoplasms, such as gastric (Yonemura et al. 1994) and pancreatic cancer (Kayahara et al. 1995), direct wall involvement is very rare and is a result of vascular propagation in periductal vessels. In this case tumor spread resembles primary ductal carcinoma. Periductal lymph node infiltration represents external compression. Very rarely, a nodal tumor deposit grows beyond the capsule of the node, directly infiltrating the wall.

Clinical Findings
The clinical presentation of patients with metastatic biliary tract disease is more or less indistinguishable from the presentation of those with primary bile duct tumors. However, the existence of a causative neoplasm makes the clinical situation clear.

CT Findings
Like the clinical picture, the morphologic pattern of biliary tract metastasis corresponds to that of ductal carcinoma, with dilated bile ducts above the level of obstruction. Interestingly, however, periductal lymphadenopathy as the underlying obstructive mechanism is somewhat better visualized on CT than is primary tumor growth. This may be due to greater tumor cell volume and the lack of desmoplastic reaction in lymph node infiltration, both of which are major factors in the malignancy of bile duct tumors.

9.3 Inflammatory Disease of the Bile Ducts

Acute cholangitis is a severe inflammatory disorder of the biliary tree. It was first summarized by Charcot by the triad of fever, obstructive jaundice, and upper quadrant abdominal pain (Charcot 1877), and is a result of biliary duct obstruction and secondary ascending bacterial infection, inducting a rise in intraductal pressure and bacterial proliferation. Ultimately, septicemia develops through intralobular entrance of bacteria into the vascular system. The major cause (80%) is biliary stone retention in the distal common bile duct. The second most frequent cause is malignant ductal stricture.

9.3.1 Primary Cholangitis

Cholangitis in its primary form is always known as "primary sclerosing cholangitis" (PSC).

General Aspects
PSC is an acalculous, non-trauma-induced chronic inflammation of the intra- and extrahepatic bile ducts of unknown origin. Microscopically it is characterized by progressive fibroplastic ductal wall changes. It is seen in up to 30% of patients with ulcerative colitis. PSC is associated with or induces periportal hepatitis, biliary duct stenosis, cholestasis, biliary cirrhosis, CCC, and Klatskin's tumor (see Sects. 9.2.2.1, 9.2.2.2).

Clinical Findings
Obviously the clinical symptoms are a direct reflection of the pathophysiologic situation. In patients with ulcerative colitis, increased bilirubin levels and liver type phosphatase are highly suggestive of the presence of PSC. In general, the clinical presentation is dominated by the onset of obstructive jaundice. Therefore, the diagnosis of PSC is established by ruling out other sources of cholangitis.

CT Findings
CT plays a very minor role in the primary diagnosis of PSC. Usually, CT is used to rule out any other reason for obstructive jaundice. However, the intrahepatic bile ducts in particular display a somewhat specific morphologic pattern on CT which is shown in Fig. 9.8. This pattern is characterized by ductal dilatation that varies in degree from segment to segment, creating a knob-like appearance of the biliary tree (Fig. 9.8). Some liver lobes may be almost completely free of ductal dilatation. This can be explained by focally increased wall sclerosis that leads to ductal shrinkage, while less affected portions of the ductal system dilate due to the rise in intraductal pressure (Ament et al. 1983; Balthazar et al. 1993).

9.3.2 Secondary Cholangitis

General Aspects
The term "secondary cholangitis" indicates that in this disease the inflammatory lesions within the biliary tree have a distinct cause. Microscopically and morphologically secondary cholangitis resembles PSC; the only difference is that it develops for a specific reason. It may develop as a result of biliary stone

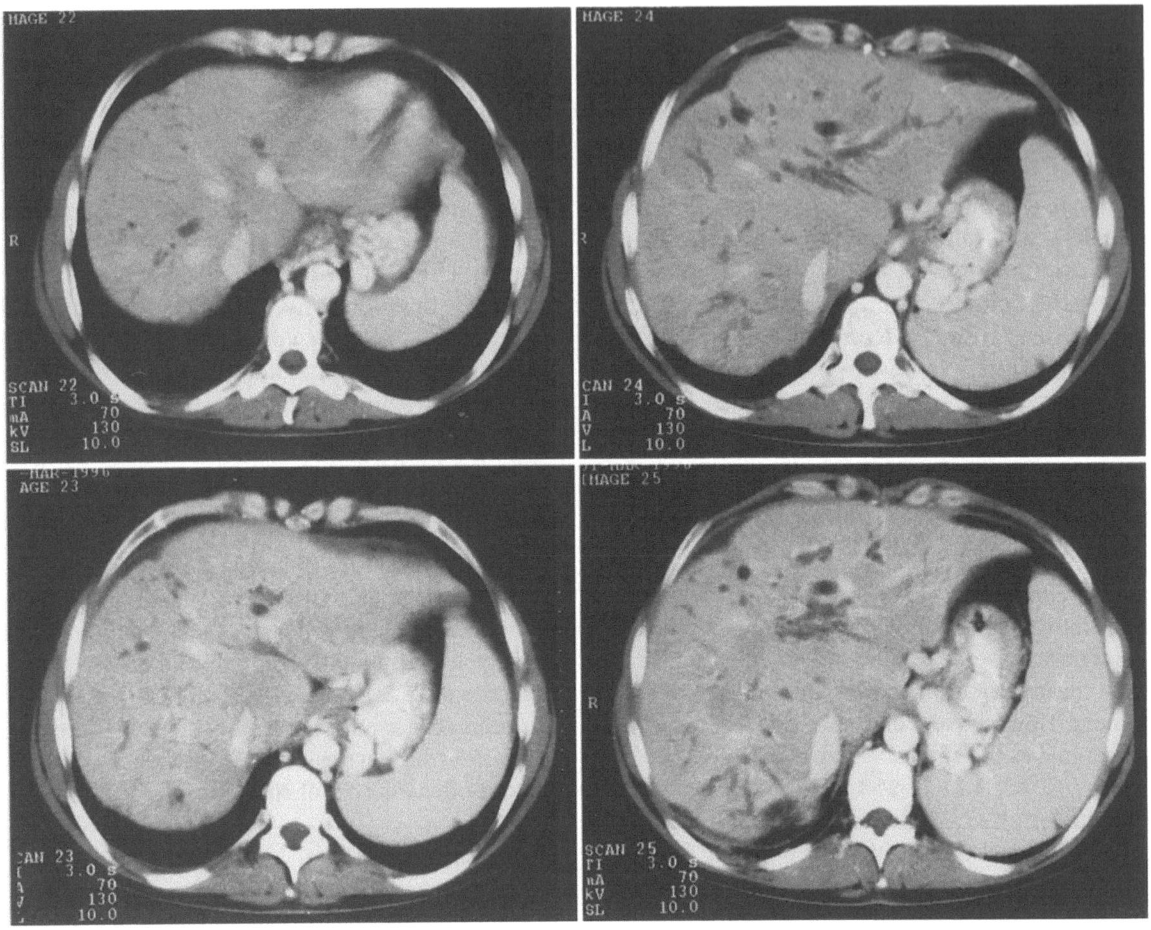

a b c d

Fig. 9.8a–d. Primary sclerosing cholangitis in contrast-enhanced scans. Four CT scans at 2-cm intervals show typical focal distribution of hyperdilating ductal structures and some liver segments to a certain extent spared ductal changes

disease, chronic bacterial infection, postoperative problems, or malignant disease.

Clinical Findings

The clinical presentation is basically determined by the underlying pathophysiologic condition. In general, recurrent obstructive jaundice associated with bouts of fever dominates the clinical picture. In 92% of patients with this disease fever is the primary symptom (Thompson et al. 1972); in 67% influenza-like symptoms associated with obstructive jaundice are reported. Charcot's triad is usually present. Laboratory findings include leukocytosis, mildly to moderately elevated serum bilirubin levels, and increased liver type phosphatase. If not treated, chronic cholangitis may ultimately induce biliary cirrhosis and portal hypertension, like PSC.

CT Findings

Again, CT is not very helpful in the primary diagnosis of this disease. Usually it is only required for detec-

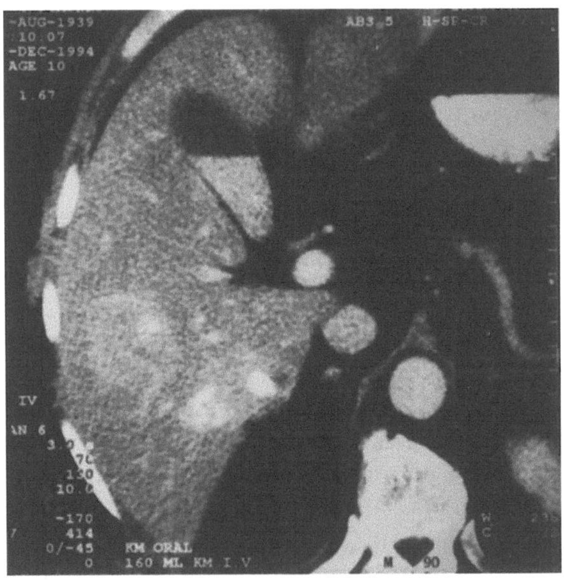

Fig. 9.9. Severe bacterial cholangitis with cholangiogenic abscesses in segment VI, significant hyperattenuation of the lesions after intravenous contrast bolus injection

tion of complications of the disorder or potential causative factors. In chronically recurrent bacterial cholangitis CT may be used to reveal secondary cystic changes resulting from microabscesses formed during the long course of the disease. However, in the diagnosis of cholangiogenic abscess CT is the method of choice. The CT morphology is very characteristic: nonenhanced scanning shows uni- or multilocular, cyst-like focal hepatic lesions in close relationship with the intrahepatic bile ducts; after intravenous contrast bolus injection a hyperattenuating ring appears at the periphery of the lesions (Fig. 9.9). Cholangiogenic abscesses may considerably vary in size.

9.3.3 Congenital Cholangitis (Caroli's Disease)

General Aspects

This disorder might be considered to be on the borderline between bile duct anomalies and bile duct inflammation. The baseline disorder is a congenital malformation of bile ducts characterized by multifocally dilated intrahepatic bile ducts. The syndrome was first described by CAROLI et al. in 1958. It is associated with ascending biliary tract infection, cholestasis, and ductal stone formation. Frequently congenital hepatic fibrosis is found. Morphologically the disorder is characterized by multiple cysts of variable sizes arranged in a chain-like pattern with normal ductal structures in between the cystically dilated segments. Microscopically the lesions are lined by cholangiocellular epithelium and are more periductal cysts than focal ductular ectasia (see below), with small-caliber communications to the ductular systems. The condition may be found throughout the entire intrahepatic biliary tree or may be restricted to lobes or segments. In the latter case some preponderance in the left-side duct system or even in individual left-side segments has been described (CAROLI 1973; CAROLI et al. 1958). In our image example, however, the condition is restricted to the right-side segments V and VI (Fig. 9.10).

Clinical Findings

As already mentioned, the disorder is present at birth. It does not become symptomatic, however, until after a long silent period, at least 5 years and sometimes 20 years. Usually, the first symptom is unexplained fever which develops secondary to bacterial infection and cholangitis. Therefore, indicator enzymes for biliary tract disease will test positive. When the syndrome is associated with congenital fibrosis, rapid development of portal hypertension is a typical clinical feature.

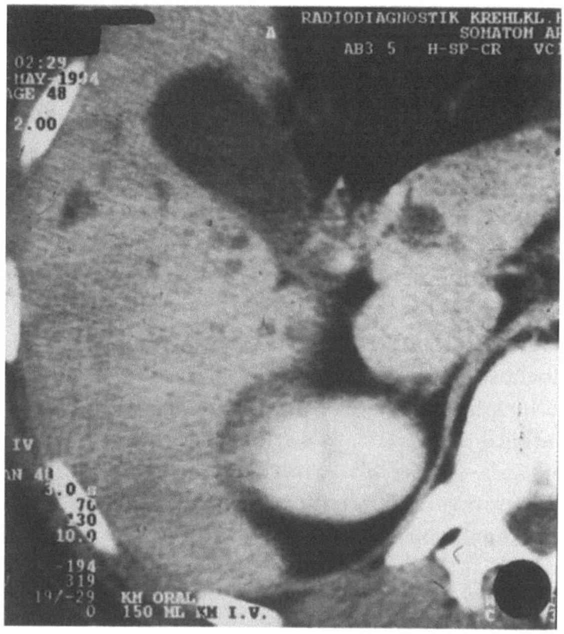

Fig. 9.10. Caroli's disease: a rare case with isolated appearance in segments I, V, and VI. Polygonal shapes and variable sizes of the multiple cystic lesions without hyperattenuation in the periphery or surrounding liver parenchyma

CT Findings

The CT morphology naturally follows from the pathologic changes in the intrahepatic bile ducts. Usually, CT depicts the multicystic pattern even on nonenhanced scanning. Particular when the cystic lesions are restricted to one lobe or to few segments, the CT morphology is highly suggestive of the diagnosis (Fig. 9.10). Tubular dilatation of segmental ducts may be found. This is more a product of the chronic concomitant cholangitis than an expression of the disorder itself. In such a case, and when the cysts are small, discrimination of Caroli's disease from PSC can be difficult. CT cholangiography may increase the accuracy of the differential diagnosis by showing the communication of the cysts with the biliary tree. When infection is present, air filling the cysts is a probable finding. Another highly specific sign is dotting of the cysts, caused by portal branches impinging on the cyst wall. In general, the fluid density of cysts in Caroli's disease is much higher than that of normal liver cysts. Sometimes intracystic calcifications may be found.

9.4 Traumatic Lesions of the Bile Ducts

General Aspects

As a major functional component of the liver, the bile ducts are prone to trauma in the same way as the vascular tree, although obviously vascular trauma results in much more acute onset of symptoms. Traumatic lesions of the bile ducts take a substantially longer time to surface with clinical symptoms. Ninety percent of traumatic bile duct lesions are iatrogenic, a term which includes a host of various underlying causative factors. The vast majority are postsurgical complications, and the biggest single offender in this group is laparoscopic gallbladder removal. Others include stenoses after conventional cholecystectomy, anastomotic strictures after choledochostomy, bilioenterostomy, and accidental lesions to the bile duct during unrelated abdominal surgical maneuvers. In a relatively small proportion of cases – 10% – biliary tract lesions may result from penetrating trauma (stitches, shot wounds) or blunt abdominal trauma.

Clinical Findings
The clinical background of the underlying causative factor dominates the symptoms. Generally, affected patients present with cholangitis and obstructive jaundice. As soon as there is extraperitoneal bile leakage, local peritonitis adds to the symptomatology. This is found in perforating and blunt liver trauma.

CT Findings
In the vast majority of cases of traumatic injury to the bile ducts the role of CT is very limited. Usually the clinical background is clear and ultrasound has already shown ductal dilatation. Only in post-trauma cases does CT have potential in demonstrating subhepatic biliomas, areas of parenchymal rupture filled with a mixture of blood, bile, and aerobilia. Now that CT cholangiography is available, however, it seems feasible that it could be used to identify a possible level of obstruction or direct trauma to the ducts.

9.5 Anomalies

9.5.1 Biliary Duct Atresia

General Aspects
Biliary duct atresia is defined as an inherited disorder of the biliary tree with missing ducts various possible degrees. A complete lack of extrahepatic ducts may be found. The pathogenetic process is still unknown; theories say that it is the result of intrauterine inflammation of the biliary tree. It is seen in one case per 12000 deliveries (Tan and Howard 1988).

Clinical Findings
Without treatment (surgery, liver transplantation), babies with biliary atresia die before the age of two. The syndrome is easily diagnosed during the neonatal period by the lack of biliary excretion.

CT Findings
CT is very rarely helpful in this disorder. Only in cases with segmental lack of bile ducts may it clarify the situation by depicting segmental liver atrophy.

9.5.2 Biliary Duct Ectasia and Choledochal Cyst

General Aspects
One percent of all benign lesions of the biliary tree are ductal cysts (Saxena et al. 1988). Eighty-two percent of these are choledochal cysts. Multiple ductal cysts are extremely rare, with an incidence of around 1% of all ductal cysts (Ono et al. 1982; Nagorney et al. 1984). In such a disorder overlap with the Caroli's syndrome might be considered. It is still unclear whether this is a duplication anomaly or a resorption cyst of the ductal glandular tissue. A very rare condition is sacciform ectasia of the common bile duct.

Clinical Findings
In the vast majority of cases patients have no specific symptoms and the lesion is diagnosed accidentally. Only in the rare case of infection of the cyst and/or intracystic stone formation may specific symptoms develop.

CT Findings
The CT findings depend largely on where the lesion is and at what level the biliary tree is affected. A ductal cyst in the liver hilum may be confused with a simple liver cyst or with the rare anomaly of a second gallbladder. When a ductal cyst is present in the distal common bile duct, CT depicts an intrapancreatic cystic mass in close relationship to the distal course of the common bile duct, which is a very specific finding. In contrast-enhanced scans, lack of hypervascularity in the cystic membrane may be used to rule out primary cystic tumor of the pancreas. By analogy to the way in which Hoglund et al. (1990) used intravenous cholangiography for demonstration of communications between the bile duct and the choledochal cyst, CT cholangiography could have a fruitful role in this rare disorder.

Table 9.1. Differential diagnosis in CT of the biliary tree

CT diagnosis	Differential diagnosis	Diagnostic clues
Cholangiocellular carcinoma	Hepatocellular carcinoma	Tumor markers (*a*-feto protein negative)
	Liver metastasis	No primary tumor
	Lymphoma	Inhomogeneity and higher vascularity
Klatskin's tumor	Gallbladder carcinoma	No hypervascularity in parenchymal phase, less mass effect
	Hilar sclerosing cholangitis	Different clinical history
Cystadenoma	Liver abscess	Smaller size, small septa
	Cystic liver metastasis	Baseline disease, septa
	Cystadenocarcinoma	No known CT clues
Common bile duct carcinoma	Ductal pancreatic carcinoma	Tumor markers (CA 19-9 negative), less invasiveness
Ampullary carcinoma	Peripapillary duodenal carcinoma	In early stage, thin-slice CT with intestinal paralysis for tumor site definition
Intrahepatic cholangitis	Liver abscess	Relationship to obstructive disease, ductal stones
Caroli's disease	Multiple liver cysts	Relationship to intrahepatic ducts, segmental involvement pattern
	Secondary cholangitis	Clinical history, segmental involvement pattern

9.6 Future Perspectives

With the advent of spiral CT and its potential for high spatial resolution in the z (longitudinal) axis (Kalendar et al. 1990), the almost dead and buried method of cholangiography was revived. With CT cholangiography, performed with intravenous administration of contrast media excreted via the biliary system and recorded on thin-slice volume scanning, 3D CT imaging of the biliary tree has become feasible (Van Beers et al. 1994). By surface rendering or maximum intensity projection the potential level of biliary obstruction can be visualized and, simultaneously, nonenhanced and contrast-enhanced scanning allow the nature of the obstruction to be defined (Goldberg 1994). In regard to restrictions on the use of CT cholangiography, the higher grades of obstructive jaundice have been mentioned (Stockberger et al. 1994). In patients in whom endoscopic retrograde cholangiopancreatography has failed it may prove very useful. However, the latest developments in magnetic resonance technology, in the form of magnetic resonance cholangiopancreatography (MRCP), would appear to be in strong competition with CT cholangiography and may make it superfluous again within a short space of time.

For distal ductal pathology, e.g., common bile duct carcinoma, ampullary carcinoma, and papillary adenoma, the relatively new "hydro-CT" technique has proved very helpful in differential diagnosis (see Figs. 9.1, 9.6, 9.7). The use of water as a negative contrast agent, high attenuation of the duodenum wall with an appropriate contrast injection protocol, and high contrast resolution with thin-slice spiral CT in combination with pharmacologic intestinal paralysis, are the key elements of this new and promising approach (Richter et al. 1996).

Table 9.1 provides a summary of current knowledge relating to the differential diagnosis of biliary tract lesions on CT.

References

Adam A, Chetty N, Roddie M, Yeung E, Benjamin IS (1991) Self expandable stainless steel endoprostheses for treatment of malignant bile duct obstruction. AJR 156:321–325

Albores-Saavedra J, Henson DE (1986) Tumors of the gallbladder and extrahepatic bile ducts. Armed Forces Institute of Pathology, Washington DC (Atlas of tumor pathology, series 2, fascicle 22)

Ament AE, Haaga JR, Wiedenmann SD, Barkmeier JD, Morrison SC (1983) Primary sclerosing cholangitis: CT findings. J Comput Assist Radiol 7:795–800

Balthazar EJ, Birnbaum BA, Naidich M (1993) Acute cholangitis: CT evaluation. J Comput Assist Radiol 17:283–289

Baron RL, Stanley RJ, Lee JKT, Koehler RE, Levitt RG (1983) Computed tomographic features of biliary obstruction. AJR 140:1173–78

Baron RL (1978) Common bile duct stones: reassessment of criteria for CT diagnosis. Radiology 162:419–424

Bismuth H, Corlette MB (1975) Intrahepatic cholangiocentric anastomosis in carcinoma of the hilus of the liver. Surg Gynecol Obstet 140:170–178

Blumgart LH, Hadjis NS, Benjamin IS (1984) Surgical approaches to cholangiocarcinoma at confluence of hepatic ducts. Lancet 1:66–70

Burhans R, Myers RT (1971) Benign neoplasms of the extrahepatic biliary ducts. Am Surg 37:161–166

Caroli J (1973) Disease of the intrahepatic biliary tree. Clinical Gastroenterol 2:147–161

Caroli J, Couinaud C, Soupault R, Porcher P, Eteve J (1958) Une affection nouvelle, sans doute congénitale, des voies biliaires. Semaine des Hopitaux de Paris 34:136–143

Charcot JM (1877) Leçons sur les maladies du foie, des voies biliares et des veins. Faculté de médecine de Paris, Bourneville et Sevestre, Paris

Dorudi S, Chapman RW, Kettlewell MG (1991) Carcinoma of the gallbladder in ulcerative colitis and primary sclerosing cholangitis. Dis Colon Rectum 34:827–828

Goldberg HI (1994) Helical cholangiography: complementary or substitute study for endoscopic retrograde cholangiography. Radiology 192:615–616

Hoglund M, Muren C, Boijsen MW (1990) Computed tomography with intravenous cholangiography contrast: a method for visualizing choledochal cysts. Eur J Radiol 10:159–161

Itai Y, Araki T, Fururi S, et al (1983) Computed tomography of primary intrahepatic biliray malignancy. Radiology 147:485–490

Jeffrey RB, Federle MP, Laing FC, Wall S, Rego J, Moss AA (1983) Computed tomograhy of choledocholithiasis. AJR 140:1179–1183

Kalendar W, Seissler W, Klotz E, Vock P (1990) Spiral volumetric CT with single-breath-hold technique, continous transport and continous scanner rotation. Radiology 176:181–183

Kayahara M, Nagakawa T, Ueno K, Ohta T, Tsukioka Y, Miyazaki I (1995) Surgical strategy for carcinoma of the pancreas head area based on clinico-pathologic analysis of nodal involvement and plexus invasion. Surgery 117:616–623

Kirschbaum JD, Kozoll DC (1941) Carcinoma of the gallbladder and extrahepatic bile ducts. Surg Gynecol Obstet 73:740–754

Klatskin G (1965) Adenocarcinoma of the hepatic duct at its bifurcation within the porta hepatis. Am J Med 38:241–256

Kuwayti K, Baggenstoss AH, Stauffer MH, Priestly JI (1957) Carcinoma of the major intrahepatic and the extrachepatic bile ducts exclusive of the papilla of Vater. Surg Gynecol Obstet 104:357–366

Levine E, Maklad NF, Wright CH, et al (1979) Computed tomographic and ultrasonic appearances of primary carcinoma of the common bile duct. Gastrointest Radiol 4:147–151

McIntyre JA, Cheng PZ (1968) Adenoma of the common bile duct causing obstructive jaundice. Can J Surg 11:215–218

Nagorney DM, McIlrath DC, Adson MA (1984) Choledochal cysts in adults: clinical management. Surgery 96:656–663

Ono J, Sakoda K, Akita H (1982) Surgical aspects of cystic dilatation of the bile duct. An anomalous junction of the pancreaticobiliary tract in adults. Ann Surg 195:203–208

Reiman TH, Balfe DM, Weyman PJ (1987) Suprapancreatic biliary obstruction: CT evaluation. Radiology 163:49–56

Richter GM, Simon C, Hoffmann V, DeBernadinis M, Seelos R, Senninger N, Kauffmann GW (1996) Hydro-Spiral CT des Pankreas im Dünnschichttechnik. Radiologe 36:397–405

Roberts-Thompson IC, Strickland RJ, Mackay IR (1973) Bile duct carcinoma in chronic ulcerative colitis. Aus N Z J Med 3:264–267

Ros PR, Buck JL, Goodman ZD, Ros AMV, Olmsted WW (1988) Intrahepatic cholangiocarcinoma: radiologic-pathologic correlation. Radiology 167:689–693

Sako S, Seitzinger GL, Garside E (1957) Carcinoma of the extrahepatic bile ducts. Review of the literature and report of six cases. Surg 41:416–417

Saxena R, Pradeep R, Chander J, Kumar P, Wig JD, Yadav RV, Kaushik SP (1988) Benign disease of the common bile duct. Br J Surg 75:803–806

Stockberger SM, Wass JL, Sherman S, Lehman GA, Kopecky KK (1994) Intravenous cholangiography with helical CT: comparison with endoscopic retrograde cholangiography. Radiology 192:675–680

Tan KC, Howare ER (1988) Surgical management of biliary atresia. Lancet 2:678–679

Thompson BW, Reed RC, White JH (1972) Sclerosing cholangitis. Arch Surg 104:460–464

Thorsen MK, Quiroz F, Lawson TL, et al (1984) Primary biliary carcinoma: CT evaluation. Radiology 152:479–483

Van Beers BE, Lacrosse M, Trigaux JP, de Canniere L, de Ronde T, Pringot J (1994) Noninvasive imaging of the biliary tree before or after laparascopic cholecystectomy: use of three-dimensional spiral CT cholangiography. AJR 162:1331–1335

Yonemura Y, Ninomiya I, Tsugawa K, Masumoto H, Takamura H, Fushida S, Yamaguchi A, Miwa K, Miyazaki I (1994) Lymph node metastasis from carcinoma of the gastric stump. Hepatogastroenterology 41:248–252

10 MRI of the Bile Ducts

C. Reinhold and P.M. Bret

CONTENTS

10.1 Introduction

Choledocholithiasis and malignant bile duct obstruction are the most common diseases involving the biliary system. A wide range of imaging techniques have been advocated for evaluating the pancreas and biliary tree. Although ultrasound (US) and computed tomography (CT) are used in the initial evaluation of patients with symptoms and signs referrable to the pancreaticobiliary system, direct opacification of the biliary tree and pancreatic duct is often needed. Current techniques for direct visualization of the biliary tree include endoscopic retrograde cholangiopancreatography (ERCP) and, to a lesser extent, percutaneous transhepatic cholangiography. Significant advantages of ERCP include the unparalleled resolution obtained with this technique and the ability to institute therapeutic measures at the time of initial diagnosis. Although generally considered a safe procedure, ERCP is associated with a morbidity and mortality of 7% and 1% respectively, as reported in a recently published multicenter trial (Lenriot et al. 1993). Important complications include the development of sepsis in the obstructed system, pancreatitis, gastric or duodenal perforation, and bleeding. Other limitations of diagnostic ERCP include the following: (1) the technique is highly operator-dependent, with unsuccessful cannulation of the common bile duct or pancreatic duct occurring in 3%–9% of cases (Lenriot et al. 1993; Assouline et al. 1993); (2) there is limited or no opacification of ducts proximal to a severe or complete obstruction; and (3) routine sedation is required.

To date, the role of magnetic resonance (MR) imaging (MRI) in diagnosing diseases of the biliary tract has been limited. However, the recent development of fast imaging sequences and MR cholangiography has generated new interest in applying these techniques to evaluating the biliary tract. Fast imaging techniques allow the upper abdomen to be scanned during a single breath-hold and are best suited to studying tissue enhancement after intravenous contrast injection. MR cholangiography, on the other hand, provides direct visualization of the biliary tract similar to endoscopic retrograde cholangiography. MR cholangiography differs from conventional MR imaging in that it allows direct visualization of the biliary tree and pancreatic duct, in a way similar to contrast cholangiogrpahy (Fig. 10.1). However, unlike direct cholangiogrpahy, it is noninvasive and does not require the administration of contrast medium. It is in this respect that MR cholangiopancreatograhy has the potential of replacing more traditional ways of imaging the pancreaticobiliary system rather than adding to the already existing battery of available imaging tests.

C. Reinhold, MD, Department of Diagnostic Radiology, Montreal General Hospital, McGill University, 1650 Cedar Avenue, Montreal, Quebec, H3G 1A4 Canada
P.M. Bret, MD, Department of Diagnostic Radiology, Montreal General Hospital, McGill University, 1650 Cedar Avenue, Montreal, Quebec, H3G 1A4, Canada

10.2 Technical Considerations

10.2.1 General Guidelines

The usual contraindications to MR imaging also apply to examinations of the biliary tract. Although the presence of surgical clips in the right upper quadrant is not a contraindication to MR imaging, surgical clips may result in signal void artifacts. An antispasmodic agent (e.g., glucagon 1 mg) is usually administered intravenously or intramuscularly at the start of the examination to decrease motion artifacts arising from bowel peristaltism. A phased array multicoil provides increased signal-to-noise (S/N) ratio and should be used when available.

10.2.2 Conventional MR Imaging

Conventional MR pulse sequences combine the use of T1- and T2-weighted sequences. Table 10.1 details our routine protocol for an examination of the biliary tract: With the use of a torso multicoil array, the increased signal of the adjacent subcutaneous fat is such that fat saturation is routinely required to reduce motion artifacts. Paramagnetic agents such as Gd-DTPA can be used to study the enhancement of the liver, the pancreas, and the wall of the bile ducts. Immediately following a bolus intravenous injection of contrast medium, several T1-weighted fast gradient-echo sequences are performed sequentially, in breath-hold, to evaluate the pattern of enhancement of a particular lesion. In order to maintain the scanning time at 15–25 s to ensure cooperation of the patient with breath-holding, an asymmetric field of view (FOV) can be used and the number of signal acquisitions reduced. It is advisable to perform the dynamic scan during one single acquisition, particularly if the images are acquired in an interleaved fashion, to avoid adjacent slices being acquired at different times after contrast injection. In patients who are unable to breath-hold, the fast gradient-echo sequences are usually of limited diagnostic value; conventional spin-echo (CSE) T1-weighted sequences acquired with respiratory ordered phase encoding and gradient moment nulling are more diagnostic. In conventional MR imaging, there is no significant benefit to acquiring coronal or oblique imaging planes, except when the portal vein is being studied during the evaluation of tumor extent.

10.2.3 MR Cholangiography

The MR cholangiopancreatographic examination requires no patient preparation and no routine administration of premedication or contrast medium injection. The technique is based on heavily T2-weighted pulse sequences which result in stationary fluids having a *high* signal intensity, while solid organs demonstrate a *low* signal intensity. In addition, flowing blood will result in little or no measurable signal. This combination provides optimal contrast between the hyperintense signal of the bile and the hypointense signal of the background. Initial MR cholangiographic studies were performed using a heavily T2-weighted gradient-echo sequence with parameters that resulted in the generation of *steady-state of free precession signals* (SSFP). Wallner et al. (1991) first applied the SSFP sequence to the evaluation of the biliary tree using a breath-hold acquisition. Morimoto et al. (1992) optimized the sequence by acquiring a three-dimensional (3D) data set which improved the contrast between the bile ducts and

Table 10.1. Conventional MR imaging protocol for gallbladder and bile ducts

Pulse sequence	Saturation	Flip angle (°)	Echo train (ms)	Echo time (ms)	Repetition time (ms)	Thickness/gap (mm)	Matrix	No. of excitations
Fast spin-echo dual echo	S/I/fat	–	8	34/112	4000	5/2.0	256	2
Spin-echo/T1	S/I	–	–	Min.	500	5/2.0	192	2
Spin-echo/T1	S/I/fat	–	–	Min.	500	5/2.0	192	2
Fast spoiled gradient-recalled	S/I/fat	70	16	4.2	150	5/2.0	128	2
Fast spoiled gradient-recalled	S/I/fat	70	16	4.2	150	5/2.0	128	2
Spin-echo/T1	S/I/fat	–	–	Min.	500	5/2.0	192	2

S/I, Superior/inferior.

surrounding tissues, thus allowing the selection of thinner slices. Several other investigators (ISHIZAKI et al. 1993; HALL-CRAGGS et al. 1993; REINHOLD et al. 1995a) have since evaluated the technique and recognized two significant limitations of the SSFP sequence: (1) the lack of routine visulization of the nondilated bile ducts or pancreatic duct (for example, distal to a site of obstruction), and (2) the inability of patients to consistently cooperate with prolonged periods of breath-hold. Although the lack of visualization of nondilated bile ducts may be due to signal loss from slow-flowing bile (WALLNER et al. 1991; ISHIZAKI et al. 1993), a more plausible explanation is the limited resolution achieved with the SSFP sequence which compromises the imaging of smaller ducts. Improving the spatial resolution of the image would further reduce the already low S/N ratio of the SSFP sequence, which requires relatively thick sections and a large FOV. A second factor contributing to the poor visualization of smaller ducts is signal loss from motion artifact. The SSFP sequence, being particularly sensitive to motion, is routinely performed during a period of apnea ranging from approximately 20 s to more than 60 s. The period of apnea required depends on the feasibility of segmenting the 3D acquisition, which is an inconstant feature on different MR systems. In addition, segmenting the acquisition often results in image degradation due to misregistration artifact.

Instead of the gradient-echo sequence described above, other investigators (REINHOLD et al. 1995a; OUTWATER 1993; TAKEHARA et al. 1994; GUIBAUD et al. 1994) have used a heavily T2-weighted two-dimensional (2D) *fast spin-echo* (FSE) sequence to generate MR cholangiopancreatograms. In this, as in the SSFP sequence, fluids with a long T2 relaxation time, such as bile, will have a very high signal intensity, whereas the surrounding liver, which has a much shorter T2 value, will generate very little signal. The FSE sequence offers several advantages over the SSFP sequence, including: (1) a higher S/N and contrast-to-noise ratio, which allows the use of thin sections even for 2D imaging; (2) diminished sensitivity to both motion artifact and slow flow; (3) decreased magnetic susceptibility effects (for example, from signal loss due to surgical clips or air in the duodenum); and (4) measurement of T2 rather than combined T2/T2* decay. In a study comparing 2D FSE and 3D SSFP pulse sequences in 26 patients (REINHOLD et al. 1995a), we have shown that the FSE sequence is significantly superior in visualizing both the dilated and nondilated biliary tree and pancreatic duct.

Several techniques have been proposed to optimize MR cholangiopancreatography. Some take advantage of recent updates in imaging software, such as the addition of gradient moment nulling to the FSE sequence, which reduces artifact from periodic motion (for example, ghosting associated with breathing motion of the gallbladder). Others use a surface coil or phased array multicoil instead of the body coil, which increases the S/N ratio and therefore improves visualization of small nondilated ducts. The addition of a fat saturation technique has also become routine, in order to improve conspicuity of the bile ducts against the surrounding intra-abdominal fat and to decrease motion artifacts assciated with the hyperintense subcutaneous fat. TAKENHARA et al. (1994) reported the results of a breath-hold 2D FSE sequence using a surface coil and a small FOV to study patients with chronic pancreatitis. In order to minimize the acquisition time, TAKEHARA et al. (1994) used an echo train length (ETL) of 32, one signal acquisition, and a matrix size of 256 × 128. To ensure cooperation with breath-holding, the acquisition was segmented with the insertion of one pause for the majority of patients. This resulted in periods of apnea ranging from 18 to 22 s. Although the use of a surface coil with a small FOV results in improved in-plane resolution, this combination would not be generally applicable to the morphotype of the Western patient population, particularly in instances where simultaneous imaging of the pancreatic duct and biliary tree is required. The routine addition of fat saturation, advocated in the majority of recent reports, is usually obtained using a chemical selective fat saturation pulse preceding the data collection. However, SHIONO and TWASAKI (1995) proposed using a fast inversion recovery sequence with a seqtion thickness of 40 mm during a single breath-hold period of 18 s. The advantage of an inversion recovery sequence is a more uniform fat suppression, as this sequence is less susceptible to field inhomogeneity. Other investigators have demonstrated the feasibility of depicting the biliary tree by using a *non-breath-hold FSE* technique, which relies on signal averaging (up to six excitations) to compensate for image degradation due to motion (REINHOLD et al. 1996; MACAULEY et al. 1995). Acquiring MR cholangiopancreatograms during quiet breathing improves the versatility of the technique: (1) patients can be imaged regardless of their ability to breath-hold, (2) segmentation of the acquisition and the potential for serious misregistration artifact is avoided, and (3) high-resolution images with prolonged imaging times become feasible. A current limitation of the 2D FSE sequence is the slice thickness, which cannot be less than 3 mm due to gradient restricitions. However, the *3D FSE* sequence allows the use of thinner slices, further improving image quality. BARISH et al. (1995) used a non-breath-hold T2-weighted, 3D multislab, turbo spin-echo sequence with respiratory triggering and fat saturation. Using a repetition time/echo time (TR/TE) of 5000 ms/240 ms, an ETL of 31, a FOV of 24 cm, a matrix of 256 × 192, and one signal acquisition, the authors were able to image a volume of 10 cm in approximately 14 min. More recently, several modifications of the *rapid acquisition by relaxation enhancement* (RARE) technique (HENNIG et al. 1986) have been proposed to obtain MR cholangiopancreatograms. LAUBENBERGER et al. (1995) studied a sequence based on a *single-shot RARE* technique, in which the spin-echoes that are generated by refocusing pulses are phase-encoded differently. With this technique, an acquisition covering the entire FOV without slice selection is possible. The acquisition is obtained during a single breath-hold of 4 s and no image postprocessing, such as maximum intensity projection (MIP), is necessary. Due to the very short acquisition time, several acquisitions in multiple angulations can be obtained within a short period of time. However, since postprocessing of the RARE MR cholangiopancreatogram is not possible, individual tomographic sections cannot be examined. The impact of this on diagnosing small stones surrounded by hyperintense bile remains to be determined. ZUO et al. (1995), using a standard 1.5-T magnet and a phased array coil, compared a 2D multislice CPMG (Carr-Purcell-Meiboom-Gill)-like RARE sequence to: (1) a thick-slice single-shot sequence similar to the one reported by LAUBENBERGER et al. (1995), but without breath-hold, and (2) a 3D-PSIF (mirrored fast imaging with steady-state precession) gradient-echo sequence. The 2D RARE sequence was performed using two sets of interleaved acquisitions (total 18 sections) with an imaging time of 11–15 s per set. ZUO et al. (1995) found that the 2D RARE sequence with MIP postprocessing provided more detail than the single-shot thick-slice technique, and was less susceptible to motion artifacts than the gradient-echo sequence. They also recommended a long effective TE, ≥650 ms, to maximize decay of the signal from the background liver tissue. The same authors (WIELOPOLSKI et al. 1995) also evaluated a 3D RARE sequence using segmented echo planar imaging readouts on a standard MR scanner. Imaging was obtained during periods of breath-hold varying from 18 to 25 s. Examinations of diagnostic quality were obtained both in healthy volunteers and patients. Optimization of this sequence will require the use of higher gradient rise times and potentially half Fourier imaging. Finally, MR cholangiopancreatography has also been performed using a half Fourier single-shot turbo spin-echo sequence (HASTE) (SANANES et al. 1995). This sequence is similar to echo planar imaging; however, it uses a single-shot rather than a multi-shot technique. The HASTE sequence acquires only one half of the lines in K-space and acquires them during a single ETL. The net effect is a marked reduction in imaging time, with one slice being acquired in less than 2 s. High gradient strengths, however, are necessary to play out the ETL of 128 in the HASTE sequence. In addition, the slice thickness is currently limited to 5 mm or greater.

Currently, we perform all MRCP with a torso multicoil array and a *2D FSE* sequence using the following parameters (REINHOLD et al. 1996): TR 8000 ms, effective TE 144 ms, ETL 32, FOV 34–48 cm (an asymmetric FOV is routinely used to decrease imaging time; however, this does not alter in-plane resolution on our system), 3 mm section thickness with no intersection gap, a matrix of 512 × 256, and four signal acquisitions (Fig. 10.1). Gradient moment nulling and fat saturation are routinely used. In addition to the usual superior and inferior saturation bands, an in-plane saturation band is placed over the left abdomen to decrease the signal arising from the splenoportal system. The sequence is nonsegmented and is acquired during quiet breathing. To minimize misregistration artifacts and optimize postprocessing, we acquire our images sequentially, without intersection spacing. Although this results in signal loss due to direct saturation of overlapping slice profiles, it does not significantly degrade image quality, and avoids the trade-offs of an interleaved sequence, i.e., an increase in imaging time and the potential for misregistration artifacts. We routinely acquire MR cholangiograms in the coronal as well as the axial plane. Images acquired in the axial plane are frequently of superior diagnostic quality, as they are plagued less by respiratory artifacts. Depending on the FOV used, a maximum of 13 sections can be obtained in 2 min 8 s. In the coronal plane, a 3/4 rectangular FOV is used and 24–26 contiguous images are obtained in 6 min 16 s. In the axial plane, a 1/2 FOV is used and 36–39 images are obtained in 6 min 25 s. These parameters provide a 100-mm thick axial slab, which routinely covers the area from the major ampulla to well above the bile duct bifurcation, and also includes the level of the pancreatic duct. The

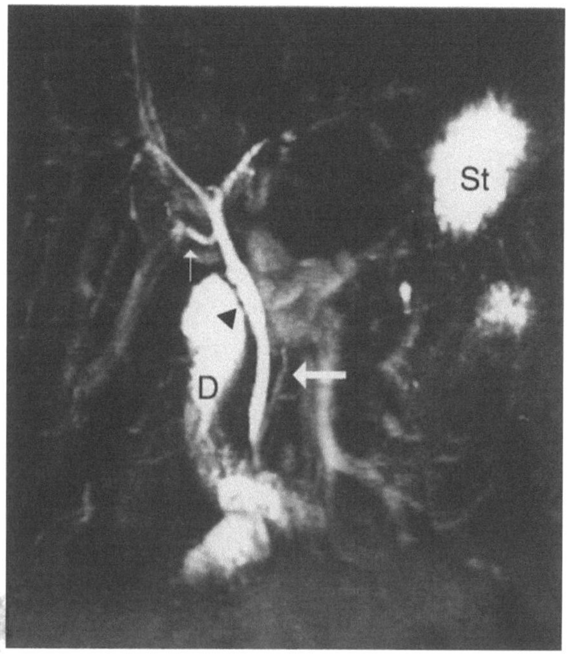

Fig. 10.1. MR cholangiography. Two-dimensional fast spin-echo (2D FSE), phased array torso coil. Coronal, targeted maximum intensity projection (MIP) reconstruction demonstrates a normal-sized bile duct (*arrow*) in a patient after cholecystectomy. The cystohepatic junction is visible (*arrowhead*) Note an aberrant right hepatic duct (*small arrows*) joining the common hepatic duct below the bile duct bifurcation. *St*, Stomach; *Du*, duodenum

coronal slab (approximately 75 mm thick) also includes most of the intrahepatic bile ducts, as well as the entire extrahepatic bile duct and pancreatic duct. The acquisition of multiple imaging planes greatly facilitates image interpretation, as rotation from one plane to another (using 3D MIP or multiplanar reconstruction) results in severe image degradation because of the anisotropic resolution inherent in the 2D sequence (increased slice thickness compared to in-plane resolution). Therefore, under certain circumstances we add to our imaging protocol an oblique plane (35°–45° to the coronal plane), that simulates a right anterior oblique projection at direct cholangiography. The routine administration of antispasmodic agents diminishes bowel artifacts associated with the prolonged imaging times.

We review all our examinations at the diagnostic console, rather than on hard-copy films. We have found that the dynamic nature of the review at the console greatly facilitates diagnosis. We primarily review individual source images as well as reformatted images, rather than relying on 3D reconstructed views. This avoids a number of potential pitfalls in diagnosis, since much of the information obtained from the individual source images is lost when the whole volume is displayed in the MIP format. This is particularly true in the diagnosis of choledocholithiasis. Common bile duct stones depicted as areas of low signal may be completely obscured by the MIP technique if they are surrounded by hyperintense bile. The 3D reconstructed images, on the other hand, are better suited to provide a complete overview of the biliary tree anatomy, which is useful for treatment planning.

To summarize, the large variety of proposed MR cholangiopancreatographic sequences and the lack of available clinical data are clear indicators that the technique is still in its infancy. Technical refinements will undoubtedly improve image quality. However, the future of MR cholangiopancreatography also depends on a variety of other factors including: (1) the ability to obtain acceptable and consistent results on existing MR systems, (2) the availability of time on MR scanners for this clinical indication, and (3) the results of comparative studies not only between MR cholangiopancreatography and ERCP, but also between MR cholangiopancreatography and new imaging techniques of the bile ducts such as CT cholangiography and endoscopic utrasound. In addition, since measurements of diagnostic accuracy merely test how a technique can perform in a controlled environment but give no indication on its clinical utility, outcome and cost-analysis studies will be needed before the role of MR cholangiopancreatography can be determined.

10.3 Normal Anatomical Structures and Variants

10.3.1 Normal Anatomical Structures

With currently available imaging protocols, the normal extrahepatic bile ducts are visualized in nearly 100% of patients (Reinhold et al. 1996). Occasionally a segment of the extrahepatic bile duct may be devoid of signal, for example in cases of pneumobilia (Fig. 10.2), in the proximity of surgical clips (Fig. 10.3), or when motion artifacts, often secondary to duodenal peristaltism, result in signal loss in the suprapancreatic segment of the common bile duct. Nondilated intrahepatic bile ducts can be followed into the outer third of the hepatic parenchyma in over 90% of cases (Macauley et al. 1995) (Fig. 10.4). The cystic duct, as well as its insertion into the com-

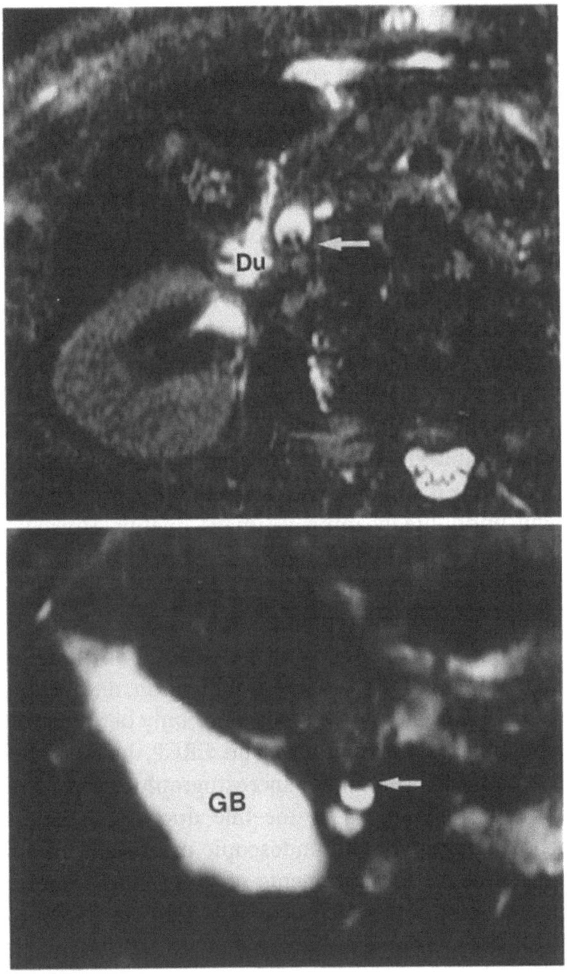

a

b

Fig. 10.2a,b. Pneumobilia versus choledocholithiasis. 2D FSE, phased array torso coil, axial source images. **a** Choledocholithiasis. A hypointense focus with a puncturate area of hyperintensity is seen within the lumen of the bile duct (*arrow*) and represents a stone in the common bile duct. Note the dependent position of the stone. **b** Pneumobilia. In a different patient, a hypointense focus is seen anteriorly in the lumen of the common bile duct (*arrow*) and represents pneumobilia. Axial sections are helpful to distinguish choledocholithiasis from pneumobilia. *GB*, Gallbladder; *Du*, duodenum

mon hepatic duct, is routinely seen (Taourel et al. 1996) (Fig. 10.5).

10.3.2 Normal Variants

It is generally accepted that the presence of an *aberrant right hepatic duct or an abnormal cysticohepatic duct junction* places the patient at a higher risk of bile duct injury during laparoscopic cholecystectomy. For some authors, identifying these variants prior to surgical dissection results in a decreased rate of bile duct injury, and justifies the routine use of

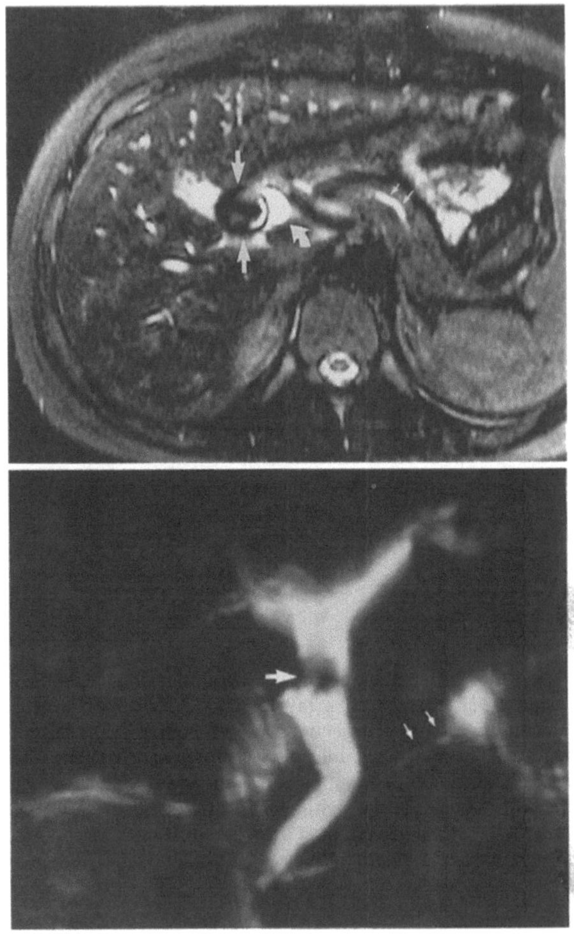

a

b

Fig. 10.3a,b. Artifact from surgical clip. 2D FSE, phased array torso coil. **a** Axial source image demonstrates a round signal void (*arrows*) in the proximity of a surgical clip and creates a pseudo narrowing on the common hepatic duct (*curved arrow*). **b** Coronal, targeted MIP reconstruction demonstrates a signal void mimicking a stone (*arrow*) at the level of the cystohepatic junction. The pancreatic duct (*small arrows*) is displayed in the region of the body and head of the pancreas

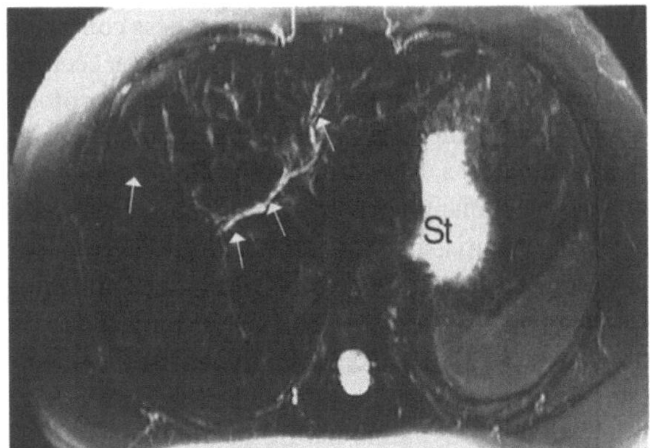

Fig. 10.4. Normal intrahepatic bile ducts. 2D FSE, phased array torso coil, axial targeted MIP reconstruction of the left lobe of the liver. The intrahepatic bile ducts can be followed into the outer third of the hepatic parenchyma (*arrows*). *St*, Stomach

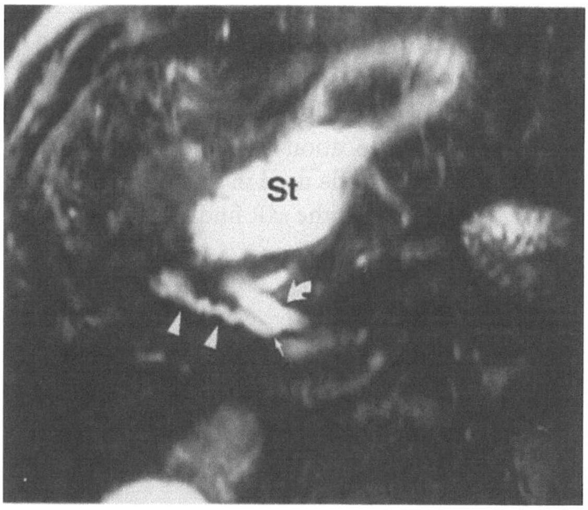

Fig. 10.5. Normal cystohepatic junction. 2D FSE, phased array torso coil. Axial source image demonstrates the cystic duct (*arrowheads*) joining the common hepatic duct (*curved arrow*) posteriorly and laterally (*small arrow*). *St*, Stomach

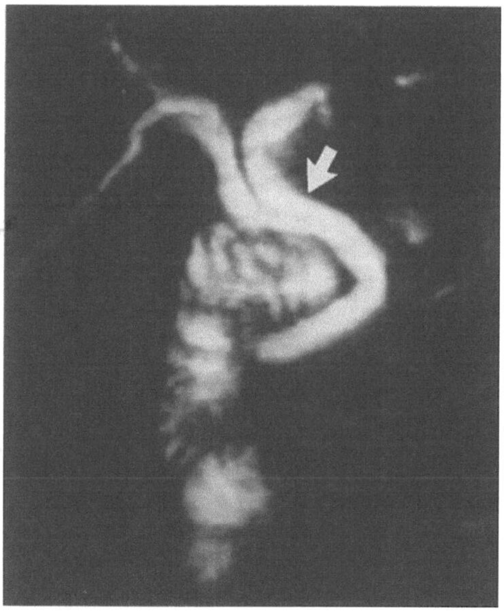

Fig. 10.6. Low bile duct bifurcation. 2D FSE, phased array torso coil. Coronal MIP reconstruction. The bile duct bifurcation is extrahepatic (*arrow*)

intraoperative cholangiography. However, the routine use of intraoperative cholangiography remains controversial, as it has also been shown that a substantial percentage of bile duct injuries occur before the intraoperative cholangiogram is actually performed. We have found that MR cholangiopancreatography can accurately detect normal anatomical variants of the bile ducts (Taourel et al. 1996) (Figs. 10.1, 10.6–10.9). The practical relevance of this remains unclear, as it is doubtful that, at this stage, the use of MR cholangiopancreatography can be advocated before every laparoscopic cholecystectomy. MR cholangiography might prove useful, however, in patients presenting

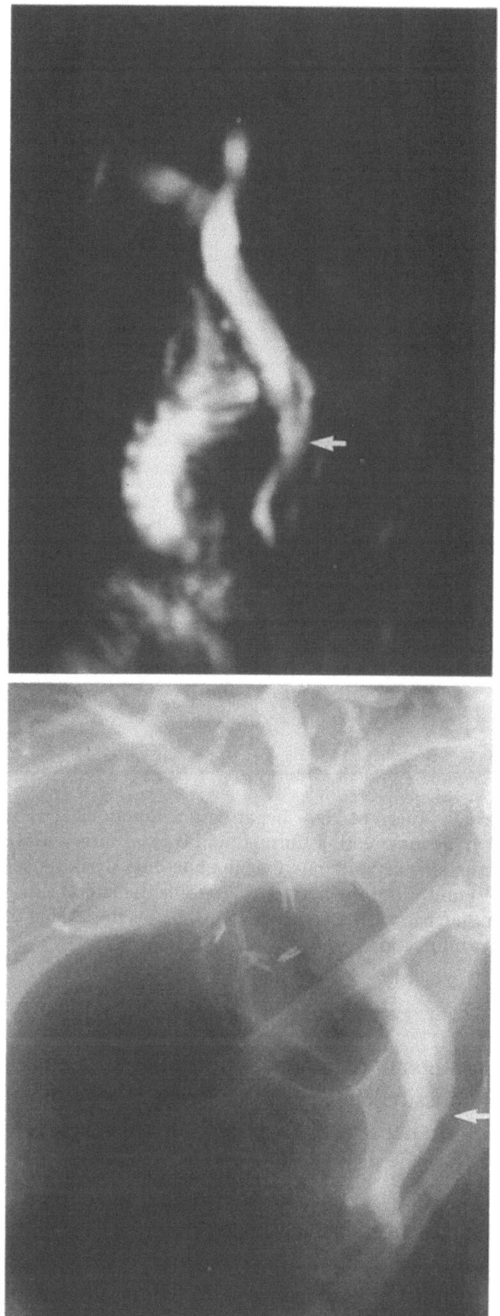

a

b

Fig. 10.7a,b. Medial and low cystic duct insertion. 2D FSE, phased array torso coil. **a** Coronal, targeted MIP reconstruction demonstrates a low cystohepatic junction on the medial side of the common bile duct (*arrow*). **b** ERCP confirms the abnormality (*arrow*)

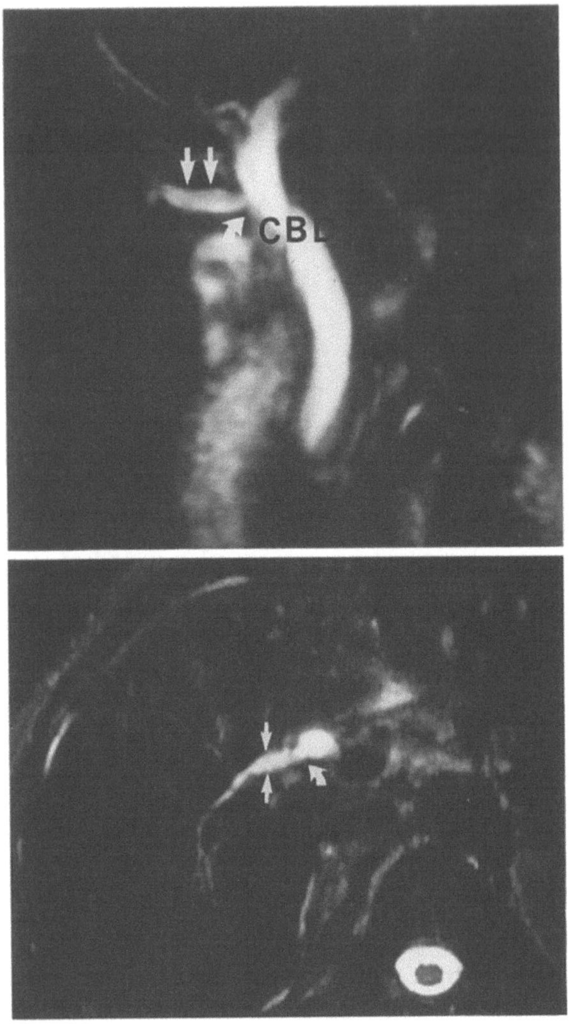

a

b

Fig. 10.8a,b. Aberrant right hepatic duct junction. 2D FSE, phased array torso coil. **a** Coronal and **b** axial targeted MIP reconstructions demonstrate that the bile duct draining segments VI and VII of the liver (*arrows*) joins the lateral side of the common hepatic duct (*curved arrow*) below the level of the bile duct bifurcation

with additional risk factors for bile duct injury, such as obesity, cholecystitis, increased fat deposits in the porta hepatis, or prior abdominal surgery.

10.3.3 Bilio-enteric Anastomosis

After a hepaticojejunostomy, ERCP is impossible since the common bile duct stump is blind-ending, and the area of the bypass cannot be cannulated endoscopically. In these instances, MR cholangiography is particularly useful to demonstrate the intrahepatic bile ducts.

10.4 Presence and Location of Bile Duct Obstruction

Ultrasound and CT are highly accurate in diagnosing the presence and location of bile duct obstruction, and MRI has been little used in this setting: In 1986, Dooms et al. reported the MR findings in 18 patients

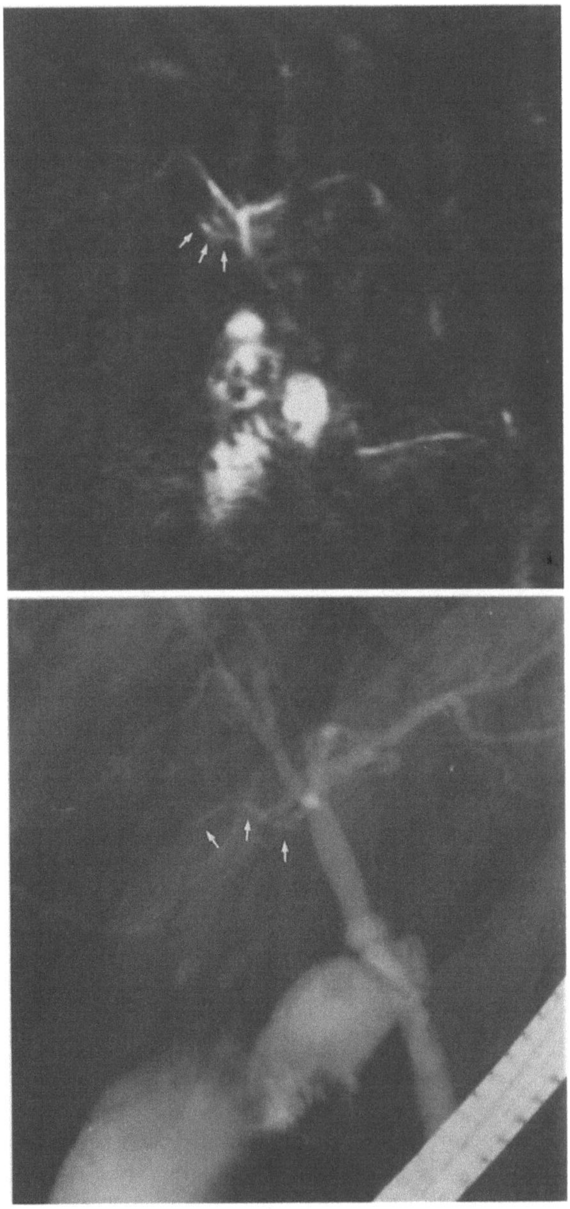

a

b

Fig. 10.9a,b. Aberrant right hepatic duct junction. 2D FSE, phased array torso coil. **a** Coronal, targeted MIP reconstruction demonstrates a small right hepatic duct (*arrows*) joining the common hepatic duct immediately below the level of the bile duct bifurcation. **b** ERCP confirms the findings (*arrows*) with a dilated biliary tract from a variety of benign and malignant causes using unenhanced T1-weighted and T2-weighted CSE sequences. Although MRI reliably demonstrated dilatation of the biliary tree, these authors concluded that the need for MRI for this purpose was limited since US and CT are less costly and have, over time, proven their value in diagnosing bile duct dilatation. However, the advent of MR cholangiography sequences has generated a new interest in MRI as a modality to provide projectional images rather than cross-sections of the bile duct: HALL-CRAGGS et al. (1993), using a 3D SSFP sequence, studied 40 patients, 37 of whom were found to have bile duct obstruction at direct cholangiography. MR cholangiopancreatography correctly diagnosed the presence of obstruction in 36 patients (97%) and identified the location of the obstruction in 33 patients (89%). However, the bile ducts distal to the site of obstruction were seen in only 11 patients. Using a breath-hold single-shot RARE sequence in 30 patients, LAUBENBERGER et al. (1995) accurately detected the presence and level of obstruction in all 25 patients with proven bile duct obstruction. The ducts distal to the site of obstruction as well as the normal bile ducts were visualized in every case. In a series of 126 patients imaged at our institution (GUIBAUD et al. 1995) with the body coil and a 2D FSE sequence, MR cholangiography correctly diagnosed 72 of 79 cases (91%) with proven bile duct obstruction and identified the level of obstruction in all 72 true-positive cases. The ducts distal to the site of obstruction were also visualized in every case. Similar results have been reported by other investigators using various modifications of the FSE sequence (MACAULEY et al. 1995; BARISH et al. 1995).

To summarize, in its current state of development MR cholangiopancreatography can diagnose the presence of bile duct obstruction in 91%–100% of cases and can determine the level of obstruction in 85%–100% of cases. The bile ducts distal to the site of obstruction are routinely identified, as well as the distance from the site of obstruction to the ampulla, which is essential for adequate treatment planning. MR cholangiopancreatography depicts the intrahepatic biliary tree more consistently than ERCP, particularly in cases of complete or high-grade obstruction. Indeed, underfilling of the intrahepatic bile ducts often occurs with ERCP, partly to minimize the risk of septic complications if immediate drainage of the obstructed system is not performed.

10.5 Etiology of Bile Duct Obstruction

10.5.1 Choledocholithiasis

Since the advent of laparoscopic cholecystectomy, there has been renewed interest in the diagnosis of choledocholithiasis. Despite technical advances in recent years, the sensitivity of US and CT in diagnosing choledocholithiasis remains low, ranging from 20% to 80% for US (PANASEN et al. 1992; RIGAUTS et al. 1992; O'CONNOR et al. 1986; CRONAN 1986; STOTT et al. 1991; WERMKE and SCHULTZ (1987); DONG and CHEN (1987) and 23% to 85% for CT (PANASEN et al. 1992; BARON 1987; TODUA et al. 1991). However, the specificity of these tests is high, being greater than 90% and 95% for US and CT respectively. Therefore, ERCP remains the standard of reference for establishing the diagnosis of common bile duct stones. In addition, ERCP has the added advantage of allowing therapy to be instituted at the time of initial diagnosis. However, as already mentioned, ERCP has a number of disadvantages: (1) it is operator-dependent and invasive, and (2) its sensitivity and specificity in the diagnosis of common bile duct stones are only 90% and 98% respectively (FREY et al. 1982). Therefore, depending on the prevalence of common bile duct stones in the population studied, MR cholangiopancreatography may be a cost-effective alternative in determining which patients would benefit from an endoscopic sphincterotomy. Thus far, little data is available regarding the accuracy of MR cholangiography in the diagnosis of common bile duct stones. ISHIZAKI et al. (1993), using a 3D SSFP sequence, correctly diagnosed all 6 cases of choledocholithiasis in 20 patients presenting with obstructive jaundice. In a nonblinded, retrospective review of 10 patients with known choledocholithiasis, we were able to demonstrate the stones in all 10 patients using a 2D FSE sequence (GUIBAUD et al. 1994). More recently, we prospectively analyzed the MR cholangiopancreatograms of 110 patients scheduled to undergo ERCP with a clinical diagnosis of bile duct obstruction, of whom 27 were proven to have choledocholithiasis. All MR cholangiopancreatograms were performed with a phased array multicoil using the parameters described above. In this series, MR cholangiopancreatography diagnosed the presence of choledocholithiasis with a sensitivity of 90%, a specificity of 100%, and an overall diagnostic accuracy of 97% (REINHOLD et al. 1995b) (Fig. 10.10–10.15). This represents a considerable improvement over the results we obtained in a previous study (sensitivity 81% and specificity 98%

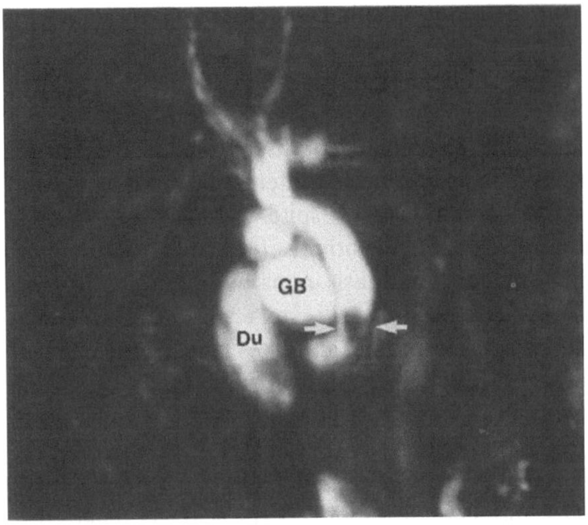

Fig. 10.10. Choledocholithiasis. 2D FSE, phased array torso coil. Coronal MIP reconstruction demonstrates a quadrangular stone (*arrow*) in the distal common bile duct. *Du*, Duodenum; *GB*, gallbladder

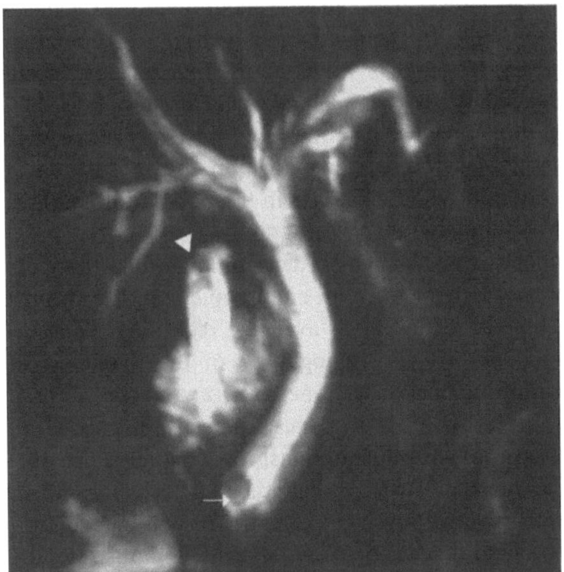

Fig. 10.11. Choledocholithiasis. 2D FSE, phased array torso coil. Coronal MIP reconstruction demonstrates an 8-mm stone (*arrow*) in the distal common bile duct. Gallstones are also present (*arrowhead*)

for diagnosing choledocholithiasis), in which the examinations were performed in the body coil and with lower-resolution imaging parameters (GUIBAUD et al. 1995). Using a 3D FSE technique with a 0.5-T MR system, LAGHI et al. (1995) obtained an 89% sensitivity and a 100% specificity in the diagnosis of choledocholithiasis in a group of 40 patients with a 45% prevalence of choledocholithiasis.

To summarize, US and CT are not sufficiently sensitive to reliably exclude the presence of stones in the common bile duct. However, because these tests are highly specific, no further imaging tests are needed when choledocholithiasis is diagnosed with US or CT. In the remaining cases, MR cholangiopancreatography, the accuracy of which is comparable to that of ERCP, can be performed first, with only the patients in whom stones are identified being referred for endoscopic sphincterotomy. We will see below that a similar algorithm can be advocated for other causes of bile duct obstruction.

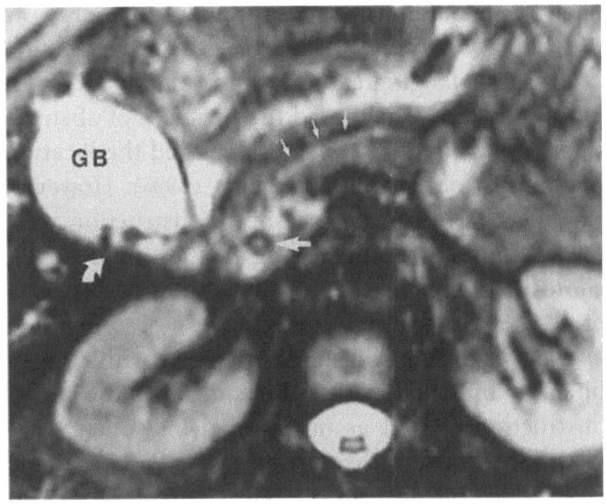

Fig. 10.12. Choledocholithiasis. 2D FSE, phased array torso coil. Axial source image demonstrates a punctuated stone (*arrow*) in the distal common bile duct. A nondilated pancreatic duct is also visible (*small arrows*), as well as small gallstones (*curved arrow*). *GB*, Gallbladder

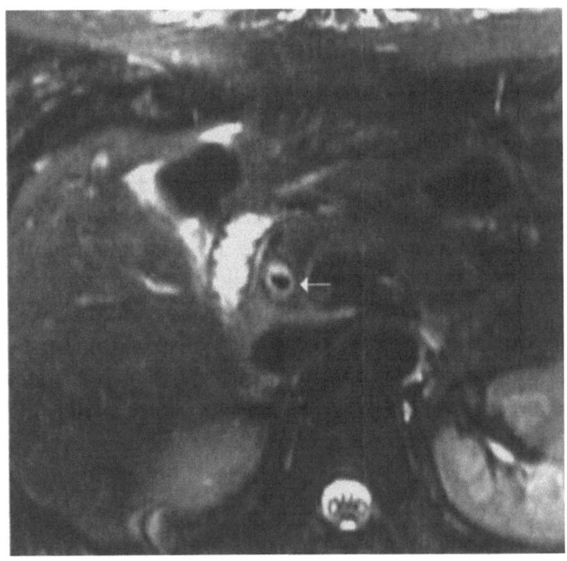

Fig. 10.13. Choledocholithiasis. 2D FSE, phased array torso coil. Axial source image demonstrates a small stone (*arrow*) in the distal common bile duct

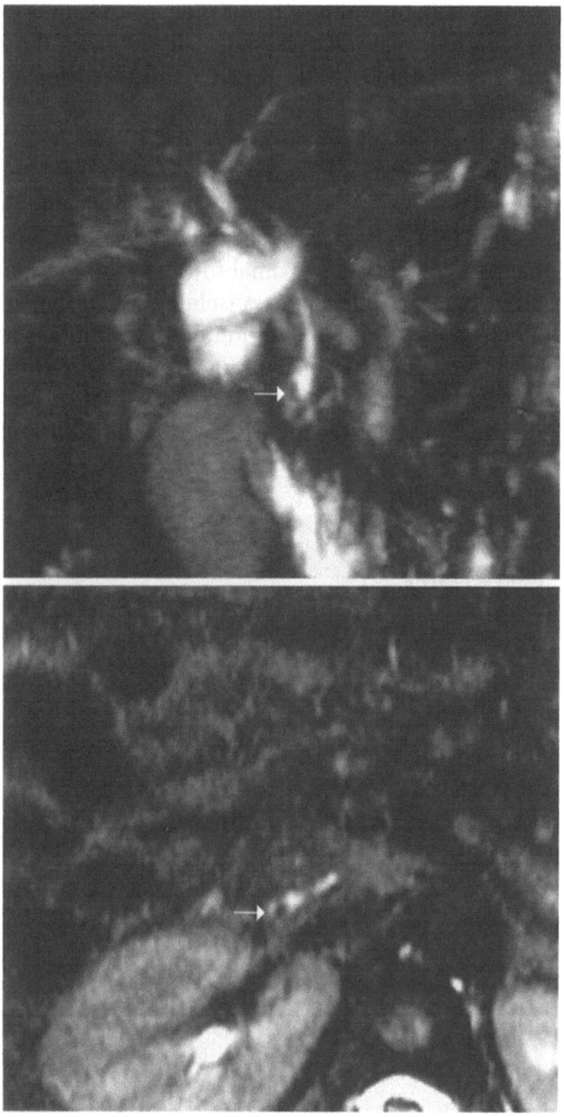

a

b

Fig. 10.14. Choledocholithiasis. 2D FSE, phased array torso coil. **a** Coronal targeted MIP reconstruction and **b** axial source image demonstrate a 3-mm stone (*arrow*) in the distal common bile duct

10.5.2 Malignant Bile Duct Obstruction

The excellent spatial resolution of ERCP usually permits a detailed analysis of the morphology of a bile or pancreatic duct stricture (i.e., the length of a stricture, the presence of asymmetry or irregularity, and the presence of tapering ends). In addition, the contrast injection during the ERCP may distend the stricture and improve the depiction of these morphological characteristics. Analysis of these features often allows a reliable distinction between benign and malignant strictures, and between stricures that are intrinsic versus those that are extrinsic. Conventional MR imaging sequences can provide information on tumor extension: in a recent study, Low et al. (1994) reported the results of a contrast-enhanced, fast multiplanar spoiled gradient-recalled sequence in 21 patients with malignant biliary obstruction. They compared the results of this technique with those obtained using FSE T2-weighted sequences and unenhanced CSE T1-weighted sequences, as well

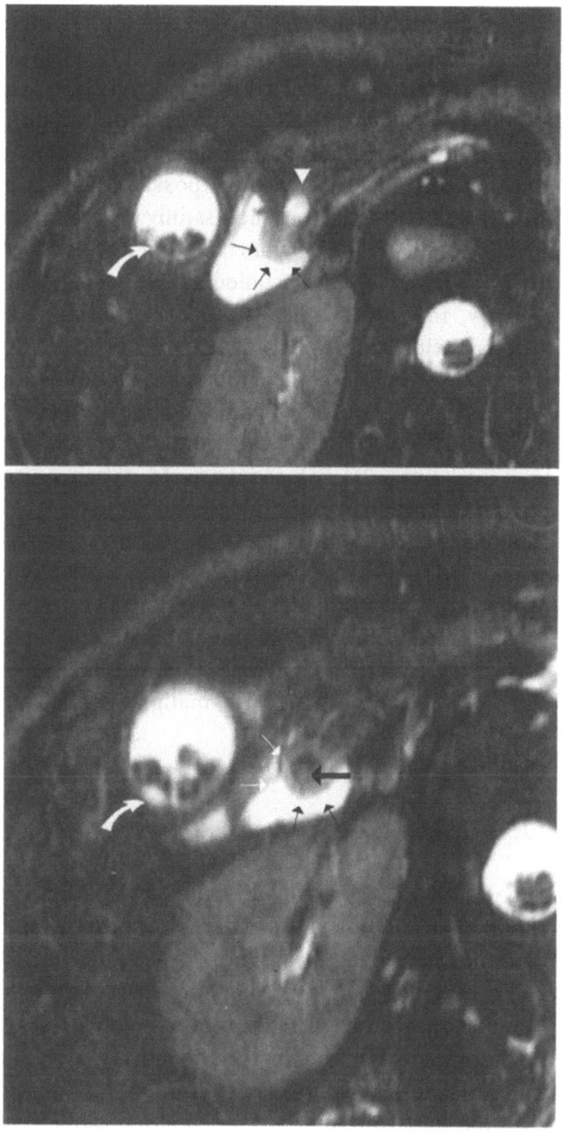

a

b

Fig. 10.15. Common bile duct stone impacted in the ampulla. 2D FSE, phased array torso coil. **a** Axial source image demonstrates a dilatation of the common bile duct (*arrowhead*) in the head of the pancreas. The cranial aspect of the ampulla is visible (*small arrows*). **b** A more caudal axial source image demonstrates a 7-mm stone (*arrow*) impacted in the enlarged ampulla (small arrows). Gallstones are seen in the dependent portion of the gallbladder (*curved arrow*) as contrast-enhanced CT. The gradient-echo sequence improved tumor conspicuity and provided additional information with respect of tumor extension when compared to the FSE T2-weighted and CSE T1-weighted sequences. However, overall, MR imaging offered no significant advantage over contrast-enhanced CT. Published studies of MR cholangiopancreatography that used the SSFP sequence reported a cutoff of the duct(s) at the site of a complete or incomplete obstruction and were therefore unable to evaluate the stricture itself, or to establish whether the narrowed segment was malignant or benign in nature (WALLNER et al. 1991; HALL-CRAGGS et al. 1993). Similarly, in a study by ISHIZAKI et al. (1993) using 3D SSFP, differentiation among lower bile duct carcinomas, pancreatic carcinomas, and metastatic disease was not possible. However, this group of investigators successfully differentiated all 14 patients with malignant bile duct obstruction from 6 patients with choledocholithiasis. More recently, improved results have been obtained in differentiating the various causes of bile duct obstruction with MR cholangiopancreatographic sequences based on the FSE technique. In a group of 79 patients with bile duct obstruction studied with the body coil and a 2D FSE sequence, 14 patients were proven to have malignant bile duct obstruction. MR cholangiopancreatography diagnosed the presence of malignant bile duct obstruction with a sensitivity of 86% and a specificity of 98% (GUIBAUD et al. 1995). In addition, MR cholangiopancreatography correctly characterized the type of neoplasm in 9 of the 14 patients (64%) with malignant bile duct obstruction.

10.5.2.1 Cholangiocarcinoma (Figs. 10.16, 10.17)

The role of imaging in the evaluation of cholangiocarcinoma is twofold: tumoral extension is determined in order to identify resectable tumors and to best plan palliative biliary drainage in patients found to have unresectable tumors. The presence of high-grade obstruction of the biliary tree and wall thickness greater than 5 mm on CT images are findings consistent with cholangiocarcinoma (SCHULTE et al. 1990). These features have also been observed on MR images (SEMELKA et al. 1992). On gadolinium-enhanced fat-suppressed MR images, cholangiocarcinoma enhances to a moderate extent, which permits better delineation of intrahepatic tumor extension than on CT images (LOW et al. 1994). Peripheral tumors appear as large intrahepatic focal mass lesions (HAMRICK-TURNER et al. 1992) after contrast enhancement. The assessment of tumor extension along the bile ducts usually requires direct cholangiography. Both ERCP and transhepatic cholangiography (THC) are of limited value in tumors that extend beyond the porta hepatis. ERCP only demonstrates the bile ducts distal to the site of obstruction, while THC usually does not opacify the distal part of the common bile duct and incompletely maps the bile ducts proximal to the lesion, particularly when multiple stenoses isolate the intrahepatic bile duct segments. In addition, both techniques result in opacification of undrained bile ducts, which exposes the patient to increased risks of septic complications. Furthermore, THC carries the additional

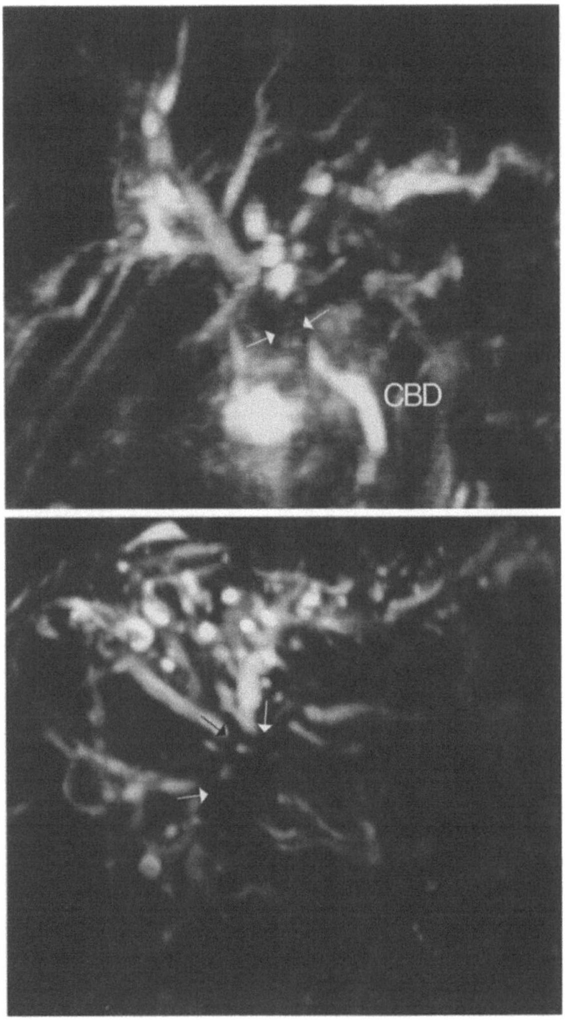

a

b

Fig. 10.16. Cholangiocarcinoma. 2D FSE, phased array torso coil. **a** Coronal, targeted MIP reconstruction demonstrates a stop (*arrow*) on the proximal common hepatic duct with proximal bile duct dilatation. The common bile duct distal to the tumor is well visualized. **b** Axial MIP reconstruction demonstrates tumoral extension along the intrahepatic branches to better advantage (*arrows*). *CBD*, Common bile duct

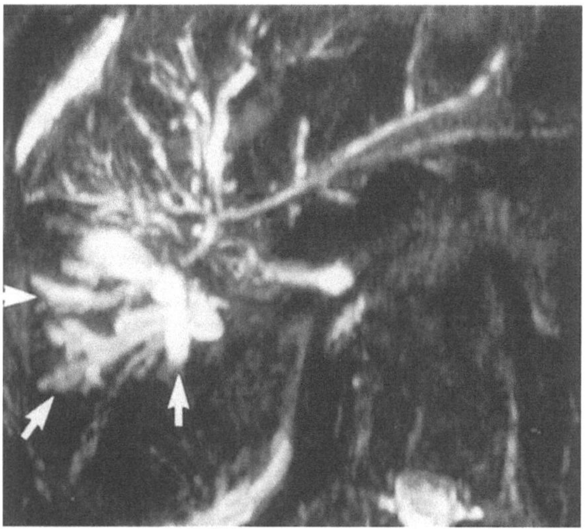

Fig. 10.17. Cholangiocarcinoma. Axial MIP reconstruction shows predominant obstruction of the right hepatic ducts (*arrows*) with associated atrophy of the right lobe of the liver. (From Reinhold et al. 1996, by permission)

risk of bile leakage if the procedure is not followed by a biliary drainage. In this clinical setting, MR cholangiopancreatography has a number of advantages over both THC and ERCP:

1. It can provide a detailed map of the biliary tree anatomy, and often provides more information than can be obtained with the combination of THC and ERCP. The coronal 3D MIP reconstructions, which simulate the images obtained at direct contrast cholangiography, greatly facilitate identification of the various intrahepatic bile duct segments. We have also found that axial, targeted, small-volume MIP reconstructions that encompass the bile duct bifurcation are particularly suited to evaluating tumor extension along the intrahepatic bile ducts.
2. Patients presenting with nonresectable tumors that are essentially asymptomatic, of patients presenting with multiple intrahepatic strictures, can be spared a biliary drainage that may be of doubtful clinical benefit.
3. Correlation of the dilated ducts on MR cholangiopancreatography with images of the surrounding hepatic parenchyma provides a means of planning optimal nonsurgical drainage in patients with complex hilar strictures, thereby avoiding drainage of atrophic segments.
4. A complete tumor staging that documents involvement of the liver, portal nodes, or the portal vein can often be obtained from the axial MR cholangiogram and by using additional conventional MR sequences (Low et al. 1994).

10.5.2.2 Pancreatic Carcinoma (Figs. 10.18, 10.19)

The majority of pancreatic carcinomas are diagnosed with cross-sectional imaging techniques and in these cases, an ERCP is not needed since it provides no additional information with respect to the resectability of the tumor. Similarly, when a carcinoma of the pancreas is diagnosed with ultrasound or CT, MR cholangiopancreatography is of limited

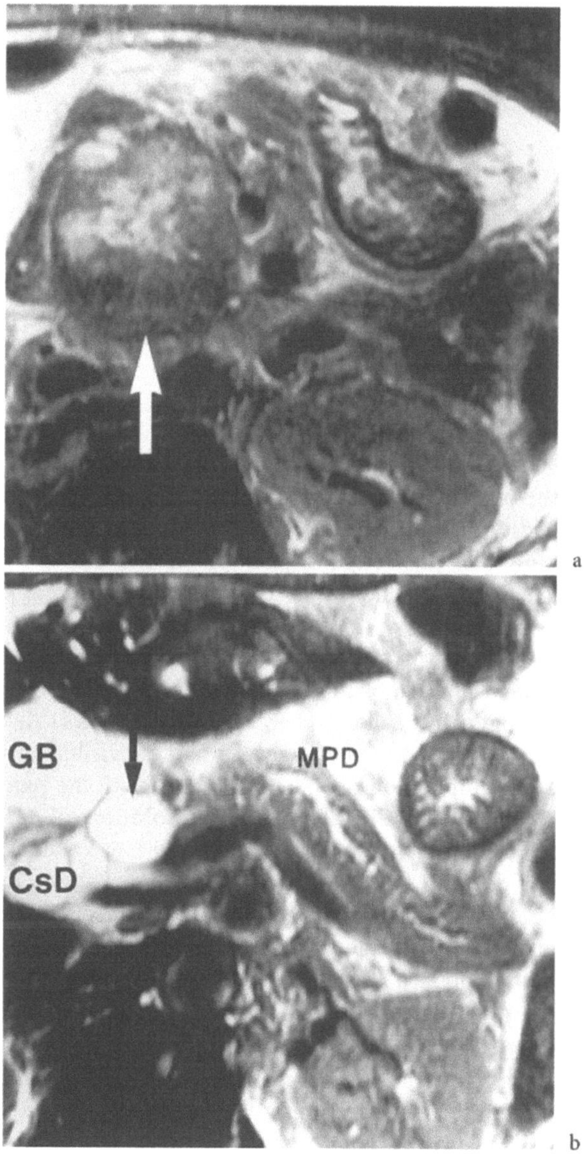

Fig. 10.18a,b. Pancreatic carcinoma. **a** Axial source image demonstrates a necrotic tumor in the head of the pancreas (*arrow*). **b** Axial source image at a more cranial level shows dilatation of the common bile duct (*arrow*) and cystic duct proximal to the tumor. Note the clear visualization of minimally dilated pancreatic side branches. *GB*, Gallbladder; *MPD*, main pancreatic duct; *CsD*, cystic duct. (From Reinhold et al. 1996, by permission)

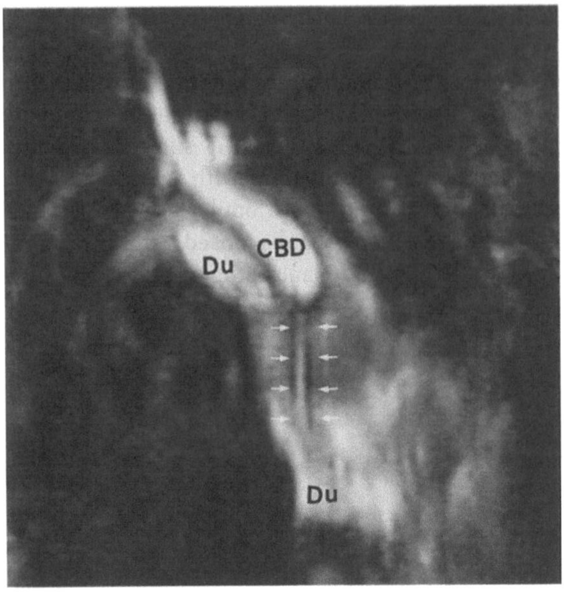

Fig. 10.19. Pancreatic carcinoma with endoscopic stent. 2D FSE, phased array torso coil. Coronal targeted MIP reconstruction demonstrates an endoscopic stent (*arrows*) placed across a tumor of the pancreas (not shown). *Du*, Duodenum; *CBD*, common bile duct

value, although it will display the ducts proximal to the obstruction more completely, and may provide additional information on the anatomy of the common bile duct (e.g., angulation of the bile duct, length of the stenosis) that may facilitate endoscopic stenting. Indeed, the majority of pancreatic carcinomas are unresectable at presentation and these patients will benefit from a palliative endoscopic biliary drainage. In small peripheral pancreatic tumors that remain undetected at US and CT and are visible only as an amputation of collateral branches of the pancreatic duct at ERCP, the low resolution of MR pancreatography may be a limitation when compared to ERCP. Conventional MR imaging may also be used as an adjunct to ultrasound and contrast-enhanced CT in the assessment of tumor extension. Fat-suppressed T1-weighted sequences are best suited to demonstrate the pancreatic tumor. Contrast injection enhances the pancreas tumor signal intensity ratio, and is helpful to evaluate the peripancreatic extension of the tumor, especially to the superior mesenteric vein.

10.5.2.3 Ampullary Carcinoma (Fig. 10.20)

Ampulllary carcinomas, especially when intraductal, remain a challenge for MR cholangiopancreatography. The main advantage of ERCP is that it provides an endoscopic view of the ampulla and allows biopsies of all suspicious areas to be performed. In a series of 6 ampullary carcinomas diagnosed in 79 patients presenting with bile duct obstruction, 2 cases of ampullary carcinoma were correctly diagnosed at MR cholangiopancreatography, 2 cases were misinterpreted as pancreatic carcinomas, 1 case

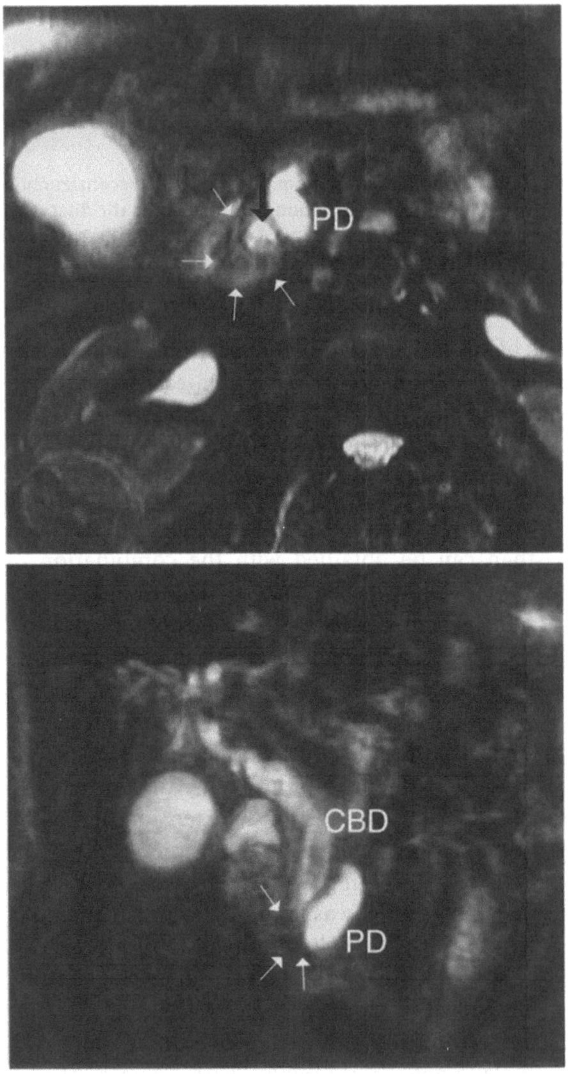

a

b

Fig. 10.20a,b. Ampullary carcinoma. 2D FSE, phased array torso coil. **a** Axial source image demonstrates a mass in the region of the ampulla (*small arrows*) bulging into the duodenum. The pancreatic duct is dilated, as well as the distal common bile duct (*arrow*), which shows presence of debris in the lumen. **b** Coronal targeted MIP reconstruction demonstrates the distal common bile duct containing debris and the obstructed pancreatic duct proximal to the ampullary tumor (*small arrows*). *PD*, Pancreatic duct; *CBD*, common bile duct

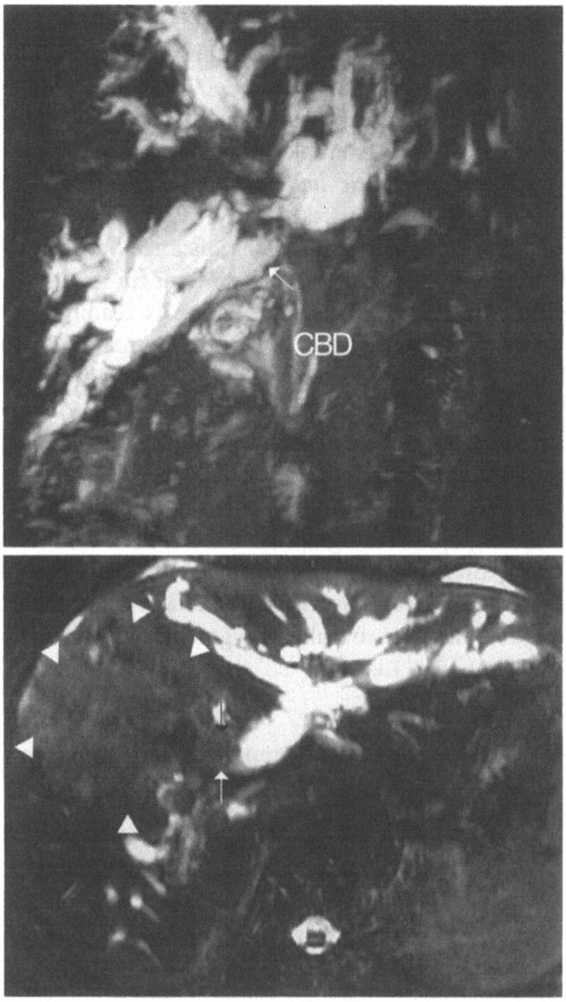

a

b

Fig. 10.21a,b. Bile duct obstruction secondary to liver metastases. 2D FSE, phased array torso coil. **a** Coronal MIP reconstruction demonstrates bile duct obstruction at the level of the porta hepatis (*arrow*) with severe intrahepatic bile duct dilatation excepted for segment VIII of the liver. The common bile duct is not dilated. **b** Axial source image demonstrates a mass in segment VIII of the liver (*arrowheads*) which extends into the lumen of the bile ducts (*arrows*) in the region of porta hepatis. *CBD*, Common bile duct

was misdiagnosed as a common bile duct stone, and the remaining case was diagnosed as a distal bile duct obstruction of undetermined cause (Guibaud et al. 1995). In addition, two benign inflammatory stenoses were misdiagnosed as ampullary tumors at MR cholangiopancreatography. In the future, improved visualization of the ampulla at MR cholangiopancreatography may help to better characterize distal bile duct obstructions, especially with respect to ampullary tumors. However, using current techniques, it has been our experience that the major papilla is only visible at MR cholangiopancreatography in approximately 40% of patients (Taourel et al. 1996).

10.5.2.4 Liver Metastases (Fig. 10.21)

In case of bile duct obstruction secondary to liver metastases, MR cholangiography allows visualization of both the site of the obstruction and the extension of the tumoral lesions.

10.5.3 Inflammatory Stenosis of the Distal Common Bile Duct

Using conventional imaging methods, a significant number of patients referred for suspected bile duct obstruction demonstrate a dilated common bile duct that tapers at the level of the ampulla without evidence of an obstructive lesion. These patients form a heterogeneous group comprising patients with postcholecystectomy common bile duct dilatation, patients with common bile duct dilatation from stones that have spontaneously passed, and patients with an inflammatory ampullary stenosis (Fig. 10.22). Therefore, it is somewhat misleading to diagnose bile duct obstruction solely on the basis of measurements of the diameter of the common bile duct, and associated relevant clinical symptomatology and laboratory abnormalities are usually required before endoscopic sphincterotomy is performed. In addition, ERCP offers the advantage over MR cholangiopancreatography that it can assess for the presence of delayed contrast emptying. The differential diagnosis of an isolated dilatation of the common bile duct at the level of the ampulla includes intraductal ampullary tumor, and ERCP remains superior to MR cholangiopancreatography in the evaluation of the ampulla, as previously discussed.

10.5.4 Bile Duct Injuries

The reported rate of bile duct injury during cholecystectomy has doubled since the advent of laparoscopic cholecystectomy. ERCP is currently the examination of choice to detect a bile leak or a bile duct injury. A frequent and benign complication of laparoscopic cholecystectomy is the presence of a cystic duct leak. While ERCP can demonstrate the actual leak, MR cholangiography shows a fluid collection in the gallbladder bed, but cannot determine whether or not the duct is actually leaking. Conversely, when bile duct injuries involve either the intra- or extrahepatic bile ducts, MR cholangiopancreatography has a number of advantages over ERCP. For example, when the injury involves the *common hepatic duct or common bile duct*, a cutoff is usually demonstrated at ERCP. However, therapeutic options depend on the distance between the proximal end of the lesion and the porta hepatis (Bismuth 1983). This information is readily available by mapping the biliary tree anatomy with MRCP. Similarly, when the injury involves one of the *intrahepatic bile ducts*, as occurs more frequently in the setting of an aberrant right hepatic bile duct, a common pitfall of ERCP is to overlook the unopacified or excluded bile duct segment. MR cholangiopancreatography on the other hand, can easily diagnose the presence of an exluded segment and demonstrate the associated dilatation of the excluded bile duct, usually in segments, VI and VII of the liver (Fig. 10.23).

a

b

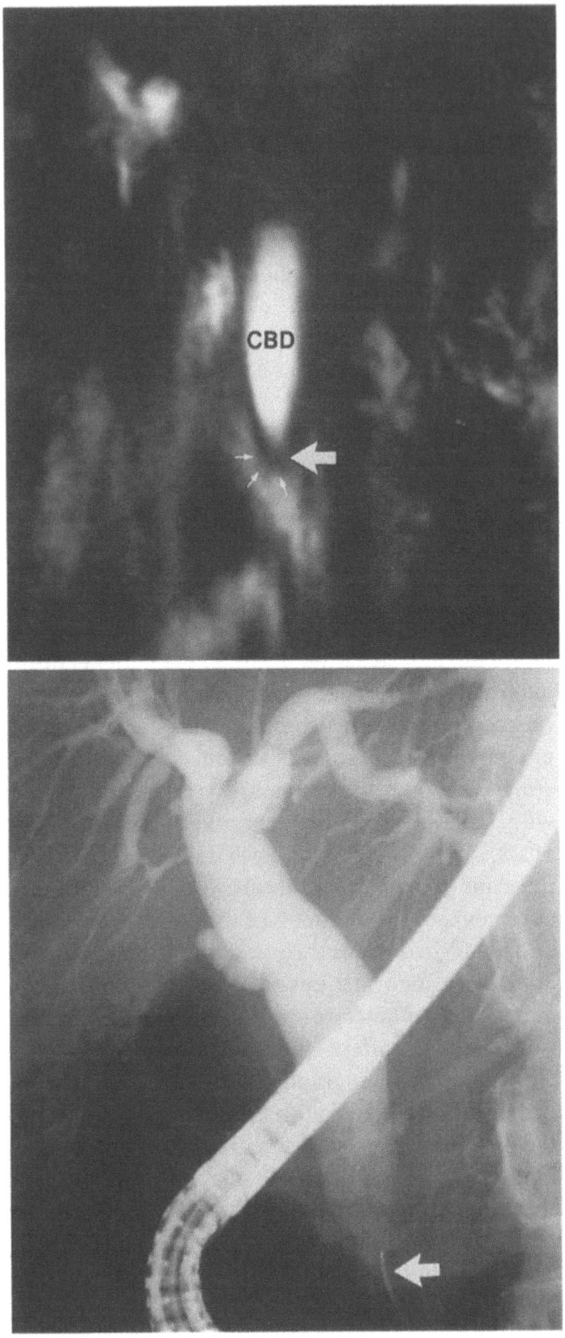

Fig. 10.22a,b. Ampullitis. 2D FSE, phased array torso coil. **a** Coronal targeted MIP reconstruction demonstrates moderate dilatation of the common bile duct (*arrow*) down to the level of the ampulla (*small arrows*), which is enlarged and hypointense. The differential diagnosis includes ampullitis or ampullary carcinoma. **b** ERCP demonstrates the dilated bile ducts down to the level of the ampulla (*arrow*). At endosocopy, the ampulla was enlarged and edematous. Endoscopic biopsy performed after endoscopic sphincterotomy confirmed the diagnosis of ampullitis. *CBD*, Common bile duct

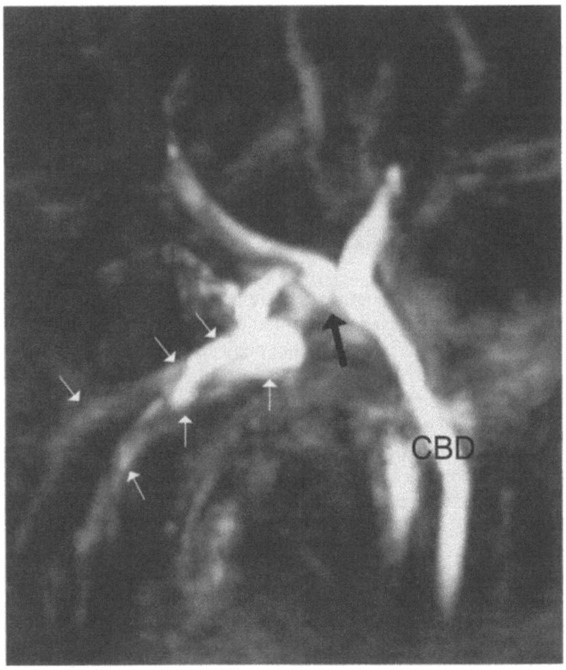

Fig. 10.23. Bile duct injury during laparoscopic cholecystectomy. 2D FSE, phased array torso coil. Coronal MIP reconstruction demonstrates that the posterior right hepatic duct (*small arrows*) is disconnected from the rest of the biliary tract (*arrow*). The duct did not opacify at ERCP, and was shown to leak into the peritoneal cavity at percutaneous transhepatic cholangiography. *CBD*, Common bile duct

To summarize, in complications of laparoscopic cholecystectomy, ERCP and MR cholangiography are complementary, and each may provide unique information relevant to the patient's management.

10.5.5 Sclerosing Cholangitis

The role of cholangiography in the evaluation of sclerosing cholangitis is to generate a detailed map of the biliary tree anatomy, including the depiction of strictures that are multifocal and involve both the extra- and intrahepatic bile ducts. ERCP is often limited by the multiplicity of the bile duct stricutres and the difficulty in opacifying the ducts proximal to a high-grade stenosis (Fig. 10.24). Conversely, the low spatial resolution of MR cholangiopancreatography limits its ability to provide a detailed analysis of the stricture morphology, especially in the setting of a cholangiocarcinoma complicating sclerosing cholangitis.

10.6 MR Cholangiography Versus Other Cholangiographic Techniques

In addition to MR cholangiopancreatography, other imaging modalities are competing with ERCP for the diagnosis of bile duct and pancreatic disease. CT cholangiography has recently been advocated for imaging the biliary tree. This technique combines thin-section helical CT and intravenous cholangiography to produce high-resolution axial images of the biliary tree, with the potential for 3D reconstruction (Klein et al. 1991; Van Beers et al. 1994). Although infrequent, reactions to the intravenous cholangiographic contrast material have been reported (Dawson et al. 1993). A successful cholangiogram cannot be obtained in patients with elevated bilirubin levels, thereby limiting its use to patients with mild obstructive jaundice. In addition, since only the biliary tree is opacified during CT cholangiography, separate evaluations of the pancreatic duct may be required. Endoscopic US is a promising new technique for the evaluation of the extrahepatic bile ducts and appears to be highly accurate in the diagnosis of choledocholithiasis (Amouyal et al. 1994). However, it is extremely operator-dependent, requires heavy sedation, and, like ERCP, can result in complications related to the endoscopic manipulation. Furthermore, due to its limited field of view (on the order of 5–7 cm from the transducer), endoscopic US is restricted to imaging the extrahepatic bile ducts and the pancreas.

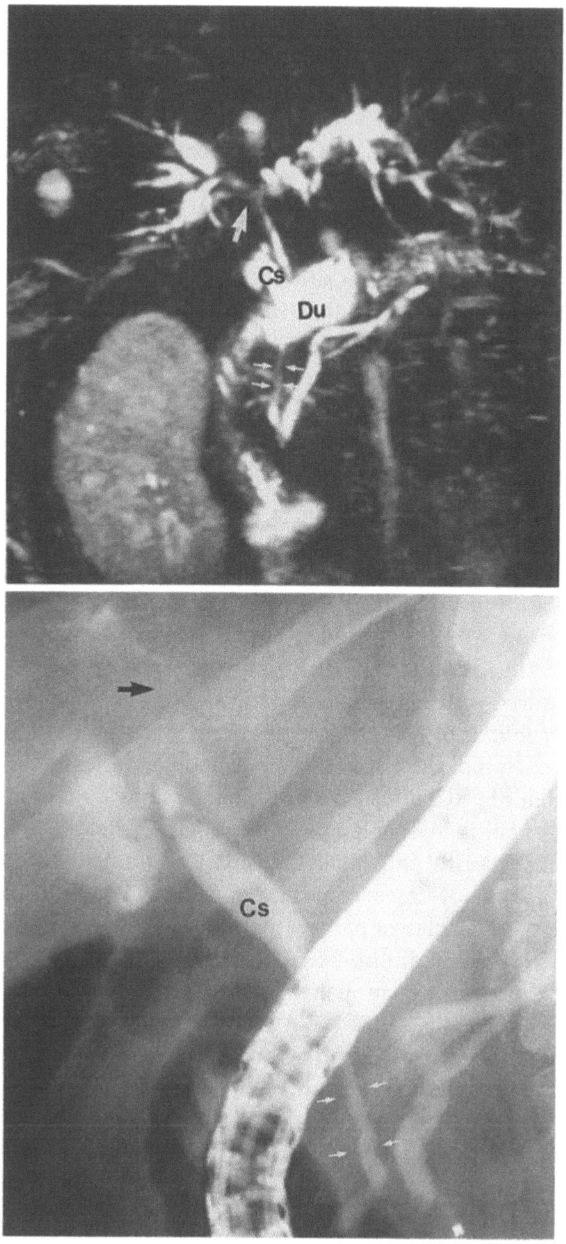

a

b

Fig. 10.24a,b. Sclerosing cholangitis. 2D FSE, phased array torso coil. **a** Coronal MIP reconstruction demonstrates moderate dilatation of the intrahepatic bile ducts with multiple strictures at the level of porta hepatis (*arrow*). There is an associated stricture of the distal common bile duct (*small arrows*). **b** ERCP demonstrates the narrowing of the distal common bile duct (*small arrows*) but fails to opacify the ducts proximal to the obstruction (*arrow*). *Cs*, Cystic duct; *Du*, duodenum

In summary, MR cholangiopancreatography should not be regarded as simply one more test in the work-up of patients with biliary or pancreatic disease. For example, when ultrasound demonstrates choledocholithiasis or a pancreatic carcinoma that is clearly unresectable, the patient will be referred for endoscopic intervention and, in these cases, there is no role for routine MR cholangiopancreatography. However, in cases where the results of US and/or CT are indeterminate, ERCP is currently the standard of reference; it is in these instances that MR cholangiopancreatography has the potential to substitute for diagnostic ERCP. With this algorithm, only patients with positive findings on MR cholangiopancreatography that require endoscopic intervention would be referred for ERCP. The remaining cases would be managed on the basis of the MR cholangiopancreatographic results. When comparing MR cholangiopancreatography with ERCP, a clear advantage of MR cholangiopancreatography is its *lack of invasiveness*. No adverse effects are encountered with MR imaging if patients are adequately screened for contraindications. On the other hand, diagnostic ERCP is associated with a 4%–7% morbidity and up to a 1% mortality (LENRIOT et al. 1993). Routine sedation is not required with MR cholangiopancreatography. The *rate of failure* of MR cholangiopancreatography (due to contraindications and claustrophobia) is less than that of ERCP. In a series of 156 patients who underwent both MR cholangiopancreatography and ERCP, unsuccessful and inadequate ERCP accounted for 8% and 6% of cases respectively, versus 4% and 1% for MR cholangiopancreatography (GUIBAUD et al. 1995). In addition, MR cholangiopancreatography is not limited in patients with altered anatomy, for example, those with a duodenal stenosis or hepatojejunostomy. MR cholangiopancreatography is *not operator-dependent* and can easily be implemented on most mid-field or high-field MR scanners. In comparison, ERCP requires highly skilled operators, and the results achieved by specialized teams are not always reproduced during multicenter trials (LENRIOT et al. 1993; ASSOULINE et al. 1993). MR cholangiopancreatography allows visualization of the biliary tree and pancreatic duct both proximal and distal to the site of obstruction. This is particularly relevant in cases of multiple bile duct strictures, where a complete cholangiogram is needed to plan therapeutic intervention. Image postprocessing, including multiplanar reformatting and 3D reconstruction, allows MR cholangiopancreatography to be viewed in any desired imaging plane. A current weakness of MR cholangiography is its relatively low spatial resolution, which limits the visualization of nondilated pancreatic duct side branches, the characterization of strictures, and the detection of calculi less than 2–3 mm in diameter (GUIBAUD et al. 1995). Signal loss from surgical clips after cholecystectomy may occur with MR cholangiopancreatography, although this rarely interferes with diagnostic accuracy. An intraductal signal void is not specific for calculi; blood clot, proteinaceous debris, or air bubbles must be considered in the differential diagnosis. However, this limitation also holds true for ERCP, where the differentiation of common bile duct stones from air bubbles is a common pitfall. A clear advantage of ERCP over MR cholangiopancreatography, however, is its ability to visualize the ampulla in every case at endoscopy and to perform biopsies and other therapeutic interventions when indicated.

10.7 Conclusions

The main advantages of MRI of the biliary tract are that it combines the benefits of a cross-sectional imaging technique, providing the same type of information as US and CT, and of a projectional modality similar to ERCP. At the current stage, it is too early to predict whether MR cholangiography will routinely substitute for diagnostic ERCP. To date, only preliminary results are available regarding the diagnostic accuracy of MR cholangiopancreatography, and larger clinical trials are needed. The future of MR cholangiopancreatography will depend not only on its diagnostic accuracy, but also on the availability of the technique and the cost at which the information is obtained. Therefore, cost-effectiveness and outcome analysis studies are urgently needed to address these issues.

References

Amouyal P, Amouyal G, Lévy P, Tuzet S, Palazzo L, Vilgrain V, Gayet B, Belghiti J, Fékété F, Bernades P (1994) Diagnosis of choledocholithiasis by endoscopic ultrasonography. Gastroenterology 106:1062–1067

Assouline Y, Liguory C, Ink O, Fritsch J, Choury A, Lefebvre J, Pelletier G, Buffet C, Etienne J (1993) Résultats actuels de la sphinctérotomie endoscopique pour lithiase de la voie biliaire principale. Gastoenterol Clin Biol 17:251–258

Barish MA, Yucel EK, Soto JA, Chuttani R, Ferrucci JT (1995) MR cholangiopancreatography: efficacy of three-dimensional turbo spin-echo technique. AJR 165:295–300

Baron RL (1987) Common bile duct stones: reassessment of criteria for CT diagnosis. Radiology 162:419–424

Bismuth H (1983) Postoperative strictures of the biliary tract. In: Blumgart LH (ed) The biliary tract. Churchill Livingstone, New York, pp 209–218

Cronan JJ (1986) US diagnosis of choledocholithiasis: a reappraisal. Radiology 161:133–134

Dawson P, Adam A, Benjamin IS (1993) Intravenous cholangiography revisited (editorial). Clin Radiol 47:223–225

Dong B, Chen M (1987) Improved sonographic visualization of choledocholithiasis. J Clin Ultrasound 15:185–190

Dooms GC, Fisher MR, Higgins CB, Hricak H, Goldberg HI, Margulis AR (1986) MR imaging of the dilated biliary tract. Radiology 158:337–341

Frey CF, Burbige EJ, Meinke WB, Pullos TG, Wong HN, Hickman DM, Belber J (1982) Endoscopic retrograde cholangiopancreatography. Am J Surg 144:109–114

Guibaud L, Bret PM, Reinhold C, Atri M, Barkun ANG (1994) Diagnosis of choledocholithiasis: value of MR cholangiography. Am J Roentgenol 163:847–850

Guibaud L, Bret PM, Reinhold C, Atri M, Barkun AN (1995) Bile duct obstruction and choledocholithiasis: diagnosis with MR cholangiography. Radiology 197:109–115

Hall-Craggs MA, Allen CM, Owens CM, Theis BA, Donald JJ, Paley M, Wilkinson ID, Chong WK, Hatfield ARW, Lees WR, Russell RCG (1993) MR cholangiography: clinical evaluation in 40 cases. Radiology 189:423–427

Hamrick-Turner J, Abbitt PL, Ros PR (1992) Intrahepatic cholangiocarcinoma: MR appearance. AJR 158:77–79

Hennig J, Friedburg H, Ströbel B (1986) Rapid nontomographic approach to MR myelography without contrast agents. J Comput Assist Tomogr 10:375–378

Ishizaki Y, Wakayama T, Okada Y, Kobayashi T (1993) Magnetic resonance cholangiography for evaluation of obstructive jaundice. Am J Gastroenterol 88:2072–2077

Klein H, Wein B, Truong S, Pfingsten F, Günther R (1991) Computed tomographic cholangiography using spiral scanning and 3D image processing. Br J Radiol 66:762–767

Laghi A, Catalano C, Broglia L, Messina A, Scipioni A, Pavone P, Passariello R (1995) Optimized 3D MR cholangiography in the evaluation of bile duct stones: superiority over endoscopic retrograde cholangiopancreatography. In: Society of Magnetic Resonance Imaging Abstract Book, p 1446. Presented at SMR Annual Meeting, August 19–25, 1995, Nice, France

Laubenberger J, Büchert M, Schneider B, Blum U, Hennig J, Langer M (1995) Breath-hold projection magnetic resonance-cholangio-pancreatography (MRCP): a new method for examination of the bile and pancreatic ducts. Magn Reson Med 33:18–23

Lenriot J, Le Neel J, Hay J, Jaeck D, Millat, B, Fagniez P (1993) Cholangio-pancréatographie rétrograde et sphinctérotomie endoscopique pour lithiase biliaire. Gastroenterol Clin Biol 17:244–250

Low RN, Sigeti JS, Francis IR, Weinman D, Bower B, Shimakawa A, Foo TK (1994) Evaluation of malignant biliary obstruction: efficacy of fast multiplanar spoiled gradient-recalled MR imaging vs spin-echo MR imaging, CT, and cholangiography. AJR 162:315–323

Macaulay SE, Schulte SJ, Sekijima JH, Obregon RG, Simon HE, Rohrmann CA Jr, Freeny PC, Schmiedl UP (1995) Evaluation of a non-breath-hold MR cholangiography technique. Radiology 196:227–232

Morimoto K, Shimoi M, Shirakawa T, Aoki Y, Choi S, Miyata Y, Hara K (1992) Biliary obstruction: evaluation with three-dimensional MR cholangiography. Radiology 183:578–580

O'Connor HJ, Hamilton I, Ellis WR, Watters J, Lintott DJ, Axon AT (1986) Ultrasound detection of choledocholithiasis: prospective comparison with ERCP in the postcholecystectomy patient. Gastrointest Radiol 11:161–164

Outwater EK (1993) MR cholangiography with a fast-spin echo sequence (abstract). JMRI 3(P):131

Panasen P, Partanen K, Pikkarainen P, Alhava E, Pirinen A, Janatuinen E (1992) Ultrasonography, CT and ERCP in the diagnosis of choledochal stones. Acta Radiol 33:53–56

Reinhold C, Guibaud L, Genin G, Bret PM (1995a) MR cholangiopancreatography: comparison between two-dimensional fast spin-echo and three-dimensional gradient-echo pulse sequences. J Magn Reson Imaging 4:379–384

Reinhold C, Taourel P, Bret PM, Barkun A, Atri M (1995b) MR imaging of choledocholithiasis using a multicoil array and high resolution imaging parameters. In: Society of Magnetic Resonance Imaging Abstract Book, p 509. Presented at SMR Annual Meeting, August 19–25, 1995, Nice, France

Reinhold C, Bret PM, Guibaud L, Barkun A, Genin G, Atri M (1996) Magnetic resonance cholangiopancreatography (MRCP): Potential clinical applications. Radiographics 16:309–320

Rigauts H, Marchal G, Van Steenbergen W, Ponette W (1992) Comparison of ultrasound and ERCP in the detection of the cause of obstructive biliary disease. Rofo Fortschr Geb Rontgenstr Neuen Bildgeb Verfahr 156:252–257

Sananes JC, Bonnet M, Lecesne R, Raymond JM, Couzigou P, Laurent F, Drouillard J (1995) Magnetic resonance cholangiography using HASTE sequence. Optimization and clinical evaluation in extrahepatic cholestasis. In: Society of Magnetic Resonance Imaging Abstract Book, p 1453. Presented at SMR Annual Meeting, August 19–25, 1995, Nice, France.

Schulte SJ, Baron RL, Teffey SA, Rohrmann CA Jr, Freency PC, Shuman WP, Foster MA (1990) CT of the extrahepatic bile ducts: wall thickness and contrast enhancement in normal and abnormal ducts. AJR 154:79–85

Semelka RC, Shoenut JP, Kroeker MA, Hricak H, Minuk GY, Yaffe CS, Micflikier AB (1992) Bile duct disease: prospective comparison of ERCP, CT, and fat suppression MRI. Gastrointest Radiol 17:347–352

Shiono T, Iwasaki N (1995) MR cholangiography with fast imaging scheme. In: Society of Magnetic Resonance Imaging Abstract Book, p 1454. Presented at SMR Annual Meeting, August 19–25, 1995, Nice, France

Stott MA, Farrand PA, Guyer PB, Dewbury KC, Browning JJ, Sutton R (1991) Ultrasound of the common bile duct in patients undergoing cholecystectomy. J Clin Ultrasound 19:73–76

Takehara Y, Ichijo K, Tooyama N, Kodaira N, Yamamoto H, Tatami M, Saito M, Watahiki H, Takahashi M (1994) Breath-hold MR cholangiopancreatography with a long-echo-train fast spin-echo sequence and a surface coil in chronic pancreatitis. Radiology 192:73–78

Taourel P, Bret PM, Reinhold C, Barkun AN, Atri M (1996) Anatomic variants of the biliary tree: Diagnosis with MR cholangiopancreatography. Radiology 199:521–527

Todua FI, Karmazanovskii GG, Vikhorev AV, Todua FI, Karmazanovskii GG, Vikhorev AV (1991) Computerized tomography of the mechanical jaundice in the involvement of the distal region of the common bile duct. Vestn Rentgenol Radiol 2:15–22

Van Beers BE, Lacrosse M, Trigaux JP, de Canniere L, De Ronde T, Pringot J (1994) Noninvasive imaging of the biliary tree before or after laparoscopic cholecystectomy: use of three-dimensional spiral CT cholangiography. Am J Roentgenol 162:1331–1335

Wallner BK, Schumacher KA, Weidenmaier W, Friedrich JM (1991) Dilated biliary tract: evaluation with MR cholangiography with a T2-weighted contrast-enhanced fast sequence. Radiology 181:805–808

Wermke W, Schultz HJ (1987) Sonographic diagnosis of bile duct calculi. Results of a prospective study of 222 cases of choledocholithiasis. Ultraschall Med 8:116–120
Wielopolski PA, Zuo C, Clouse M, Buff B (1995) Breath-hold 3D cholangiography using RARE and segmented echo planar imaging readouts. In: Society of Magnetic Resonance Imaging Abstract Book, p 1448. Presented at SMR Annual Meeting, August 19–25, 1995, Nice, France
Zuo C, Buff B, Wieloposki P, Clouse M (1995) MR cholangiography with fast imaging scheme. In: Society of Magnetic Resonance Imaging Abstract Book, p 1445. Presented at SMR Annual Meeting, August 19–25, 1995, Nice, France

11 Endoluminal Ultrasonography of the Biliary Tree

H.-J. BRAMBS

CONTENTS

11.1 Introduction

With the miniaturization of ultrasound probes, it has become possible to perform endoluminal ultrasonography via catheter in various tubular structures. Early experiences with intravascular ultrasonography have shown that valuable information about vascular pathology may be obtained in this way. Intravascular ultrasonography depicts vascular stenoses and arterial plaques in great detail. It is an accurate method of evaluating the success of balloon angioplasty and is more sensitive than angiography in detecting postangioplasty intimal flaps (TOBIS et al. 1988; ISNER et al. 1990).

In a comparable way to endoscopic ultrasonography (EUS), these devices allow assessment of the urinary tract (GOLDBERG et al. 1991), the female reproductive organs (GOLDBERG et al. 1990), the trachea and the bronchial tree (HÜRTER and HANRATH 1990), the esophagus (LIU et al. 1993), and the pancreas (YASUDA et al. 1992). Intraluminal ultrasonography opens up a wide range of diagnostic and therapeutic possibilities in the bile ducts (ENGSTRÖM and WIECHEL 1990; VAN SONNENBERG et al. 1992; YASUDA et al. 1992; BRAMBS and RIEBER 1993; WAGNER et al. 1995).

H.-J. BRAMBS, MD, Prof. Dr., Department of Radiology, University Hospital, Steinhövelstr. 9, 89075 Ulm, Germany

11.2 Technique

Several endoluminal ultrasound instruments and a variety of catheters are available. Two types of transducers have been developed, one using mechanically rotated transducer elements and the other electronically switched phased arrays. Mechanically driven systems have a simpler transducer design, better signal processing, and superior image quality than phased array systems (CROWLEY et al. 1989).

The most frequently used system (Boston Scientific, in association with the supporting ultrasound unit from Diasonics or Hewlett Packard) consists of a single element crystal oscillating at 12.5 or 20 MHz mounted at the end of a flexible shaft. The transducer is rotated at 1800 rpm within a guiding 6-F or 9-F plastic catheter 100 cm in length. Real-time cross-sectional images at 10° from perpendicular are obtained by 360° rotation. Sterile water provides acoustic coupling between the transducer and the catheter. These catheters are either blunt-tipped or contain guidewires, which are useful for advancing into small lumina.

The catheter is introduced percutaneously through a transhepatic 7- or 8-F sheath directly into the bile duct. The position of the transducer is monitored both by ultrasound and by intermittent fluoroscopy after injection of contrast material through the sheath (Fig. 11.1). Recently, longer ultrasound probes (Olympus) have been developed which allow transpapillary insertion via the working channel of a duodenoscope (YASUDA et al. 1992).

The depth of penetration approaches 2–2.5 cm with 12.5-Mhz transducers and 1.5–2 cm with 20-Mhz transducers (Table 11.1), yielding effective fields of view of 5 cm or 4 cm, respectively. Because of the high frequency, axial accuracy of between 0.1 mm (20 MHz) and 0.3 mm (12.5 MHz) is theoretically possible. Thus, detection of areas of focal wall thickening, irregularity, and distortion of normal architecture is possible.

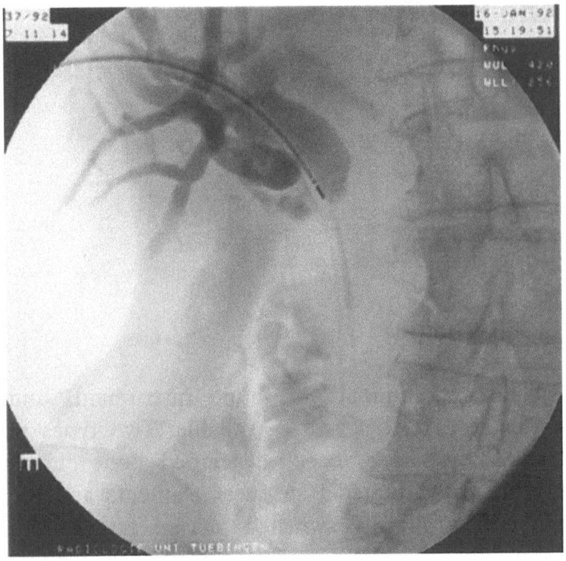

Fig. 11.1. Fluoroscopy shows the ultrasound probe to be positioned close to the proximal margin of a hilar cholangiocarcinoma

Table 11.1. Endoluminal ultrasonography: technical data

	Frequency	
	12.5 MHz	20 MHz
Penetration depth	~2 cm	~1.5 cm
Axial resolution[a]	0.3 mm	0.1 mm
Lateral resolution[a]	0.45 mm	0.3 mm

[a]Arithmetical values.

11.3 The Normal Bile Duct

After entry into the papilla has been fluoroscopically confirmed, the ultrasound investigation should be started within the wide, fluid-filled lumen of the duodenum. The folds and various layers of the bowel wall can be precisely identified as in endoscopic ultrasonography. The papilla is a protruding structure consisting of relatively echogenic tissue (Fig. 11.2). During retraction of the probe, the fluid-filled lumen of the distal bile duct, the bile duct wall, and the adjacent part of the pancreatic head can be identified.

In individuals with normal anatomy the branching of large and small ducts is visualized, including the main right and left hepatic duct confluences and the entrance of the cystic duct into the common duct. Portal vein branches and the hepatic artery are well seen in their characteristic position adjacent to the bile duct. The periductal structures including the lymph nodes can be identified and are clearly seen within a radius of about 20–30 mm. However, the topographic relationship of the bile duct to the adjacent vessels is quite variable, and frequently the vessels can be evaluated only in the middle and proximal parts of the common bile duct, especially

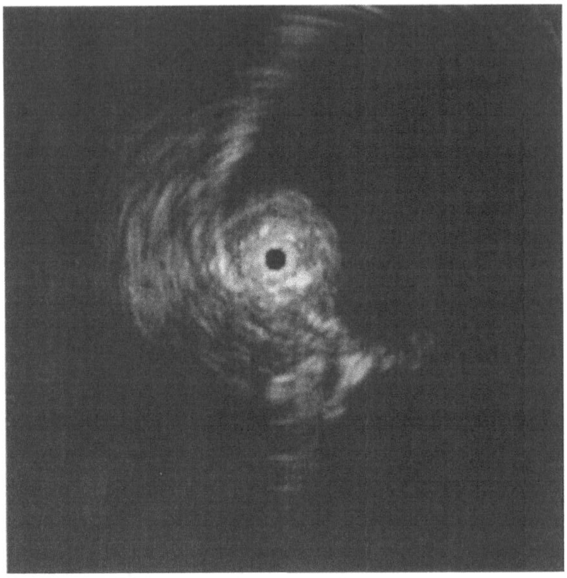

Fig. 11.2. Papilla protruding into the lumen of the fluid-filled duodenum

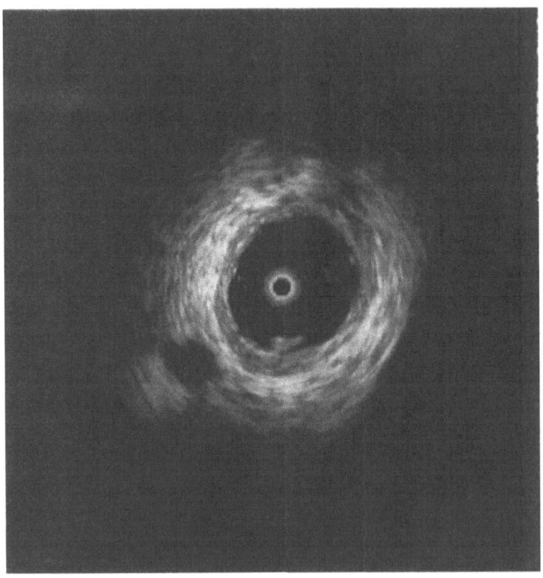

Fig. 11.3. A normal hypoechoic bile duct wall which is well defined against the surrounding tissue. The small anechoic spot adjacent to the bile duct corresponds the hepatic artery

when high-frequency transducers are used. Within the hilum the liver tissue can be identified and delineated; however, intrahepatically the margins of the biliary ducts are ill-defined.

Anatomically, the bile duct wall has a single-layered epithelium and a thick lamina propria containing some mucoid glands. The 1- to 2-mm-thick bile duct wall lacks a distinct muscle layer and muscle cells are scattered irregularly within the wall. Because the normal bile duct wall consists of this kind of fibromuscular layer, it appears sonographically fairly uniform (Fig. 11.3). Despite the histologically dense fibromuscular layer, the wall is well defined against the surrounding echogenic connective tissue, and thus assessment of tumor infiltration of the connective tissue is possible.

11.4 Benign Stricture of the Bile Duct

The bile duct wall is usually thickened in benign strictures, but the thickening is concentric, less pronounced, and the architecture less distorted than in wall thickening due to bile duct carcinoma (Fig. 11.4).

In sclerosing cholangitis the bile duct wall is thickened with onion-shaped wall layers of variable echogenicity (Fig. 11.5). Sometimes cystlike structures within the bile duct wall can be identified, corresponding with intramural diverticula (Fig. 11.6). Frequently this disease is associated with enlarged lymph nodes along the hepaticoduodenal ligament. The irregular architecture of the bile duct wall makes differentiation from a tumor growing within an area of sclerosing cholangitis extremely difficult, but endoluminal ultrasonography may at least be helpful in indicating the region where intraluminal biopsies should be performed.

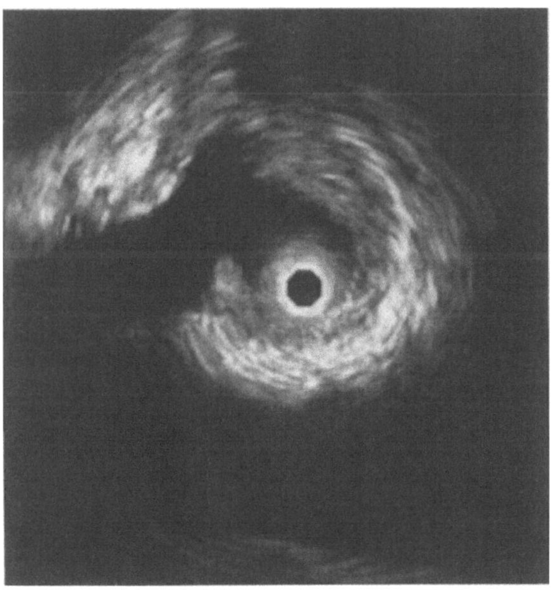

Fig. 11.4. Benign stricture of the common hepatic duct following laparoscopic cholecystectomy. The ultrasound probe is positioned at the level of the bifurcation

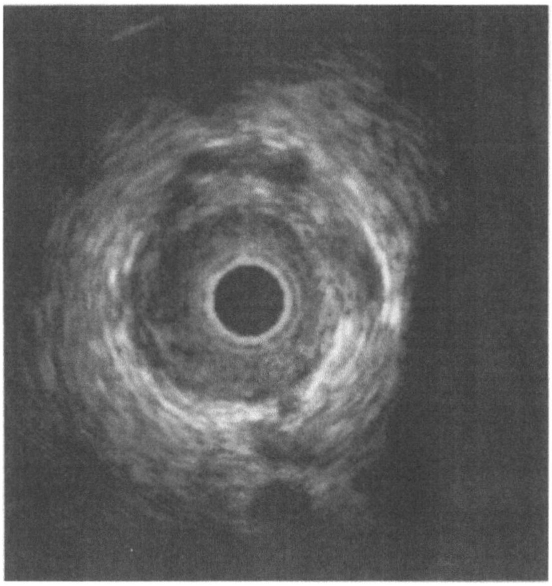

Fig. 11.5. Eccentric thickening consisting of different wall layers in primary sclerosing cholangitis

11.5 Malignant Stricture of the Bile Duct

The most striking signs of malignancy are luminal narrowing due to thickening of the bile duct wall (Fig. 11.7), infiltration of the surrounding connective tissue and adjacent vessels (Fig. 11.8), and enlarged lymph nodes (Fig. 11.9). In bile duct cancer the thickening is frequently irregular, nodular, and eccentric with distortion of the normal wall architecture.

Both transmural extent (T stage) and the longitudinal spread can be determined by endoluminal ultrasonography. The longitudinal extent is important for assessing resectability, especially in cases where the tumor is at the bifurcation. This ability to determine the tumor volume by longitudinal and axial extent can be used in the planning of intraluminal brachytherapy (Minsky et al. 1992).

Tumor staging is according to the TNM system. Differentiating between stages T1a and T1b is not

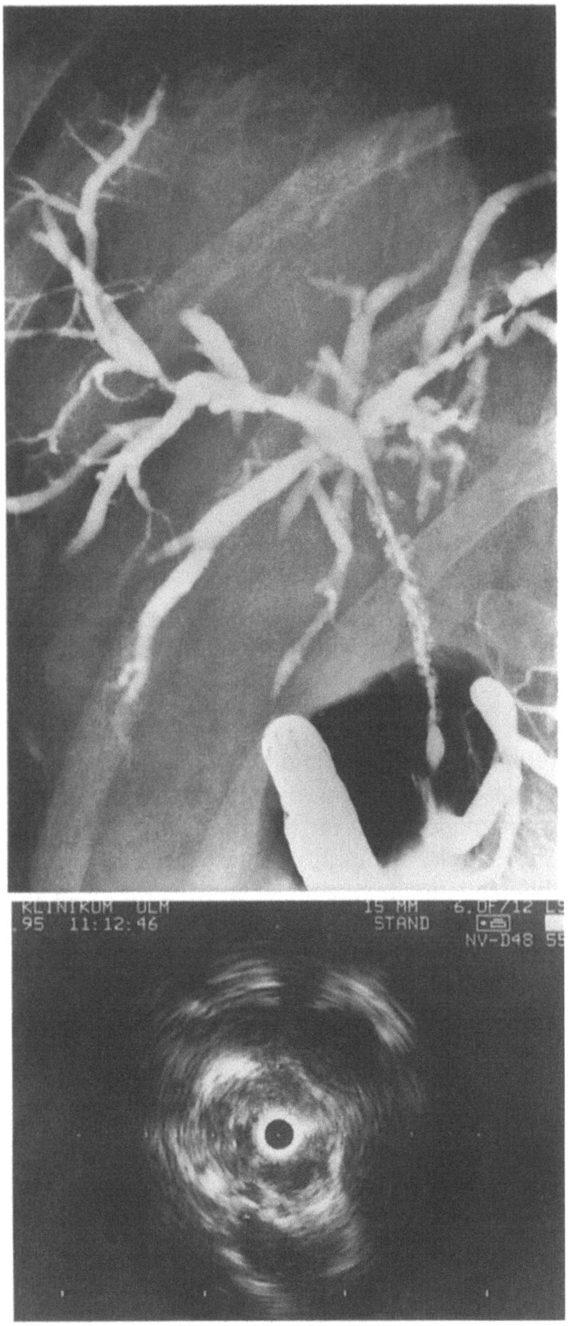

a

b

Fig. 11.6a,b. Primary sclerosing cholangitis. **a** Intramural diverticular outpouchings. **b** Cyst-like lesions within the thickened bile duct wall

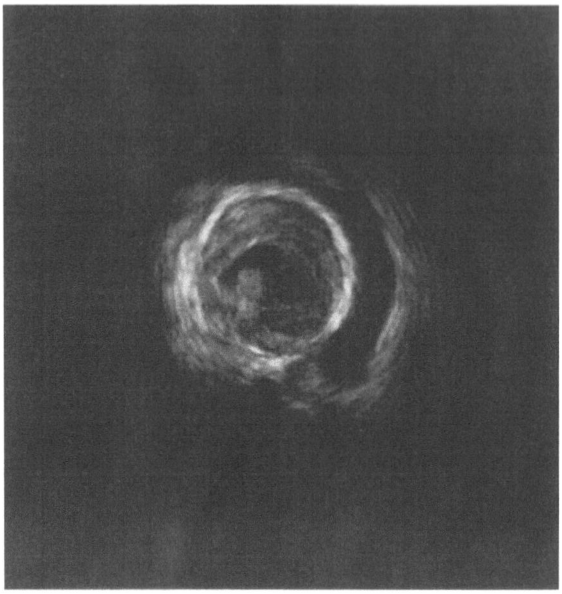

Fig. 11.7. Concentric well-defined thickening of the bile duct wall in a case of proximal bile duct carcinoma

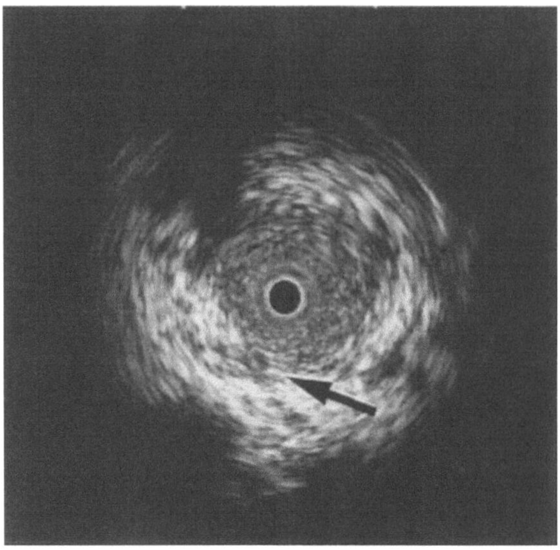

Fig. 11.8. Bile duct carcinoma: irregular thickening of the bile duct wall with signs of infiltration into the adjacent connective tissue (*arrow*)

possible because the bile duct wall has no distinct muscle layer. Thus, the stages are:

T1: Tumor limited to the bile duct wall (mucosa and fibromuscular layer)

T2: Tumor invasion into the periductal connective tissue. Tumor growth interrupts the outer margin of the bile duct

T3: Tumor infiltration into adjacent structures such as major blood vessels, liver, pancreas, gallbladder, stomach, duodenum, or colon

N0/N1: Absence/presence of lymph node metastases

In a prospective study of a small group of patients with primary bile duct carcinoma, endoluminal ultrasonography allowed correct T staging in all

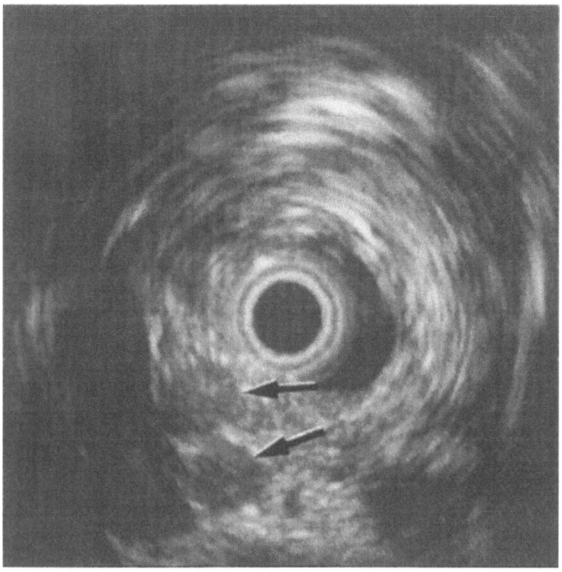

Fig. 11.9. Bile duct carcinoma: enlarged lymph nodes within the hepatoduodenal ligament (*arrows*)

cases (WAGNER et al. 1995). Due to its limited penetration, the authors noted that it needed to be combined with other, conventional imaging methods.

The recognition of malignant involvement of lymph nodes by conventional ultrasonography and computed tomography (CT) relies on size alone (DORFMAN et al. 1991). Therefore, metastases in normal-sized lymph nodes frequently escape detection during these diagnostic imaging procedures (RÖSCH et al. 1992). For lymph node evaluation using endoluminal ultrasound transducers, a frequency of 12.5 MHz is required, because 20-MHz transducers have not yielded satisfactory results in depicting the anatomy of the hepatoduodenal ligament, owing to their inadequate depth of penetration (BRAMBS 1993).

The use of sonographic criteria to improve the accuracy of endoluminal ultrasonography in detecting lymph node metastases was proposed by TIO et al. (1990), and our findings using catheter-based endoluminal ultrasonography lend support to the theory that a detailed analysis of lymph node sonomorphology may improve results. Endoluminal ultrasonography may prove helpful in identifying normal-sized lymph nodes involved in cancer by allowing analysis of the echogenicity, contour, and distinctness of boundaries (DUDA et al., to be published). A lymph node can be classified as malignant when two of the following three sonographic criteria are met: hypoechogenicity, a round configuration, and conspicuous margins.

Endoluminal ultrasonography is a sensitive diagnostic technique. The accuracies of endoluminal ultrasonography and of CT for diagnosing metastatic involvement of lymph nodes in the hilum of the liver, the hepaticoduodenal ligament, and the region around the pancreatic head are 83% and 28%, respectively (DUDA et al., to be published). Sensitivity rates for the assessment of lymph node metastases in pancreatic cancer have been reported to range from 72% to 74% for endoscopic ultrasonography (TIO et al. 1990; RÖSCH et al. 1992). Specificity, however, is low because of the difficulty in distinguishing between metastatic and nonmetastatic lymph node abnormalities.

11.6 Bile Duct Stones

Bile duct stones appear as highly echogenic foci with acoustic shadowing (Fig. 11.10), whereas debris within the bile ducts or the gallbladder is noted as moderately echogenic material without posterior shadowing (VAN SONNENBERG et al. 1992; THOMSON et al. 1993). Compared with fluoroscopy after injection of contrast material, endoluminal ultrasonography offers the possibility to distinguish stones from debris, clot, or air bubbles within the bile ducts and the gallbladder. Endoluminal ultrasonography could be helpful in various interventional procedures to ascertain response to stone disintegration by laser (Fig. 11.10), electrohydraulic shock waves,

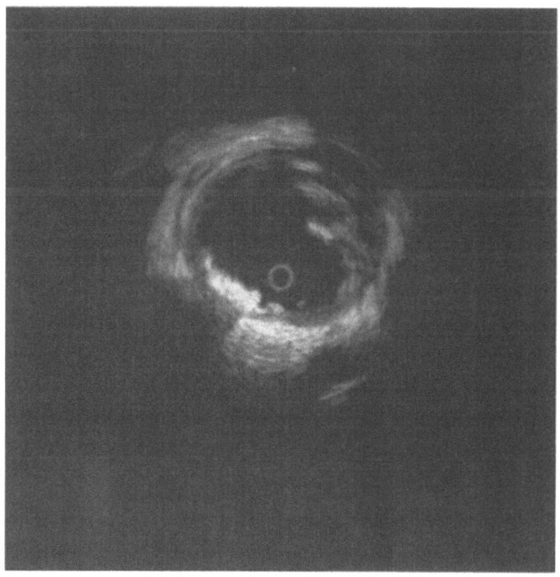

Fig. 11.10. Stone fragments shown within the common bile duct following laser lithotripsy

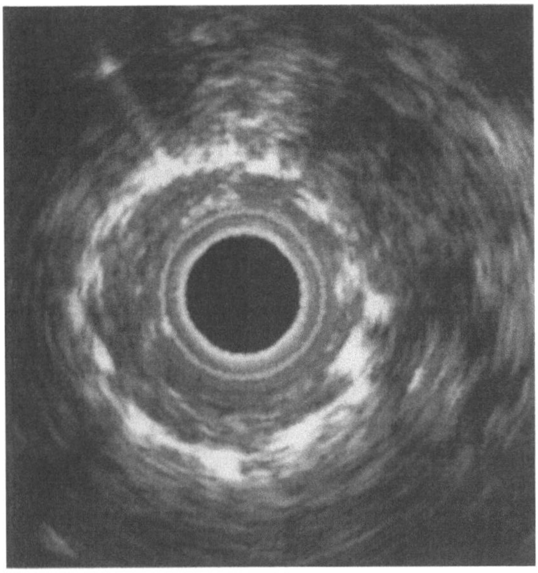

Fig. 11.11. Stent occlusion due to tumor ingrowth in a case of cholangiocarcinoma

or solvents such as MTBE (methyl-tert-butyl-ether). It can also be used to evaluate the common duct pathology during laparoscopic cholecystectomy (Ascher et al. 1992; Thomson et al. 1993).

11.7 Bile Duct Stents

Metallic stents are quite echogenic with small posterior shadows. In cases of stent occlusion the cause of obstruction such as debris, stones, or tumor ingrowth (Fig. 11.11) and overgrowth can be evaluated. The effect of tumor ablation using laser or thermal probes can be monitored by endoluminal ultrasonography.

11.8 Conclusions

Endoluminal ultrasonography is an attractive method for detailed assessment of bile duct disease. In most cases, benign strictures can be distinguished from stenoses caused by tumors. However, such a diagnosis should be confirmed by intraductal biopsy, and endoluminal ultrasonography could help to determine the exact place where the specimen should be obtained.

In malignant disease, endoluminal ultrasonography facilitates the difficult staging of bile duct tumors and provides information on longitudinal and axial tumor extent, infiltration of adjacent connective tissue and vessels, and lymph node enlargement. Preliminary studies have shown that endoluminal ultrasonography is an accurate method for T and N staging of bile duct tumors. However, further studies need to be performed to assess the impact of this method, especially in the difficult cases of tumors at the hilum, when precise diagnosis of liver infiltration is required.

This technique might be useful in combination with stone fragmentation and removal procedures, and can be used in association with laparoscopic cholecystectomy.

Disadvantages of endoluminal ultrasonography include its invasive character and the difficulties of anatomic orientation owing to the limited field of view. For this reason, it needs to be used in combination with conventional diagnostic procedures. The procedure is simple and devoid of risk, but requires some further refinement in order to attain more widespread use for different clinical problems.

References

Ascher SM, Evans SRT, Goldberg JA, Garra BS, Benjamin SB, Davros WJ, Zeman RK (1992) Intraoperative bile duct sonography during laparoscopic cholecystectomy: experience with a 12.5-MHz catheter-based US probe. Radiology 185:493–496

Brambs HJ (1993) Intraduktaler Ultraschall der Gallenwege. Radiologe 33:385–390

Brambs HJ, Rieber A (1993) Perkutaner intraduktaler Ultraschall der Gallenwege. Bildgebung 60:83–87

Crowley RJ, von Behren PL, Couvillon LA, et al (1989) Optimized ultrasound imaging catheters for use in the vascular system. Int J Card Imaging 4:145–151

Dorfman RE, Alpern MB, Gross BH, Sandler MA (1991) Upper abdominal lymph nodes: criteria for normal size determined with CT. Radiology 180:319–322

Duda SH, Huppert PE, Schott U, Brambs HJ, Claussen CD (to be published) Percutaneous transhepatic intraductal biliary sonography for lymph node staging at 12.5 MHz in malignant bile duct obstruction. J Vasc Interv Radiol

Engström CF, Wiechel KL (1990) Endoluminal ultrasound of the bile ducts. Surg Endosc 4:187–190

Goldberg BB, Liu JB, Merton DA, Kurtz AB (1990) Endoluminal US: experiments with nonvascular uses in animals. Radiology 175:39–43

Goldberg BB, Bagley D, Liu JB, Merton DA, Alexander A, Kurtz AB (1991) Endoluminal sonography of the urinary tract: preliminary observations. Am J Roentgenol 156:99–103

Hürter T, Hanrath P (1990) Endobronchiale Sonographie zur Diagnostik pulmonaler und mediastinaler Tumoren. Dtsch Med Wochenschr 115:1899

Isner JM, Rosenfield K, Losordo DW, et al (1990) Percutaneous intravascular US as adjunct to catheter-based interventions: preliminary experience in patients with peripheral vascular disease. Radiology 175:61-70

Liu JB, Miller LS, Feld RI, Barbarevech CA, Needleman L, Goldberg BB (1993) Gastric and esophageal varices: 20 MHz transnasal endoluminal US. Radiology 187:363-366

Minsky B, Botet J, Gerdes H, Lightdale C (1992) Ultrasound directed extrahepatic bile duct intraluminal brachytherapy. Int J Radiat Oncol Biol Phys 23:165-167

Rösch T, Braig C, Gain T, et al (1992) Staging of pancreatic and ampullary carcinoma by endoscopic ultrasonography: comparison with conventional sonography, computed tomography, and angiography. Gastroenterology 102:188-199

Thomson H, Kisslo K, Farouk M, Chung K, Saperstein LA, Meyers C (1993) Technique of intraluminal ultrasonography during laparoscopic cholecystectomy. Am J Surg 165:265-269

Tio TL, Tytgat GN, Cikot RJ, Houthoff HJ, Sars PR (1990) Ampullopancreatic carcinoma: preoperative TNM classification with endoscopy. Radiology 175:455-461

Tobis JM, Mallery JA, Gessert J, et al (1988) Intravascular ultrasound visualization before and after balloon angioplasty. Circulation 78:84 (abstract)

Van Sonnenberg E, D'Agostino HB, Sanchez RL, Goodacre BB, Esch OG, Easter DE, Gosink BB (1992) Percutaneous intraluminal US in the gallbladder and the bile ducts. Radiology 182:693-696

Wagner HJ, Hoppe M, Klose KJ (1995) Perkutane intraluminale Ultraschalluntersuchung der Gallenwege in der Diagnostik maligner biliärer Obstruktionen. Dtsch Med Wochenschr 120:472-477

Yasuda K, Mukai H, Nakajima M, Kawai K (1992) Clinical application of ultrasonic probes in the biliary and pancreatic duct. Endoscopy 24[Suppl 1]:370-375

12 Laparoscopic Ultrasonography in Benign and Malignant Biliary Diseases

M. Bezzi, O.J.M. van Delden, R. Merlino, and J.W.A.J. Reeders

CONTENTS

12.1 Introduction

In the mid 1980s, intraoperative ultrasonography (US) during open surgery was reported to be an effective procedure for the detection and the staging of benign and malignant disease of the gallbladder and bile ducts. However, only a few surgical groups proposed its routine use during biliary tract surgery (Jakimovicz et al. 1987).

M. Bezzi, MD, Assistant Professor, Department of Radiology, University of Rome "La Sapienza", Policlinico Umberto I, 00161 Rome, Italy

O.J.M. van Delden, MD, Department of Gastrointestinal Radiology and Hepato-pancreato-biliary Imaging, Academic Medical Center, Meibergdreef 9, 1105 AZ Amsterdam, The Netherlands

R. Merlino, MD, Department of Radiology, University of Rome "La Sapienza", Policlinico Umberto I, 00161 Rome, Italy

J.W.A.J. Reeders, MD PhD, Head, Department of Gastrointestinal Radiology and Hepato-pancreato-biliary Imaging, Academic Medical Center, Meibergdreef 9, 1105 AZ Amsterdam, The Netherlands

The recent introduction and rapid development of laparoscopic surgical techniques has prompted a new interest in intraoperative US, and high-frequency ultrasonographic probes have been designed to be passed through laparoscopic ports to assist the surgeon in evaluating the biliary tract during laparoscopic surgery or diagnostic laparoscopy. With laparoscopic US, high-resolution images can be obtained of the biliary ducts, the gallbladder, the liver parenchyma, and the pancreas without the image degradation caused by overlying bowel gas or a thick abdominal wall. Several authors are now reporting their preliminary results on the use of laparoscopic US in the detection of biliary stones, and the consensus is that the technique is at least as accurate as laparoscopic cholangiography.

In addition, laparoscopy has been long known as a method of staging intra-abdominal malignancy. The technique has a high accuracy in detecting peritoneal deposits and small superficial liver metastases (Cuschieri et al. 1978; Warshaw et al. 1990), but its efficacy is limited by the inability to reliably assess lesions deeply located in the liver parenchyma or in the retroperitoneum. These limitations can be overcome by the use of laparoscopic US, and this imaging technique is increasingly being used in staging malignancies involving the biliary tree and other abdominal organs.

12.2 Equipment, Examination Technique, and Anatomy

12.2.1 Laparoscopy

Laparoscopy and laparoscopic surgery are performed under general anesthesia. According to the laparoscopic or surgical technique, two to four trocars can be used. Access for the US probe to the peritoneal cavity is through 10- or 11-mm ports which are usually placed at the umbilical and subcostal sites; these are the same ports as give access to the video camera.

The peritoneal surface and the liver can be inspected visually with the exception of the most cranial part of the hepatic dome. The gallbladder and the hepatoduodenal ligament are visualized by raising the edge of the right hepatic lobe. In the case of a tumor in the pancreatic head area, special attention should be paid to the Treitz ligament and the transverse mesocolon for signs of tumor ingrowth (Bemelman et al. 1995). Suspected metastatic liver lesions, peritoneal deposits, and lymph nodes can if necessary undergo needle or excisional biopsy under direct laparoscopic sight and ultrasonographic guidance. Ascites can be aspirated for cytological examination and isotonic saline solution can be instilled into the peritoneal cavity for cytological sampling.

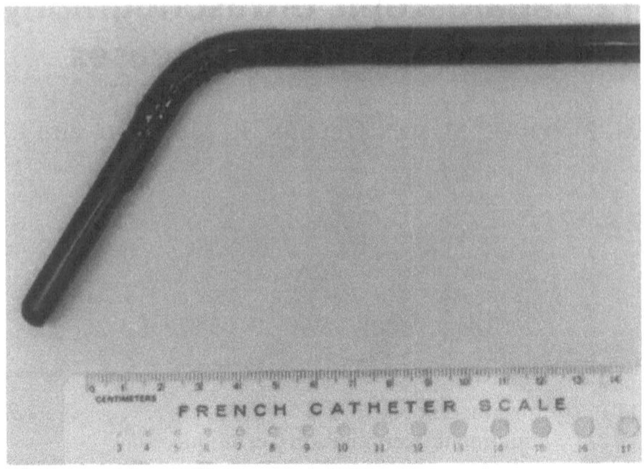

Fig. 12.1. Probe for laparoscopic ultrasonography. The probe shown is a model with flexible tip (±90° in-plane flexibility), equipped with a side-view 7.5-MHz linear array transducer (Esaote Biomedica, Genua, Italy). (From Bezzi et al. 1995, with permission)

12.2.2 Laparoscopic Probes and Ultrasound Scanners

The probes used in laparoscopic US essentially consist of a transducer on the tip of a long shaft. The shaft is usually 10 mm or less in diameter, to be introduced through the surgical accesses, and 35 to 45 cm in length, to permit a wide examination field from any single port. Transducers are available in the 5- to 10-MHz frequency range, with either a linear or a curved array. They are usually mounted in side-view fashion on a rigid or deflectable tip (Fig. 12.1). Probes with a flexible tip give more versatility in scanning less accessible regions, such as the dome of the liver and the hepatoduodenal ligament. 7.5-MHz transducers provide excellent image resolution, particularly when contact scanning is used on the hepatoduodenal ligament; the penetration depth, however, is somewhat limited and this can be a problem when examining a fatty infiltrated liver, the omentum, or the retroperitoneal structures. An ideal solution would be a transducer whose frequency can be switched from 5 to 7.5 MHz according to the area of interest. Some manufacturers also provide sector scanners with end or side viewing; these have the advantage of a small contact surface which requires a limited acoustic window, but are characterized by a small near field and reduced near-field resolution.

The probes are connected to a US apparatus that is generally a compact, mobile, real-time B-mode system providing high-quality imaging. Doppler options should be always included for intraoperative examinations. Color Doppler imaging is presently provided only in more expensive equipment seldom used in the operating room.

The choice of procedure for probe sterilization depends on how often the probe is used and also on the hospital sterilization policy. Sterilization may be either by gas sterilization using ethylene oxide or by cold sterilization using several disinfectant glutaraldehyde solutions. Gas sterilization is preferable to achieve high levels of virucidal and bactericidal effectivity. A cycle with ethylene oxide, however, takes 12–18 h, and this time constraint prevents constant availability of the instrument. Cold sterilization with glutaraldehyde is therefore sometimes preferred since it is more practical; the procedure is both bactericidal and virucidal, and sterility is usually achieved after at least 20 min immersion, depending on the solution used. Most probe cables are waterproof and can be immersed in the bactericidal solution; reference should be always made, however, to the manufacturer's instructions. The cable and the connector to the US machine, if not sterile, can be covered with a sterile plastic cover.

12.2.3 Examination Technique and Normal Anatomy

The examination is usually performed by contact scanning using the liquid film of the peritoneum as a coupling agent. The stomach is emptied to avoid air-related artefacts. Surgical dissection, if necessary, is kept to a minimum, as the presence of gas in the tissues caused by electrocoagulation may interfere with subsequent imaging.

The examination technique is somewhat different for linear array and sector probes. Here we mainly describe the technique used with a linear array transducer (Bezzi et al. 1995).

Transverse and oblique scans of the *gallbladder* and *liver hilum* are obtained by introducing the probe through the 10-mm subxiphoid cannula. The hepatic parenchyma of segment IV is used as an acoustic window. This allows good visualization of

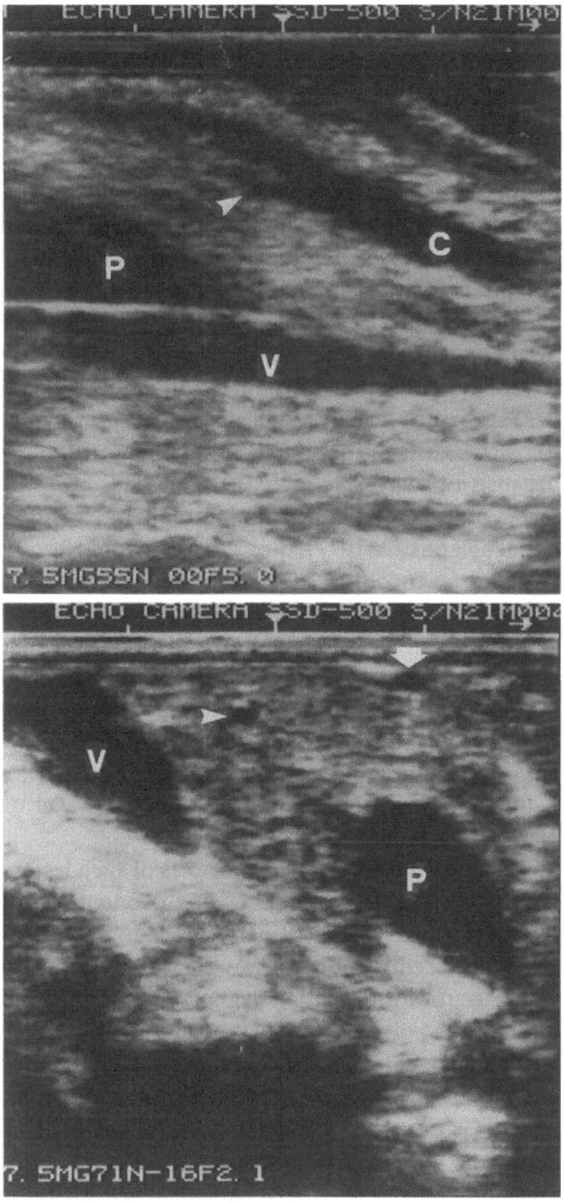

a

b

Fig. 12.2a,b. Normal anatomy of the common bile duct. **a** In a longitudinal view at the level of the hepatoduodenal ligament, the confluence of the cystic duct is demonstrated (*arrowhead*). **b** In a transverse scan at the level of the pancreatic head, the common bile duct (*arrowhead*) and the gastroduodenal artery (*arrow*) are seen in their short axis. *V*, Inferior vena cava; *P*, portal vein; *C*, common bile duct. (From Bezzi et al. 1995, with permission)

the intrahepatic bile duct, ductal confluence, and proximal *common bile duct* (CBD) (Fig. 12.2). Study of the gallbladder is not possible when it is emptied accidentally during surgical manipulation or when the bile is aspirated by the surgeon (e.g., in case of hydrops). The surgeon then raises the gallbladder anteriorly; the probe is placed directly onto the hepatoduodenal ligament and gently slid toward the duodenum and the pancreatic head. These steps allows investigation of the CBD in its entire length, down to the region of the papilla (Fig. 12.2). The most difficult part of the the examination is the identification of the papilla, since images are disturbed by the duodenal contents.

The approach through the umbilical port is used to examine the gallbladder and the bile ducts in sagittal planes, which also permit identification of the cystic duct confluence (Fig. 12.2). Alternate use of the umbilical and subxiphoid approaches will allow imaging of organs in different scanning planes, in the same way as in conventional US. This is especially important when determining the relationships between a tumor and adjacent vascular structures.

The various structures are recognized by their appearance and by their anatomical location. A bile duct endoprosthesis, if present, is recognized as a double hyperechoic line, and this can be used to rapidly locate the CBD. Air in the CBD from a previous sphincterotomy may prevent adequate evaluation of ductal lumen. Apart from their anatomic location, arteries can be recognized by their pulsation, and large veins by internal echoes due to slowly flowing blood. Color Doppler and pulsed Doppler make the identification obvious. During the examination, one realizes that the pressure exerted to achieve probe contact may compress and flatten the veins and the bile ducts, but not the arteries. This flattening should be taken into account, particularly when measuring the CBD.

The *liver* should be scanned from the superior and from the posterior surface. The dome can be imaged easily with a flexible probe. When using a rigid probe, the dome is more difficult to study because of lack of adequate contact between the straight probe and the curving liver surface. This can be easily overcome by instilling 1-3 l isotonic saline into the right subphrenic region and scanning the highest part of the organ through the fluid.

The laparoscopic US examination is safe and is not difficult to perform. Interpretation of the images may take some time and there is a distinct learning curve. The examination and image interpretation is done by surgeons in some centers; in both our institutions it is performed by a team consisting of a laparoscopic surgeon and a radiologist trained in intraoperative US. The study is usually recorded on videotape, and hard copies are also made of the most significant images.

12.3 Laparoscopic US in Laparoscopic Cholecystectomy and Lithiasis

M. Bezzi and R. Merlino

12.3.1 Gallbladder

At laparoscopic US the appearance of gallbladder abnormalities is similar to the usual findings on transabdominal US, although with a clearer definition of details.

In patients undergoing laparoscopic cholecystectomy for gallstones, laparoscopic US may be used to confirm the diagnosis, if preoperatively uncertain, and to document the size of the stone(s) and the presence of associated microcalculi and sludge. The thickness of the gallbladder wall may be evaluated and wall inflammation and edema characteristically recognized by the presence of alternating hyperechoic and hypoechoic layers (Fig. 12.3).

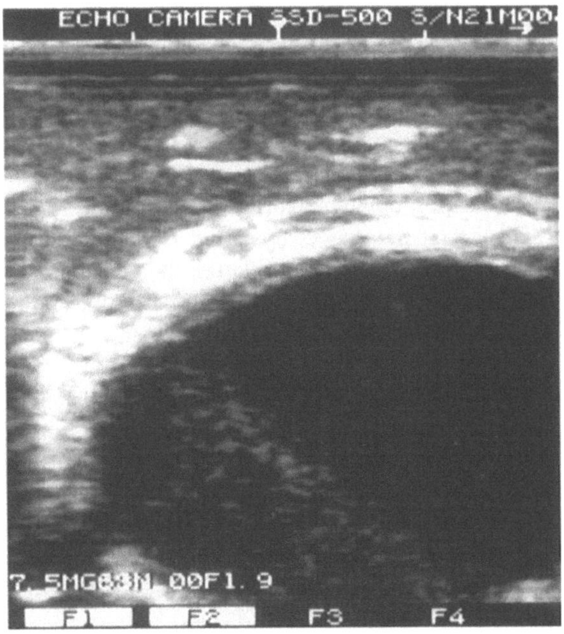

Fig. 12.3. Acute cholecystitis. The image shows a transverse scan of the gallbladder obtained through the "window" of the liver parenchyma. The typical edema of the wall is shown by the characteristic appearance of alternating hypoechoic and hyperehcoic layers. Biliary sludge is also seen in the dependent portion of the gallbladder. (From Bezzi et al. 1995, with permission)

Laparoscopic US is also valuable in the assessment of the gallbladder wall in patients who are undergoing laparoscopic surgery for gallbladder polyps or in whom a pathological condition of the gallbladder wall is suspected or discovered incidentally.

Polypoid lesions of the gallbladder are commonly classified into tumors and pseudotumors (Christensen and Kamal 1970). Tumors include malignant lesions such as adenocarcinoma and metastases, and benign lesions those such as adenoma, leiomyoma, granular cell tumor, and lipoma. Pseudotumors include cholesterol polyps, inflammatory polyps, adenomatous hyperplasia, and heterotopia. It is generally accepted that cholesterol polyps never degenerate, whereas adenoma and adenocarcinoma are considered respectively to be premalignant and malignant lesions.

In a patient with a polypoid lesion of the gallbladder, surgery is usually indicated because of symptoms, because the lesion is larger than 10 mm, or because a progressive increase in size has been detected on US. Preoperative and intraoperative differential diagnosis between the various polypoid lesions that may be encountered is useful since it may affect surgical strategy. Cholesterol polyps, by far the most common, can be safely removed by laparoscopic cholecystectomy, while premalignant and malignant lesions are best treated by open surgery, since unsuspected gallbladder cancer removed by laparoscopic cholecystectomy may result in spread to the peritoneum or within the abdominal wall (Wade et al. 1994).

Preoperative differential diagnosis of such lesions is difficult. Transabdominal US is the only imaging test performed before surgery but it is not highly reliable. Recently, Sugiyama et al. (1995) reported more accurate differential diagnosis based on evaluation of the polyp echotexture by endoscopic US.

At the Department of Radiology, "La Sapienza" University in Rome, we perfomed preoperative US and laparoscopic US in patients with polypoid lesions who underwent laparoscopic cholecystectomy, and studied 105 polyps in 84 patients. For each lesion we recorded the size, morphology, echotexture, and the presence or absence of a pedicle. In addition, for each polyp we calculated the height, the width, and the ratio of height to width, thus performing a geometric quantitative analysis (Bezzi et al. 1996). Laparoscopic US was in all cases more effective than preoperative US in visualizing the lesion and its pedicle, and in assessing the real diameters and orientation of the lesion in respect to the gallbladder wall. In our series all the cholesterol polyps had a height greater than the width, with a height/width ratio >1.15, whereas the malignant and premalignant lesions showed a tendency to "flat" growth, parallel to the wall, with a height/width ratio <0.70 (Figs. 12.4, 12.5). In addition, cholesterol polyps were often attached to the underlying mucosa by a fragile pedicle (86% of cases), which was best visualized during laparoscopic US (Fig. 12.4). Our analysis also demonstrated that multiple polyps were more frequently benign ($p < 0.05$), and that age over 55 years and the presence of gallstones were more often associated with malignant lesions ($p < 0.001$ and $p < 0.005$ respectively). Differences in echotexture were not relevant to the differential diagnosis.

If these observations are confirmed by other studies, it will be possible safely to use laparoscopic US to assess the nature of a suspected gallbladder wall lesion during laparoscopic cholecystectomy.

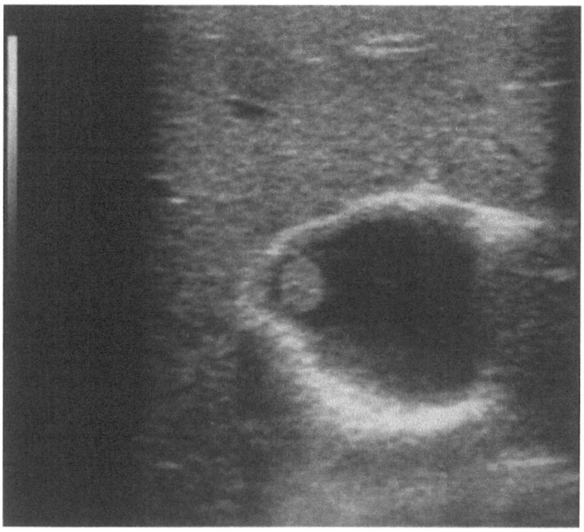

Fig. 12.4. Cholesterol polyp of the gallbladder. The lesion is attached to the gallbladder wall by a thin pedicle; the wall is thickened due to cholesterolosis. The height/width ratio of the lesion (see text) in this case was 1.43. Histologic study after laparoscopic cholecystectomy confirmed the nature of the polyp

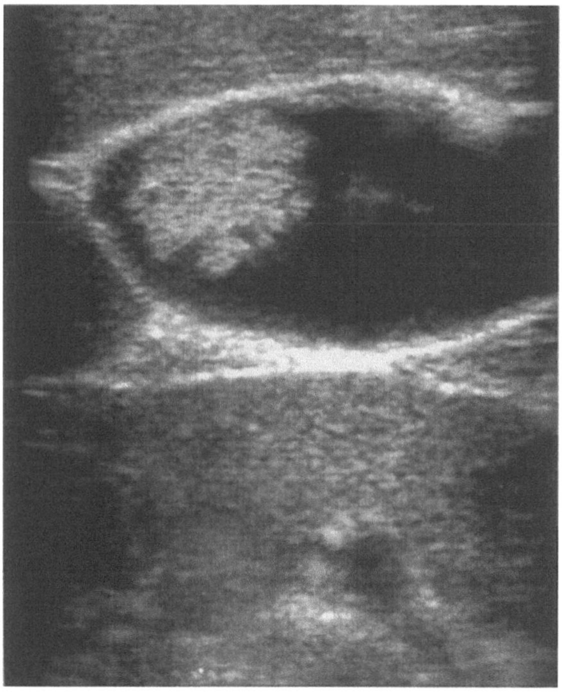

Fig. 12.5. Adenoma of the gallbladder. The polypoid lesion is sessile, with no appreciable pedicle. The height/width ratio was 0.68. Histology confirmed the adenomatous nature of the lesion

12.3.2 Bile Ducts

The role of operative cholangiography (OCG) during laparoscopic cholecystectomy, and whether it should be performed selectively or routinely, is a highly debated issue. Its routine use is advocated by some authors in order to rule out iatrogenic lesions of the bile duct, to evaluate possible anatomical variations of the biliary tree, and to verify the presence of CBD stones. Others support a more selective use of OCG based on preoperative clinical, radiological, and biochemical considerations.

Laparoscopic US has been proposed as an alternative to OCG in the study of the biliary tree, particularly in regard to the issue of residual CBD stones. The appearance of CBD stones is identical to that of gallbladder calculi (Fig. 12.6). Large stones are more obvious, while the combined use of longitudinal and transverse views will allow identification of stones no larger than a few millimeters (Fig. 12.6b). Thickening of the CBD, due to inflammation, can be seen on US images in cases of longstanding lithiasis (Fig. 12.6a).

Preliminary experiences in the use of laparoscopic US to detect CBD stones have been published by several authors. Greig and coworkers (Greig et al. 1994), using a 7.5-MHz linear array probe, examined the CBD of 54 patients undergoing laparoscopic cholecystectomy. They found that laparoscopic US had a sensitivity of 71% and a specificity of 96% for CBD stones. In the same study OCG had a sensitivity of 83% and a specificity of 95%. Röthlin et al. performed laparoscopic US and OCG (using conventional static film) on 100 consecutive patients during laparoscopic cholecystectomy. They reported a sen-

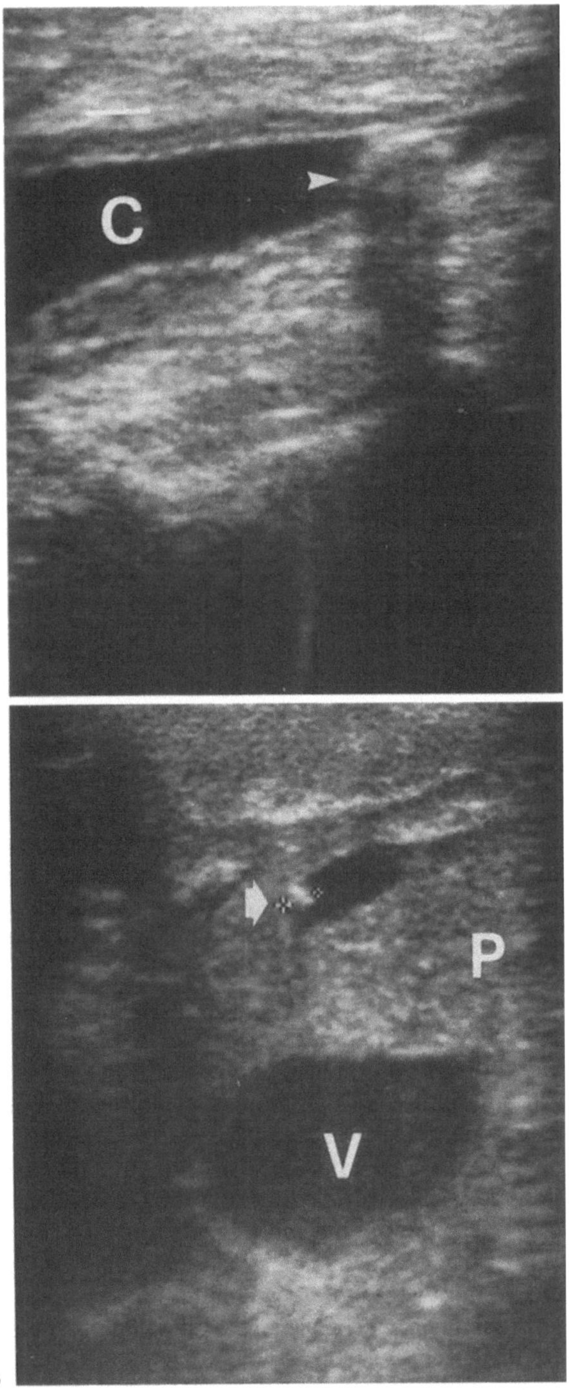

a
b

Fig. 12.6a,b. Common bile duct stones. a Longitudinal scan. A 7-mm stone is seen in the distal common bile duct (*c*), with thickened wall due to inflammation. The typical acoustic shadowing can be well appreciated. b Transverse scan in another patient. A 3-mm stone is demonstrated along the medial wall of the common bile duct. *P*, Pancreatic head; *V*, inferior vena cava

sitivity of 100% and a specificity of 99% for CBD stones using laparoscopic US, both superior to OCG (Röthlin et al. 1994). In another study conducted by Barteau and coworkers on 125 patients, OCG (using digital videofluoroscopy) was slightly superior to laparoscopic US; it was more sensitive (92.8% vs 71.4%) but less specific (76.2% vs 100%) for choledocholithiasis than was US (Barteau et al. 1995). Similar results have been reported by other authors (John et al. 1994a; Teefey et al. 1995; Yamashita et al. 1993).

The general consensus among the surgeons who use both techniques, is that laparoscopic US is at least as accurate as OCG in diagnosing choledocholithiasis during laparoscopic cholecystectomy. Both techniques have a definite learning curve, but once skills have been developed, laparoscopic US requires less time than OCG. On the other hand, OCG provides a precise map of the entire anatomy of the biliary tree, unattainable by laparoscopic US. This advantage of OCG is frequently cited by some authors, who maintain that the primary objective of biliary imaging during laparoscopic cholecystectomy is anatomic definition, so that the incidence of iatrogenic injuries to the bile ducts can be reduced. The issue is far from being resolved. Whether laparoscopic US is superior to laparoscopic OCG or whether the two techniques are complementary will be probably answered by more experience in larger groups of patients.

12.4 Laparoscopic US in Staging of Bile Duct Malignancies

O.J.M. van Delden and J.W.A.J. Reeders

Laparoscopic US can be used, together with laparoscopy, in staging malignancies involving the biliary tree. A thorough study in a patient examined for such condition should include the liver, the bile duct, the pancreas, the regional lymph nodes, and the vascular structures.

With laparoscopic US, a liver lesion as small as few millimeters can be visualized. Liver metastases of tumors involving the bile ducts are usually hypoechoic or isoechoic compared to normal liver parenchyma, but can occasionally be hyperechoic (Fig. 12.7). Cysts can easily be distinguished from solid lesions. Small hemangiomas are usually homogeneously hyperechoic and can in many cases be differentiated from metastases by compressing them with the US probe. The fact that hemangiomas are

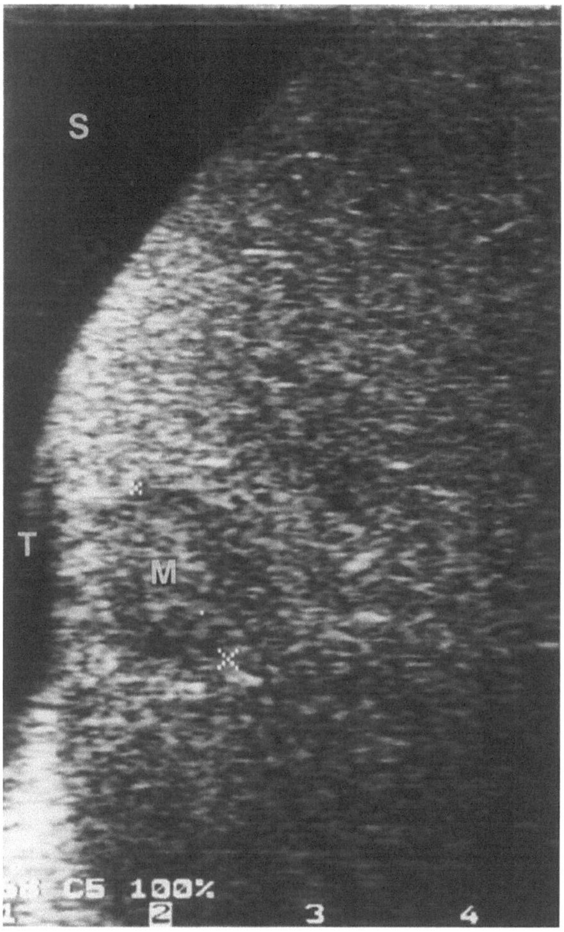

Fig. 12.7. Liver metastasis (*M*), almost isoechoic to the hepatic parenchyma, located cranially in the right liver lobe. Good visualization after instillation of isotonic saline (*S*) into the right subphrenic space. *T*, Ligamentum triangulare

compressible is known from intraoperative US and can be used in laparoscopic US as well (PAUL et al. 1994). Suspected intrahepatic lesions can be biopsied percutaneously under combined laparoscopic and laparoscopic US guidance.

The intrahepatic bile ducts, the hilar confluence, and the proximal CBD can best be imaged through the liver, placing the probe on segmente IV. The middle and distal CBD are visualized with the probe directly placed on the hepatoduodenal ligament and pancreatic head. Duct dilation, localized bile duct tumors or diffuse wall thickening, concrements, and sludge can be well documented.

The pancreatic head can be found by following the CBD distally. Pancreatic duct dilatation can be easily recognized.

Distal cholangiocarcinomas, pancreatic adenocarcinomas, and carcinomas of the papilla of Vater are almost always hypoechoic compared to normal pancreatic parenchyma (Figs. 12.8, 12.9). Since inflamed pancreatic tissue is also hypoechoic compared to normal pancreatic parenchyma, it can in our experience be difficult to differentiate focal obstructive pancreatitis from a malignant neoplasm.

Lymph nodes in the hepatoduodenal ligament and at the celiac axis can be well imaged. US features of benign hyperplastic lymph nodes are a longitudinal shape and the presence of a hyperechoic center, representing hilar fat. US features of lymph nodes with metastatic disease are a more rounded shape, a hypoechoic appearance, and loss of the hyperechoic hilum. Since there is considerable overlap between these findings, no certain differentation is possible on US (HULSMANS et al. 1992). In most patients with bile duct obstruction, enlarged lymph nodes are encountered in the hepatoduodenal ligament, in many cases with a benign - inflammatory - appearance. Enlarged lymph nodes on the celiac axis should be biopsied, since metastatic nodes in that site are con-

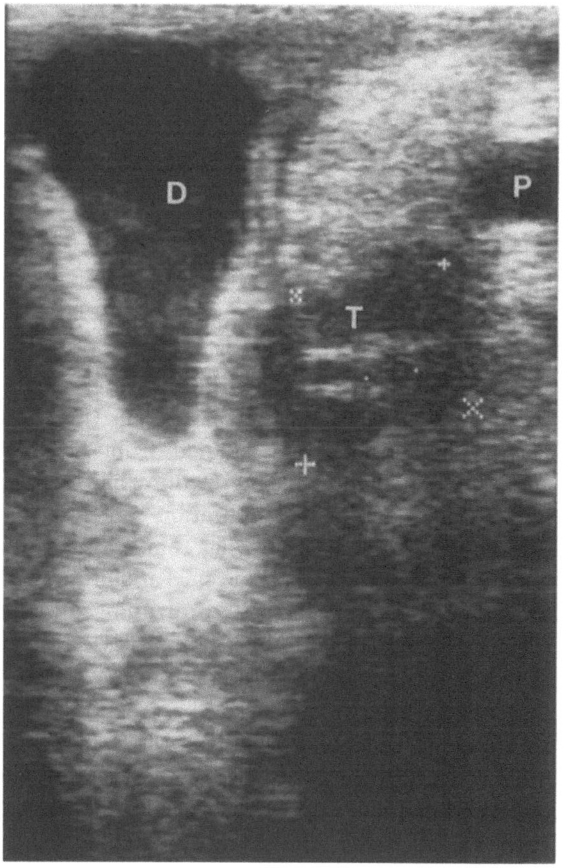

Fig. 12.8. Laparoscopic sonogram of a small distal cholangiocarcinoma (*T*) in the pancreatic head. The double hyperechoic line inside the tumor represents the bile duct endoprosthesis. There is no tumor infiltration of the portal vein (*P*). *D*, Fluid-filled duodenum

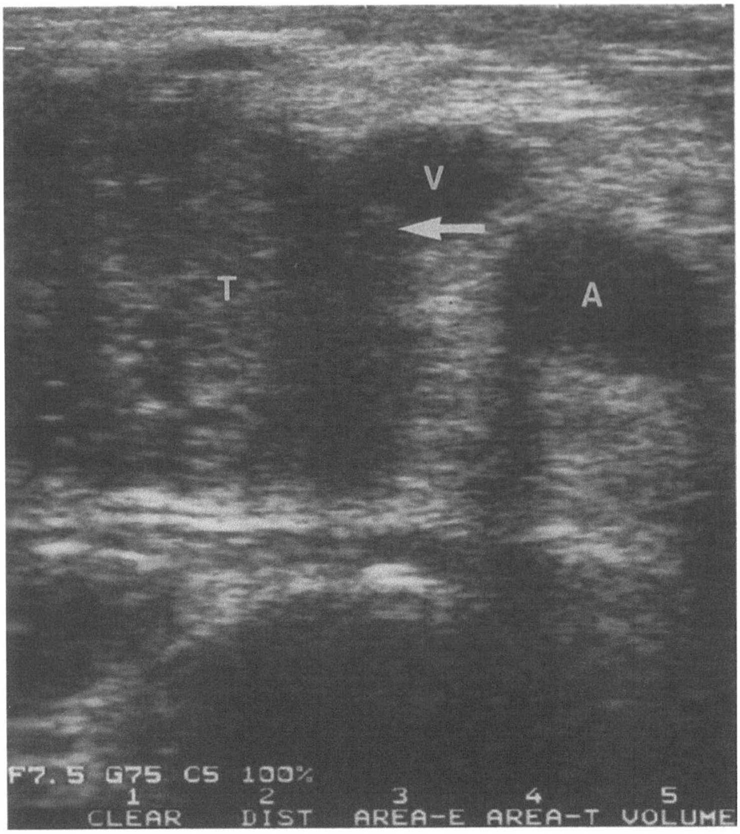

Fig. 12.9. Pancreatic head carcinoma (*T*) infiltrating the superior mesenteric vein (*V*). The tumor protrudes into the vascular lumen (*arrow*). The superior mesenteric artery (*A*) is free of tumor

sidered a contraindication for curative tumor resection in most centers.

The splenic vein, the superior mesenteric vein, the portal vein with its intrahepatic branches, the celiac axis, and the hepatic and superior mesenteric arteries can be well followed with laparoscopic US. These vessels should be closely studied from different angles to evaluate possible ingrowth by bile duct tumors. Signs of tumor infiltration are vascular thrombosis, luminal narrowing, and loss of the hyperechoic interface between vessel and tumor with or without the tumor actually protruding into the vascular lumen (Fig. 12.9). Collateral vessels or an aberrant hepatic artery arising from the superior mesenteric artery can also be identified by laparoscopic US. Color Doppler probes may help quickly to differentiate between vessels and bile ducts and may show up vascular occlusion.

12.4.1 Distal Bile Duct Malignancies

The main distal bile duct malignancies that are eligible for preoperative laparoscopic staging are pancreatic ductal adenocarcinoma, distal cholangiocarcinoma, and carcinoma of the ampulla of Vater. Liver metastases, peritoneal metastatic implants, and metastases to distant lymph nodes are frequently present at the time of detection of the primary tumor, and each of them precludes curative resection. Local tumor extension outside the pancreatic head or into portal or superior mesenteric vessels is in many centers considered a contraindication for resective therapy.

Small liver metastases and subtle vascular tumor infiltration or encasement may be missed in a significant number of patients using the currently available noninvasive imaging techniques such as transabdominal US, computed tomography (CT), and magnetic resonance imaging (MRI). Peritoneal tumor dissemination will be missed in most cases, unless ascites is present.

The first preliminary report of the diagnosis of pancreatic cancer by laparoscopic US was published in 1984 by Okita et al. Cuesta et al. reported a small series in 1993, in which no additional information was obtained by laparoscopic US in patients with cancer of the pancreatic head region.

A larger series reported by BEMELMAN et al. (1995) clearly showed that additional information may be provided by laparoscopic US patients with malignant distal bile duct obstruction. Combined laparoscopy and laparoscopic US revealed metastatic disease in 16 out of 72 patients who were defined as having potentially resectable tumors after noninvasive preoperative staging. Liver metastases were detected in 13 patients. In 6 of these, metastases were seen only by laparoscopic US, and in 3 they were seen by both laparoscopy and laparoscopic US. Laparoscopic US showed a high specificity and positive predictive value in evaluating overall unresectability (96% and 97% respectively) and unresectability due to vascular tumor ingrowth (96% and 93% respectively).

JOHN et al. (1995) also report that this technique added to the information provided by diagnostic laparoscopy in a series of 40 patients. Laparoscopic US demonstrated factors confirming an unresectable tumor in 23 patients (59%), provided staging information in addition to that of laparoscopy alone in 20 patients (53%), and changed the decision regarding tumor resectability in 10 patients (25%). Laparoscopy combined with laparoscopic US was more specific and accurate in predicting tumor resectability than laparoscopy alone (88% and 89% versus 50% and 65% respectively).

In a series by VAN DELDEN et al. (1996) some very small bile duct tumors measuring between 1 and 2 cm were imaged, which could not be seen with US and CT.

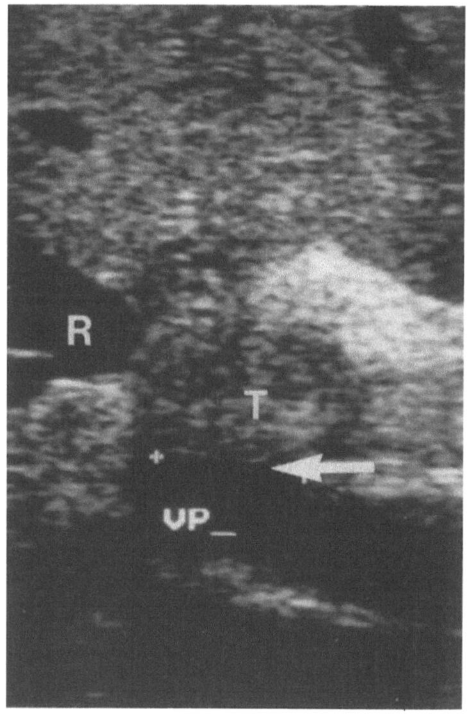

Fig. 12.10. Bifurcation tumor (*T*, Klatskin's tumor) infiltrating the portal vein (*VP*). There is a loss of hyperechoic interface between the tumor and the vascular lumen (*arrow*). The tumor extends into the dilated right hepatic duct (*R*)

12.4.2 Proximal Bile Duct Malignancies

Proximal bile duct malignancies which can be staged laparoscopically are gallbladder cancer and carcinoma of the proximal CBD or bifurcation (Klatskin's tumor). Gallbladder cancer may in many cases be unresectable at the time of detection, due to local invasion of the liver or hepatoduodenal ligament or because of metastatic spread to the liver or distant lymph nodes. These features can be well studied by laparoscopic US.

In cancer of the proximal CBD or bifurcation, distant tumor spread is less frequently seen. In these cases it is important preoperatively to rule out tumor ingrowth into the portal vein, hepatic artery, or liver, and to assess tumor extension into left and right hepatic ducts and segmental intrahepatic biliary radicles (Fig. 12.10).

At present, only GOUMA et al. have reported on this patient group. Out of 24 patients with a malignant proximal bile duct obstruction, laparotomy could be avoided by laparoscopy combined with laparoscopic US in 13 patients (54%) (GOUMA et al. 1996).

12.4.3 Intrahepatic Bile Duct Malignancies and Liver Tumors

Laparoscopic staging of intrahepatic bile duct tumors is similar to staging of other malignant primary intrahepatic tumors and potentially resectable metastases from colorectal carcinoma. Intraoperative US is an established technique in imaging of these tumors, for localizing a known tumor and defining its relationship with vascular structures as well as detecting additional occult intrahepatic lesions. Laparoscopic US of the liver, which is the most easily accessible organ for this technique, has the same possibilities.

In a series of 19 patients with primary or secondary hepatic tumors reported by CUESTA et al. (1993), the combination of laparoscopy and laparoscopic US provided new information leading to a change in surgical approach in 16 patients, avoiding laparotomy in 11 patients. In these 16 patients, the additional information was obtained by laparoscopy alone in 2 patients, by laparoscopy and laparoscopic US combined in 9 patients, and by laparoscopic US alone in 5 patients. In a similar series by GOUMA et al. (1996), laparoscopy and laparoscopic US proved unresectable disease due to new liver metastases or extrahepatic disease in 8 of 18 patients after extensive noninvasive preoperative imaging. In a series of 43 patients reported by JOHN et al. (1994b), additional information compared to laparoscopy alone was obtained by laparoscopy and laparoscopic US combined in 14 patients (33%). The rate of resectability of liver tumors was significantly higher after laparoscopy and laparoscopic US combined (91%) than after laparoscopy alone (58%).

12.4.4 Future Prospects

12.4.4.1 Technical Improvements

Ultrasound equipment is continually improving, and this will also apply to laparoscopic US imaging. Probes with a flexible tip may prove useful in scanning difficult areas, such as the liver dome, and color Doppler facilities may be useful for distinguishing between vessels and bile ducts, investigating vessel patency, and possibly even characterizing liver lesions (e.g., distinguishing malignant nodules from hemangiomas).

A very important step forward in laparoscopic US will be the development of specially designed puncture probes that allow accurate cytological puncture of small, deep-seated liver lesions by guiding the biopsy needle. With currently available probes it is often extremely difficult to obtain adequate biopsies from such lesions.

12.4.4.2 Indications

Laparoscopic US as a technique for staging intrahepatic tumors is probably here to stay. It has the same potentials as intraoperative US, which has established itself by proving its value in comparative studies with such techniques as dynamic contrast-enhanced CT and CT arterioportography. Studies by CUESTA et al. (1993), GOUMA et al. (1996) and JOHN et al. (1994b) confirm this expectation.

There are already a few prospective studies (JOHN et al. 1995; BEMELMAN et al. 1995) that indicate an additional value of combined laparoscopy and laparoscopic US in staging of distal bile duct malignancies. However, prospective studies comparing this technique with state-of-the-art helical/dynamic contrast-enhanced CT, new MRI techniques, and transabdominal US including duplex Doppler of the portal and superior mesenteric vessels are required to define the precise role of laparoscopic US in these patients. The main potential benefits of laparoscopic US are in increased small lesion detection, improved characterization of liver lesions and enlarged lymph nodes, and exclusion of vascular tumor ingrowth.

At present the role of laparoscopic US in staging of proximal bile duct malignancies is still unclear. The potential benefit in gallbladder cancer might be increased detection of metastatic disease, whereas in bifurcation tumors local staging could be improved.

It is to be expected that in the coming years the potentials and benefits of laparoscopy combined with laparoscopic US will be more extensively studied and it will become clear whether the technique can play a significant role in the staging of bile duct malignancies. This will be dependent not only on the technical possibilities of laparoscopic US, but also on developments in noninvasive imaging modalities and treatment strategies (e.g., surgical bypass versus endoscopic stenting in unresectable bile duct malignancies).

References

Barteau JA, Castro M, Arregui ME, Tetik C (1995) A comparison of intraoperative ultrasound versus cholangiography in the evaluation of the common bile duct during laparoscopic cholecystectomy. Surg Endosc 9:490–496

Bemelman WA, De Wit LTh, Van Delden OM, Smits NJ, Obertop H, Rauws EAJ, Gouma DJ (1995) Diagnostic laparoscopy combined with laparoscopic ultrasonography in staging of cancer of the pancreatic head region. Brit J Surg 82:820–824

Bezzi M, Merlino R, Orsi F, Di Nardo R, Silecchia G, Basso N, Passariello R, Rossi P (1995) Laparoscopic sonography during abdominal laparoscopic surgery: technique and imaging findings. AJR 165:1193–1198

Bezzi M, Merlino R, Penna A, Spaziani E, Silecchia G, Raparelli L, Tretola V, Basso N (1996) Ultrasonography of polypoid lesions of the gallbladder: diagnosis and indications for laparoscopic cholecystectomy (abstract). Surg Endosc 10:195(S92)

Christensen AH, Kamal GI (1970) Benign tumors and pseudotumors of the gallbladder. Arch Pathol 90:423–432

Cuesta MA, Meijer S, Borgstein PJ, Sibinga Mulder L, Sikkenk AC (1993) Laparoscopic ultrasonography for hepatobiliary and pancreatic malignancy. Br J Surg 80:1571–1574

Cuschieri A, Hall AW, Clark J (1978) Value of laparoscopy in the diagnosis and treatment of pancreatic carcinoma. Gut 19:672–677

Gouma DJ, De Wit LT, Nieveen van Dijkum E, Van Delden OM, Bemelman WA, Rauws EAJ, Van Lanschot JJB, Obertop H (1996) Laparoscopic ultrasonography for staging of GI malignancy. Scand J Gastroenterol 31:(in press)

Greig JD, John TG, Mahadaven M, Garden OJ (1994) Laparoscopic ultrasonography in the evaluation of the biliary tree during laparoscopic cholecystectomy. Br J Surg 81:1202-1206

Hulsmans FJ, Bosma A, Mulder PJJ, Reeders JWAJ, Tytgat GNJ (1992) Perirectal lymph nodes in rectal cancer: in vitro correlation of sonographic parameters and histopathologic findings. Radiology 184:553-560

Jakimowicz JJ (1993) Intraoperative ultrasonography during minimal access surgery. J R Col Surg Edinb 38:231-238

Jakimowicz JJ, Rutten H, Jürgens PJ, Carol EJ (1987) Comparison of operative ultrasonography and radiography in screening of the common bile duct for calculi. World J Surg 11:628-634

John TG, Banting SW, Pye S, Paterson-Brown S, Garden OJ (1994a) Preliminary experience with intracorporeal laparoscopic ultrasonography using a sector scanning probe: a prospective comparison with intraoperative cholangiography in the detection of choledocholithiasis. Surg Endosc 8:1176-1181

John TG, Greig JD, Crosbie JL (1994b) Superior staging of liver tumors with laparoscopy and laparoscopic ultrasound. Ann Surg 220:711-719

John TG, Greig JD, Carter DC, Garden OJ (1995) Carcinoma of the pancreatic head and periampullary region. Tumor staging with laparoscopy and laparoscopic ultrasound. Ann Surg 221:156-164

Okita K, Kodama T, Oda M, Takemoto T (1984) Laparoscopic ultrasonography: diagnosis of liver and pancreatic cancer. Scand J Gastroenterol 19(Suppl 94):91-100

Paul MA, Sibinga Mulder L, Cuesta MA, Sikkenk AC, Lyesen GKS, Meijer S (1994) Impact of intraoperative ultrasonography on treatment strategy for colorectal cancer. Br J Surg 81:1660-1663

Röthlin MA, Schlumpf R, Largiadèr F (1994) Laparoscopic sonography. An alternative to routine intraoperative cholongiography? Arch Surg 129:694-700

Sugiyama M, Atomi Y, Kuroda A, Muto T, Wada N (1995) Large cholesterol polyps of the gallbladder: diagnosis by means of US and endoscopic US. Radiology 196:493-497

Teefly SA, Soper NJ, Middleton WD, Balfe DM, Brink JA, Strasberg SM, Callery M (1995) Imaging of the common bile duct during laparoscopic cholecystectomy: sonography versus videofluoroscopic cholangiography. AJR 165:847-851

van Delden OM, Smits NJ, Bemelman WA, De Wit LT, Gouma DJ, Reeders JWAJ (1996) Comparison of laparoscopic and transabdominal ultrasonography in staging cancer of the pancreatic head region. J Ultrasound Med 16:207-212

Wade TP, Comitalo JB, Andrus CH (1994) Laparoscopic cancer surgery. Lessons from gallbladder cancer. Surg Endosc 8:698-701

Warshaw AL, Gu Z, Wittenberg J, Waltmann AC (1990) Preoperative staging and assessment of resectability of pancreatic cancer. Arch Surg 125:230-233

Yamashita Y, Kurohiji T, Hayashi J, Kimitsuki H, Hiraki M, Kakegawa T (1993) Intraoperative ultrasonography during laparoscopic cholecystectomy. Surg Laparosc Endosc 3:167-171

13 Nuclear Medicine in Acute and Chronic Disorders of the Hepatobiliary Tree

D. Fink-Bennett and H. Balon

CONTENTS

13.1 Acute Cholecystitis

13.1.1 Introduction

It is estimated that approximately 10%–15% of Americans have gallstones. Of these 20 million Americans (16 million female and 4 million male), 20%–25% will develop acute cholecystitis and 20%–35% will develop symptomatic chronic cholecystitis, while the remainder remain asymptomatic (Anderson 1971; Hardy 1988).

Gallstones occur more frequently in women than in men, since estrogens, progesterone, and pregnancy increase the production of lithogenic bile and the formation of nucleation sites – factors that permit or promote gallstone formation. In addition, estrogens, progesterone, and pregnancy cause decreased gallbladder motility, which contributes to gallstone formation. Obesity and rapid weight loss also increase cholesterol supersaturation and decrease gallbladder motility. These conditions are not unique to women but occur more commonly in them. In fact, of the one million newly diagnosed patients with cholelithiasis every year, the majority are female (Anderson 1971; Hardy 1988; Anonymous 1993).

Gallstones are comprised principally of cholesterol, although pigmented stones do occur. Gallstones increase in size over a 2- to 3-year period, then stabilize. Most are small (less than 2 cm in diameter). Unless obstruction of the cystic duct occurs, their presence may go undetected. If, however, a stone becomes impacted in the cystic duct, acute cholecystitis develops and can be life-threatening (Anonymous 1993).

In 95% of patients with an acutely inflamed hemorrhagic gallbladder, one or more gallstones are responsible for the cystic duct obstruction. In the remaining 5% of patients with acute cholecystitis, the obstruction is caused by inflammation, edema, gallbladder mucus, or a tumor (Anderson 1971; Hardy 1988). The pathogenesis of acute cholecystitis is still not completely understood, but acute cholecystitis is believed to occur when the obstructed gallbladder becomes distended and its mucosa irritated and inflamed by retained lithogenic bile. Vascular congestion, mural edema, leukocytic infiltration, and ultimately hemorrhagic necrosis of the gallbladder wall ensue (Hardy 1988).

The signs and symptoms of acute cholecystitis include severe epigastric or right upper quadrant pain, fever, nausea, and vomiting. A mild to moderate leukocytosis and abnormal liver function values are also usually present. If the pain radiates to the back in the region of the scapula or there is rebound tenderness, the clinical diagnosis of acute cholecystitis is strengthened (Rhoads et al. 1970).

If a stone becomes impacted in the common bile duct, hyperbilirubinemia will develop. A dilated common bile duct is identified sonographically 3 days after its obstruction.

Choledocholithiasis is not uncommon. It is estimated that 8%–15% of patients under the age of 60 and 15%–60% of patients over the age of 60 will have stones in their common bile duct in conjunction with one in their cystic duct (Anonymous 1993).

D. Fink-Bennett, MD, Department of Nuclear Medicine, William Beaumont Hospital, 3601 West 13 Mile Road, Royal Oak, MI 48073-6769, USA
H. Balon, MD, Department of Nuclear Medicine, William Beaumont Hospital, 3601 West 13 Mile Road, Royal Oak, MI 48073-6769, USA

The diagnosis of acute cholecystitis is not always an easy one, as other conditions (e.g., coronary artery disease, chronic passive congestion of the liver, intestinal obstruction, renal colic, herpes zoster, and phlegmonous gastritis) can have a similar presentation (RHOADS et al. 1970).

Most patients with acute cholecystitis do not require urgent surgery. Surgery is usually delayed for up to 72 h in order to stabilize the patient and begin antibiotic therapy. Elderly patients and up to 10%–15% of patients with a history of acute cholecystitis which resolved spontaneously, do, however, require early surgery, as gangrenous changes with or without perforation, gallstone ileus, or cholangitis occur in up to 25% of these patients if the proper diagnosis is not made and early treatment not instituted (ANDERSON 1971; HARDY 1988; ANONYMOUS 1983; RHOADS et al. 1970). Since history, physical examination, and laboratory tests alone cannot always establish the diagnosis, numerous imaging modalities have evolved to confirm the clinical suspicion of acute cholecystitis. These include hepatobiliary scintigraphy, real-time ultrasonography, and computed tomography of the gallbladder.

13.1.2 Radiopharmaceuticals and Their Uptake

Hetapobiliary scintigraphy is performed following the intravenous administration of a technetium-99m-labeled iminodiacetic acid (IDA) derivative. There are two IDA derivatives that are commercially available: disofenin (Hepatolite; Dupont-Merck) and mebrofenin (Choletec; Bracco). Both compounds are extracted by the hepatocytes, secreted into the bile canaliculi, and transported via the intra- and extrahepatic biliary ducts into the gallbladder and small bowel.

The hepatic uptake of IDA derivatives is accomplished by the same carrier-mediated, non-sodium-dependent organic anion pathway as the uptake of bilirubin. The degree of hepatic uptake of these compounds and their rate of excretion is dependent upon their chemical structure. Approximately 85% of disofenin and 98% of mebrofenin is extracted by the hepatocytes. The clearance rate of mebrofenin is 6–28 min, that of disofenin is 6–32 min (KRISHNAMURTHY and KRISHNAMURTHY 1988).

Hepatic extraction of IDA derivatives decreases with increasing serum bilirubin levels, and thus in hyperbilirubinemia with serum concentrations greater than 8 mg/dl, common bile duct and gallbladder visualization is achieved better with mebrofenin than with disofenin. Mebrofenin is useful in patients with bilirubin levels up to 30 mg/dl. In the absence of hyperbilirubinemia, either agent provides excellent visualization of the hepatobiliary tree. Mebrofenin does, however, have a rapid biliary-to-bowel transit, a characteristic that must be taken into consideration when evaluating patients with suspected acute cholecystitis: If too little tracer remains within the intrahepatic biliary tree at 1 h, none will be available to enter the gallbladder following morphine administration or upon delayed imaging (FREITAS 1982; DATZ 1988; FINK-BENNETT and BALON 1993).

13.1.3 Procedure

13.1.3.1 Conventional Hepatobiliary Scintigraphy

Conventional hepatobiliary scintigraphy is performed after a 2- to 4-h fast and following the intravenous administration of 5 mCi technetium-99m disofenin or 5 mCi technetium-99m mebrofenin. If a patient has fasted for more than 24–48 h, he or she should be pretreated with a synthetic octapeptide of cholecystokinin (CCK; sincalide, Kinevac). The usual dose of sincalide is 0.01–0.02 μg/kg i.v. 30–60 min before radiotracer administration. Sincalide should be injected slowly over a 3–5 min period. This will minimize abdominal cramping, one of the potential side effects of CCK administration (ZEISSMAN et al. 1992). An anterior one-million-count large-field-of-view gamma camera image or 500 000-count small-field-of-view gamma camera image of the liver and biliary tree is then obtained with a low-energy all-purpose collimator and a 20% window centered on the 140-keV technetium-99m photopeak at 2 min aftered injection. Images are then obtained every 10 min for the same time as was required to obtain the initial anterior image. Right lateral views are obtained at 30 and 60 min after radiotracer administration. Oblique views are obtained, if necessary, to distinguish gallbladder from small bowel activity. Continuous dynamic computer acquisition at a rate of one frame per minute is also recommended. This will permit the determination of a hepatic extraction fraction (HEF) as well as allow the study to be viewed in cinematic mode if so desired.

In order for hepatobiliary imaging to achieve its high degree of accuracy in the detection of acute cholecystitis, images have to be obtained up to 4 h following radiotracer administration. If imaging is not continued beyond 1 h, the specificity of the procedure is reduced from 93%–96% to 80%–88%

(FREITAS et al. 1980; FREITAS 1981; WEISSMAN 1981). This is because as many as 20% of patients with a patent cystic duct require up to 4h for visualization of their gallbladder because of increased intraluminal pressure within the gallbladder from viscous or stagnant bile, stones, or sludge (WEISSMAN 1981).

To shorten imaging time, patients clinically suspected of acute cholecystitis can either be premedicated with sincalide (0.01–0.02 μg/kg) or be given morphine sulfate if the gallbladder does not visualize within 60 min of imaging and if radiotracer is seen within the small intestine (EIKMAN et al. 1975; CHOY et al. 1984).

Pretreatment with the octapeptide of CCK causes the gallbladder to contract, empty, and subsequently relax, thereby creating an optimal state for radiotracer accumulation, provided the cystic duct is patent. Thus, the potential source of increased resistance to bile flow (viscous bile, stones, sludge, etc.) is eliminated and delaying imaging becomes unnecessary. In fact, this state of low resistance to bile flow accounts for preferential gallbladder filling and delayed biliary-to-bowel transit in some patients pretreated with CCK.

The sensitivity and specificity of pretreatment CCK cholescintigraphy in the diagnosis of acute cholecystitis are identical to those of conventional hepatobiliary scintigraphy. Why then do most laboratories not use CCK premedication in all patients undergoing hepatobiliary scintigraphy for the detection of acute cholecystitis? The reason is that FREEMAN et al. (1981) were concerned that patients with chronic cholecystitis would not be distinguished from those without cholecystitis; patients would be unnecessarily exposed to a pharmacological agent that, although it has not been associated with significant side effects, can cause nausea, cramping, and, in some individuals, exacerbation of gallbladder pain; and, occasionally, patients with acute cholecystitis who had delayed gallbladder visualization would be totally missed – a situation that occurred in 5/143 patients with proven acute cholecystitis (FREEMAN et al. 1981). In addition, FREEMAN et al. were concerned that if delayed views were not obtained, the incidental findings thus obtained of malrotation, enterogastric reflux, and masses displacing or inflammatory processes affecting the small bowel would go undetected (FREEMAN et al. 1981).

The use of pretreatment CCK to shorten the time required to establish the diagnosis of acute cholecystitis is now rarely, if ever, considered, since morphine-augmented cholescintigraphy is equally accurate, rapid, and without the potential pitfalls of CCK pretreatment.

13.1.3.2 Morphine Augmentation

Morphine sulfate enhances sphincter of Oddi tone. This results in increased pressure within the lumen of the common bile duct. The pressure within the common bile duct following the intravenous administration of as little as 0.04 mg/kg morphine sulfate is increased by approximately 50% (from 12.7 ± 2.4 to $20 \pm 2.4\,cmH_2O$). It is this pharmacologically induced increased pressure that diverts bile away from the sphincter of Oddi into a functionally obstructed gallbladder if the cystic duct is patient (CHOY et al. 1984; FINK-BENNETT et al. 1991b; KESLAR and TURBINER 1987; KIM et al. 1986; VASQUES et al. 1988).

Morphine-augmented cholescintigraphy is performed in the same way as conventional hepatobiliary scintigraphy except that morphine sulfate (0.04 mg/kg diluted in 10 ml saline) is administered intravenously over 3 min when there is nonvisualization of the gallbladder at 60 min after technetium-99m IDA administration, provided radiotracer is demonstrated within the small bowel. Five-minute serial postmorphine images are then obtained for 30 min for the same time as the premorphine images. Persistent nonvisualization of the gallbladder is indicative of acute cholecystitis, whereas chronic cholecystitis is present if the gallbladder visualizes following morphine administration. By administering morphine at 60 min instead of 30–40 min after technetium-99m IDA administration, the diagnosis of chronic cholecystitis is not sacrificed, the diagnosis of acute cholecystitis is accurately made, and imaging time is reduced from 4 to 1.5 h (CHOY et al. 1984; KIM et al. 1990; FINK-BENNETT et al. 1991b; KESLAR and TURBINER 1987; VASQUES et al. 1988; FINK-BENNETT and BALON 1993).

13.1.4 Interpretation

A normal hepatobiliary scan is one in which the gallbladder and small bowel are visualized within 1 h of the intravenous administration of technetium-99m IDA (Fig. 13.1).

In patients in whom a conventional hepatobiliary scan is being performed for suspected acute cholecystitis, one of six scan patterns is usually seen:

1. Gallbladder visualization within 1 h following radiotracer administration

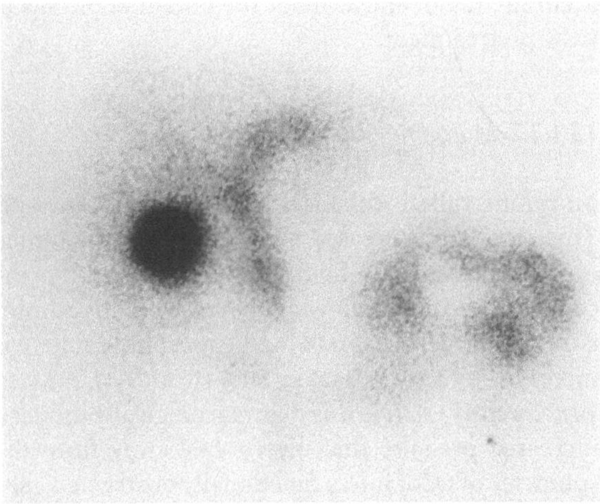

Fig. 13.1. Normal hepatobiliary scan. The gallbladder, biliary radicals, and small bowel are visualized within 60 min following the intravenous administration of technetium-99m iminodiacetic acid (IDA)

2. Nonvisualization of the gallbladder up to 4 h after radiotracer administration or up to 30 min after morphine administration (Fig. 13.2)
3. Delayed gallbladder visualization between 1 and 4 h after radiotracer or within 30 min of morphine administration (Figs. 13.3, 13.4)
4. Persistent (up to 24 h) nonvisualization of the intra- and extrahepatic biliary tree in a patient with a normal hepatic extraction efficiency (Fig. 13.15)
5. Preferential gallbladder filling with delayed (>60 min) visualization of the small bowel (Fig. 13.6)
6. Presence of a nubbin of activity adjacent to the common bile duct and medial to the gallbladder fossa which persists for up to 4 h after radiotracer administration ("dilated cystic duct sign"; Fig. 13.7)

Failure to visualize the gallbladder up to 4 h after radiotracer administration in conjunction with a normal technetium-99m IDA hepatic extraction efficiency and normal biliary-to-bowel transit indicates cystic duct obstruction and, in an appropriate clinical setting, acute cholecystitis (WEISSMAN 1981) (Fig. 13.2). If a band or rim of increased activity is located adjacent to the gallbladder fossa, a "rim sign" is present. This increased pericholecystic hepatic activity (PCHA or rim sign) is not only indicative of acute cholecystitis, but often is a manifestation of the inflammatory changes resulting from a gangrenous gallbladder wall (BUSHNELL et al. 1986) (Fig. 13.8).

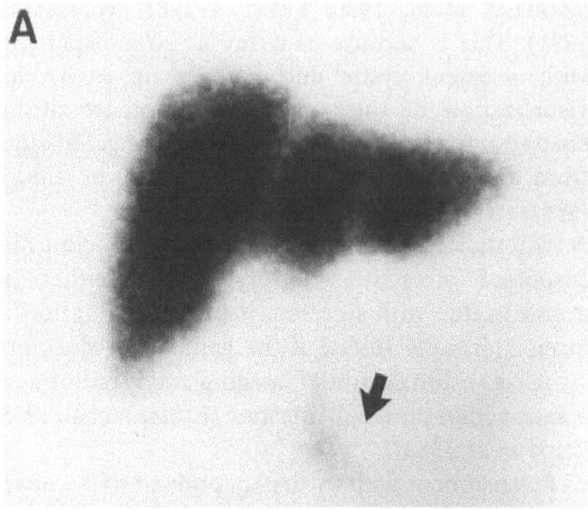

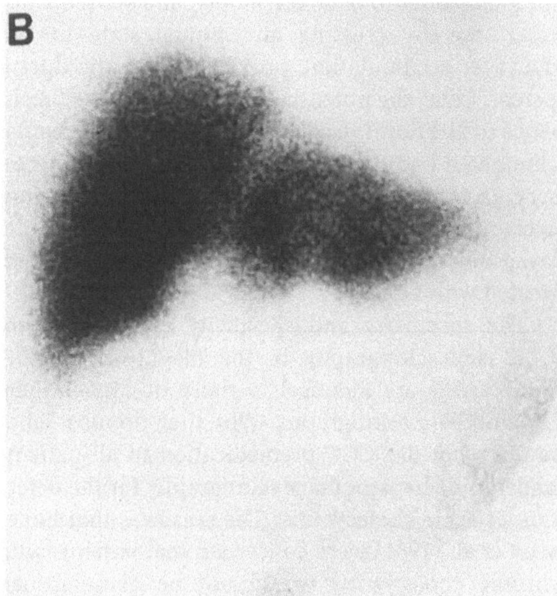

Fig. 13.2a,b. Acute cholecystitis with choledocheolithiasis. **a** Sixty-minute anterior hepatobiliary scan reveals nonvisualization of the gallbladder and spartan radiotracer in the small bowel (*arrow*). **b** Three-hour delayed anterior hepatobiliary scan reveals persistent nonvisualization of the gallbladder with continued spartan bowel activity in a patient with acute cholecystitis and choledocholithiasis

Delayed gallbladder visualization at between 1 and 4 h is most often the result of increased intraluminal gallbladder pressure resulting from viscous bile or sludge in patients with chronic cholecystitis (Fig. 13.3). Delayed gallbladder visualization, however, may also be seen in patients with hepatocellular disease. The slow biliary transit resulting from reduced hepatic extraction and excretion of technetium-99m IDA delays gallbladder

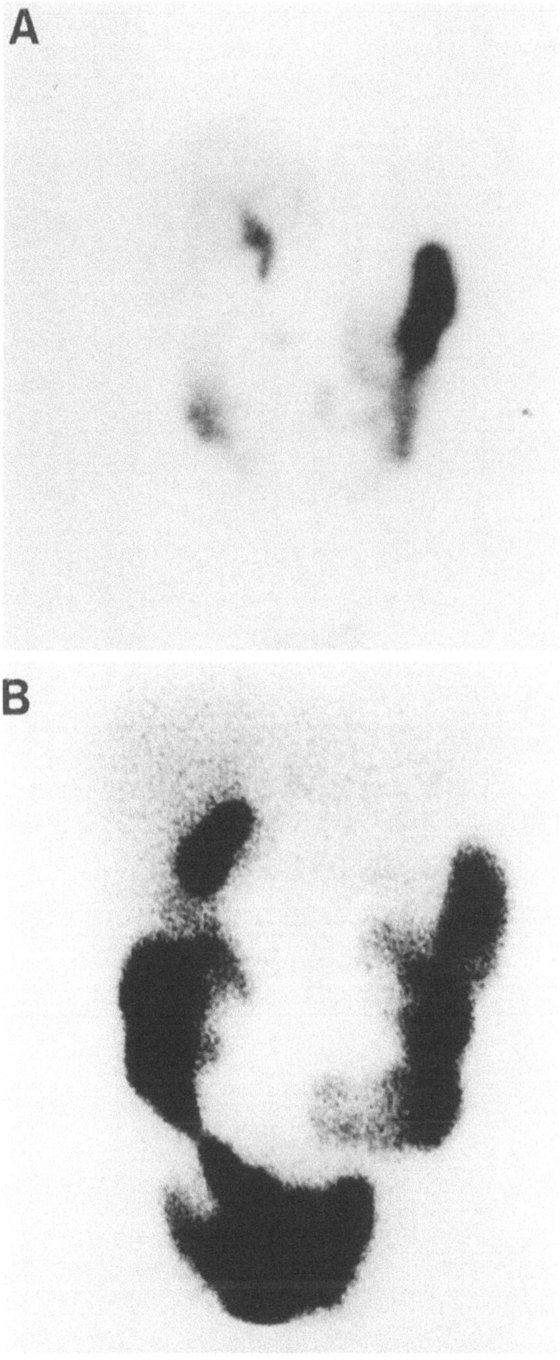

Fig. 13.3a,b. Chronic cholecystitis. **a** Anterior hepatobiliary scan at 60 min demonstrates nonvisualization of the gallbladder. **b** Three-hour delayed hepatobiliary scan reveals radiotracer in a chronically inflamed gallbladder

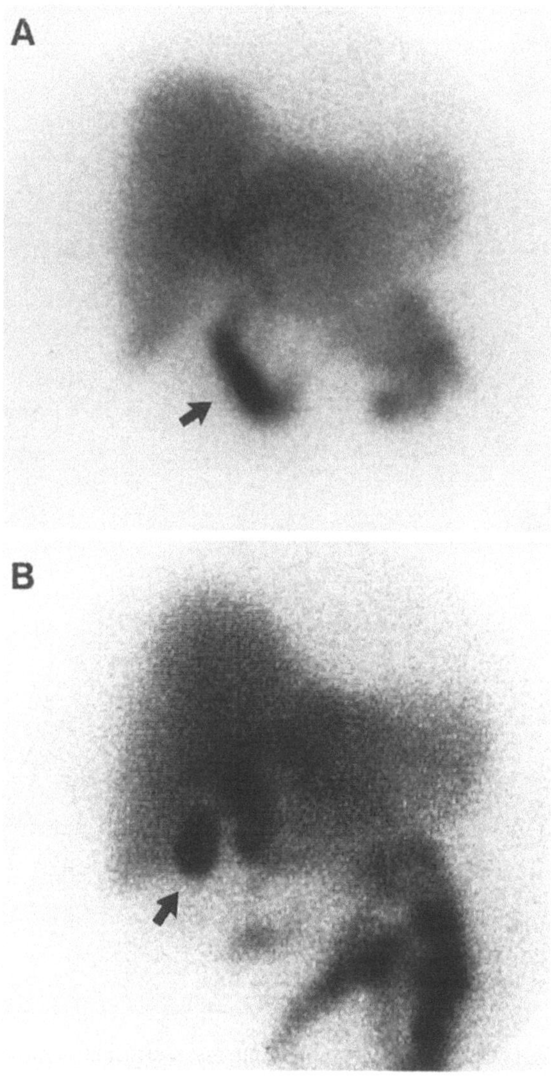

Fig. 13.4a,b. Chronic cholecystitis. **a** Anterior 60-min hepatobiliary scan reveals nonvisualization of the gallbladder. Radiotracer is seen within the duodenal loop (*arrow*) and small bowel. **b** Anterior hepatobiliary scan 15 min after morphine administration reveals radiotracer in a small, contracted, chronically inflamed, sludge-filled gallbladder (*arrow*)

visualization and does not necessarily imply chronic cholecystitis (Freitas 1982; Weissman 1981).

Nonvisualization of the intrahepatic and extrahepatic biliary tree at 6–24 h after technetium-99m IDA administration in patients with normal liver function is indicative of high-grade or total common bile duct obstruction. The presence or absence of a concomitant cystic duct obstruction and acute cholecystitis cannot, however, be determined when this scintigraphic patern is seen. The reason for this is that choledocholithiasis in conjunction with acute cholecystitis is reported to have occurred in 14%–93% of patients in whom only liver activity was demonstrable at 24 h (Freitas 1982) (Fig. 13.5). In situations where radiotracer is identified in the bowel between 6 and 24 h after IDA administration,

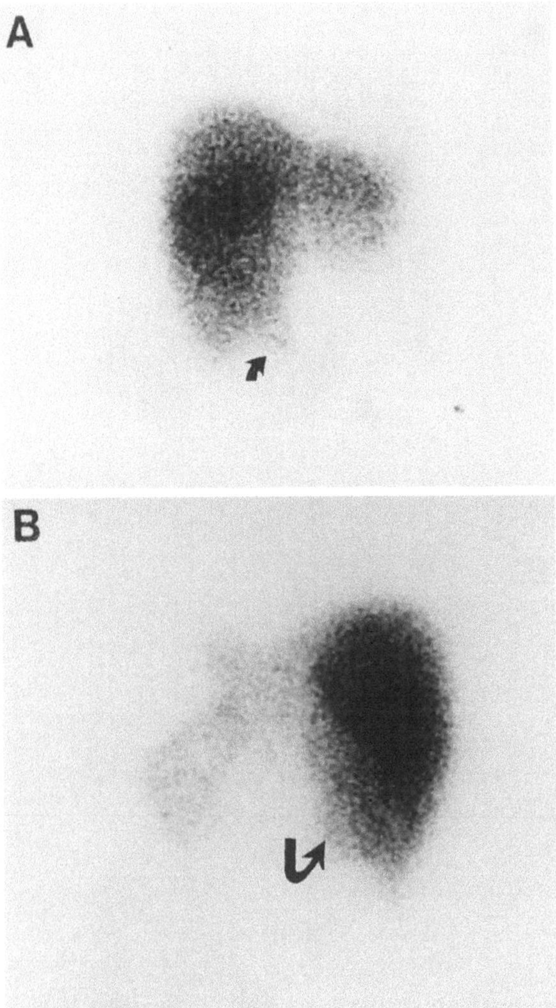

Fig. 13.5a,b. Complete common bile duct obstruction. **a** Anterior 24-h hepatobiliary scan demonstrates changes due to high-grade or complete common bile duct obstruction, i.e., visualization of the liver but no biliary tree or small bowel activity. The activity located inferior to the right hepatic lobe resides in the kidney (*arrow*). **b** Posterior 24-h hepatobiliary scan confirms that the activity below the right lobe of the liver is located within the kidney (*arrow*)

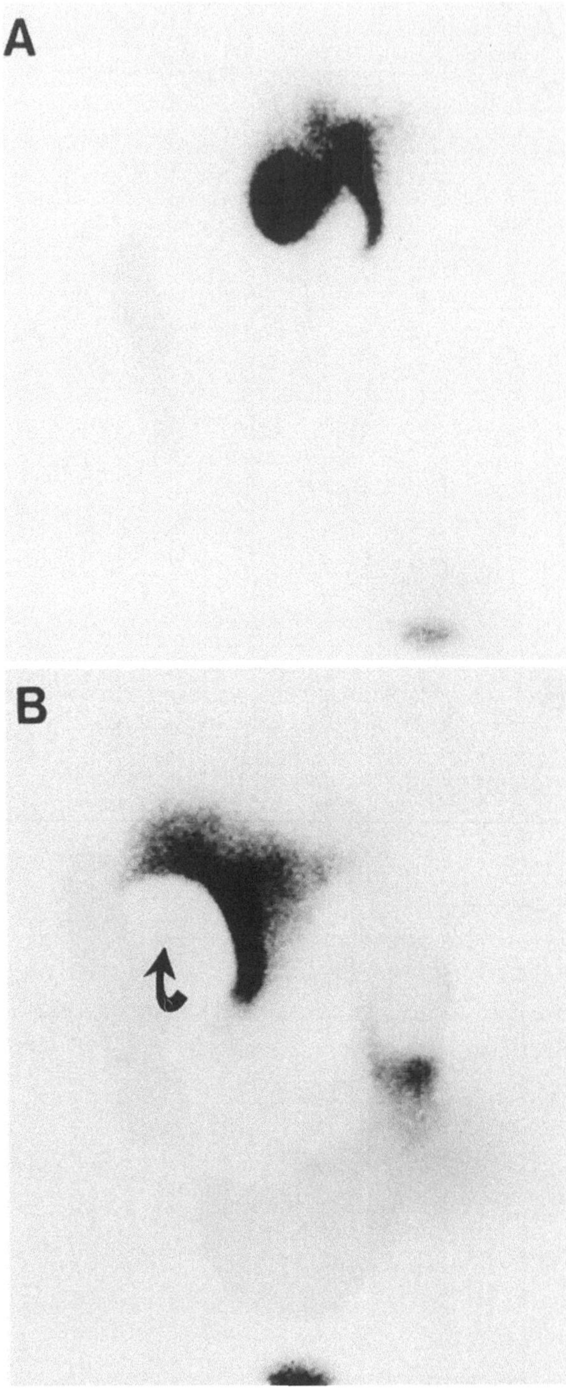

Fig. 13.6a,b. Preferential gallbladder fillng. **a** Fifty-minute anterior hepatobiliary scan reveals preferential gallbladder filling in a patient pretreated with cholecystokinin (CCK). **b** The gallbladder-shielded (*arrow*) 65-min anterior image reveals some activity in the small bowel

endoscopic retrograde cholangiopancreatography (ERCP) is often required to determine the underlying etiology of the partial occlusion of the common bile duct (FREITAS 1982; WEISSMAN 1981).

The presence of a nubbin of activity located adjacent to the common bile duct and medial to the gallbladder fossa persistent up to 4 h after radiotracer administration is a manifestation of a "dilated cystic duct sign" and acute cholecystitis. The nubbin of activity represents radiotracer within a patient cystic duct proximal to its site of obstruction, e.g., by calculus (COLEMAN et al. 1984) (Fig. 13.7). Preferential gallbladder filling (i.e., gallbladder visualization within 1 h but bowel visualization delayed beyond 60 min) (Fig. 13.6) occurs when the common bile duct is functionally obstructed as a result of increased tone of the sphincter of Oddi (e.g., sphincter

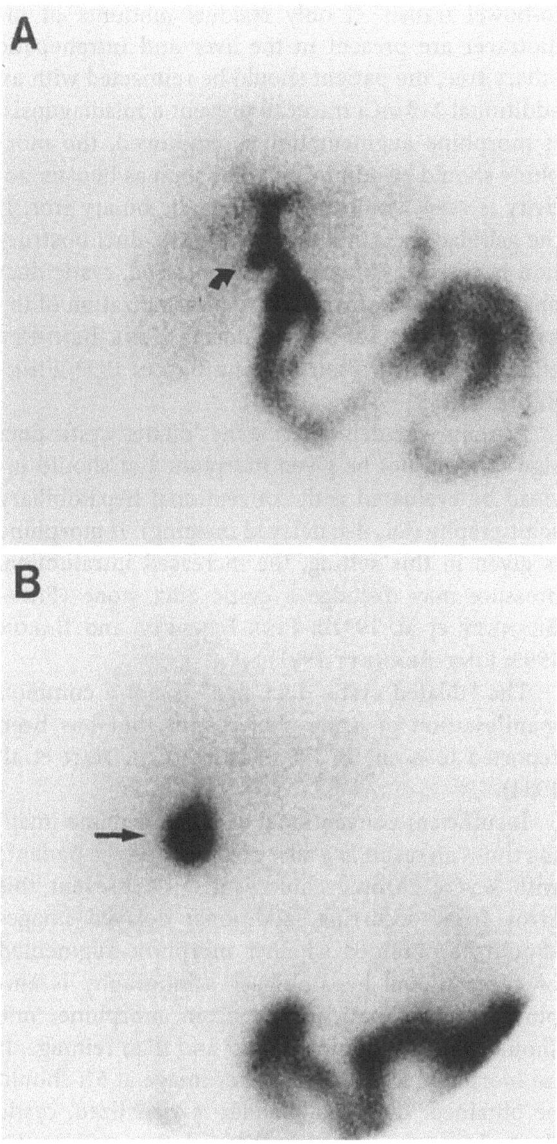

Fig. 13.7a,b. Dilated cystic duct sign. **a** Anterior 60-min hepatobiliary scan reveals a focus of technetium-99m IDA adjacent to the common bile duct (*arrow*). **b** Three-hour delayed scan reveals a persistent nubbin of increased radiotracer accummulation adjacent to the common bile duct. It represents radiotracer within a dilated cystic duct (*arrow*) proximal to its site of obstruction: the dilated cystic duct sign

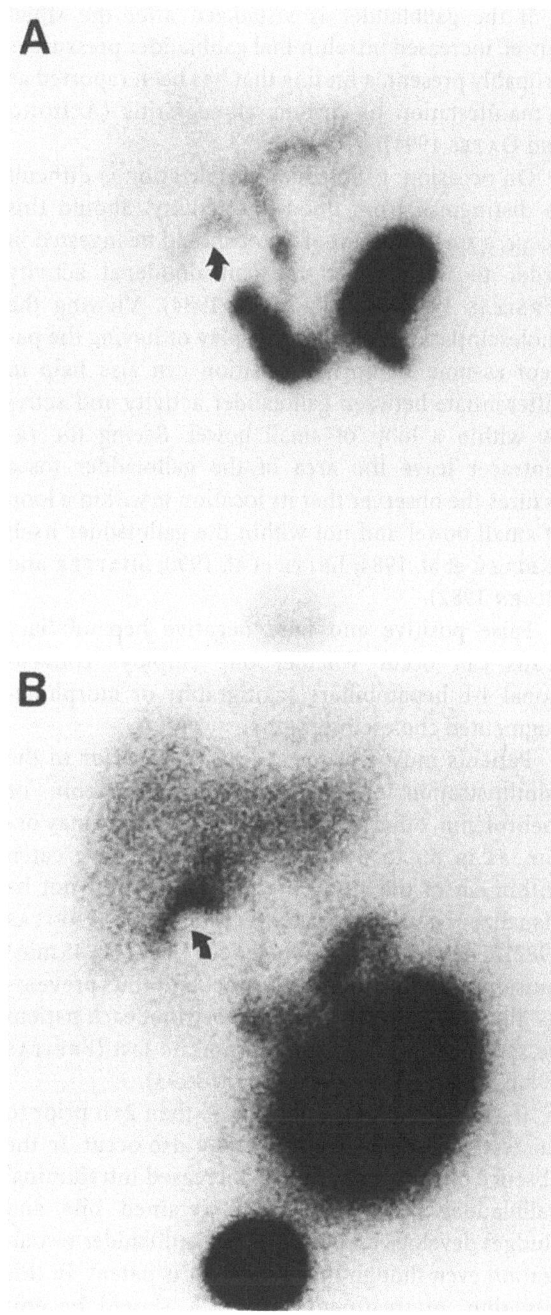

Fig. 13.8a,b. Acute cholecystitis (rim sign). **a** Sixty-minute anterior hepatobiliary scan reveals no radiotracer within the gallbladder. A band of radiotracer (*arrow*), however, is present superior to the gallbladder fossa: the rim sign. **b** Thirty minutes after morphine administration there is persistent nonvisualization of the gallbladder and increased pericholecystic activity (*arrow*) above the gallbladder fossa, where an acutely inflamed gangrenous gallbladder resided

of Oddi dyskinesia, effect of opioids). Increased pressure within the common bile duct can also occur in ampullitis resulting from repeated passage of small stones through the ampulla of Vater in patients with chronic calculous cholecystitis. Over 50% of patients pretreated with CCK will also demonstrate preferential gallbladder filling, as there is less resistance to bile flow into the gallbladder than through the sphincter of Oddi following CCK administration (WEISSMAN 1981; KIM et al. 1990).

If the gallbladder is visualized after the small bowel, increased intraluminal gallbladder pressure is probably present, a finding that has been reported as a manifestation of chronic cholecystitis (ACHONG and OATES 1994).

On occasion, gallbladder visualization is difficult to distinguish from duodenal activity. Should this occur, a small amount of water should be ingested in order to wash away transient duodenal activity (FREITAS 1982; KELLER et al. 1984). Viewing the cholescintiscan in dynamic display or having the patient assume an upright position can also help to differentiate between gallbladder activity and activity within a loop of small bowel. Seeing the radiotracer leave the area of the gallbladder fossa assures the observer that its location is within a loop of small bowel and not within the gallbladder itself (KELLER et al. 1984; LETTE et al. 1990; SHAFFER and OLSEN 1982).

False positive and false negative hepatobiliary scans can occur whether one employs conventional 4-h hepatobiliary scintigraphy or morphine-augmented cholescintigraphy.

Patients must fast for at least 2–4 h prior to the administration of technetium-99m disofenin or mebrofenin, otherwise a false positive study may occur, as in up to 64% of patients who have eaten within 4 h of the study the gallbladder will not be visualized despite a patent cystic duct (FREITAS 1982). The presence of endogenous CCK ($T_{1/2}$ 45 min) causes the gallbladder to contract and thus prevents its filling. It is therefore mandatory that each patient be screened to assure an appropriate fast (FREITAS 1982; FINK-BENNETT and BALON 1993).

If a patient has fasted for more than 24 h prior to the test, a false positive study may also occur. In the absence of endogenous CCK, increased intraluminal gallbladder pressure (due to retained bile and sludge) develops and can prevent gallbladder visualization even though the cystic duct is patent. In this situation, pretreatment with CCK should be employed (see above). CCK will cause the gallbladder to contract, eject the retained bile and sludge, and thus eliminate the source of increased resistance to bile flow (WEISSMAN 1981; EIKMAN et al. 1975; FINK-BENNETT 1985, 1991).

Care must always be taken to ensure that an adequate amount of radiotracer is present within the hepatobiliary tree to permit gallbladder visualization after morphine administration or following a 4-h delay. If this is not done, a false positive study will occur. This is especially true when technetium-99m mebrofenin is used, as it has a very rapid biliary-to-bowel transit. If only residual amounts of radiotracer are present in the liver and intrahepatic biliary tree, the patient should be reinjected with an additional 2–3 mCi tracer to prevent a misdiagnosis. If morphine augmentation is employed, the morphine should be administered as soon as booster activity is seen within the intrahepatic biliary tree. If the gallbladder is still not seen, cystic duct obstruction is present. If morphine is not used, cystic duct obstruction is confirmed by non-visualization of the gallbladder 4 h after booster dosing (FINK-BENNETT et al. 1991a; FINK-BENNETT and BALON 1993; FINK-BENNETT 1991).

Patients who demonstrate the "dilated cystic duct sign" should not be given morphine but should instead be evaluated with conventional hepatobiliary scintigraphy (i.e., 4-h delayed imaging). If morphine is given in this setting, the increased intraluminal pressure may dislodge a cystic duct stone (FINK-BENNETT et al. 1991b; FINK-BENNETT and BALON 1993; FINK-BENNETT 1991).

The "dilated cystic duct sign" is not a common manifestation of acute cholecystitis, but has been reported to occur in 7% of cases (COLEMAN et al. 1984).

Insufficient conventional or postmorphine imaging time can result in a false positive study in patients with severe chronic cholecystitis. To prevent this error from occurring, additional delayed images should be obtained whether morphine-augmented or conventional hepatobiliary scintigraphy is employed. If the patient was given morphine, one should wait an additional hour and then reimage. If no morphine was used, another image at 5 h should be obtained. If the gallbladder is visualized, cystic duct obstruction is excluded and the presence of chronic cholecystitis confirmed. Delayed imaging should also be performed when it cannot be determined with certainty whether a small focus of activity in the gallbladder fossa represents a small contracted gallbladder or a part of the small bowel. By obtaining further delayed images, the nature of this finding can be ascertained (bowel activity will dissipate or change in position) (FREITAS 1982; KESLAR and TURBINER 1987; FINK-BENNETT and BALON 1993; FINK-BENNETT 1991).

Other causes of a false positive hepatobiliary scan include severe hepatocellular disease, sepsis, total parental hyperalimentation, and prolonged fasting. In these situations, the use of real-time ultrasonography of the gallbladder, computed tomography, and indium-111 white blood cell scintigraphy should help to prevent a misdiagnosis (FINK-BENNETT and

BALON 1993; FINK-BENNETT 1991; FINK-BENNETT et al. 1991a; DATZ 1986).

False negative scans can also occur, particularly in patients with acute acalculous cholecystitis, when the intraluminal pressure within the edematous cystic duct and gallbladder is insufficient to prevent radiotracer from entering it. In this situation, and in any situation where there is a strong pretest suspicion of acute cholecystitis but a negative hepatobiliary scan, real-time ultrasonography of the gallbladder, gallbladder computed tomography and/or indium white blood cell scintigraphy should be employed. In the absence of hypoalbuminemia, cirrhosis, or ascites, the pathognomonic (strict criteria) real-time sonographic findings of acute cholecystitis are the presence of pericholecystic fluid and gallbladder wall edema. Using liberal criteria, the sonographic findings are those indicative of, but not pathognomonic for, acute cholecystitis. These include cholelithiasis, a thick gallbladder wall, nonshadowing echoes, and the sonographic Murphy's sign.

The sensitivity of cholescintigraphy in detecting acute cholecystitis is greater than 95%. Using liberal sonographic criteria the sensitivity is only 80%–86%, and it is even lower (24%) using the strict sonographic criteria (FINK-BENNETT et al. 1985a).

The characteristic findings of acute cholecystitis in computed tomography are gas bubbles within the gallbladder wall (KANE et al. 1983; MOSS et al. 1983). The sensitivity of computed tomography is similar to that of real-time ultrasonography.

The specificity of hepatobiliary imaging in excluding acute cholecystitis is also superior to those of real-time ultrasonography and computed tomography. The specificity of real-time ultrasonography ranges from 60% to 64%, whereas that of hepatobiliary imaging is 93%–96%. Gallbladder computed tomography is similar in specificity to real-time ultrasonography, both modalities being subject to and limited by a high false positive rate in the presence of sludge, stones, or a thick gallbladder wall; in fact, in several studies, more than one-third of patients with acute epigastric and right upper quadrant pain would have been misdiagnosed as having acute cholecystitis had they only undergone gallbladder sonography (MOSS et al. 1983; SAMUELS et al. 1983; FREITAS 1982).

The positive predictive value of hepatobiliary scintigraphy for acute cholecystitis is as high as 92.1%, whereas that of real-time ultrasonography is only 40%–50%. Cholescintigraphy's negative predictive value is 99% (SAMUELS et al. 1983; FREITAS et al. 1980, 1983).

Hepatobiliary imaging is also a highly accurate modality for detecting common bile duct obstruction. Its sensitivity for total or high-grade common bile duct obstruction is over 95%, and it can detect it immediately on stone impaction. On the other hand, it takes more than 72 h from the time of impaction before ductal dilatation is demonstrable sonographically (WEISSMAN 1981; KAPLUN et al. 1985; MILLER et al. 1984).

Indium-111 white blood cell gallbladder scintigraphy is performed using indium-111-labeled autologous white blood cells. White blood cells are a mixed population and are labeled using the method described in the Amersham package insert for indium-111 oxyquinoline solution for the radiolabeling of leukocytes except that acid citrate dextrose is used as an anticoagulant, hetastarch is employed in all sedimentation steps, and centrifugation is performed at 300 $g \times 5$ min. Anterior images of the gallbladder are obtained at 6 and 24 h after the intravenous administration of 500 μCi labeled leukocytes. A medium-energy collimator with a 20% window centered on the 175- and 247-keV indium-111 photopeaks is used. Twenty-four-hour delayed images are required, for without them the sensitivity of this procedure is reduced from 85% to 66% (FINK-BENNETT et al. 1991a). Accumulation of labeled white cells within the gallbladder wall indicates acute

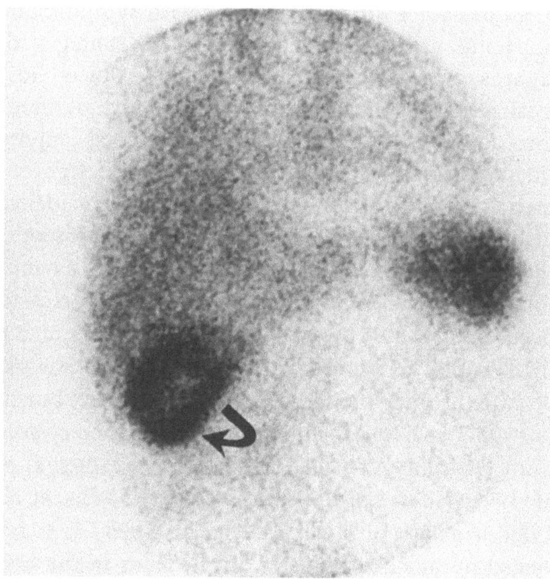

Fig. 13.9. Acute cholecystitis. The 24-h anterior indium-111 scan of the abdomen reveals increased indium-111-labeled white blood cells within an acutely inflamed hemorrhagic gallbladder wall (*arrow*)

cholecystitis (Fink-Bennett 1991; Fink-Bennett et al. 1991a; Datz 1988) (Fig. 13.9).

It should be remembered that gallbladder scintigraphy with indium-111-labeled white blood cells is not dependent on intraluminal gallbladder pressure and labeled leukocytes are therefore an ideal imaging agent for diagnosing acute cholecystitis in patients in whom a potential false positive or false negative hepatobiliary scan is clinically suspected.

Fortunately, acute acalculous cholecystitis accounts for only 5%–6% of all patients with acute cholecystitis. In addition, the majority of patients with acute acalculous cholecystitis do have sufficiently increased intraluminal gallbladder and cystic duct pressure to prevent radiotracer from entering the gallbladder (Freitas 1982).

Even though many false positive and false negative conventional and morphine-augmented hepatobiliary scans do occur, the accuracy of cholescintigraphy for the detection of acute cholecystitis still remains superior to that of any other imaging modality.

13.2 Bile Leak

Laparoscopic cholecystectomy was first introduced in 1987 in France (Cuschieri et al. 1991; Dubois et al. 1991) and has since become widely accepted as an alternative to open cholecystectomy and the treatment of choice for most patients with symptomatic nonacute gallbladder disease. The advantages of laparoscopic cholecystectomy include shorter hospitalization (median hospital stay: 1 day) and, therefore, lower costs, earlier return to normal activity (median: 8 days), reduced morbidity, and better cosmetic results (Donohue et al. 1992). The incidence of bile leaks, however, is higher after laparoscopic than after open cholecystectomy; small asymptomatic leaks are very common and are reported to occur in up to 31% of patients. Larger leaks requiring intervention are reported in approximately 0.3%–4% of patients after the laparoscopic procedure, but in only 0.07%–0.2% of patients after open cholecystectomy (Andren-Sandberg et al. 1991; Cuschieri et al. 1991; Hawasli and Lloyd 1991; Rayter et al. 1989). Analysis of the experience at Mayo Clinic revealed no bile duct injuries or bile leaks in the first 200 laparoscopic cholecystectomies, while Meyers et al. (1991) reported a 0.5% prevalence of bile duct injuries in their series of over 1500 laparoscopic cholecystectomies (Donohue et al. 1992; Meyers et al. 1991). The complications of bile leak (bile peritonitis, abscess, sepsis) can be life-threatening, so early diagnosis and management are critical.

The cause of postoperative bile leak include damage to small bile ducts during gallbladder dissection, injury to the hepatic or common bile duct, and a slipped or dislodged clip or suture. Less common, nonoperative causes include blunt or penetrating abdominal trauma and gallbladder perforation with or without fistula formation as a result of acute inflammation.

The clinical signs and symptoms of bile leak (abdominal pain, fever, hyperbilirubinemia, elevated alkaline phosphatase, leukocytosis) are nonspecific. While ultrasound and computed tomography can document the presence of an abdominal fluid collection, these modalities cannot differentiate a bile leak from a postoperative seroma, hematoma, or lymphocele (Walker et al. 1992).

Hepatobiliary scintigraphy is a simple, noninvasive procedure that can be used to detect and localize a suspected bile leak. It can demonstrate the contiguity or lack of it of the biliary tree with the fluid collection seen on computed tomography or ultrasonography. It also helps to determine the need for an intervention: if bile flows preferentially into the small bowel, the leak will probably resolve spontaneously, whereas if the preferential pathway of flow is into the peritoneal cavity, the leak requires intervention (Gentili et al. 1993). Treatment options include surgery, percutaneous transhepatic biliary drainage, and endoscopic sphincterotomy with stent placement.

A bile leak is manifested scintigraphically as a collection of activity outside the liver, hepatic or common bile duct, or small bowel (Fig. 13.10). Various patterns of bile leakage have been described:

- Extravasation in the perihepatic, perisplenic, or subdiaphragmatic space (Siddiqui et al. 1986)
- Bile ascites (Nagle et al. 1985; Siddiqui et al. 1986)
- Activity in the gallbladder fossa (Shih et al. 1993) (this may give rise to a false negative study by simulating a normal gallbladder)
- Activity in the right or left paracolic gutter or the pelvis
- Activity around the superior or posterior edge of the liver or in the porta hepatis

In a patient suspected of having a bile leak it is important to obtain delayed images if the leak is not

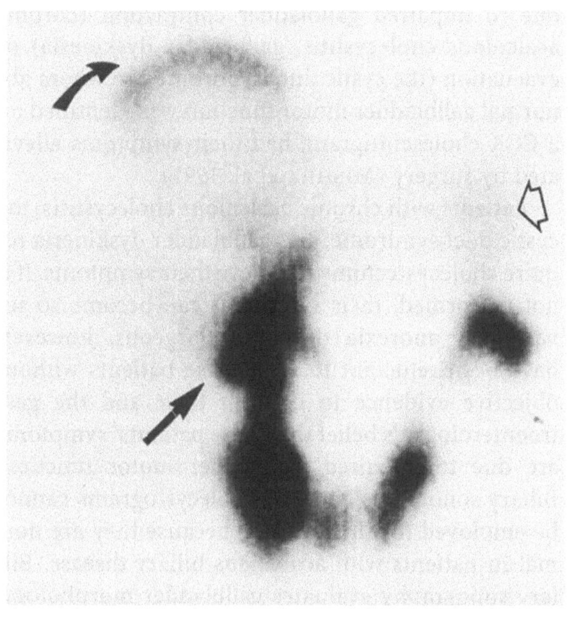

Fig. 13.10. Bile leak. The 2-h anterior hepatobiliary scan of a patient who had a laparoscopic cholecystectomy reveals radiotracer emanating from a partially ligated cystic duct into the gallbladder fossa (*arrow*) and then up and over the dome of the liver. Enterogastric reflux is also present (*open arrow*)

demonstrated within the 1st h of imaging. A small or slow leak may not be visualized for several hours, or it may be obscured by activity in normal structures and thus only become apparent after the activity in them clears. It is also helpful to use continuous digital acquisition when obtaining these studies and then to view the images in a dynamic (cinematic) display mode to detect small leaks and to distinguish them from other structures (BALON and NAGLE 1992). It is also important to know the true anatomic location of the gallbladder and common bile duct from correlation with other modalities (computed tomography, ultrasonography), to help distinguish, for example, a loculated leak from the gallbladder.

Bile ascites may be difficult to recognize due to its relatively faint intensity; images of the lower abdomen and pelvis should be obtained whenever a leak is suspected. Bile ascites will manifest itself as diffuse mild activity throughout the abdomen, interspersed with irregular photon-deficient areas representing bowel. To further confirm the presence of bile ascites, paracentesis may be performed or fluid from drainage catheters aspirated and imaged or counted in a well counter (CARICHNER and NAGLE 1987; NAGLE et al. 1985).

13.3 Chronic Acalculous Biliary Disease

CCK cholescintigraphy can be employed for objective confirmation of the clinical diagnosis of chronic acalculous biliary disease.

The chronic acalculous disorders of the biliary tree are chronic acalculous cholecystitis, the cystic duct syndrome, biliary dyskinesia, and sphincter of Oddi dyskinesia.

Chronic acalculous cholecystitis is a disorder that is histologically characterized by hypertrophy of the gallbladder wall (>1.5–2.0 mm), the presence of Rokitansky-Aschoff sinuses, foamy macrophages filling the tips of mucosal folds, hypertrophy of the muscularis propriae with or without a monocellular infiltrate, and/or the presence of yellow papillary nodules (cholesterolosis) (ANDERSON 1971; HARRISON et al. 1966; CECIL 1979).

Clinically, chronic acalculous cholecystitis is manifested by recurrent postprandial right upper quadrant pain and biliary colic. These symptoms result from the inability of the chronically inflamed gallbladder wall to respond appropriately to endogenous CCK (ANDERSON 1971; HARRISON et al. 1966; CECIL 1979).

Chronic acalculous cholecystitis is reported to occur more frequently in men than in women. It is precipitated by conditions that result in gallbladder atony and biliary stasis. Thus, patients who are critically ill, burn victims, and postoperative patients in whom prolonged hyperalimentation is required are prime candidates for developing a chronic inflammation of their atonic, non-CCK-stimulated gallbladder wall (ANDERSON 1971; HARRISON et al. 1966; CECIL 1979).

The cystic duct syndrome is another of the acalculous disorders of the hepatobiliary tree. It too is clinically manifested by recurrent postprandial right upper quadrant pain and biliary colic, not as a result of a chronically inflamed gallbladder wall, but from a partially obstructed cystic duct. The cystic duct is narrowed either by inflammatory changes within its wall, by adhesions constricting it, or by a kink in it (FINK-BENNETT et al. 1985b; COZZOLINO et al. 1963).

The cystic duct syndrome occurs in patients with normal gallbladder function as well as in patients with impaired gallbladder contraction. European and South American synonyms for cystic duct syndrome are infundibular cervicocystic dyskinesia, mechanical dyskinesia, cystic cholecystic syndrome, and gallbladder siphopathy (FINK-BENNETT et al.

1985b; Cozzolino et al. 1963; McFarland and Currin 1965; Camishion and Goldstein 1967).

Gallbladder dyskinesia is a condition that occurs when CCK receptor sites or neurotransmitter-mediated contractor receptor cells located within the gallbladder wall are abnormally or nonuniformly distributed throughout it. Recurrent postprandial right upper quadrant pain and biliary colic occur as these CCK receptors and/or neurotransmitter-mediated contractor receptor sites respond abnormally to endogenous CCK. Instead of causing a uniform contraction of the gallbladder, a disorganized one occurs. Increased intraluminal gallbladder pressure results, followed by the characteristic symptoms of gallbladder disease, i.e., post-prandial recurrent right upper quadrant pain, flatulence, epigastric fullness, fatty food intolerance, heartburn, and abdominal fullness (Fink-Bennett 1991; Fink-Bennett et al. 1991c; Bolen and Javitt 1982; Harvey and Oliver 1980).

Sphincter of Oddi dyskinesia is caused by a paradoxical response to endogenous CCK or by spontaneous spasms of the sphincter of Oddi. Clinically, the symptoms created by increased pressure with the common bile duct following an inappropriate CCK-mediated or spontaneous sphincter of Oddi contraction are similar to those that occur from impaired gallbladder contraction (chronic acalculous cholecystitis, gallbladder dyskinesia) or evacuation (the cystic duct syndrome). The diagnosis of sphincter of Oddi dyskinesia, however, is much more difficult to make. It usually requires CCK-augmented ERCP manometric determinations (Lechin et al. 1978; Hogan et al. 1982; Sostre et al. 1992). Sphincter of Oddi dyskinesia is the commonest cause of the postcholecystectomy syndrome (Sostre et al. 1992).

CCK cholescintigraphy is a very accurate, noninvasive test that can be used to objectively confirm the clinical diagnosis of symptomatic chronic acalculous cholecystitis, the cystic duct syndrome, and gallbladder dyskinesia. Its sensitivity and specificity in identifying patients with impaired gallbladder contraction or evacuation range from 89% to 100% and from 80% to 100%, respectively (Newman et al. 1983; Pickelman et al. 1985; Fink-Bennett et al. 1991c; Halverson et al. 1992; Yap et al. 1991; Zech et al. 1991; Brugge et al. 1986; Fink-Bennett et al. 1985b). In addition, CCK cholescintigraphy can predict with a similar degree of accuracy which patient with abnormal gallbladder motility will benefit from cholecystectomy. Ninety-five percent of patients whose symptoms were clinically believed to be due to impaired gallbladder contraction (chronic acalculous cholecystitis, gallbladder dyskinesia) or evacuation (the cystic duct syndrome), in whom abnormal gallbladder motor function was identified on a CCK cholescintigram, had their symptoms alleviated by surgery (Mishra et al. 1991).

Patients with chronic acalculous cholecystitis, the cystic duct syndrome, and gallbladder dyskinesia require cholecystectomy to relieve their symptoms. If it not performed, their symptoms can become so severe that anorexia develops. Surgeons, however, have been reluctant to treat these patients without objective evidence to confirm their and the gastroenterologist's belief that these patients' symptoms are due to impaired gallbladder motor function. Biliary sonograms and oral cholecystograms cannot be employed for this purpose because they are normal in patients with acalculous biliary disease. Biliary sonography evaluates gallbladder morphology, oral cholecystography the ability of the gallbladder to concentrate bile – neither of which is abnormal in patients with chronic acalculous biliary disease. CCK cholescintigraphy, however, can confirm the clinical diagnosis of acalculous biliary disease because it evaluates the gallbladder's ability to contract and eject bile, the impairment of which is the underlying etiology of chronic acalculous cholecystitis, gallbladder dyskinesia, and the cystic duct syndrome.

It must be remembered that CCK cholescintigraphy is a confirmatory test to be employed only in patients who have been thoroughly screened to make sure that their symptoms are not a manifestation of another disorder with symptoms that mimic gallbladder disease, such as Crohn's disease, irritable bowel syndrome, bile gastritis, colitis, lactose intolerance, ulcer disease, etc. If, and only if, it is performed in this selected population, the efficacy and prognostic value of CCK cholescintigraphy will be maintained. Removing a gallbladder that is chronically inflamed or functionally impaired in a patient whose symptoms are due to a disease other than intrinsic gallbladder disease will serve no purpose. Remember, acalculous biliary disease, like the calculous form, can be asymptomatic.

The hypothesis upon which CCK cholescintigraphy is based is that individuals with a partially obstructed, chronically inflamed, or functionally impaired gallbladder with respond differently to CCK than individuals with normal gallbladder contraction and evacuation, i.e., will have a reduced maximal gallbladder ejection fraction response to CCK of less than 35% (Krishnamurthy et al. 1981, 1993; Bobba et al. 1984).

Endogenous and exogenous CCK have identical effects on the gastrointestinal and biliary system. Both cause the gallbladder to contract, the sphincter of Oddi to relax, enhance pyloric sphincter tone, enhance large and small bowel motility, increase bile production, and increase the secretion of pancreatic enzymes and enterokinase (HARVEY and OLIVER 1980; MORLEY 1982).

Endogenous CCK is a 33-amino-acid polypeptide chain that is produced by the duodenal mucosa in response to fats, amino acids, and small polypeptides. The portion of CCK that causes the gallbladder to contract, the sphincter of Oddi to relax, that enhances pyloric tone, enhances small bowel motility, and increases the secretion of pancreatic enzymes and enterokinase resides in the last eight amino acids of the 33-amino-acid chain. It is these last eight amino acids that comprise the "cholecystokinetic" portion of CCK.

CCK was discovered by IVY and OLDEBERT in 1928. It was isolated by JOBES and MUTT from the porcine gastrointestinal tract in 1961. CCK can be purchased in its entirety as a 33-amino-acid polypeptide chain from the Karolinska Institute, Stockholm, Sweden, or from Boots Ltd. in Nottingham, England, under the respective appellations of cholecystokinin and pancreozymin. The cholecystokinetic portion of CCK, sincalide (Kinevac), can be purchased from E.R. Squibb & Son, Princeton, New Jersey. It is this agent that is routinely used in conjunction with a hepatobiliary scan to perform a CCK cholescintigram in the United States of America.

CCK cholescintigramas are performed after an overnight fast and following the intravenous administration of either 5 mCi technetium-99m disofenin (Hepatolite) or 5 mCi technetium-99m mebrofenin (Choletec).

Anterior 500 000–1 000 000-count hepatobiliary images are obtained at 10-min intervals for 1 h or until the gallbladder is maximally filled. Maximal gallbladder filling is deemed present when little to no activity is present within the major hepatic radicals, most within the gallbladder.

The hepatobiliary images are obtained utilizing a large-field-of-view gamma camera and a low-energy all-purpose collimator with a 20% window centered at the 140-keV technetium-99m photopeak.

When the gallbladder is maximally filled and bowel activity seen, 0.01–0.02 μg/kg sincalide is administered. Sincalide is administered intravenously for a 3-min period. CCK can be administered manually by dilution in 10 ml normal saline or can be given via an infusion pump. However, and most importantly, it must never be administered as a bolus injection. A bolus could cause spasm of the neck of the gallbladder, resulting in a spuriously reduced gallbladder ejection fraction.

Following completion of the 3-min CCK infusion, anterior post-CCK analogue hepatobiliary images are obtained every 5 min for 20 minutes. Post-CCK analogue images are obtained for equal time intervals, not counts, and the time for each image is determined by the number of seconds required to obtain a pre-CCK anterior biliary scintiscan. Gallbladder ejection fractions are determined from data simultaneously acquired on a computer at one frame per minute for 20 min. The data are stored on a 64 × 64 × 16 computer matrix. Acquisition is begun 1 min before and continues for 20 min after the 3-min CCK infusion. Ejection fractions are determined manually by assigning areas of interest around the gallbladder and an adjacent background area on the pre-CCK and the 5-, 10-, 15-, and 20-min post-CCK digital images. The background area (region of interest, ROI) is selected adjacent and to the right of the gallbladder with a width measuring approximately 3–4 pixels and a height equal to the gallbladder's length. Background activity is subtracted from both pre- and post-CCK images. The total counts and the number of pixels in each ROI are determined and the gallbladder ejection fraction calculated according to the following formula:

$$\mathrm{GBEF_i} = \frac{(\text{net pre-CCK GB counts}) - (\text{net post-CCK GB counts}_i)}{\text{Net pre-CCK GB counts}}$$

where GBEF is the gallbladder ejection fraction, i is the time after-CCK administration and net GB counts = total GB counts – (background counts/pixel × number of GB pixels).

Gallbladder motor function can also be assessed by evaluating the gallbladder ejection fraction response to a 45-min continuous infusion of CCK. If this method is employed, a gallbladder ejection fraction above 40% is deemed normal (YAP et al. 1991). Three-minute and 45-min infusion cholescintigraphy are similar in efficacy (FINK-BENNETT et al. 1991c; YAP et al. 1991; BRUGGE et al. 1986). Both accurately identify patients with abnormal gallbladder motility as the result of a chronically inflamed, partially obstructed, or dyskinetic gallbladder (Figs. 13.11, 13.12).

Whether one performs a 3- or a 45-min infusion cholescintigram, normal biliary-to-bowel transit must be present. If it is not, a potential false positive

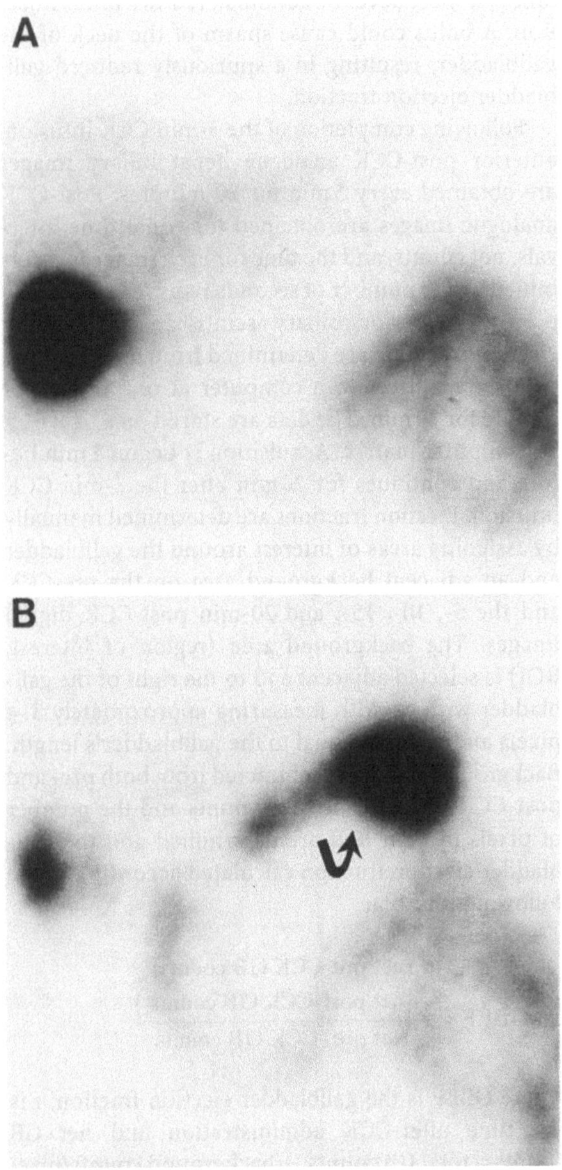

Fig. 13.11a,b. Normal gallbladder ejection fraction in a patient with bile gastritis. Pre-CCK 60-min anterior hepatobiliary scan demonstrates the gallbladder to be maximally filled. Little radiotracer resides within the liver and biliary radicals. Radiotracer is seen in the small bowel. **b** Twenty minutes after CCK administration, anterior hepatobiliary scan demonstrates a normal gallbladder ejection fraction in response. Enterogastric reflux is seen after CCK administration (*arrow*) in this patient with bile gastritis

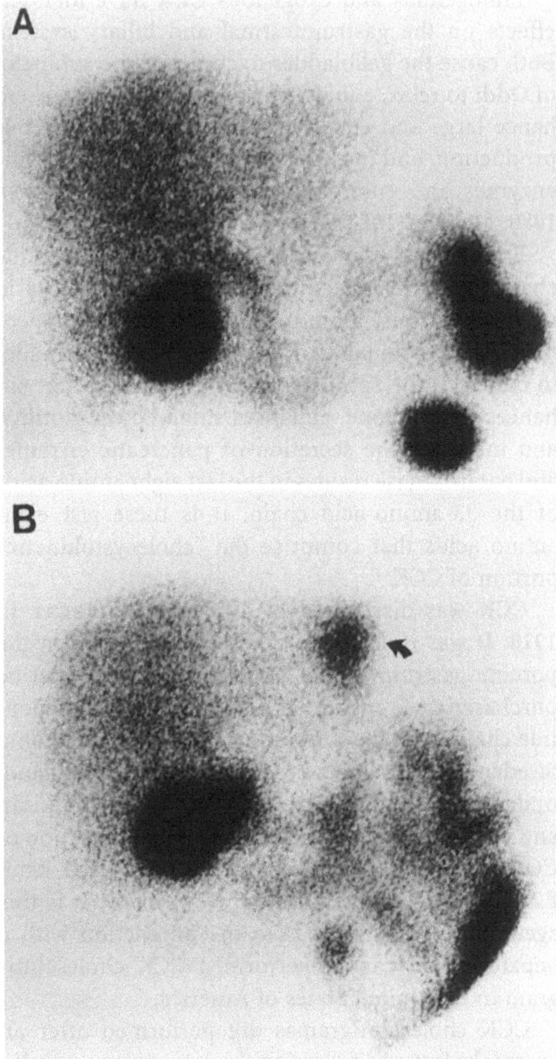

Fig. 13.12a,b. Chronic acalculous cholecystitis. **a** Pre-CCK anterior hepatobiliary scan reveals a maximally filled gallbladder and normal biliary-to-bowel transit. **b** Twenty minutes after CCK administration, hepatobiliary scan reveals reduced gallbladder motor function as manifested by a maximal gallbladder ejection fraction response of less than 35%. The calculated gallbladder ejection fraction was 20%. Enterogastric refulx is also demonstrated (*arrow*)

interpretation may occur, since the etiology of a reduced gallbladder ejection fraction response to CCK in a patient with delayed biliary-to-bowel transit is either due to reduced gallbladder motor function or to the normal gallbladder's inability to contract against increased pressure at the sphincter of Oddi.

Sphincter of Oddi dyskinesia demonstrates itself scintigraphically by a paradoxical response of the sphincter of Oddi to CCK. Scintigraphically, this is manifested by dilatation of the common bile duct after CCK administration, i.e., the dilated common bile duct sign (Fig. 13.13). Delayed biliary-to-bowel transit is also present if the basal pressure of the sphincter of Oddi is elevated (DeRidder and Fink-Bennett 1984).

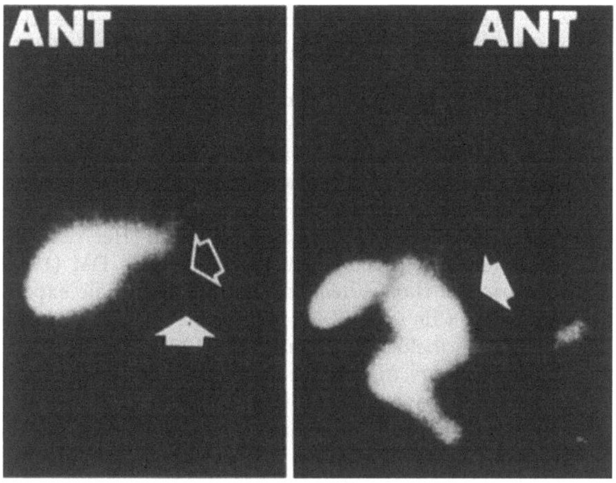

a,b

Fig. 13.13a,b. Sphincter of Oddi dyskinesia. **a** Seventy-five-minute anterior hepatobiliary scan reveals normal gallbladder visualization and a delay in biliary-to-bowel transit. The common bile duct is barely visible (*open arrow*) up to the level of the sphincter of Oddi (*closed arrow*). **b** The anterior hepatobiliary scan 5 min after CCK administration demonstrates a normal gallbladder ejection fraction response to CCK but a paradoxical response of the sphincter of Oddi (*arrow*): the dilated common bile duct sign

13.4 Postcholecystectomy Syndrome

The postcholecystectomy syndrome is defined as recurrent or persistent biliary colic following cholecystectomy. It occurs in approximately 45 000 patients per year who have undergone cholecystectomy in 6200 of whom an abnormal sphincter of Oddi is the cause (STEINBERG 1988; BAR-MEIR et al. 1984). The sphincter is either partially occluded by a stone, narrowed from recurrent episodes of ampullitis, or functionally impaired as the result of spontaneous sphincter of Oddi spasms or a dyskinetic sphincter.

Until now, only an ERCP could determine whether an abnormality of the sphincter of Oddi was the underlying etiology of a case of postcholecystectomy syndrome, and a mean of 7.2 ERCPs had to be performed to identify one impaired sphincter (SOSTRE et al. 1992). Now it appears that patients with a potentially impaired sphincter of Oddi can have it evaluated noninvasively by generating a technetium-99m IDA "scintigraphic score." To do this, patients are pretreated with CCK 15 min prior to the intravenous administration of 5 mCi technetium-99m IDA. Anterior hepatobiliary scans are then obtained over a 60-min period. Simultaneously, a dynamic study is acquired at a rate of one frame per minute and stored on a computer. ROIs are placed over the liver and common bile duct, from which a time activity curve is created. From this, the percentage of common bile duct emptying is computed and the time to peak hepatic uptake determined. The percentage of common bile duct (CBD) emptying is calculated from the following formula:

$$\begin{matrix}\%\ \text{CBD} \\ \text{emptying}\end{matrix} = \frac{(\text{peak CBD count}) - (\text{CBD counts at 60 min})}{\text{peak CBD counts}} \times 100.$$

These values, along with qualitative assessments obtained from the 60-min analogue hepatobiliary images, permit the recognition of patients whose postcholecystectomy pain is due to sphincter of Oddi dysfunction or disease. The qualitative assessments determined from the analogue hepatobiliary images include: the time at which the intrahepatic biliary tree is first visualized, the presence or absence of a prominent intrahepatic biliary tree, the time at which bowel activity is first identified, and the creation of a common-bile-duct-to-liver ratio. The common-bile-duct-to-liver ratio is obtained by visually comparing the intensity of uptake of technetium-99m IDA within the common bile duct at 15 and 60 min to the intensity of uptake within the liver parenchyma. A value is then assigned to each qualitative and quantitative determinant (Table 13.1). From this, a

Table 13.1. Criteria for scoring scintigrams

	Score
1. Peak time	
Less than 10 min	0
10 min or more	1
2. Time to biliary visualization	
Less than 15 min	0
15 min or more	
3. Prominence of biliary tree	
Not prominent	0
Prominent major intrahepatic ducts	1
Prominent small intrahepatic ducts	2
4. Time to bowel visualization	
Less than 15 min	0
15–30 min	1
More than 30 min	2
5. CBD emptying	
More than 50%	0
Less than 50%	1
No change	2
Shows increasing activity	3
6. CBD-to-liver ratio	
$CBD_{60} \leq liver_{60}$	0
CBD_{60} higher than $liver_{60}$ but lower than $liver_{15}$	1
CBD_{60} higher than $liver_{60}$ and equal to $liver_{15}$	2
CBD_{60} higher than both $liver_{60}$ and $liver_{15}$	3

CBD, common bile duct.

scintigraphic score is calculated. Scores of 0–4 are indicative of normal sphincter of Oddi function. Scores between 5 and 12 reflect an impaired sphincter and thus the need for an ERCP to determine the underlying etiology of its abnormality, i.e., whether it is a functional or a mechanical one.

In a pilot study conducted on 26 symptomatic and 14 asymptomatic postcholecystectomy patients, the sensitivity and specificity of this non-invasive test for identifying a sphincter of Oddi abnormality was 100% (Sostre et al. 1992).

References

Achong DM, Oates E (1994) A reversed sequence of gallbladder and small bowel visualization during cholescintigraphy: its relationship to chronic cholecystitis. Clin Nucl Med 19:89–92

Anderson WHD (1971) Pathology. Mosby, St Louis, pp 1261–1265

Andren-Sandberg A, Johansson S, Bengmark S (1991) Accidental lesion of the common bile duct at cholecystectomy. Ann Surg 201:452–455

Anonymous (1993) Gallstones and laparoscopic cholecystectomy. JAMA 29:1018–1024

Balon HR, Nagle CE (1992) Computer-assisted imaging and display for biliary leak detection. Clin Nucl Med 17:336–337

Bar-Meir S, Halpern Z, Bardon E, et al (1984) Frequency of papillary stenosis among cholecystectomized patients. Hepatology 4:328–330

Bobba VR, Krishnamurthy GT, Kingson E, et al (1984) Gallbladder dynamics induced by a fatty meal in normal subjects and patients with gallstones (concise communication). J Nucl Med 25:21–24

Bolen G, Javitt NB (1982) Biliary dyskinesia: mechanisms and management. Hosp Pract 17:115–130

Brugge W, Brand D, Atkins H, Lane B, Abel W (1986) Gallbladder dyskinesia in chronic acalculous cholecystitis. Dig Dis Sci 31:461–467

Bushnell DL, Perlman SC, Wilson MA, Polycr RD (1986) The rim sign: association with acute cholecystitis. J Nucl Med 27:353–356

Camishion RD, Goldstein F (1967) Partial, noncalculous cystic duct obstruction (cystic duct syndrome). Surg Clin North Am 47:1107–1114

Carichner SL, Nagle CE (1987) Hepatobiliary scintigraphy in patients with bile leaks. J Nucl Med Technol 15:182–186

Cecil B (1979) Textbook of medicine. Saunders, Philadelphia, pp 1624–1628

Choy D, Shi EC, McLean RG, Hoschl R, Murry IPC, Ham JM (1984) Cholescintigraphy in acute cholecystitis: use of intravenous morphine. Radiology 151:203–207

Coleman RE, Freitas JE, Fink-Bennett D, Bree RL (1984) The dilated cystic duct sign - a potential cause of false negative cholescintigraphy. Clin Nucl Med 9:134–136

Cozzolino HJ, Goldstein F, Greening RR, Wirts CW (1963) The cystic duct syndrome. JAMA 185:100–104

Cuschieri A, Dubois F, Mouiel J, Mouret P, Becker H, Buess G, Trede M, Troidl H (1991) The European experience with laparoscopic cholecystectomy. Am J Surg 161:385–387

Datz FL (1986) Utility of indium-111-labeled leukocyte imaging in acute acalculous cholecystitis. AJR 147:813–814

Datz F (1988) Handbook in radiology, nuclear medicine. Mosby Yearbook, St Louis, pp 154–155

DeRidder P, Fink-Bennett D (1984) The dilated common duct sign: a potential indicator of a sphincter of oddi dyskinesia. Clin Nucl Med 9:262–263

Donohue JH, Farnell MB, Grant CS, Van Heerden JA, Wahlstrom HE, Sarr MG, Weaver AL, Ilstup DM (1992) Laparoscopic cholecystectomy; early Mayo Clinic experience. Mayo Clin Proc 67:449–455

Dubois F, Berthelot G, Levard H (1991) Laparoscopic cholecystectomy: historic perspective and personal experience. Surg Laparosc Endosc 1:52–57

Eikman EA, Cameron JL, Coleman M, Natarajan TK, Dugal P, Wagner HN Jr (1975) A test for patency of cystic duct in acute cholecystitis. Ann Intern Med 82:318–322

Fink-Bennett D (1985) The role of cholecystogogues in the evaluation of biliary tract disorders. In: Freeman LM, Weissman HS (eds) Nuclear medicine annual. Raven, New York

Fink-Bennett D (1991) Augmented cholescintigraphy: its role in detecting acute and chronic disorders of the hepatobiliary tree. Semin Nucl Med 21:128–139

Fink-Bennett D, Balon H (1993) The role of morphine-augmented cholescintigraphy in the detection of acute cholecystitis. Clin Nucl Med 18:891–897

Fink-Bennett D, Freitas JE, Ripley SD, Bree RL (1985a) The sensitivity of hepatobiliary imaging and real-time ultrasonogrphy in the detection of acute cholecystitis. Arch Surg 120:904–906

Fink-Bennett D, DeRidder P, Kolozsi W, Gordon RM, Rapp J (1985b) Cholecystokinin cholescintigraphic findings in the cystic duct syndrome. J Nucl Med 26:1123–1128

Fink-Bennett D, Clark K, Tsai D, Nuechterlein P (1991a) Indium-111 WBC imaging in acute cholecystitis. J Nucl Med 32:803–804

Fink-Bennett D, Balon H, Robbins T, Tsai D (1991b) Morphine augmented cholescintigraphy: its efficacy in detecting acute cholecystitis. J Nucl Med 32:1231–1233

Fink-Bennett D, DeRidder P, Kolozsi W, Gordon R, Jaros R (1991c) Cholecystokinin cholescintigraphy: detection of abnormal gallbladder motor function in patients with chronic acalculous gallbladder disease. J Nucl Med 32: 1695–1699

Freeman L, Sugarman L, Weissman H (1981) Role of cholecystokinetic agents in 99m-Tc-IDA cholescintigraphy. Semin Nucl Med 11:186–193

Freitas JE (1982) Cholescintigraphy in acute and chronic cholecystitis. Semin Nucl Med 12:18–26

Freitas JE, Rajinder M, Gulati MD (1980) Rapid evaluation of acute abdominal pain by hepatobiliary scanning. JAMA 244:1585–1587

Freitas JE, Coleman RE, Nagle CE, Bree RL, Krewer KD (1983) Influence of scan pathologic criteria on the specificity of cholescintigraphy (concise communication). J Nucl Med 24:876–879

Gentili A, Gilkeson RC, Adler LP (1993) Scintigraphic detection of bile leaks after laparoscopic cholecystectomy. Clin Nucl Med 18:1–6

Halverson JD, Garner BA, Siegel BA, Alexander R, Edundowicz SA, Sampbell W, Miller JE (1992) The use of hepatobiliary scintigraphy in patients with acalculous biliary colic. Arch Intern Med 152:1305–1307

Hardy JD (1988) Hardy's textbook of surgery, 2nd edn. Lippincott, Philadelphia, pp 677–690

Harrison TR, Adams RD, Bennett IL, et al (1966) Principles of internal medicine. McGraw-Hill, New York, pp 1088–1093
Harvey RF, Oliver JM (1980) Cholecystokinins and the gallbladder. Gastroenterology 78:1117–1119
Hawasli A, Lloyd LR (1991) Laparoscopic cholecystectomy: the learning curve: report of 50 patients. Am J Surg 57:542–545
Hogan W, Green J, Dodds I, et al (1982) Paradoxical response to cholecystokinin (CCK-OP) in patients with suspected sphincter-of-Oddi dysfunction (abstract). Gastroenterology 82:1085
Ivy AC, Oldbert E (1928) A hormone mechanism for gallbladder contraction and evacuation. Am J Physiol 86:599–613
Jopes JE, Mutt V (1961) The gastrointestinal hormones, secretin and cholecystokinin-pancreomyzin. Ann Intern Med 55:395–405
Kane RA, Costello P, Duszlak E (1983) Computed tomography in acute cholecystitis: new observations. AJR 141:697–701
Kaplun L, Weissman HS, Rosenblatt RR, Freeman LM (1985) The early diagnosis of common bile duct obstruction using cholescintigraphy. JAMA 254:2431–2434
Keller IA, Weissman HS, Kaplan LL, et al (1984) The use of water ingestion to distinguish the gallbladder and duodenum on cholescintigrams. Radiology 152:151
Keslar PJ, Turbiner EH (1987) Hepatobiliary imaging and the use of intravenous morphine. Clin Nucl Med 12:592–596
Kim EE, Pjura G, Lowry P, Nguyen M, Pollack M (1986) Morphine-augmented cholescintigraphy in the diagnosis of acute cholecystitis. AJR 147:1177–1179
Kim CK, et al (1990) Delayed biliary-to-bowel transit in cholescintigraphy after cholecystokinin treatment. Radiology 176:533–556
Krishnamurthy S, Krishnamurthy K (1988) Quantitative assessment of hepatobiliary diseases with Tc99m-IDA scintigraphy. In: Freeman LM, Weissman HS (eds) Nuclear Medicine Annual. Raven, New York, pp 309–313
Krishnamurthy GT, Bobba VR, Kingston E (1981) Radionuclide ejection fraction: a technique for quantitative analysis of motor function of the human gallbladder. Gastroenterology 80:482–490
Krishnamurthy GT, Bobba VR, McConnell D, et al (1983) Quantitative biliary dynamics: introduction of a new noninvasive scintigraphic technique. J Nucl Med 24:217–3
Lechin F, Van Der Dijs B, Bentolila A, Pena F (1978) Adrenergic influences on the gallbladder emptying. Am J Gastroenterol 69:662–668
Lette J, Morin M, Heyen F, Paquet A, Levasseur A (1990) Standing views to differentiate gallbladder or bile leak from duodenal activity on cholescintigrams. Clin Nucl Med 15:231–236
McFarland JO, Currin J (1965) Cholecystokinin and the cystic duct syndrome. Clinical experience in a community hospital. Am J Gastroenterol 515–522
Meyers WC, Branum GD, Farduk M, et al (1991) A prospective analysis of 1518 laparoscopic cholecystectomies. N Engl J Med 324:1073–1078
Miller DR, Egbert RM, Braunstein P (1984) Comparison of ultrasound and hepatobiliary imaging in the early detection of acute total common bile duct obstruction. Arch Surg 119:1233–1237
Mishra DC, Blooson GB, Fink-Bennett DF, Glover JJ (1991) Results of surgical therapy for biliary dyskinesia. Arch Surg 26:957–959
Morley JE (1982) The ascent of cholecystokinin (CCK) - from gut to brain. Life Sci 30:479–493
Moss AA, Gamsu G, Genant HK (1983) Computed tomography of the body. Saunders, Philadelphia, pp 14–18, 691–697
Nagle CE, Fink-Bennett D, Freitas JE (1985) Bile ascites in adults: diagnosis using hepatobiliary scintigraphy. Clin Nucl Med 10:403–405
Newman P, Browne MK, Mowat W (1983) A simple technique for quantitative cholecystokinin-HIDA scanning. Br J Radiol 56:500–502
Pickelman J, Peiss R, Henkin R, et al (1985) The role of sincalide cholescintigraphy in the evaluation of patients with acalculous gallbladder disease. Arch Surg 120:693–697
Rayter Z, Tonge C, Bennett CE, Robinson PS, Thomas MH (1989) Bile leaks after simple cholecystectomy. Br J Surg 76:1046–1048
Rhoads JE, Allen JG, Harkins HN, Moyer CA (1970) Surgery: principles and practice. Saunders, Philadelphia
Samuels BI, Freitas JE, Bree RL, Schwab RE, Heller ST (1983) A comparison of radionuclide hepatobiliary imaging and real-time ultrasound for the detection of acute cholecystitis. Radiology 147:207–210
Shaffer PB, Olsen JO (1982) Differentiation of the gallbladder from the duodenum on cholescintigrams by dynamic display. Radiology 145:217
Shih WJ, Magoun S, Mills BJ, Pulmano C (1993) Bile leak from gallbladder perforation mimicking bowel activity and a false-negative result in a morphine-augmented cholescintigraphy. J Nucl Med 33:131–133
Siddiqui AR, Ellis JH, Madura JA (1986) Different patterns for bile leakage following cholecystectomy demonstrated by hepatobiliary imaging. Clin Nucl Med 11:751–753
Sostre S, Kalloo A, Speigler E, Camaigo E, Wagner H (1992) A noninvasive test of sphincter of Oddi dysfunction in post cholecystectomy patients: the scintigraphic score. J Nucl Med 33:1216–1222
Steinberg WM (1988) Sphincter of Oddi dysfunction: a clinical controversy. Gastroenteroloy 95:1409–1415
Vasques TE, Greenspan G, Evans DG, Halpern SE, Ashburn WL (1988) Clinical efficacy of intravenous morphine administration in hepatobiliary imaging for acute cholecystitis. Clin Nucl Med 13:4–6
Walker AT, Shapiro AW, Brooks DC, Braver JM, Tumeh SS (1992) Bile duct disruption and biloma after laparoscopic cholecystectomy: imaging evaluation. AJR 158:785–789
Weissman H (1981) The clinical role of technetium-99m iminodiacetic acid cholescintigraphy. In: Freeman LM, Weissman HS (eds) Nuclear medicine annual. Raven, New York, pp 35–90
Yap L, Wycherley AG, Morphett AD, Toouli J (1991) Acalculous biliary pain: cholecystectomy alleviates symptoms in patients with abnormal cholescintigraphy. Gastroenterology 101:786–793
Zech ER, Simmons LB, Kendrick RR, Soballe PW, Olcese JAM, Goff WB, II, Lawrence DP, DeWeese RA (1991) Cholecystokinin enhanced hepatobiliary scanning with ejection fraction calculation as an indicator of disease of the gallbladder. Surg Gynecol Obstet 172:21–22
Zeissman H, Fahey F, Hilson, D (1992) Calculation of a gallbladder ejection fraction: advantage of continuous sincalide infusion over the three-minute infusion method. J Nucl Med 33:537–541

14 Diagnostic Imaging of the Gallbladder and Hepatobiliary System in Children

D.C.B. Redd and K.E. Fellows

CONTENTS

14.1 Diagnostic Imaging Techniques

Selection of the appropriate imaging techniques for evaluation of the gallbladder, biliary tract, and liver in the pediatric patient depends upon the specific clinical problem at hand. Many situations require two or more imaging methods to sufficiently narrow the diagnostic possibilities. To determine the optimal combination of studies, direct communication between the referring physician and radiologist is essential.

D.C.B. Redd, MD, Department of Radiology, 3rd Floor, Children's Hospital of Philadelphia, 34th Street and Civic Center Blvd., Philadelphia, PA 19104, USA

K.E. Fellows, MD, Department of Radiology, 3rd Floor, Children's Hospital of Philadelphia, 34th Street and Civic Center Blvd., Philadelphia, PA 19104, USA

14.1.1 Ultrasonography

Ultrasound is particularly useful to evaluate the biliary tree and gallbladder in the pediatric population. Ultrasonography is an ideal diagnostic modality in this age group as it does not utilize ionizing radiation, has relatively low cost, and the equipment is portable. It is ideally suited as a screen for hepatic disease and for the detection of ascites. Doppler ultrasonography is a useful, noninvasive technique for monitoring flow within the hepatic vasculature, including the mesenteric and portal veins. The patency of vascular shunts may also be monitored following surgery or percutaneous intervention (transjugular intrahepatic portosystemic shunting; Fig. 14.1).

Normal liver parenchyma has a uniform echotexture with major branches of the hepatic and portal venous systems clearly discernible. Ultrasound is highly accurate in distinguishing hepatocellular from obstructive causes of jaundice. The normal common bile duct is regularly visualized in children, measuring only 2–3 mm in diameter; intrahepatic bile ducts are not normally identified. The normal gallbladder should be visible in virtually every fasting patient; failure to demonstrate this structure has a high correlation with gallbladder disease. Ultrasonography is usually the first imaging study to be performed in patients with hepatic dysfunction following orthotopic liver transplantation (OLT); this examination is useful in detecting biliary obstruction, bile leak, or vascular compromise.

Diffuse hepatic parenchymal abnormalities such as chronic hepatitis, cirrhosis, or metabolic storage diseases may uniformly increase hepatic echotexture. Metastatic disease has a wide spectrum of ultrasonographic appearance, varying from highly echogenic to isoechoic. Hepatocellular carcinoma, hemangioendothelioma, hepatoblastoma, and hemangioma appear as localized regions of increased echogenicity and may be associated with hepatomegaly. Cystic lesions are readily identifiable, even if the cysts are small.

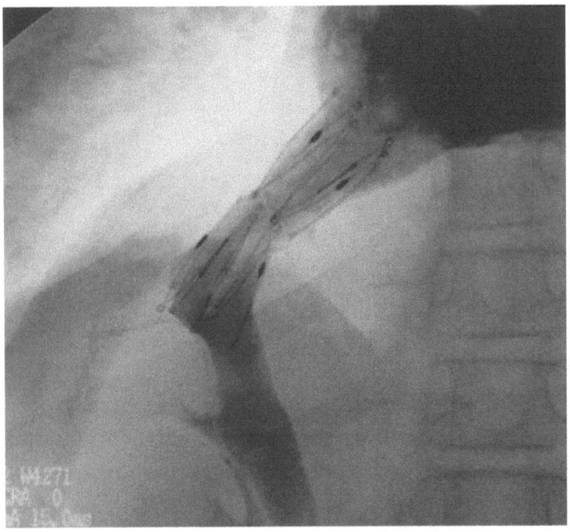

a

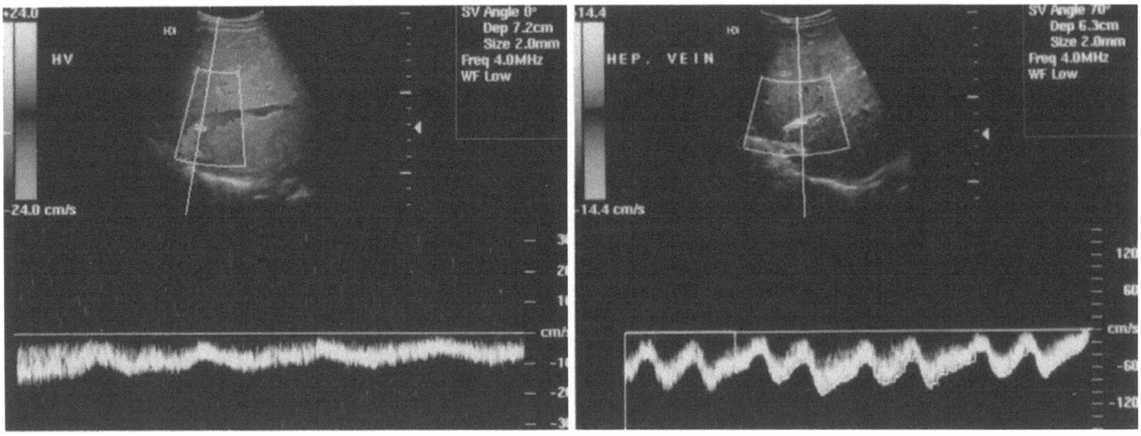

b

Fig. 14.1a,b. An articulated Gianturco-Rösch Z stent was deployed for treatment of refractory stenosis of the suprahepatic inferior vena cava in a child who had an orthotopic liver transplant. After stent placement there was restoration of antegrade flow within the inferior vena cava (**a**). **b** Prior to stent deployment Doppler ultrasound (*left*) demonstrated aphasic flow in the hepatic veins with a mean velocity of approximately 8 cm/s. After stent placement (*right*), there was restoration of phasic flow in the hepatic veins with mean flow velocity increased to approximately 50 cm/s

14.1.2 Computed Tomography

Today computed tomographic (CT) imaging is widely available and allows precise anatomic information related to tissue density. The radiation received during a CT examination is similar to that in other diagnostic procedures (e.g., barium enema). Spiral acquisition techniques with bolus contrast injection allow dynamic evaluation of pertinent vascular structures such as the aorta and inferior vena cava, as well as of mesenteric and portal venous structures. Disadvantages of CT include its expense, the potential requirement for sedation or anesthesia, and the total cumulative radiation dose, especially if multiple follow-up studies are required.

14.1.3 Magnetic Resonance Imaging

Sensitive tissue differentiation is achieved by Magnetic resonance imaging (MRI), based upon variations in proton density, T1 and T2 relaxation times, chemical shift, and blood flow. MRI offers the advantages of multiplanar imaging, superior contrast resolution, and lack of ionizing radiation. Hepatic mass lesions may be detected, including hepatic vascular lesions such as hemangioma, with greater contrast sensitivity than with CT. The MRI appearance of a hepatic mass may be sufficiently characteristic to diagnose malignancy. Blood flow can be assessed without the need for the injection of contrast material. Like CT, MRI cannot reliably distinguish primary from metastatic disease, and abscess may be difficult to distinguish from necrotic tumor.

MRI is useful in quantifying hepatic fat content and for monitoring deposition of iron, copper, and other metabolites. While the gallbladder and biliary tree may be satisfactorily imaged by MR cholangiography, there is no apparent advantage over other techniques as current imaging sequences depend on breath-hold techniques, a disadvantage in the pediatric population. Limitations of MRI include cost, long acquisition sequences which result in blurred images, and the requirement for sedation or anesthesia in certain patients.

14.1.4 Radionuclide Imaging

A variety of radioisotopes are used to study the structure and function of the hepatobiliary system. These examinations are safe, noninvasive, usually simple, and with few exceptions require no patient preparation. Radiopharmaceuticals are physiologically innocuous and do not produce toxic, allergic, or adverse pharmacologic effects, and do not result in hemodynamic or osmotic overload. Absorbed radiation dose varies with examination, age, weight, and physiologic condition, and is proportional to the administered dose. When kept to the minimum amount required for a technically acceptable examination, the absorbed dose may be equal to or significantly lower than that from an equivalent X-ray examination.

Radioisotopes may be selected to be taken up by hepatic parenchymal cells, Kupffer cells, or neoplastic or inflammatory cells. Intravenously injected N-substituted iminodiacetic acid (IDA) compounds (50 μCi/kg; minimum 1 mCi) are quickly extracted from the blood by hepatocytes, excreted unchanged through the biliary ducts, concentrated by the gallbladder, and excreted through the common bile duct into the intestinal tract. Most of the radiotracer is normally accumulated in the liver within 5 min following injection; the gallbladder is normally visualized within 10–15 min with radiotracer appearing in proximal small bowel within 30 min. Approximately 5%–15% of IDA compounds undergo renal excretion.

Biliary imaging studies may be useful in evaluating prolonged neonatal jaundice, suspected biliary tract anomaly, or acute cholecystitis. Dynamic imaging permits definitive diagnosis of vascular masses, such as hemangioendothelioma. Disadvantages of radionuclide imaging are low spatial resolution and poor anatomic detail. As a result, ultrasonography, CT, and MRI have replaced it in the evaluation of most hepatobiliary disorders.

14.1.5 Percutaneous Transhepatic Cholangiography

Percutaneous transhepatic cholangiography (PTC) offers precise visualization of the intra- and extrahepatic biliary trees via direct percutaneous injection of iodinated contrast through a small-caliber needle. Fine-needle techniques permit opacification of the intrahepatic biliary tree with low morbidity (Mueller et al. 1982). With proper technique, this procedure is safe, reliable, and diagnostically accurate in infants and children. When the intrahepatic bile ducts are dilated, success approaches 100%. When ductal dilatation is absent, success rates are considerably lower (50%–75%). In this setting (e.g., primary sclerosing cholangitis or biliary atresia), direct cholangiography may be performed by percutaneous puncture of the gallbladder (Fig. 14.2). As in adults, complications of PTC occur infrequently, but include hemorrhage, bile peritonitis, and sepsis. PTC also allows entry into the biliary tree for transhepatic interventional procedures.

PTC in children typically requires only conscious sedation, although general anesthesia may be necessary in selected patients. Initial duct puncture using a fine needle (25–21 G) may be optimized by use of

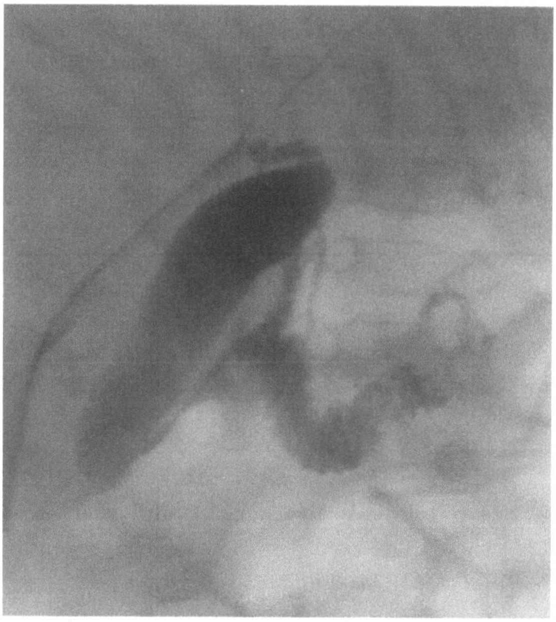

Fig. 14.2. A 6-week-old infant on hyperalimentation was evaluated for prolonged neonatal jaundice after necrotizing enterocolitis. Radionuclide imaging showed no excretion of radiotracer on delayed images. Percutaneous transhepatic cholecystogram performed to rule out biliary atresia shows a patent extrahepatic biliary tree with reflux of contrast into central right and left intrahepatic ducts. Percutaneous cholecystography carries lower morbidity than laparoscopy and is equally valuable in evaluating the patency of the extrahepatic biliary tree. Here a 3-French catheter was inserted into the gallbladder; extravasated contrast stains the undersurface of the liver edge and was tolerated without treatment

ultrasound guidance. A right transcostal approach is used for right and a subxiphoid midline approach for left duct puncture. As there is occasionally poor correlation between the caliber of the bile ducts and the degree of obstruction, manometric techniques may be employed to determine the functional significance of findings (van Sonnenberg et al. 1983).

14.1.6 Endoscopic Retrograde Cholangiopancreatography

In endoscopic retrograde cholangiopancreatography (ERCP), a fiberoptic endoscope is used to visualize and cannulate the ampulla of Vater for selective injection of iodinated contrast into the common bile and pancreatic ducts. The success of ERCP in the pediatric population is reported to range between 90%–100% in experienced hands; individual technical success reflects a combination of patient size and anatomy and operator experience (Allendorph et al. 1987; Putnam et al. 1991; Guelrud et al. 1992). Unlike PTC, this technique does not depend upon dilatation of the biliary tract. Periprocedural complications of pediatric upper endoscopy occur in less than 2% of patients and include pancreatitis and cholangitis (Ament et al. 1988).

The procedure selected for direct cholangiography depends upon the clinical situation and the availability of a skilled operator, either an endoscopist or an interventional radiologist. Although both techniques allow the biliary system to be visualized in many situations, ERCP may be preferable when the bile ducts are not dilated, when the gallbladder is not accessible, or when the presence of an ampullary, pancreatic, or distal bile duct lesion is suspected.

14.2 Hepatobiliary Diseases of Childhood

14.2.1 Congenital Anomalies

At surgery, congenital anomalies of the biliary tract and gallbladder may be encountered in up to 1.6–10% of patients (Puchetti et al. 1976; Berci 1992). Gallbladder agenesis is rare (an incidence of 0.04%–0.07% in autopsy series; Puente and Bannura 1983). Gallbladder hypoplasia also may be observed, congenital or associated with cystic fibrosis. Ectopia of the gallbladder occurs adjacent to the left lobe of the liver, less commonly intrahepatically or retrohepatically. Longitudinal and transverse septations of the gallbladder may also be identified (Tan et al. 1993). Biliary tract anomalies may include aberrant drainage of either the intrahepatic or cystic ducts (Charels and Kloppel 1989). Intrahepatic ducts may have anomalous insertion into the common hepatic duct, common bile duct, cystic duct, or gallbladder. Duplications of the common duct and cystic duct have been noted (Knechtle and Filston 1990), as has aberrant drainage of the biliary ductal system (Walia et al. 1986; Picardi et al. 1989; Giuly et al. 1992; Wong 1994). Congenital bronchobiliary fistula may cause respiratory distress in infants presenting with bilious sputum or pneumobilia (Iuchi et al. 1983).

Atretic or hypoplastic segments of the extrahepatic and major intrahepatic biliary branches are the most common biliary anomalies of clinical relevance in infants who present clinically with obstructive jaundice and acholic stools. Biliary hypoplasia is caused by a developmental hepatocellular or bile canalicular defect; this results histologically in absence or reduction of the number of bile ductules. The intrahepatic biliary tree is gracile on direct cholangiography in biliary hypoplasia. Two forms have been described: isolated or syndromic biliary hypoplasia and arteriohepatic dysplasia (Alagille's syndrome).

Alagille's syndrome is an uncommon syndrome characterized by typical facial features, pulmonary artery stenosis, butterfly vertebrae, and progressive neonatal jaundice. The appearance on hepatobiliary scintigraphy is similar to that seen in biliary atresia; thus, the diagnosis may require surgical exploration (Summerville et al. 1988). The differential diagnosis includes choledochal cyst, extrahepatic biliary obstruction, and other intrahepatic cholestatic syndromes.

14.2.2 Congenital Biliary Atresia

The single most common cause of neonatal cholestasis, congenital biliary atresia is characterized by obliteration of the lumen of the common bile duct. The etiology is unclear. Obliterative proliferation by the biliary ductal epithelium may be biochemical, viral, inflammatory, or ischemic in origin. There is no gender predilection. Fifteen percent of patients have trisomy 18 or the noncardiac polysplenia syndrome. The gallbladder is preserved in approximately 12%–25% of patients. The common hepatic duct and intrahepatic biliary tree are in continuity in approximately 10% of patients, allowing surgical correction by primary Roux-en-Y hepaticojejunostomy anastomosis (DEHNER 1987). However, the majority of infants with biliary atresia have disruption of the intrahepatic biliary tree, requiring an alternative surgical treatment, hepatoportoenterostomy (KASAI procedure) to be used (KASAI et al. 1968).

The KASAI procedure is successful as a temporizing measure in up to 90% of patients when performed within the first 2 months of life; if surgery is delayed beyond 3 months, success is poor (LILLY 1975). Thus, a timely work-up of neonatal jaundice is imperative (ABRAMSON et al. 1982). Following Kasai palliation, if intrahepatic biliary drainage is incomplete, most patients develop chronic cholangitis, extensive perihepatic fibrosis, secondary biliary cirrhosis, and portal hypertension, ultimately requiring OLT for cure (LAURENT et al. 1990).

Proliferation of bile ducts and periportal fibrosis can be detected early in infants with biliary atresia by ultrasonography displaying increased periportal echogenicity (BRUN et al. 1985). A cystic structure at the porta hepatis, representing a remnant of the common bile duct, may uncommonly (<10%) be observed (BRUNELLE et al. 1994). On PTC the intrahepatic biliary tree has a tangled, beaded appearance (Fig. 14.3) (CHAUMONT et al. 1982).

Nuclear scintigraphy using ^{99m}Tc-labeled IDA derivatives provides a reliable method for differentiating biliary atresia from neonatal hepatitis. Phenobarbital (5 mg/kg per day) administered orally for at least 5 days prior to the examination significantly increases the accuracy of the examination by inducing hepatic microsomal enzyme activity. Demonstration of radiotracer within the gastrointestinal tract, with or without visualization of gallbladder, indicates patency of the extrahepatic biliary tree. However, decreased hepatic extraction with nonvisualization of the gastrointestinal tract may also reflect the severity of hepatocellular disease and contribute to a false positive finding.

14.2.3 Congenital Hepatic Fibrosis

Congenital hepatic fibrosis is an autosomal recessive disease characterized by an increased number of ectatic, distorted terminal bile ductules which are usually too small to be recognized radiographically. Extensive periportal fibrosis results in increased echogenicity on ultrasonography and contributes to chronic cholangitis and portal hypertension. Hepatomegaly is usually present. Liver biopsy may be required to confirm the diagnosis, but, due to hetero-

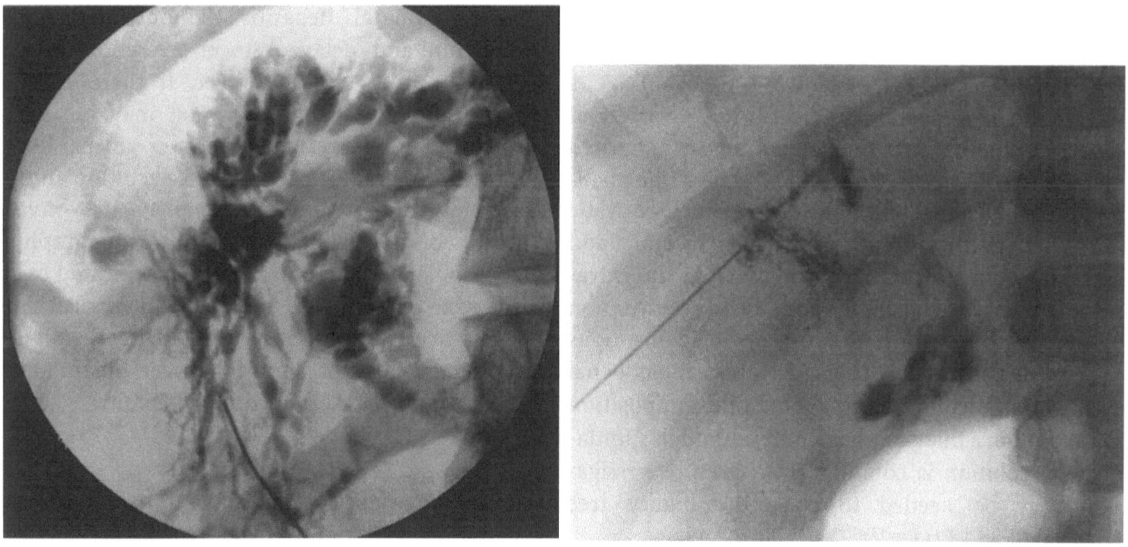
a b

Fig. 14.3. **a** Percutaneous transhepatic cholangiogram with external biliary drainage performed in a 19-year-old with a history of congenital biliary atresia, treated with Kasai hepatoportoenterostomy at age 6 weeks. Left duct puncture with a 22G Chiba needle shows the marked irregularity of the intrahepatic biliary tree which has multiple regions of saccular dilatation, interspersed between narrowed segments of intrahepatic bile duct. This represents secondary (acquired) sclerosing cholangitis. **b** A 3.5-year-old boy with a history of congenital biliary atresia, treated with Kasai hepatoportoenterostomy in the neonatal period. Percutaneous transhepatic cholangiogram (25G spinal needle) performed to evaluate persistently elevated liver function tests shows small neoductules which drain poorly across the biliary/enteric anastomosis geneous involvement, may be falsely negative. Prognosis is poor, with death from variceal bleeding or hepatic failure often occurring in childhood. Hepatocellular and cholangiocellular carcinoma may be seen as late complications of this disease.

14.2.4 Simple Hepatic Cysts

Simple hepatic cysts are fluid-filled masses lined by bile duct epithelium which do not communicate with the biliary tree. They may present at any age, have a predilection for females, and are usually asymptomatic unless complicated by hemorrhage, infection, or compressive biliary obstruction. On ultrasonography hepatic cysts are typically unilocular and anechoic. CT shows them to have thin walls and fluid content of water density; if aspirated, clear, straw-colored fluid is obtained. If multiple and small, they may mimic solid lesions. On sulfur colloid and IDA scans they represent photopenic lesions. On MRI these lesions have low signal intensity on T1- and high signal intensity on T2-weighted images. Treatment is required only if symptomatic; many hepatic cysts are amenable to percutaneous drainage or sclerosis (Montorsi et al. 1994; van Sonnenberg et al. 1994; Yamada et al. 1994).

14.2.5 Neonatal Jaundice

Persistence of jaundice beyond 3–4 weeks of age is referred to as neonatal jaundice. It has a variety of causes including infection, acquired or inherited metabolic conditions, and biliary tract abnormalities. Inherited metabolic abnormalities include cystic fibrosis, a_1-antitrypsin deficiency, galactosemia, and hereditary tyrosinemia. Abnormalities of the extrahepatic biliary tree include spontaneous perforation of the bile duct which presents with an insidious onset of bilious ascites (Fig. 14.4), and choledochal cyst. Conditions leading to altered bile composition may result in a bile plug syndrome; although spontaneous resolution is common, intraoperative irrigation may be needed to clear the biliary tree (Holland and Lilly 1992).

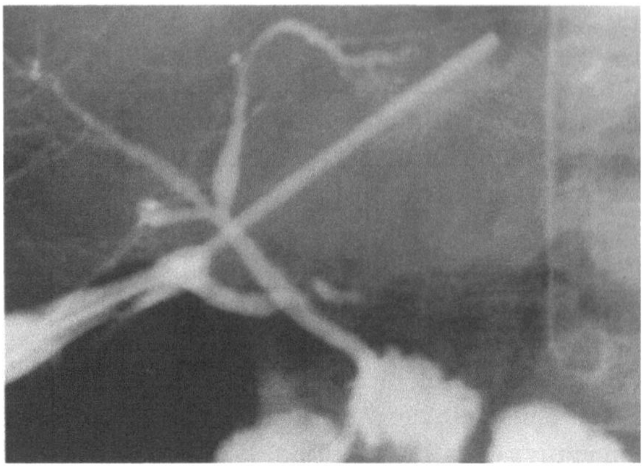

Fig. 14.4. Spontaneous perforation of the common bile duct in a 2-month-old boy presenting with bilious ascites and jaundice. Follow-up cholecystogram is performed through a surgically placed cholecystostomy tube. Focal dehiscence in the common duct opposite the cystic duct insertion healed following biliary diversion

Biliary atresia and neonatal hepatitis account for over two-thirds of the cases of persistent neonatal jaundice. Diagnostic workup of the neonate should be initiated if conjugated hyperbilirubinemia persists beyond 2 weeks of age or exceeds 2 mg/dl. In addition, episodes of intermittent jaundice should be investigated at any age to exclude obstructive lesions of the biliary tract. Appropriate surgical intervention can preserve satisfactory hepatic function in nearly all cases of extrahepatic obstruction (Ryckman and Noseworthy 1987).

The sonographic features of both neonatal hepatitis and biliary atresia are nonspecific. Postphenobarbital 99mTc-IDA cholescintigraphy is the best imaging study to rule out biliary atresia (Majd et al. 1981). Neonatal hepatitis demonstrates normal extraction and delayed tracer excretion into the gastrointestinal tract; however, if there is severe hepatocellular damage, excretion may be delayed or absent. In biliary atresia, there is normal hepatic extraction of radiotracer but no excretion into the gastrointestinal tract (Kirks et al. 1984). Nonvisualization of the gallbladder by ultrasonography or hepatobiliary scintigraphy is nonspecific and cannot be used to discriminate medical from surgical causes of jaundice (Burton et al. 1990). Direct cholangiography may help distinguish between biliary atresia, biliary hypoplasia, and neonatal hepatitis.

14.2.6 Primary Sclerosing Cholangitis

Primary sclerosing cholangitis (PSC) is an insidiously progressive disease characterized by fibrosis and inflammation of the intra- and extrahepatic bile ducts. It is an immunoregulatory abnormality (either primary or acquired) with elevated levels of circulating antinuclear and smooth muscle autoantibodies (Classen et al. 1987; el-Shabrawi et al. 1987;

ZAULI et al. 1987). A reduction in total number of circulating T cells and an increase in B cells correlates with the histologic stage (LINDOR et al. 1987). An association with HLA phenotypes B8 and DR3 also has been noted. Viral infections (e.g., rheovirus type 3) are recognized to induce obliterative cholangitis in childhood, but their role in PSC remains undefined.

PSC may present at any age; in a review of 56 children with PSC, first symptoms occurred at a mean age of 3.7 years, one-quarter presenting with neonatal jaundice (DEBRAY et al. 1994). Median survival time was approximately 10 years from onset of clinical symptoms. On cholangiography, the intrahepatic bile ducts were abnormal in all patients; extrahepatic involvement was present in two-thirds. Isolated intrahepatic involvement was present in less than one-quarter of cases, and the cystic duct was involved in approximately 20%.

Histologic findings included neoductular proliferation, generalized thickening of the submucosal and subserosal layers of the biliary tree, and extensive portal fibrosis with obliterative fibrosis of the biliary radicles (DEBRAY et al. 1994). When isolated to a single region of the biliary tree, prognosis appears to be better than if the disease is either diffuse or multifocal (SOKAL et al. 1990b). There is slow progression from an asymptomatic stage to cirrhosis and portal hypertension. Premature death usually results from liver failure, or as a consequence of portal hypertension with bleeding gastroesophageal varices. Bile duct cancer appears to be a frequent complication of long-standing PSC (WIESNER et al. 1985).

On cholangiography, there is marked irregularity of the biliary tree with multifocal strictures having a predilection for branch points (Figs. 14.5, 14.6). Skip lesions representing uninvolved ductal segments may be interspersed with areas of saccular dilatation. Small bile duct diverticula or pseudodiverticula are pathognomonic. Involvement of the extrahepatic biliary tree may be prominent (Fig. 14.7); the gallbladder is uncommonly involved. Differential diagnosis includes primary biliary cirrhosis, ascending cholangitis, and sclerosing cholangiocarcinoma.

There is increased echogenicity of portal triads on ultrasonography, caused by periportal fibrosis. On CT imaging, one sees nodularity of the extrahepatic biliary tree, duct wall thickening, and mural contrast enhancement. Lack of biliary ductal dilatation does not rule out biliary obstruction. Retention of radiotracer in the intrahepatic biliary tree, marked prolongation of hepatic clearance, and possible nonvisualization of the gallbladder (30%) are noted on hepatobiliary scintigraphy.

PSC may be highly associated (50% of patients) with chronic inflammatory bowel disease, especially ulcerative colitis, and also has been associated with AIDS. Absence of such associated disorders (e.g., inflammatory bowel disease or autoimmunity) suggests a better prognosis.

Therapy with cholestyramine may help control symptoms of pruritus. Antibiotics are useful for controlling cholangitis. Glucocorticoids are not efficacious. Surgical or percutaneous intervention may be required to relieve extrahepatic obstruction in selected patients. Hepatic transplantation may be considered for the treatment of end-stage PSC.

Because transplantation is an option, palliative surgical biliary drainage and portal decompression procedures should be avoided, so that hepatic transplantation may be performed without increased operative risk.

Disease entities similar to PSC include idiopathic sclerosing cholangitis, a neonatal form of the disease also characterized by small, beaded intrahepatic bile ducts which appear in an irregular network. This

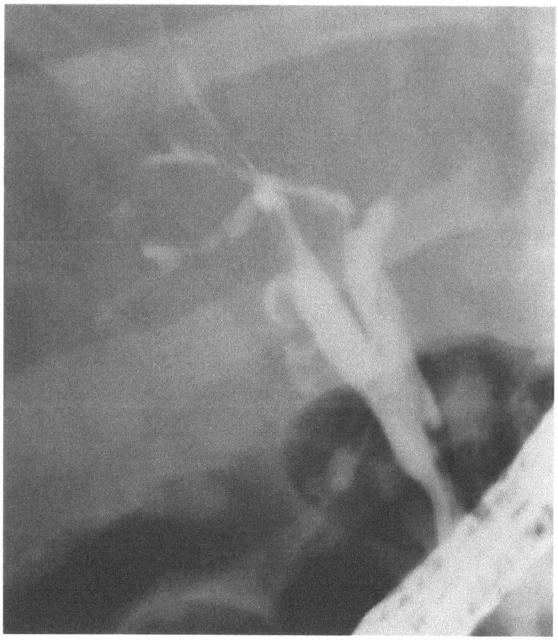

Fig. 14.5. Sclerosing cholangitis diagnosed by ERCP in a 10-year-old girl with Crohn's disease with diffuse intrahepatic involvement and dilatation of the extrahepatic biliary tree. This patient presented with acute right upper quadrant pain. Ultrasonography showed a dilated common bile duct (7 mm) with sludge in the gallbladder. DISIDA showed no evidence of biliary obstruction

appearance is similar to that in biliary atresia, except that the extrahepatic biliary tree is always patent. Secondary sclerosing cholangitis may develop as a long-term complication of choledocholithiasis, cholangiocarcinoma, traumatic or iatrogenic biliary tract injury, or due to other chronic inflammatory processes. Histiocytosis X has cholangiographic features similar to PSC, including stenoses of the central intrahepatic bile ducts, but uncommonly involves the extrahepatic biliary tree (Fig. 14.8).

a

b

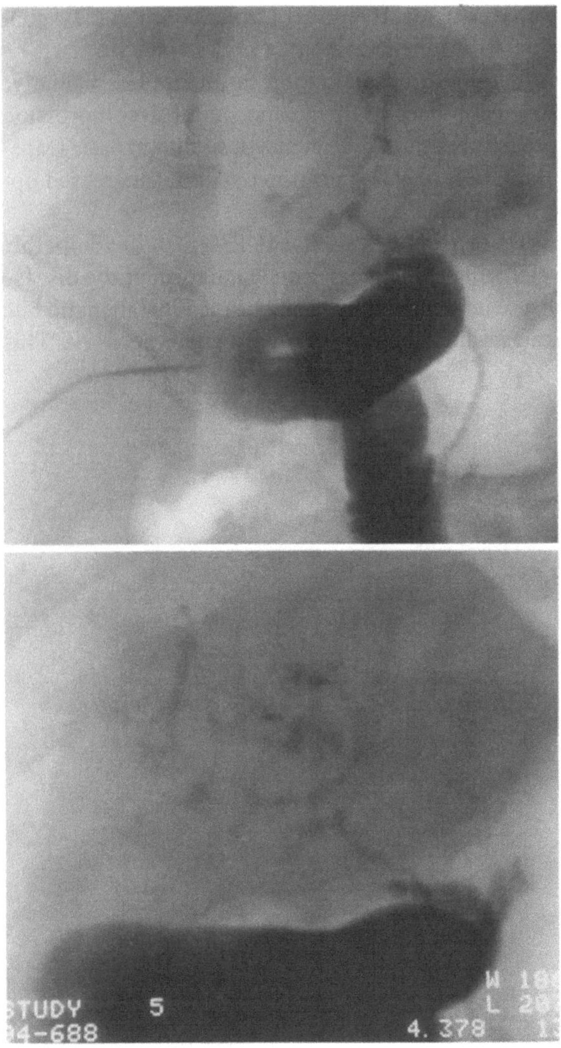

Fig. 14.6a,b. Sclerosing cholangitis diagnosed following liver biopsy in a 9-month-old boy presenting with elevated liver function studies. Following ultrasound-guided transhepatic puncture of the gallbladder, a 3-French catheter was introduced for cholecystography. Diffuse involvement of the intrahepatic biliary tree is evident; differential diagnosis includes biliary hypoplasia and idiopathic sclerosing cholangitis. Contrast freely drains through the extrahepatic biliary tree into the bowel

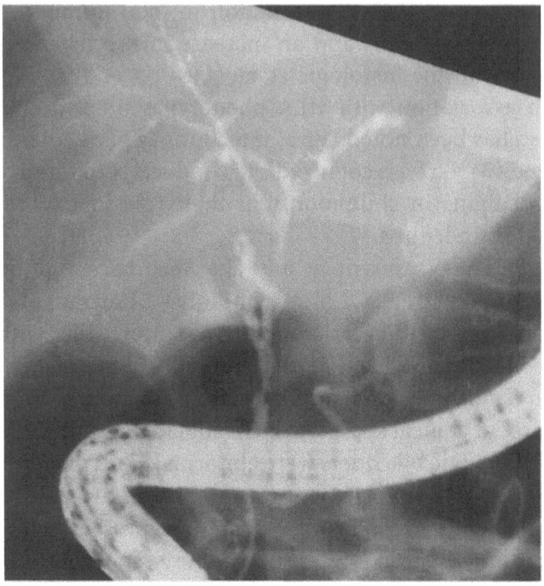

Fig. 14.7. Primary sclerosing cholangitis diagnosed by ERCP in an 8-year-old girl presenting with liver dysfunction and no other pertinent clinical history. Cholangiography shows diffuse, characteristic "aneurysms" of both the intra- and extrahepatic biliary tree

14.2.7 Biliary Cirrhosis

Prolonged obstruction of the biliary system leads to biliary cirrhosis, an acquired injury resulting from replacement of normal hepatic parenchyma by regenerating nodules and surrounding fibrous tissue. Primary biliary cirrhosis (PBC) and secondary biliary cirrhosis are separate histopathological entities with respect to etiology, but both result in a similar clinical presentation in which there is impairment of biliary excretion, destruction of hepatic parenchyma, and progressive periportal fibrosis. Hepatocellular loss results in impaired synthetic function; periportal fibrosis results in occlusion of the postsinusoidal venules with resultant portal hypertension. Clinical findings of PBC are similar to those of PSC, and include hepatosplenomegaly. On cholangiography, the intrahepatic ducts are attenuated and irregular; unlike PSC, the extrahepatic biliary tree is not involved, thus distinguishing this entity. Macronodular cirrhosis may be suggested by ultrasonography; liver biopsy is required to identify micronodular cirrhosis.

14.2.8 Chronic Granulomatous Disease of Childhood

Chronic granulomatous disease of childhood is a systemic disease cause by recessive, sex-linked poly-

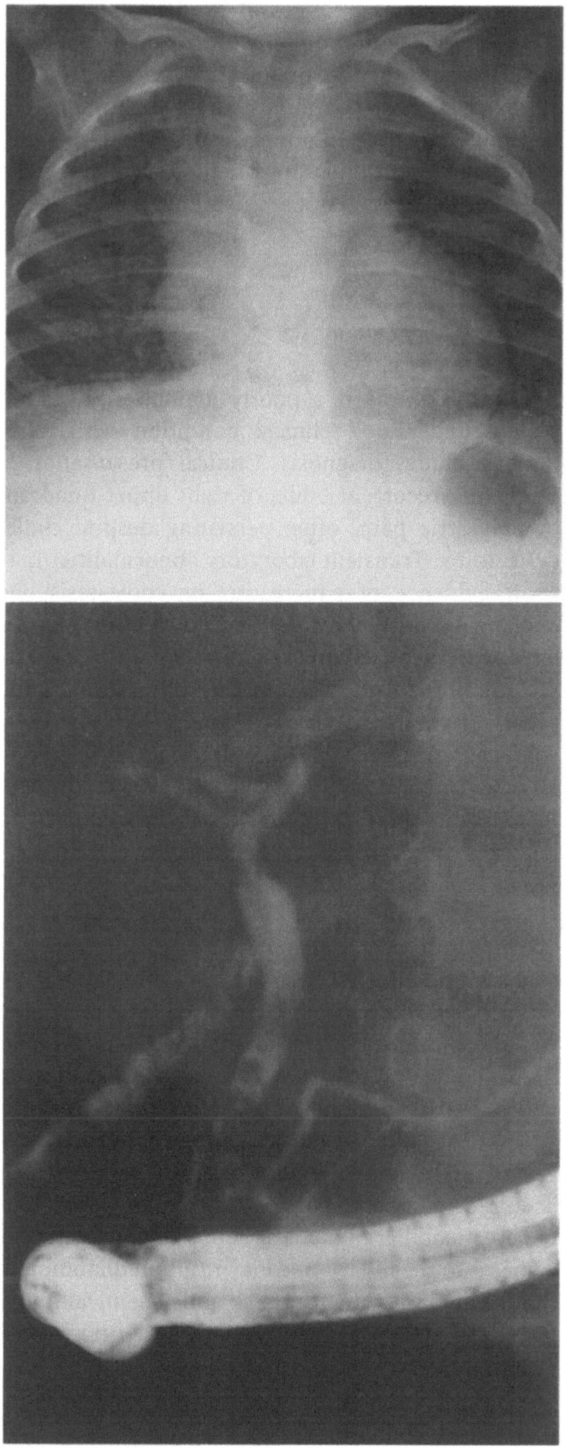

a

b

Fig. 14.8a,b. Sclerosing cholangitis diagnosed by ERCP in a 2-year-old boy presenting with hepatosplenomegaly and dermatitis. a Chest X-ray shows diffuse interstitial involvement consistent with histiocytosis X. A skeletal series demonstrated lytic bone changes in the calvarium and mandible. ERCP was performed to evaluate progressive liver dysfunction and shows cutoff of the central intrahepatic bile ducts (b). The patient's clinical status rapidly deteriorated; the child died despite hepatic transplantation

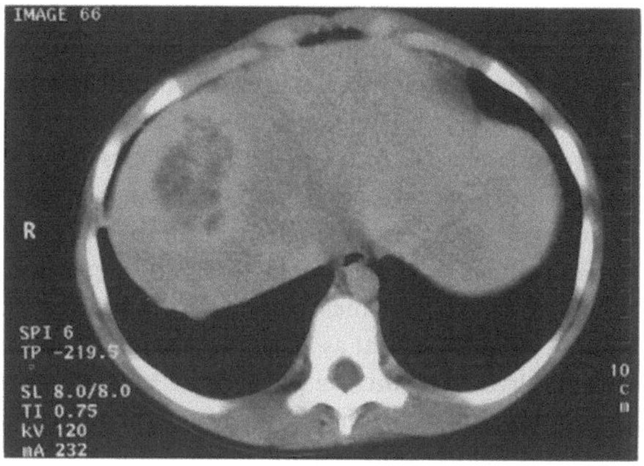

Fig. 14.9. Liver abscesses involving both left and right hepatic lobes in a 6-year-old boy with chronic granulomatous disease of childhood. Repeat admissions for fever of unknown origin resulted in the diagnosis of hepatic abscesses, which responded poorly to both intravenous antibiotics and surgical debridement

morphonuclear leukocyte dysfunction in which there is a defect in microbicidal activity, causing prolonged intracellular survival of phagocytized bacteria. Histologically there is evidence of chronic infection, with granuloma formation, caseation, and suppuration in many organs, most commonly liver and lung. The disease has a childhood onset with a male predilection. Chronic recurrent infection may lead to lymphadenitis which may involve the porta hepatis. There may be associated hepatosplenomegaly, hepatic abscesses, liver calcifications, and ascites (Fig. 14.9).

14.2.9 Byler's Disease

Familial progressive intrahepatic cholestasis, known as Byler's disease, is an autosomal recessive disease with rapidly progressive hepatic insufficiency. This disease presents in infancy and must be suspected in the infant with cholestasis and normal serum level of γ-glutamyl transpeptidase. Liver biopsy shows marked fibrosis and micronodular cirrhosis with little regeneration. On cholangiography the intra- and extrahepatic biliary trees are normal.

14.2.10 Hepatobiliary Masses

It is distinctly uncommon to encounter hepatobiliary masses in the neonate, but those that do occur include hepatic hemangioendothelioma, solitary liver cyst, hepatoblastoma, choledochal cyst, and gallbladder hydrops. Clinical presentation may include an asymptomatic abdominal mass, hepatomegaly, or symptoms related to hepatocellular dysfunction. Approximately 5% of all pediatric tumors beyond the newborn period are hepatobiliary in origin. A majority (~65%) are malignant tumors of either hepatocellular (e.g., hepatoblastoma and hepatocellular carcinoma) or mesenchymal origin. Hepatic tumors usually present as an upper abdominal mass without liver dysfunction. The right lobe is more commonly involved than the left; approximately one-third involve both hepatic lobes.

Hepatoblastomas are more common in children and usually occur in patients under 3 years of age. There is a slight male predominance. Hepatoblastomas are usually echogenic on ultrasound. Hepatocellular carcinomas are encountered more frequently after age 5 years, and are more likely superimposed upon preexisting liver disease (e.g., chronic hepatitis or cirrhosis). Hepatomegaly is present in less than half of patients. The location and extent of tumor is more precisely defined by CT or MR; prognosis depends upon suitability for surgical resection. Arteriography may be required preoperatively to define vascular anatomy.

Hemangioendotheliomas are usually diffuse, multiple lesions that involve the entire liver and present before 6 months of age. Most patients have hepatomegaly or an abdominal mass (Keslar et al. 1993). This condition, which has a predilection for females, may frequently be associated with cardiomegaly and congestive failure due to arteriovenous shunting. Approximately 50% of these infants have cutaneous hemangiomas. Treatment is directed according to the clinical situation. Most hemangioendotheliomas spontaneously involute; rare life-threatening complications include thrombocytopenia and disseminated intravascular coagulation (Wong and Masel 1995). Treatment with steroids or transcatheter embolization may be required for the management of refractory congestive failure. The main role of ultrasonography is in following the response to therapy. CT usually shows a low-attenuation solid lesion with peripheral enhancement (Mahboubi et al. 1987). Central enhancement is often lacking, except in smaller lesions. These lesions usually have high signal intensity on T2-weighted MR images (Keslar et al. 1993). Arteriography is indicated only if embolization, vascular ligation, or resection is contemplated.

Embryonal rhabdomyosarcoma (sarcoma botryoides) represents a bile duct malignancy originating in the submucosal tissues of the extrahepatic bile ducts, especially the common bile duct and gallbladder. There is no gender predilection; the age group 1–10 years is most commonly affected. Prognosis is poor due to local tumor spread. Both ultrasonography and cholangiography may show one or more intraluminal mass lesions which have soft tissue attenuation on CT.

14.2.11 Biliary Dyskinesia

Biliary dyskinesia is a poorly defined and incompletely understood clinical condition which frequently eludes diagnosis. Clinical presentation is frequently recurrent bouts of right upper quadrant or epigastric pain, often persisting despite cholecystectomy. Transient laboratory abnormalities may show evidence of pancreatitis or cholestasis. On cholangiography, the common bile duct may be dilated with delay in contrast emptying. Etiology may include fibrosis or spasm of the sphincter of Oddi; biliary manometry shows elevation of the basal sphincter pressures (vanSonnenberg et al. 1983). Histologic abnormalities are not always present. Endoscopic sphincterotomy has been reported to decrease sphincter of Oddi pressures for up to 2 years.

14.2.12 Choledochal Cysts

An uncommon type of hepatobiliary mass in children and young adults is the choledochal cyst, in which there is focal dilatation of the biliary tree (Fig. 14.10). The etiology may be either congenital (reported as early as the 15th week of gestation), or acquired in nature, possibly due to an anomalous pancreaticobiliary junction. Reflux of pancreatic enzymes into the biliary tree may result in dilatation of the common bile duct from chronic chemical injury (Babbitt 1969; Crittenden and McKinley 1985; O'Neill 1992).

Clinical presentation includes the triad of jaundice (70%), abdominal mass (60%), and abdominal pain (50%); this triad is complete in a minority of cases (<25%) (Crittenden and McKinley 1985; Grosfeld et al. 1994). One-half present by age 10 years; only 5% are symptomatic before 6 months. Ultrasonography shows a cystic mass in the porta hepatis, separate from the gallbladder, in contiguity with distended biliary ducts. Radionuclide scintigraphy shows normal hepatic extraction of ra-

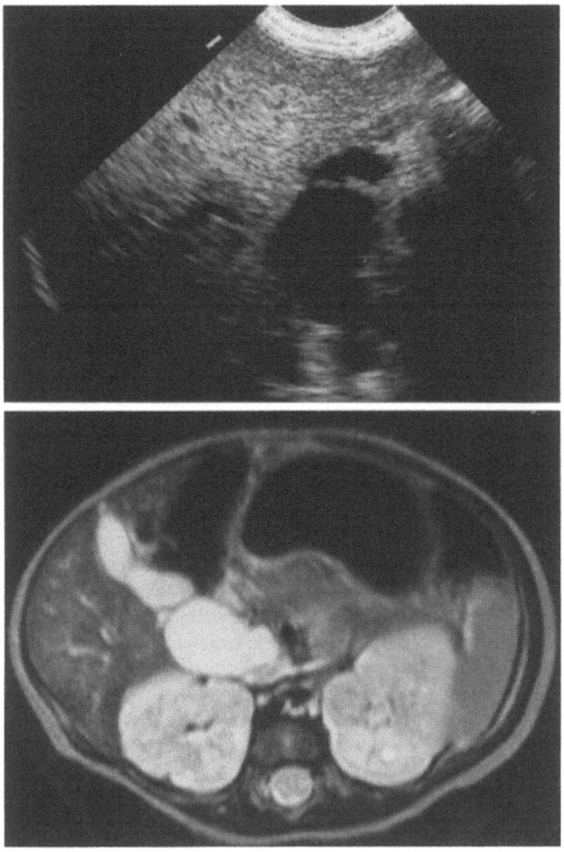

a

b

Fig. 14.10a,b. Focal dilatation of the extrahepatic biliary tree is seen both on ultrasound (**a**) and MRI (**b**) in this 6-month-old girl who was diagnosed at age 3 months to have polycystic kidney disease. A cystic lesion in the region of the common bile duct was noted on ultrasound. Preoperative workup included MRI in which the gallbladder and choledochal cyst have high signal intensity on this T2-weighted sequence

diotracer with excretion into the biliary system and eventual accumulation in the cyst. There is a female predilection. Complications include cholangitis, complete obstruction of distal common bile duct by stones, biliary cirrhosis, cyst rupture with bile peritonitis, and bile duct dysplasia or malignancy.

14.2.13 Caroli's Disease

The characteristic feature of Caroli's disease is cystic, nonobstructive dilatation of the intrahepatic bile ducts, which are diffusely involved. Frequently, ectasia of extrahepatic biliary tree is also observed. It may be associated with congenital hepatic fibrosis and recessive polycystic kidney disease. In a few cases it is discovered in association with a choledochal cyst. On cholangiography there is segmental saccular cystic dilatation of major intrahepatic bile ducts which converge toward the porta hepatis. Involvement may either be segmental or diffuse and extend to the periphery of the liver. CT and ultrasonography demonstrate branching intrahepatic ducts with focal areas of ectasia. On radionuclide IDA imaging studies the dilated intrahepatic ducts are initially cold, but later fill as radiotracer is excreted. Bile stasis predisposes these patients to calculus formation, secondary pyogenic cholangitis with resultant intrahepatic abscess formation, and cholangiocarcinoma. Liver function is usually not impaired; death is usually due to sepsis.

14.2.14 Choledocholithiasis

The occurrence of stones in the common bile duct is rare in children, but it is being recognized more frequently with the widespread use of ultrasound. Stones can be secondary to a structural anomaly of the duct (e.g., choledochal cysts, or sclerosing cholangitis) or due to a metabolic disturbance (e.g., hemolysis or sickle cell anemia). In older children hemolytic anemia represents the most common cause of choledocholithiasis. Interventional radiological techniques are occasionally used postoperatively in children, the aim of the procedure being to flush the biliary tree in order to push residual stone debris into the bowel. Following percutaneous cholangiography, an external drainage catheter may be left in the biliary tree or gallbladder and this access used for biliary stone removal.

14.3 Pediatric Liver Transplantation

The most common indication for transplantation in the pediatric population is biliary atresia. Although the survival rates for liver transplantation continue to improve, significant postoperative morbidity remains. Older children usually receive a whole liver with reconstruction of the hepatic artery, hepatic veins, and portal veins as well as a choledochal choledochostomy with a T-tube, or, less often, a Roux-en-Y choledochal jejunostomy (Colonna et al. 1992). Limited availability of suitably sized, matched-donor whole organs has been a major cause of morbidity and mortality in infants and smaller children. In the younger child reduced-size grafts, using either the entire left lobe or left lateral segment, may be used (but also require a Roux-en-Y choledochal jejunostomy biliary reconstruction). Reduced-size hepatic allografts result in an alteration of the position and number of hepatic vessels and bile ducts, leading to the existence of a cut liver surface (CARON et al. 1992). Additionally, the hepatic artery is smaller and undergoes a significant change in diameter at the surgical anastomosis, which makes it more prone to thrombosis (SOKAL et al. 1990a). Reduced-size liver transplants have been successful and have resulted in decreased patient mortality (BROELSCH et al. 1988).

Despite immunosuppression, homograft rejection still occurs in a majority of patients within 1–6 weeks of transplantation (Fig. 14.11). Early laboratory changes seen in rejection include leukocytosis and elevation in serum bilirubin, aminotransferase, and alkaline phosphatase activity. These may be followed clinically by fever, right upper quadrant tenderness, diarrhea, and ascites, with progressive deterioration of liver function. Biliary complications occur in 20%–40% of cases and may be encountered early (1–2 days) or late (6 months) after liver transplantation (LETOURNEAU and CASTANEDA-ZUNIGA 1990; PARIENTE et al. 1991; PECLET et al. 1994). The incidence of biliary complications is similar for full and reduced-size grafts (HEFFRON et al. 1992). Biliary complications seldom result in death or graft loss unless there is associated sepsis or bile duct necrosis (SHENG et al. 1994).

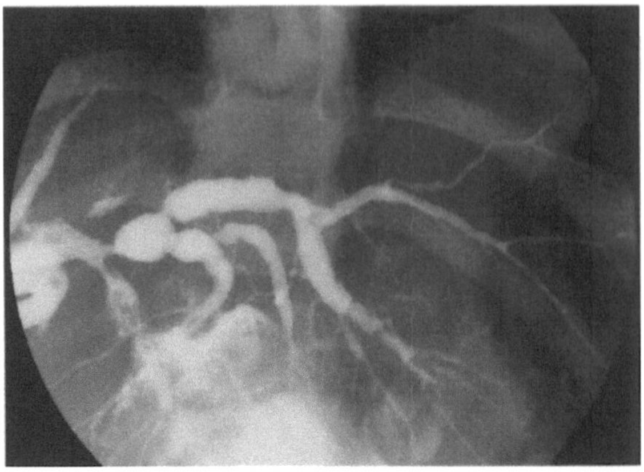

Fig. 14.12. Hepatic artery thrombosis. A percutaneous transhepatic biliary drainage performed in a 3-year-old girl who had biliary atresia, who underwent transplantation within the first few months of life. A whole liver was transplanted; the patient has recurrent cholangitis requiring antibiotic therapy. Percutaneous balloon dilatation with internal/external stent placement was performed for strictures of the central left and right bile ducts and common hepatic duct. Here, tube cholangiography was performed to evaluate results of internal stenting. There is persistent narrowing of the left bile duct at the confluence with the right. Diffuse dilatation of the central left bile ducts contributes to bile duct stasis. Peripheral branches of the biliary tree show diffuse narrowing

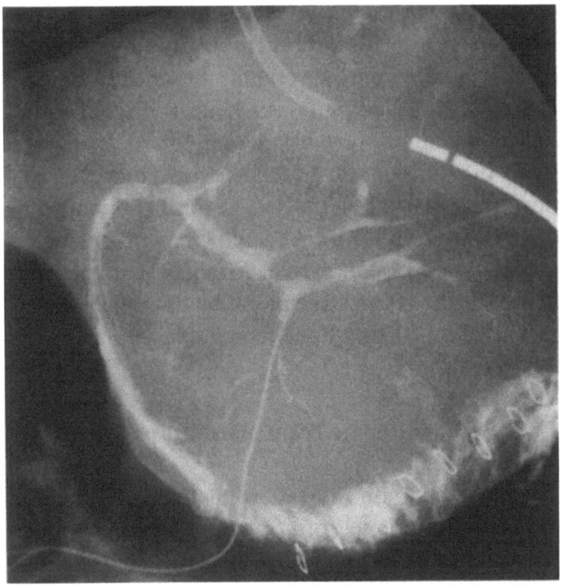

Fig. 14.11. Acute rejection. A percutaneous transhepatic cholangiogram performed in a 6-year-old boy with a history of Alagille's syndrome. The patient received his second hepatic allograft at age 6, approximately 2 months prior to this examination. A left lateral segment reduced-size graft was performed, and the patient developed early rejection. PTC was performed to evaluate elevated bilirubin and other laboratory abnormalities. The study was performed via a 3-French catheter in the biliary tree. This demonstrates irregular dilatation of the central left bile duct with a patent biliary-enteric anastomosis. Diffuse narrowing of intrahepatic branches is consistent with rejection. The patient died subsequently of disseminated intravascular coagulation and intracranial hemorrhage

The major cause of biliary complications is hepatic artery thrombosis, which causes ischemia and necrosis of the biliary tract. Diffuse ischemic biliary ductal injury may result from improper homograft preservation during the cold ischemia time prior to transplantation (LI et al. 1992). Ischemic strictures that develop early are associated with a poor prognosis (SANCHEZ-URDAZPAL et al. 1993a,b). Necrosis resulting from bile duct ischemia may occur within any portion of the biliary tract and lead to bile leakage, stenosis, or dilatation (Fig. 14.12). On cholangiography, dilated intrahepatic bile ducts with shaggy borders and/or large intraluminal filling defects suggest epithelial necrosis due to hepatic artery thrombosis (Fig. 14.13a–c). Prognosis is often poor, many patients requiring retransplantation. Nonanastomotic biliary strictures may also be related to

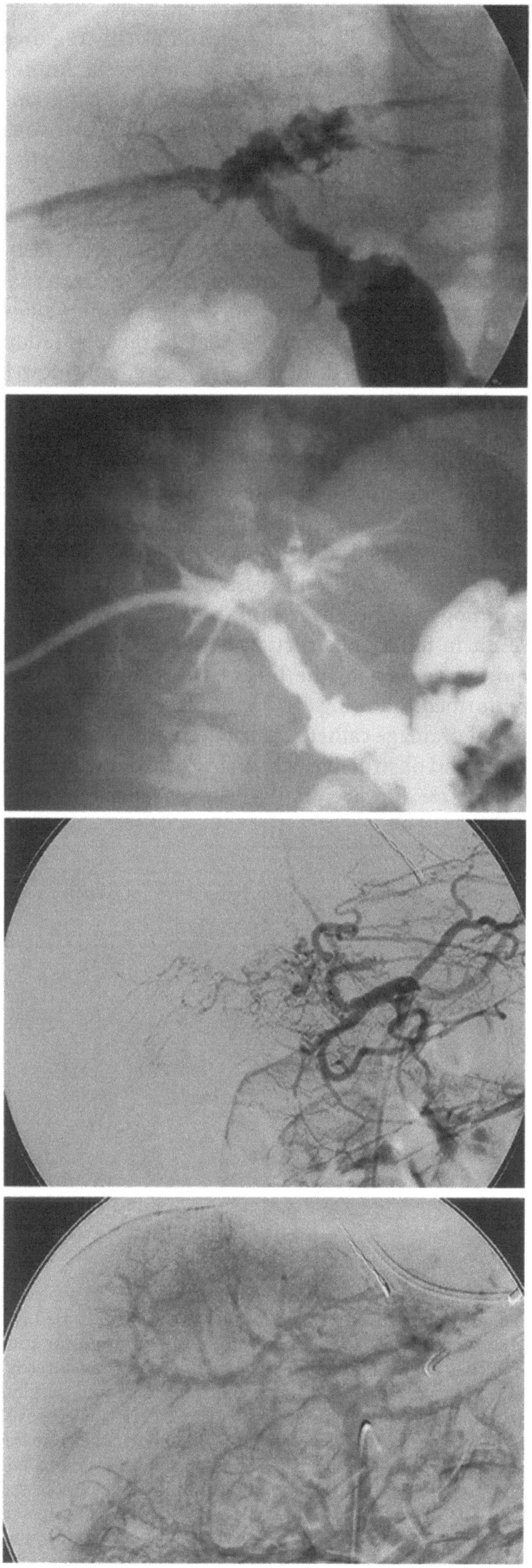

a
b
c
d

ductopenic cellular rejection, cytomegalovirus infection, ABO blood incapability, and prolonged cold ischemic time prior to allograft transplantation (SANCHEZ-URDAZPAL et al. 1992). Peripheral biliary strictures are more common in patients with the pretransplantation diagnosis of sclerosing cholangitis (MCDONALD et al. 1991). Biliary-enteric anastomotic strictures may be successfully treated with percutaneous transluminal angioplasty which may avoid or delay surgical intervention (Fig. 14.14).

Chronic rejection is associated with progressive cholestasis, bile duct proliferation, focal parenchymal necrosis, mononuclear infiltration, and fibrosis; these findings are similar to chronic viral hepatitis, which may be difficult to differentiate. Rejection may also cause biliary-enteric anastomotic strictures.

Laboratory studies may suggest, but do not diagnose these biliary complications. Scintigraphy, ultrasonography, and CT in various combinations can be used to screen for biliary obstruction or leak. Direct cholangiography is necessary to define the cause and location of most lesions and to allow a treatment plan to be formulated (CAMPBELL et al. 1994). Liver biopsy may be required in selected cases to demonstrate intrahepatic cholestasis.

PTC is the most frequent invasive postoperative procedure performed in pediatric liver transplant patients. Indications include laboratory abnormalities, suspected bile leaks, symptoms of biliary obstruction or leakage, cholangitis, sepsis, and decreased excretion on IDA scintigraphy (ZAJKO et al. 1987). The intrahepatic ducts are usually not dilated. PTC is successful in approximately 75% of cases without ductal dilatation; however, when the bile

Fig. 14.13a–d. Hepatic artery thrombosis. A percutaneous transhepatic biliary drainage was performed to evaluate an 11-month-old girl who at age 1 month received a whole liver allograft for biliary atresia. Recurrent bouts of cholangitis required antibiotic therapy; PTC was performed for cholangitis. A right transcostal approach was utilized and internal/external drainage performed after dilatation of the left hepatic duct was demonstrated. Shaggy borders with intraluminal debris suggested hepatic artery thrombosis (**a**), but Doppler ultrasound showed pulsatile flow within hepatic arterial branches. A follow-up tube cholangiogram (**b**) 3 months later showed partial resolution of the changes in the left intrahepatic bile duct, which is now smoothly dilated and has pseudodiverticula. Visceral arteriography was later performed to evaluate continuing laboratory abnormalities and demonstrated proximal obstruction of the left and right hepatic arteries; these reconstitute more peripherally through serpiginous collaterals (**c**). Additionally, a large segment of the dome of the liver receives only portal venous flow (**d**)

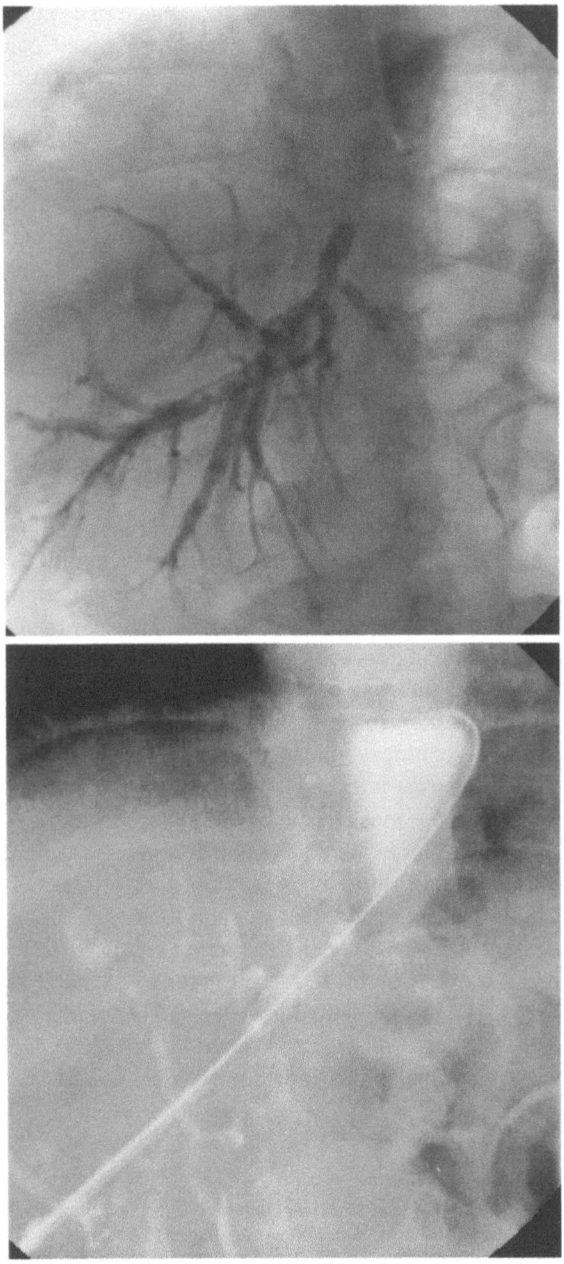

a

b

Fig. 14.14a,b. Biliary-enteric anastomotic stricture treated by balloon angioplasty after gaining access to the biliary tree by PTC in a 5-year-old boy. The patient received a reduced-size hepatic allograft (left lateral segment) at age 3 years and has continued to do well clinically. Follow-up laboratory studies detected mild, but progressive increase in γ-glutamyl transpeptidase. Near-complete obstruction was demonstrated at the biliary-enteric anastomosis (**a**). After crossing into the Roux-en-Y with a guidewire, balloon angioplasty (**b**) was performed and an internal/external stent placed for 3 months

ducts are dilated success approaches 100%. PTC can be performed easily in children and usually requires only conscious sedation and local anesthesia. Antibiotic coverage is essential, and coagulation parameters should be normalized if possible. Percutaneous biliary balloon dilation is at least transiently effective in treating most strictures; however, if extensive or recurrent, surgical revision or retransplant may be required (MacDonald 1991). Percutaneous techniques for dilating strictures (internal/external stenting) do prolong graft survival and often allow smaller patients to grow so that retransplantation may be performed more easily and with lower morbidity.

Bile leaks may result either from oozing at the cut surface of the liver, in proximity to the biliary-enteric anastomotic site, or from unincorporated or aberrant bile ducts (Fig. 14.15). Bile leaks are usually effectually treated by internal/external drainage (Lopez et al. 1992). Inspissated bile (sludge) and postbiopsy hemorrhage into the ductal system may result in biliary obstruction or sepsis, and may require irrigation of the biliary tree. This is usually performed following PTC with placement of an external drainage catheter. The biliary tree may be irrigated and obstructive debris cleared from within the biliary tree; however, if this debris is tenacious, mac-

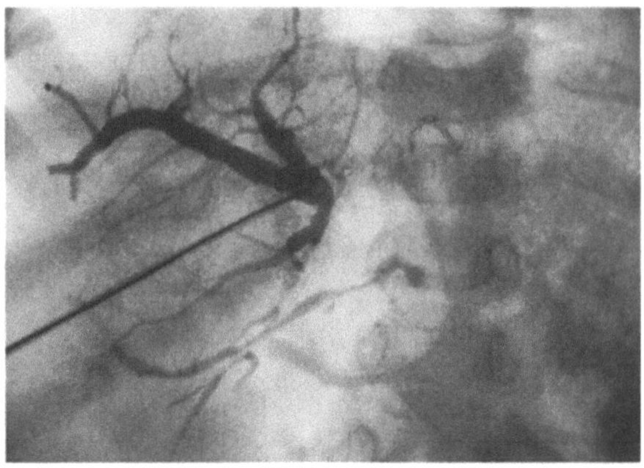

Fig. 14.15. Biliary-enteric anastomotic stricture in a 2-year-old girl who received a reduced-size hepatic allograft (left lateral segment). Transection of the superior and inferior branches of the left bile duct at organ harvest was unrecognized, and resulted in a postoperative biloma and surgical revision for creation of a second biliary-enteric anastomosis. PTC was performed for elevated γ-glutamyl transpeptidase; serum bilirubin was normal. The larger, more cephalad biliary-enteric anastomosis is occluded and has been accessed with a 0.018-inch (0.45-mm) guidewire and 3-French catheter. The smaller, more caudal branch of the left bile duct brains freely into the Roux-en-Y limb. The patient was treated successfully with internal/external stenting for 3 months

eration with a guidewire or stone-basket may be required to assist in its removal.

References

Abramson S, Treves S, Teele R (1982) The infant with possible biliary atresia: evaluation by ultrasound and nuclear medicine. Pediatr Radiol 12:1-5

Allendorph M, Werlin S, Geenen J, Hogan W, Venu R, Stewart E, Blank E (1987) Endoscopic retrograde cholangiopancreatography in children. J Pediatr 110:206-211

Ament M, Berquist W, Vargas J, Perisic V (1988) Fiberoptic upper intestinal endoscopy in infants and children. Pediatr Clin North Am 35:141-155

Babbitt DP (1969) Congenital choledochal cysts: new etiological concept based on anomalous relationships of the common bile duct and pancreatic bulb (multilingual). Ann Radiol (Paris) 12:231-240

Berci G (1992) Biliary ductal anatomy and anomalies. The role of intraoperative cholangiography during laparoscopic cholecystectomy. Surg Clin North Am 72:1069-1075

Broelsch CE, Emond JC, Thistlethwaite JR, Whitington PF, Zucker AR, Baker AL, Aran PF, Rouch DA, Lichtor JL (1988) Liver transplantation, including the concept of reduced-size liver transplants in children. Ann Surg 208:410-420

Brun P, Gauthier F, Boucher D, Brunelle F (1985) Ultrasound findings in biliary atresia in children. Ann Radiol 28:259-263

Brunelle F, Pariente D, Chaumont P (1994) Bile ducts. In: Liver disease in children: an atlas of angiography and cholangiography. Springer, London Berlin Heidelberg pp 95-118

Burton EM, Babcock DS, Heubi JE, Gelfand MJ (1990) Neonatal jaundice: clinical and ultrasonographic findings. South Med J 83:294-302

Campbell WL, Sheng R, Zajko AB, Abu-Elmaged K, Demetris AJ (1994) Intrahepatic biliary strictures after liver transplantation. Radiology 191:735-740

Caron KH, Strife JL, Babcock DS, Ryckman FC (1992) Left-lobe hepatic transplants: spectrum of normal imaging findings. AJR 159:497-501

Charels K, Kloppel G (1989) The bile duct system and its anatomical variations. Endoscopy 1:300-308

Chaumont P, Martin N, Riou J, Brunelle F (1982) Percutaneous transhepatic cholangiography in extrahepatic biliary duct atresia in children. Ann Radiol 25:94-100

Classen M, Gotze H, Richter HJ, Bender S (1987) Primary sclerosing cholangitis in children. J Pediatr Gastroenterol Nutr 6:197-202

Colonna J II, Shaked A, Gomes AS, Colquhoun SD, Jurim O, McDiarmid SV, Millis JM, Goldstein LI, Busuttil RW (1992) Biliary strictures complicating liver transplantation. Incidence, pathogenesis, management, and outcome. Ann Surg 216:344-350

Crittenden SL, McKinley MJ (1985) Choledochal cyst - clinical features and classification. Am J Gastroenterol 80:643-647

Debray D, Pariente D, Urvoas E, Hadchouel M, Bernard O (1994) Sclerosing cholangitis in children. J Pediatr 124:49-56

Dehner LP (1987) Liver, gallbladder, and extrahepatic biliary tract. In: Dehner LP (ed) Pediatric surgical pathology, 2nd edn. Williams and Wilkins, Baltimore, p 433

El-Shabrawi M, Wilkinson ML, Portmann B, Mieli-Vergani G, Chong SK, Williams R, Mowat AP (1987) Primary sclerosing cholangitis in childhood. Gastroenterology 92:1226-1235

Giuly J, Francois GF, Reynaud B (1992) Canaux biliaires aberrants et canaux cysto-hépatiques. Essai de synthèse (Aberrant biliary ducts and cysto-hepatic ducts. A comprehensive study.) (in French). Chirurgie 118:360-369

Grosfeld JL, Rescorla FJ, Skinner MA, West KW, Scherer L III (1994) The spectrum of biliary tract disorders in infants and children. Experience with 300 cases. Arch Surg 129:513-518

Guelrud M, Mendoza S, Jaen D, Plaz J, Machuca J, Torres P (1992) ERCP and endoscopic sphincterotomy in infants and children with jaundice due to common bile duct stones. Gastrointest Endosc 38:450-453

Heffron TG, Emond JC, Whitington PF, Thistlethwaite J Jr, Stevens L, Piper J, Whitington S, Broelsch CE (1992) Biliary complications in pediatric liver transplantation. A comparison of reduced-size and whole grafts. Transplantation 53:391-395

Holland RM, Lilly JR (1992) Surgical jaundice in infants: other than biliary atresia. (Review.) Semin Pediatr Surg 1:125-129

Iuchi K, Kusaka Y, Yamamoto S, Mori T, Ri T, Nakamoto K, Nakamura K, Hashimoto S, Nagaoka Y, Kawahara M (1983) Case of bronchobiliary fistula due to congenital anomalies of the liver and biliary tract (in Japanese). Nippon Kyobu Shikkan Gakkai Zasshi 21:282-287

Kasai M, Kimura S, Asakura Y (1968) Surgical treatment of biliary atresia. J Pediatr Surg 3:665-675

Keslar PJ, Buck JL, Selby DM (1993) From the archives of the AFIP. Infantile hemangioendothelioma of the liver revisited. (Review.) Radiographics 13(3):657-670.

Kirks DR, Coleman RE, Filston HC, Rosenberg ER, Merten DF (1984) An imaging approach to persistent neonatal jaundice. Am J Roentgenol 142:461-465

Knechtle SJ, Filston HC (1990) Anomalous biliary ducts associated with duodenal atresia. (Review.) J Pediatr Surg 25:1266-1269

Laurent J, Gauthier F, Bernard O, Hadchouel M, Odievre M, Valayer J, Alagille D (1990) Long-term outcome after surgery for biliary atresia. Study of 40 patients surviving for more than 10 years. Gastroenterology 99:1793-1797

Letourneau JG, Castañeda-Zuñiga WR (1990) The role of radiology in the diagnosis and treatment of biliary complications after liver transplantation. Cardiovasc Intervent Radiol 13:278-282

Li S, Stratta RJ, Langnas AN, Wood RP, Marujo W, Shaw B Jr (1992) Diffuse biliary tract injury after orthotopic liver transplantation. Am J Surg 164:536-540

Lilly JR (1975) The Japanese operation for biliary atresia: remedy or mischief? (Review.) Pediatrics 55:12-19

Lindor KD, Wiesner RH, Katzmann JA, LaRusso NF, Beaver SJ (1987) Lymphocyte subsets in primary sclerosing cholangitis. Dig Dis Sci 32:720-725

Lopez RR, Benner KG, Ivancev K, Keeffe EB, Deveney CW, Pinson CW (1992) Management of biliary complications after liver transplantation. Am J Surg 163:519-524

MacDonald CA (1991) Biliary atresia. J Pediatr Nurs 6:374-383

Mahboubi S, Sunaryo FP, Glassman MS, Patel K (1987) Computed tomography, management, and follow-up in infantile hemangioendothelioma of the liver in infants and children. J Comput Assist Tomogr 11:370-375

Majd M, Reba R, Altman R (1981) Effect of phenobarbital on ^{99m}Tc-IDA scintigraphy in the evaluation of neonatal jaundice. Semin Nucl Med 11:194-204

McDonald V, Matalon TA, Patel SK, Brunner MC, Sankary H, Foster P, Williams J (1991) Biliary strictures in hepatic transplantation. J Vasc Intervent Radiol 2:533–538
Montorsi M, Torzilli G, Fumagalli U, Bona S, Rostai R, De Simone M, Rovati V, Mosca F, Filice C (1994) Percutaneous alcohol sclerotherapy of simple hepatic cysts. Results from a multicentre survey in Italy. HPB Surg 8:89–94
Mueller PR, van Sonnenberg E, Simeone JF (1982) Fine-needle transhepatic cholangiography. Indications and usefulness. Ann Intern Med 97:567–572
O'Neill J Jr (1992) Choledochal cyst. (Review.) Curr Prob Surg 29:361–410
Pariente D, Bihet MH, Tammam S, Riou JY, Bernard O, Devictor D, Gauthier F, Houssin D, Chaumont P (1991) Biliary complications after transplantation in children: role of imaging modalities. Pediatr Radiol 21:175–178
Peclet MH, Ryckman FC, Pedersen SH, Dittrich VS, Heubi JE, Farrell M, Balistreri WF, Ziegler MM (1994) The spectrum of bile duct complications in pediatric liver transplantation. J Pediatr Surg 29:214–219
Picardi N, Monti M, Pasta V, Nudo R, Costantini D, Cassano C (1989) Rara malformazione delle vie biliari di interesse chirurgico. (Rare malformation of the bile ducts of surgical interest.) (In Italian.) Ann Ital Chir 60:399–403
Puchetti V, Modena S, Abrescia F, Canton C, Carolo F (1976) La nostra esperienza in tema di anomalie congenite delle vie biliari. [Our experience with congenital anomalies of the bile ducts (with presentation of 13 clinical cases).] (In Italian.) Chir Ital 28:632–664
Puente SG, Bannura GC (1983) Radiological anatomy of the biliary tract: variations and congenital abnormalities. World J Surg 7:271–276
Putnam PE, Kocoshis SA, Orenstein SR, Schade RR (1991) Pediatric endoscopic retrograde cholangiopancreatography. Am J Gastroenterol 86:824–830
Ryckman FC, Noseworthy J (1987) Neonatal cholestatic conditions requiring surgical reconstruction. (Review.) Semin Liver Dis 7:134–154
Sanchez-Urdazpal L, Gores GJ, Ward EM, Maus TP, Wahlstrom HE, Moore SB, Wiesner RH, Krom RA (1992) Ischemic-type biliary complications after orthotopic liver transplantation. Hepatology 16:49–53
Sanchez-Urdazpal L, Gores GJ, Ward EM, Hay E, Buckel EG, Wiesner RH, Krom RA (1993a) Clinical outcome of ischemic-type biliary complications after liver transplantation. Transplant Proc 25:1107–1109
Sanchez-Urdazpal L, Gores GJ, Ward EM, Maus TP, Buckel EG, Steers JL, Wiesner RH, Krom RA (1993b) Diagnostic features and clinical outcome of ischemic-type biliary complications after liver transplantation. Hepatology 17:605–609
Sheng R, Sammon JK, Zajko AB, Campbell WL (1994) Bile leak after hepatic transplantation: cholangiographic features, prevalence, and clinical outcome. Radiology 192:413–416
Sokal EM, Veyckemans F, de Ville de Goyet J, Moulin D, Van Hoorebeeck N, Alberti D, Buts JP, Rahier J, Van Obbergh L, Clapuyt P, Carlier M, Claus D, Latinne D, de Hemptinne B, Otte JB (1990a) Liver transplantation in children less than 1 year of age. J Pediatr 117:205–210
Sokal EM, de Ville de Goyet J, Buts JP, Habets S, Gosseye S, Clapuyt P, Claus D, Otte JB (1990b) Unifocal stricture of the common bile duct in two children: a localized form of primary sclerosing cholangitis. J Pediatr Gastroenterol Nutr 11:268–274
Summerville DA, Marks M, Treves ST (1988) Hepatobiliary scintigraphy in arteriohepatic dysplasia (Alagille's syndrome). A report of two cases. Pediatr Radiol 18:32–34
Tan CE, Howard ER, Driver M, Murray-Lyon IM (1993) Non-communicating multiseptate gall bladder and choledochal cyst: a case report and review of publications. (Review.) Gut 34:853–856
Van Sonnenberg E, Ferrucci J Jr, Neff CC, Mueller PR, Simeone JF, Wittenberg J (1983) Biliary pressure: manometric and perfusion studies at percutaneous transhepatic cholangiography and percutaneous biliary drainage. Radiology 148:41–50
Van Sonnenberg E, Wroblicka JT, D'Agostino HB, Mathieson JR, Casola G, O'Laoide R, Cooperberg PL (1994) Symptomatic hepatic cysts: percutaneous drainage and sclerosis. Radiology 190:387–392
Walia HS, Abraham TK, Baraka A (1986) Gall-bladder interposition: a rare anomaly of the extrahepatic ducts. Int Surg 71:117–121
Wiesner RH, Ludwig J, LaRusso NF, MacCarty RL (1985) Diagnosis and treatment of primary sclerosing cholangitis. (Review.) Semin Liver Dis 5:241–253
Wong DC, Masel JP (1995) Infantile hepatic haemangioendothelioma. Australas Radiol 39:140–144
Wong DK (1994) The accessory bile duct of Luschka and bile leakage in laparoscopic cholecystectomy. Hawaii Med J 53:164–165
Yamada N, Shinzawa H, Ukai K, Makino N, Matsuhashi T, Wakabayashi H, Togashi H, Takahashi T (1994) Treatment of symptomatic hepatic cysts by percutaneous instillation of minocycline hydrochloride. Dig Dis Sci 39:2503–2509
Zajko AB, Bron KM, Campbell WL, Behal R, Van Thiel DH, Starzl TE (1987) Percutaneous transhepatic cholangiography and biliary drainage after liver transplantation: a five-year experience. Gastrointest Radiol 12:137–143
Zauli D, Schrumpf E, Crespi C, Cassani F, Fausa O, Aadland E (1987) An autoantibody profile in primary sclerosing cholangitis. J Hepatol 5:14–18

Interventional Radiology
Benign Diseases

15 Percutaneous Cholecystostomy for Treatment of Benign Gallbladder Disease

R.L. VOGELZANG

CONTENTS

15.1 Introduction

Until about 10 years ago, the gallbladder was generally off limits to interventional radiologists. Despite its prominent and superficial position in the right upper quadrant of the abdomen, puncture and/or catheterization of the gallbladder was felt to entail an excessive risk of bile leakage and peritonitis. Sporadic reports of successful catheterization were published in the latter portion of the 1970s but it was not until 1982 that SHAVER et al. established that percutaneous drainage of the gallbladder could be accomplished (ELYADERANI et al. 1979; SHAVER et al. 1982; AMBERG and CHUN 1981). Since that time, a large number of reports have documented the safety and utility of percutaneous gallbladder catheterization (HAWKINS 1985; EGGERMONT et al. 1985; MCGAHAN and LINDFORS 1989; TEPLICK et al. 1990; LO et al. 1995; BOLAND et al. 1993, 1994a; BROWNING et al. 1993). These efforts initially focused on the use of percutaneous cholecystostomy for decompression of acute cholecystitis, but the scope of utilization quickly broadened to attempts to treat cholelithiasis, with the use of a large array of techniques and devices. For the most part those efforts, although initially successful, have fallen by the wayside as a result of the parallel development of laparoscopic cholecystectomy. Despite the lack of broad use of percutaneous gallstone extraction and/or dissolution, percutaneous cholecystostomy is now an accepted interventional technique that is performed throughout the world.

In spite of the fact that most authors to date have reported high technical success rates, catheterization of and subsequent intervention in the gallbladder can be problematic, occasionally even disastrous, because of the peculiar nature of the gallbladder, which is a thin-walled, pliable, mobile, and distensible organ that has inconstant and variable attachments to the liver. In this chapter we will concentrate on the technical, anatomic, and clinical aspects of gallbladder catheterization.

R.L. VOGELZANG, MD, Professor of Radiology, Northwestern University Medical School, and Chief, Vascular and Interventional Radiology, Northwestern Memorial Hospital, 710 North Fairbanks Coust, Chicago, IL 60611 USA

15.2 Gallbladder Disease: Scope of the Problem

Gallbladder disease (primarily related to the development of gallstones) is extremely common and affects a large percentage of the population of most countries of the world. In the United States, 20 million people have gallbladder disease and close to a million individuals develop new gallstones every year. Gallstones are also more common in the elderly (GLENN and DILLON 1980; SCHOENFELD and LACHIN 1981; RANOFSKY 1978; MCGAHAN 1990). The clinical implications and complications of cholelithiasis and associated acute or chronic cholelithiasis are substantial. They include choledocholithiasis, gallstone pancreatitis, and acute cholecystitis with gallbladder perforation. Acute abdominal pain and sepsis related to cholecystitis are also common reasons for admission to a hospital. In the group of patients who are already hospitalized, acalculous cholecystitis is a distinctive variant of cholecystitis which tends to affect the acutely ill patient in the intensive care unit. Finally, gallbladder carcinoma is one of the most feared complications of untreated chronic gallstone disease.

Cholecystectomy is the preferred treatment for gallbladder disease, with laparoscopic cholecystectomy having almost completely replaced open cholecystectomy in the surgical armamentarium. This change in surgical technique has sharply reduced postoperative morbidity for the majority of patients. Around 500 000 cholecystectomies are performed annually in the United States (RANOFSKY 1978). Despite the widespread use of and low morbidity and mortality associated with cholecystectomy, a cohort of patients still exists for whom surgical removal of the gallbladder remains a procedure with excessive morbidity and/or mortality. Patients in this risky category include those with systemic disease and/or comorbidities such as cardiac, renal, or pulmonary disease. Other patients in whom cholecystectomy is poorly tolerated are those with systemic malignancies, those with so-called "hostile" abdomen as a result of multiple previous surgical operations, and patients who have recently undergone major surgical procedures in the chest or abdomen. For these patients, percutaneous cholecystostomy is an ideal acute solution. It is a rapid and simple procedure requiring only local anesthesia for decompression of the gallbladder in acute cholecystitis and critically ill patients. It offers a safe alternative to cholecystectomy and now functions as the replacement for surgical cholecystostomy in patients who are not good surgical candidates. In these situations, catheter drainage of the gallbladder can be life saving as a temporizing procedure that allows the patient's condition to improve until elective cholecystectomy can be safely performed later.

15.3 Indications for Percutaneous Cholecystostomy

The current indications for this procedure include: (1) treatment of acute cholecystitis, including empyema and perforation, (2) gallbladder access for stone dissolution, fragmentation, and removal, (3) drainage of common bile obstruction, and, rarely, (4) performance of diagnostic cholangiography (LO et al. 1995; BURTON et al. 1993; VOGELZANG and NEMCEK 1988; GERVAIS and MUELLER 1996).

15.3.1 Acute Cholecystitis

Drainage of acute cholecystitis remains the most commonly performed interventional gallbladder procedure. Virtually all patients with acute cholecystitis referred for the procedure will be critically ill and at high risk for significant surgical morbidity and mortality if cholecystectomy is performed. In the past, operative tube cholecystostomy, often performed with the patient under local anesthesia, was the temporizing procedure of choice to allow these patients to weather the course of a concomitant illness that contraindicated more extensive surgery (SKILLINGS et al. 1980). Despite the use of this less morbid procedure, these patients still had very high rates of in-hospital mortality, ranging from 23% to 59%, which reflected the severity of other coexistent diseases (LO et al. 1995).

Acute cholecystitis can be a very difficult disease to diagnose in critically ill patients who are in the hospital. These patients often do not show typical symptoms of acute gallbladder conditions and their clinical signs may be misleading due to recent trauma, surgery, and pain medications or the use of antibiotics. Noninvasive imaging like ultrasound and/or hepatobiliary scanning is often used to confirm clinical suspicions, but these techniques have some limitations. Biliary scintigraphy, for example, is generally very accurate in the diagnosis of acute cholecystitis, but it may be falsely positive in patients who are fasting or who have diseases or intercurrent illness which leads to decreased bile production and/or stasis (LO et al. 1995; KAFF et al. 1983; LARSEN et al. 1982).

Percutaneous gallbladder catheterization is now the replacement for operative placement of a gallbladder tube. A number of reports have documented the reliability and efficacy of this technique (HAWKINS 1985; EGGERMONT et al. 1985; MCGAHAN and LINDFORS 1989; TEPLICK et al. 1990; LO et al. 1995; BOLAND et al. 1993, 1994a; BROWNING et al. 1993; LEE et al. 1991). Recently, we reported on the use of this technique for the treatment of acute calculous and acalculous cholecystitis. The results of our study mirror to a large degree the overall patient population with which interventional radiologists are now confronted. We performed ultrasound-guided cholecystostomy in 58 consecutive hospitalized patients with suspected cholecystitis who were not surgical candidates. Twenty-eight were thought to have calculous disease and 30 were believed to have acalculous cholecystitis. The gallbladder was successfully catheterized in all 58 patients. Forty-eight of 58 patients (83%) had a final diagnosis of acute cholecystitis. Clinical benefit was seen in 93% of patients with gallstones and in 80% of patients with acalculous cholecystitis (Fig. 15.1). The six patients who did not respond had extensive transmural inflammation and/or gangrene of the gallbladder wall (Fig. 15.2) (Lo et al. 1995). Our experience points out what others have also found: acute cholecystitis in these patients is a difficult diagnosis to make, and a good percentage of patients who undergo gallbladder drainage do *not* have cholecystitis (Browning et al. 1993; Boland et al. 1994a; Lee et al. 1991). In our series 17% of our population fell into this category. Furthermore, some patients will have extensive gallbladder necrosis and will not respond to simple decompression of the gallbladder. These patients must be operated upon.

15.3.2 Gallbladder Access for Stone Removal

Access to the gallbladder for subsequent removal of stones is now an infrequently performed procedure. There was an early wave of enthusiasm for dissolution (with methyl *tert* butyl ether), pulverizing (with a rotary device modeled after a rotation atherectomy catheter or direct contact lithotripsy), or simple fragmentation and extraction of gallstones (May and Thistle 1986; Laffey and Margin 1986; Lindberg et al. 1993) (Fig. 15.3). Still other investigators vigorously pursued attempts at percutaneous gallbladder ablation with a variety of agents (Yedlicka et al.

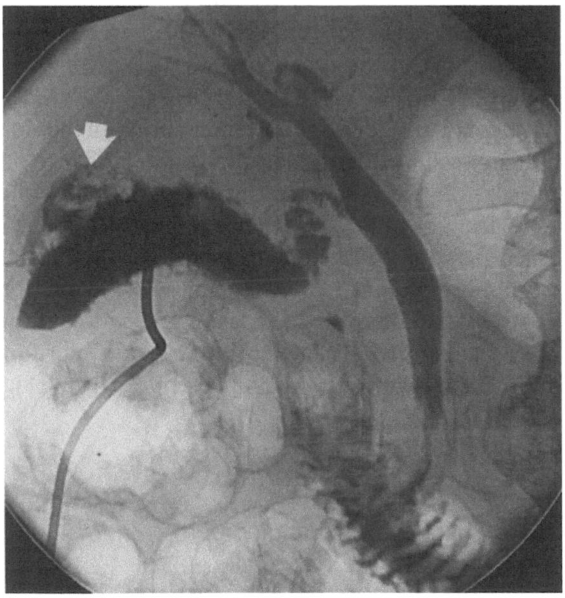

Fig. 15.1. Percutaneous cholecystostomy for drainage of acute cholecystitis with perforation. Cholangiogram 3 days after placement of cholecystostomy demonstrates irregularity of gallbladder wall, as well as extravasation of contrast from the site of perforation (*allow*). The patient improved markedly after drainage and was able to undergo successful open cholecystectomy under improved clinical conditions

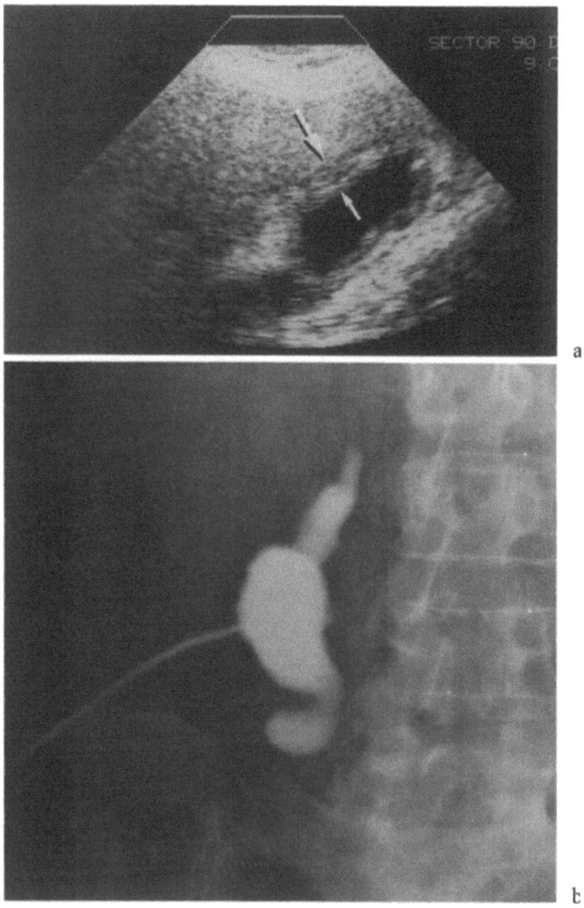

Fig. 15.2a,b. Gangrenous cholecystitis. **a** This 54-year-old patient with right upper quadrant pain of 5 days' duration underwent percutaneous cholecystostomy. Note the markedly thickened edematous gallbladder wall (*arrows*). Despite obvious ultrasound and nuclear medicine findings of acute cholecystitis, the patient did not respond to percutaneous drainage. She was operated upon and gangrenous gallbladder wall was found. In patients who are strongly felt to have cholecystitis and who do not respond to percutaneous cholecystostomy, the possibility of gangrenous cholecystitis or gallbladder wall necrosis should be considered. **b** Cholangiogram 24 h after drainage of gallbladder in the same patient. The gallbladder remains somewhat thick-walled and no cystic duct filling is seen. The patient had not improved and underwent surgery, when gangrenous gallbladder wall was found

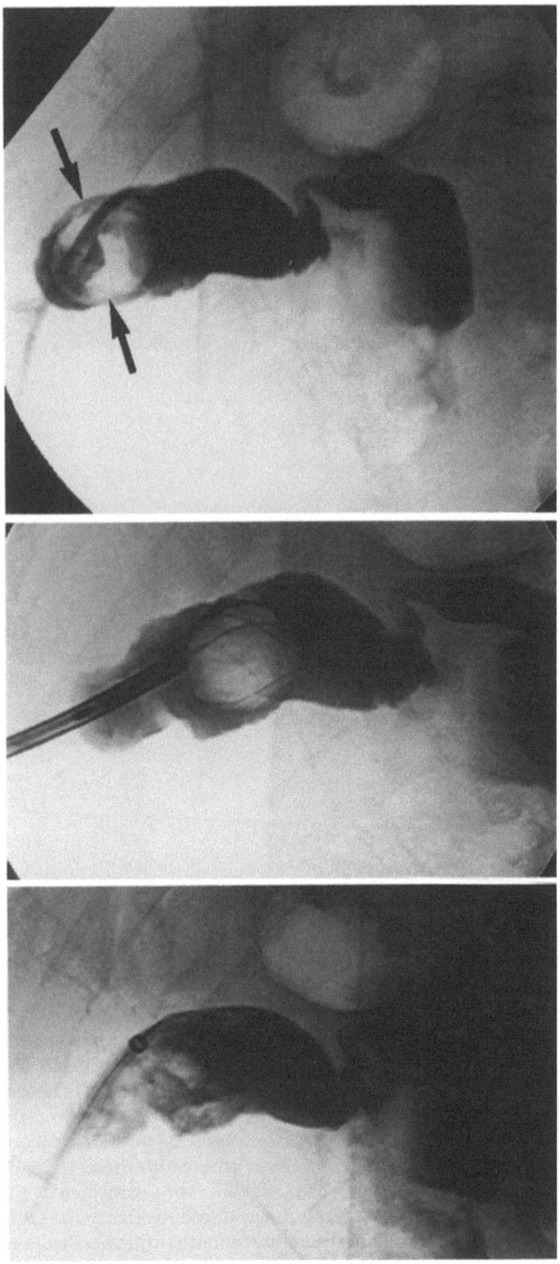

a b c

Fig. 15.3a–c. Mechanical fragmentation and extraction of gallstones. **a** Large solitary gallstone (*arrows*) in the gallbladder 2 weeks after percutaneous drainage for acute episode of cholecystitis. **b** A large sheath was placed and a stone-crushing basket was used to ensnare the stone. The stone was fragmented. **c** Following extraction of large fragments, some smaller fragments remain in the fundus of the gallbladder. These were effectively removed

1993; Becker et al. 1988; Getrajdman et al. 1985; Becker and Burhenne 1991). Some of these attempts were successful in humans (Becker and Burhenne 1991), but the parallel development of laparoscopic cholecystectomy essentially preempted the widespread use of these techniques. Several groups, however, continue to use percutaneous gallstone extraction selectively to good effect. Picus et al. have demonstrated the long-term efficacy of their approach which uses percutaneous access to the gallbladder and tract dilatation with direct endoscopic fragmentation and extraction of stones. They have used the technique successfully in over 100 patients and continue to perform the procedure on a selected group of high-risk individuals (Picus et al. 1989; Burton et al. 1993) (Fig. 15.4). Other authors also continue to use gallstone extraction techniques selectively (Boland et al. 1994b).

15.3.3 Drainage of Common Duct Obstruction

The majority of distal common bile duct obstructions are now managed by endoscopic means; however, in patients with sepsis and distal common bile duct obstruction who are candidates for percutaneous drainage, we have substituted the safer, quicker, and easier performance of a percutaneous cholecystostomy for transhepatic drainage, which we believe is more morbid. The technique is particularly helpful in patients who have minimal biliary dilatation, but its use is not confined to them. In our experience, most of these patients have sepsis or other conditions that do not permit transfer to the radiology department. Bedside placement of a catheter into a dilated gallbladder under ultrasound guidance often permits life-saving treatment of a patient too ill to tolerate any other procedure (Fig. 15.5). We also place percutaneous cholecystostomy catheters for temporary drainage of malignant distal common duct obstruction prior to surgical procedures if the operating surgeon wishes to lower serum bilirubin and/or improve liver function preoperatively (Fig. 15.6).

15.3.4 Diagnostic Cholangiography

Diagnostic cholangiography can also be performed by injection of contrast into the gallbladder. We and others have described the performance of cholangiography as a diagnostic study or as a maneuver preparatory to performing percutaneous biliary drainage, in which placement of a catheter into a dilated gallbladder allows controlled ductal opacification. This technique is rarely used.

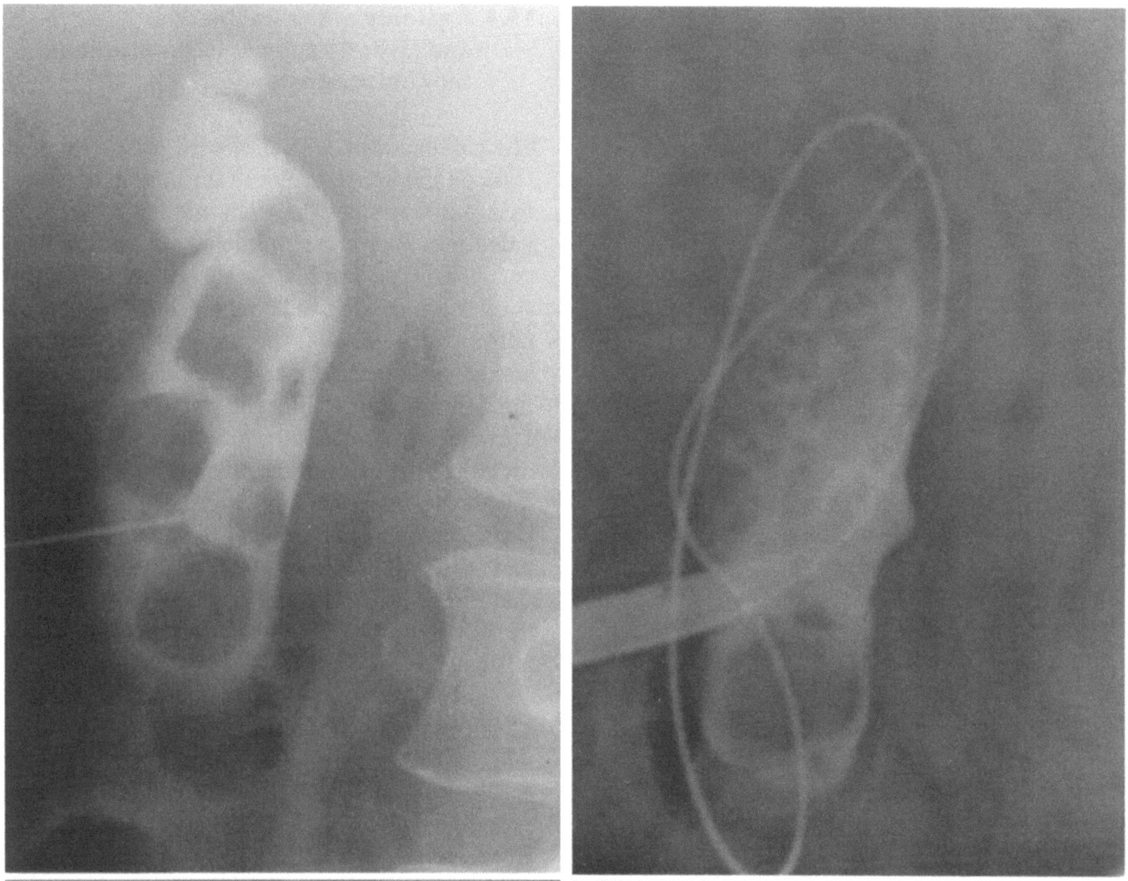

a c

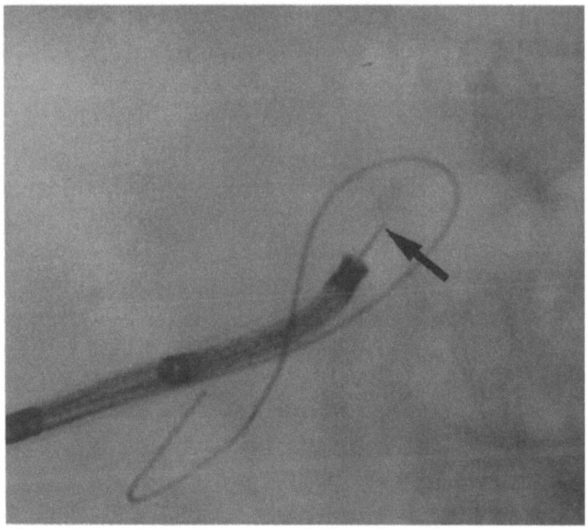

b

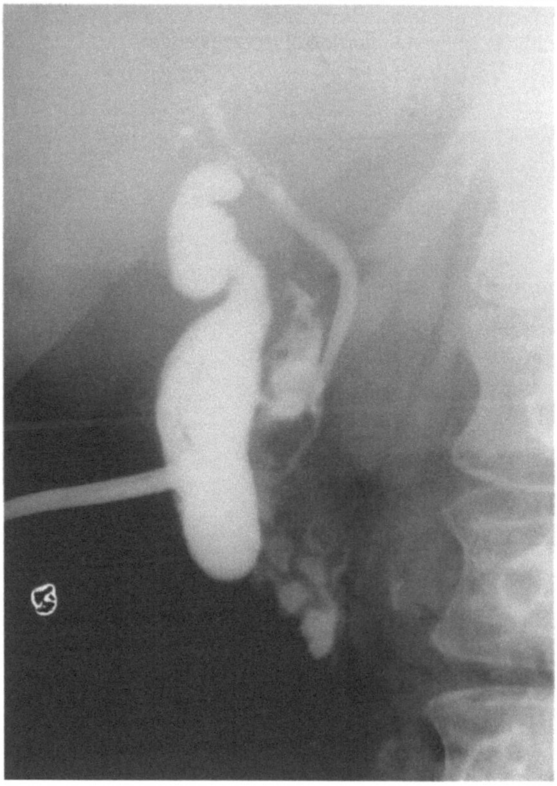

d

Fig. 15.4a–d. Percutaneous gallstone lithotripsy. Treatment of large gallstones with cholecystostomy and percutaneous endoscopic gallstone lithotripsy in an elderly patient unable to undergo cholecystectomy. **a** Initial cholecystostomy for gallbladder access showing multiple large faceted stones. **b** After tract dilatation and insertion of biliary endoscope. Spot film showing lithotrite probe (*arrow*) emerging from the endoscope in contact with a stone. **c** During the procedure multiple small stone fragments from lithotripsy are visible. A single large stone remains in the fundus. **d** Completion cholangiogram after stone removal. (Courtesy Daniel Picus, M.D., Mallinckrodt Institute of Radiology, Washington University School of Medicine)

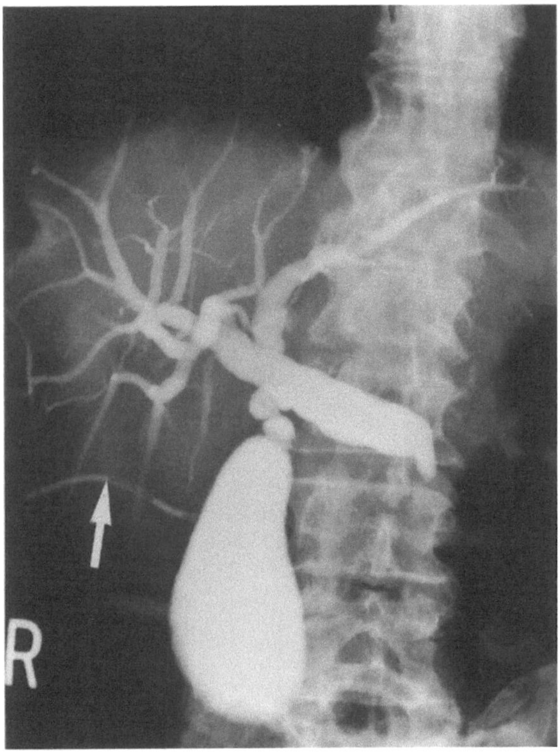

Fig. 15.5. Drainage of the gallbladder for relief of malignant bile duct obstruction. A 57-year-old patient with pancreatic carcinoma with malignant biliary obstruction. The gallbladder was used to decompress the biliary tract preoperatively before resection of the pancreatic mass. Note good position of the catheter in the mid portion of the gallbladder

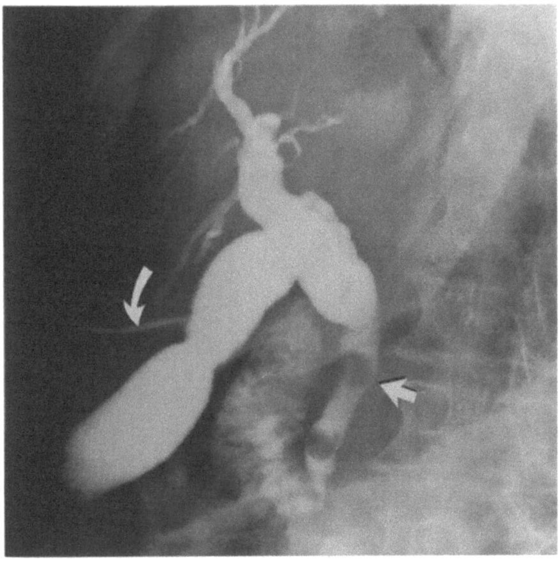

Fig. 15.6. The use of percutaneous cholecystostomy for treatment of choledocholithiasis. Two stones in the distal common bile duct (*straight arrow*) cause biliary obstruction and sepsis. Placement of a percutaneous cholecystostomy catheter (*curved arrow*) alleviated the acute septic episode

15.4 Anatomy of the Gallbladder and Liver Attachments: Implications for Cholecystostomy

The gallbladder is a pear-shaped organ lying in the fossa on the undersurface of the liver in line with the fissure that divides the right and left lobes. It is divided into four areas: fundus, body, infundibulum, and neck. The fundus usually extends slightly beyond the liver edge and contains most of the smooth muscle in the organ. The body is the main storage area and has elastic fibers. The neck continues from the infundibulum into the small, tortuous cystic duct, whose point of entry into the extrahepatic duct divides the common hepatic duct from the common bile duct. The blood supply to the gallbladder is the cystic artery, which usually arises from the right hepatic artery. Venous return is via small, short veins that enter the liver through the bare area directly and through the cystic vein which drains into the right portal vein (McNulty 1990; Berk 1983; Schwartz 1979).

The cystic duct, the connection to the biliary system, usually enters the common hepatic duct at an acute angle, but variations in its course which change the point at which the cystic duct inserts are common and include a shorter or absent duct and a duct that runs parallel to the common duct and inserts as far distally as the duodenum. Insertion on the left side of the common hepatic duct is also seen. Anomaly of the gallbladder in number, form, and/or position is of great importance for the performance of percutaneous procedures. Very rarely, the gallbladder may be duplicated, absent, left-sided, or retrodisplaced. More commonly seen are the variations in position, which mainly result from different attachments of the organ to the liver. It is the latter that most concern the interventional radiologist in his or her attempt to catheterize the organ either transhepatically or transperitoneally. This thin, muscular sac also varies considerably in size, shape, and position depending on the activity of the gastrointestinal tract, the degree of filling, the amount of distention of the transverse colon adjacent to it, and the body habitus of the individual.

Normally, the superior aspect of the neck and body of the gallbladder is firmly attached to the serosa of the undersurface of the liver by the peritoneum. This extraperitoneal plane of fixation between the gallbladder and liver is known as the bare area. The size and shape of the bare area are quite variable, however, and an increase in the percentage of the gallbladder circumference that is covered or invested by the peritoneum will result in progressively more gallbladder motility or floppiness. The gallbladder may also be suspended from the liver by a complete mesentery, or only the neck may be attached, with the fundus and body hanging freely. This very mobile gallbladder (it is said to occur in about 5% of patients), may result in torsion and, from an interventional radiologist's point of view, may cause considerable problems as the gallbladder is very difficult to puncture since it moves away from an advancing catheter or guidewire.

It would seem then that such a highly mobile organ should be entered through a point of maximum extraperitoneal fixation to stabilize the deformable structure during catheter and guidewire passage and prevent leakage of bile into the peritoneum. Transhepatic puncture of the gallbladder has accordingly been the most widely recommended technique for performance of percutaneous cholecystostomy (Lo et al. 1995; Vogelzang and Nemcek 1988; Picus et al. 1989), although others have used transperitoneal puncture, including Garber et al., who recently reported on a large series of patients with no more complications than in other series where transhepatic placement was performed (Garber et al. 1994). Those authors who advocate transperitoneal puncture cite avoidance of injury to the hepatic parenchyma as a major advantage, which may be true: in our series one patient died as a result of a major hepatic bleeding episode from inadvertent puncture of an aberrant hepatic capsular artery (Lo et al. 1995). At present, however, the most that can be said is that acceptable results have been reported with both techniques and, to our knowledge, no se-

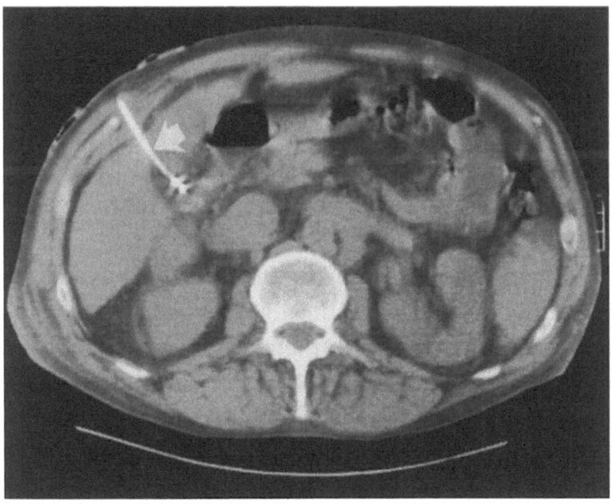

Fig. 15.7. Transhepatic puncture of the gallbladder. Computed tomogram showing percutaneous cholecystostomy catheter (*arrow*) passing through the liver into the gallbladder

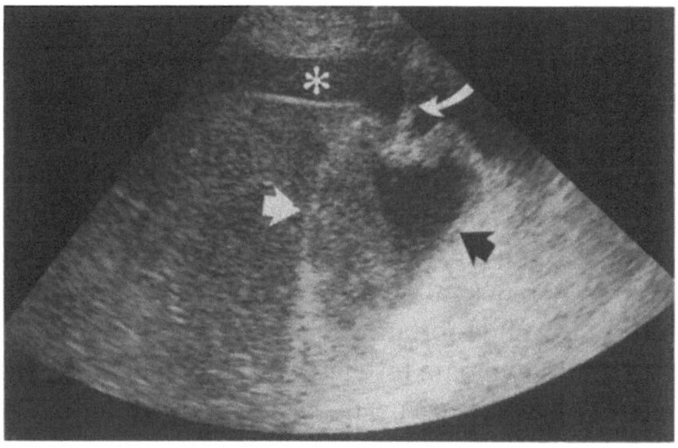

Fig. 15.8. Transperitoneal puncture of the gallbladder. Necrotic gallbladder (*straight arrows*) was actively leaking bile, producing ascites (*asterisk*). Direct puncture of the fundus was necessary because of the anatomic position of the gallbladder. An 18-gauge needle (*curved arrow*) indents the fundus of the gallbladder. The gallbladder was drained successfully without incident

ries of patients or work yet exists to document the superiority of either approach.

In an attempt to take this question further and answer it one way or the other, we studied the anatomy of the liver and gallbladder attachments in 19 cadavers and found that the bare area could not be reliably punctured using ultrasound guidance. Of 19 catheters placed under ultrasound guidance transhepatically into the gallbladder, only 8 (42%) traversed the bare area and the other 58% entered the free peritonealized gallbladder wall immediately adjacent to the serosal attachment. We concluded that the bare area was not reliably imaged or punctured, and that transhepatic catheterization seemed to facilitate but in no way insured puncture of the gallbladder across the bare area, due to the variability in the gallbladder and liver attachment (Nemcek et al. 1991). Also, the studies previously alluded to seemed to indicate that free wall puncture is not more morbid than transhepatic puncture. We ourselves, however, continue to prefer the transhepatic approach, because transhepatic catheterization does have the advantage that the liver provides catheter stability (Fig. 15.7). We do not completely eschew transperitoneal puncture, however; we use it when we think it offers the best way to access the gallbladder and/or solve the problem (Fig. 15.8).

15.5 Technique of Percutaneous Cholecystostomy

Patient preparation for percutaneous cholecystostomy depends on the indication for the procedure.

For the most part, the patients we now see are very ill and frequently are in the intensive care unit. Many have significant problems such as sepsis, hemodynamic instability, cardiac, and/or pulmonary conditions which add to the risk of the procedure. If the procedure is indicated, we make every effort to optimize the patient's condition before proceeding, but we rarely refuse to perform the procedure since the morbidity and mortality are very low.

In the candidate for elective cholecystostomy (an increasingly rare event) the usual preprocedural testing is performed.

15.5.1 Imaging Guidance

Real-time ultrasound, in our opinion, is mandatory for the performance of all percutaneous gallbladder procedures. We like to use adjunctive fluoroscopy to permit manipulation of catheters and guidewires, but X-ray guidance is not essential, as evidenced by the large number of successful gallbladder procedures performed at the bedside by ourselves and others. Real-time ultrasound guidance (which *is* absolutely necessary) allows accurate localization and assessment of the position of the liver and gallbladder so that appropriate transhepatic or transperitoneal approaches can be made (Fig. 15.9). Placement of the catheter under ultrasound control can be done with a minimum of manipulation. We believe strongly that the use of ultrasound for this procedure is mandatory. We do not perform percutaneous gallbladder drainage without real-time ultrasound guidance, even when the gallbladder is still opacified from previous percutaneous or endoscopic procedures.

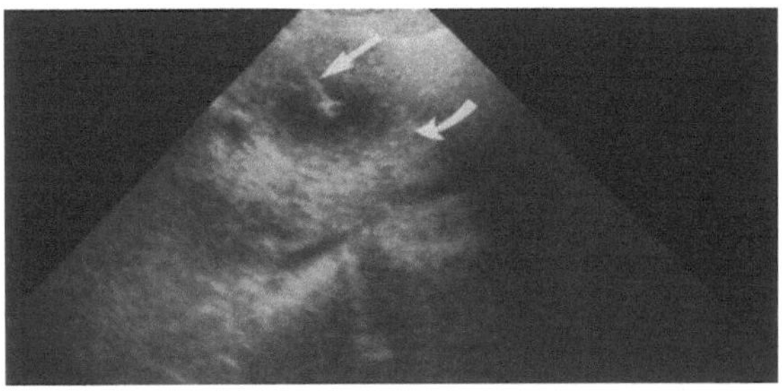

Fig. 15.9. Real-time ultrasound guidance for gallbladder puncture. Echogenic needle (*straight arrow*) enters a markedly thickened edematous gallbladder (*curved arrow*). Ultrasound guidance is essential for ascertaining when the catheter is correctly placed in the gallbladder

15.5.2 Pain Management

Pain management is an important aspect of any procedure. The pain caused by percutaneous cholecystostomy is usually well managed by intravenous sedation and regional anesthetic techniques, such as local anesthesia and/or the use of intercostal blocks.

15.5.3 Access Techniques and Catheter Systems

Catheterization of the gallbladder can be accomplished by a variety of techniques. Many people use the trocar technique, but the majority of practicing interventionalists (including ourselves) continue to use standard or modified guidewire exchange methods. Either technique is perfectly acceptable and the experience and skill of the operator is the primary determinant of which technique should be used. Some of the equipment available for percutaneous entry into the gallbladder includes trocar sets in which the catheter and its inner stiff sharpened stylet are directly placed into the gallbladder. Once the device lies within the lumen, the stylet is removed and the catheter advanced over the fixed, rigid cannula (McGahan 1988). Other devices include removable hub needles and so called "one-stick" and access systems (van Sonnenberg et al. 1986, 1990). These widely used systems allow entry of the catheter into the target with a 20- or 21-gauge needle. Through this thin needle (which can be placed into the gallbladder with a minimum of trauma), a small wire (0.018 inch diameter) is advanced into the organ and the needle removed. Conversion to a standard size wire is accomplished by the use of a special catheter. Finally, a standard 18- or 19-gauge needle can be placed into the gallbladder, through which a heavy duty guidewire can be directly placed. We actually prefer this method as it is cheaper, simpler, and quicker than any other. The larger needle is also much easier to see with ultrasound than smaller needles.

A large range of drainage systems are available, including catheters specifically designed for gallbladder use. Many types of catheters can be used, but the most important feature a catheter should have is a Cope loop or some other self-retaining design (McGahan 1988; Andrews and Hawkins 1984). The use of self-retaining catheters is important in the gallbladder since catheter dislodgment tends to occur at a higher rate here than in other organs, probably because of excessive respiratory excursion of the liver and gallbladder. Accidental dislodgment can lead to significant bile leaks and/or hepatic bleeding and should be avoided whenever possible by good catheter fixation.

15.5.4 Performance of Cholecystostomy

After selection and anesthetization of the skin entry site, the area and ultrasound transducer are prepared and sterilely draped. Under real-time guidance, the needle is advanced through the anterior wall with a short jab. This motion is often necessary because normal gallbladder wall pliancy or diffuse inflammatory thickening of the gallbladder wall can impede needle penetration. After entry of the gallbladder, needle position within the structure is confirmed, bile aspirated, and a guidewire advanced into the lumen. After tract dilatation, an 8- or 10-French self-retaining catheter is placed within the gallbladder and all bile aspirated. Catheter position may be secondarily confirmed by injection of a small (less than 10 ml) amount of contrast agent. At the bedside the catheter can usually be seen with ultrasound or a small amount of air can be injected to confirm the position. The catheter is placed to gravity drainage without further manipulation and bile sent for appropriate culture and gram stain.

15.5.5 Complications and Technical Problems

Problems happen during the procedure despite the best efforts of the interventionalist. Most are uncommon and can be prevented with good technique, but even then difficulties can occur. There can be failure to penetrate the gallbladder wall because of marked inflammatory thickening or rapid movement of the gallbladder away from a catheter needle or guidewire. Recognition of the problem as it occurs is paramount to avoid guidewire buckling and dislodgment. Fluoroscopy plays a vital role here since constant visualization of the guidewire and/or catheter can prevent the buckling which may occur as an attempt is made to pass through the gallbladder wall (Fig. 15.10). If buckling occurs, a large loop of catheter and/or guidewire may form and the examiner will be surprised to find the catheter dislodged and lying completely free in the peritoneum. Occasionally, the gallbladder may be so difficult to catheterize that the only catheter which can be passed into the lumen may be a small 4- or 5-French angiographic catheter. When this happens, the catheter should be left in place for 24–48 h, after which tract formation will allow placement of a definitive tube.

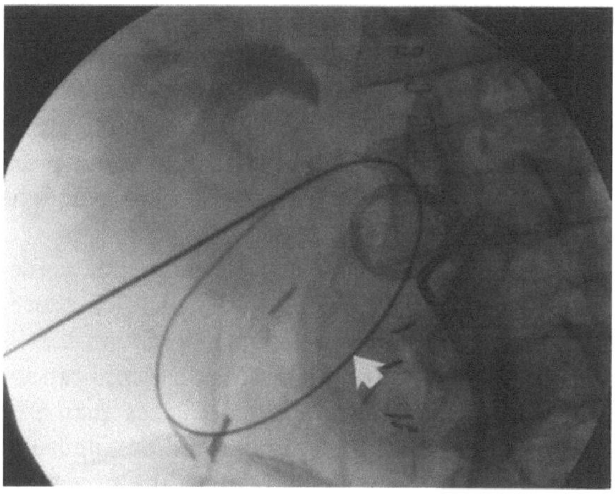

Fig. 15.10. Utility of fluoroscopic guidance. Spot film from fluoroscopic procedure showing guidewire (*arrow*) coiled within the gallbladder prior to tract dilatation. Fluoroscopy enables visualization of the guidewire to prevent buckling and guidewire dislodgment which can occur during catheterization of thick-walled gallbladders

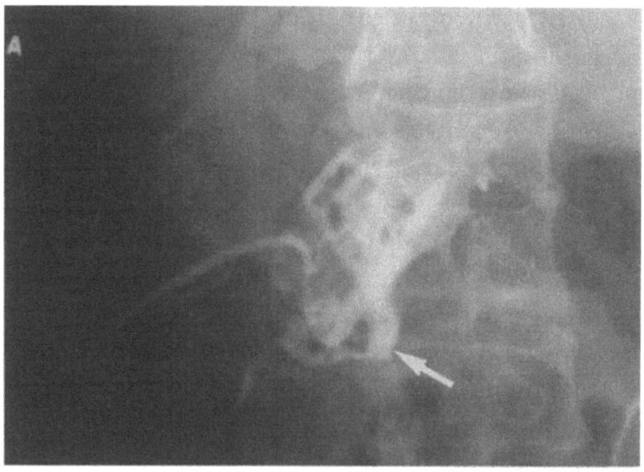

Fig. 15.11. Complication of cholecystostomy. Perforation of gallbladder wall opposite the insertion site with a small amount of contrast extravasation around the edge of the liver (*arrow*). The patient had pain as a result of this event. The catheter was repositioned without incident

Other problems include perforation of the opposite wall of the gallbladder, which is usually handled by simple withdrawal of the catheter (Fig. 15.11), or occasionally the gallbladder may be completely decompressed, leaving a totally collapsed and leaking gallbladder that cannot be recatheterized. In such cases, we have been able to prevent bile peritonitis by leaving a drainage catheter in the subhepatic space to aspirate and drain the leak.

Some authors have reported severe vasovagal reactions to occur during percutaneous cholecystostomy (VAN SONNENBERG et al. 1990). We have not encountered this complication in over 200 patients, but continue to be aware of the potential for its occurrence. The prophylactic administration of atropine sulfate is not a routine part of our practice, but the drug certainly is available in the interventional suite.

Most reports indicate that percutaneous cholecystostomy is a very safe and well tolerated procedure, even in the very ill patient population that is best served by this technique. Major complications have consistently been reported in less than 5%; most of these are septic episodes and/or bile leaks. Some deaths directly attributable to the procedure have been seen, but percutaneous catheterization of the gallbladder has clearly been shown to be safe.

15.5.6 Postprocedural Care

In patients with acute inflammatory conditions (who constitute the majority of our patients), catheter manipulation or injection is avoided for at least 24–48 h until the patient improves clinically. After 1 or 2 days, the irregular edematous gallbladder wall often reverts to a more normal appearance and cystic duct patency is reestablished (Figs. 15.12, 15.13). The gallbladder must also show signs that it functions adequately and can empty in response to the usual gastrointestinal stimuli. In general, we will clamp the catheter for about 24 h after the patient is eating; lack of symptoms and a nondistended gallbladder demonstrate that the catheter can be removed. Prior to removal we strongly recommend that assessment of the catheter tract be done by injection of contrast, making what we call a "tractogram," to be sure that leakage of bile into the peritoneum will not occur after catheter removal (Fig. 15.14).

In patients with gallstone disease, the catheter should not be withdrawn or removed unless the stones are percutaneously removed or the cholecystectomy is done. Nowadays, the gallbladder is removed in the vast majority of cases, although percutaneous gallstone removal can be effective in managing these patients (BURTON et al. 1993). In desperately ill patients unable to have their gallbladders removed, cholecystostomy can serve as a means of external drainage for an indefinite period of time. In our clinical series of 48 patients with cholecystitis, 18 underwent cholecystectomy, 25 recovered and had their catheters removed, and 5 died of other causes with the catheters in place (Lo et al. 1995).

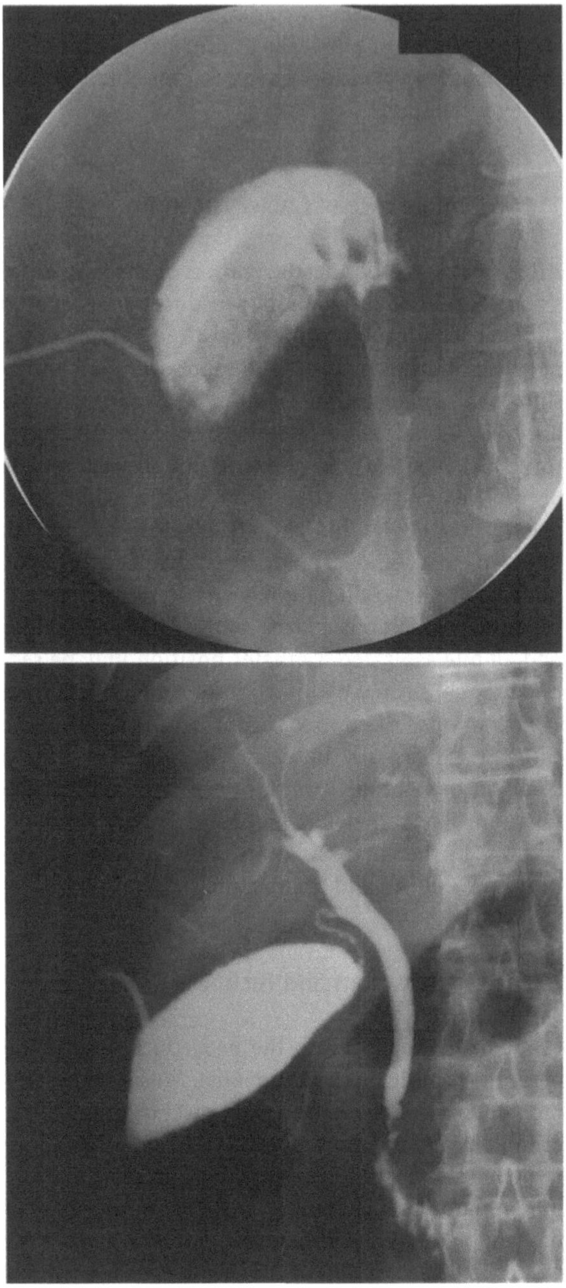
a
b

Fig. 15.12a,b. Typical appearance after percutaneous cholecystostomy. **a** Immediately following drainage, the gallbladder shows edematous wall and thickening. **b** Follow-up cholangiography in another patient after drainage of acute acalculous cholecystitis shows typical appearance of a relatively normal appearing gallbladder and widely patent biliary ducts

Other authors have reported similar results (HAWKINS 1985; EGGERMONT et al. 1985; McGAHAN and LINDFORS 1989; TEPLICK et al. 1990; Lo et al. 1995; BOLAND et al. 1993, 1994a; BROWNING et al. 1993).

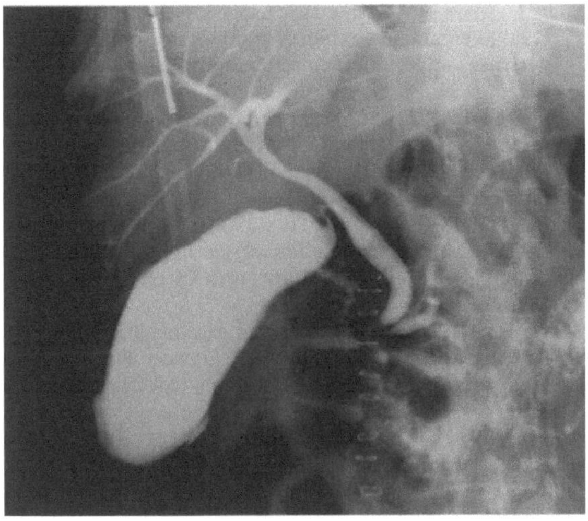

Fig. 15.13. Normal cholangiogram 4 days after percutaneous cholecystostomy for acute acalculous cholecystitis in a postoperative patient. There is filling of the gallbladder and bile ducts with flow into the duodenum

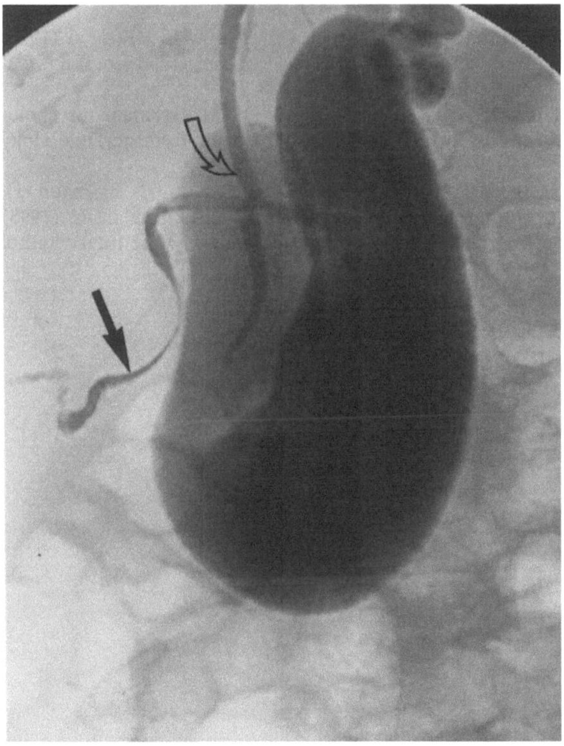

Fig. 15.14. "Tractography" showing intact tract (*straight arrow*) into gallbladder without leakage into the peritoneum, indicating it is safe to withdraw the catheter. Note that the tract crosses a biliary radical (*curved arrow*). This occurred without clinical sequelae

15.6 Conclusion

Percutaneous cholecystostomy is a valuable interventional radiology procedure that has proven to be of great benefit in a variety of clinical circumstances. While the initial enthusiasm for the use of gallbladder access for stone removal has waned considerably, the acutely ill patient with acalculous or calculous cholecystitis continues to be an important beneficiary of percutaneous gallbladder drainage.

References

Amberg JR, Chun G (1981) Transcystic duct treatment of common bile stones. Gastrointest Radiol 6:361–362

Andrews RC, Hawkins IF (1984) The Hawkins needle-gauge system for percutaneous catheterization. I. Instrumentation and procedure. AJR 142:1191–1195

Becker CD, Burhenne HJ (1991) Percutaneous ablation of the cystic duct and the gallbladder: experimental and early clinical results. Semin Roentgenol 26:259–271

Becker CD, Quenville NF, Burhenne JH (1988) Long-term occlusion of porcine cystic duct by means of endoluminal radio-frequency electrocoagulation. Radiology 167:63–68

Berk RN (1983) Radiology of the gallbladder. In: Margulis AR, Burhenne HJ (eds) Alimentary tract radiology, 3rd edn. Mosby, St Louis, pp 1434–1460

Boland GW, Lee MJ, Dawson SL, et al (1993) Percutaneous cholecystostomy for acute acalculous cholecystitis in a critically ill patient. AJR 160:871

Boland GW, Lee MJ, Leung J, et al (1994a) Percutaneous cholecystostomy in critically ill patients: early response and final outcome in 82 patients. AJR 163:339

Boland GW, Lee MJ, Mueller PR, et al (1994b) Gallstones in critically ill patients with acute calculous cholecystitis treated by percutaneous cholecystostomy: nonsurgical therapeutic options. AJR 162:1101–1109

Browning PD, McGahan JP, Gerscovich EO (1993) Percutaneous cholecystostomy for suspected acute cholecystitis in the hospitalized patient. J Vasc Interv Radiol 4:531

Burton KE, Picus D, Hicks ME, Darcy MD, Vesely TM, Kleinhoffer MA, Aliperti GP, Edmundowiz SA (1993) Fragmentation of biliary calculi in 71 patients by use of intracorporeal electrohydraulic lithotripsy. J Vasc Interv Radiol 4:251–258

Eggermont AM, Laméris JS, Jeekel J (1985) Ultrasound-guided percutaneous transhepatic cholecystostomy for acute acalculous cholecystitis. Arch Surg 120:1354–1356

Elyaderani M, Gabriele OF (1979) Percutaneous cholecystostomy and cholangiography in patients with obstructive jaundice. Radiology 130:601–602

Garber SJ, Mathieson JR, Cooperberg PL, MacFarlane JK (1994) Percutaneous cholecystostomy: safety of the transperitoneal route. J Vasc Interv Radiol 5:295

Gervais DA, Mueller PR (1996) Percutaneous cholecystostomy. Semin Intervent Radiol 13:35–43

Getrajdman GI, O'Toole K, LoGerfo P, Laffey KJ, Martin EC (1985) Transcatheter sclerosis of the gallbladder in rabbits. A preliminary study. Invest Radiol 20:393–398

Glenn F, Dillon LD (1980) Developing trends in acute cholecystitis and choledocholithiasis. Surg Gynecol Obstet 151:528–532

Hawkins IF Jr (1985) Percutaneous cholecystostomy. Semin Intervent Radiol 2:97–103

Kaff V, Froelich JW, Thrall JH (1983) Predictive value of an abnormal hepatobiliary scan in patients with severe intercurrent illness. Radiology 146:191–194

Laffey KJ, Margin EC (1986) Percutaneous removal of large gallstones. Gastrointest Radiol 11:165–168

Larsen MJ, Kingensmith WC III, Kuni CC (1982) Radionuclide hepatobiliary imaging: nonvisualization of the gallbladder secondary to prolonged fasting. J Nucl Med 23:1003–1005

Lee MJ, Saini S, Brink JA, et al (1991) Treatment of critically ill patients with sepsis of unknown cause: value of percutaneous cholecystostomy. AJR 156:1163–1166

Lindberg CG, Jeppsson B, Lundsted C, et al (1993) Percutaneous rotational lithotripsy of gallbladder stone: clinical results with Rotolith lithotriptor. Acta Radiol 34:273

Lo LD, Vogelzang RL, Braun MA, Nemcek AA Jr (1995) Percutaneous cholecystostomy for the diagnosis and treatment of acute calculous and acalculous cholecystitis. J Vasc Interv Radiol 6:629

May GR, Thistle JL (1986) Percutaneous cholecystostomy for gallstone dissolution by methyl *tert*-butyl ether. Radiology 161:90

McGahan JP (1988) A new catheter design for percutaneous cholecystostomy. Radiology 166:49–52

McGahan JP (1990) The gallbladder. In: McGahan JP (ed) Interventional ultrasound. Williams & Wilkins, Baltimore, pp 159–169

McGahan JP, Lindfors KK (1989) Percutaneous cholecystostomy: an alternative to surgical cholecystostomy for acute cholecystitis. Radiology 173:481–485

McNulty JG (1990) Interventional radiology of the gallbladder: percutaneous cholecystostomy. Springer, Berlin Heidelberg New York

Nemcek Jr AA, Bernstein J, Vogelzang RL (1991) Percutaneous cholecystostomy: does transhepatic puncture preclude a transperitoneal puncture route? J Vasc Interv Radiol 2:545–547

Picus D, Marx MV, Hicks ME, Lange EV, Edmundowicz SA (1989) Percutaneous cholecystolithotomy: preliminary experience and technical considerations. Radiology 173:487–491

Ranofsky AL (1978) Surgical operations in short-stay hospitals: 1975. National Center for Health Statistics, Hyattsville, MD (Vital and Health Statistics, series 13, no 34, DHEW; publication # pHS 78-178S)

Schoenfield LJ, Lachin JM (1981) Chenodiol (chenodeoxycholic acid) for dissolution of gallstones: the National Cooperative Gallstone Study. Ann Intern Med 95:257–282

Schwartz SI (1979) Gallbladder and extrahepatic biliary system. In: Schwartz SI, Shires GT, Spencer FC, Storer EH (eds) Principles of surgery, 3rd edn. McGraw-Hill, New York, pp 1317–1352

Shaver RW, Hawkins Jr IF, Soong J (1982) Percutaneous cholecystostomy. AJR 138:1133–1136

Skillings JC, Kurnai C, Hinshau JR (1980) Cholecystostomy: a place in modern biliary surgery? Am J Surg 139:865–869

Teplick SK, Brandon JC, Wolferth CC, et al (1990) Percutaneous interventional gallbladder procedures: personal experience and literature review. Gastrointest Radiol 15: 133–136

Van Sonnenberg E, d'Agostino HB, Casola G, Varmey RR, Ainge G (1990) Interventional radiology in the gallbladder: diagnosis, drainage, dissolution, and management of stones. Radiology 174:1–6

Van Sonnenberg E, Wittich GR, Schiffman HR, et al (1986) Percutaneous drainage access: simplified coaxial technique. Radiology 159:266–268

Vogelzang RL, Nemcek Jr. AA (1988) Percutaneous cholecystostomy: diagnostic and therapeutic efficacy. Radiology 168:29–34

Yedlicka JW Jr, Coleman CC, Peterson C, Bjarnaon H, Hunter DW, Casteñeda-Zuñiga WR, Amplatz K (1993) Thermal ablation of the gallbladder. J Vasc Interv Radiol 4:367

16 Benign Biliary Diseases: Surgical Approach

V. SPERANZA and D. LOMANTO

CONTENTS

16.1 Choledocholithiasis

16.1.1 Conventional Surgery

Surgical treatment of extrahepatic bile duct lithiasis has been established since 1882, when Karl Langenbuch performed the first cholecystectomy, while Ludwig Courvoisier in Basel in 1890 was the first to remove stones from the common bile duct (CBD) by choledochotomy. Two subsequent major innovations which led to marked improvement in diagnosis and treatment were the introduction of cholangiography by Mirizzi in 1931 and of choledochoscopy by Bakes in 1923. These procedures have enabled a reduction from 8%–10% to 2%–4% in the incidence of false negative findings, while intraoperative cholangiography has allowed the assessment of the anatomy of the biliary tree. Anatomical anomalies were detected in 15%–20% of cases, thus preventing iatrogenic lesions as well as showing the presence of impalpable or clinically silent stones in 10%–20% of cases.

Subsequently, and until the 1990s, only surgery was considered for the treatment of this disease, with two different approaches. Some suggested the removal of stones by "ideal choledochotomy" (feasible only in 5%–10% of cases) or with the positioning, in 60%–80% of cases, of an external biliary drainage (transcystic or with T-tube); others favored transduodenal sphincterotomy.

In our experience, when the diagnosis of CBD stones has been established during cholecystectomy (10%–12% of cases), the best procedure to remove the stones and to prevent recurrence is transduodenal papillostomy for a proper biliary drainage procedure (SPERANZA et al. 1982; SPERANZA 1991). For transduodenal papillostomy, particular precautions are required, such as a minimal duodenotomy after a wide Kocher maneuver, section of the papilla of Vater by a diathermic knife large enough to allow passage of a Randall forceps into the bile duct without damaging the sphincter of the CBD (partial inferior sphincterotomy), and suturing of incision margins using three or four 000 polyglactin 910 stitches with an atraumatic needle only, in order to prevent hemorrhage and stenosis. Removal of stones, lavage with soft probes, and the use of à choledochoscope to assess clearance are the next steps. These are the technical details that we recommend to achieve a successful papillostomy. In a study carried out on 25000 patients operated on for papillostomy, the results demonstrated the validity of this procedure (NEGRO et al. 1984). In 96% of patients it was possible to remove the stones, with less than 8% morbidity and 1.9% mortality, and with a mean hospital stay of 12 ± 3 days. These results support the opinion that careful technique is mandatory to avoid the major complications attributed in the past to papillostomy.

Choledochoduodenostomy is an alternative drainage procedure which has some disadvantages, such as resulting in ascending cholangitis due to duodenal reflux, stenosis of the anastomosis, and

V. SPERANZA, Professor, Head, Clinica Chirurgica II, University of Rome "La Sapienza", Policlinico Umberto I, Viale del Policlinico, 155, 00161 Rome, Italy
D. LOMANTO MD, PhD, Assistant Professor, Clinica Chirurgica II, University of Rome "La Sapienza," Policlinico Umberto I, Viale del Policlinico 155, 00161 Rome, Italy

"sump syndrome," which favors bacterial overgrowth. It may be indicated after failure of sphincterotomy with stenosis or in patients with chronic pancreatitis.

Bilio-enteric anastomosis, such as choledochojejunostomy or hepaticojejunostomy, is indicated in no more than 15% of cases, usually in subjects with a markedly enlarged CBD, multiple lithiasis or large stones, and in elderly patients or patients with iatrogenic stricture (STEFANINI et al. 1975). Overall mortality in the now well-known series of these surgical procedures reported by STEFANINI et al. (1975) ranged from 3.7% to 6.6% and was related to age, associated disease, and surgical timing. Morbidity was 8%–10%, with a mean hospital stay of 17 ± 5 days. Residual lithiasis was around 1%–3%.

Finally, it should be stressed that the above findings relate to patients who underwent surgery for gallstones; in patients who have had a previous cholecystectomy but have bile duct stones, endoscopic sphincterotomy (ES) is currently the treatment of choice. In particular cases, such as patients with gallbladder and bile duct stones in whom severe jaundice or acute cholangitis or gallstone pancreatitis are the main clinical features, ES is recommended to drain and clear the biliary tree. In elderly or debilitated patients with symptomatic choledocholithiasis, too, ERCP/ES and CBD stone extraction may be adequate without cholecystectomy; in fact less than 20% of such patients require cholecystectomy during prolonged follow-up. (LEUSCHNER and SEIFERT 1991).

16.1.2 Approaches in the Laparoscopic Era

Biliary lithiasis was for a long time and still is most commonly treated by conventional surgery, with a well-standardized technique and intraoperative examinations which in most studies yielded similar result in terms of mortality, morbidity, and residual lithiasis. In recent years, with the introduction of laparoscopic surgery, the treatment of biliary lithiasis has been stimulated by the advances in the field of laparoscopic optics and the miniaturized camera. Operative laparoscopy, which has been used in gynecology for over 20 years, was proposed worldwide for the standard treatment of gallstones only in 1987, when Mouret performed the first laparoscopic cholecystectomy, was enthusiastically accepted by surgeons, and met with great interest from patients (BERCI et al. 1991). The indications for laparoscopic cholecystectomy are similar to those for conventional surgery (NIH CONSENSUS CONFERENCE 1993). Absolute contraindications are: generalized peritonitis, septic shock from cholangitis, acute severe pancreatitis, terminal cirrhosis with portal hypertension, severe coagulopathy unresponsive to medical treatment, the proven presence of gallbladder cancer, cholecystoenteric fistulae, and the third trimester of pregnancy because of the risk of uterine injury during the various procedures. Relative contraindications are acute cholecystitis, acute remittent biliary pancreatitis, previous surgery to the upper abdomen, and symptomatic lithiasis during the second trimester of pregnancy. Obviously, these conditions require an experienced surgeon, able to tackle more complex situations.

The risks and complications related to laparoscopic cholecystectomy are practically the same as those of laparotomy, although a higher incidence (albeit not significantly so) of iatrogenic lesions to bile ducts during laparoscopic surgery has been observed: 0.2%–0.6% vs 0.1%–0.2% in conventional surgery. General complications should be distinguished from those specifically related to the procedure. Of the latter, those related to the creation of pneumoperitoneum have an incidence of 0.06%. Vascular injury, usually caused by the insertion of Veress needles and of trocars, has an incidence of 0.25%, while intestinal lesions occur in 0.14% of cases. Hospital stay after laparoscopic surgery is definitely shorter (1–2 days vs 5 days after conventional surgery), and health service costs are similar or slightly lower than those of conventional surgery. Recovery takes 1–2 weeks, compared to the decidedly longer period of 3–6 weeks, with a marked difference in terms of prompt restoration of the patient's ability to return to work (Table 16.1). The mortality associated with laparoscopic cholecystectomy is similar to that of open cholecystectomy, which is around 0%–1%.

The results and advantages of laparoscopic cholecystectomy are by now well established and accepted (PETERS et al. 1991; PERISSAT 1993; GADACZ 1993), and today over 95% of patients with cholelithiasis are treated with this method. Other diseases, especially those most frequently associated with cholelithiasis or detected during this type of surgery (e.g., inguinal hernia, ovarian cyst), and especially choledocholithiasis, present in 3%–33% of cases of cholelithiasis (ARNOLD 1970; GLENN and MCSHERRY 1975), have been considered for laparoscopic treatment. In particular, the assessment and management of choledocholithiasis has involved all those who deal with laparoscopic biliary surgery. Although still in its initial phase, laparoscopic CBD exploration is gaining ground in cases of associated choledocholithiasis, following the introduction of new equipment and better surgical procedures. The longer life expectancy and the increased popularity of laparoscopic cholecystectomy have led to a higher incidence of associated choledocholithiasis being reported today in patients undergoing cholecystectomy, between 8% and 15% in patients younger than 60 years and from 15% to 60% in patients older than 60 (NIH CONSENSUS CONFERENCE 1993). One may say that a growing number of patients with choledocholithiasis will have to be treated in the near future.

16.1.2.1 *Predictive Criteria of Choledocholithiasis*

With the widespread use of laparoscopic cholecystectomy, preoperative diagnosis of choledocholithiasis has assumed great significance. All patients undergo blood chemistry tests to assess liver function, such as determination of bilirubin, transaminase, alkaline phosphatase, and LDH levels, and imaging procedures such as hepatobiliary ultrasonography and perfusional cholangiography to allow diagnosis, while ERCP is performed when CBD stones are suspected. It has been shown that an increase in the level of any one enzyme carries a 17%–22% predictivity of the presence of CBD stones, and there was a two- to three-fold increase if two, three, or more values were 50%–60% increased above the normal range (DEL SANTO et al. 1985; HAUER-JENSEN et al. 1985). These findings were confirmed by the fact that in a series of patients undergoing laparoscopic cholecystectomy, only 4% showed the presence of choledocholithiasis in spite of normal blood chemistry (PETELIN 1991). As for diagnostic imaging, abdominal ultrasonography has shown 55% sensitivity to the presence of CBD stones, as confirmed at surgery. In intravenous preoperative cholangiography the biliary tree opacified in only 60%–70% of cases. In a prospective study of 113 patients at low risk for choledocholithiasis, intravenous cholangiography was shown to be effective in the diagnosis of this disease only in 1.5% of cases (PATEL et al. 1993). In a number of studies, these predictive factors were analyzed separately or correlated, with unsatisfactory results (BARKUN et al. 1994). In their predictive model, BARKUN et al. (1994) showed that when one or none of the factors under consideration was present, the probability of associated choledocholithiasis was in the range of 19%–38%, while it reached 49%–94% when three or four of the factors were present. Based on predictive clinical factors, blood chemistry, and imaging findings, ERCP was reserved for patients with CBD dilatation and high levels of three or more specific enzymes which indicate a high probability of choledocholithiasis; however, the advent of MRI cholangiography has modified this practice. By contrast, 4% of patients in whom lithiasis was not suspected on the basis of blood chemistry were shown by intraoperative cholangiography to have stones (HUNTER 1992).

It may be concluded from the above that the choice between an accurate, more invasive preoperative diagnostic procedure and the routine use of

Table 16.1. Laparoscopic cholecystectomy: review of the literature

	No. of patients	Success rate (%)	IOC (%)	Complications (%)	Lesions to the CBD (%)	Mortality (%)	Hospital stay (days)	Return to work (days)
DEZIEL et al. 1993	77604	98.8	?	2	0.59	0.04	?	?
LARSON et al. 1992	1983	95.5	37.6	2.1	0.25	0.1	?	?
SOUTHERN SURGEONS CLUB (1991)	1518	95.3	29.2	5.1	0.5	0.07	1.2	<10
BAIRD et al. 1992	800	97.7	14	3.1	0	0.13	0.89	8.4
VOYLES et al. 1991	453	95	9.3	1.3	0	0	?	?
AIRAN et al. 1992	418	95	90	?	0.2	?	1	?
LILLEMOE et al. 1992	400	96	2	9.3	0.5	0	1	?
BAILEY et al. 1991	375	95	37.6	3.5	0.3	0.3	1.3	7
Total	83531	95–99	2–90	1.3–9.3	0–0.59	0–0.3	0.89–1.3	7–10

IOC, Intraoperative cholangiography.

intraoperative cholangiography depends on the level of preoperative clinical suspicion, on the surgeon's experience, and on the availability of equipment.

16.1.2.2 Combinations of Endoscopic and Laparoscopic Approaches

A number of prospective studies are in progress to assess the risk and success of the various combinations of minimally invasive approaches to treat patients with choledocholithiasis and to define their sequencing. At present, the therapeutic possibilities with endoscopy or laparoscopic surgery are as follows:

- Preoperative ERCP/ES and laparoscopic cholecystectomy
- Laparoscopic cholecystectomy and intraoperative ERCP/ES
- Laparoscopic cholecystectomy and postoperative ERCP/ES
- Laparoscopic CBD exploration

Preoperative ERCP and Laparoscopic Cholecystectomy

Since the early 1980s, treatment of choledocholithiasis was exclusively surgical, with surgical technique, pre- and intraoperative examinations, and expected results all well established. However, with the introduction of endoscopic sphincterotomy (ES) and then of laparoscopic surgery in the 1990s, a complete minimally invasive approach to treating choledocholithiasis was sought. Sequential treatment by ES and laparoscopic surgery soon replaced open surgery, because feasibility was high, results satisfactory, and patient compliance good. However, the results of the three randomized trials carried out by Neoptolemos et al. (1987) Stain et al. (1991) and Van Stiegmann et al. (1992) should be kept in mind. In these studies the combined treatment (conventional cholecystectomy and ES) was compared with conventional surgery in a population of patients with a preoperative diagnosis of choledocholithiasis (Neoptolemos et al. 1987; Stain et al. 1991) and in patients in whom it was suspected. Stain et al. and Van Stiegmann et al. showed that there is no definite advantage of one treatment over the other. On the other hand, Neoptolemos et al. observed a slightly higher incidence of complications in the group of patients undergoing the combined treatment. This was due to ERCP, no allowance being made for the lack of real usefulness of the procedure and the incidence of failure of ES. However, 40%–60% of patients have an "unnecessary" ERCP prior to laparoscopic cholecystectomy.

We will now analyze the findings of the numerous reports on the sequential combined treatment (Table 16.2). Most authors now agree that ERCP should be carried out in a selected group of patients, namely those with preoperative sonographic evidence of choledocholithiasis (100% diagnostic accuracy on ERCP) or a dilated CBD (47% accuracy on ERCP). Surick et al. (1993) have reported 28% accuracy of positive sonographic features, while stones were present in the CBD in 53% of patients with altered liver enzymes, in 60% with cholangitis, and in 87.5% with jaundice. Results of ES were 74% successful in terms of CBD clearance. Reported failures related to cases of multiple lithiasis or paravaterian diverticula.

Table 16.2. Results of preoperative ERCP/ES and laparoscopic cholecystectomy in cases of suspected choledocholithiasis: review of the literature

	Presence of stones suspected (no. of cases)	Stones confirmed on ERCP (no. of cases)	Clearance achieved (no. of cases)	Clearance achieved (%)
Arregui et al. 1992	29	11	5	45
Baird et al. 1992	28	11	9	82
Ellul et al. 1992	48	18	17	94
Graves et al. 1991	11	3	3	100
Larson et al. 1992	65	20	18	90
Southern Surgeons Club 1991	63	10	9	90
Stockman & Soper 1991	16	8	8	100
Vanneman et al. 1992	30	11	9	82
Graham et al. 1993	41	18	18	100
Total	331	110	96	87

ERCP, Endoscopic retrograde cholangiopancreatography; ES, endoscopic sphincterotomy.

ERCP/ES carried a 5% morbidity. In about 8% of cases the procedure had to be repeated. Mortality was 0%. A study by GRAHAM et al. (1993) reported 88% feasibility of ERCP, with diagnosis of choledocholithiasis in 50% of cases. Morbidity was 11%. Analysis of the various predictive factors showed that visualization of CBD stones on ultrasonography plays a major role, since this diagnosis was confirmed by ERCP in 100% of cases. Simple dilatation was predictive of lithiasis in only 71% of cases, and altered enzyme levels in 40%. FRANCESCHI et al. (1993), in a report on 401 patients, further confirmed the good correlation between sonographic suspicion of CBD stones and proven presence of stones on ERCP ($p < 0.001$). ERCP has detected the presence of stones in about one-third of suspected cases (range: 27%–50%). CBD clearance was achieved in 90% (range: 45%–100%) of patients, with 0%–9% morbidity. The combination of these two minimally invasive procedures has been proposed as a valid alternative to conventional surgery, although there has been much criticism and controversy about this.

Laparoscopic Cholecystectomy and Intraoperative ERCP/ES

On the basis of the results of sequential ERCP/ES ad LC and of criticisms of this type of approach, some authors have opted for an intraoperative treatment of CBD stones during laparoscopic cholecystectomy in spite of doubts about the feasibility of ERCP under these conditions. In a recent study (FIOCCA et al. 1994) where sequential treatment was performed, the authors reported a failure rate of ERCP in 36% of cases, which led them subsequently to opt for intraoperative treatment of CBD stones. This involved intraoperative cholangiography based on the clinical and instrumental findings, and if stones were diagnosed, intraoperative endoscopic treatment was performed. The preoperative predictive criteria were applied and 32 patients were selected for the treatment. Patients with a CBD over 12 mm in diameter with stones larger than 10 mm or more than three stones diagnosed on ultrasonography were excluded and underwent preoperative ERCP/ES. Of 14 patients (43%) undergoing intraoperative cholangiography, CBD clearance was achieved in 12 (86%), while in the two in whom it failed, a nasobiliary catheter was positioned and ERCP/ES was successfully performed the day after. A mean of 22 min was required for intraoperative ERCP/ES (range: 10–70 min). No particular technical difficulty was caused by the patient's position or pneumoperitoneum. Intraluminal air insufflation after endoscopy did not hinder the laparoscopic cholecystectomy. As for morbidity, there was a slight increase in amylase levels, which reverted to normal after 24 h. All patients were discharged 48 h after combined intraoperative treatment. In this limited experience, although the number of patients does not allow statistical significance, the results seem encouraging in terms of both positive outcome and cost-effectiveness, suggesting that 35%–40% of the useless ERCPs reported in the world literature could be prevented.

Laparoscopic Cholecystectomy and Postoperative ERCP/ES

In a study by GRAHAM et al. (1993), 95% postoperative feasibility of ERCP/ES is reported in a group of 22 patients with choledocholithiasis, in 6 or whom the diagnosis was established on intraoperative cholangiography, while in 16 it was suspected on the basis of signs of symptoms suggestive of residual lithiasis. Stones were confirmed on ERCP in 11 patients (52%), 5 of whom were among the 6 patients with positive cholargiographic findings and only 6 among the 16 with a clinically suspected diagnosis. CBD clearance was achieved in all patients with CBD stones. In three patients (27%) several attempts were necessary to remove the stones endoscopically. Overall morbidity was 4%. Other studies (Table 16.3) have confirmed CBD clearance in 97% of cases (range: 67–100%), with morbidity varying from 0 to 9%. In fact, this approach appears risky, mainly because with two procedures the risk of complications doubles, but also because postoperative ES may be unsuccessful. For this reason, until the results of prospective studies are published, it is better to reserve this combined approach for high-risk or elderly patients.

Laparoscopic CBD Exploration

With the widespread use of laparoscopic cholecystectomy as the gold standard of therapy for cholelithiasis, treatment of stones in the CBD was initially based on the combination of two minimally invasive procedures, laparoscopic cholecystectomy and ERCP/ES, the latter being performed before, during, or after surgery. Results, as shown by the present continuing search for a valid alternative, were not entirely favorable. So far, within the various trends no one particular approach has taken the lead. At the same time, with the improvement and the availability of new, more sophisticated surgical

Table 16.3. Results of laparoscopic cholecystectomy and postoperative ERCP/ES in choledocholithiasis: review of the literature

	Stones present (no. of cases)	Clearance achieved	
		(no. of cases)	(%)
ALIPERTI et al. 1990	4	4	100
ARREGUI et al. 1992	5	5	100
BAIRD et al. 1992	6	4	66.6
GRAVES et al. 1991	3	3	100
LARSON et al. 1992	4	4	100
MANOUKIAN et al. 1992	22	22	100
PETELIN 1991	5	4	80
PRUITT et al. 1991	38	37	97.3
PUTNAM et al. 1991	13	13	100
QUATTLEBAUM & FLANDERS 1991	6	6	100
SOUTHERN SURGEONS CLUB 1991	10	10	100
GRAHAM et al. 1993	11	11	100
Total	127	123	96.8

equipment, and the greater manual skill acquired by individual operators, techniques of laparoscopic CBD exploration have advanced markedly.

CBD exploration has revived the role of intraoperative cholangiography as the major tool in screening for extrahepatic bile duct lithiasis (SACKIER et al. 1991; BERCI et al. 1991; FLOWERS et al. 1992; LEZOCHE et al. 1994). In this area, too, there have been some who favor routine use of intraoperative cholangiography and others who argue for its selective use on the basis of the predictive factors already mentioned. However, this method, criticized (initially because of technical problems) after the introduction of laparoscopic cholecystectomy, is now regaining its earlier importance in the intraoperative diagnosis and treatment of choledocholithiasis. With technological improvements, ultrasonography has also found a place in intraoperative diagnosis during laparoscopic surgery. With the high-frequency (7- to 10-MH2) probes fitted for use through trocars, adequate screening of parenchymal structures as well as of the CBD is feasible. This procedure has clear advantages, such as noninvasiveness, lack of requirement of contrast, and high resolution, which bypass the intrinsic limitations of intraoperative cholangiography, such as the difficulty of visualizing small stones, especially in areas of contrast medium stagnation. The single flaw is that, the examination being static, the functional information gained with cholangiography is lost. Therefore, in our opinion, ultrasonography is only a complement to intraoperative cholangiography (CARDONE et al. 1994).

When stones are detected in the CBD during laparoscopic cholecystectomy, it can be explored and cleared either by a transcystic route or by choledochotomy. The feasibility and selection of the procedure depend on several factors such as the size of the stone or stones, the anatomy of the cystic duct, the equipment available, and the skill and experience of the surgeon. The cystic duct most suitable for transcystic removal of stones is the wide, short one that inserts laterally on the CBD. Unfortunately, this is present in only 17% of cases (SCHWARTZ 1989), while there are a number of anatomic variations such as the spiral and/or small cystic duct with medial or left lateral insertion, where exploration is unpredictable. Transcystic exploration (PETELIN 1993; LEZOCHE et al. 1994) is feasible in about 90% of cases and is successful in about 90% when choledochoscopy is employed. Stones are removed using a Dormia basket (3–6 F) alone in about 50%–60% cases. If stones are less than 8 mm in diameter, they can be advanced into the duodenum with a balloon catheter and pneumatic divulsion of the papilla of Vater performed with a high-pressure balloon catheter. For very large stones, various alternatives have been proposed (LEZOCHE et al. 1994). It is possible to use laser dye-pulsed, electrohydraulic lithotripsy or the alexandrite laser to reduce the size of the stone. Results published to date report a 73%–90% rate of successful transcystic CBD clearance, with a hospital stay of 1–2 days and return to work within 7–10 days (Table 16.4). In the 10% of cases where clearance fails, because of impacted stones, anatomic anomalies, or intrahepatic stones,

Table 16.4. Results of laparoscopic common bile duct exploration for biliary lithiasis: review of the literature

Authors	Laparoscopic cholecystectomy (*n*)	Lithiasis at IOC (*n*)	Lithiasis at IOC (%)	Clearance achieved laparoscopically (%)	Conversion to open surgery (*n*)	Postoperative ES or PTC (*n*)
Jacobs et al. 1991	200	8	4	88	0	1
Hunter 1992	252	20	8	85	3	0
Spaw et al. 1991	500	29	6	35	3	16
Sackier et al. 1991	516	35	7	?	8	5
Carroll et al. 1992	555	57	10	98	1	?
Petelin 1993	1000	95	9	76	5	18
Lezoche et al. (1996)	1059	120	11.3	92.5	4	5

choledochotomy is required. Retained stones occur in about 2%–4% of cases, an incidence equal to that in other forms of CBD exploration, including ES. However, it can be stated that choledochotomy should be performed in all cases where a long, narrow, tortuous cystic duct with a medial or distal insertion, or a stone 10 mm or more in diameter, make transcystic exploration unfeasible or unsuccessful. Through a 10-mm transverse choledochotomy stones are removed with a Fogarty catheter or a Dormia basket. This can be done blindly or with digital fluoroscopic guidance. Then the CBD is explored with a 6-mm choledochoscope with an operative channel that permits the use of a Fogarty catheter, Dormia basket, or laser probes to complete clearance. The choledochotomy can be closed with absorbable monofilament sutures (PDS or Maxon). If a T-tube is placed, the CBD is sutured around the drainage with a continuous absorbable 4-0 monofilament suture (Lezoche et al. 1994). Published data indicate a hospital stay of 4–7 days, return to work within 14–30 days, and 10.4% morbidity. CBD clearance is successful in 70%–96% of cases according to the various authors, with a 2%–20% conversion rate. ERCP/ES or postoperative interventional radiologic approaches are required in 2%–7% of cases in the various series.

To conclude, laparoscopic or combined-technique CBD exploration allows the removal of bile duct stones and results in reduced morbidity, hospital stay, and direct and total costs of treatment. Open bile duct exploration has still a place and is indicated in difficult cases and patients who require biliary bypass.

16.2 Acute Cholangitis

In spite of the improvements in diagnosis and therapy, acute cholangitis remains a serious illness with a mortality rate of 10%–20% (Kadakia 1993). In 1877 Charcot provided the first description of the characteristic symptom triad (fever, jaundice, and abdominal pain); in 1959 Reynolds and Dargan added shock and central nervous depression to the triad called it a "pentad." Two factors are necessary for the onset of clinical symptoms: the presence of significant numbers of bacteria and elevated intraductal pressure that causes the passage of endotoxins and bacteria into the blood. Obstruction of the biliary tree plays an important role in the pathogenesis, and CBD stones seem to be the most common cause of this. Less common causes are carcinoma of the pancreas, of the CBD, or of the ampulla of Vater, or benign strictures; adenopathy, pancreatitis, and parasitic infections are rare causes. A recent study (Bilbao et al. 1976) has shown that an increased number of nonoperative biliary manipulations – often performed in patients with malignant strictures – such as percutaneous transhepatic drainage, placement of an endoprosthesis, or endoscopic retrograde cholangiography, results in cholangitis (9%–11%). Once the diagnosis is suspected, the initial treatment is antibiotic therapy (taking account of the antibacterial spectrum, biliary excretion, and serum and tissue levels, toxicity, and cost of the antibiotics) and general supportive measures (fluids, nasogastric decompression, steroids in patients with septic shock, etc.). The majority of patients usually respond to antibiotic therapy, but in those who do not, biliary decompression is mandatory. Percutaneous transhepatic biliary drainage (PTBD) is a valid alternative, with initial external drainage (90% success rate) followed if necessary by placement of an internal prosthesis after resolution of the sepsis (success rate 60%–78%). Endoscopic retrograde cholangiography can play a part in both diagnosis and treatment, with a decompression success rate of about 96%. By this technique it is possible to remove stones or to position a nasobiliary tube in patients with a benign or malignant stricture. Morbidity ranges between 5% and 28%, with 2%–7% mortality.

Surgery is indicated after initial decompression or when medical and decompressive therapy have failed (KADAKIA 1993). CBD exploration and T-tube positioning are performed for cholangitis due to bile duct stones, while a bilio-enteric bypass is performed in cases of malignant stricture. Mortality is higher (4%–40%), with morbidity at 40%–57%. In difficult cases, as in severely ill patients with toxic cholangitis, surgical exploration should be very careful in palpation fo the liver (hepatic abscesses) and where there is a benign or malignant stricture a biliary stent or T-tube should be placed above the obstruction and an elective procedure perfomed some weeks later (PITT and CAMERON 1987). As for elective surgery after a successful biliary decompression, a range of operative choices can be considered. In our experience, transduodenal sphincteroplasty seems the best choice, since sphincteroplasty provides good biliary drainage and allows the passage of stones, and carries low mortality (<2%) and morbidity rates (<8%). Choledochoduodenostomy is indicated in patients with multiple or impacted duct stones or distal bile duct stenosis. It is contraindicated in patients with pancreatitis, and its role is disputed in cases of malignant obstruction. Choledochoduodenostomy carries a mortality rate of 0–8%, with morbidity ranging from 10%–34%.

16.3 Biliary Pancreatitis

While about 5% of patients with bile duct stones develop acute pancreatitis, 50% of patients with acute pancreatitis have gallstones. After the initial hypothesis that a stone impacted in the papilla of Vater caused pancreatitis due to bile reflux in Wirsung's duct, the more convincing and accepted opinion now is that biliary pancreatitis is due to transient blockage of the papilla by the migrating stone. This is of cardinal importance, together with several local factors such as stone size, density, and shape, which first may allow passage through the cystic duct and then impact at the papilla. In conclusion, the characteristics of the gallstone and the motor function of the biliary tract seem to play a crucial role in the pathogenesis of biliary pancreatitis.

As in cholangitis, supportive measures (fluids, analgesia, antibiotics, nasogastric tube, etc.) are the first step in the management of biliary pancreatitis; the timing and extent of additional nonoperative or operative measures are controversial. ERCP with endoscopic sphincterotomy, if bile duct stones are present, is mandatory in cases of severe biliary pancreatitis, but its role is debatable in mild pancreatitis. In fact, the results and complications of ERCP/ES and of operation are similar. The timing of surgery (early, i.e., <48 h, versus delayed) for severe or mild pancreatitis seem to have different effects on morbidity and mortality; in mild pancreatitis timing made no difference, whereas in severe biliary pancreatitis an early operation increased both mortality and morbidity (83% morbidity with early surgery, 18% with delayed surgery). The surgical options start from simple cholecystectomy with intraoperative cholangiography to decide whether to perform bile duct exploration or routine transduodenal sphincteroplasty for biliary and pancreatic drainage and to treat bile duct stones. These options are suitable especially in patients with mild pancreatitis after an acute attack.

In patients with severe pancreatitis, early ES, restoring ampullary potency, improves outcome in terms of complications and the need for urgent surgery. Indications for surgery are based on clinical status in patients with renal failure, sepsis, or imaging features diagnostic of necrosis. Total pancreatectomy or partial resection is recommended in severely ill patients who do not respond to alternative treatments. Mortality is high (40%–60%), and necrosectomy with debridement of necrotic tissue (mortality 3%–33%) seems to be the best option, especially if combined with peritoneal lavage (mortality down to 8%) (MARSHALL 1993).

16.4 Sclerosing Cholangitis

Sclerosing cholangitis is a rare (prevalence in the USA is 1–4 cases/100000 head of population), chronic cholestatic disease of all or parts of the biliary tree, characterized by inflammation causing obliterative, fibrous narrowing of intrahepatic and extrahepatic ducts. The etiology of sclerosing cholangitis is still unknown, but the most widely accepted hypotheses are of bacterial or viral infection and impaired immune response, as suggested by the association of sclerosing cholangitis with other immunologic diseases (WILLIAMS and SCHOETZ 1981). With the advent of techniques for imaging the biliary tree, more cases of sclerosing cholangitis were diagnosed before surgical exploration, causing a major change in the surgical approach to this disease. In the past (1960–1970) patients were subjected to cholecystectomy, dilatation with Bake's dilators, and drainage by T-tube, but results were not satisfactory (52% 5-year survival rate). In more recent years new approaches have been follow (biliary-enteric bypass), with a significant improvement in terms of survival (80% at 5 years) and quality of life. The present management of sclerosing cholangitis is mainly based on the signs and symptoms (jaundice and recurrent fever or bilirubin levels persistently exceeding 5 mg/dl), the cholangiographic pattern, and the presence of cirrhosis (WIESNER 1994). Currently, surgical treatment of sclerosing cholangitis includes procedures to relieve biliary obstruction by biliary-enteric bypass with or without stenting and liver transplantation, indicated in patients not older than 60 years with significant esophageal varices, repeated cholangitis with persistently elevated bilirubin levels, and partial loss of liver function. The results of liver transplantation are so encouraging that it is now an important therapeutic intervention for patients with primary sclerosing cholangitis complicated by cirrhosis or recurrent bouts of cholangitis: mortality and morbidity rates have rapidly decreased in recent years, while recurrence of sclerosing cholangitis after transplantation is not yet well documented (HARRISON and MCMASTER 1994).

16.5 Benign Tumors

Benign tumors are a rare occurrence in the gallbladder or bile ducts: the incidence of benign tumors in the gallbladder has been calculated to be 3% of all cholecystectomies performed, and the incidence of benign tumors in the bile ducts is even lower (0.1% of all operations on the biliary tract). Consequently, prospective studies are difficult and many aspects, especially those relating to malignant transformation of the tumors, are still under debate (NARHWOLD 1987). The symptoms of benign tumors are often similar to those of other disease such as lithiasis or malignant tumors; in a certain percentage of cases ultrasonography, CT, and cholangiography may suggest the presence of a tumor, but only an endoscopic biopsy (possible only in periampullary tumors) can assess it as benign. In most cases, especially of gallbladder tumors, diagnosis is usually intraoperative (it is always very important to open the gallbladder immediately after extraction), and pathologists must distinguish benign tumors from benign pseudotumors (adenomatoid hyperplasia, gastric mucosa heteroplasia).

Cholecystectomy is the surgical treatment for benign tumors of the gallbladder. A conservative approach is justified only in patients in poor general condition who are high-risk candidates for surgery. Knowledge of the location, extent, and histological type (surgical biopsy) of the lesion are necessary to decide the operative approach to tumors of the biliary tree. In the case of benign tumors, resection of the tract involved should be performed with a biliary end-to-end over a T-tube anastomosis or an end-to-side biliary-enteric anastomosis with duodenum or jejunum (10% mortality). In the case of periampullary tumors, local excision should be performed through a duodenotomy whenever possible, otherwise complete resection is possible only by a pancreaticoduodenectomy. In elderly patients or high-risk surgical candidates a safe alternative is a biliary-enteric bypass, leaving the tumor in place – a procedure that exposes the patients to the possibility of malignant transformation of the tumor.

16.6 Benign Biliary Strictures

Benign biliary stenosis is the result of surgical injury, damage from CBD stones, stricture of the papillary region, toxic or ischemic lesion of the hepatic artery, or primary infection, such as in primary or sclerosing cholangitis. It may be also due to external penetrating or blunt trauma (WAY 1987; LILLEMOE et al. 1990). We discuss biliary duct injuries in this section only because the others are treated in other sections. Morbidity is high and generally clinical manifestations at onset consist of episodes of cholangitis; if they are untreated, secondary biliary cirrhosis or even hepatic failure may follow. The major cause of bile duct strictures is biliary injuries inflicted during cholecystectomy, usually at the level of the cystic junction and hepatic bifurcation. These accidents occur during a routine procedure that is associated with low mortality and morbidity, in young patients (30–50 years), in whom gallstone disease is more frequent, and cause external biliary fistula or jaundice. The underlying cause of biliary stricture has not changed much from the early years of this century to the development of laparoscopic technique, but its diagnosis and treatment have changed greatly, paralleling the profound changes in the diagnosis and treatment of biliary disease. The increase in biliary injuries that has come with the advent of laparoscopic cholecystectomy (0.1% in open surgery vs 0.3% for the laparoscopic procedure seems to be related to the surgeon's experience and to the method used in performing the cholecystectomy (MARTIN and ROSSI 1994). Furthermore, we should remember that residents are no longer familiar with open cholecystectomy and the hilus of the liver. Accidents have been more common during operations performed by unsupervised residents than during those performed by surgeons who have completed their training.

If a biliary injury is diagnosed intraoperatively during laparoscopic surgery, conversion and immediate repair are almost always mandatory. Iatrogenic biliary accidents can be divided into three types: leaks, transections, and strictures (GEENEN 1995). Leakage is due to inadequate closure of the cystic stump and clinical manifestations are present immediately after surgery, with abdominal pain and fever. If a drainage tube has been placed during the operation, an external collection of bile is observed, or an abdominal scan (ultrasound or CT) will often show a bile collection located in a subhepatic area or Morrison's pouch, or a biloma. In most of these cases conservative management is indicated, as leaks tend to close spontaneously. Sometimes it is necessary to improve the bile outflow using a bile duct stent, nasobiliary drainage, or endoscopic sphincterotomy; only a few patients need relaparotomy to suture the injury. In the case of traumatic transection, too, a number of radiologic and endoscopic alternatives to surgery are available today, but they fail in a significant percentage of cases. Thus an operation is required for suture of the transected duct, or, more often, a bilio-enteric anastomosis such as choledochojejunostomy or hepaticojejunostomy. Occasionally, it is possible to perform end-to-end repair with a T-tube. Transection or clipping of a major bile duct (often the right or central hepatic duct results in lost continuity of the duct, which is not amenable to any alternative endoscopic or radiologic approach and must always be managed by an experienced hepatobiliary surgeon. This injury generally occurs at the time of operation, and various studies have shown in these cases intraoperative cholangiography is very rare. In the majority of cases the injury is not recognized intraoperatively, and the clinical manifestations the day after surgery show its severity.

The problem is different when the bile ducts have been injured but not transected during surgery, and become strictured during the healing process due to ischemic factors. Patients develop symptoms such as cholangitis, jaundice, and elevation of hepatic enzymes months or years after surgery. Subsequent imaging procedures (CT or ultrasound) show dilation of the intrahepatic duct and ERCP confirms the injury. In such cases, lesions are localized at the level of the common hepatic duct, close to the insertion of the cystic duct. The injury may be treated by endoscopic or radiologic maneuvers if the duct is not transected. The site of the injury is an important factor in deciding treatment. Lesions of the lower or mid hepatic or common bile duct are easily managed by endoscopy or surgery, while strictures of the hilum are a challenge. In the latter type of injury, intrahepatic cholangiojejunostomy, often in combination with the preoperative radiologic approach, is the best option for complex strictures or stenosis of earlier bilio-enteric anastomosis. Finally, surgery is the last chance in benign postoperative strictures if endoscopic dilatation and stenting fail after 1 year, while if benign stricture is due to recurrence of chronic pancreatitis, bypass surgery is mandatory (ROSSI and TSAO 1994).

16.7 Papillary Stenosis

Stenosis of the papilla of Vater due to some kind of stricture in the papillary region is a very interesting problem and implies a partial reduction in the outflow of bilio-pancreatic secretions. The occurrence of papillary stenosis has often been denied, especially by authors in the English-speaking countries, in contrast to European-Latin-American opinion (SPERANZA 1988). The clinical features of papillary stenosis are major episodes of biliary colic, precipitated by a fatty meal or analgesics containing codeine, with an increase in serum bilirubin or, more frequently, serum alkaline phosphatase. In rare cases, patients present with recurrent symptoms of abdominal pain associated with hyperamylasemia, showing recurrent idiopathic pancreatitis. The majority of these patients have already undergone gallstone surgery and show significant CBD dilation (>10 mm) on intravenous cholangiography, ERCP, or intraperative cholangiography, in the absence of CBD stones.

The crucial point is how to diagnose papillary stenosis. In the past, it was diagnosed radiologically by the presence of CBD dilation, delayed ductal emptying, and narrowing of the papillary region, confirmed by the inability of the surgeon to advance different probes through the papilla into the duodenum. These criteria resulted in a high number of false positive diagnoses and an overestimation of the prevalence of the disease. In fact, to establish a correct diagnosis we need more precise and objective diagnostic tools. For this purpose, we have adopted routine intraoperative flow manometry (von Brucke's cholangiometer) with video fluoroscopy. Perfusion of contrast medium is at a constant pressure of 30 cm H_2O. Using this instrument it is possible to evaluate flow rates through the papilla and the residual pressure. Moreover, we obtain better quality operative cholangiography, and overall accuracy is enhanced. On the basis of our data, we assume as suggestive of papillary stenosis a flow rate of less than 12 ml/min associated with a residual pressure of more than 15 cm of H_2O. Further diagnostic improvement has been achieved by combining flow manometry with pharmacodynamic tests (cerulein, a CCK analog that relaxes the sphincter of Oddi under physiologic conditions) in order to distinguish patients with stable and persistent papillary stenosis from those with a transient functional abnormality. There is no doubt that direct manometric recording of the sphincter of Oddi by cannulation of the papilla with a miniaturized catheters is the most recent reliable method of confirming papillary stenosis. It can be performed either during ERCP or intraoperatively. Intraluminal pressures are transmitted to an external transducer and the duodenal pressure, the basal pressure of the sphincter of Oddi, the amplitude and frequency of contraction, the direction of propagation, and the response to administration of cerulein are recorded on a multichannel polygraph [De Masi et al. 1984]. We classify sphincter of Oddi motility disorder, on the basis of direct sphincter of Oddi manometry, into two categories: stenosis and dyskinesia. Abnormal elevated basal pressure of the sphincter, combined with absence or disorganization of peristaltic activity, suggests organic stenosis of the sphincter, whereas recording of excess retrograde contractions, rapid contraction frequency, and a paradoxical response to CCK suggest dyskinesia of the sphincter. A third important factor by which to diagnose papillary stenosis is the pathological findings. We have detected different histological aspects in patients with manometric abnormalities, finding a slight degree of cholesterolosis in some patients or, more often, pseudopolyps of various appearance as an expression of hyperplasia. Cystic dilation of submucosal glands or muscle hypertrophy, or evidence of mild inflammation or dense chronic infiltrate through all layers of the papilla, up to marked fibrosis, have also been observed. In conclusion, we are able to diagnose papillary stenosis only if all three types of findings – radiologic, manometric, and histologic – are present. On the basis of these criteria, papillary stenosis is present in 4.5% of patients undergoing biliary surgery.

The best treatment in our opinion, advocated by most authors to facilitate biliary drainage, is transduodenal sphincteroplasty (already described in Sect. 16.1.1). This produces good long-term results with a low morbidity rate. Endoscopic sphincterotomy is a questionable choice because of the frequent difficulty of cannulating the common bile duct and the higher incidence of restenosis. Side-to-side choledochoduodenostomy is favored by some surgeons, because it is relatively simple, but long-term complications such as cholangitis, pain, and stenosis of the anastomoses are frequent. Moreover, with this procedure drainage of the main pancreatic duct is not facilitated.

16.8 Miscellaneous Conditions

16.8.1 Recurrent Pyogenic Cholangitis

Recurrent pyogenic cholangitis, endemic in Eastern countries, is an infection of the biliary tree caused by parasitic infestation or by enteric pathogens (see Chap. 22). Although the whole hepatobiliary system may be involved, most lesions are localized to the intrahepatic ducts, which undergo significant fibrosis with stenosis or dilatation. Moreover, the infection of the bile promotes the formation and growth of multiple stones which contain infected bile and mucus. The management of recurrent pyogenic cholangitis includes the treatment of the infection and of the stones. Cholecystectomy, if the patient is in fair general condition, is mandatory. In order to obtain complete clearance of the biliary tree, sphincteroplasty or choledochotomy can be performed, draining the biliary system with a T-tube. If these procedures fail, choledochojejunostomy is indicated, even though results are poor in 15%–20% of cases.

16.8.2 Ascariasis

Ascariasis is a parasitic disease particularly common in Africa and Central and South America. In 40% of cases it involves the biliary tract, passing through the duodenum and causing ductal infection and obstruction. A medical approach is justified in most cases, but in the presence of persistant disease or complications, such as hepatic abscesses, cholecystitis, peritonitis, and prolonged jaundice, a surgical approach is justified, especially if radiologic and endoscopic measures fail. Surgical treatment consists in exploration of the biliary tract, removal of the parasites by a choledochotomy, washing of the biliary tree with a vermifuge, and drainage with a T-tube. For patients who need a more aggressive approach, bilio-enteric anastomosis has also been described in literature. Of course, in the presence of a hepatic abscess the collection should be drained.

16.8.3 Congenital Biliary Atresia

Congenital biliary atresia is a rare (1 every 10000 births) obliterative process that may involve the whole biliary tract, with complete destruction of the ducts due to an unknown cause. The prognosis of patients with congenital biliary atresia was very poor until recently, but the advent of portoenterostomy, proposed by Kasai in the 1970s, changed the results of surgery significantly, achieving a 55% success rate in the cure of these patients. This rate has further increased with the increasing practice of liver transplantation, which is indicated as first treatment or after failure of a portoenterostomy (about 80% of patients alive at 1 year after transplantation).

References

Airan M, Appel M, Berci G, et al (1992) Retrospective and prospective multi-institutional laparoscopic cholecystectomy study organized by the Society of American Gastrointestinal Endoscopic Surgeons. Surg Endosc 6:169–176

Aliperti G, Edmundowitz SA, Soper NJ (1990) Early experience with combined endoscopic sphincterotomy and laparoscopic cholecystectomy in patients with choledocholithiasis. Am J Gastroenterol 85:1245

Arnold DJ (1970) 28621 cholecystectomies in Ohio: results of a survey in Ohio hospitals by the Gallbladder Survey Committee, Ohio Chapter, American College of Surgeons. Am J Surg 119:714–717

Arregui ME, Davis CJ, Arkush AM, Nagan RF (1992) Laparoscopic cholecystectomy combined with endoscopic sphincterotomy and stone extraction or laparoscopic choledochoscopy and electrohydraulic lithotripsy for management of cholelithiasis and choledocholithiasis. Surg Endosc 6:10–15

Bailey RW, Zucker KA, Flowers JL, Scovill WA, Graham SM, Imbembo AL (1991) Laparoscopic cholecystectomy: experience with 375 consecutive patients. Ann Surg 214:531–541

Baird DR, Wilson JP, Mason EM, et al (1992) An early review of 800 laparoscopic cholecystectomies at university affiliated community teaching hospitals. Am Surg 58:206–210

Barkun AN, Barkun JS, Fried GM, et al (1994) Useful predictors of bile duct stones in patients undergoing laparoscopic cholecystectomy. Ann Surg 220:32–39

Berci G, Sackier JM, Paz-Partlow M (1991) Routine or selected intraoperative cholangiography during laparoscopic cholecystectomy? Am J Surg 161:355–360

Berci G, Sackier JM (1991) Laparoscopic cholecystectomy. In: Berci G (ed) Laparoscopic surgery. Lippincott, Philadelphia, pp 284–319. (Problems in general Surgery, vol 8, no 3)

Bilbao MK, Dotter CT, Lee TG, et al (1976) Complications of ERCP: a study of 10000 cases. Gastroenterology 70:314–318

Carroll BJ, Phillips EH, Daykhovsky L, et al (1992) Laparoscopic choledochoscopy: an effective approach to the common duct. J Laparoendosc Surg 2:15–21

Cardone G, Di Girolamo M, Lomanto D, et al (1994) Ruolo dell'ecografia intraoperatoria nella colecistectomia laparoscopica. Radiol Med (Torino) 88:3:233–237

De Masi E, Corazziari E, Habib FI, et al (1984) Manometric study of the sphincter of Oddi in patients with and without common bile duct stones. Gut 25:275–278

Del Santo P, Kazarian KK, Rogers JF, et al (1985) Prediction of operative cholangiography in patients undergoing elective cholecystectomy with routine liver function chemistries. Surgery 98:7–11

Deziel DJ, Millikan KW, Economan SG, et al (1993) Complications of laparoscopic cholecystectomy: a national survey of 4292 hospitals and an analysis of 77604 cases. Am J Surg 165:9–14

Ellul JPM, Wilkinson ML, McColl I, Dowling RH (1992) A predictive ERCP study of patients with gallbladder stones (GBS) and probably choledocholithiasis-predictive factors. Gastrointest Endosc 38:266

Fiocca F, Grasso E, Basso N, et al (1994) Laparoscopic cholecystectomy and intraoperative ERCP with endoscopic sphincterotomy. The new gold standard. In: Monduzzi (ed) 14th World Congress Collegium Internationale Chirurgiae Digestivae. Los Angeles, 29 September–1 October, pp 451–454

Flowers JL, Zucker KA, Scovill WA, Imbembo AL, Bailey RW (1992) Laparoscopic cholangiography: results and indications. Ann Surg 215:209–216

Franceschi D, Brandt C, Margolin D, et al (1993) The management of common bile duct stones in patients undergoing laparoscopic cholecystectomy. Am Surg 59(8):525–532

Gadacz TR (1993) US experience with laparoscopic cholecystectomy. Am J Surg 165:450–454

Geenen JE (1995) Management of benign biliary strictures. In: Proceedings of ASGE Clinical Symposium. Digestive Disease Week, San Diego, 17 May 1995, pp 42–44

Glenn F, McSherry CK (1975) Calculous biliary tract disease. In: Ravitch MM (ed) Current problems in surgery. Yearbook Medical Publishers, Chicago pp 1–38

Graham SM, Flowers JL, Scott TR, et al (1993) Laparoscopic cholecystectomy and common bile duct stones. Ann Surg 218:61–67

Graves HA, Ballinger JF, Anderson WJ (1991) Appraisal of laparoscopic cholecystectomy. Ann Surg 213:655–664

Harrison J, McMaster P (1994) The role of orthotopic liver transplantation in the management of sclerosing cholangitis. Hepatology 20:14–19

Hauer-Jensen M, Karesen R, Nygaard K, et al (1985) Predictive ability of choledocholithiasis indicators. Ann Surg 202:64–68

Hitch DC, Shikes RH, Lilly JR (1979) Determinants of survival after Kasai's operation for biliary atresia using actuarial analysis. J Pediatr Surg 14:310–315

Hunter JG (1992) Laparoscopic transcystic common bile duct exploration. Am J Surg 163:53–58

Jacobs M, Verdeja JC, Goldstein HS (1991) Laparoscopic choledocholithotomy. J Laparoendosc Surg 1:79–82
Kadakia SC (1993) Biliary tract emergencies. Acute cholecystitis, acute cholangitis and acute pancreatitis. Med Clin North Am 77:1015–1036
Larson GM, Vitale GC, Casey J, et al (1992) Multipractice analysis of laparoscopic cholecystectomy in 1983 patients. Am J Surg 163:221–226
Leuschner U, Seifert E (1991) The role of endoscopy in the treatment of gallstones. In: Speranza V, Barbara L (eds) Changing concepts in biliary stone management. Lippincott, Philadelphia, pp 582–586 (Problems in general surgery, vol 8, no 4)
Lezoche E, Paganini A, Guerrieri M, Carlei F, Lomanto D, Sottili M, Nardovino M (1994) Techniques and results of routine dynamic cholangiography during 528 consecutive laparoscopic cholecystectomies. Surg Endosc 8:1443–1447
Lezoche E, Paganini AM, Carlei F, Feliciotti F, Lomanto D, Guerrieri M, Nardovino M, Sottili M (1996) Laparoscopic treatment of gallbladder and common bile duct stones: a prospective study on 120 unselected, consecutive cases. World J Surg (in press)
Lillemoe KD, Pitt HA, Cameron JL (1990) Post-operative bile duct strictures. Surg Clin North Am 70:1355–1380.
Lillemoe KD, Yeo CJ, Talamini MA, et al (1992) Selective cholangiography: current role in laparoscopic cholecystectomy. Ann Surg 215:669–676
Lomanto D, Pavone P, Fiocca F, Catalano C, Laghi A, Nardovino M, Dalsasso G, Zerba Meli E, De Luca A, Gracovazzo F, Salvio A, Lezoche E (1996) MR cholangiography (MRCP): predictive value in assessing CBD stones before cholecystectomy. Fourth International Congress of the European Association for Endoscopic Surgery, 23–26 June 1996, Trondheim, Norway (abstract book)
Manoukian AV, Schmalz MJ, Geenen JE, Yenu RP, Johnson GL (1992) Post-laparoscopic cholecystectomy problems: "minimally invasive" ERCP therapy. Gastrointest Endosc 38:250
Marshall JB (1993) Acute pancreatitis, a review with an emphasis on new development. Arch Intern Med 153(10):1185–1198
Martin RF, Rossi RL (1994) Bile duct injuries: spectrum, mechanism of injury and their prevention. Surg Clin North Am 74(4):781–804
Nahrwold DL (1987) Benign tumors and pseudotumors of the biliary tract. In: Way LW, Pellegrini CA (eds) Surgery of the gallbladder and bile ducts. Saunders, Philadelphia, pp 459–470
National Institute of Health (1993) Consensus development conference statement on gallstones and laparoscopic cholecystectomy. Am J Surg 165:390–398
Negro P, Tuscano D, Flati G, et al (1984) Le risque operatoire de la sphincterotomie oddienne: résultats d'une enquête internationale (25541 cas). J Chir (Paris) 121: 133–136
Neoptolemos JP, Carr-Locke DL, Fossard DP (1987) Prospective randomized study of preoperative endoscopic sphincterotomy versus surgery alone for common bile duct stones. Br Med J 294:470–474
Patel JC, McInnes GC, Bagley JS, et al (1993) The role of intravenous cholangiography in pre-operative assessment for laparoscopic cholecystectomy. Br J Radiol 66:1125–1127
Perissat J (1993) Laparoscopic cholecystectomy: the european experience. Am J Surg 165:444–449
Petelin JB (1991) Laparoscopic approach to common bile duct pathology. Surg Laparosc Endosc 1:33–41
Petelin JB (1993) Laparoscopic approach to common duct pathology. Am J Surg 165:487–491
Peters JH, Ellison EC, Iunes JT, et al (1991) Safety and efficacy of laparoscopic cholecystectomy; a prospective analysis of 100 initial patients. Ann Surg 213:3–7
Pitt HA, Cameron JL (1987) Acute cholangitis. In: Way LW, Pellegrini CA (eds) Surgery of the gallbladder and bile ducts. Saunders Philadelphia, pp 295–310
Pruitt RE, Bailey AH, Foust TW, Olsen DO, Spaw A, Reddick EJ (1991) Endoscopic retrograde cholangiography with sphincterotomy and common bile duct stone extraction combined with laparoscopic laser cholecystectomy: our initial experience. Gastrointest Endosc 37:286
Putnam WS, Wegley SJ, Rosen SN, Lewis ST (1991) The impact of laparoscopic cholecystectomy on ERCP in a community hospital. Gastrointest Endosc 37:246
Quattlebaum JK, Flanders HD (1991) Laparoscopic treatment of common bile duct stones. Surg Laparosc Endosc 1:26–32
Sackier JM, Berci G, Phillips E, Carroll B, Shapiro S, Paz-Partlow M (1991) The role of cholangiography in laparoscopic cholecystectomy. Arch Surg 126:1021–1026
Reynolds BM, Dargan EL (1959) Acute obstructive cholangitis: a distinct clinical syndrome. Ann Surg 150:299–304
Rossi RL, Tsao JI (1994) Biliary reconstruction. Surg Clin North Am 74:825–844.
Sackier JM, Berci G, Phillips E, Carroll B, Shapiro S, Paz-Partlow M (1991) The role of cholangiography in laparoscopic cholecystectomy. Arch Surg 126:1021–1026
Schwartz S (1989) Principles of surgery. McGraw-Hill, New York
Soper NJ, Dunnegan DL (1992) Routine vs selective intraoperative cholangiography during laparoscopic cholecystectomy. World J Surg 16:1133–1140
Southern Surgeons Club (1991) A prospective analysis of 1518 laparoscopic cholecystectomies performed by Southern US surgeons. N Engl J Med 324:1073–1078
Spaw AT, Reddick EJ, Olsen DO (1991) Laparoscopic laser cholecystectomy: analysis of 500 procedures. Surg Laparosc Endosc 1:2–7
Speranza V (1988) Papillary stenosis: fact or fiction? Ital J Surg Sci 18:401–406
Speranza V (1991) Surgical treatment of extrahepatic stones. In: Speranza V, Barbara L (eds) Changing concepts in biliary stone management. Lippincott, pp 582–586 (Problems in general surgery, vol 8, no 4)
Speranza V, Lezoche E, Minervini S, Carlei F, Basso N, Simi M (1982) Transduodenal papillostomy as a routine procedure in managing choledocholithiasis. Arch Surg 117:875
Stain SC, Cohen H, Tsuishoysha M, et al (1991) Choledocholithiasis: endoscopic sphincterotomy or common bile duct exploration. Ann Surg 213:627–634
Stefanini P, Carboni M, Patrassi N, Basoli A, De Bemardinis G, Negro P (1975) Roux-en-Y hepaticojejunostomy: a reappraisal of its indications and results. Ann Surg 181(2):213–219
Stockmann PT, Soper NJ (1991) Early results of laparoscopic cholecystectomy at a teaching institution. Perspect Gen Surg 2:1–19
Surick B, Washington M, Ghazi A (1993) Endoscopic retrograde cholangiopancreatography in conjunction with laparoscopic cholecystectomy. Surg Endosc 7:388–392
Vanneman W, Kingsbury R, Duberman E, Lee M (1992) When is ERCP indicated before laparoscopic cholecystectomy? Gastrointest Endosc 38:265

Van Stiegmann GV, Goff JS, Mansour A, et al (1992) Precholecystectomy endoscopic cholangiography and stone removal is not superior to cholecystectomy, cholangiography and common bile duct exploration. Am J Surg 163:227–230
Voyles CR, Petro AB, Meena AL, Haick AJ, Khoury AM (1991) A practical approach to laparoscopic cholecystectomy. Am J Surg 161:365–370
Way LW (1987) Biliary stricture. In: Way LW, Pellegrini CA (eds) Surgery of the gallbladder and bile ducts. Saunders, Philadelphia, pp 419–436
Wiesner RH (1994) Currents concepts in primary sclerosing cholangitis. Mayo Clin Proc 69:969–982
Williams LF, Schoetz DJ (1981) Primary sclerosing cholangitis. Surg Clin North Am 61:951–963

17A Percutaneous Treatment of Benign Biliary Stenoses and Injuries

M. Rossi, M. Bezzi, F. Maccioni, F.M. Arata, G. Bonomo, and P. Rossi

CONTENTS

17A.1 Introduction

Benign biliary strictures present different clinical problems to malignant biliary strictures, because the life expectancy of the patients is much longer and is not tumor-related. Benign biliary strictures are less common than malignant biliary strictures, so the number of cases in reported series is always rather limited (Pellegrini et al. 1984; Davids et al. 1993; Lillemoe et al. 1992). In the majority of cases benign biliary stenosis follows surgical complications and affects people of working age. Patient numbers and the duration of invalidity are therefore very important.

M. Rossi, MD, Department of Radiology, University of Rome "La Sapienza," Policlinico Umberto I, 00161 Rome, Italy
M. Bezzi, MD, Department of Radiology, University of Rome "La Sapienza," Policlinico Umberto I, 00161 Rome, Italy
F. Maccioni, MD, Department of Radiology, University of Rome "La Sapienza," Policlinico Umberto I, 00161 Rome, Italy
F.M. Arata, MD, Department of Radiology, University of Rome "La Sapienza," Policlinico Umberto I, 00161 Rome, Italy
G. Bonomo, MD, Department of Radiology, University of Rome "La Sapienza," Policlinico Umberto I, 00161 Rome, Italy
P. Rossi, MD, Department of Radiology, University of Rome "La Sapienza," Policlinico Umberto I, 00161 Rome, Italy

Surgical repair is usually the first choice of treatment for benign biliary stenosis; however, the incidence of recurrence increases with the number of surgical procedures performed (Pitt et al. 1982). Interventional radiology techniques proposed by Burhenne and Molnar (Burhenne 1975; Burhenne and Morris 1980; Molnar and Stockum 1978) may offer many solutions to this serious problem; the acquired experience has shown quite satisfactory results.

In this chapter we consider the types of injuries, their imaging, and their possible management with interventional radiology techniques, discussing also the surgical and endoscopic alternatives. We have not included primary sclerosing cholangitis in the discussion, because usually patients with this disease do not respond well to any surgical or percutaneous therapy and are probable candidates for liver transplantation.

17A.2 Etiology of Benign Biliary Strictures

Benign biliary strictures are a result of iatrogenic damage in 90%–95% of cases, not only after open or laparoscopic cholecystectomy, but also after gastric or hepatic resection, portacaval shunt, or biliary-enteric anastomosis (Andrén-Sandberg et al. 1985a, 1985b; Lindenauer 1973). Bile duct strictures can also follow liver transplantation (Petersen et al. 1996; Rossi et al. 1995). Other causes of stenosis are blunt or penetrating trauma, inflammation associated with lithiasis, chronic pancreatitis, primary sclerosing cholangitis, and stenosis of Oddi's sphincter. Many factors, including inadequate exposure, congenital anomalies, obesity, acute pre-existing cholecystitis, and bleeding during the operation, are associated with bile duct injury during cholecystectomy (Lillemoe et al. 1992).

Early reports on laparoscopic cholecystectomy suggest that this new procedure has been associated with a higher incidence of bile duct injury than open surgery (Zucker et al. 1991). Laparoscopic technique has gained popularity only during the last few years and many surgeons are in the learning phase. In addition, thermic injuries to the ductal wall, produced by the propagation of heat during electrocutting or electrocoagulation, may produce a progressive stenosis, usually localized at the bifurcation.

In recent years the importance of ischemia of the bile duct in the formation of postoperative benign biliary stenosis has also been emphasized. Extensive dissection of the bile duct during surgery may damage the major arteries to the proximal portion of the bile duct (Terblanche et al. 1983).

Repeated surgical attempts to repair strictures at the site of a biliary-enteric anastomosis are associated with an increased incidence of recurrences (Pitt et al. 1982; Warren et al. 1971).

17A.3 Clinical Presentation

The clinical presentation of a patient with a benign biliary stenosis may be early or late.

Early presentation is usually caused by an acute obstruction of the bile duct and becomes clinically evident with increasing jaundice and alterations of liver function tests a few days after surgery. Fever and infection may not be present, but may appear later. Sometimes bile leaks into the peritoneal cavity and may accumulate in a sterile or infected biloma.

Late presentation occurs from few months to several years after surgery, usually with cholangitis, septic fever, and jaundice. Sometimes in later presentations stenoses are associated with bile stasis and stone formation. If the strictures involve one duct only (right or left), the patient may have symptoms related to cholangitis but not jaundice. If diagnosis is delayed, the patient may present with advanced biliary cirrhosis and portal hypertension. However, regardless of the timing of diagnosis, the most common clinical presentation of patients with postoperative bile duct strictures appears to be cholangitis and jaundice (Lillemoe et al. 1992).

17A.4 Imaging Studies

Abdominal ultrasound and computed tomography (CT) play the initial role in the evaluation of patients with benign biliary strictures. Biliary tree dilatation, the level of obstruction, and fluid collections are usually easily detected by these two imaging modalities.

In patients who present a long time after surgery, CT can also help to rule out biliary, pancreatic, or other extrinsic neoplasms as causes of jaundice.

Magnetic resonance (MR) cholangiography is the newest technique of imaging the biliary tree and brings interesting advantages. It is a noninvasive technique which enables three-dimensional visualization of the intra- and extrahepatic ducts from different angles of view.

Percutaneous transhepatic cholangiography (PTC) is a minimally invasive technique performable in every fluoroscopy room using easily available and inexpensive material. This procedure is generally more reliable than endoscopic retrograde cholangiopancreatography (ERCP), and it can be followed by placement of an external-internal transhepatic drainage catheter to decompress the biliary system, to assist surgical reconstruction, and as an access for nonoperative dilatation.

ERCP has been almost completely replaced by MR cholangiography for diagnostic purposes and its role is now confined to interventional therapies. Furthermore, ERCP cannot be easily performed in patients with biliary-jejunal anastomosis.

In many cases, especially when the hepatic duct bifurcation is involved, or in the presence of a biliary-enteric anastomosis, it may be quite difficult to distinguish cholangiocarcinoma from benign strictures on ultrasonography or CT. Even the cholangiographic appearance may be ambiguous. In such cases, multiple brushing or, preferably, cholangioscopically guided biopsy may give the definitive solution to the diagnostic problem (Rabiovitz et al. 1990; Rossi et al. 1996).

17A.5 Percutaneous Treatment

During the last 15 years percutaneous treatment of benign biliary strictures with balloon dilatation and stenting has been widely accepted as a valid therapeutic alternative to surgery, because of its clinical efficacy and low complication rate. The reported success rates are comparable to those of surgery, with lower morbidity and a minimal incidence of procedure-related deaths (Citron and Martin 1991; Moore et al. 1987; Mueller et al. 1986; Williams et al. 1987; Rossi et al. 1990). This procedure has been termed "cholangioplasty" by analogy to "angioplasty."

17A.5.1 Indications

The main indications for balloon dilatation are:

1. High patient risk for surgery (age, poor general condition, multiorgan failure, etc.)
2. Multiple failure of previous surgical repairs
3. Purulent cholangitis with or without intrahepatic stones
4. Inclusion in a randomized trial
5. Patient refusal of surgical intervention

Percutaneous biliary dilatation may be performed in any patient with benign strictures of any etiology, whether single or multiple. The best response to percutaneous dilation is generally obtained in patients with a common bile duct stricture, although those with anastomotic or intrahepatic ductal stenoses without a large inflammatory component also respond fairly well. Balloon dilation is not effective in primary inflammatory stricture of the biliary ducts, such as sclerosing cholangitis, because of the high incidence of restenosis.

Our technique may be summed up as follows:

- Biliary drainage
- Passing through the stenotic tract
- Balloon dilatation
- Catheter stenting and follow-up
- Catheter removal

17A.5.2 Concomitant Medical Treatment

All patients receive intravenous antibiotics at least 12 h before percutaneous treatment. Since every attempt at dilation of biliary strictures is very painful for the patient, deep sedation and anesthesiologic assistance, especially during the dilation, are necessary. General anesthesia is often preferable. All procedures are done under continuous EKG and oxygen saturation monitoring.

17A.5.3 Approach

Biliary drainage is performed following the standard technique; however, some consideration is necessary in the choice of the best percutaneous access.

The right transhepatic approach is safer for the radiologist because it does not involve direct exposure of the hands to the X-ray beam and it is most commonly used. However the left bile duct approach is also frequently performed, and in order to avoid direct exposure of the operator's hands we suggest to using ultrasound guidance. The most difficult step in performing a percutaneous treatment is passing through the stricture, and sometimes the left transhepatic access may be more advantageous for this. MR cholangiography or preliminary PTC, including oblique views, give the necessary information about the three-dimensional morphology of the dilated biliary tree and allow the best approach to be selected. A good reason for preferring the left approach is that after dilatation a plastic stent usually needs to be left in place for long time, and for this the anterior subxiphoid position is better tolerated by the patient.

17A.5.4 Crossing the Stenosis

Whatever the entrance site, the second step is to cross the stricture. A steerable wire is used to negotiate the curves and bends of the bile ducts, which are usually dilated. When the wire reaches the obstruction, the needle sheath is advanced and an attempt is made to cross the obstruction. In our daily experience, J-shaped hydrophilic guidewires (Terumo Corporation, Japan) allow the stenosis to be crossed in almost all cases. This type of wire is less traumatic to the patient and easier to handle than other steerable wire types.

Obstructions of the biliary tree are easier to cross than choledochojejunostomy obstructions, because of the intense fibrosis and often total obstruction of the lumen at anastomotic sites. The retrograde percutaneous transjejunal biliary approach is a useful alternative for crossing a biliary-enteric anastomosis and for performing multiple dilations avoiding multiple duct punctures. (Martin et al. 1989; Perry et al. 1995; Russell et al. 1986).

When pyogenic cholangitis is detected, intrabiliary manipulation must be minimized; an external catheter is positioned for drainage for a few days and dilatation postponed until stabilization occurs.

17A.5.5 Balloon Dilatation

The maneuvers described above are difficult and the balloon catheters are the same as those employed for angioplasty. They can support a pressure of up to 17 atm and possess an extremely high radial force, able to force tight strictures. Once the inflation starts, the profile of the balloon is characterized by a "waist"

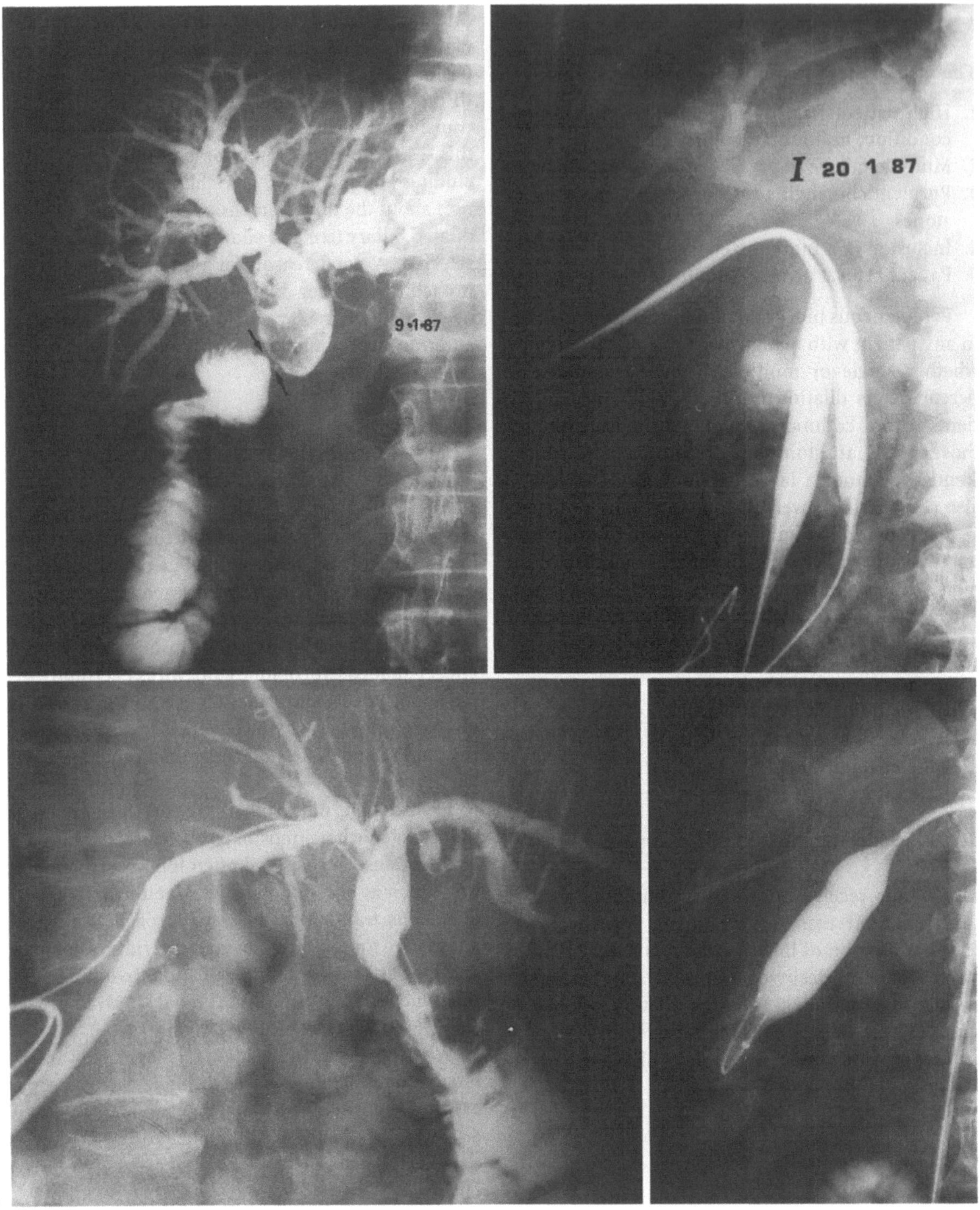

a b c d

Fig. 17A.1a–f. A 55-year-old man was admitted to December 1986 for septic fever. After surgical repair of a biliary-enteric anastomosis a drainage catheter was placed through a right-side approach and left in place for 2 weeks. The patient returned in January 1987 without fever. **a** Cholangiography confirmed the presence of a tight stenosis with a large stone at the biliaryenteric anastomosis. **b** An 8-mm balloon and a 10-mm balloon were simultaneously inflated at the anastomosis. Soon after this procedure the left bile duct was also catheterized. The cheek cholangiogram performed 2 days later showed no evidence of stones. **c** First cholangiographic follow-up at 1 month showing a good response to treatment and reduced bile duct dilation. **d** A recurrent stenosis was treated with an 18-mm balloon. A catheter was left in place for 6 months and then removed. **e** Two years later, in February 1989, the patient returned with cholangitis. On the suspicion of restenosis, percutaneous cholangioscopy was performed, demonstrating good passage of contrast material into the jejunal loop and diffuse signs of cholangitis. **f** In 1993 a CT examination of the abdomen was performed to evaluate a cancer of the stomach. No recurrent symptoms of biliary pathology occurred

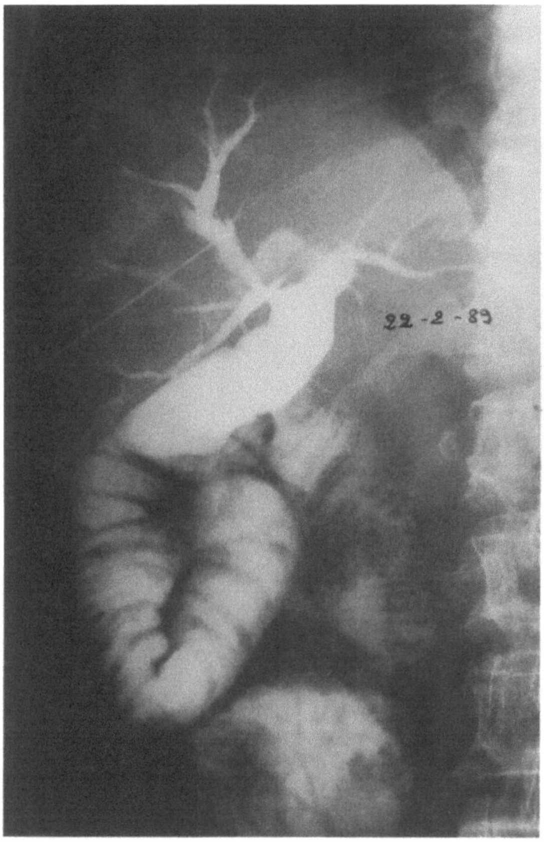

e

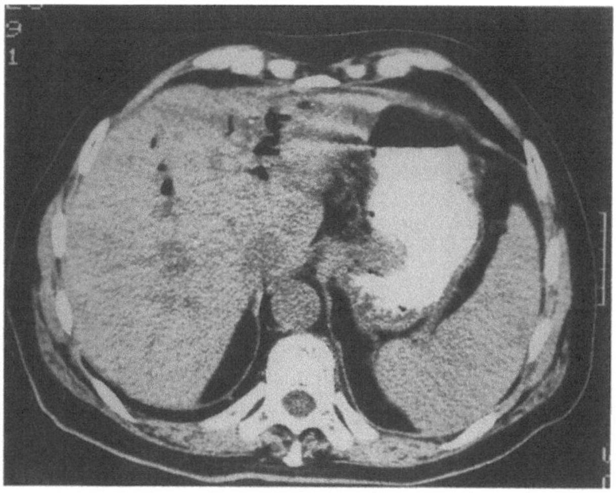

f

Fig. 17A.1e,f

at the stricture site that usually disappears as the pressure increases and the stricture gives way. In soft strictures, this may take only minimal inflation for several seconds.

The time of inflation may vary from person to person. The balloon is sometimes inflated for up to 2–3 min and then the procedure repeated several times in different sessions.

The size of the balloon is related to the type and location of the stenosis. Dilation is usually progressive in order to avoid laceration of the duct and bleeding. The most common size is a 10-mm balloon, but for intrahepatic ducts a 6- to 8-mm balloon may also be used. For biliary-enteric anastomoses the balloon is larger, varying from 10 to 20 mm in diameter (Fig. 17A.1b).

In patients who have undergone multiple surgery and have rigid fibrotic strictures, a larger balloon diameter with high-pressure inflation for a longer time may be necessary (Fig. 17A.1d).

Elastic recoil of the stricture is the most difficult situation and requires overdilation to adequately stretch the elastic fibers.

When biliary lithiasis is associated with the stricture (Fig. 17A.1a), the stones are fragmented and pushed into the bowel after dilation of the stenosis. If the stones are too large and cannot be crushed with a balloon or a basket, they are fragmented with an Electrohydraulic Lithotriptor and pushed with a balloon into the bowel, often after endoscopic sphincterectomy. Our success rate with this procedure has been 90% (Rossi et al. 1996).

17A.5.6 Stenting

After dilation, to keep a lumen during the healing process and provide access to the biliary tree if restenosis occurs, placement of an indwelling stent is advocated for a period ranging from 2 to 12 months, according to different authors' experiences. Generally, large-bore catheters are preferred, ranging from 10 to 20 F in size (Fig. 17A.2b).

The role of stent placement after dilation remains controversial because, while it may improve long-

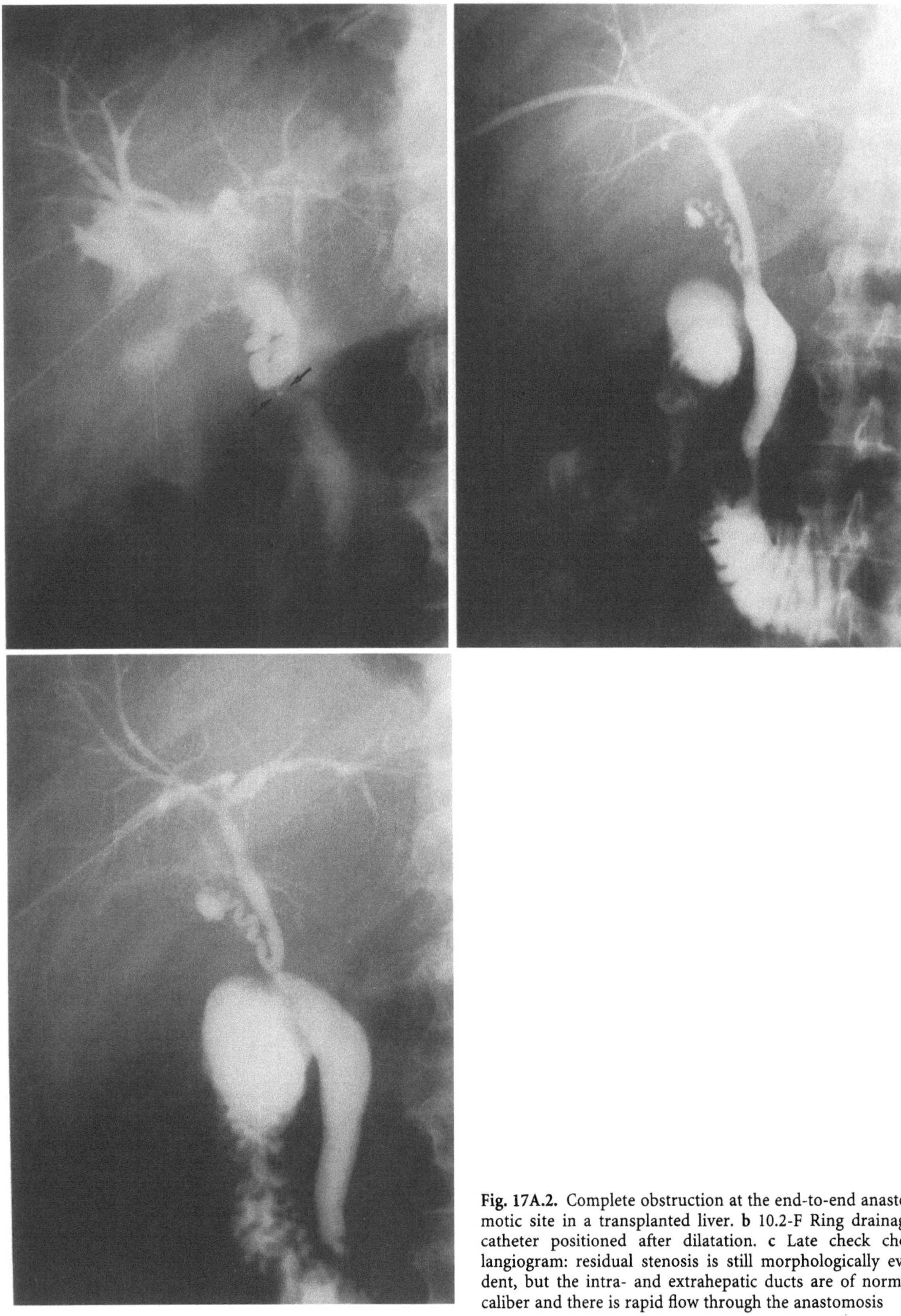

a b c

Fig. 17A.2. Complete obstruction at the end-to-end anastomotic site in a transplanted liver. **b** 10.2-F Ring drainage catheter positioned after dilatation. **c** Late check cholangiogram: residual stenosis is still morphologically evident, but the intra- and extrahepatic ducts are of normal caliber and there is rapid flow through the anastomosis term success, the tube itself may stimulate inflammatory reaction and fibrosis (Morrison et al. 1990; Mueller et al. 1986). Some authors prefer just to dilate and then to remove everything from the biliary tree if cholangioscopy has shown that there is no debris, erosion, or surface irregularity in the stenotic tract (Mueller et al. 1986).

The evaluation of the results before stent removal is usually based on cholangiographic-morphologic criteria. A decreased or normal caliber of the biliary ducts, disappearance of signs of cholangitis, and an absence of "filling defects" in the stenotic tract all indicate a good response to the treatment (Figs. 17A.1e, 17A.2b). We usually leave a small 5-F catheter above the dilated tract for 15–20 days to see the modifications of the bile duct during this time. If a residual stenosis is present, redilation can be performed in the same session. Multiple redilatations may be required in difficult cases.

Some Authors use a biliary manometric perfusion test as predictor of long-term success after treatment (Savader et al. 1994). The intrabiliary pressure is measured during mechanical injection of a saline solution at various rates (2–20 ml/min) through a catheter above the stricture. If the pressure does not exceed 20 cmH_2O, the patients are considered to have passed the biliary manometric perfusion test. In the study by Savader et al., biliary manometric perfusion test and clinical trial with a catheter above the stricture showed equal predictive value regarding the success or failure of the treatment.

17A.5.7 Complications

Minor complications after biliary dilatation such as transient fever or bleeding are common (Moore et al. 1987; Perry et al. 1995; Rossi et al. 1990; Williams et al. 1987).

Major complications that have been encountered are: septic shock (1.2%, Rossi et al. 1990), septicemia (5%, Mueller et al. 1986, and Perry et al. 1995; 22%, Williams et al. 1987), and hemorrhage requiring transfusion or hepatic artery embolization (4%, Rossi et al. 1990; 11% Williams et al. 1987). Pancreatitis (1/74 cases) and catheter-related duodenal perforation (1/74 cases) have been reported as occasional complications (Williams et al. 1987).

Episodes of cholangitis can occur while the stent is in place (1/17 cases) or after it has been removed (2/17 cases), and have been treated with intravenously administered antibiotics (Citron and Martin 1991).

Procedure-related death is extremely rare, occurring in 0.3% of cases in our experience. The primary morbidity and mortality of the procedure appear mostly related to transhepatic access and catheter permanency rather than to the balloon dilatation itself.

In recurrent cases, after several attempts at surgical repair or balloon dilatation with long-term catheter stenting, the therapeutic options are limited. Nevertheless, these patients need to be adequately treated, because of the potential of benign disease to progress toward biliary cirrhosis, secondary sclerosing cholangitis, and possibly even malignancy. In these cases, selected patients may be considered as potential candidates for placement of a metallic stent.

17A.5.8 Results

Recurrence of benign stenosis after cholangioplasty may be due to chronic inflammation or possibly to poor stricture compliance owing to an elastic component in the scar tissue. The common difficulty in treating biliary strictures is that this is a fibrosing disease, the course of which is interrupted but not stopped, and the strictures tend to recur (Morrison et al. 1990).

In most series success is defined as the absence of symptoms following stent removal and normal bilirubin and alkaline phosphatase levels. However, the results vary greatly from center to center depending on the size of the patient series, the type and location of the strictures, the presence of sclerosing cholangitis, and the length of follow-up. Success rates ranging from 70% to 100% at a mean follow-up of 23–59 months (range 6–78 months) have been reported (Citron and Martin 1991; Moore et al. 1987; Mueller et al. 1986; Pitt et al. 1989; Rossi et al. 1990; Vogel et al. 1985; Williams et al. 1987). Lower success rates – 55% – are reported in series with longer follow-up times (5–7.5 years) (Pitt et al. 1989; Jan et al. 1994).

At our institution 243 patients have been treated between 1983 and 1996 for benign biliary strictures. Cumulative primary patencies of 70% at 60 months and 70% at 100 months have been obtained. Metallic stents were employed in only 20 cases, 17 of them in 1988–1989, and the remaining 3 in the last 6 years. Secondary, "assisted" patency was 82% at 60 months and 75% at 140 months.

17A.6 Therapeutic Alternatives

17A.6.1 Surgery

Primary repair of bile duct strictures can be performed with a large number of different surgical techniques:

- End-to-end repair
- Roux-en-Y hepaticojejunostomy or choledochojejunostomy
- Choledochoduodenostomy
- Mucosal grafting

The choice of technique depends upon the extent and location of the stricture, the experience of the surgeon, and the timing of the intervention.

Fundamental principles for successful surgical repair are: (1) exposure of healthy proximal bile duct, (2) creation of a tension-free anastomosis, and (3) creation of a direct biliary-enteric, mucosa-to-mucosa, anastomosis. Choledochocholedochal, end-to-end, anastomosis or choledochoduodenostomy can rarely be accomplished, due to the invariable loss of duct length resulting from fibrosis associated with the injury, and to the consequent technical difficulty in obtaining a tension-free anastomosis. Thus, in almost all cases a Roux-en-Y hepaticojejunostomy is the preferred choice. These anastomoses can be created either between the common hepatic duct or the left hepatic duct or both the right and the left and the jejunum, depending on the level of the obstruction and whether the bifurcation is involved or not.

The role of postoperative biliary stenting and the length of time that the stent should be in place, which ranges from several weeks to 2 years, remain controversial (Pitt et al. 1989; Lillemoe et al. 1992). The likelihood of stenting being of benefit in a wide mucosa-to-mucosa anastomosis with normal duct tissue (Bismuth type I) is relatively small. However, a long-term postoperative stent may keep a tenuous anastomosis through scar tissue open, and prevents late fibrosis and restricture.

In the last 10 years most surgical series have shown lower mortality than in the previous 10 years. Surgical mortality is now less than 5%. Unfavorable factors are biliary tract infection and underlying liver disease, which can increase the mortality up to 30%. General or specific postoperative morbidity is around 20%–30% (Blumgart et al. 1984; Innes et al. 1988; Lillemoe et al. 1992; Pitt et al. 1982, 1989).

Long-term results indicate a 70%–90% success rate after surgery, with follow-up periods ranging from 57 to 133 months (Lillemoe et al. 1992). Although two-thirds of restenoses occur within 2 years, and 90% within 7 years from previous repair, long-term follow-up is very important in analyzing the results, since restenoses or complications such as stone formation can occur up to 20 years after the initial procedure.

17A.6.2 Endoscopic Dilatation

Endoscopic biliary dilatation begins with an ERCP and endoscopic sphincterotomy. The stricture is traversed in a retrograde fashion with an atraumatic guidewire and sequential dilatation with 6- to 10-mm balloon catheters is performed. In most cases an endoprosthesis is left in place for at least 6 months following dilatation. Success rates range from 55% to 88% at a follow-up of 6–48 months (Davids et al. 1993; Smith et al. 1995).

This procedure is not free of minor (5%–10%) or major (1%–2%) complications, most of them associated with ERCP procedures (Smith et al. 1995). Major complications include common bile duct perforation, bleeding, pancreatitis or sepsis, and respiratory arrest (Foutsch and Sivak 1985; Smith et al. 1995).

The main limitation of this technique is that it is possible only in patients with primary bile duct stricture or a choledochocoduodenal anastomosis.

References

Andrén-Sandberg A, Alinder G, Bengmark S (1985a) Accidental lesions of the common bile duct at cholecystectomy: pre- and perioperative factors of importance. Ann Surg 201:328–332

Andrén-Sandberg A, Johansson S, Bengmark S (1985b) Accidental lesions of the common bile duct at cholecystectomy: II. Results of treatment. Ann Surg 201:452–455

Blumgart LH, Kelley CJ, Benjamin IS (1984) Benign bile duct strictures following cholecystectomy: critical factors in management. Br J Surg 71:836–843

Burhenne HJ (1975) Dilatation of biliary tract strictures: a new roentgenologic technique. Radiol Clin 44:153–159

Burhenne HJ, Morris DC (1980) Biliary stricture dilatation: use of the Gruntzig balloon catheter. J Can Assoc Radiol 31:196–197

Citron SJ, Martin LG (1991) Benign biliary strictures: treatment with percutaneous cholangioplasty. Radiology 178: 339–341

Davids PHP, Tanka AKF, Rauws EAJ, et al (1993) Benign biliary strictures. Ann Surg 217:237–243

Foutsch PG, Sivak MV Jr (1985) Therapeutic endoscopic balloon dilatation of the extrahepatic biliary ducts. Am J Gastroenterol 80:575–580

Innes JT, Ferrara JJ, Carey LC (1988) Biliary reconstruction without transanastomotic stent. Am Surg 54:27–30
Jan YY, Chen MF, Hung CF (1994) Balloon dilatation of intrahepatic duct and biliary-enteric anastomosis strictures. Int Surg 79:103–105
Lillemoe KD, Pitt HA, Cameron JL (1992) Current management of benign bile duct strictures. Adv Surg 25:119–173
Lindenauer SM (1973) Surgical treatment of bile duct strictures. Surgery 73:875–880
Martin EC, Laffey KJ, Bixon R (1989) Percutaneous transjejunal approaches to the biliary system. Radiology 172:1031–1034
Molnar W, Stockum AE (1978) Transhepatic dilatation of choledochoenterostomy strictures. Radiology 129:59–64
Moore AV, Illescas FF, Mills SR, et al (1987) Percutaneous dilation of benign biliary strictures. Radiology 163:629–634
Morrison CM, Lee MJ, Saini S, Brink JA, Mueller PR (1990) Percutaneous balloon dilatation of benign biliary strictures. Radiol Clin North Am 28:1191–1201
Mueller PR, Van Sonnenberg E, Ferrucci JT, et al (1986) Biliary strictures dilatation: multicenter review of clinical management in 73 patients. Radiology 106:17–22
Pellegrini CA, Jean Thomas M, Way LW (1984) Recurrent biliary stricture. Am J Surg 147:175–180
Perry LJ, Stokes KR, David Lewis W, Jenkins RL, Clouse ME (1995) Biliary intervention by means of percutaneous puncture of the antecolic jejunal loop. Radiology 195:163–167
Petersen BD, Maxfield SR, Ivancev K, Uchida BT, Rabkin JM, Rosch J (1996) Biliary strictures in hepatic transplantation: treatment with self-expanding Z stents. J Vasc Interv Radiol 7:221–228
Pitt HA, Miyamoto T, Parapatis SK, Tompkins RK, Longmire WP Jr (1982) Factors influencing outcome in patients with postoperative biliary strictures. Am J Surg 144:14–19
Pitt HA, Kaufman SL, Coleman J, et al (1989) Benign postoperative biliary strictures: operate or dilate? Ann Surg 210:417–427
Rabiovitz M, Zajko AB, Hassanein T, et al (1990) Diagnostic value of brush cytology in the diagnosis of bile duct carcinoma: a study in 65 patients with bile duct stictures. Hepatology 12:747–752
Rossi M, Salvatori FM, Ingianna D, Greco M, Iappelli M, Rossi P (1995) Non-vascular interventional radiology for complications following orthotopic liver transplantation. Clinical-radiological correlation and technical considerations. Radiol Med 90:291–297
Rossi P, Salvatori FM, Bezzi M, Maccioni F, Porcaro ML, Ricci P (1990) Percutaneous treatment management of benign biliary strictures with balloon dilation and self-expanding metallic stents. Cardiovasc Intervent Radiol 13:231–239
Rossi P, Bezzi M, Fiocca F, Salvatori FM, Grasso E, Speranza V (1996) Percutaneous biliary endoscopy. Semin Interv Radiol 13:185–193
Russell E, Irizarry JM, Huber JS, et al (1986) Percutaneous transjejunal biliary dilatation: alternate management for benign strictures. Radiology 159:209–214
Savader SJ, Cameron LJ, Pitt HA (1994) Biliary manometry versus clinical trial: value as predictors of success after treatment of biliary tract strictures. J Vasc Interv Radiol 5:757–763
Smith MT, Sherman S, Lehman GA (1995) Endoscopic management of benign strictures of the biliary tree. Endoscopy 27:253–266
Terblanche J, Allison HF, Northover JMA (1983) An ischemic basis for biliary strictures. Surgery 94:52–57
Vogel SB, Howard RJ, Caridi J, Hawkins IF (1985) Evaluation of percutaneous transhepatic balloon dilatation of benign biliary strictures in high-risk patients. Am J Surg 149:73–79
Warren KW, Mountain JC, Midell AI (1971) Management of strictures of the biliary tract. Surg Clin North Am 51:711–730
Williams HJ, Bender CE, May GR (1987) Benign postoperative biliary strictures: dilatation with fluoroscopic guidance. Radiology 163:629–634
Zucker KA, Bailey RW, Gadacz TW, et al (1991) Laparoscopic guided cholecystectomy: a plea for cautious enthusiasm. Am J Surg 161:36–44

17B Use of Metallic Stents in Benign Biliary Strictures: Mid- and Long-Term Results

F. Maccioni, M. Bezzi, M. Rossi, H.-J. Brambs, A. Rieber, and P. Rossi

CONTENTS

17B.1 Introduction

In the last 7 years biliary metallic stents of various shapes and materials have been widely and successfully used to relieve malignant biliary strictures. They seem to present several advantages over conventional plastic stents and will probably come to replace them in the future, especially in the form of the covered metallic stent (Mueller et al. 1985; Yoshioka et al. 1990; Lammer 1990; Lee et al. 1991; Adam et al. 1991; Nicholson and Royston 1993; Roeren et al. 1990; Rossi 1996).

At the beginning there was some hesitation about placing metallic stents in benign biliary strictures, since no information was available on their long-term patency, the most important factor to consider in patients with a long life expectancy (Carrasco et al. 1985; Alvarado et al. 1989; Coons et al. 1989; Dick et al. 1989; Rossi et al. 1990a,b; Mueller et al. 1990). In patients with malignant strictures a 1- or 2-year patency is more than satisfactory, since this covers the life expectancy, but in patients with benign disease the stent needs to remain patent for several decades. In fact, once positioned, stents cannot be removed, since the ductal epithelium progressively grows over the metallic struts until they finally become an integral part of the ductal wall. If occlusion occurs, surgical reconstruction is possible if the stent is in the common bile duct, but not if it has been placed at the level of intrahepatic ducts.

A note of awareness is therefore mandatory in regard to using metallic stents for benign strictures. If any percutaneous attempts at stent recanalization should fail, and surgical repair is not feasible, the only solution to stent occlusion is liver transplantation.

For these reasons, when metallic stents became available for clinical use, approximately 7 years ago, we decided that in patients with benign stenoses they had to be considered the very last resort to recanalize the biliary tree (Rossi et al. 1990a,b).

17B.2 General Indications for Metallic Stent Placement

Among patients with benign biliary strictures, in our opinion the only candidates for placement of metallic stents are those *patients with recurrent strictures after failure of several surgical and/or percutaneous treatments, who are no longer candidates for reconstructive biliary surgery or have refused additional surgery*. In these selected cases, metallic stents represent the only way to recanalize the biliary tree before liver transplant.

Surgery is still considered the best treatment for patients affected by recurrent benign strictures, with

F. Maccioni, MD, Department of Radiology, University of Rome "La Sapienza," Policlinico Umberto I, Rome, Italy
M. Bezzi, MD, Department of Radiology, University of Rome "La Sapienza," Policlinico Umberto I, Rome, Italy
M. Rossi, MD, Department of Radiology, University of Rome "La Sapienza," Policlinico Umberto I, Rome, Italy
H.-J Brambs, MD, Radiologische Klinik und Poliklinik, Ulm, Germany
A. Rieber, MD, Radiologische Klinik und Poliklinik, Ulm, Germany
P. Rossi, MD, Department of Radiology, University of Rome "La Sapienza," Policlinico Umberto I, Rome, Italy

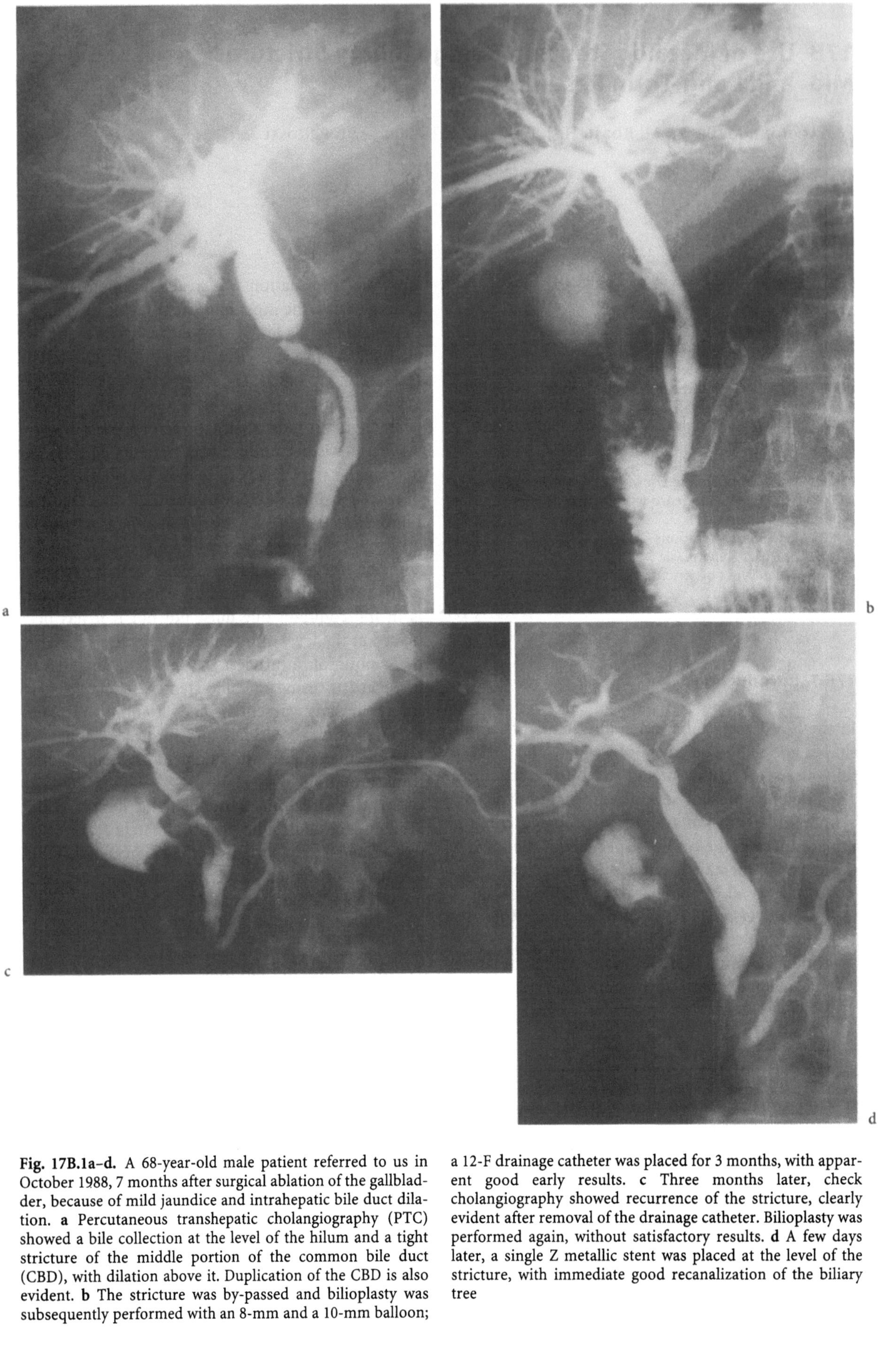

Fig. 17B.1a–d. A 68-year-old male patient referred to us in October 1988, 7 months after surgical ablation of the gallbladder, because of mild jaundice and intrahepatic bile duct dilation. **a** Percutaneous transhepatic cholangiography (PTC) showed a bile collection at the level of the hilum and a tight stricture of the middle portion of the common bile duct (CBD), with dilation above it. Duplication of the CBD is also evident. **b** The stricture was by-passed and bilioplasty was subsequently performed with an 8-mm and a 10-mm balloon; a 12-F drainage catheter was placed for 3 months, with apparent good early results. **c** Three months later, check cholangiography showed recurrence of the stricture, clearly evident after removal of the drainage catheter. Bilioplasty was performed again, without satisfactory results. **d** A few days later, a single Z metallic stent was placed at the level of the stricture, with immediate good recanalization of the biliary tree a primary success rate of 70%–90% (Warren et al. 1971; Bolton et al. 1980; Glenn 1978; Way et al. 1981). However, if strictures recur after initial surgery, 22%–36% of patients undergoing further surgery will have recurrent strictures after each procedure. In fact, after two or more interventions, the presence of exuberant inflammatory fibrotic tissue at the anastomotic site and the progressive shortening of the main bile duct no longer favor effective surgical reconstruction. The risk of surgical intervention in these cases is also increased by biliary cirrhosis, the incidence of which depends on the duration of the obstruction (Pitt et al. 1982; Blumgart et al. 1984).

Cholangioplasty can represent a safe and effective alternative to surgery in patients affected by recurrent benign biliary stenoses (Mueller et al. 1986; Moore et al. 1987; Willliams et al. 1987; Citron and Martin 1991; Rossi et al. 1990a,b) (Fig. 17B.1a,b). Percutaneous balloon dilation of benign strictures is associated with a low risk of morbidity and mortality and high long-term success rates: in our experience, in a group of 243 benign stenoses treated between 1983 and 1996, the primary patency rate is 70% at 7-year follow up, and the assisted patency rate is 76% at more than 10 years (Bezzi et al. 1995; Rossi 1996b). Other authors have reported a mid-term 3-year patency rate of 67% for anastomotic and 76% for iatrogenic strictures (Mueller et al. 1986). Unfortunately, however, strictures recur after cholangioplasty as well after surgery; in our experience, approximately 80% of recurrences became clinically evident within 3 years from the end of the treatment (Fig. 17B.1c).

In patients who are no longer candidates for reconstructive biliary surgery, when repeated balloon dilatations and prolonged catheter stenting have failed, metallic stent implantation should be considered (Fig. 17B.1d). The fact is that for recurrent benign strictures there are limited therapeutic options. The use of plastic endoprostheses, which tend to occlude within 6–8 months with sludge, is unsatisfactory in patients with a normal life expectancy and many of whom are young. Repeated surgical repair or a long-term indwelling catheter are generally distressing and undesirable without offering any guarantee of permanent patency. If untreated, recurrent benign stenoses may eventually evolve toward biliary cirrhosis and secondary sclerosing cholangitis – serious events if one considers that the majority of these strictures are the consequence of a simple cholecystectomy. Metallic stents may represent a permanent solution for recurrent benign strictures, because they can provide structural support of the bile duct, preventing elastic recoil of the lumen and possible repeat stenosis (Adam 1994; Coons 1992; Bezzi et al. 1991; Maccioni et al. 1992, 1993). Metallic stents also produce a larger degree of dilation than any other endoprosthesis, thus ensuring long-term drainage and a better quality of life for the patient.

17B.3 What Type of Metallic Stent?

Presently, there are several different types of metallic endoprostheses commercially available in Europe for use in the biliary tree: the Gianturco Z stent (Cook, Bloomington, Indiana), the Wallstent (Medinvent, Lausanne, Switzerland), Strecker's nitinol stent (Elastalloy, Meditech, USA – for experimental use only), the Palmaz stent (Johnson and Johnson, Warren, New Jersey), nitinol Angiomed stents (Instent, Angiomed, Karlsruhe, Germany). All metallic stents have certain similarities, in that they are constructed of a thin wire that is shaped into different specific designs. They may be basically divided in two groups: balloon-expandable (Strecker's Tantalum and Palmaz stents) and self-expanding metallic stents (Wallstents, Z stents and Strecker's and Angiomed nitinol stents); all of them are made of a fine, very flexible wire mesh, except the Gianturco stent, which is more rigid because it is made of a stainless steel wire arranged in a zigzag pattern. Recent developments include stents covered with silicone, throughout their length or only in the middle part, to avoid tumor ingrowth between the struts of the stent when used to treat malignant stricture (Roeren et al. 1990).

On the basis of our experience, we and other authors (Coons 1992; Maccioni et al. 1992; Petersen et al. 1996; Uflaker 1996) suggest the use of longer, fine-mesh, flexible stents – such as Wallstents or Strecker's nitinol stents – for malignant strictures, whereas we recommend the use of the shorter stents – such as Z and Palmaz stents – for benign strictures, for several reasons. Firstly, the stent should contain as little metal as possible, to allow total incorporation into the ductal wall without significant adverse reaction. Benign stenoses are generally web-like and shorter than malignant ones and are well treated with shorter rather than longer stents. The space between structs can be large because no tumor ingrowth is expected. Finally, unsatisfactory results have been reported with the use of longer, fine-mesh metallic stents, such as Wallstents, to release benign strictures (Kugler et al. 1996).

The most significant advantage that all metallic stents have over conventional stents is their large internal diameter, which varies from 8 12 mm (26–34 F), considerably greater than the diameter of the largest conventional plastic devices. Experimental work by Rey et al. (1985) has proved that patency is directly related to stent diameter; the severalfold greater diameter of the metallic stent should therefore promote long-term patency and drainage. The relationship between internal and external diameter (stent profile) is significantly lower in metallic stents than in plastic endoprostheses. Moreover, the small surface area of the metallic structure offers less chance for bile sludge accummulation.

All these structural characteristics, associated with good radial strength, are responsible for a better patency rate of metallic stents versus plastic ones; in malignant stenoses an occlusion rate of 3% was reported for metallic stents versus 16% for plastic endoprostheses (Lammer 1990). In benign stenoses, however, the most important technical feature of a metallic stent is its high radial expanding strength, to resist the circumferential pressure of the scar tissue (Uflaker 1996).

Because of their design, all metallic stents, whether self-expanding or balloon-expandable, will be incorporated into the bile duct mucosa within 1 week or more after deployment; consequently they cannot be removed, either percutaneously or endoscopically (Alvarado et al. 1989; Adam 1994; Coons 1992; Bezzi et al. 1991; Maccioni et al. 1992, 1993). However, an occluded stent, although not removable, can usually be crossed with a guidewire and subsequently recanalized. Ways to recanalize an obstructed stent include balloon dilatation, electrocutting, or placement of an additional metallic stent ("stent within a stent").

17B.4 Technical Features of Metallic Stents Used in Benign Strictures

Gianturco ZigZag Stent

The Z stents are made of a stainless steel wire, 0.08–0.1 mm in diameter, according to a modified Gianturco design (Uchida et al. 1988). The wire is bent in a zigzag and tied with a nylon suture that passes through an eyelet at the end of each bend to form a cylindrical structure. The stents are self-expanding and when completely expanded have a diameter of 8–12 mm and a length of 1.5 cm; when compressed, they can be introduced through an 8.5- or 10-F delivery catheter; finally, two or three stents can be linked together head to tail to dilate long segments.

When placing a Z stent, an essential point is to center the stent in the midpoint of the stenosis. Once extruded from the delivery system, Z stents expand to their full size in 12–36 h. Any interventional maneuver through the stent during this period should be avoided, to prevent dislodgment or malpositioning.

After stent deployment, a 5-F angiographic catheter is left above the stricture to allow flushing, being removed when stent expansion is considered satisfactory.

Palmaz Stent

The Palmaz stent is a balloon-expandable stent made of stainless steel 0.015 mm thick and dilating to a diameter of 8–10 mm (30 F). The length of the Palmaz stent is approximately 3 cm and it is introduced through a 9-F sheath. It is easy to position, due to excellent radiopacity associated with predictable shortening. The stent is mounted on its own balloon catheter.

Once positioned at the level of the stricture, the balloon is inflated, causing the stent to expand completely; the balloon is then deflated and carefully withdrawn.

17B.5 Major Issues in the Early Clinical Application of Metallic Stents in Benign Strictures

The action of the metallic stent on the biliary wall was unknown and was the major concern in our clinical trial (Rossi et al. 1990a,b; Maccioni et al. 1992). For proper evaluation of definitive results and any late complications, a long follow-up period of 5–10 years is necessary, since late recurrences after surgical repair can occur over 10 years after the event. Another intrinsic problem with metallic stents was that, once positioned, they cannot be removed and in most cases prevent further surgery on the biliary tree. Furthermore, no information was available regarding the possibility of a percutaneous reintervention on the metallic stent if it became occluded.

Therefore, to obtain as much information as possible, we implanted most of the stents within 1 year (1988–1989) and planned careful monitoring of our patients in the following years, by laboratory tests and imaging procedures such as ultrasound, abdominal plain films, and CT; percutaneous

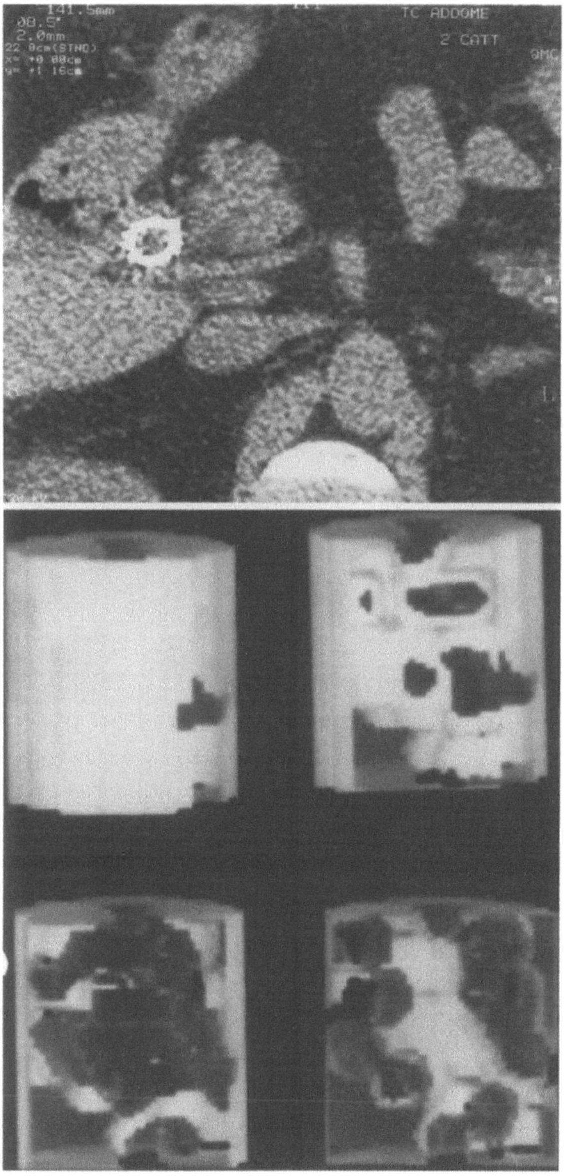
a
b

Fig. 17B.2a,b. Same patient as Fig. 17B.1. In the period following Z stent placement the patient remained asymptomatic, with laboratory test values within the norm. CT performed 3 years after stent placement (a) showed good stent expansion, with an inner low density suggestive of the presence of air or bile, although at some levels a higher density was evident within the stent, as also showed by three-dimensional reconstructions (b; see *upper* images showing reconstruction at the mid level of the stent)

cholangiography was performed in cases of recurrence only (Rossi et al. 1991a,b; Maccioni et al. 1992) (Figs. 17B.2, 17B.3a).

At almost 7 years follow-up, we have achieved enough experience to discuss the possible causes of occlusion or procedural failure, and to define the strictures best suited to treatment with these stents.

17B.6 Evidence of the European Experience

17B.6.1 Long-Term Follow-Up with Z Stents

Our series consisted of 20 patients treated with metallic stents between 1988 and 1994, at the "La Sapienza" University Hospitals of Rome. Eighteen patients were treated between 1988 and 1989, one patient in 1993, and another in 1994. The group as a whole will be regarded as having "long-term" results (78 months average follow-up).

All patients had a stent placed for recurrent post-surgical stricture: in nine cases post-cholecystectomy CBD stricture, in nine biliary-enteric anastomotic stenosis, and in two post-liver-transplant biliary-enteric stricture.

We used a total of 29 single or double zigzag stents: 23 were Gianturco Z stents (Cook, Bloomington, IN, USA), 6 were prototype Z stents (Angiomed, Karlsruhe, Germany), and one a Wallstent (Medinvent, Lausanne, Switzerland).

At an average follow-up of nearly 7 years (71–84 months, average 78 months), 11 out 20 patients (55%) had complete diappearance of symptoms and had not needed any additional procedure after stent placement (3 of them died with the stent still patent, and 8 are alive in good condition). Nine out 20 (45%) did well initially, but lately had recurrence of symptoms and were subsequently retreated. In this group of patients with recurrence, 5 were successfully retreated and are now asymptomatic, 1 patient underwent surgical revision of the anastomotic stricture, while 3 patients were retreated but subsequently died due to severe underlying diseases (lymphoma, biliary cirrhosis).

The cause of recurrence was late stent migration after 7 and 60 months, respectively, in 2 patients, and late stent occlusion after 15–60 months (average 35 months) in 7 cases, due to either hyperplastic mucosa and or stone formation (Fig. 17B.3a–d).

The long-term primary patency rate is 55% (11/20), while the assisted patency rate is 75% (15/20).

If we consider long-term results according to the type and level of the stricture treated, in the group with single CBD strictures (9 patients), only 1 stent occluded, whereas in the group with anastomotic strictures (11 patients), 8 stents occluded: 3 patients in this group are asymptomatic, 5 had recurrence and were retreated because of stent occlusion (3 patients) or stent migration (2 patients), and 3 died, all of them with stent malfunction or occlusion.

The modalities of reintervention for stent occlusion were: balloon dilation of the stent ($n = 7$), ex-

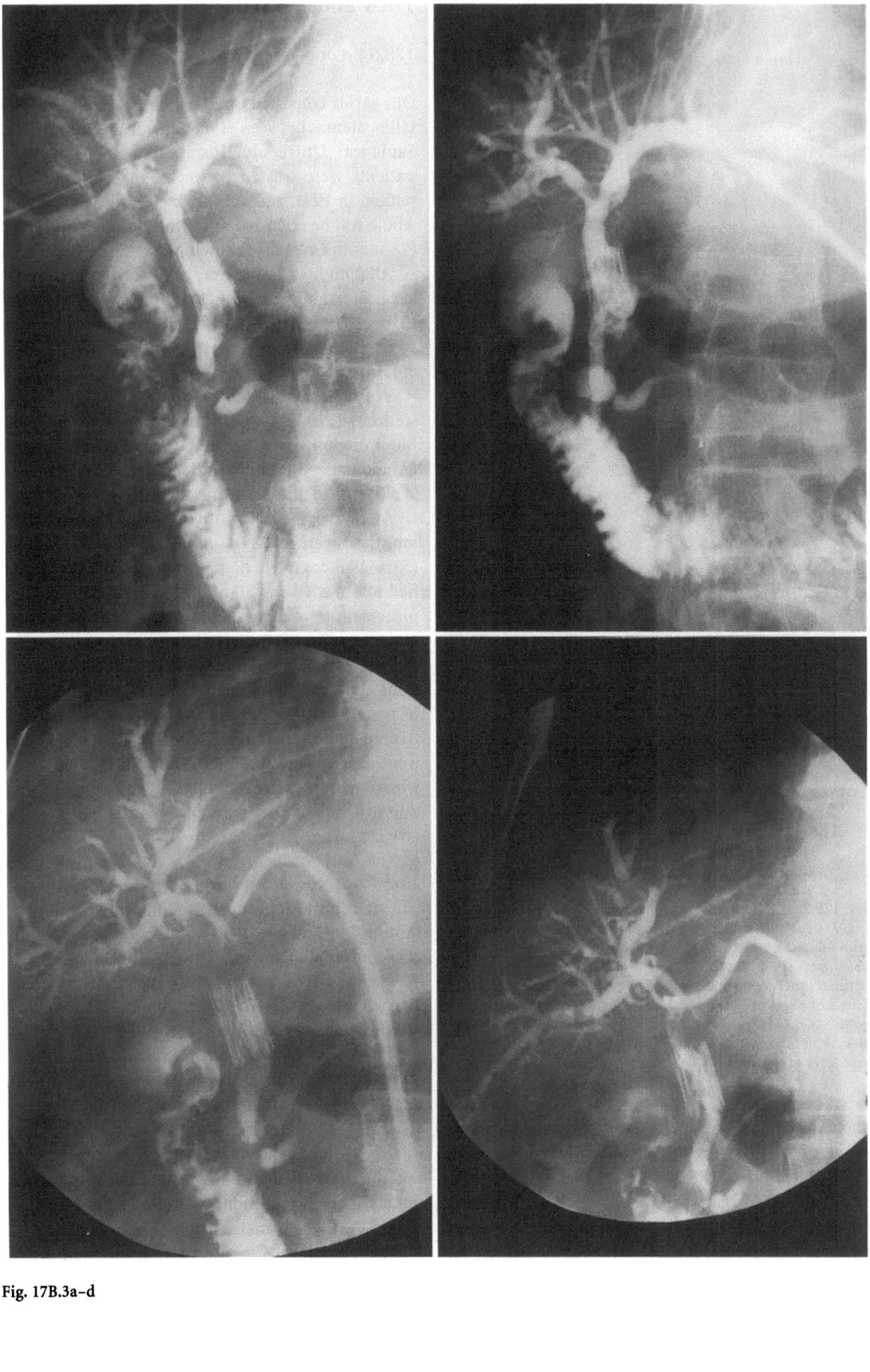

a b c d

Fig. 17B.3a–d

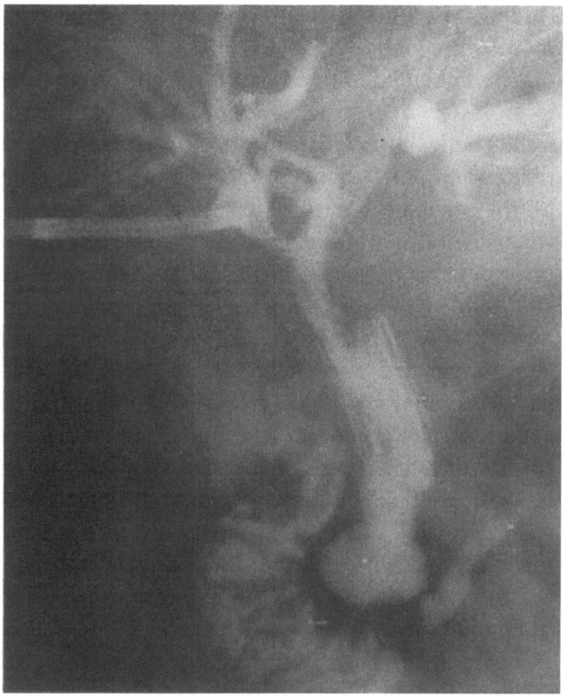

e

Fig. 17B.3a–e. Same patient as Fig. 17B.1. He remained asymptomatic for approximately 5 years after Z stent placement. In March 1994 he had an episode of fever and cholangitis with a rise in liver enzymes, treated with antibiotics. In the following year the patient had other similar episodes, until in October 1995 he was readmitted to hospital for worsening of symptoms. **a** PTC showed mild duct dilation above the stent, several small stones within the right ducts, and several filling defects within the stent itself. **b** A left drainage catherer was percutaneously placed. Through the left approach, a cholangioscope was percutaneously introduced, revealing abundant granulation tissue and a stone within the stent (**c**) and a tight stricture at the origin of the right hepatic ducts, with several stones above it (**d**). The metallic stent was subsequently dilated and the stone was fragmented and removed with the cholangioscope. **e** Through a right percutanous approach, another drainage catheter was placed. Cholangioscopic removal and fragmentation of the remaining right intrahepatic stones was then performed

traction of stones ($n = 3$), electrocoagulation ($n = 1$), and placement of other stents ($n = 1$).

17B.6.2 Mid-Term Experience with Palmaz Stents

The second series consisted of 25 patients treated with metallic stents between 1991 and 1995 at the University of Ulm (Brambs and Rieber, unpublished data), who are regarded as having "mid-term" results (23 months average follow-up). In this group of 25 patients, 34 single or double Palmaz stents were implanted.

At 3–66 months after stent placement (average 18 months) 20/25 patients are asymptomatic (80%).

Five of 25 (20%) suffered recurrence of symptoms after 1.5–48 months (mean 24 months) and were retreated; 4 are still asymptomatic, whereas 1 patient died.

Causes of occlusion were stricture above the stent, stent migration, intrahepatic stones, and in two cases hyperplastic mucosa. The modalities of reintervention for stent occlusion were: stent balloon dilation ($n = 3$), extraction of stones ($n = 2$), electrocoagulation ($n = 1$), and placement of other stents ($n = 1$).

The mid-term primary patency rate in this group is 80%, while the assisted patency rate is 96% (24/25).

17B.7 Complications

Early complications include stent misplacement or dislodgement, and are easily managed with placement of an additional stent. They only happened in the learning phase of our experience.

17B.7.1 Late Stent Migrations

Late stent migrations may represent a possible cause of procedural failure and recurrence (Maccioni et al. 1992). In our experience this happened three times, with individual metallic stents placed at the level of biliary-enteric anastomoses; two of the three patients had recurrent jaundice and were eventually retreated. An interesting point is that the third patient with late stent dislodgement did not suffer any recurrence of symptoms after stent migration and still remains asymptomatic; we believe that the temporary presence of the stent in situ could have had a therapeutic effect, confirming the potential usefulness of retrievable metallic stents, still to be developed.

In agreement with other authors, we now believe that double Z stents are more adequate to treat anastomotic strictures and could prevent this complication.

17B.7.2 Mucosal Hyperplasia

There has been some controversy about the possible presence of hyperplastic mucosa within the struts of

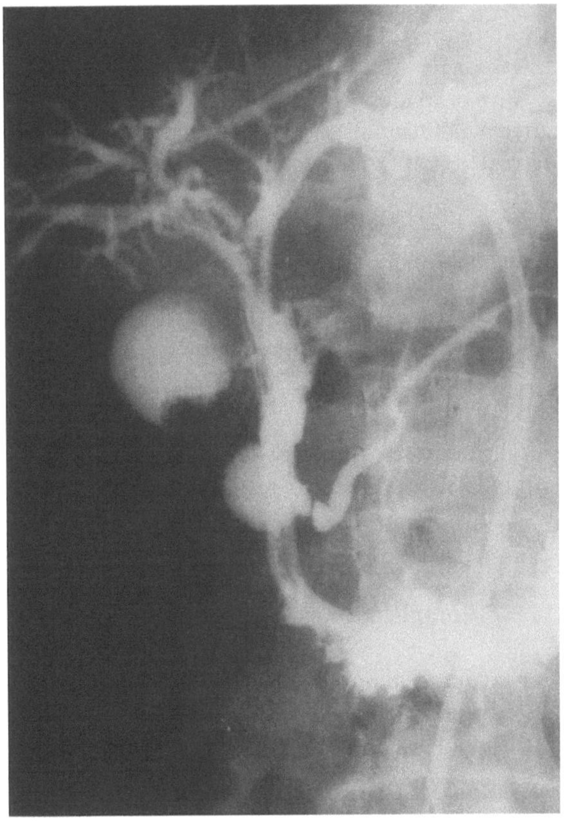

Fig. 17B.4. Same patient as Fig. 17B.1. In February 1996, thc last check cholangiogram showed complete removal of the intrahepatic and CBD stones, with resolution of the right hepatic stricture and patency of the metallic stent. Both catheters were therefore removed, and the patient is asymptomatic at 3 months follow-up

metallic stents. Preliminary studies on animal models (Carrasco et al. 1985; Alvarado et al. 1989) and subsequent early clinical studies (Maccioni et al. 1991) reported the possibility of a papillary hyperplastic reaction of he ductal mucosa between the struts of the stent, causing partial obstruction within the first 6 months after stent placement.

Actually, late stent occlusion by tissue ingrowth represents the main cause of recurrence in most reported series (Coons 1989, 1992; Rossi et al. 1990; Maccioni et al. 1992; Bezzi et al. 1995; Petersen et al. 1996). In our series, in all six patients with recurrent jaundice a percutaneous cholangiogram revealed almost complete stent obstruction, associated with stone formation in four cases (Fig. 17B.3a–d); subsequent biopsies obtained through cholangioscopes or by brushing confirmed the presence of hyperplastic epithelium and/or granulation tissue within the stent. In all cases hyperplastic tissue occluded the stents late, after an average time of 35 months.

This phenomenon it is likely to be present, to a lesser degree, in asymptomatic patients as well. The lack of symptoms in most of the patients treated may be explained by the large final caliber of the metallic stent, which is almost six times wider than the common bile duct. We believe that in these patients tissue ingrowth reaches a certain point and than stabilizes, allowing a physiological bile flow.

Although the possibility of late tissue ingrowth has been definitely proved by our and other authors' experience, we do not think that it should contraindicate the use of metallic stents in benign lesions, since all patients with partial or complete stent occlusion due to tissue ingrowth were satisfactorily treated with balloon dilation, electrothermal ablation, or stone removal, and are now asymptomatic (Figs. 17B.3c–e, 17B.4).

None of the patients in our or other series had serious procedue-related complications. In our experience, surgical revision after stent placement was needed in one case only, although collaboration with the endoscopist is often necessary to treat stent occlusion due to either tissue ingrowth or stoncs.

17B.8 What Types of Stricture Recur?

Reviewing our experience, we have observed that all patients with complete stent occlusion had a *biliary-enteric anastomosis* and preexisting *diffuse cholangitis* associated with *intrahepatic strictures*, which may explain the presence of abundant granulation tissue. We presently consider secondary sclerosing cholangitis to be a relative contraindication to the use of metallic stents, especially if they are to be used to correct intrahepatic ductal stenoses, because of the high occlusion rate. Recently, Coons (1992) has reported poor long-term results as well, although patients with primary sclerosing cholangitis in his series did somewhat better.

On the other hand, the long-term results obtained in patients with *single latrogenic common bile duct strictures* should be considered extremely satisfactory, with a reported long-term success rate of nearly 90%. *Single anastomotic strictures* not associated with intrahepatic stenoses can be satisfactorily treated with metallic stents as well, although associated with a higher risk of late stent dislodgement and late stent occlusion by tissue ingrowth.

17B.9 Conclusion

In conclusion, recurrence of biliary strictures after repeated balloon dilation still represents the main indication for the use of metallic stents in patients with benign disease. In this group of patients metallic stents should be considered as extremely effective therapeutic devices, representing the last resort for recanalization of the biliary tree without a permanent drainage catheter. The 7-year long-term results are more than satisfactory, with a primary success rate of 55% and a secondary success rate of 75%.

Irrespective of these positive results, however, we still believe, that metallic stents should never be placed primarily, but only after failure of an initial trial with repeated balloon angioplasty, which still remains a less invasive, less expensive, and more widely available procedure. Balloon dilatation and metallic stent placement are not in competition with each other but complementary to each other, and the association of the two may significantly raise the overall success rate of the percutaneous radiological management of benign biliary strictures.

References

Adam A (1994) Metallic biliary endoprostheses. Cardiovasc Intervent Radiol 17:127–132

Adam A, Chetty N, Roddie M, Yeung E, Benjamin IS (1991) Self-expandable stainless steel endoprostheses for treatment of malignant bile duct obstruction. AJR Am J Roentgenol 156:321–325

Alvarado R, Palmaz J, Garcia OJ, et al (1989) Evaluation of polymer-coated balloon-expandable stents in bile ducts. Radiology 170:975–978

Bezzi M, Salvatori FM, Maccioni F, Ricci P, Rossi P (1991) Biliary metallic stents in benign strictures. Semin Interv Radiol 8:321–329

Bezzi M, Bonomo G, Salvatori FM, Orsi F, Maccioni F, Rossi P (1995) Ten years follow-up of percutaneous management of benign biliary strictures: how successful are we? (Abstract.) Radiology 197(P):241

Blumgart LH, Kelley CJ, Benjamin IS (1984) Benign bile duct strictures following cholecystectomy: critical factors in management. Br J Surg 71:836–843

Bolton JS, Braasch JW, Rossi RL (1980) Management of benign biliary strictures. Surg Clin North Am 60:313–332

Carrasco CH, Wallace S, Charnsangavej C, et al (1985) Expandable biliary endoprostheses: an experimental study. AJR 145:1279–1281

Citron SJ, Martin LG (1991) Benign biliary strictures: treatment with percutaneous cholangioplasty. Radiology 178:339–341

Coons HG (1989) Self-expanding stainless steel biliary stents. Radiology 170:979–983

Coons H (1992) Metallic stents for the treatment of biliary obstruction: a report of 100 cases. Cardiovasc Intervent Radiol 15:367–374

Dick R, Gillams A, Dooley JS, Hobbs KEF (1989) Stainless steel mesh stents for biliary strictures. Intervent Radiol 4:95–98

Dick BW, Gordon RL, La Berge JM, Doherty MM, Ring EJ (1990) Percutaneous transhepatic placement of biliary endoprostheses: results in 100 consecutive patients. J Vasc Interv Radiol 1:97–100

Gillams A, Dick R, Dooley JS, et al (1990) Self-expandable stainless steel braided endoprosthesis for biliary strictures. Radiology 174:137–140

Glenn F (1978) Iatrogenic injuries to the biliary ductal system. Surg Gynecol Obstet 146:430–434

Kugler C, Hausegger KA, Uggowitze MM, Karaic R, Klein GE, Maurer M (1996) Benign biliary strictures: treatment with Wallstents. J Vasc Interv Radiol 7:292 (Scientific program, Society of Cardiovascular and Interventional Radiology, 21st Annual Meeting, March 1996)

Lammer J (1990) Biliary endoprostheses: plastic versus metal stents. Radiol Clin North Am 28:1211–1222

Lammer J, et al (1992) Cardiovascular and Interventional Radiological Society of Europe (CIRSE). Annual Meeting, Barcelona

Lee JS, Stoker J, Nijs HGT, Zonderland HM, et al (1991) Malignant biliary obstruction: percutaneous use of self-expandable stents. Radiology 179:703–707

Maccioni F, Ricci P, Gandini R, Rossi P (1991) Mucosal hyperplasia: a factor conditioning long-term patency of biliary metallic stents. 77th Meeting of the Radiological Society of North America, Chicago, 1991

Maccioni F, Rossi M, Salvatori FM, Ricci P, Bezzi M, Rossi P (1992) Metallic stents in benign biliary strictures: three year follow-up. Cardiovase Intervent Radiol 15:6

Maccioni F, Bezzi M, Gandini R, Rossi M, Ricci P, Broglia L, Salvatori FM, Rossi P (1993) Metallic stents in benign biliary stenosis. A four-year follow-up. Radiol Med (Torino) 86:294–301

Moore AV, Illescas FF, Mills SR, et al (1987) Percutaneous dilation of benign biliary strictures. Radiology 163:625–628

Mueller PR (1991) Metallic endoprostheses: boon or bust? Radiology 179:603–605

Mueller PR, Ferrucci JT Jr, Teplick SK, et al (1985) Biliary endoprosthesis: analysis of complications in 113 patients. Radiology 156:637–639

Mueller PR, Van Sonnenberg E, Ferrucci JT, et al (1986) Biliary stricture dilatation: multicenter review of clinical management in 73 patients. Radiology 106:17–22

Mueller PR, Tegmeyer CJ, Saini S, et al (1990) Metallic biliary stents: early experience (abstract). Radiology 177:138

Nicholson AA, Royston CM (1993) Palliation of inoperable biliary obstruction with self-expanding metal endoprostheses; a review of 77 patients. Clin Radiol 47:245–250

Petersen BD, Maxfield SR, Ivancev K, Uchida BT, Rabkin JM, Rösch J (1996) Biliary strictures in hepatic transplantation: treatment with self-expanding Z stents. J Vasc Interv Radiol 7:221–228

Pitt HA, Miyamoto T, Parapatis SK, Tompkins RK, Longmire WP Jr (1982) Factors influencing outcome in patients with postoperative biliary strictures. Am J Surg 144: 14–19

Rey JF, Marpetit D, Greff M (1985) Experimental study of biliary endoprostheses efficiency. Endoscopy 17:145–148

Roeren T, Brambs HJ, Ritcher GM, Kauffmann GW (1990) Coated balloon-expandable stent for percutaneous treat-

ment of malignant biliary obstruction (abstract). Radiology 177:238–239
Rossi P (1996a) Covered stents in malignant billiary obstruction. 25th Annual Meeting of the Japanese Society of Angiography and Interventional Radiology, 23–26 April 1996
Rossi P (1996b) Percutanous treatment of benign biliary strictures. Ten year follow up. 25th Annual Meeting of the Japanese Society of Angiography and Interventional Radiology, 23–26 April 1996
Rossi P, Bezzi M, Salvatori FM, Maccioni F, Porcaro ML (1990a) Recurrent benign biliary strictures: management with self-expanding metallic stents. Radiology 175:661–665
Rossi P, Salvatori FM, Bezzi M, Maccioni F, Porcaro ML, Ricci P (1990b) Percutaneous management of benign biliary strictures with balloon dilation and self-expanding metallic stents. Cardiovasc Intervent Radiol 13:231–239
Salomonowitz EK, Antonucci F, Heer M, et al (1992) Biliary obstruction: treatment with self-expanding endoprostheses. Cardiovasc Intervent Radiol 3:365–370
Uchida BT, Putnam JS, Rosh J (1988) Modification of Gianturco expandable wire stents. AJR 150:1185–1187
Uflacker R (1996) Interventions in benign biliary disease. J Vasc Interv Radiol 7:225–228 (Scientific program, Society of Cardiovascular and Interventional Radiology, 21st Annual Meeting, March 1996)
Warren KW, Mountain JC, Midell AI (1971) Management of strictures of the biliary tract. Surg Clin North Am 51:711–730
Way LW, Bernhoft RA, Thomas MJ (1981) Biliary strictures. Surg Clin North Am 61:963–969
Williams HJ, Bender CE, May GR (1987) Benign postoperative biliary strictures: dilation with fluoroscopic guidance. Radiology 163:629–634
Yoshioka T, Sakaguchi H, Yoshimura H, et al (1990) Expandable metallic biliary endoprostheses: preliminary clinical evaluation. Radiology 177:253–257

17C Percutaneous Biliary Endoscopy: Technique and Clinical Applications

P. Rossi, M. Bezzi, F.M. Salvatori, and F. Fiocca

CONTENTS

17C.1 Introduction

Percutaneous cholangioscopy consists in the direct visualization of the biliary tree - for both diagnostic and therapeutic purposes - using a flexible endoscope that can be introduced percutaneously, either through a transhepatic tract created during biliary drainage or through a surgical T-tube tract. The technique was initially used in the early 1970s by some Japanese surgeons (Takada et al. 1974; Yamakawa et al. 1976) and then taken up around the world in the late 1980s and early 1990s by radiologists and surgeons (Picus et al. 1989; Nimura et al. 1989; Bonnel et al. 1991; Picus 1995; Rossi et al. 1996). More recently, many surgeons have begun to use a fine-caliber endoscope during laparascopic operations, to visualize the intrahepatic ducts or for therapeutic maneuvers (Lezoche and Paganini 1995).

P. Rossi, MD, Department of Radiology, University of Rome "La Sapienza," Policlinico Umberto I, 00161 Rome, Italy
M. Bezzi, MD, Department of Radiology, University of Rome "La Sapienza," Policlinico Umberto I, 00161 Rome, Italy
F.M. Salvatori, MD, Department of Radiology, University of Rome "La Sapienza," Policlinico Umberto I, 00161 Rome, Italy
F. Fiocca, MD, Department of Surgery, University of Rome "La Sapienza," Policlinico Umberto I, 00161 Rome, Italy

17C.2 Cholangioscopes

The characteristics of a cholangioscope for the percutaneous approach are the following:

- Small caliber (less invasive)
- Steerability
- Working channel of adequate caliber to permit the introduction of baskets, biopsy forceps, and lithotripsy probes

Their small size has many advantages - like flexibility and ease of introduction - but, compared with larger instruments (8 mm), they have poorer image resolution.

Larger cholangioscopes provide higher-resolution images because of the larger number of optical fibers; however, they require a larger transhepatic tract, and often do not permit exploration of the subsegmental biliary radicles. Smaller scopes have fewer light fibers, which has a negative effect on image quality, but they can be more easily advanced into the peripheral biliary tree, and through a smaller percutaneous entry.

Miniature cholangioscopes (<1 mm in diameter) have been recently developed for the pancreatic and biliary system (Tajiri et al. 1993; Riemann and Kohler 1993). These scopes, however, have a limited observation field and their image quality becomes suboptimal in enlarged ducts; in addition, they do not have an operating channel for biopsy and/or lithotripsy. The clinical use of these scopes is mainly limited to visual inspection of the ducts for diagnostic purposes.

Several fine-caliber cholangioscopes, ranging in size between 1.9 and 2.8 mm, are also commercially available (Guenther et al. 1990). Their primary advantage is that they can be used through smaller tracts, thus extending the range of indications for percutaneous cholangioscopy. Some of these fine-caliber scopes, however, are not steerable, and this makes manipulation within the biliary tree difficult. These scopes have no instrument channel at all, or only a small working channel, and are used for visual inspection and for limited therapeutic procedures.

Two types of flexible cholangioscopes are currently used at our institution. The first one introduced into our practice is 4.9 mm (15 F) in diameter with a 2.2-mm working channel (Olympus, model CHF-P20, Olympus Optical Co. GmbH, Hamburg, Germany). The working channel allows the use of a large variety of operative instruments, up to a size of 5–6 F. The total usable length of the scope is 67 cm. The lighting system has two light guides, while the optical system is characterized by forward viewing with a 120° field of view and a depth of field between 3 and 50 mm; this depth of field permits the exploration of highly dilated ducts. The distal end of the scope can be deflected by means of a control lever (the range of bending is 160° up and 120° down).

The second scope currently in use in our department is a ureteroscope (Olympus, model URF-P). It is 3.9 mm (12 F) in size, with a 1.2-mm instrument channel. This channel permits the introduction of instruments up to 3 F. The lighting system has only one light guide, and this gives an image quality inferior to the larger 4.9-mm scope. The optical system is characterized by forward viewing with a 90° field of view and a depth of field similar to the CHF P20 scope. The distal end of the instrument can also be bent (180° up and 100° down).

The light source for both instruments is a xenon light source. The eyepiece of both cholangioscopes may be attached to a video camera, directly or by means of an adaptor. Digital charge-coupled device (CCD) video cameras provide endoscopic images with better resolution than conventional video cameras. The use of a video camera has the main advantage that the operator is not obliged to look directly into the cholangioscope; this allows a second operator and other people in the room to follow the procedure.

17C.2.1 Cholangioscope Sterilization

Complete sterilization of the instruments is of vital importance, considering how many patients have viral hepatitis. Special care should therefore be taken in cleaning the working channel and the instrument tip, as well as careful removal of all debris.

Either gas sterilization using ethylene oxide or cold sterilization using several disinfectant glutaraldehyde solutions may be used. Gas sterilization is preferable in order to achieve high levels of virucidal and bactericidal efficiency. A cycle with ethylene oxide takes 12–18 h; this time constraint limits the availability of the instrument. Cold sterilization with glurtaraldehyde is both bactericidal and virucidal, and is usually complete after few hours of immersion, depending on the solution used. Most modern cholangioscopes are waterproof and can therefore be totally immersed to achieve sterility. In the case of nonimmersible fiberscopes, only partial sterility is obtained since only the instrument shaft can be soaked, while the optic plug and the camera adaptor should not be exposed to the disinfecting fluid. If complete asepsis is desired, these portions of the instrument can be covered with a sterile plastic cover.

The problem of complete sterilization is common to all endoscopists using the same instruments to examine many patients every day.

17C.3 Technique

Any visual exploration of the biliary tree with a cholangioscope requires a direct approach and adequate tract maturation and dilation. The area where the percutaneous cholangioscopy is to be done is be considered when deciding the approach. The left ductal approach allows more direct access to the common bile duct; in addition, it causes less pain during dilation due to the absence of an intercostal tract. If the biliary drain, however, has been already placed through the right ducts and the tract is mature, percutaneous cholangioscopy of the common bile duct can be done from this approach and there is no need for a second approach from the left. When the lesion to be treated is located in the intrahepatic biliary tree, either right or left, a contralateral approach is usually preferred. This usually allows exploration of all the segmental branches of one lobe from the other side; this is particularly useful in case of intrahepatic stones and/or strictures (Fig. 17C.1).

Tract dilation should be done progressively in order to avoid severe complications as reported by Bonnel et al. (1991); this author reports severe complications in 22% of a group of 50 patients treated with percutaneous cholangioscopy for bile duct stones, and a mortality of 8% when percutaneous cholangioscopy was performed within 3 days after the initial percutaneous drainage.

The introduction of the scope can be performed either through a peel-away Teflon sheath or directly through the cutaneous biliary fistula. Larger dilation

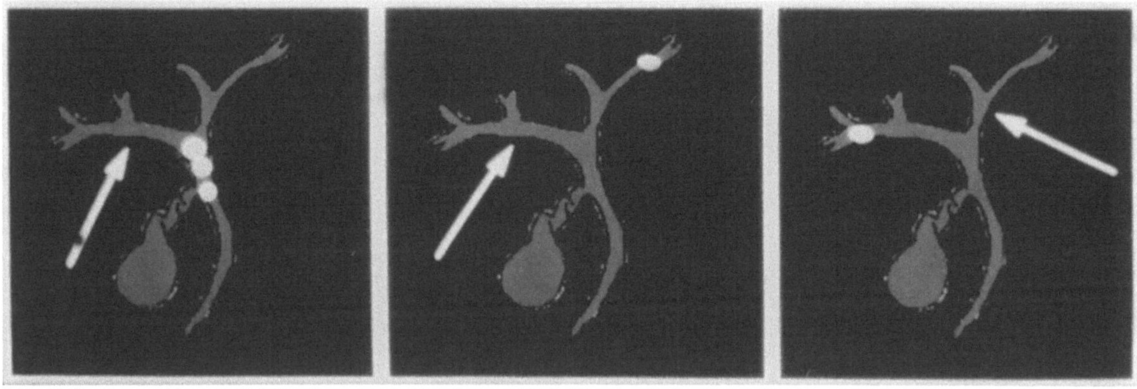

Fig. 17C.1. Selection of left or right percutaneous approach according to the location of the biliary stones

of the tract is, of course, required if a peel-away sheath is used. When we use small cholangioscopes (3.9 mm), a 15-F peel-away sheath or tract dilation to 12 F is required. The 4.9-mm instrument can be advanced through a 18-F sheath or through a cutaneous fistula dilated to 14 or 15 F. The use of a peel-away sheath requires a further 2-F dilation of the tract. When the tract is mature, there is no need for the peel-away sheath (Neuhaus et al. 1993; Picus 1995). Many operators suggest tract maturation of 7–10 days to reduce the complications due to the mechanical introduction of the instrument, and very often we perform percutaneous cholangioscopy on an out-patient basis when the tract is completely mature.

17C.3.1 Patient Sedation and Treatment

All patients receive intravenous antibiotics at least 12 h before percutaneous cholangioscopy. The skin is infiltrated with local anesthesia. Intravenous analgesia and sedation are commonly used; usually fentanyl (Fentanest; Farmitalia, Italy) and midazolam (Versed; Roche, Nutley, NJ) are given. These drugs are administered directly by the interventional radiologist, assisted by a nurse. When general anesthesia or deep sedation is necessary, this is accomplished by the anesthesiologist. All procedures are done under continuous EKG and oxygen saturation monitoring.

17C.3.2 Cholangioscopy

During the procedure, a clear view is guaranteed by irrigation, which provides for lumen distension and bile wash-out. Usually, gravity irrigation is performed; forceful flushing is, however, sometimes necessary when instruments are advanced through the working channel, since this maneuver reduces the free lumen available for irrigation. It is very important to remember that continuous irrigation may cause overdistention of the bile ducts, resulting in pain and biliary sepsis, diarrhea, and fluid overload. These conditions can be prevented by providing adequate outflow of irrigation fluid, or by placing a nasoduodenal or nasojejunal tube.

The instrument can be advanced under direct vision or over a guidewire into the biliary tree; a combination of shaft rotation and tip deflection usually allows entry into the different areas to be explored. Injection of contrast medium into the biliary tree under fluoroscopy is helpful when one wants to be sure of the exact position and orientation of the scope or when one finds difficulties in advancing the instrument into the desired duct. If a bilateral approach has already been established, for whatever reason, grasping a contralateral guide wire with the scope may help in entering angulated ducts or in negotiating difficult strictures (Fig. 17C.2).

17C.4 Clinical Applications

The procedure indications are different depending on whether a lesion is benign or malignant. For benign lesions the indications are:

1. Visual exploration of the biliary duct
2. Percutaneous treatment of biliary stones with or without lithotripsy (Fig. 17C.2)

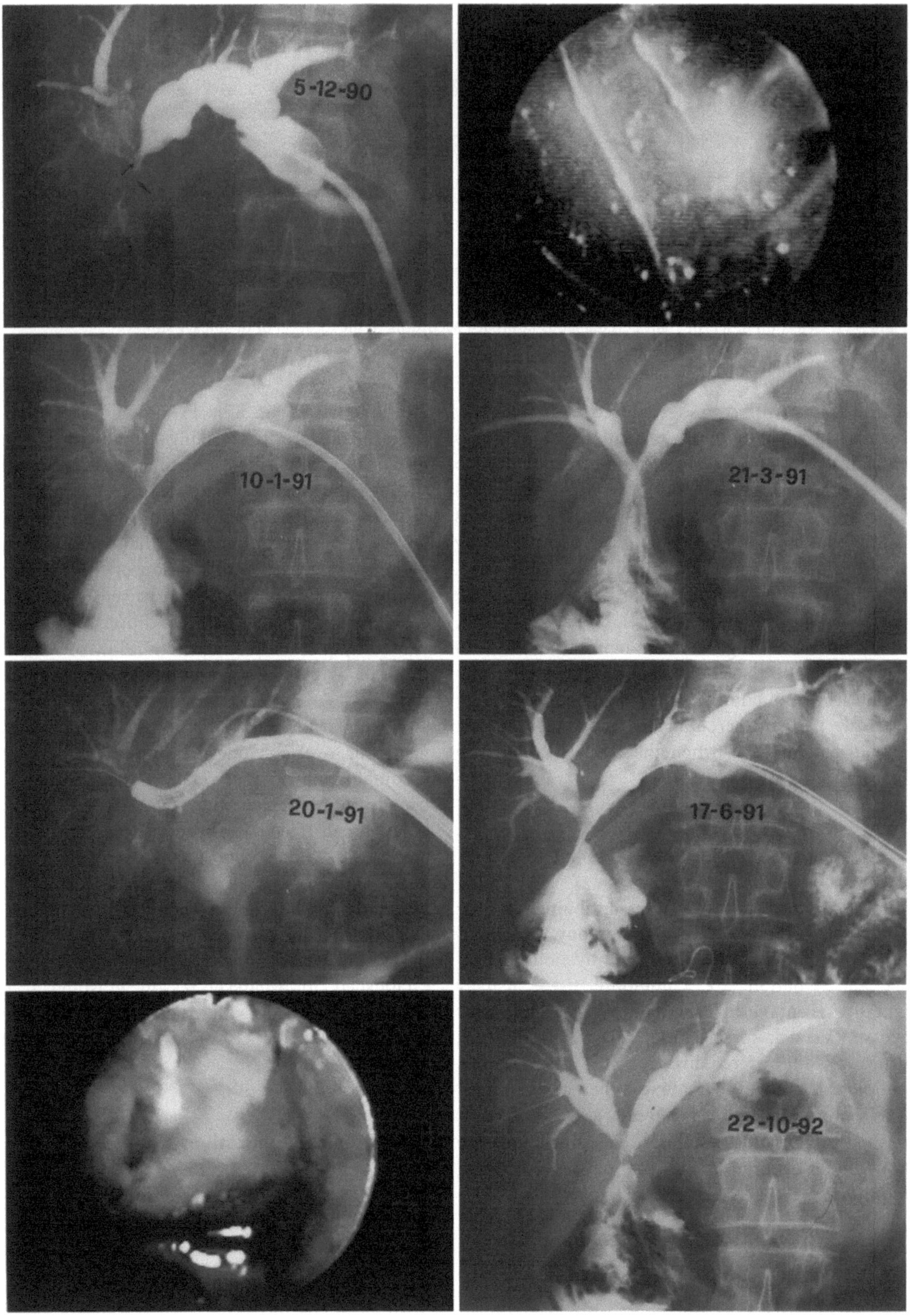

a e b f c g d h

3. Evaluation of a stenotic area to rule out inflammations or ulcerations (Figs. 17C.3, 17C.4)
4. Support in crossing very tight stenoses
5. Biopsy for the differential diagnosis of benign versus malignant lesions
6. Tissue removal where there is abundant tissue growth at the level of injured ducts
7. Removal of clips or sutures (Fig. 17C.5)

The indications for malignant lesions are completely different:

1. Biopsy of the tumor for the identification of its histological type
2. Evaluation of tumor extension (Fig. 17C.6)
3. Identification of the superficial spread of the tumor, which cannot be seen with cholangiography or with other techniques, to decide on operability: granular mucosal surface or tortuous vessels
4. Various

17C.4.1 Biliary Stones

Percutaneous cholangioscopy finds its most common indication in the removal and fragmentation by lithotripsy of biliary stones. It is usually performed when removal of the stones is not possible by the retrograde endoscopic approach or by conventional transhepatic techniques performed under fluoroscopic guidance.

Several methods of intracorporeal lithotripsy are available; they include laser, ultrasonic, and electrohydraulic lithotripsy. Electrohydraulic lithotripsy is the preferred method for lithotripsy at our institution. This technique was first introduced in the biliary tree by Burhenne (1975) and performed under fluoroscopy. With electrohydraulic lithotripsy, the lithotripsy is achieved by emission of shock waves caused by a spark discharge from the tip of a bipolar coaxial electrode probe (Fig. 17C.2e). The spark is generated by a electrohydraulic generator (Lithotron Walz EL-23; Olympus GmbH, Hamburg, Germany). The energy of the shock waves is absorbed by the stone, causing it to fragment. For energy transmission it is critical that there is direct contact between the probe and the stone, without interposition of air; the best contact is provided by water. Contact between the activated probe and the bile duct wall may result in wall damage with hemobilia, perforation, and, possibly, stricture formation. Direct vision is essential to monitor correct positioning of the probe. The electrohydraulic lithotripsy method is also very rapid and efficient, allowing stone fragmentation and clearance in one to two sessions, even in the case of large impacted stones.

Percutaneous cholangioscopy is also useful in determining whether a filling defect demonstrated at cholangiography is a stone; other causes of filling defects, such as bubbles of air, mucus, and blood clots, can be easily differentiated from stones.

At our institution, percutaneous cholangioscopy for the treatment of biliary stones is usually performed in selected cases when stone clearance is not possible with standard techniques. Electrohydraulic lithotripsy was needed for larger stones or when the stone is in a peripheral duct and cannot be displaced, and required more than one session for complete stone fragmentation and clearance of debris (average 1.8 sessions). Each electrohydraulic lithotripsy session required 20–40 min percutaneous

Fig. 17C.2a–h. Percutaneous cholangioscopy with lithotripsy of biliary stones in a 32-year-old patient who had undergone cholecystectomy and hepatojejunostomy 3 years previously and was referred for pyogenic cholangitis and history of recurrent cholangitis. **a** A left percutaneous biliary drainage is in place. Cholangiography shows a 1.5-cm stone in the left hepatic duct and a 2-cm stone wedged into the right hepatic duct associated with a tight stenosis of the biliary-enteric anastomosis. After dilating the anastomosis with a 2-cm balloon catheter, the stone in the left hepatic duct was pushed into the jejunal loop, but 24 h later massive gastrointestinal bleeding occurred, which was successfully embolized. A control cholangiogram performed 3 weeks later showed the left duct to be decompressed, the right duct unchanged, and the anastomosis still stenotic. **b** Control cholangiogram 5 weeks later demonstrates a considerable improvement of the biliary condition, although the right duct is occupied by stones. **c** Ten days later, a right percutaneous biliary drainage was performed and an 8-mm cholangioscope was advanced from the left biliary duct into the right one, using the *rendezvous* technique. The cholangioscope was pulled into the right duct. Electrohydraulic lithotripsy was then performed, clearing the right duct of stones. **d** Good visualization of the stone in the duct. **e** Electrohydraulic lithotripsy in progress: note the sparkling and the fragmentation of the calculi. **f** Two months afterwards, the control cholangiogram showed patency of the biliary-enteric anastomosis and of the intrahepatic bile ducts with complete removal of stones. The biliary drainages were left in place. **g** Five months after lithotripsy, control cholangiography showed patency of the intrahepatic bile ducts and the anastomosis with no evidence of stones. Both catheters were removed. **h** Sixteen months after **g**, the patient returned for follow-up and a percutaneous cholangiogram through a right intracostal approach showed mild narrowing of the anastomosis secondary to mucosal hypertrophy, as documented by repeat percutaneous cholangioscopy. No stones were seen; there are signs of secondary cholangitis on the right. The patient is still well at 5-year follow-up

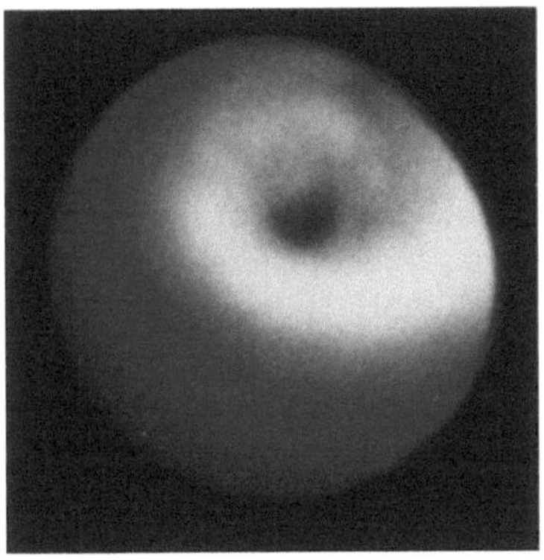

Fig. 17C.3. Smooth surface at the stenosis of a biliary duct. No signs of inflammation are present. (Courtesy Prof. G. Gandini, Novara, Italy)

cholangioscopy time. All stone fragments were flushed into the duodenum or jejunum, leaving all patients free of stone. Recurrent or residual intrahepatic calculi occurred in 13% of our patients after an average follow-up interval of 24 months.

Stone removal performed with the aid of percutaneous cholangioscopy is usually a procedure with a high success rate. Picus (1995) reported on 11 patients treated with electrohydraulic lithotripsy performed under percutaneous cholangioscopic guidance. The stones were successfully fragmented and removed without complications in all cases. Mo et al. (1988) report a 100% success rate in a group of ten patients treated by a similar technique. Bonnel et al. (1991) achieved complete stone clearance in 92% of a group of 50 patients, with a mortality of 8%. A slightly lower success rate is reported by Jeng et al. (1989), who were able to obtain complete stone fragmentation in 78% of 50 patients.

Long-term results depend on the presence of associated biliary disease. Jan and Chen (1995) reported on 48 patients with intrahepatic stones followed for a period of 4–10 years. The overall rate of stone recurrence was 40%; stones, however, recurred only in 16/31 patients with recurrent strictures and did not recur in patients who were stricture-free ($p < 0.05$).

Our success rate in the complete clearing of stones is approximately 87%, since very often after complete dilation in high-risk patients we leave some small stones if they are difficult to remove.

Bleeding occurred in 4% of our patients; however, it was not directly related to tract dilation or the introduction of the cholangioscope, but was usually secondary to a long-term indwelling 10-F drainage catheter.

17C.4.2 Benign Biliary Strictures

In patients with benign biliary strictures, percutaneous cholangioscopy is mainly used to negotiate difficult strictures and to obtain biopsy samples of suspect ductal lesions.

Strictures which are difficult to cross are not rarely encountered during routine percutaneous biliary drainage. However, in eccentric stenosis direct cholangioscopic vision may facilitate passage of a guidewire through the lesion (Nimura et al. 1989). In a series of 101 patients reported by Neuhaus et al. (1993), difficult bile duct strictures which were not catheterized by retrograde or percutaneous techniques were present in 15 patients. Percutaneous cholangioscopy allowed cannulation of the stenotic duct in 14 of these 15 cases.

If benign stenosis on cholangiographic examination presents a smooth surface without ulcerations (Fig. 17C.3), an inflammatory component can be ruled out and a "single-shot" dilation can be attempted without leaving a stent (Mueller et al. 1986). If ulcerations are present (Fig. 17C.4), placement of a stent after balloon dilation is mandatory.

Strictures associated with stones often present particular technical problems. If the stones are located peripheral to an intrahepatic stricture it is impossible to reach them with the cholangioscope. In this case it may be necessary to use a balloon dilation or a *rendezvous* technique. The approach for percutaneous cholangioscopy is established through the ducts contralateral to the stricture (Fig. 17C.1), while the ducts peripheral to the strictures are punctured with standard percutaneous methods. A guidewire is passed through the stenosis and is grasped by the

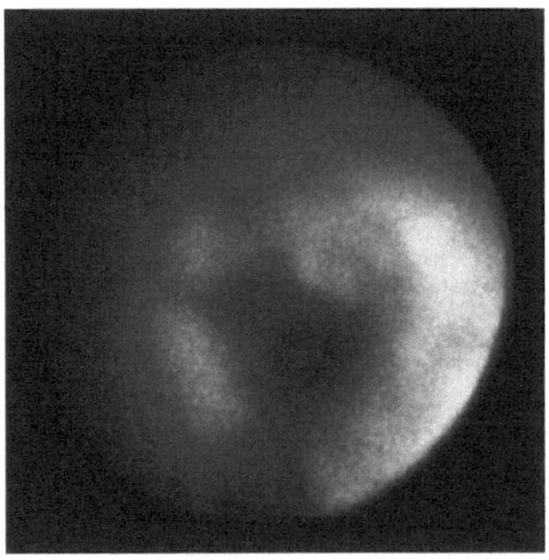

Fig. 17C.4. Biliary stenosis with signs of inflammatory lesions characterized by superficial ulcerations. (Courtesy Prof. G. Gandini, Novara, Italy)

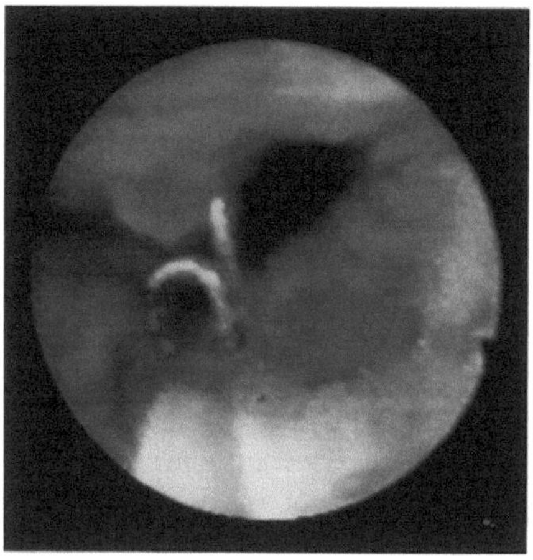

Fig. 17C.5. Metallic clip protruding into the lumen at the site of biliary-enteric anastomosis and producing a narrowing of the lumen. The metallic clip was removed pulling on a guidewire looped around an eyelet of the metallic suture (*green line*)

endoscope (Fig. 17C.2c). The stenosis is dilated and the guidewire is pulled back, together with the endoscope, which is now in an adequate position for stone treatment (MAETANI et al. 1993).

Occasionally, metallic staples or suture material (Fig. 17C.5) will be encountered protruding into the bile duct lumen, usually at the level of biliary-enteric anastomoses, thus preventing complete balloon dilation of a stricture.

The last problem in benign lesions is the differential diagnosis between malignant and benign disease. Sometimes, patients are treated for many months or years for benign biliary strictures and then a biopsy reveals the presence of a neoplastic growth. In this case, we do not know whether the benign lesion – in the continuous presence of a catheter – degenerates into a neoplastic lesion, or if a slow-growing lesion was misdiagnosed from the start as benign.

Bile duct wall biopsy is particularly useful in patients with sclerosing cholangitis, to rule out the presence of cholangiocarcinoma, which may grow slowly and not be readily apparent on the cholangiograms (VENBRUX et al. 1991).

17C.4.3 Malignant Biliary Strictures

In cases of malignant biliary obstruction the main role of percutaneous cholangioscopy is to obtain definitive tissue diagnosis of malignancy (Fig. 17C.6).

Cholangioscopy to differentiate biliary strictures from polyps has been mainly practiced by Japanese authors (YAMASE et al. 1988; SATO et al. 1994). Cholangioscopy is also very useful for aimed biopsies since you can see the lesion and take a fragment from the most suspicious area. NIMURA (1993) reported a series of 257 biopsies performed for suspected malignant biliary obstruction with a positive biopsy rate of 81% in the entire series. In this group of patients, cholangiocarcinomas were easily diagnosed with a positive biopsy (96% cancerous cells in the bioptic specimens) because they were clearly visible. Negative biopsies, on the other hand, were obtained in diffusely infiltrating bile duct carcinoma in which the cancer was not present in the stenotic segment.

Biopsy samples are usually taken in areas of mucosal irregularity or where dilated vessels, so-called "tumor vessels," are seen on the mucosal surface (NIMURA et al. 1989).

NIMURA et al. (1990) report a very useful combination of cholangiography and cholangioscopy to determine the extent of tumors in the intra- and extrahepatic biliary tree. This technique allowed the correct preoperative staging of bile duct malignancies in a large series of 501 patients with hepatobiliary and pancreatic malignancies.

Cholangioscopic signs indicative of tumor diffusion are the presence of dilated and tortuous vessels on the ductal wall and the presence of a papillary or granular mucosal surface. Vascular dilation is more frequently associated with invasive mucosal carcinoma, while the papillary granular appearance was found specifically in noninvasive mucosal carcinoma, which diffuses by superficial spreading (SATO et al. 1994). In the experience of NIMURA (1993), improved staging obtained by percutaneous transhepatic cholangiography and percutaneous cholangioscopy resulted in a high rate of resectability and in considerably improved long-term survival. Such thorough staging, however, is necessary only when the surgeon plans an extended surgical ap-

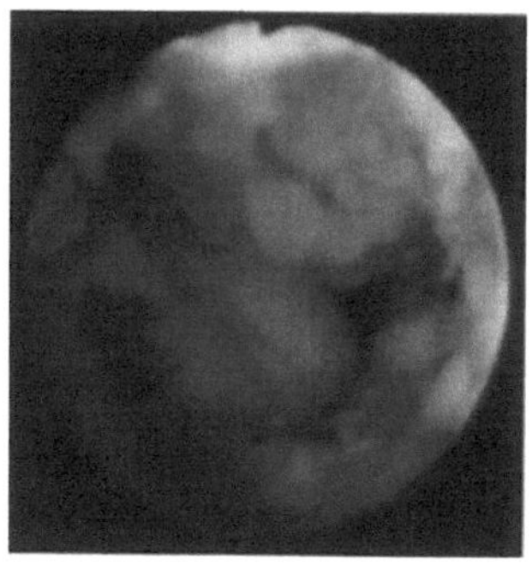

Fig. 17C.6. Neoplastic invasion of a biliary duct with polypoid masses proach to the liver hilum, which includes caudate lobe resection and, often, resection of other segments as well. This approach, more used in Japan, may not be commonly followed by Western surgeons.

KUBOTA et al. (1993) also described the use of cholangioscopy for the positioning of a single Y-shaped stent introduced through a single percutaneous approach. At our institution, percutaneous cholangioscopy was also employed for immediate control of covered stents at 1–2 days and for biopsy in occluded covered stents (ROSSI et al. 1996; BEZZI et a. 1996). Very recently, we carried out a clinical trial in 21 patients with malignant obstructive jaundice using polyurethane-covered Wallstents (Schneider, Lausanne, Switzerland).

17C.5 Complications

Complications of percutaneous cholangioscopy can be divided into two groups: those related to the percutaneous approach and those directly related to the procedure.

Hemorrhage into the biliary ducts is the most serious complication related to the approach and is usually caused by tract dilation. In our series of 97 cholangioscopic procedures severe hemobilia occurred in four patients (4.1%). In no case did the bleeding occur during manipulation of the instrument, but either before or several days after the procedure, probably due to erosion caused by a large, 10-F Ring drainage catheter; the reason for using such a catheter is that it allows easy passage of the 3.9-mm cholangioscope without any further dilation. All patients underwent hepatic arteriography followed by successful embolization of the bleeding site; no blood transfusion was necessary in any case and there were no deaths or any need for emergency surgery. BONNELL et al. (1991) reported massive hemobilia in 6/50 patients (12%), which was fatal in two patients despite embolization. This type of complication can be minimized – but not avoided – by accessing the biliary tree in a peripheral duct and by dilating the tract for at least 4–8 days after the initial drainage. In other reports the incidence of severe hemobilia ranged from none to 10% (JENG et al. 1989).

Minor bleeding, which usually stops spontaneously after the procedure is interrupted, may be seen during percutaneous cholangioscopy and may be caused by the tip of the scope damaging the ductal wall or by electrohydraulic lithotripsy. Often minor bleeding occurs in the presence of severe ductal strictures, ductal angulation, or periductal hypervascularity surrounding areas of chronic inflammation. The incidence of such minor hemorrhage varies greatly; in our series of 97 percutaneous cholangioscopies we saw it in 15 patients (15.5%) and managed it conservatively with success.

Laceration of the bile duct is another potential complication of percutaneous cholangioscopy and may be caused by incautions advancement of the scope into the ducts without adequate vision. If the ducts are lacerated, the procedure is stopped and a drainage catheter is left in place. We encountered this complications in two cases (2%).

Other minor complications include transient bacteremia, vagal reactions, nausea, fever (in 90%–95% of cases for 1–2 days), and diarrhea due to excessive outflow of water and contrast medium into the bowel.

17C.6 Conclusions

Percutaneous cholangioscopy has proven to be a therapeutic diagnostic technique for biliary disease. We have found, in our experience, that its widest use is in benign lesions associated with lithiasis. Using the percutaneous approach combined with stone fragmentation by laser or electrohydraulic lithotripsy has made stone removal easier than other by established techniques, such as surgery or retrograde endoscopic stone fragmentation and removal. Moreover, percutaneous cholangioscopy is also a reliable tool for the differential diagnosis of bile duct lesions. In the staging of bile duct malignancy it may also play an indispensable part when extended hilar surgery is planned.

The interventional radiologist is in a unique position to take advantage of this relatively new but well-established procedure, being able to combine in the same percutaneous access both fluoroscopic and endoscopic techniques for the management of biliary tract disease.

References

Bezzi M, Panzetti C, Bonomo G, Pedicini V, Salvatori FM, Rossi P (1996) Plastic covered Wallstents in malignant biliary obstruction: evaluation in seventeen patients. Abstract book of the Society of Cardiovascular and Interventional Radiology 21st Scientific Meeting, p 115

Bonnel DH, Liguory CE, Cornud FE, Lefebvre JFP (1991) Common bile duct and intrahepatic stones: results of

transhepatic electrohydraulic lithotripsy in 50 patients. Radiology 180:345–348
Burhenne HJ (1975) Electrohydraulic fragmentation of retained common duct stones. Radiology 117:721–722
Guenther RW, Vorwerk DV, Klose KJ, et al (1990) Fine caliber cholangioscopy. Radiol Clin N Am 28:1171–1183
Jan YY, Chen MF (1995) Percutaneous transhepatic cholangioscopic lithotomy for hepatolithiasis: long-term results. Gastrointest Endosc 42:94–96
Jeng KS, Chiang HJ, Shih SC (1989) Limitations of percutaneous transhepatic cholangioscopy in the removal of complicated biliary calculi. World J Surg 13:603–610
Kubota Y, Nakahashi Y, Nakatani S (1993) Bilateral internal drainage of hilar cholangiocarcinoma with self-expanding metallic stents via a single percutaneous tract. Gastroenterology 104:A367
Lezoche E, Paganini AM (1995) Single-stage laparoscopic treatment of gallstones and common bile duct stones in 120 unselected, consecutive patients. Surg Endosc 9:1070–1075
Maetani I, Hoshi H, Ohashi S, et al (1993) Cholangioscopic extraction of intrahepatic stones associated with biliary strictures using a rendez-vous technique. Endoscopy 25:303–306
Mo LR, Hwang MH, Yeuh SK, Yang JC, Lin C (1988) Percutaneous transhepatic choledochoscopic electrohydraulic lithotripsy (PTCS-EHL) of common bile duct stones. Gastrointest Endosc 34:122–125
Mueller PR, Van Sonnenberg E, Ferrucci JT, et al (1986) Biliary strictures dilation: multicenter review of clinical management in 73 patients. Radiology 160:17–22
Neuhaus H, Hoffmann W, Classen M (1993) Nutzen and Risiken der perkutanen transhepatischen Cholangioskopie. Dtsch Med Wochenschr 118:574–581
Nimura Y (1993) Staging of biliary carcinoma: Cholangiography and Cholangioscopy. Endoscopy 25:76–80
Nimura Y, Kamiya J, Hayakawa N, Shionoya S (1989) Cholangioscopic differentiation of biliary strictures and polyps. Endoscopy 21:351–356
Nimura Y, Hayakawa N, Kamiya J, et al (1990) Hepatic segmentectomy with caudate lobe resection for bile duct carcinoma of the hepatic hilus. World J Surg 14:535
Picus D (1995) Percutaneous biliary endoscopy. J Vasc Interv Radiol 6:303–310
Picus D, Weyman PJ, Marx MV (1989) Role of percutaneous intracorporeal electrohydraulic lithotripsy in the treatment of biliary tract calculi. Radiology 170:989–993
Riemann JF, Kohler B (1993) Endoscopy of the pancreatic duct: value of different endoscopes types. Gastrointest Endosc 39:367–370
Rossi P, Bezzi M, Fiocca F, Salvatori FM, Grasso E, Speranza V (1996) Percutaneous cholangioscopy. Semin Interv Radiol 13:185–193
Sato M, Maetani I, Ohashi S, et al (1994) Relationship between percutaneous transhepatic cholangioscopy findings and pattern of carcinomatous spread in the bile duct. Diagn Therap Endosc 1:45–50
Tajiri H, Kobayashi M, Niwa H, et al (1993) Clinical application of an ultrathin pancreatoscope using a sequential video converter. Gastrointest Endosc 39:371–374
Takada T, Suzuki S, Nakamura K, et al (1974) Studies in percutaneous biliary tract endoscopy. Gastroenterol Endosc 16:106–111
Venbrux AC, Robbins KV, Savader SJ, et al (1991) Endoscopy as an adjuvant to biliary radiologic intevention. Radiology 180:355–361
Yamakawa T, Mieno K, Noguchi T, Shikata J (1976) An improved choledochofiberscope and non-surgical removal of retained biliary calculi under direct visual control. Gastrointest Endosc 22:160–164
Yamase H, Nimura Y, Hayakawa N, et al (1988) Differential diagnosis on stenosis of distal bile duct by percutaneous transhepatic cholangioscopy (PTCS). Gastroenterol Endosc 30:1175–1182

18 Role of Endoscopy in the Treatment of Benign Biliary Disease

G. Costamagna

CONTENTS

18.1 Introduction

Therapeutic endoscopy has shown a tremendous impact on biliopancreatic diseases in the last two decades. The first landmark of this evolution was set in the early 1970s by Classen in Germany and Kawai in Japan, who performed endoscopic sphincterotomy to extract bile duct stones. Continuing technical improvements in endoscopes and accessories along with the growth of clinical experience all over the world are still constantly pushing back the frontiers of the applicability of nonoperative endoscopic methods. The arsenal of endoscopic treatment for benign biliary disease has been increasing and now includes biliary sphincterotomy and sphincteroplasty, stone extraction with or without lithotripsy, drain and stent placement, and balloon dilation of strictures.

Nonoperative treatment is justified when a precise and definite diagnosis can be established without laparotomy: very close interaction between radiologists and endoscopists is therefore essential in order to collect all the clinical information and results of imaging studies (computed tomography, ultrasonography, and magnetic resonance) and endoscopic retrograde cholangiopancreatography (ERCP) prior to a therapeutic decision. Since ERCP still yields the most detailed information about the morphology of the biliary duct, the diagnostic phase of ERCP should always be carefully carried out; as a rule, attempts at endoscopic treatment without proper visualization and radiologic documentation of biliary ducts should be avoided. ERCP is routinely performed under conscious sedation and only rarely under general anesthesia. Deep sedation with propofol is becoming more and more popular because it gives to the patient complete amnesia regarding the procedure, but it requires the presence of an anesthesiologist.

ERCP is best performed in a dedicated X-ray room located in or near the radiology department, to improve interactivity with other imaging techniques. High-quality X-ray equipment should be available.

Various diagnostic and therapeutic side-viewing duodenoscopes are available with different-sized channels. The current tendency, except in special situations (e.g., duodenal strictures, pediatric patients, etc.), is to use large-channel (3.7–4.5 mm) therapeutic duodenoscopes, in order to be able to cope with all possible circumstances during the procedure. A wide choice of ancillary equipment, including catheters, guidewires, sphincterotomes, nasobiliary drains and stents, balloons catheters, and Dormia baskets, should be immediately available in dependence on the problems presented by each individual patient. Special devices such as a lithotriptor, needle knives, injection needle, foreign body forceps, snares, and stent retrievers should be present for special situations and immediate management of some complications.

Therapeutic ERCP should at all times be considered part of an integrated care system in which continuous interdisciplinary discussion and interaction with radiologists and surgeons is mandatory. Thera-

G. Costamagna, MD, Department of Surgery, Catholic University School of Medicine, Largo Gemelli 8, 00168 Rome, Italy peutic ERCP is a part of the modern team approach to biliary diseases.

18.2 Bile Duct Stones

Gallstones may pass from the gallbladder into the common bile duct (CBD) or may form primarily in the duct. About 15% of patients with gallbladder calculi also have stones in the CBD and, conversely, 90% of patients with CBD stones also have gallbladder calculi (SOWBERG et al. 1993).

Therapeutic ERCP was first employed for patients with residual and/or recurrent CBD stones after cholecystectomy and patients with CBD and gallbladder stones who presented a high surgical risk. With the advent of laparoscopic cholecystectomy the endoscopic approach has become the treatment of choice for virtually all patients with CBD stones and an endoscopically accessible papilla, with few exceptions (e.g., giant CBD stones, Mirizzi's syndrome, etc.) (COTTON 1993) (Fig. 18.1). Future developments in the laparoscopic approach to CBD stones may well redefine the role of endoscopic extraction (PETELIN 1993).

The presence of CBD stones is almost always suspected clinically on the basis of biliary colic with elevated liver function test values, jaundice, cholangitis, or pancreatitis. Imaging studies may be misleading if ducts are not dilated (CLAIR et al. 1993).

18.2.1 Diagnostic ERCP

A high-quality cholangiogram is essential when looking for CBD stones. Even if CBD stones have been detected prior to ERCP, the endoscopic and

a,b

c,d

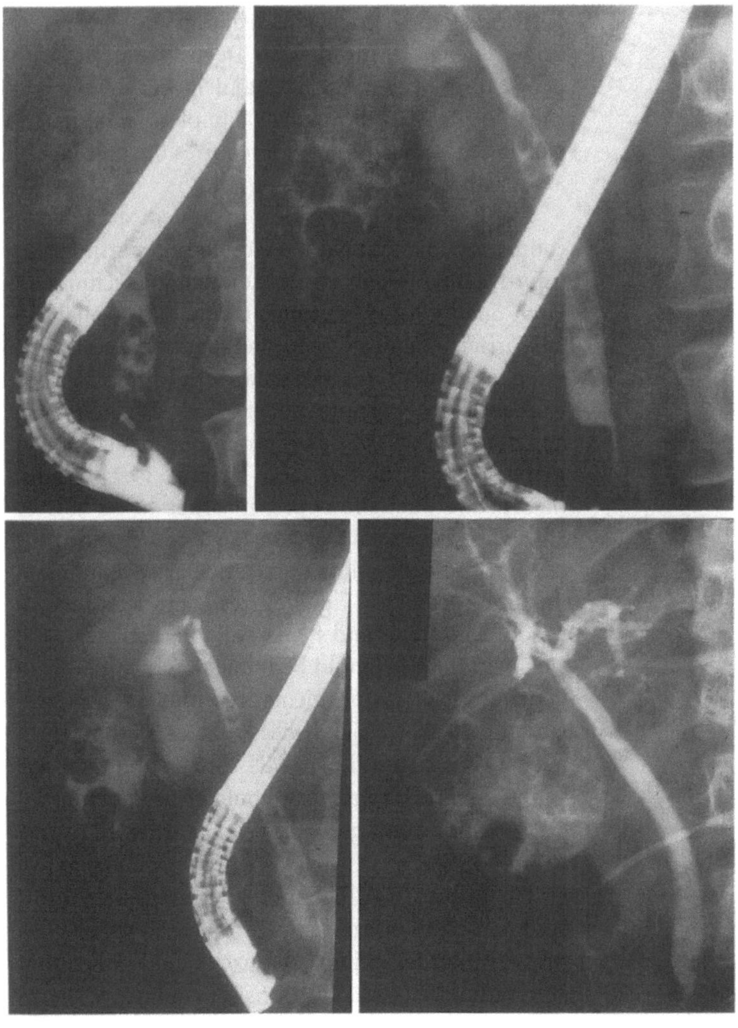

Fig. 18.1. **a** Early filling cholangiogram in a young patient with common bile duct (CBD) stones. **b** Deep cannulation of the CBD and filling of the gallbladder, which also contains multiple stones. **c** A long-nose sphincterotome has been inserted into the CBD. **d** After endoscopic sphincterotomy and stone extraction, a nasobiliary drain has been inserted in order to perform a check cholangiography the following day during laparoscopic cholecystectomy

radiologic technique should always follow the same method. Before attempting cannulation of the CBD, the papilla should be carefully examined in order to identify direct or indirect signs of pathology (peripapillary fistulae, intra-ampullary or exophytic tumor, distorsion of the ampullary axis, etc.). To avoid injecting air into the ducts the catheter should be preflushed with contrast medium. Initially, we prefer to obtain opacification of the biliary duct by injecting contrast medium with the tip of the cannula just touching the papillary orifice, before deep cannulation is achieved. This "trick" is essential when looking for microlithiasis, which may be easily overlooked; often, only in "early filling" radiographs, with contrast medium flowing back in the CBD, is it possible to show stones of less than 5 mm in diameter.

Production of a complete cholangiogram is the next step towards a precise diagnosis. This is better obtained after selective deep cannulaion of the CBD. Care must be taken to depict all the biliary branches in order to rule out intrahepatic stones. Changing the patient's position may increase the radiological accuracy. In expert hands, the sensibility of diagnostic ERCP for detecting CBD stones approaches 100%.

18.2.2 Therapeutic ERCP

After the diagnosis of CBD stones has been confirmed, the next step is to perform a sphincterotomy to enable stone extraction. In few instances, and only for small stones, however, extraction may be possible through an intact sphincter with or without pharmacological relaxation (e.g., sublingual or drip infusion of nitrate) (Ibuki et al. 1992). Extraction of stones after pneumatic dilation of the papilla (sphincteroplasty) is another alternative to endoscopic sphincterotomy which has been recently investigated (see below).

Endoscopic sphincterotomy is performed using a specialized catheter with an active wire protruding from its distal portion for a distance of 2–3 cm. By tightening the wire the tip of the catheter is bowed. Many types of sphincterotome are commercially available today, differing mainly as to the length of the exposed diathermy wire (we prefer wires of 20 mm), the length of the distal "nose," which acts as a stabilizer once deeply inserted into the CBD (the "nose" should be at least 15–20 mm long), and their ability to be used over a guidewire, which enhances positioning and stability in the ducts. Wireguided sphincterotomy is at the moment the technique preferred by most endoscopists, especially in situations where initial deep cannulation with the diagnostic catheter has been difficult.

When deep cannulation fails, precut techniques, which involve cutting the roof of the papilla with a needle knife or with a "distal" papillotome to gain access to the biliary duct, may be used. However, these potentially risky techniques should only be employed when the presence of a biliary lesion requiring endoscopic treatment has already been ascertained, and then only by experienced endoscopists (Shakoor and Geenen 1992). Needle-knife precut papillotomy is most useful and generally safe in cases of impacted stone in the ampulla, which may prevent deep cannulation of the CBD (Siegel 1992).

As a rule, the length of the sphincterotomy should be tailored to the size of the CBD irrespective of the size of the stones to be removed, since small sphincterotomies carry the risk of stricture and of stone recurrence.

18.2.3 Complications of Endoscopic Sphincterotomy

Initially, endoscopic sphincterotomy was utilized in older, high-risk patients judged to be unfit for surgery. With the broadening of indications, it is now performed in patients who could undergo operation with reasonable safety. This has focused attention on the risks of sphincterotomy and on management of its complications.

Most published series show similar figures for complications and mortality; significant complications occur in about 10% of patients, with an overall hospital mortality of approximately 1.5%. In 1990 Cotton and other selected experts (Cotton et al. 1991) organized a consensus conference on complications of endoscopic sphincterotomy and their management, reporting complication and mortality rates respectively at 8.2% and 1.3% of 7729 sphincterotomies. At our center, 1120 endoscopic sphincterotomies were performed for various indications in a 5-year period. Complications related to the procedure occurred in 5.1%, and related mortality was 0.2%.

Before discussing the individual complications, it must be emphasized that there are some general factors believed to increase the risk of complications: these include the patient's general state of health, the indication for the examination, anatomical factors, and the skill of the endoscopist:

1. Age, unstable cardiorespiratory status, renal insufficiency, malignancies, sepsis, malnutrition, and chronic liver disease can all affect patient outcome. Rarely have endoscopists reported the actual impact of these factors individually.
2. Complications appear to be more frequent in patients with sphincter of Oddi dysfunction (SHERMAN et al. 1991), and bleeding is more likely to occur during a sphincterotomy for papillary tumors.
3. It seems that patients with Billroth II stomach and peripapillary diverticula do not have an increased risk of complications, probably because these patients are usually referred to experts. Many authors have reported complications to be more common in patients with non-dilated bile ducts (SHERMAN et al. 1991).
4. Recently pre-cutting has come into widespread use, especially to achieve a difficult deep cannulation. In expert hands, and when used only for specific indications, the risk associated with precutting does not appear to be greater than that of conventional sphincterotomy, but there are alarming reports of series with increasing rates of complications (COTTON 1989).
5. The skill of endoscopist is probably very important, but it is not easy to document. As one reads the literature on endoscopic sphincterotomy and its complications, it becomes clear that despite the now universal performance of this procedure, the morbidity and mortality associated with sphincterotomy have remained constant at the level they were initially when it was performed by only handful of experts.

Despite all these considerations, complications do occur following endoscopic sphincterotomy and the endoscopist should be familiar with them in order to recognize and manage them.

18.2.3.1 Bleeding

Bleeding is the most common complication of sphincterotomy, but it can be controlled in the majority of cases. Large series indicate an overall incidence of 2.5%–5% with a mortality of about 0.3%. In our experience, significant bleeding occurred in 21 cases (1.9%). In 17 of these (80.9%) the bleeding stopped spontaneously or by coagulation with the sphincterotome; in 3 patients (14.3%) it was successfully managed with the injection of adrenaline 1:10000 at the site of bleeding. One patient (4.9%) with severe recurrent bleeding underwent surgery. Bleeding occurred in 16 out of 633 of our patients treated for CBD stones: 13 out of 609 undergoing standard sphincterotomy (2.1%) and 3 out of 24 undergoing precutting ($p = 0.02$). Therefore, in our experience, precutting is a risk factor for bleeding in patients treated for CBD stones.

Comparisons of reported incidences are difficult because the definitions of complications are rarely similar. Bleeding is often seen endoscopically at the time of sphincterotomy but it does not produce symptoms. In the Middlesex series of 1000 sphincterotomies (VAIRA et al. 1989), 39 patients bled (3.9%); only 6 (0.6%) received transfusions and only 1 (0.1%) needed surgery. Generally hemorrhage occurs at the moment when the sphincter is sectioned, and it is usually slight, but it can be delayed for hours or even days. In these cases, it is important to confirm the source of bleeding endoscopically.

Coagulation status should be checked routinely and improved by appropriate treatment where possible. Clotting may be impaired in patients with chronic liver disease; aspirin and nonsteroidal anti-inflammatory drugs are known to affect platelet function.

Bleeding does not appear to occur more frequently in patients with peripapillary diverticula. The risk seems to be higher when sphincterotomy is performed for papillary tumors, in patients with Oddi sphincter dysfunction and/or with a nondilated CBD, and when a previous sphincterotomy is enlarged within a few days or weeks.

It should be stressed that most bleeding incidents are related to technical mistakes. Most experts emphasize that the sphincterotomy must be performed sectioning the papilla slowly, pausing at each millimeter of cut to see whether coagulation is needed at its edges ("step by step" sphincterotomy). Bleeding may be due to a rapid uncontrolled incision ("zipper" sphincterotomy).

Serious bleeding probably occurs because of variants in vascular anatomy. Attempts to document aberrant vessels have been made by endoscopic application of a Doppler probe directly on the papilla, but this maneuver cannot be proposed routinely.

In most cases bleeding stops spontaneously or is managed successfully by standard conservative measures. Endoscopic and angiographic management are alternatives to the surgical intervention. When sphincterotomy results in severe arterial bleeding, local treatment is impossible because the endoscopic field becomes rapidly obscured by blood. However, endoscopic measures can be utilized when bleeding is less severe. If the sphincterotome is still in place, it can be used to coagulate the specific source of bleeding. This method is not always effective and most authors prefer to inject 1:10 000 adrenaline and/or sclerosing agents with a sclerotherapy needle into the cut margins of the sphincterotomy. Care should be taken to avoid injecting close to the pancreatic orifice, since the resulting edema may lead to pancreatitis.

Endoscopic balloon tamponade has also been recommended. The inflated balloon is placed on the bleeding source and is compressed between the sphincterotomy and the tip of the endoscope.

When hemostasis is obtained, it is mandatory to end the procedure by inserting a nasobiliary drain in order to avoid the risk of cholangitis secondary to impacted clot in the lower CBD.

If all endoscopic measures fail or if bleeding is arterial, angiographic treatment should be proposed, but only where high-quality experts are rapidly available.

Surgery must be considered when all nonsurgical methods have failed or when endoscopy has failed to deal with the initial problem, e.g., retained stones (Cotton et al. 1991; Oujaudé et al. 1994).

18.2.3.2 Pancreatitis

Pancreatitis may occur after diagnostic ERCP, but sphincterotomy seems to increase this risk. An increase in pancreatic enzymes in both serum and urine is common after ERCP, but it is generally asymptomatic and returns to normal within 24–48 h after the procedure. It is our policy not to order monitoring of amylase levels routinely after ERCP. Diagnosis of pancreatitis must include clinical criteria such as abdominal pain, abdominal distension, nausea, vomiting, leukocytosis, and fever (Cotton et al. 1991).

The overall incidence of pancreatitis is 1%–3.3% of patients, and mortality is 0.2%. Surgery is required in 0.1%. In our series 7 patients (0.6%) developed clinical acute pancreatitis and all were managed conservatively.

There are many factors which may increase the risk of pancreatitis (Chen et al. 1994). Excessive use of coagulation current during sphincterotomy may compromise the pancreatic orifice by thermal injury or edema. That is why some authors recommend using pure cutting diathermic current for the sectioning of the first millimeters, just to divide the two orifices, and then using blended diathermic current to complete the sphincterotomy. Precutting may increase the incidence of pancreatitis.

Trauma to the papilla and repeated injections of contrast medium are probably additional factors, especially when difficulties are experienced in achieving deep cannulation of the bile duct.

Unfortunately, clinical studies utilizing protease inhibitors or antisecretory drugs prophylactically have not shown convincing benefit. A recent Italian multicenter study has shown that high-dose gabexate mesylate, a protease inhibitor, is able to significantly reduce post-ERCP and post-endoscopic sphincterotomy hyperamylasemia and clinical pancreatitis when given prophylactically (Cavallini et al. 1996).

The management of pancreatitis after sphincterotomy is not different from that of pancreatitis due to diagnostic ERCP. It is important to rule out retroduodenal perforation with a plain film and CT scan, especially if the patient is acutely ill within 24 h after the procedure. Patients are kept without oral intake, with intravenous fluids and nasogastric suction. With conservative measures, most patients recover completely within few days. Some patients with severe pancreatitis require longer hospitalization with parenteral nutrition. The need to cover routinely with broad spectrum antibiotics is still debatable. When sepsis develops, it may be helpful to obtain culture samples from the pancreas by percutaneous aspiration to plan appropriate antibiotic therapy. If infection is confirmed, this is usually an indication for surgery. If infection is not confirmed, it is very difficult to indicate the timing and the indications for surgery.

18.2.3.3 Perforation

Retroduodenal perforation has been reported to occur in about 1.3% of sphincterotomies, with a mortality of 0.2%. Intraperitoneal perforation is very rare. In our series, retroduodenal perforations occurred in 5 of 1120 sphincterotomies (0.4%). All were successfully treated conservatively, giving the patients nothing by mouth, nasobiliary drainage, nasogastric suction, antisecretory drugs, and antibiotics. Perforation is easy to recognize during the procedure by radiological observation of air or contrast in the retroperitoneum. The kidney may be outlined and the margin of the psoas obscured by air. Frequently, however, perforation is not diagnosed at the time of the sphincterotomy and is not suspected until a few hours after the procedure. The patient develops pain, fever, rarely subcutaneous emphysema. Small localized perforations are sometimes difficult to detect (no air on plain films of the abdomen) and the clinical syndrome may resemble acute pancreatitis. An early CT scan and pancreatic enzyme values are usually discriminatory. Perforation usually occurs when the incision is directed outside the usual recommended sector (11–1 o'clock). The risk of perforation is higher in patients with Oddi sphincter dysfunction and a small-diameter CBD, but not in patients with peripapillary diverticula or in patients with a Billroth II stomach. In expert hands, precutting seems not to increase the risk of perforation.

If a retroperitoneal leak is suspected or proved, further leakage can be minimized by insertion of a nasobiliary drain (to reduce retroperitoneal contamination), by nasogastric suction, appropriate intravenous fluid therapy, antisecretory drugs (pancreatic and gastric), antibiotics, and giving nothing by mouth. Most patients have been treated conservatively with satisfactory results. Sepsis is usually a bad prognostic sign.

A major consideration in deciding for or against early surgery is the status of the CBD and its drainage. If perforation is recognized when there are still stones or obstruction, surgery is usually indicated to complete the therapeutic procedure and to assure effective drainage of the duct after the endoscopist has failed to complete the treatment endoscopically. Sealing of the perforation can be revealed by serial radiographic studies (CT scan, Gastrografin swallow, nasobiliary cholangiogram), and the patient's status will indicate the progress of the retroperitoneal infection.

Both medical and operative treatment appear more successful when initiated early after sphincterotomy.

18.2.3.4 Sepsis

The possibility of iatrogenic infection in the pancreas and in the biliary ducts has been stressed by many authors. However, most septic complications occur from bacteria already present in the biliary tree. Effective drainage is the key to preventing biliary sepsis. This can be achieved by sphincterotomy, stone extraction, and stenting in cases of stricture. A nasobiliary drain should be inserted when all the stones are removed or when the patient has been subjected to a long and difficult examination. Cremer recommends prophylactic antibiotics in patients with cholestasis before undergoing ERCP. Sepsis is reported in 1% of the patients in most series.

Among 633 patients treated for CBD stones at our center, sepsis occurred in 14 (2.2%); 8 had severe cholangitis and 6 cholecystitis. Cholangitis occurred in two cases after precut papillotomy, in eight cases because of retained stones, and in one case because of bad management of the nasobiliary drainage. Seven patients were successfully treated with endoscopic and conservative management. One died of septic shock. All the cases of cholecystitis occurred in gallbladders with stones. The patients were operated on. One patient with a lymphoma died a few days after the cholecystectomy. The figures for the incidence of sepsis reported in the literature depend on the definition used and on whether failed procedures are included in the analysis.

Recognition of sepsis is usually very easy. It is very important to remember that cholecystitis can occur after ERCP in patients with the gallbladder in situ, especially in the presence of stones. Ultrasonography may be helpful in the differential diagnosis. The principles of management are the use of parenteral antibiotics (after blood cultures) and prompt drainage of the biliary duct. This may require an urgent repeat endoscopic examination or percutaneous drainage.

18.2.4 Stone Extraction

Small stones, under 10 mm diameter, may pass the endoscopically performed sphincterotomy, but active extraction is today recommended by all the experts. Two devices are usually employed to extract stones: the Dormia basket and the Fogarty balloon catheter (Fig. 18.2). The latter is particularly useful when biliary sludge or multiple small stones are present. Wire-guided extraction devices are now also commercially available: selective cannulation of the intrahepatic branches with the guidewire allows easier extraction of intrahepatic stones. Overall duct clearance may be expected in about 90% of the patients undergoing successful endoscopic sphincterotomy and standard extraction techniques.

Several methods of lithotripsy are available for stones that are unextractable with conventional devices (Binmoeller et al. 1993).

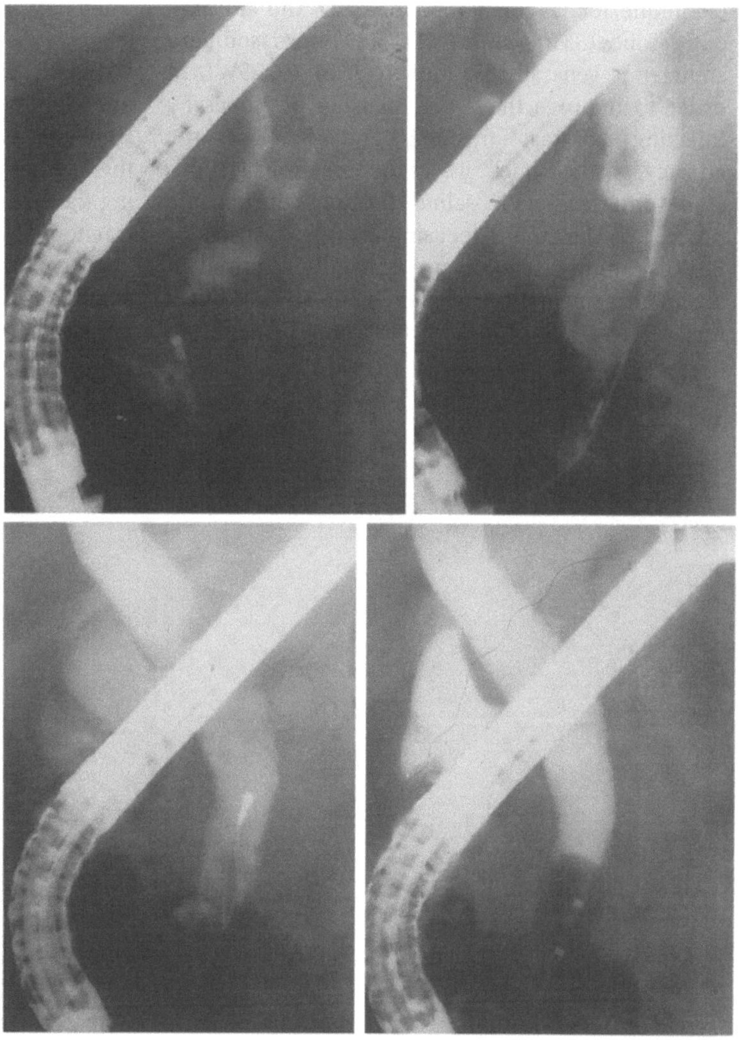

a b c d

Fig. 18.2. **a** Early filling cholangiogram showing multiple CBD stones in a cholecystectomized patient. **b** After deep cannulation of the bile duct a wire-guided sphincterotome is positioned. **c** Extraction of stones with Dormia basket after completion of endoscopic sphincterotomy. **d** Further extraction of biliary sludge with Fogarty ballon and final check cholangiogram showing an alithiasic bile duct

18.2.4.1 Lithotripsy

When conventional extraction devices fail to remove CBD stones, various lithotripsy techniques may be applied.

Mechanical lithotripsy is the simplest method of crushing stones that are too big to extract through a standard sphincterotomy. The stone is captured in a wire basket and then crushed by pulling the basket into a metal sheath. Two methods are currently available: the extraendoscopic lithotriptor (Soehendra type, Wilson Cook), in which the metal sheath is advanced over the basket wire once the endoscope has been removed, and the intraendoscopic device (Olympus), which works through the operative channel of the duodenoscope.

Inability to capture the stone is the most frequent cause of failure of mechanical lithotripsy (SHAW et al. 1993). We have recently reviewed 106 patients in whom mechanical lithotripsy had been attempted to identify variables predictive of the positive or negative outcome of the procedure. In both univariate and multivariate analysis, only stone size proved to be predictive of the outcome of lithotripsy (diameter of grasped vs not grasped stones: 21.7 ± 6.7 mm vs 28.3 ± 10.4 mm; $p < 0.001$). Overall, the procedure was both safe (2.8% specific complications) and effective (83.9% stone clearance) (CIPOLLETTA et al. 1995).

When mechanical lithotripsy is unsuccessful, various nonsurgical alternatives may be employed. Intracorporeal electrohydraulic or pulsed dye laser lithotripsy via a cholangioscope is an effective but demanding technique because of its high cost and need for local expertise. Patients with large unretrievable stones are more and more often referred to extracorporeal shock wave lithotripsy (ESWL) units: ESWL is a safe and effective technique for fragmentation of CBD and intrahepatic stones (Sauerbruch et al. 1992). A nasobiliary catheter is generally positioned prior to ESWL in order to improve the X-ray focusing of stones by injecting contrast medium and increase the efficacy of the shock waves by creating a fluid-stone interface with saline flushing during the procedure. Ultrasound focusing may also be used for ESWL. Multiple sessions are generally required to achieve fragmentation of giant CBD stones, but a final success rate of at least 80% may be expected when ESWL is performed by experienced (dedicated) teams.

Finally, in old and high-risk patients, placement of endobiliary stents to avoid impaction of retained stones in association with long-term oral bile acid therapy has also been shown to be effective in reducing the size of stones, allowing their subsequent extraction (Johnson et al. 1993).

In conclusion, the management of difficult CBD stones should depend on the local facilities and expertise (Devière and Cremer 1994).

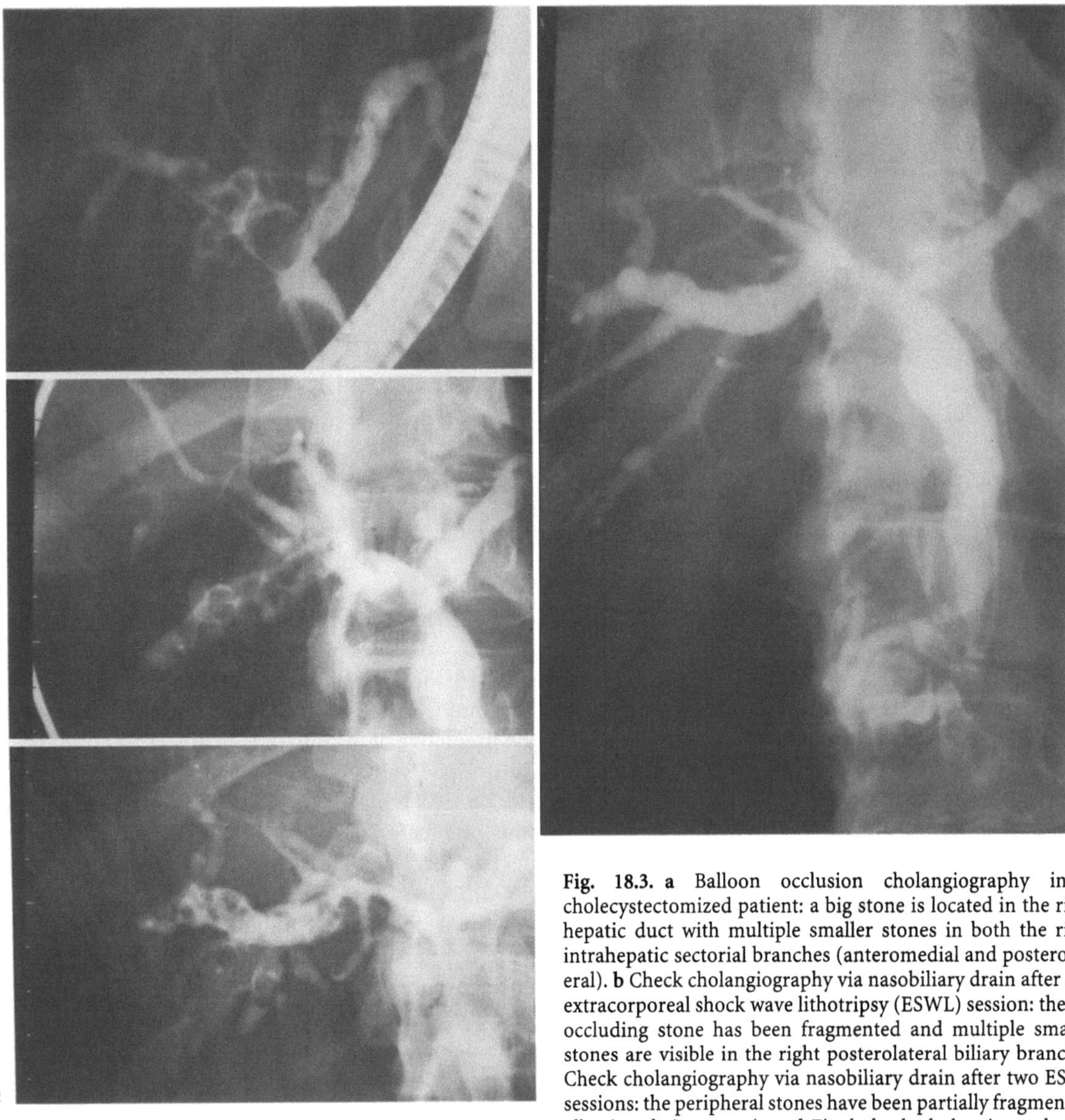

a b c d

Fig. 18.3. a Balloon occlusion cholangiography in a cholecystectomized patient: a big stone is located in the right hepatic duct with multiple smaller stones in both the right intrahepatic sectorial branches (anteromedial and posterolateral). **b** Check cholangiography via nasobiliary drain after one extracorporeal shock wave lithotripsy (ESWL) session: the big occluding stone has been fragmented and multiple smaller stones are visible in the right posterolateral biliary branch. **c** Check cholangiography via nasobiliary drain after two ESWL sessions: the peripheral stones have been partially fragmented, allowing their extraction. **d** Final check cholangiography via nasobiliary drain after stones and fragments extraction: the entire biliary tree is free of stones

18.2.4.2 Intrahepatic Stones

Wire-guided extraction devices, i.e., Dormia basket and Fogarty balloon, permit selective access to intrahepatic branches: therefore clearing stones in these difficult areas may often be tried by endoscopic means. However, when access to peripheral branches is impaired by a stricture, intrahepatic stones are better dealt with by percutaneous interventional techniques, or by resectional hepatic surgery if the affected parenchyma is atrophic. Intracorporeal electrohydraulic lithotripsy under percutaneous cholangioscopic guidance has also been described in an Asiatic series (Sheen-Chen et al. 1993). ESWL may be extremely useful to fragment stones located more proximally, closer to the main confluence: in fact removal of the obstructing stone often facilitates clearance of smaller ductal stones and of distally located biliary sludge.

Multiple sessions are usually necessary to complete clearance of stones from the intrahepatic duct (Fig. 18.3).

18.2.5 Biliary Sphincteroplasty

Endoscopic balloon dilatation of the sphincter of Oddi prior to stone extraction has been reported as an alternative to sphincterotomy with a view to avoiding post-endoscopic sphincterotomy complications and preserving sphincter function (McMathuna et al. 1993). Several studies comparing endoscopic sphincterotomy to sphincteroplasty are ongoing. Available data at the moment show that balloon sphincteroplasty is effective and safe when used for stones of less than 10 mm diameter. In a prospective trial (Bergman et al. 1994) balloon sphincteroplasty allowed bile duct clearance in 85% of the patients, with a complication rate comparable to that of sphincterotomy. Endoscopic balloon dilatation may become an alternative to sphincterotomy, especially in younger patients with small stones and non-dilated biliary ducts.

18.3 Bile Duct Injuries

18.3.1 Biliary Fistulas

Biliary fistula may occur after any kind of operation involving the biliary tract as a result of inadvertent biliary injury or leakage from the cystic duct stump or from the choledochotomy closure.

Fistulas often do not heal spontaneously because of persisting distal obstruction due to stones, strictures, or sometimes only because of a competent sphincter of Oddi. Many reports have shown that endoscopic decompression of the CBD is highly effective in healing biliary leaks (Davids et al. 1992; Foutch et al. 1993; Foco et al. 1992). The aim of endoscopic therapy is to lower the resistance to biliodeudenal flow in order to decrease bile flow through the fistulous tract and to allow its healing. Controversies still exist concerning the choice of the ideal technique: internal or external drainage with stent or nasobiliary tube with or without concomitant endoscopic sphincterotomy have all been used successfully to lower the intrabiliary pressure and induce healing of the fistulous tract. Prospective randomized studies are therefore needed to clarify which strategy is the best suited to endoscopic treatment of biliary leaks. To date there is general agreement that the endoscopic approach should represent the first diagnostic and therapeutic step in all instances of suspected biliary leak, even in patients who have undergone liver transplantation (Sherman et al. 1993).

18.3.2 Biliary Strictures

The majority of benign strictures of the bile ducts occur as a result of surgical intervention, especially cholecystectomy. If biliary continuity is not completely interrupted, endoscopic management may be undertaken: complete transection should be surgically repaired immediately. The aim of endoscopic therapy is to reestablish long-term patency of the CBD by dilation and stenting. A detailed cholangiogram is mandatory to outline the precise morphology of the entire biliary tree and to define any associated stones, leaks, and fistulas, then endoscopic sphincterotomy is routinely performed to get easier access to the bile ducts in view of repeated endoscopic maneuvers.

Traversing the stricture with a guidewire is often more difficult in benign than in malignant strictures: a broad choice of various guidewires (straight, J-tip, Z-tip, hydrophilic, torquable) should be always available to overcome technical difficulties.

Benign strictures may be initially dilated with bougies (7–10 F), but balloon dilation is often required to achieve adequate calibration before stenting. High-pressure (10–12 atm) hydrostatic balloons are now available for endoscopic use: care should be taken in order to avoid overdistension and

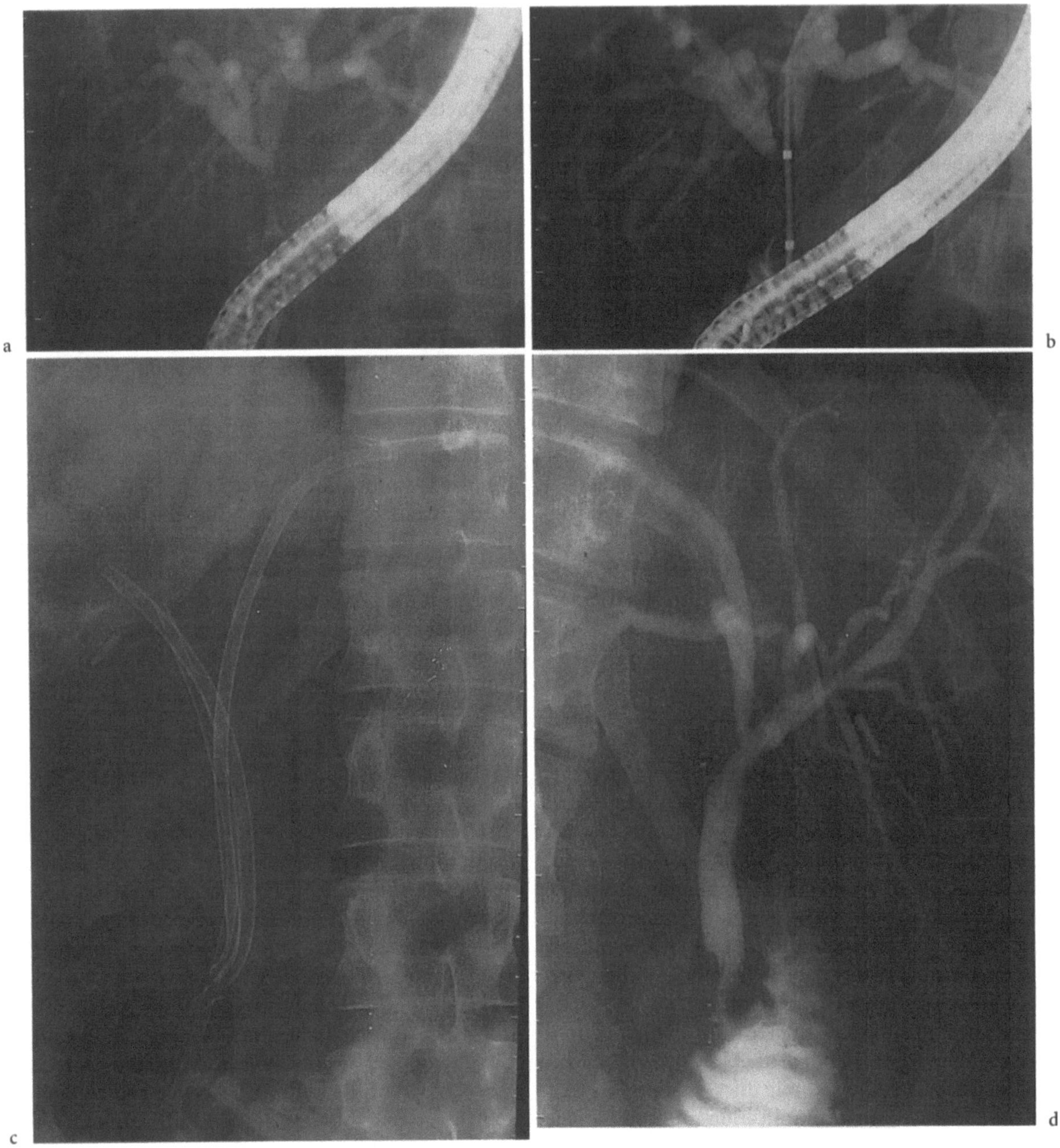

Fig. 18.4. **a** Balloon occlusion cholangiography after endoscopic sphincterotomy: post-cholecystectomy stricture of the CBD reaching the main confluence. **b** Pneumatic dilation of the stricture with a high-pressure hydrostatic balloon. **c** Three large-bore 10-F stents have been inserted to dilate the stricture. The stents are routinely charged every 3–4 months over 1 year. **d** Final result of endoscopic management after 1 year of stenting: the stricture is no longer visible and the intrahepatic branches are of normal caliber

risk of perforation. The current policy of most experienced endoscopic teams is to stent strictures after initial dilation: 10-F plastic (polyethylene or Teflon) straight (Amsterdam-type) stents are generally used. Stent exchange is scheduled at 3- to 4-month intervals over 1 year: at each exchange, at least one additional stent is inserted, the ideal being at least two 10-F stents, left in place for a minimum of 3 months (Davids et al. 1993; Smith et al. 1995). As a rule all stents are removed after 1 year and the patient is followed up clinically and with laboratory tests (Fig. 18.4). If symptoms of relapse and cholestasis occur, morphologic evaluation using ultrasound and magnetic resonance cholangiography may be useful to

Table 18.1. Endoscopic management of postoperative biliary strictures

	No. of postop strictures	Primary treatment method	Associated features	Technicial success	Average follow-up after treatment (months)	Response to treatment	Complications of treatment
DAVIDS et al. (1993) (Amsterdam)	80	Stenting (2–10 F)	Bile leaks 20 Stones 17	83% of total 94% of those Attempted	42	excellent or good (83%)	Early (<30 d) 10.6% Late >30 d 34.8%
WALDEN et al. (1993) (Toronto)	32 in 25 patients	Stenting (7, 9.5, or 11 F)	Bile leaks 5 Stones 8	85%	65	excellent 59% good 14% poor 27%	16%
GEENEN et al. (1989) (Racine)	18	Dilation/ stent	-	92%	48	good 78% fair 17% poor 4%	8.6%

(Modified from SMITH et al. 1995).

identify recurrent stricture and to plan further treatment; a new endoscopic approach may be undertaken if the stricture is easy to redilate, but surgery should be advocated for multiple recurrences in good-risk patients. Comparable recurrence rates of 17% each have been reported in endoscopically and surgically treated patients respectively in a recent series (DAVIDS et al. 1993).

Endoscopically placed self-expandable metal mesh stents have also been used in benign strictures, but preliminary experiences have been disappointing and their use should at present be restricted to malignant strictures (COSTAMAGNA et al. 1992).

Details of short-term and long-term results of endoscopic management of postoperative biliary strictures are given in Table 18.1.

References

Bergman JJGHM, Rauws EAJ, Tytgat GNJ, Huibregtse K (1994) A prospective randomized trial comparing endoscopic sphincterotomy (EST) with endoscopic balloon dilation (EBD) for removal of common bile stones (CBDS); initial report. Gastrointest Endosc 40:P99

Binmoeller KF, Bruckner M, Thonke F, Soehendra N (1993) Treatment of difficult bile duct stones using mechanical, electrohydraulic and extracorporeal shock wave lithotripsy. Endoscopy 25:201–206

Cavallini G, Tittobello A, Frulloni L, et al (1996) Gabexate mesylate in the prevention of pancreatic damage induced by ERCP. A multicenter double-blind randomized study. N Engl J Med (in press)

Chen YK, Foliente RL, Santoro MJ, et al (1994) Endoscopic sphincterotomy-induced pancreatitis: increased risk associated with non-dilated bile ducts and sphincter of Oddi dysfunction. Am J Gastroenterol 89:327–333

Cipolletta L, Costamagna G, Bianco MA, et al (1995) Mechanical lithotripsy of bile duct stones: a score for predicting the outcome. Endoscopy 27:S15, 402

Clair DG, Carr-Locke DL, Becker JM, et al (1993) Routine cholangiography is not warranted during laparoscopic cholecystectomy. Arch Surg 128:551–555

Costamagna G, Perri V, Mutignani M, et al (1992) Long-term results of endoscopic self-expanding metal stents in benign biliary strictures. Gastroenterology 102:A309

Cotton PB (1989) Pre-cut papillotomy: a risky technique for experts only. Gastrointest Endosc 35:578–579

Cotton PB (1993) Endoscopic retrograde cholangiopancreatography and laparoscopic cholecystectomy. Am J Surg 165:474–478

Cotton PB, Lehman G, Vennes J, et al (1991) Endoscopic sphincterotomy complications and their management: an attempt at consensus. Gastrointest Endosc 37:383–393

Davids PHP, Rauws EAJ, Tytgat GNJ, et al (1992) Postoperative bile leakage: endoscopic management. Gut 33:118–122

Davids PHP, Tanka AKF, Rauws EAJ, et al (1993) Benign biliary strictures. Surgery or endoscopy? Ann Surg 217:237–243

Devière J, Cremer M (1994) Biliary endoscopy. In: LaMont JT, Afdhal NH (eds) Biliary tract. Current Science, Philadelphia, pp 567–572

Foco A, Garbarini A, Franchello A, et al (1992) Management of postoperative bile leakage with endoscopic sphincterotomy (EST) and a nasobiliary drain (NBD). Hepato-gastroenterology 39:301–303

Foutch PG, Harlan JR, Hoefer M (1993) Endoscopic therapy for patients with a postoperative biliary leakage. Gastrointest Endosc 39:416–421

Geenen DJ, Geenen JE, Hogan WJ, et al (1989) Endoscopic therapy for benign bile duct strictures. Gastrointest Endosc 35:367–371

Ibuki Y, Kudo M, Todo A (1992) Endoscopic retrograde extraction of common bile duct stones with drip infusion of isosorbide dinitrate. Gastrointest Endosc 38:178–180

Johnson GK, Geenen JE, Venu RP, et al (1993) Treatment of non-extractable common bile duct stones with combination ursodeoxycholic acid plus endoprostheses. Gastrointest Endosc 39:528-532

McMathuna P, White P, Lennon J (1993) Balloon sphincterotomy for removal of bile duct stones: an alternative to papillotomy? Gastroenterology 104:369A

Oujaudé J, Pelletier G, Fritsch J, et al (1994) Management of clinically relevant bleeding following endoscopic sphincterotomy. Endoscopy 26:217-220

Petelin JB (1993) Laparoscopic approach to common duct pathology. Am J Surg 165:487-491

Sauerbruch T, Holl J, Sackmann M, et al (1992) Fragmentation of bile duct stones by extracorporeal shockwave lithotripsy: a five-year experience. Hepatology 15:208-214

Shakoor T, Geenen JE (1992) Pre-cut papillotomy. Gastrointest Endosc 38:623-627

Shaw MJ, Mackie RD, Moore JP, et al (1993) Results of a multicenter trial using a mechanical lithotriptor for the treatment of large bile duct sones. Am J Gastroenterol 88:730-733

Sheen-Chen S, Chon F, Lee C, et al (1993) The management of complicated hepatolithiasis with intrahepatic biliary stricture by the combination of T-tube tract dilation and endoscopic electrohydraulic lithotripsy. Gastrointest Endosc 39:168-171

Sherman S, Ruffolo TA, Hawes RH, et al (1991) Complications of endoscopic sphincterotomy. A prospective series with emphasis on the increased risk associated with sphincter of Oddi dysfunction and non dilated bile ducts. Gastroenterology 101:1068-1075

Sherman S, Shaked A, Cryer HM, Goldstein LI, Busuttil RW (1993) Endoscopic management of biliary fistulas complicating liver transplantation and other hepatobiliary operations. Ann Surg 218:167-175

Siegel JH (1992) Endoscopic retrograde cholangiopancreatography. Raven New Yrok, p 193.

Smith MT, Sherman S, Lehman GA (1995) Endoscopic management of benign strictures of the biliary tree. Endoscopy 27:253-266

Sowberg KA, Way LW, Sleisinger MH (1993) Complications of gallstone disease, 5th edn. Saunders, Philadelphia, p 1805

Vaira D, Ainley C, Williams S, et al (1989) Endoscopic sphincterotomy in 1000 consecutive patients. Lancet 2:431-434

Walden D, Raijman I, Frichs E, et al (1993) Long-term follow-up of endoscopic stenting (ES) for benign postoperative bile duct strictures (BPBDS). Gastrointest Endosc 39:335-339

19 Benign Biliary Disease: Management of Bile Duct Stones

G. Gandini, M.C. Cassinis, P. Fonio, J. Maass, M. Natrella, and D. Righi

CONTENTS

19.1 Introduction

Biliary lithiasis is a widespread disease in the industrialized countries and is one of the most common reasons for abdominal surgery (Barbara 1984). Autopsy series and recent studies based upon clinical and instrumental investigations in selected groups or entire populations have revealed that the prevalence of the disease is much higher in females, and in both sexes rises with age (Bennion et al. 1979; Capocaccia and Ricci 1985; Covarrubias et al. 1983; Morino et al. 1988). In South-East Asia, stone disease characteristically involves the intrahepatic biliary tree and the common bile duct, sparing the gallbladder being often unaffected, and stones are of a mixed or inflammatory type (containing calcium bilirubinate and cholesterol). Although the exact etiology is still unknown, bile duct infection is believed to play a major role, either provoking lithiasis directly or causing duct strictures with subsequent formation of calculi. In Caucasians, on the other hand, more than 80% of stones are gallbladder or common bile duct stones composed of cholesterol; only a small proportion are found in the intrahepatic ducts (Maki et al. 1982).

The clinical introduction of laparoscopic cholecystectomy has hastened the decline of nonsurgical techniques in the treatment of gallbladder lithiasis to the point that the role of interventional radiology in this field is now extremely limited. The primary limitation is that only cholesterol stones can be successfully treated by means of interventional radiological procedures.

Chemical stone dissolution with methyl-*ter*-butyl-ether (MTBE) via a percutaneous cholecystostomy catheter has nearly been abandoned, even though encouraging results have been reported by others (Allen et al. 1985) and by ourselves (Gandini et al. 1988a). This may be due to the low tolerance by patients of this quite invasive technique and the long duration of treatment, which takes several hours.

Extracorporeal shock wave lithotripsy (ESWL) of gallbladder calculi has shown clear advantages over chemical litholysis, especially since the introduction of second- and third-generation lithotriptors, which allow treatment to be performed on an outpatient basis. However, its indications are limited: only patients having less than three stones with a combined diameter less than 3 cm, a functioning gallbladder, and who present a high surgical risk or refuse surgery, are potential candidates. Even with these restricted indications, however, the results reported in

G. Gandini, Professor, Instituto di Radiologia Diagnostica e Interventistica, Univ. di Torino, Osp. Maggiore, 28100 Novara, Italy
M.C. Cassinis, MD, Instituto di Radiologia Diagnostica e Interventistica, Univ. di Torino, Osp. Maggiore, 28100 Novara, Italy
P. Fonio, MD, Instituto di Radiologia Diagnostica e Interventistica, Univ. di Torino, Osp. Maggiore, 28100 Novara, Italy
J. Maass, MD, Instituto di Radiologia Diagnostica e Interventistica, Univ. di Torino, Osp. Maggiore, 28100 Novara, Italy
M. Natrella, MD, Instituto di Radiologia Diagnostica e Interventistica, Univ. di Torino, Osp. Maggiore, 28100 Novara, Italy
D. Righi, MD, Instituto di Radiologia dell'Univ. di Torino, Osp. S. Giovanni Battista, 10126 Torino, Italy

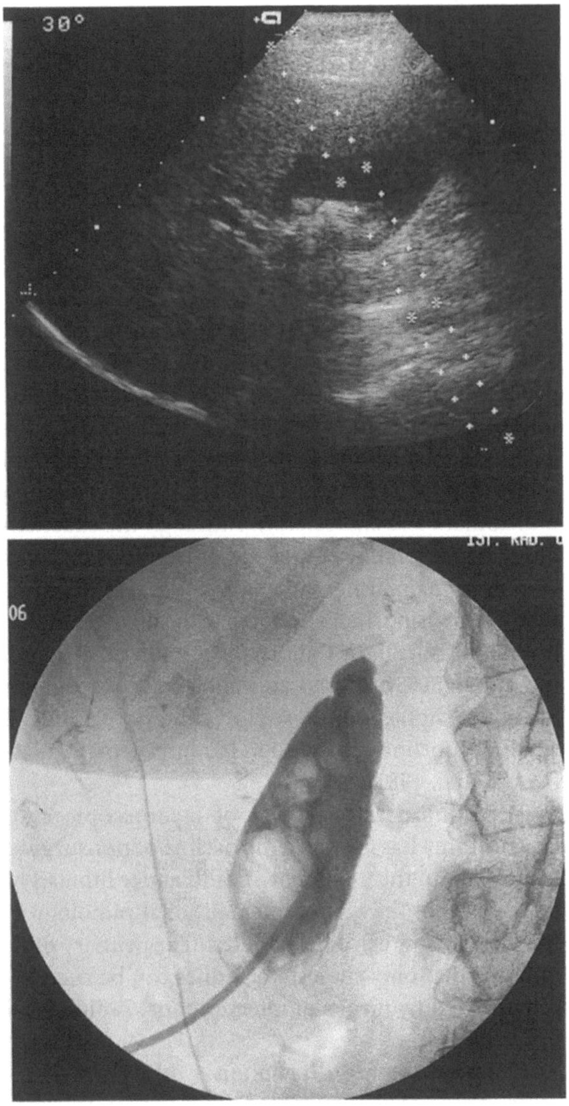

a

b

Fig. 19.1. a Ultrasound targeting of gallbladder empyema prior to percutaneous cholecystostomy. The entrance site is marked on the skin. **b** Percutaneous cholecystostomy is performed using a pig-tail catheter (7 F)

literature are quite discordant. Our own experience of gallbladder stones treated using a third-generation lithotriptor has been generally disappointing.

Interventional radiology is thus regarded as retaining an important role in gallbladder lithiasis exclusively in cases of acute empyema treatable by percutaneous emergency cholecystostomy: this maneuver is easily performed under ultrasound guidance, has a high success and a low complication rate, and consistently allows resolution of the acute inflammation so that surgery can be delayed and performed as an elective procedure (Fig. 19.1) (Fonio et al. 1990; Ort et al. 1990). Thus, this chapter considers only lithiasis of the bile ducts, in which the multidisciplinary therapeutic approach involves interventional radiology as a fundamental tool.

The classification of bile duct lithiasis employed in this chapter is based exclusively on the anatomic site of lithiasis. Three major groups of bile duct lithiasis are distinguished:

1. *Extrahepatic or common bile duct lithiasis.* This group is subdivided into residual post-cholecystectomy lithiasis (patients with a T tube or a transcystic tube), recurrent lithiasis (with onset at least after 1 year after cholecystectomy), and bile duct lithiasis associated with gallbladder lithiasis.
2. *Intrahepatic bile duct lithiasis.* Within this group a distinction is made between patients with an intact common bile duct and those who have undergone a bilioenteric anastomosis.
3. *Intra- and extrahepatic bile duct lithiasis.* This group relates to the distinct entity of massive lithiasis involving more than four segmental ducts.

Our choice of this classification is based upon our experience, which began in 1982 at the University of Turin and was continued from 1991 onwards at the Radiology Department in Novara. During this period, 293 patients with bile duct lithiasis were treated by percutaneous techniques. Analysis of this series demonstrates how the choice of therapeutic approach and treatment technique depend fundamentally on the location of the stones and on other factors such as the presence of a T tube, a bilioenteric anastomosis, or massive lithiasis. Using this classification, the 293 patients are grouped as follows:

1. Extrahepatic lithiasis: 121 patients (56 with a normal gallbladder or recurrence after cholecystectomy and 65 with a T tube)
2. Intrahepatic lithiasis: 66 patients (21 with an intact common bile duct and 45 with a bilioenteric anastomosis)
3. Combined intra- and extrahepatic lithiasis: 106 patients (52 with an intact common bile duct, 27 with a bilioenteric anastomosis, and 27 with massive lithiasis)

19.2 Diagnostic and Therapeutic Approach

Cholestasis is a constant symptom in bile duct lithiasis, except in those rare cases in which very small gallstones pass into the duodenum without causing any obstruction. When cholestasis is diagnosed by means of laboratory tests (alkaline phosphatase and γ-glutamyltranspeptidase, which are the earliest and most sensitive indicators), the first instrumental examination to perform is ultrasonography, in order to distinguish lithiasis from other types of biliary obstruction and to locate the calculi in the biliary tree. Moreover, ultrasonography allows the best percutaneous approach to the bile ducts to be chosen, and at least in selected cases it should be repeated in the interventional radiology suite and the access marked directly on the skin immediately before the procedure is started.

When intrahepatic lithiasis is diagnosed or suspected on the basis of ultrasound, CT of the liver should always be performed for better demonstration of the number and site of stones and the degree of dilation of the ducts involved (Fig. 19.2). The ability of CT to show the density of stones, and consequently their chemical composition, may have important implications for the subsequent choice between different therapeutic options.

The final step in the diagnostic algorithm of bile duct lithiasis is direct cholangiography (endoscopic retrograde cholangio-pancreatography, ERCP), T-tube or transcystic cholangiography, or percutaneous transhepatic cholangiography, PTC), which at the same time constitutes the first step of any minimally invasive intervention. In very selected cases of bile duct lithiasis, percutaneous cholangioscopy with a small-caliber fiberscope may be useful to obtain a chemical analysis of the stones (cholesterol stones are yellow and pigment stones are black), and to guide contact lithotripsy or intraluminal ultrasonography, which may be essential in the diagnosis of rare cases of submucosal bile duct lithiasis (Gandini et al. 1988b). Thus, direct cholangiography represents the link between diagnosis and therapy and may be followed by biliary drainage maneuvers. In the case of ERCP, if lithotripsy is not completed in a single session or if a biliary obstruction persists, a nasobiliary tube must be left in place. If complete clearance of the common bile duct is obtained, especially after concomitant endoscopic sphincterotomy, placement of a drainage tube can be obmitted.

With T-tube cholangiography, too, if the lithiasis has resolved and no obstruction is present, the cath-

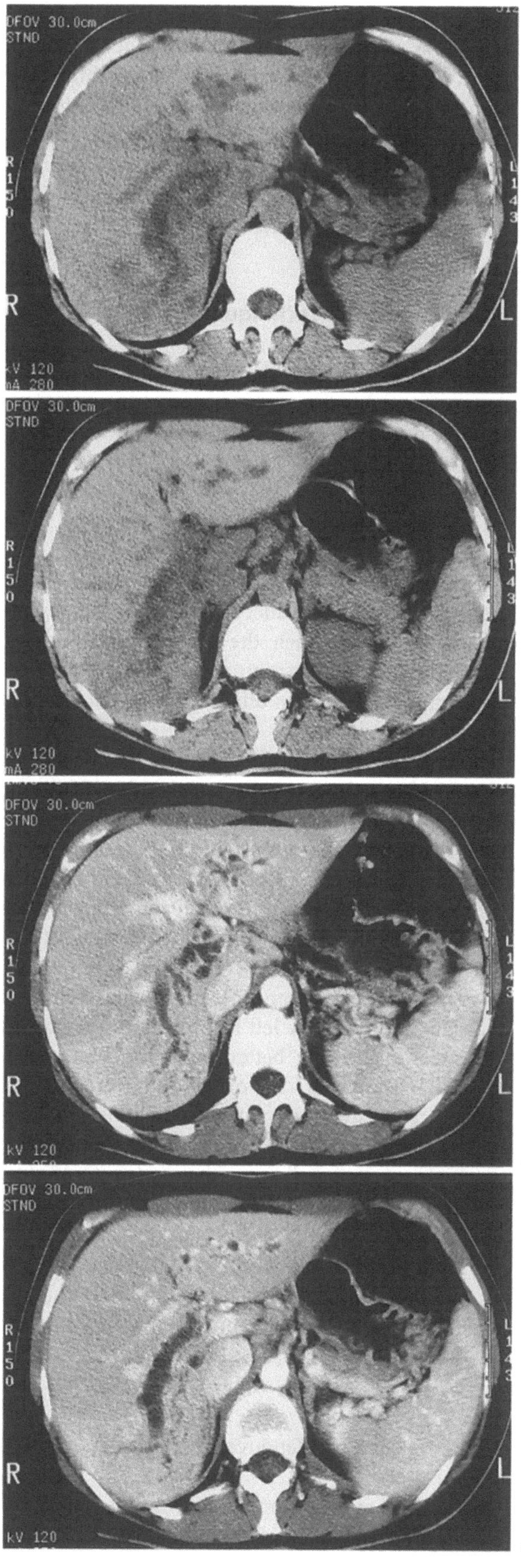

a b c d

Fig. 19.2a–d. CT scans without (**a,b**) and with (**c,d**) contrast enhancement: dilation of the right (sixth and seventh segments) and left ductal system (first, second, and third segments). The ducts of the fourth, fifth, and eighth segments are normal eter can theoretically be removed; however, it is usual practice to leave a biliary drainage catheter in place, which may be external (with its tip in the common bile duct or an intrahepatic duct) or external-internal (with its tip in the bowel loop). After PTC, placement of a transhepatic drainage catheter is essential; again, it may be external or external-internal (Ring and Kerlan 1984).

19.3 Therapeutic Techniques

This chapter describes the various therapeutic options offered by the percutaneous approach, irrespective of the site of lithiasis.

19.3.1 Percutaneous Transhepatic Bilioplasty

For percutaneous lithotripsy of whatever kind to succeed, any existing bile duct stricture has to be eliminated. Moreover, when the common bile duct is intact and endoscopic sphincterotomy has not been performed first, balloon dilation of the papilla must precede lithotripsy.

Bilioplasty is performed using angioplasty balloon catheters. In most cases the caliber of the inflated balloon is between 8 and 12 mm, but smaller balloons may be useful in the treatment of intrahepatic strictures, and bigger balloon catheters with an outer diameter of up to 20 mm are generally employed in the dilation of bilioenteric strictures. The simultaneous dilation technique, using two expandable 10-mm balloons inserted respectively from the right and from the left hepatic duct, may be advantageous, allowing better modeling of the narrowed anastomosis. In order to overcome the resistance of stenoses, particularly those that are iatrogenic or anastomotic in origin, it is essential to use high-pressure balloon catheters (up to 18 atm), which are able to dilate even the tightest strictures (Gandini et al. 1990).

In our personal series of over 300 patients treated by percutaneous transhepatic bilioplasty, the primary success rate was 94% and the patency rate at 5 years was 75% (Righi et al. 1990).

19.3.2 Mechanical Maneuvers

Small calculi may be eliminated from the biliary tree by simple flushing using large quantities of saline, which is infused through biliary drainage catheters. Saline flushing using the Broxo-Jet – a high-frequency pulsation hydrojet suited to routine use in dental hygiene – directly connected with the drainage catheters, has proved very effective. With this technique, necrotic material and fibrin deposits, so characteristic in cholangitis, which often accompanies intrahepatic lithiasis, are eliminated together with small stones and debris. Flushing with the hydrojet, however, is contraindicated in cases of suppurative cholangitis, because of the risk of septicemic spread. In these cases, low-pressure flushing with saline every 8–12 h, preceded by aspiration of the infected bile, is helpful (Cesarani et al. 1988).

If repeated flushing is insufficient to remove stones, venous occlusion balloon catheters may be employed. For this, the balloon is inflated with air or diluted contrast medium distally to the stones to be removed. It is essential to use heavy-duty torque-control guidewires (Amplatz Super Stiff type), in order to be able to exercise forceful pressure or traction on the calculi. During the passage of the stone through the biliary system, the outer diameter of the balloon has to be continuously adjusted to the ductal caliber, in order to avoid trapping of the stone between the balloon and the duct wall (Juliani and Gandini 1988).

Instead of venous occlusion balloon catheters, Dormia baskets can be employed, which allow the entrapped stones to be advanced through the papilla into the bowel loop. In addition, the basket has the advantage of a mechanical lithotriptor, and so bigger stones, once caught, can be crushed by forceful pulling of the basket against the tip of the sheath (Mazzariello 1976).

19.3.3 Extracorporeal Shock Wave Lithotripsy

The extracorporeal shock wave lithotriptor can be employed for the fragmentation of large gallstones when percutaneous lithotripsy techniques fail. In theory, the indications for ESWL of bile duct and gallbladder stones are the same. In practice, the indications for ESWL of bile duct stones are less restricted and complete fragmentation is not always necessary, since residual fragments can be eliminated by subsequent mechanical maneuvers.

Targeting is radiographic, by the injection of contrast medium through the biliary drainage catheter. Several treatment sessions (3–6), each giving 2500–3000 shocks with a power of 18–26 kV (using the Medas Lithoring Multione unit), are often required to crush a biliary stone (Burhenne et al. 1989).

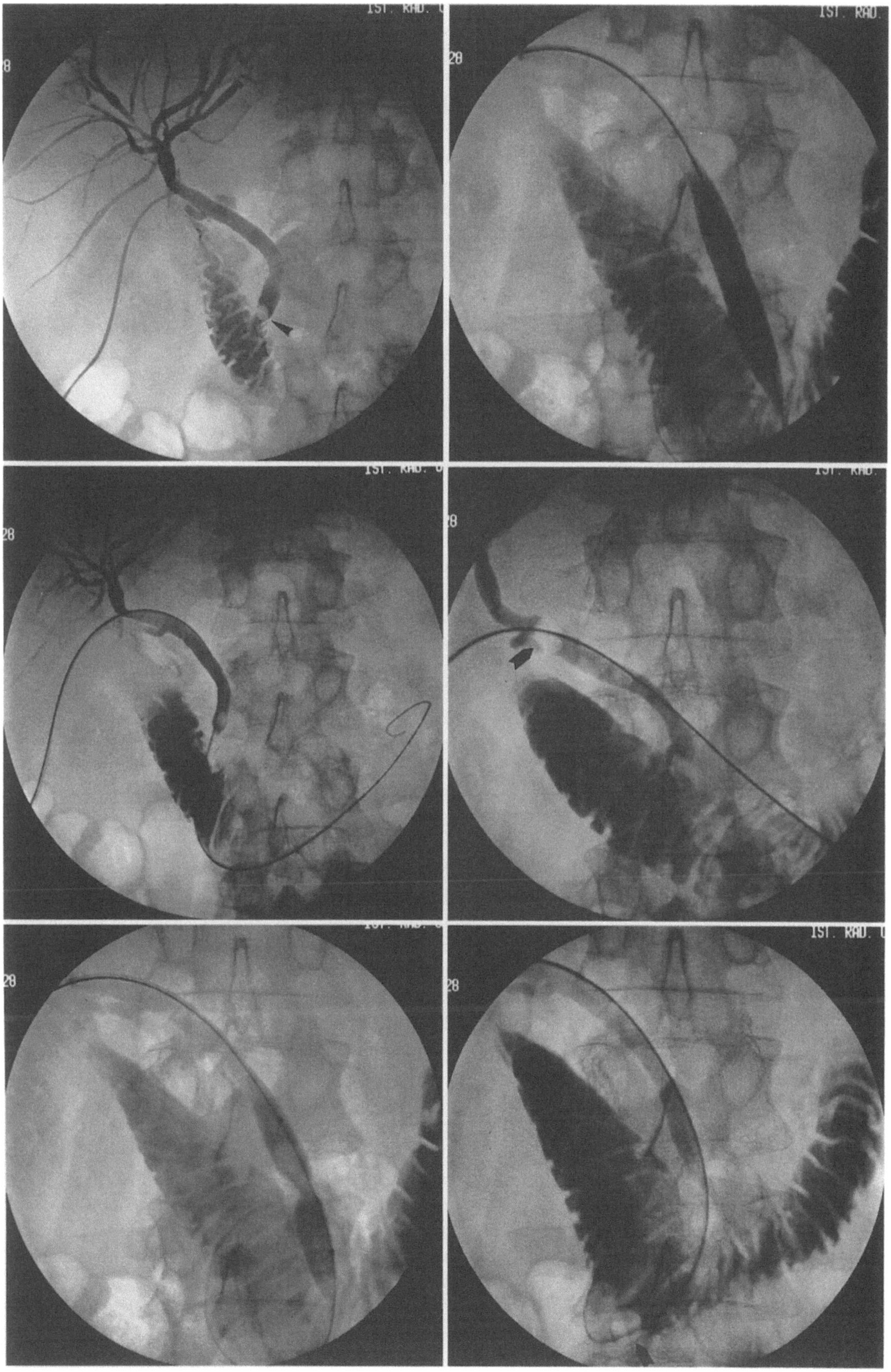

a b c d e f

Fig. 19.3a–f

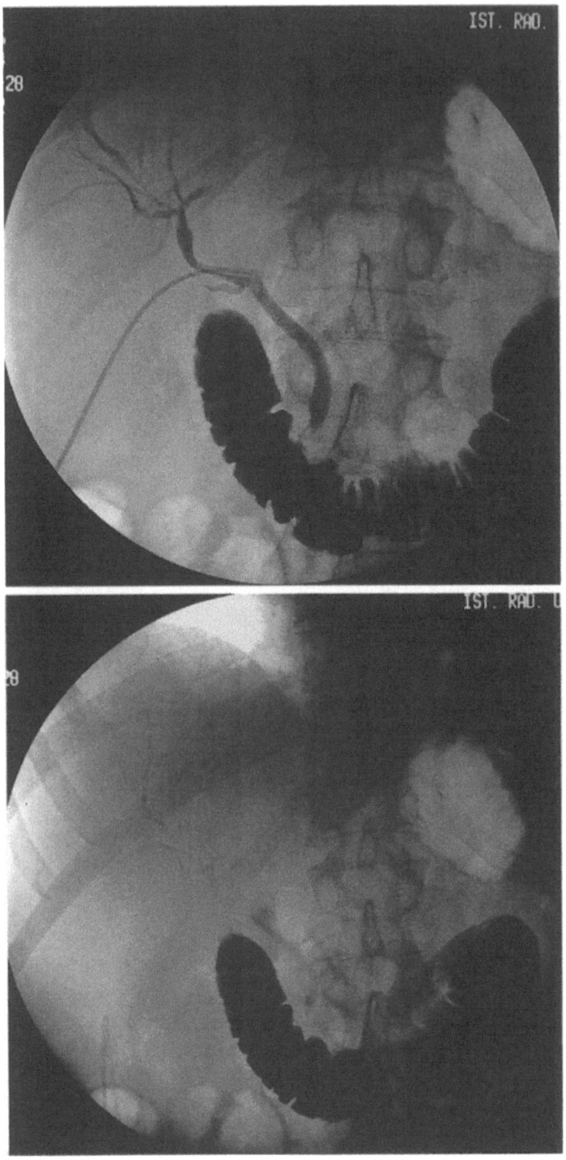

g

h

Fig. 19.3. **a** Transcystic cholangiography: residual radiolucent stone in the common bile duct above the papilla (*arrow*). **b** The transcystic catheter is replaced over a guidewire which is advanced into the duodenum. **c,d** Dilation of the papilla with a 1-cm-diameter balloon catheter. **e,f** The residual stone is removed by means of a venous occlusion balloon catheter (*arrow*). **g** Control cholangiogram: the common bile duct is free of stones. **h** Rapid emptying of the common bile duct after removal of the transcystic catheter

19.3.4 Contact Lithotripsy

As an alternative to ESWL or when extracorporeal lithotripsy fails, large stones, or those not treatable with mechanical devices because of their position, can be crushed by means of contact lithotripsy. The flexible electrohydraulic sounds are equipped with two coaxial electrodes at the tip; a high-tension discharge produces a spark at the bipolar electrode, with concurrent evaporation of the surrounding liquid and generation of a high-amplitude, low-frequency shock wave strong enough to crush a gallstone. The sound has to be directed under cholangioscopic control to ensure that it is pointed against the stone and not against the duct wall, which would be damaged by the shock wave (Juliani et al. 1988).

19.4 Choice of Therapeutic Technique

The various therapeutic options for lithiasis in different locations are as follows:

19.4.1 Extrahepatic Lithiasis

19.4.1.1 Residual Lithiasis

In patients with residual lithiasis, direct cholangiography is performed through the T tube or transcystic catheter; dilute contrast medium (20 mg% iodine) is used in order to visualize even small filling defects. It is also necessary to obtain opacification of the intrahepatic ducts, varying the patient's position, in order to exclude migration of calculi at this level. The surgical biliary tube is then removed over a stiff guidewire (Amplatz type), since unlike with percutaneous maneuvers the route of the catheter is "free" in the peritoneum, and thus the use of soft guidewires would increase the risk of loosing the access.

The sheath of an introducer is then placed in the biliary tree in order to advance a security guidewire into the duodenum or into the intrahepatic ducts, using preshaped catheters. The second option may be advantageous in cases in which during the preceding cholangiography, complete opacification of the intrahepatic ducts was not achieved and thus the contrast medium can be injected directly into the biliary tree (Fig. 19.3).

Only if the percutaneous approach fails is endoscopic extraction of stones from the common bile duct justified (Burhenne 1980).

In our series we obtained complete success in 65 out of 65 patients (100%); however, in one case it was necessary to convert the endoscopic to a transhepatic access, and in another case the percutaneous maneuvers required the collaboration of an endoscopist as sphincterotomy was necessary.

Only one complication occurred (1.5%): fissuring of the common bile duct at the site of entrance of the

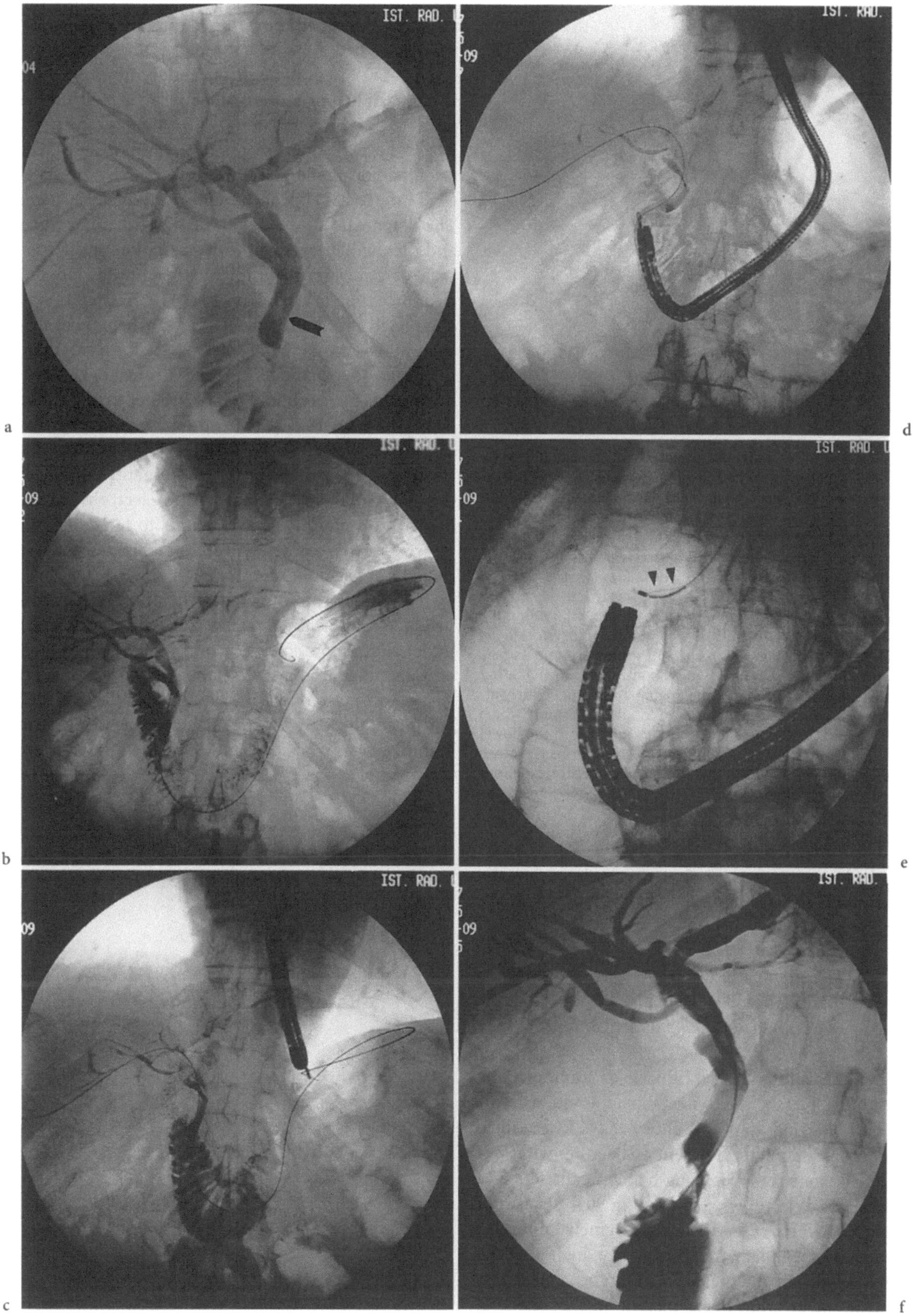

Fig. 19.4a–f

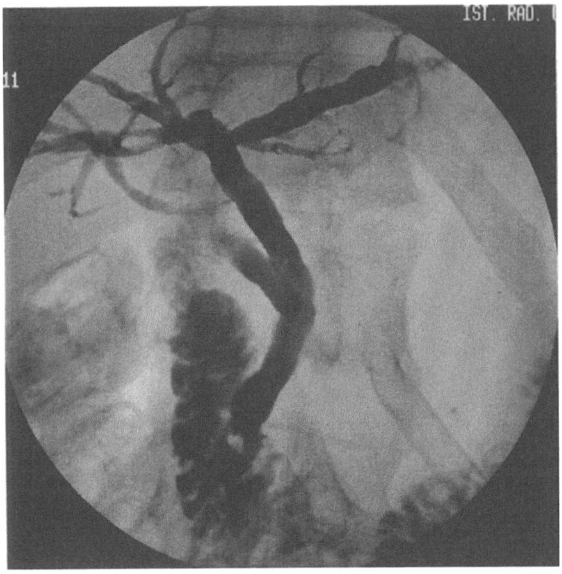

g

Fig. 19.4. **a** Percutaneous transhepatic cholangiography (PTC) of a patient who had previously undergone Billroth II gastric resection: small radiolucent gallstone in the common bile duct (*arrow*) and cicatricial stricture of the papilla. The intrahepatic bile ducts are not dilated. **b** The stricture of the papilla is passed and the guidewire advanced through the bowel loop to the gastric stump. **c** The guidewire is clamped by the biopsy forceps of a gastroscope previously introduced into the gastric stump. **d** The guidewire is withdrawn, pulling the endoscope up to the papilla. **e** A common sphincterotome is inserted percutaneously (*arrows*) and sphincterotomy is performed under endoscopic vision for control of the orientation of the instrument. **f** Control cholangiography immediately after the procedure demonstrates a wide sphincterotomy which allows rapid passage of the contrast medium into the bowel loop. **g** Control cholangiography after 48 h confirms clearance of the common bile duct; the stone has been already eliminated via the sphincterotomy

surgical tube. It was resolved by maintaining the transcystic catheter in place for 1 week.

19.4.1.2 Recurrent Lithiasis or Common Bile Duct Lithiasis Associated with Gallbladder Lithiasis

In patients with recurrent lithiasis or commons bile duct lithiasis associated with gallbladder lithiasis, direct cholangiography of choice is ERCP, since endoscopic sphincterotomy in association with mechanical stone removal using balloon catheters and Dormia baskets allows resolution in 90% of cases of lithiasis. In these cases, too, it is extremely important to opacify the biliary tree completely, including the intrahepatic bile ducts. This goal can be achieved by varying the position of the patient and by injecting the contrast medium under pressure, occluding the common bile duct above the papilla by means of a venous occlusion balloon catheter.

In about 5% of patients the endoscopic approach is unfeasible (e.g., due to sequelae of Billroth II gastrectomy or paravaterian diverticula), and PTC has to be performed to remove common bile duct calculi anterogradely. PTC is performed by the standard intercostal approach to the right ductal system. The diagnostic study includes the entire biliary tree and the papillary region, with evaluation of the passage of contrast medium into the bowel and the clearance time of the biliary tree. The examination is completed by the placement of a transhepatic drainage catheter, which may be external or external-internal. The diameter of the drainage catheter during this phase should not be greater than 7 F, but it may be increased at the end of the following session, 5–7 days after the first cholangiography, according to the caliber of the instruments employed (balloon catheters, cholangioscope, electrohydraulic lithotriptor, etc.). Recently, thanks to technological improvements, catheters with a diameter greater than 9 F are seldom required.

If a sphincterotomy is required, it may be percutaneous. However, the maneuver has to be performed under endoscopic control, requiring the cooperation of a gastroenterologist. In patients who have had a gastrectomy, a percutaneously inserted guidewire is advanced through the papilla towards the gastric stump, where it may be clamped by the foreign-body

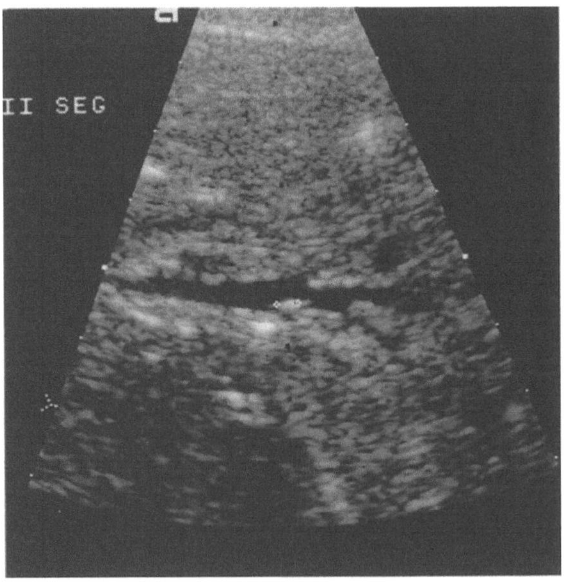

Fig. 19.5. Ultrasound targeting of the third segmental duct, which is dilated and contains a small stone

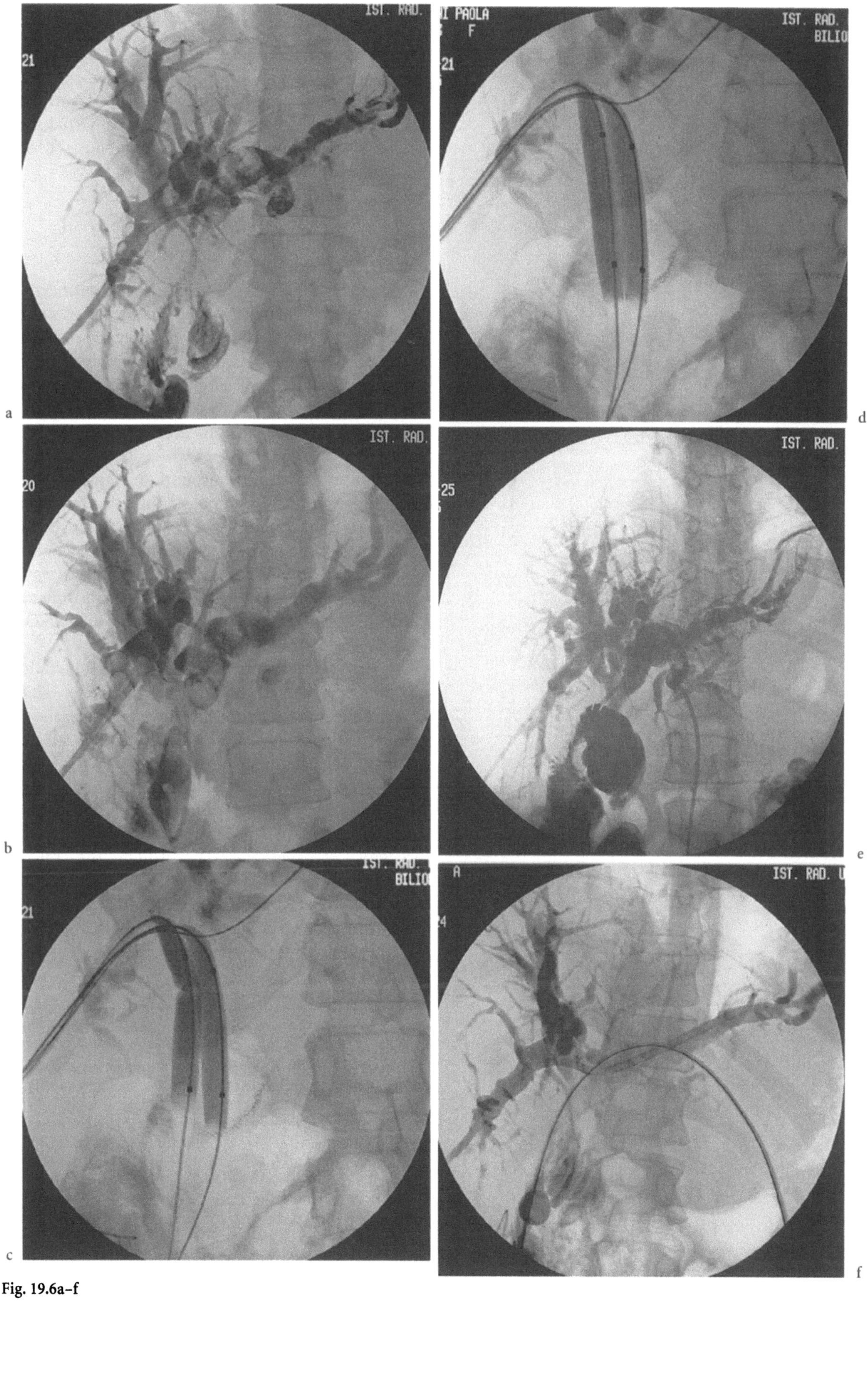
IST. RAD.
21
a
PAOLA
F
IST. RAD.
BILIO
21
d
IST. RAD.
20
b
IST. RAD.
25
e
IST. RAD.
BILIO
21
c
A
IST. RAD. U
24
f

Fig. 19.6a–f

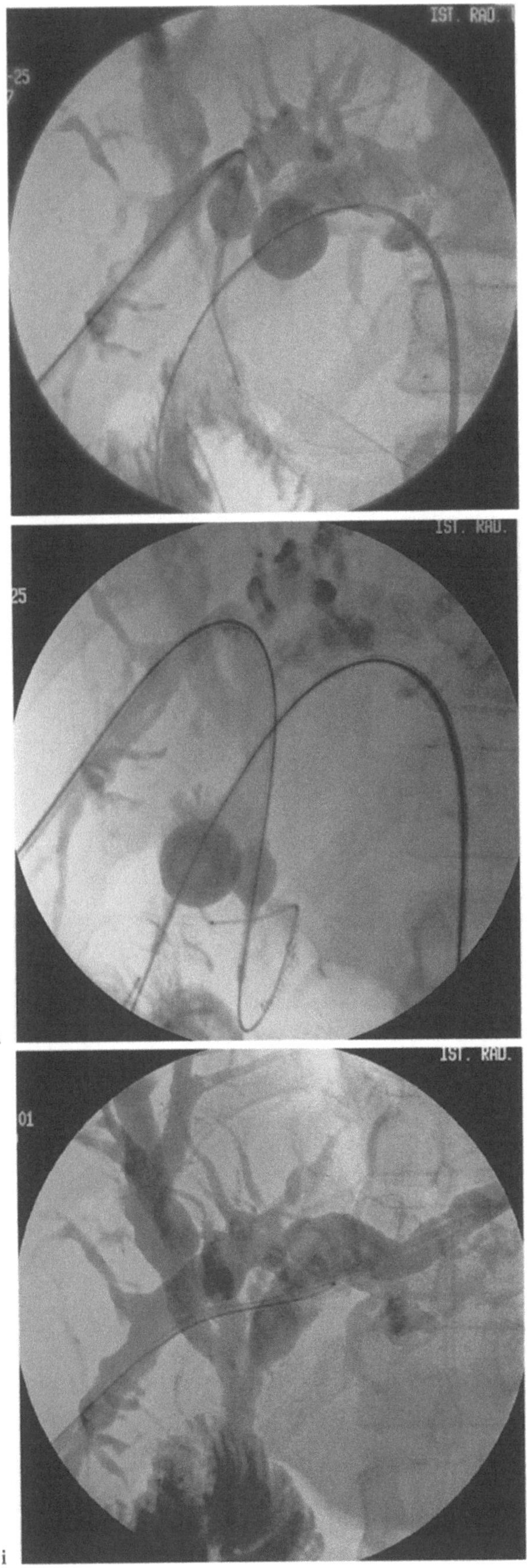

g

h

i

forceps of the fiberscope, which is then drawn by the guidewire to the papilla (Fig. 19.4) (Gandini et al. 1980). In this group of patients we obtained complete success in 51 of 56 patients (91.1%) and partial success in 3 of 56 (5.3%). There was one failure (1.8%) and one death (mortality 1.8%).

19.4.2 Intrahepatic Lithiasis

In patients with intrahepatic lithiasis the approach of choice is percutaneous transhepatic; in patients with a bilioenteric anastomosis it is the only approach. In these cases, as mentioned above, preinterventional diagnostic evaluation is essential in order to choose the right number and routes of percutaneous access, on which depends the success or otherwise of treatment. In regard to this, we again stress the importance of an ultrasound examination in the interventional radiology suite immediately before starting the intervention, with patient already in the position required for percutaneous cholangiography. This is especially important in cases in which the scheduled percutaneous access is unusual, difficult, or has to be directed to the only dilated duct (Fig. 19.5).

During percutaneous cholangiography, the filling of the biliary tree has to be complete, since nonvisualization of ducts may indicate the presence of obstructing stones at the ostium. On this subject, it may be advantageous to perform CT cholangiography 24–48 h after the percutaneous cholangiography, injecting dilute contrast medium (3 mg% iodine, the quantity evaluated during the previous cholangiography) through the biliary drainage catheter(s), and using the spiral scanning technique (with 7-mm slices and a pitch of 1).

Percutaneous cholangiography is also essential for the evaluation of strictures, including of seg-

◄

Fig. 19.6. a PTC demonstrates numerous calculi in the intrahepatic bile ducts above a tight stricture of the bilioenteric anastomosis. An external drainage catheter is positioned with its tip in the left ductal system. **b** By means of hydrophilic guidewires the stenotic anastomosis is passed and an external-internal biliary drainage catheter left in place for a few days in order to resolve the cholangitis. **c** Dilatation of the bilioenteric anastomosis is performed, using two 1-cm-diameter balloon catheters. **d** The stricture yields completely at an inflation pressure greater than 15 atm. **e** A percutaneous approach to the left ductal system is performed to achieve elimination of the larger calculi. **f–h** Numerous stone removal maneuvers are performed using two venous occlusion balloon catheters simultaneously. **i** After repeated flushing with the Broxo-Jet, complete clearance of the biliary tree is achieved mental ducts, which are often associated with intrahepatic lithiasis, such as in acute cholangitis and secondary sclerosing cholangitis. The presence of one or more of these factors affects the timing, duration and method of subsequent treatment.

As mentioned above, transhepatic cholangiography must be followed by the placement of a biliary drainage catheter, which may be external or external-internal (across the papilla or trans-anastomotic). If intrahepatic stones are located above a strictured bilioenteric anastomosis, which is the most common cause, the stenosis should be passed and if possible dilated during the first session. Flushing with saline then allows elimination of at least the smaller calculi or stone fragments. In many cases it is easier to pass the anastomotic stricture by a left epigastric approach.

The caliber of the drainage catheters positioned after the cholangiography may be from 7 to 9 F, depending on whether an bilioenteric anastomosis has been dilated. In the positioning of drainage catheters it must be kept in mind that obstructed ducts and those containing infected bile have to be drained separately.

During the resolution of intrahepatic lithiasis, progressive reduction in the size of the catheter prior to its definitive removal may be helpful, while if more than one catheter has been placed, they should be removed during subsequent sessions, at 24-h intervals (Fig. 19.6).

19.4.3 Intra- and Extrahepatic Lithiasis

In patients with intra- and extrahepatic lithiasis a combined endoscopic-radiologic approach is the best proposition. In the most simple cases the cooperation of an endoscopist is the only adjunctive intervention required: previous endoscopic

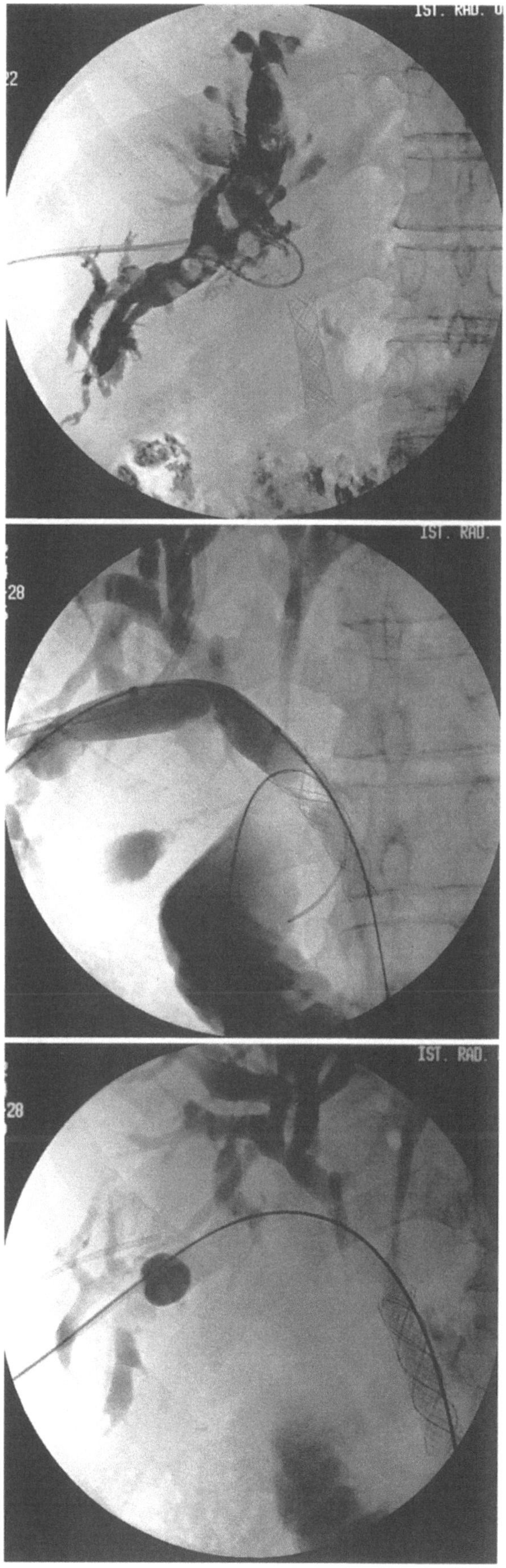

a

b

c

Fig. 19.7. **a** In this case of combined intra- and extrahepatic lithiasis, the left intrahepatic ducts and the common bile duct have been cleared endoscopically and a self-expandable metallic stent has been placed; the endoscopic approach to the right ductal system was unfeasible. PTC was therefore performed, demonstrating numerous calculi obstructing the right hepatic duct. **b** Percutaneous bilioplasty of the right hepatic duct is performed, using a 1-cm balloon catheter. **c,d** The first stones are removed by means of venous occlusion balloon catheters. **e** Cholangiogram reveals a residual stricture of the left hepatic duct; a percutaneous approach to this duct is therefore established. **f** Bilioplasty of the left hepatic duct is performed, using a 1-cm balloon catheter. **g,h** Lithotripsy is completed by repeated flushing and simultaneous passage of two venous occlusion balloon catheters. **i** Control cholangiogram demonstrates complete resolution of lithiasis

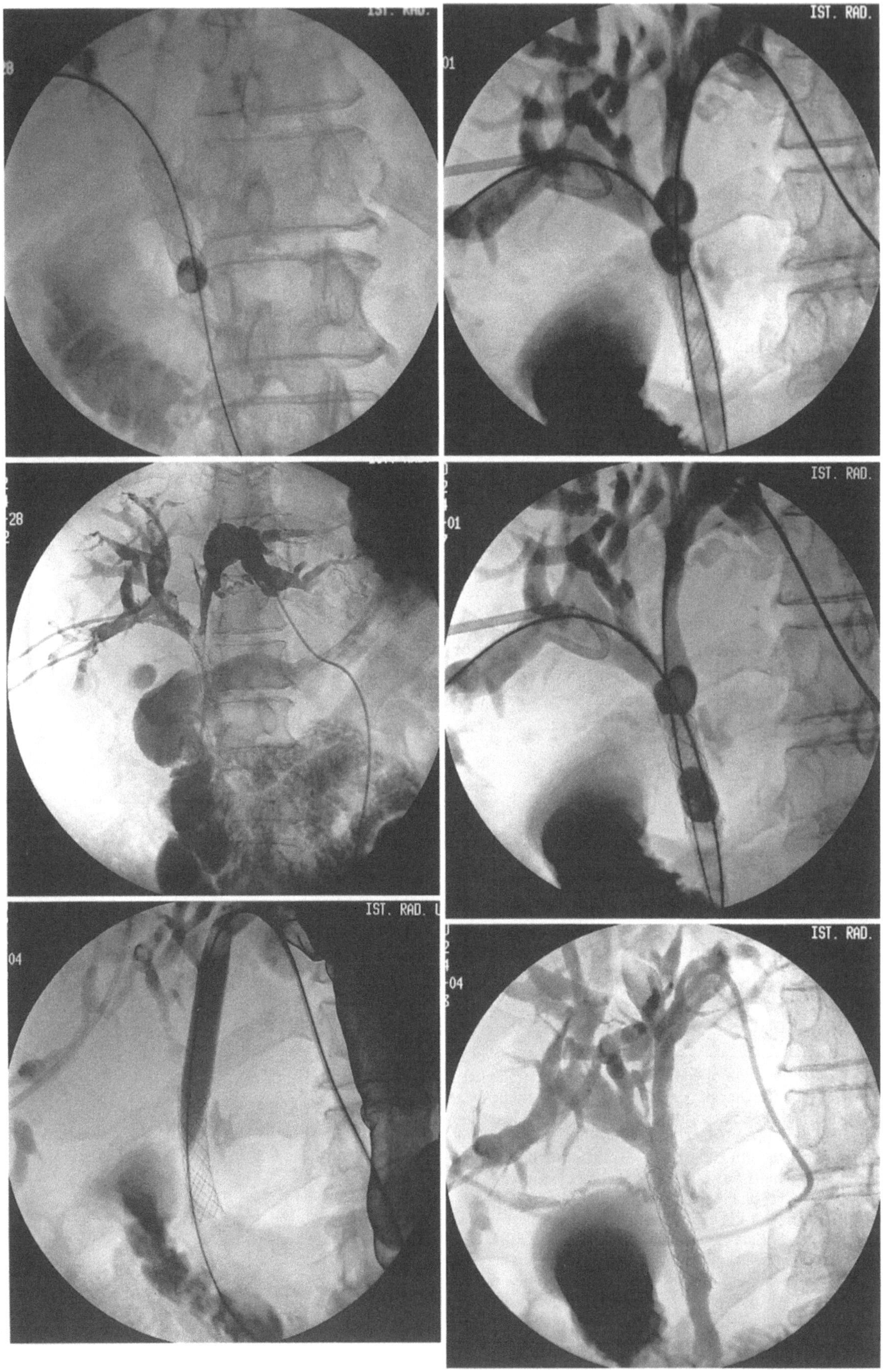

d e f g h i

Fig. 19.7d–i

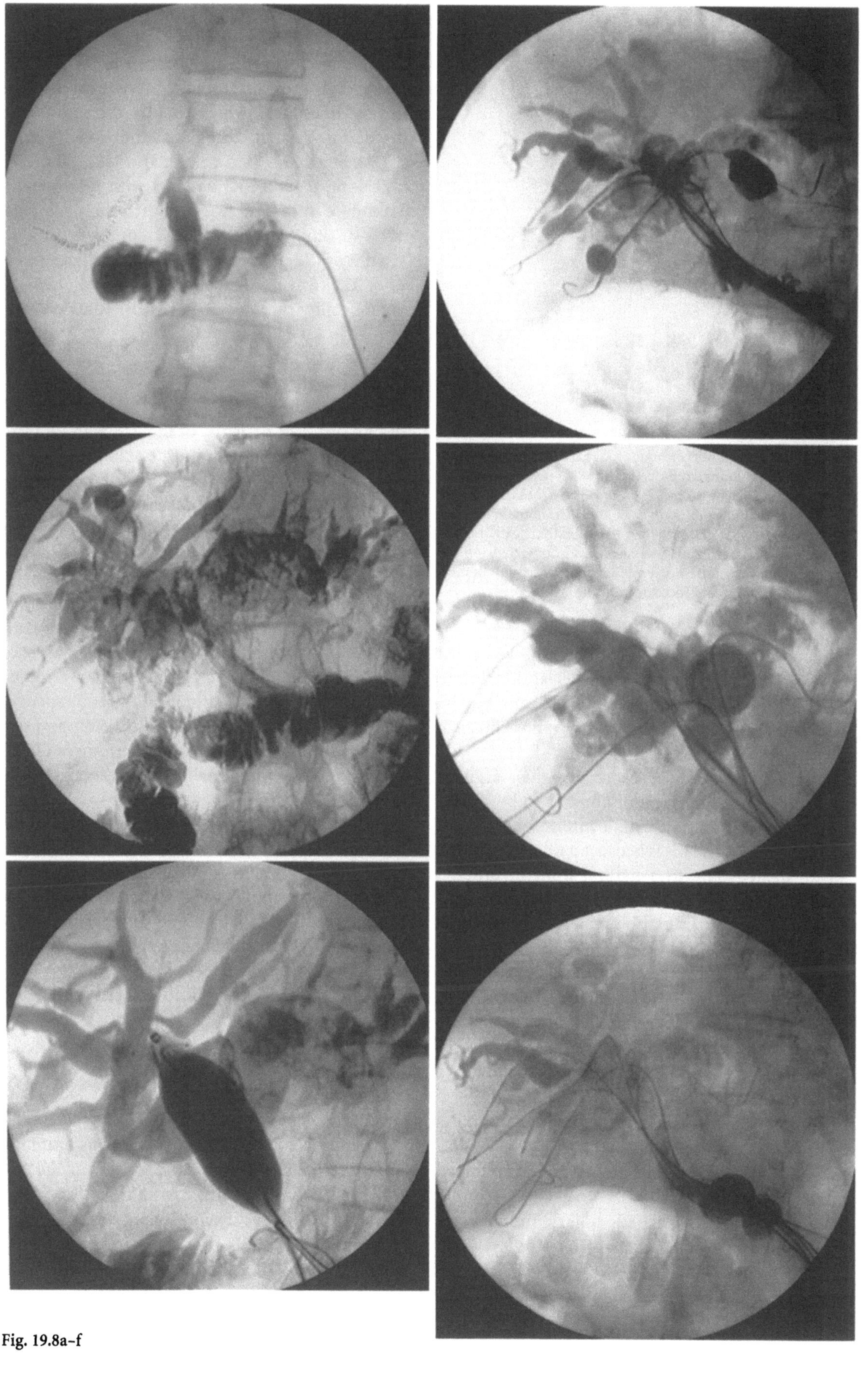

a b c d e f

Fig. 19.8a–f

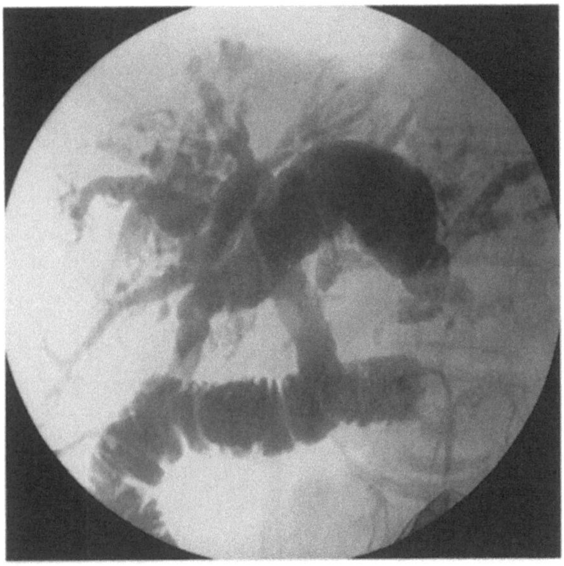
g

Fig. 19.8. a After intervention of hepatico-jejuno-cutaneostomy, the blind loop, fixed under the skin, is percutaneously catheterized. The anastomosis is wide. b Cholangiography demonstrates numerous stones in the intrahepatic ducts. c The anastomosis is dilated with a 2-cm balloon catheter. d–f Stone removal is performed, using five venous occlusion balloon catheters simultaneously. g Final cholangiogram demonstrates almost complete resolution of lithiasis

sphincterotomy allows stones to be removed retrogradely from the common bile duct and facilitates antegrade lithotripsy of intrahepatic calculi, bearing in mind what was said above as to the choice of percutaneous accesses (Fig. 19.7).

In more complex cases, especially when large calculi are present, a surgical option has to be considered, which consists in a bilioenteric anastomosis and stone removal. This is because a normal bile duct can be dilated only to a natural limit in order to eliminate a stone transhepatically or endoscopically, while an anastomosis can be greatly overdistended (Choi et al. 1982).

When intra- and extrahepatic lithiasis recurs after a "low" bilioenteric anastomosis, the endoscopic approach is naturally impossible. This situation, however, is not substantially different to that of intrahepatic lithiasis in the presence of a bilioenteric anastomosis, apart from the fact that the stump of the common bile duct has to be taken into account in the diagnosis and treatment.

Finally, in the particular case of massive lithiasis, surgery is essential and must precede every other maneuver except transhepatic biliary drainage performed to resolve cholangitis. The surgical intervention includes open bile duct revision and performance of a hepatico-jejuno-cutaneostomy; the blind loop fixed subcutaneously can be punctured under ultrasound guidance or with fluoroscopy, if metallic sutures are employed, in order to create a retrograde transanastomotic access to the intrahepatic bile ducts, which enables the radiologist or the endoscopist to complete lithotripsy (Fig. 19.8) (Chao et al. 1987; Morino et al. 1988).

Taking together all patients with intrahepatic or combined intra- and extrahepatic lithiasis, including massive lithiasis, we achieved complete success in 156 out of 172 patients (90.7%) and partial success in 11 of 172 (6.4%). Two failures occurred (1.2%), and the mortality rate was 1.7% (Righi et al. 1992).

19.5 Complications

In our series of patients, of the 228 who underwent transhepatic maneuvers (i.e., excluding the group with residual lithiasis), 44 suffered complications (19.3%). These were: 6 cases (13.6%) of acute pancreatitis requiring prolonged hospital stay; 1 case (2.3%) of acute renal failure, which resolved with dialysis; 11 (25%) biliomas, treated percutaneously under ultrasound or CT guidance; 3 (6.8%) intracapsular hematomas, which resolved spontaneously but required a prolonged hospital stay; 6 (13.6%) intraparenchymal hematomas treated by angiographic embolization of pseudoaneurysms (one patient required blood transfusion and observation in the intensive care unit); 7 cases (16%) of fissuring of the common bile duct, resolved by maintaining the biliary drainage catheter in place for 3–7 days; 3 (6.8%) hepatic abscesses, treated by means of ultrasound- or CT-guided drainage; 1 (2.3%) splenic abscess requiring CT-guided drainage; and 6 cases (13.6%) of hemobilia, 3 of which were treated by plasma substitutes, 2 by blood transfusion, and 1 by blood transfusion and angiographic embolization of a pseudoaneurysm.

None of these complications required surgical reintervention.

19.6 Conclusion

The approach to bile duct lithiasis should be multidisciplinary: interventional radiology is in some cases complementary to endoscopy or surgery, while in other cases it represents a valid alternative.

In residual lithiasis after cholecystectomy with a T tube in place, it is the therapeutic approach of choice: as mentioned above, a surgically positioned T tube is an ideal access to the common bile duct, allowing easy removal of residual stones.

In intrahepatic lithiasis, too, the percutaneous approach is the first choice. Here, accurate diagnostic evaluation beforehand is essential to identify the ducts involved for subsequent catheterization.

Beyond the classification of bile duct lithiasis based on topographic criteria, which we employ for comprehensive exposition of the subject, obviously every single patient has to be considered an individual case: for each, the best therapeutic approach should be discussed by surgeon, endoscopist, and radiologist together, cooperating as a team with respect for the specific competences and skills of each member. In this regard, it would be desirable to create increasingly specialized centers for the treatment of biliary pathology, in which the expertise of the operators involved and the technological progress in materials and equipment would guarantee constantly improving results.

References

Allen MJ, Borodj TJ, Bugliosi TF, et al (1985) Cholelitholysis using MTBE. Gastroenterology 88:122–125

Barbara L for the "Progetto Sirmione" (1984) Epidemiology of gallstone disease: the Sirmione study. In: Capocaccia L, Ricci G, Angelico F, Angelico M, Attili AF (eds) Epidemiology and prevention of gallstone disease. MTP, Lancaster

Bennion L, Knowler W, Mott D, et al (1979) Development of lithogenic bile during puberty in Pima Indians. N Engl J Med 300:873–876

Burhenne HJ (1980) Percutaneous extraction of retained biliary tract stones: 661 patients. AJR 134:888–898

Burhenne HJ, Becker CD, Malone DE, et al (1989) Biliary lithotripsy: early observations in 106 patients. Radiology 171:363–367

Capocaccia L, Ricci G (1985) Epidemiology of gallstone disease. Ital J Gastroenterol 17:215–218

Cesarani F, Gandini G, Righi D, et al (1988) L'importanza dell'idrogetto pulsato ad alta frequenza nel trattamento della calcolosi delle vie biliari. Radiol Med (Torino) 76:453–457

Chao Z, Tian F, Gao B, et al (1987) Diagnosis and management of intrahepatic retained stones through a subcutaneously placed afferent loop of Roux en Y choledochojejunostomy. Clin Med J 100(7):523–526

Choi TK, Wong J, Ong GB (1982) The surgical management of primary intrahepatic stones. Br J Surg 69:86–90

Covarrubias C, Valdivieso V, Nervi F (1983) Epidemiology of gallstone disease in Chile. In: Capocaccia L, Ricci G, Angelico F et al. (eds) Epidemiology and prevention of gallstone disease. MTP, Lancaster

Fonio P, Maisano U, Giacometti R, et al (1990) La colecistostomia percutanea: indicazioni e tecnica, Atti XXXIV Congresso Nazionale SIRM, Torino, p 108

Gandini G, Zanon E, Righi D, et al (1980) La sfinterotomia percutanea transepatica. Radiol Med (Torino) 90:893–897

Gandini G, Bonardi L, Cesarani F, et al (1988a) The use of methyl-ter-butyl-ether (MTBE) in the therapy of biliary cholesterinic lithiasis. J Intervent Radiol 3:150–157

Gandini G, Cesarani F, Juliani E, et al (1988b) Percutaneous transhepatic cholangioscopy with a 2.8 mm fiberscope. Endoscopy 30:114–117

Gandini G, Righi D, Regge D, et al (1990) Percutaneous removal of biliary stones. Cardiovasc Intervent Radiol 13:245–251

Juliani E, Righi D, Cesarani F, et al (1988) Sull'impiego di un litotritore elettroidraulico nel trattamento percutaneo della calcolosi delle vie biliari. Esperienza clinica preliminare in 4 casi. Radiol Med (Torino) 76:448–452

Juliani G, Gandini G (1988) La radiologia interventistica nelle malattie non neoplastiche delle vie biliari. Rapporto su 93 casi personali. Radiol Med (Torino) 76:448–452

Maki T, Matsushiro T, Suzuki N (1982) Classification of the nomenclature of pigment gallstones. Am J Surg 144:303–305

Mazzariello R (1976) Residual biliry tract stones: nonoperative treatment of 570 patients. Surg Annu 8:113–114

Morino F, Fronda GR, Toppino M, et al (1988) Risoluzione di litiasi intraepatica massiva mediante epatico-digiuno-cutaneo-stomia. Chirurgia 1:153–155

Ort K, Haas S, Oettinger W, et al (1990) Das Risiko der elektiven Cholezystektomie. Z Gastroenterol 28:616–620

Righi D, Fonio P, Cristina MC, et al (1990) La bilioplastica percutanea transepatica: risultati a distanza. Radiol Med (Torino) 80:492–500

Righi D, Fonio P, Gandini G, et al (1992) Trattamento percutaneo della litiasi dei dotti biliari. Esperienza personale nei primi 150 casi. Radiol Med (Torino) 83:526–534

Ring EJ, Kerlan RK (1984) Interventional biliary radiology. AJR 142:31–34

20 Role of Interventional Radiology in Laparoscopic Injuries of the Bile Duct*

A.L. Zerbey, S.L. Dawson, and P.R. Mueller

CONTENTS

20.1 Introduction

In the years since its introduction in 1987, laparoscopic cholecystectomy has gained widespread acceptance among both surgeons and their patients. With this increased use has come an increased incidence of associated injuries. This chapter discusses the interventional radiological management of bile duct injuries due to laparoscopic surgery, both as primary treatment of an injury and as a temporizing measure in anticipation of future open surgical repair.

Approximately 500 000 cholecystectomies are performed annually in the United States (Crist and Gadacz 1993; Davidoff et al. 1992; Donohue et al. 1992; Graber et al. 1992; Larson et al. 1992; Ress et al. 1993; Southern Surgical Club 1991). The majority of these procedures are now being performed using laparoscopic techniques. At the Mayo Clinic, for example, 80% of cholecystectomies for uncomplicated cholecystitis are performed using laparoscopic techniques (Donohue et al. 1992; Larson et al. 1992; Ress et al. 1993).

Part of the appeal of laparoscopic cholecystectomy is the assumption that the apparently simpler technique leads to fewer complications. This does not necessarily happen. While the true incidence of injury during laparoscopic cholecystectomy is unknown, reported incidences of bile duct injury in several single-institution experiences range from 0.2% to 0.8% (Crist and Gadacz 1993; Davidoff et al. 1992; Donohue et al. 1992; Graber et al. 1992; Larson et al. 1992; Ress et al. 1993; Southern Surgical Club 1991). Indeed, because of the increase in the number of cholecystectomies performed, the absolute number of complications occurring may actually have increased. In an attempt to analyze the source of complications associated with laparoscopic cholecystectomy, the Southern Surgical Club prospectively analyzed 1518 cases of laparoscopic cholecystectomy. They noted that bile duct injury occurred during laparoscopic cholecystectomy in an average of 2.2% of patients of a surgeon's initial 13 procedures. This rate declined to 0.1% following the thirteenth procedure (Southern Surgical Club 1991). This strongly suggests a correlation between operator experience and the rate of biliary injury. In a similar study of 1983 laparoscopic cholecystectomies, 3 of 5 bile duct injuries occurred during a surgeon's first 10 laparoscopic cholecystectomies. This also suggests a correlation between operator inexperience and bile duct injury (Larson et al. 1992).

Mortality rates for laparoscopic cholecystectomy are in the range of 0–0.9%; these are comparable with the mortality rates of 0–0.5% following open cholecystectomy (Crist and Gadacz 1993). Still, a number of deaths have been reported as directly related to technical complications of laparoscopic surgery. A national survey of 77 604

A.L. Zerbey, MD, Department of Radiology, Massachusetts General Hospital, Harvard Medical School, 32 Fruit Street, Boston, MA 02114, USA
S.L. Dawson, MD, Department of Radiology, Massachusetts General Hospital, Harvard Medical School, 32 Fruit Street, Boston, MA 02114, USA
P.R. Mueller, MD, Department of Radiology, Massachusetts General Hospital, Harvard Medical School, 32 Fruit Street, Boston, MA 02114, USA

* Reprinted in part with the permission of Thieme Medical Publishers, Inc., from *Semin Intervent Radiol* (1996)

laparoscopic cholecystectomies showed that slightly more than half of the total number of deaths (18/33, 54%) were related to technical complications occurring during the procedure (CRIST and GADACZ 1993). By contrast, deaths following traditional cholecystectomy usually occur as a result of underlying disease and related complications (CRIST and GADACZ 1993).

20.2 Types of Bile Duct Injury

20.2.1 Ligation and Transection

Laparoscopic cholecystectomy is performed in a confined space with limited exposure and visualization of the involved structures. This makes identification of structures to be ligated and transected difficult, and contributes to the incidence of bile duct injury during cholecystectomy. This incidence is made greater by the presence of biliary duct anomalies (BERCI 1992) (Figs. 20.1, 20.2).

Traditional teaching holds that in 75% of cases the cystic duct enters the common bile duct laterally and at an angle. Anomalous insertions, entering the common bile duct in parallel, posterior, or spiral fashion, are thought to be far less common. However, BERCI (1992) found in a series of several thousand cholangiograms that the classic lateral insertion was present in only 17% of cases, and that in the overwhelming majority the cystic duct drained into the common bile duct in a spiral form (35%), posteriorly (41%), or in parallel (7%) fashion. These anomalous insertions of the common bile duct can lead to misidentification during cholecystectomy. Similarly, aberrant drainage of right hepatic ducts into the gallbladder and small ducts from the gallbladder bed (ducts of Luschka) can be interrupted during surgery, resulting in a biliary leak (Figs. 20.1, 20.2).

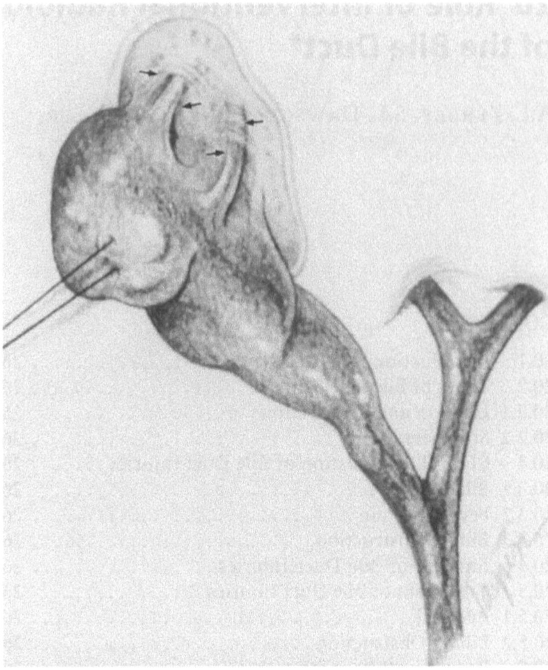

Fig. 20.2. Schematic drawing demonstrating ducts of Luschka extending from the gallbladder fundus into the hepatic bed (*arrows*). Upon removal of the gallbladder these ducts may leak if not seen at surgery and appropriately sutured

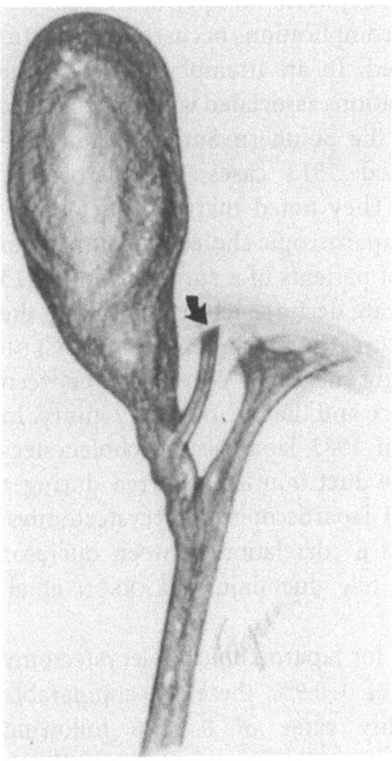

Fig. 20.1. Schematic drawing of an accessory intrahepatic duct draining into the cystic duct (*arrow*). At laparoscopic surgery this may be difficult to visualize and may cause a postoperative biloma if the duct is cut

Anomalous coursing of the extrahepatic bile ducts can also lead to confusion. For example, an aberrant right hepatic duct can travel posterior to the gallbladder and be misidentified at surgery. Inadvertent clipping of this duct will lead to segmental intrahepatic biliary obstruction; inadvertent transection of the duct will cause bile leakage (Fig. 20.3).

In the normal patient, the most common injury involves confusion of the common hepatic duct and common bile duct with the cystic duct. Ligation and transection of the common hepatic duct and common bile duct, with injury to the right hepatic artery coursing behind the common bile duct, occur as a result of this misidentification. The unfortunate results of this include biliary obstruction and/or bile leak and sacrifice of the right hepatic artery. A variant of this classic injury includes placement of proximal clips on the cystic duct and distal clips on the common duct. Division of the cystic duct between these two sites results in common bile duct obstruction below the cystic duct with decompression of the biliary system into the peritoneum via the open cystic duct stump. Another variation on this theme is seen in patients with a low bifurcation of the common hepatic duct. In these cases the right hepatic duct can be mistaken for the cystic duct; clipping of the duct leads to right hepatic biliary obstruction and/or leak.

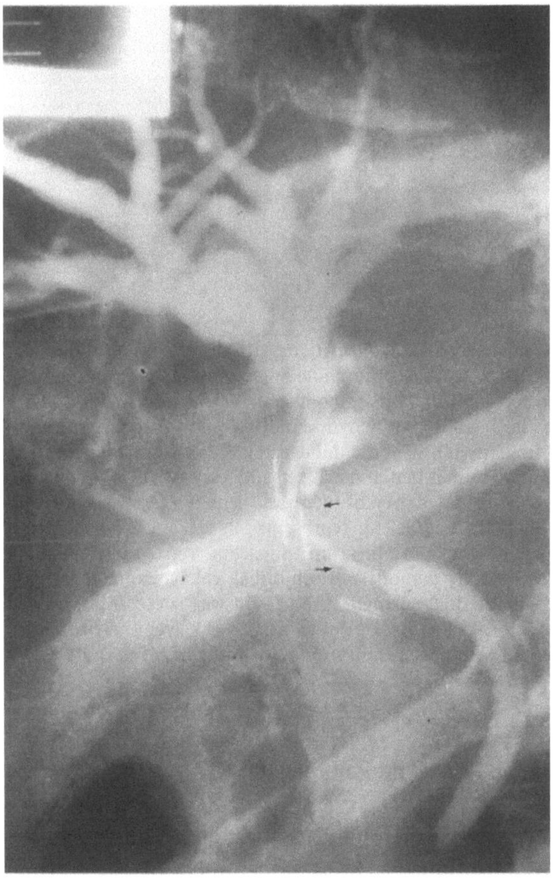

Fig. 20.3. Transhepatic cholangiogram of a 47-year-old woman with abdominal pain and elevated bilirubin level 10 days after laparoscopic cholecystectomy. Intrahepatic ducts are dilated and there are multiple surgical clips which were placed during the laparoscopic procedure. There is a long stricture (*arrows*) of the mid common bile duct with a normal duct distally. This injury is typical of thermal injury caused by laser electrocautery during laparoscopic surgery

Less common complications of laparoscopic cholecystectomy include incomplete resection of the gallbladder with a chronic sustained leak, and leakage of bile from the cholecystectomy bed via the intrahepatic ducts of Luschka (MARTIN and ROSSI 1994; DAWSON and MUELLER 1994; KOZAREK 1994; LEE et al. 1991; RAY et al. 1993). Both of these conditions result in the formation of bilomas; the latter will tend to close spontaneously while the former tend to continue to drain. Cystic duct stump leaks in otherwise uncomplicated cholecystectomies can also occur as a result of the cystic duct clip falling off after surgery. This leads to leakage of bile from the liver via the cystic duct, and can lead to bilomas. If these are small they can be cured by endoscopic drainage (Fig. 20.4) or a combination of percutaneous biloma drainage and endoscopy. Other complications, including hemorrhage, wound infection, and bile duct injury repaired at the time of surgery, fall within the purview of the surgeon and are not discussed in this chapter.

20.2.2 Strictures

Bile duct strictures are another complication of laparoscopic cholecystectomy (DAWSON and MUELLER 1994; KOZAREK 1994; LEE et al. 1991; RAY et al. 1993; STOKES 1994; TREROTOLA et al. 1992; VAN SONNENBERG et al. 1990, 1993; WRIGHT et al. 1993). They can be caused by thermal damage from both laser and electrocautery, or occur as a late complication of vascular denudation of the involved segment of duct.

Strictures of varying degree were found in 68% (13 of 19 patients) treated at the Mayo Clinic for complications of laparoscopic surgery (DONOHUE et al. 1992). Late onset of jaundice is typical of this injury, presenting weeks to months after the initial injury (Fig. 20.5).

20.3 Clinical Presentation of Bile Duct Injuries

Patients with ductal injuries unrecognized at surgery can present hours to days following the injury. The patient's symptoms can give clues to the underlying abnormality.

20.3.1 Bile Leak

Bile leaks present in two ways (DAWSON and MUELLER 1994; KOZAREK 1994; LEE et al. 1991; RAY

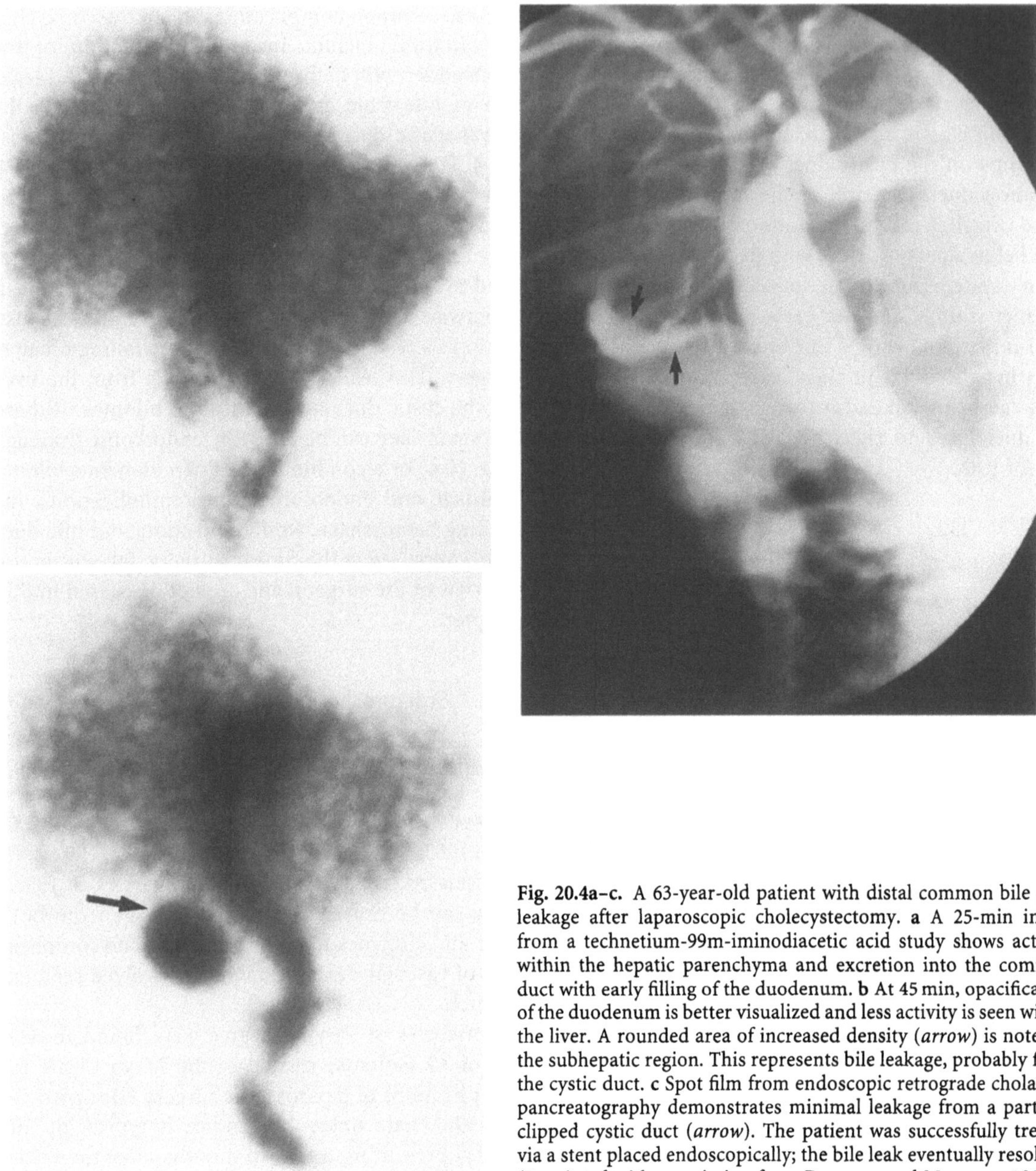

a b c

Fig. 20.4a–c. A 63-year-old patient with distal common bile duct leakage after laparoscopic cholecystectomy. **a** A 25-min image from a technetium-99m-iminodiacetic acid study shows activity within the hepatic parenchyma and excretion into the common duct with early filling of the duodenum. **b** At 45 min, opacification of the duodenum is better visualized and less activity is seen within the liver. A rounded area of increased density (*arrow*) is noted in the subhepatic region. This represents bile leakage, probably from the cystic duct. **c** Spot film from endoscopic retrograde cholangio pancreatography demonstrates minimal leakage from a partially clipped cystic duct (*arrow*). The patient was successfully treated via a stent placed endoscopically; the bile leak eventually resolved. (Reprinted with permission from Dawson and Mueller 1994)

et al. 1993; Stokes 1994; Trerotola et al. 1992; van Sonnenberg et al. 1990, 1993; Wright et al. 1993). Bile peritonitis presents as increasing and sometimes severe abdominal pain, often referred to the right shoulder. This appears in the postoperative period, often as quickly as hours following surgery. A subhepatic or subphrenic abscess can mimic bile peritonitis, but usually takes days to develop. By contrast, bile ascites is often asymptomatic, announcing itself over the course of days or weeks by the patient's increasing girth.

20.3.2 Fever and Pain

Fever is a potentially useful sign of postoperative complications, as it is uncommon in the normal laparoscopic cholecystectomy patient. Only 4% of patients at the Mayo Clinic who underwent an uncomplicated cholecystectomy developed fever in the first several days after surgery (Larson et al. 1992).

Similarly, pain can be a sign of complication in laparoscopic cholecystectomy. Laparoscopic cholecystectomy is usually marked by rapidly decreasing

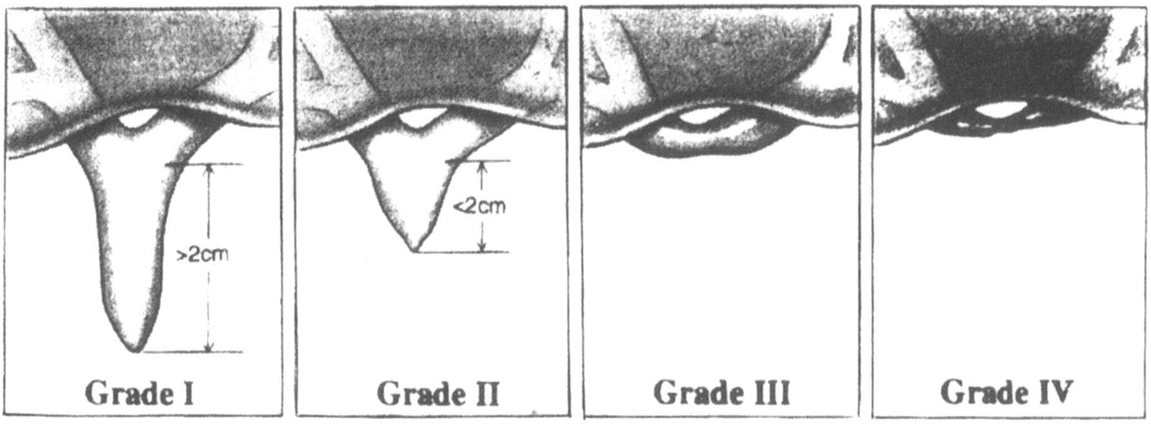

Fig. 20.5. Bismuth classification of bile duct strictures as described by BLUMGART et al. (1984). This classification can be applied to laparoscopic injuries of the bile duct. It can be important to recognize the different grades (I-IV) in order to perform the correct intervention. (Reprinted with permission from MARTIN and ROSSI 1994)

postoperative pain. The presence of protracted pain, usually localized to the right upper quadrant, or of severe or increasing pain is a sign of a serious problem. Further investigation to evaluate for complications is warranted in these cases.

20.3.3 Biliary Obstruction

Increasing jaundice is the hallmark of biliary ductal obstruction (DAWSON and MUELLER 1994; KOZAREK 1994; LEE et al. 1991; RAY et al. 1993; STOKES 1994; TREROTOLA et al. 1992; VAN SONNENBERG et al. 1990, 1993; WRIGHT et al. 1993). This will appear days to weeks after surgery. Laboratory evaluation of liver enzymes, alkaline phosphatase, and the bilirubin levels can provide objective evidence of biliary obstruction. Any of the above symptom complexes can present with fever.

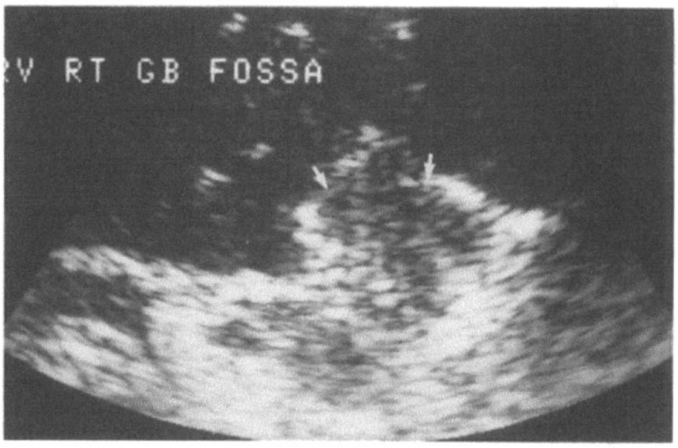

Fig. 20.6. Typical appearance of a gallbladder fossa after successful laparoscopic surgery. Right upper quadrant ultrasonogram demonstrates echogenic collection in the gallbladder fossa (*arrows*). It is not uncommon to have a small collection of fluid or a hematoma within the fossa. These invariably regress and do not require treatment

20.4 Imaging of Bile Duct Injuries

Patients suspected of post-laparoscopic-cholecystectomy injuries who are not immediate surgical candidates are best served by an initial abdominal ultrasound examination (Fig. 20.6) (DAWSON and MUELLER 1994; LEE et al. 1991; RAY et al. 1993; STOKES 1994; TREROTOLA et al. 1992; VAN SONNENBERG et al. 1990, 1993; WRIGHT et al. 1993). This permits evaluation of the liver for intrahepatic biliary dilatation due to bile duct ligation or stricturing, and evaluation of the abdomen for fluid due to bile leaks, bile ascites, bleeding, and abscess. In this way the major complications of laparoscopic bile duct injuries can be quickly and inexpensively screened for and assessed. This examination also permits transabdominal fluid aspiration to be sent for microbiological, chemical, and hematological analysis.

CT scanning can provide similar information and has the advantage of viewing a more complete area of the abdomen as well as visualizing areas difficult to see by routine ultrasonography, such as the pelvis and the retroperitoneum. Both ultrasound and CT afford the opportunity to perform therapeutic aspiration or drainage immediately an abnormal area is visualized (Fig. 20.7).

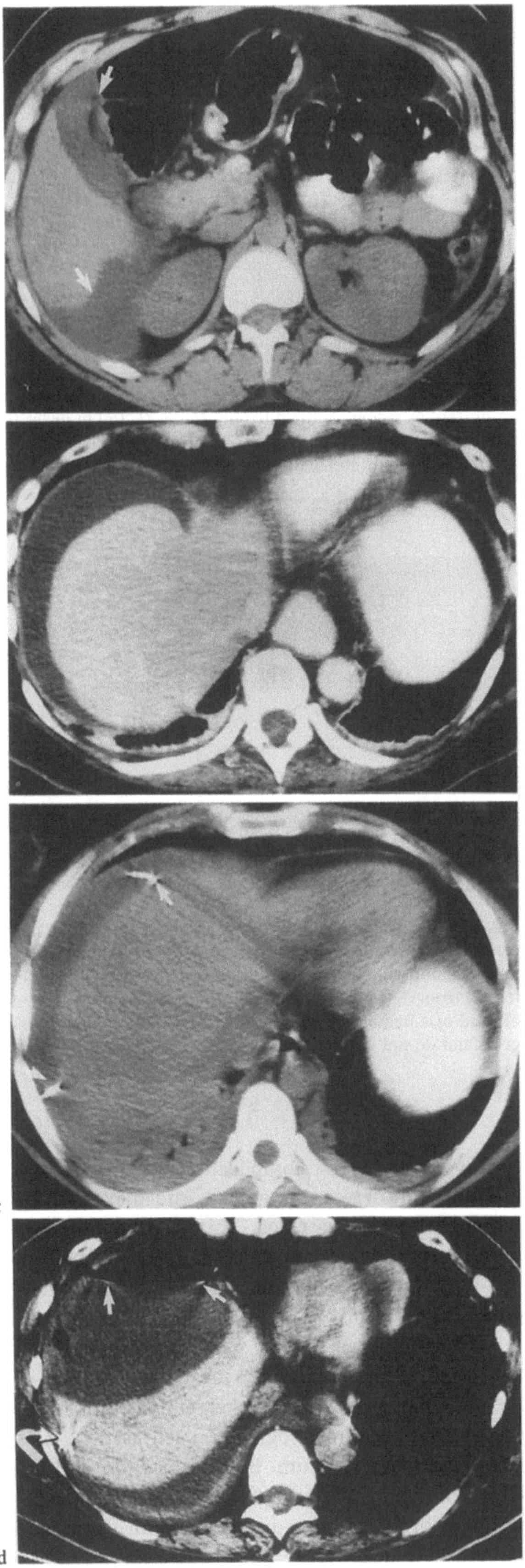

a b c d

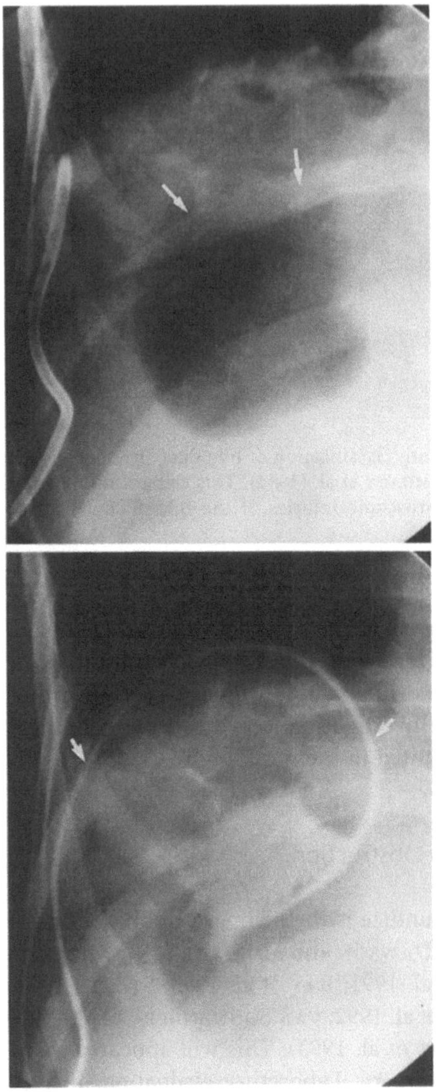

e f

Fig. 20.7a–f. CT scan obtained 1 week after a laparoscopic gallbladder removal. **a** Fluid is seen in the gallbladder fossa (pseudogallbladder sign) (*arrow*). There is also fluid in the Morison's pouch in the subhepatic space (*arrow*). **b** The fluid extended superiorly into the subphrenic space around the liver. **c** A guidewire and catheter were placed via CT guidance (*arrows*) and 400 ml bilious material was removed. The catheter was left in place for 3 weeks. The patient did well initially. **d** CT scan 10 days after the catheter was placed. The patient had spiked a fever to 103° after several days of drainage. The CT examination demonstrates a large air-fluid level over the dome of the liver (*arrows*). The catheter (*curved arrow*) is positioned inferior to the collection. **e** Plain film demonstrates the position of the air-filled abscess (*arrows*) medial and anterior to the catheter. **f** Under fluoroscopic guidance the lateral catheter was manipulated into the collection and a large catheter was placed (*arrows*). The patient responded to treatment and after 10 more days of drainage the loculated biloma was drained

We do not routinely use nuclear medicine studies to evaluate the injured liver and biliary ductal system (Dawson and Mueller 1994). Although the study can show a bile leak or obstruction of the biliary system, it does not afford the anatomic detail necessary for further treatment planning. However, in some cases it can serve as a initial screening device to delineate a bile leak (Fig. 20.4).

Endoscopic retrograde cholangiography (ERC) is a valuable tool in evaluating and treating the injured but unobstructed biliary system (Kozarek 1994). Retrograde injection can demonstrate an area of leak or obstruction along the course of the bile duct and/or the cystic duct. In cases of small leaks from the cystic duct stump, the ducts of Luschka, or even the common bile duct, endoscopic placement of a stent in the common bile duct can provide temporizing decompression and allow the leak to heal (Fig. 20.4). Subsequent removal of the stent is easily performed at a later date using endoscopic techniques. A caveat in the use of ERC is that obstructed proximal biliary segments may not be imaged by this technique and details of anatomy may be lacking.

Transhepatic cholangiography (THC) combined with ultrasound imaging may be a valuable adjunct in patients with bile duct injury (Trerotola et al. 1992; van Sonnenberg et al. 1990, 1993; Wright et al. 1993). Transhepatic cholangiography may be performed in patients showing biliary obstruction on the initial ultrasound or CT scan. Ultrasound guidance is helpful in selected cases. These are usually patients who have a dilated left biliary duct where the duct is anatomically lying in a horizontal course. A needle can then be directed into a peripheral duct via a subxyphoid approach. THC is essential in patients with complete biliary obstruction shown on ERCP. THC can demonstrate the complete anatomy, the segmental anatomy, and whether specific segments are separately blocked; this may aid in directing percutaneous biliary drainage (PBD) or even surgery (Fig. 20.8).

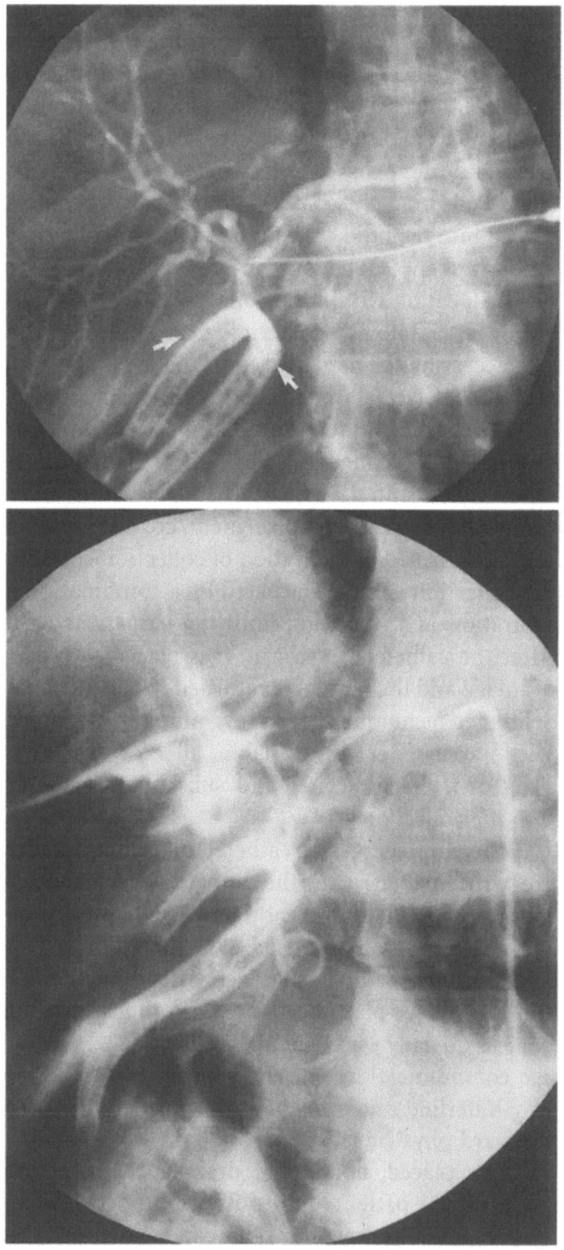

a

b

Fig. 20.8a,b. Transhepatic cholangiogram on a completely severed duct as a complication of laparoscopic surgery. **a** Initial cholangiogram demonstrates only mildly dilated right and left hepatic ducts. **a** A surgical drain (*arrows*) is seen in the region of the porta. The patient was essentially decompressing her biliary system through the drain. **b** Bilateral percutaneous biliary drainage catheters were placed with difficulty. The patient eventually went to surgery for a choledochojejunostomy

20.5 Treatment of Bile Duct Injuries

Bile duct injuries following laparoscopic cholecystectomy include the two major groups discussed above - bile leak and biliary obstruction. While both may occur in the same patient, they will be discussed separately.

20.5.1 Bile Leak

Our technique for treating bile collections includes a preliminary ultrasound or CT examination to define the location and extent of the collection. Using either technique, an access route to the collection is chosen that does not transgress vital structures. Once an appropriate access route has been chosen, the skin entrance site is prepared with iodine sterilization, cutaneous and deep local anesthesia, a generous skin nick, and blunt dissection of the subcutaneous tissues. Following this, an 18- to 22-gauge needle is placed into the fluid collection. Aspiration of a small amount of the collection contents permits proof of needle location within the collection as well as providing a sample for bacteriological analysis. After safe access has been gained to the collection, a catheter can be placed into the abnormality using a trocar or Seldinger technique (Fig. 20.8).

Once in the collection, the catheter is fed off the inner stylet/cannula assembly and into the collection. Following catheter placement, correct positioning is confirmed by aspiration of collection contents; the collection is then evacuated by a combination of aspiration and irrigation until the irrigant is clear. After the catheter is sewn in, it is placed to gravity drainage and the patient is returned to the ward. For Seldinger technique there are a number of "single-stick" access devices available, all of which allow placement of guidewires and subsequent catheters into the collection.

Following successful catheter placement, orders are written to irrigate the catheter every shift, to monitor output of the catheter, and to notify our service in case of problems with the catheter. The interventional radiology service visits all patients in whom we have placed catheters or other indwelling devices on daily ward rounds. In this way we encourage communication with the clinical teams and tacitly underline our availability to meet the needs of the ward physicians.

Once placed, abdominal drainage catheters remain in the patient until outputs drop below approximately 10 ml per day. This can mean protracted catheter placement times in patients who have persistent bile leaks. Persistent bile leaks are usually simple to recognize clinically. Although an abscess from an infected biloma may yield only pus on initial aspiration, if a bile leak is present, the character and the amount of drainage will change over time. For example, a simple collection of infected bile or hematoma will show steadily decreasing amounts of drainage over a 24- to 72-h period. An infected biloma which still communicates with the bile duct will demonstrate an increase in catheter drainage after the first 24–72 h and a change in appearance to yellow or greenish biliary fluid. If the leak is small and the bile duct is unobstructed, this will eventually heal. Most bile leaks heal within 10 days to 3 weeks unless there is a distal obstruction. For example, if the distal bile duct has been ligated or clipped, the bile leak will continue. Still, in these patients surgical repair is easier if the infected biliary collection has been drained. In those patients in whom a long-term catheter is necessary, nursing assistance can be provided for the patient at home, permitting the patient to leave the hospital while the biliary duct injury heals. The patient can then return to the Radiology Department at a later date for removal of the catheter.

20.5.2 Biliary Obstruction

Biliary obstruction due to inadvertent bile duct ligation usually manifests in the near postoperative period with jaundice, increasing pain, and, occasionally, fever (Dawson and Mueller 1994; Kozarek 1994; Lee et al. 1991; Ray et al. 1993; Stokes 1994; Trerotola et al. 1992; van Sonnenberg et al. 1990, 1993; Wright et al. 1993). As stated above, ultrasound imaging usually provides anatomic information as to the level and location of the obstruction. In some patients the surgeon may simply want the anatomy demonstrated so he or she can perform a primary hepatico- or choledochojejunostomy. A THC will then confirm the ultrasound findings and provide anatomic detail (Fig. 20.8). In other circumstances percutaneous drainage is required. Using the cholangiogram as a guide, the goal in cases of both partial and presumed complete obstruction is to re-establish continuity of the biliary system across the obstructed segment. The indications for a THC or PBD are listed in Table 20.1.

In most fresh cases of biliary injury a surgical repair will be attempted. However, in cases where there have been multiple previous surgical repairs a percutaneous approach may be safer for the patient. Such patients often have multiple adhesions and scars, and delicate surgery in these cases is difficult.

Table 20.1. Indications for transhepatic cholangiography (THC)/percutaneous biliary drainage (PBD)

- Anatomic detail prior to surgery (THC)
- Temporizing measure in a presurgical candidate who is septic or medically unstable (PBD)
- Temporizing measure to guide surgeon to appropriate intrahepatic/extrahepatic ducts for a surgical repair (PBD)
- Prior to percutaneous biliary dilatation of a previously failed/restrictured surgical anastomosis (PBD)

The technique of PBD is fairly standard and has been described. Using the preliminary THC as a guide, we use a variety of sheath/catheter systems to place a guidewire, usually 0.038 in. (0.95 mm) in diameter, to a peripheral biliary duct above the level of obstruction. Peripheral placement of the working wire permits adequate purchase for the catheter that will be placed later.

The technique for crossing an obstructed duct segment involves gentle probing at the site of the obstruction. The obstruction may be marked by clips that are causing the obstruction. The intention is to seek out whatever small orifice may be present without causing any reactions which might occlude the narrowed lumen. We have found hydrophilic-coated guidewires to be especially effective in this regard.

The best technique to use in searching for the residual lumen is a gentle, persistent tapping motion with the guidewire at the obstruction. This permits exploration of multiple possible lumen orifices while limiting the risk of edema, bleeding, or other reactive changes which would tend to occlude the orifice of the stenosed ductal segment. In performing this maneuver, it is often helpful to use an angled-tip guidewire introduced through an angled-tipped catheter which has been advanced almost to the level of obstruction. Using this combination, the guidewire tip can be rotated through 360° and presented to the narrowed lumen from a variety of directions. The catheter can also be used to perform intermittent injections of contrast material to help define the lumen.

In cases in which we are successful in placing a guidewire across the stenotic segment of biliary duct, the ultimate goal is to place a biliary drainage tube through the stenosed segment, across the papilla of Vater, and into the duodenum. Doing this serves two functions: it provides internal/external drainage across the stenosed segment(s), and provides landmarks for the surgeon at operation should this become necessary (Table 20.1).

In severely damaged systems the obstructed biliary segment cannot be crossed using percutaneous techniques. In these cases, external drainage catheters should be placed in the obstructed biliary ductal segments with their tips at the level of the obstruction. This is usually marked by the obstructing surgical clips. In the case of stricturing secondary to thermal injury, where a clip may not be present, the catheter tip should be placed as close as possible to the most distal point of guidewire passage. These catheters provide two functions. First, they permit external drainage of the obstructed biliary segments. Second, they provide a tactile and visual reference point for the surgeon at operation. This is very helpful to the surgeon, as in many cases the surgical bed will be inflamed and the native anatomy difficult to identify (Table 20.1). Placement of these catheters can shorten 5- to 6-h procedures to as little as 1–2 h (Martin and Rossi 1994).

20.5.3 Recurrent Biliary Strictures

Although laparoscopic bile duct injury followed by surgical repair is a relatively new phenomenon, some general remarks can be directed to the long-term radiological care of these injuries. Recurrent biliary strictures occurring after surgical repair are often referred to the radiologist and are difficult to treat, because (1) the intrahepatic ducts are often only minimally dilated; (2) the surgical anastomoses tend to be "high", often subcostal in location; (3) there may be segmental duct narrowing requiring more than single catheter placement. Nevertheless, the interventional approach to these patients remains the same as has been described for those with any biliary stricture. Once access to the surgical anastomosis has been gained, biliary dilatation is performed. Controversy remains over the size of balloon, the number of dilatations, the length and time of the dilatations, and the use of long-term catheter placement after dilatation (Lee et al. 1991). In our institution we use large balloons (10–12 mm); we usually perform dilatation on the patient's first visit to the Department if we can achieve biliary drainage without complications. The balloon is inflated several times for 1–3 min. Evaluation of success is by injection and fluoroscopic observation rather than quantitative pressures. We also have a drainage catheter placed through the anastomosis for 1–3 weeks after dilatation. No large series with long-term follow-up has been reported from any institution because of the small numbers of cases. However, our initial results have compared favorably with those published on open cholecystectomy series.

We do not routinely balloon-dilate the stenosed biliary ductal segments with an intention of cure. Initial enthusiasm for the technique (Lee et al. 1991) has been tempered by the realization that long-term restenosis rates approach 30%–50% (depending on the institution). While arguments can be made for either balloon dilation or primary reanastomosis, surgical therapy appears to provide an attractive alternative to balloon dilation and may prove to be the

Table 20.2. Summary of experience to date with metallic stents in recurrent biliary strictures. (Reproduced from RING 1994, by permission of the author)

Authors	Year	No. of patients	Follow-up (months)	Patency (%)
Irving	1989	11	6–21	72.7
Coons	1989	15	6–14	86.6
Rossi	1990	17	4–12	82.4
Coons	1992	40	12	87
Ring	1994	15	44	80

most logical choice for initial therapy, with percutaneous management of recurrent postoperative strictures being available for subsequent use if necessary.

Metallic Stents. The use of metallic stents in benign disease remains controversial. However, there has been a growing interest in placing them in these types of patients particularly if they have had multiple percutaneous dilatation procedures over a short period of time (S. WALL, personal communication); RING and WALL have described the use of open, shorter Gianturco stents as opposed to the Wallstent in these types of patients. It is too early to define long-term results; however, they appear favorable (Table 20.2).

20.6 Summary

Laparoscopic surgical techniques offer substantial benefits when compared to open surgery. They are, however, not without risk. In the case of bile leaks and bile duct injuries, interventional radiology provides a spectrum of proven techniques for abscess drainage and, in some cases, treatment of biliary ductal injuries. In cases of biliary duct injuries destined for surgery, interventional radiology provides temporizing measures and anatomic delineation valuable to the operating surgeon. In other types of laparoscopic injuries, interventional radiology provides the capacity to treat abdominal abscesses. As the numbers and types of laparoscopic surgical procedures performed increase, and as injuries inevitably increase with them, the interventional radiologists will be a valuable adjunct to the surgical team in managing the laparoscopically injured patient.

References

Berci G (1992) Biliary ductal anatomy and anomalies: the role of intraoperative cholangiography during laparoscopic cholecystectomy. Surg Clin North Am 72:1069–1075

Blumgart LH, Kelley CJ, Benjamin IS (1984) Benign bile duct stricture following cholecystectomy: critical factors in management. Br J Surg 17:836–843

Crist DW, Gadacz TR (1993) Complications of laparoscopic surgery. Surg Clin North Am 73:265–289

Davidoff AM, Pappas TN, Murray EA, et al (1992) Mechanisms of major biliary injury during laparoscopic cholecystectomy. Ann Surg 215:196–202

Dawson SL, Mueller PR (1994) Interventional radiology in the management of bile duct injuries. Surg Clin North Am 74:865–874

Donohue J, Farnell M, Grant C, et al (1992) Laparoscopic cholecystectomy: the early Mayo Clinic experience. Mayo Clinic Proc 67:449–455

Graber JN, Schulta LS, Pietrafitta JJ, et al (1992) Complications of laparoscopic cholecystectomy: a prospective review of an initial 100 consecutive cases. Lasers Surg Med 12:92–97

Kozarek RA (1994) Endoscopic techniques in management of biliary tract injuries. Surg Clin North Am 74:883–893

Larson G, Vitale G, Casey J, et al (1992) Multipractice analysis of laparoscopic cholecystectomy in 1983 patients. Am J Surg 163:221–226

Lee MJ, Mueller PR, Saini S, Hahn PF, Dawson SL (1991) Percutaneous dilatation of benign biliary strictures: single-session therapy with general anesthesia. Am J Radiol 157:1263–1266

Martin RF, Rossi RL (1994) Bile duct injuries: spectrum, mechanisms of injury and their preventions. Surg Clin North Am 74:781–803

Ray CE, Hibbeln JR, Wilbur AC (1993) Complications after laparoscopic cholecystectomy: imaging findings. AJR 160:1029–1032

Ress AM, Sarr MG, Nagorney DM, Farnell MB, Donohue JH, McIlrath D (1993) Spectrum and management of major complications of laparoscopic cholecystectomy. Am J Surg 165:655–662

Ring EJ (1994) Biliary strictures: dilation and stenting. Presented at the Second International Symposium on Cardiovascular are Interventional Radiology, Boston, 3–7 October 1994

Southern Surgical Club (1991) A prospective analysis of 1518 laparoscopic cholecystectomies. N Engl J Med 324:1074–1078

Stokes KR (1994) Commentary: interventional radiology in the management of bile duct injuries. Surg Clin North Am 74:879–881

Trerotola SO, Savader SJ, Lund GB, et al (1992) Biliary tract complications following laparoscopic cholecystectomy: imaging and intervention. Radiology 184:195–200

Van Sonnenberg E, Casola G, Wittich GR, et al (1990) The role of interventional radiology for complications of cholecystectomy. Surgery 107:632–638

Van Sonnenberg E, D'Agostino HB, Easter DW, et al (1993) Complications of laparoscopic cholecystectomy: coordinated radiologic and surgical management in 21 patients. Radiology 188:399–404

Wright TB, Bertino RB, Bishop AF, et al (1993) Complications of laparoscopic cholecystectomy and their interventional radiologic management. Radiographics 13:119–128

21 Biliary Complications After Orthotopic Liver Transplantation: Imaging and Intervention

J.M. LLERENA and W.R. CASTAÑEDA-ZUÑIGA

CONTENTS

21.1 Introduction

Orthotopic liver transplantation is now an accepted and successful mode of treatment for patients with various forms of end-stage liver disease. Graft survival has improved significantly in the last decade, with a 5-year survival estimated at 65%–78% (BELLE et al. 1993). As hepatic transplantation has become more common and successful, radiologists have been challenged not only with the early diagnosis of graft dysfunction, but also with the treatment of complications using interventional radiologic techniques. We will discuss the role of the noninvasive and invasive imaging techniques in the detection and management of biliary complications after orthotopic hepatic transplantation.

21.1.1 Indications for Liver Transplantation

The accepted indications for liver transplantation have not changed significantly in recent years. Among children, biliary atresia (52%) remains the most common indication for hepatic transplantation, followed by acute fulminant hepatic failure (11%), α_1-antitrypsin deficiency (9%), cryptogenic cirrhosis (6%), and others (22%) (WESTRA et al. 1993). Among adults, cirrhosis (60%) as an indication is followed by primary cholestatic liver disease (22%), fulminant hepatic failure (6%), hepatic malignancy (5%), and others (7%) (BELLE et al. 1994).

21.1.2 Pretransplantation Noninvasive and Invasive Imaging Evaluation

A detailed discussion of the role of noninvasive and invasive imaging modalities before liver transplantation is beyond the scope of this chapter. However, it is important to remember the key information with which the radiologist has to provide the surgeon prior to surgical intervention which includes a calculation of the liver volume, an evaluation of the size and patency of the portal vein, superior mesenteric vein, hepatic veins, hepatic artery, inferior vena cava (IVC), and portosystemic shunts, and an anatomical evaluation after checking for liver masses and extrahepatic disease.

Although there is no single ideal imaging technique for the preoperative evaluation of liver transplant candidates, a recent review by REDVANLY et al. (1995) has recommended computed tomography (CT) as the primary and most cost-effective imaging modality for the preoperative evaluation of these patients. Dynamic bolus CT, preferably with spiral technique, has proven to be a good technique to calculate liver volumes and to evaluate for liver masses and extrahepatic disease. CT is also adequate for evaluating the vascular anatomy, especially when three-dimensional data is reformatted as a CT angiogram. In this context, other imaging modalities, such as color Doppler ultrasonography, magnetic resonance imaging, angiography, and percutaneous transhepatic cholangiography, may be used only as secondary targeted examinations for solving specific questions not answered by CT.

J.M. LLERENA, MD, Assistant Professor, Louisiana State University, University Medical Center, Department of Radiology, 2390 West Congress Street, Lafayette, LA 70506, USA
W.R. CASTAÑEDA-ZUÑIGA, MD, Professor and Chairman, Department of Radiology, Louisiana State University Medical Center, 1542 Tulane Avenue, New Orleans, LA 70112, USA

21.1.3 Surgical Anatomy in Liver Transplantation

Surgical techniques have been reviewed by LERUT et al. (1988) in 393 consecutive grafts. In routine orthotopic hepatic transplantation, the donor liver is placed within the recipient's hepatic fossa and five end-to-end anastomoses are performed: four vascular anastomoses (hepatic artery, extrahepatic portal vein, suprahepatic and infrahepatic IVC) and one biliary anastomosis.

The hepatic artery anastomosis in adults frequently uses a Carrel patch (donor celiac axis with small aortic patch) anastomosed end-to-end with the recipient hepatic or celiac arteries. On the other hand, children usually undergo end-to-end anastomoses between the donor and recipient hepatic or celiac vessels. Alternatively, a donor iliac artery is anastomosed inferiorly to the recipient infrarenal aorta and superiorly to the donor Carrel patch. In donor livers with dual blood supply, a single donor vessel is created prior to anastomosis with the recipient hepatic or celiac arteries.

Revascularization of the portal vein and IVC is routinely performed with end-to-end anastomoses between the donor and recipient vessels. In patients with recipient portal vein thrombosis, a donor common iliac vein is anastomosed inferiorly to the recipient infrapancreatic superior mesenteric vein and superiorly to the donor extrahepatic portal vein. Patients with large discrepancy between the donor and recipient IVC require an end-to-side (piggy-back) caval anastomosis.

The routine biliary anastomosis in adult transplantation is an end-to-end choledocho-choledochostomy, stented externally with a T-tube for 2–3 months. Adults with diseased extrahepatic biliary ducts and pediatric patients routinely undergo end-to-end choledocho-jejunostomy using a Roux-en-Y with internal stenting. A choledocho-cholecysto-choledochostomy has been used only in Europe with variable results (EVANS et al. 1990).

21.2 Biliary Complications

Biliary complications remain a major cause of morbidity and mortality in liver transplant recipients. The largest series in the literature show biliary tree complications occurring in 13%–25% of patients (KUO et al. 1994; PECLET et al. 1994; PARIENTE et al. 1991; LETOURNEAU and CASTAÑEDA-ZUÑIGA 1990). Leak is the most common biliary complication during the first weeks after transplantation; on the other

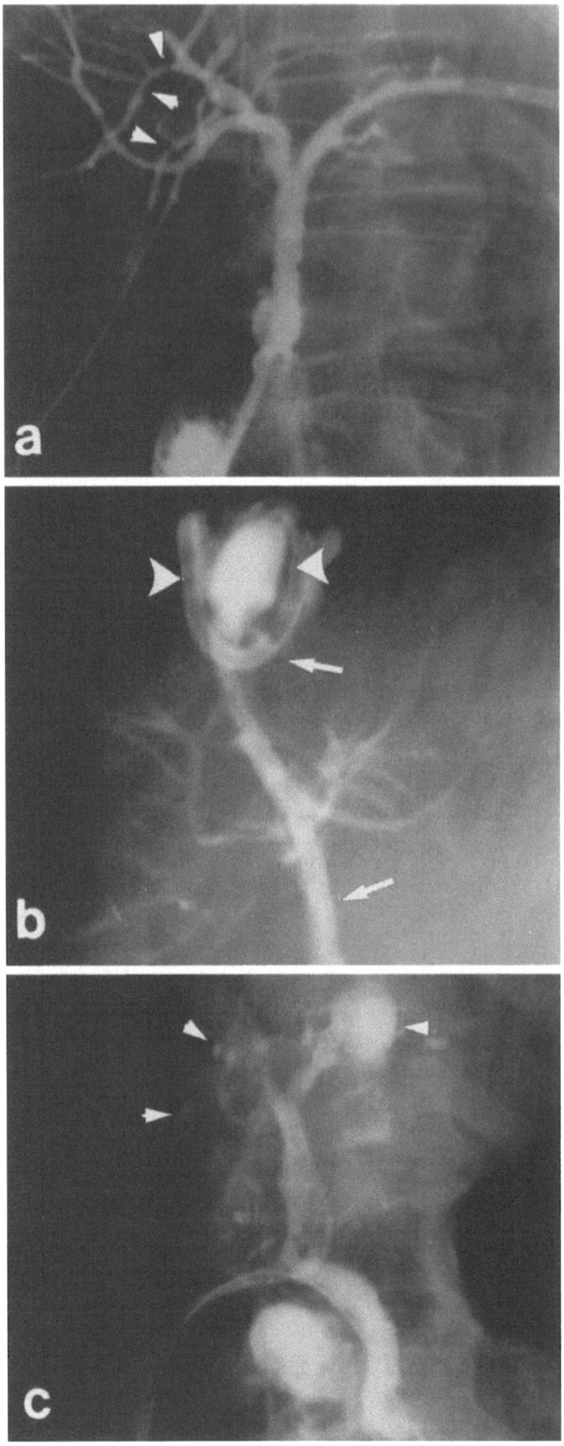

Fig. 21.1a–c. Biliary complications in patients with hepatic artery thrombosis. **a** Nonanastomotic intrahepatic biliary strictures (*arrowheads*). Cholangiography was performed through an internal-external biliary catheter. **b** Non-anastomotic biloma (*arrowheads*). Patient was treated conservatively with retrograde biliary drainage (*arrows*). **c** Intrahepatic abscesses complicating nonanastomotic bilomas (*arrowheads*). This patient required retransplantation

hand, after the first postoperative month biliary strictures predominate. The close relation between biliary complications and ischemia has been emphasized by several authors (HOLLAND et al. 1991; SANCHEZ-URDAZPAL et al. 1993). Cholangiographic abnormalities have been reported in 84% of patients with proven complete or partial occlusion of the hepatic artery (ZAJKO et al. 1987). Most ischemic complications are nonanastomotic leaks, strictures, and abscesses (Fig. 21.1).

Cholangiography is the imaging modality of choice for the early diagnosis of biliary complications after liver transplantation. This examination is usually performed by injecting contrast medium through an existing T tube or by direct puncture of the biliary tree during percutaneous transhepatic cholangiography; occasionally, the endoscopic retrograde approach has been used (O'CONNOR et al. 1991). Cholangiography has been successfully compared to other imaging modalities, such as ultra sonography and hepatic scintigraphy. ZEMEL et al. (1988), and ZAJKO et al. (1988) found that ultra sonography had sensitivities only in the range of 46%–54% in the evaluation of 41 and 50 liver transplant patients respectively with various biliary complications confirmed by cholangiography. Cholangiography is also more specific and accurate than hepatic scintigraphy (ANSELMI et al. 1990). Although CT has been used in the evaluation of biliary complications after liver transplantation, the sensitivity, specificity, and accuracy of this imaging modality have not been compared with those of cholangiography, the gold standard (LETOURNEAU et al. 1987).

21.2.1 Biliary Strictures

Obstruction of the biliary tree has been reported in 5% of transplant patients and commonly occurs after the first postoperative month (LERUT et al. 1987). Early biliary obstruction is rare and usually related to technical difficulties while performing the anastomosis, problems with the T tube or internal surgical stent, biliary sludge-cast formation, and/or extrinsic compression. Other causes of biliary obstruction, such as lithiasis or mucocele of the allograft cystic duct remnant, are quite rare and occur in the late postoperative period. Obstruction of the biliary tree is best diagnosed by cholangiography, since that real-time ultra sonography may be falsely negative in patients with early biliary obstruction (ZEMEL et al. 1988).

The most common causes of obstruction in the liver transplant population are anastomotic and nonanastomotic biliary strictures. Anastomotic strictures are usually related to postoperative fibrosis, while nonanastomotic stenoses have been related to a variety of conditions including hepatic artery thrombosis and stenosis, prolonged cold ischemia, chronic rejection, ABO blood group incompatibility, infection, and recurrent primary disease. Ischemic strictures are common among patients with hepatic artery thrombosis and/or stenosis; this is related to the fact that the hepatic artery provides the only blood supply to the graft's biliary tree. Early appearance of ischemic strictures has been associated with poor graft survival (SANCHEZ-URDAZPAL et al. 1992).

The initial treatment of choice in liver transplant recipients with biliary strictures is percutaneous biliary drainage with subsequent balloon dilation. The technique of biliary drainage in hepatic allografts is similar to that used in the nontransplant population. Puncture of the biliary tree is performed using the right midaxillary or left epigastric approaches. Some authors prefer to minimize biliary trauma and only place an external drainage catheter in the initial session. In the following days, the stenotic area is crossed with a guidewire for placement of an internal-external biliary catheter that has sideholes above and below the area of obstruction. Other authors favor crossing the obstruction and placing the internal-external drainage tube during the initial session.

Balloon dilation of strictures is performed after adequate biliary drainage has been established. Most patients require several sessions with progressive increases in balloon size. The internal-external catheter is replaced after dilation and removed only after adequate resolution of the stricture. ZAJKO et al. (1995) have recently evaluated the long-term results of balloon dilation of biliary strictures in 72 patients. Fifty-six of these patients had anastomotic narrowing, and the remaining 16 had nonanastomotic strictures. These authors found overall patency rates of 81% and 70% after 6 months and 6 years respectively, with better long-term results in the nonanastomotic strictures.

Balloon dilation has caused few complications and is considered the treatment of choice in liver transplant recipients with biliary strictures. We have used a different approach for performing repeated balloon dilations in liver transplant patients with choledocho-jejunal anastomoses (Fig. 21.2). In our series, five patients with recurrent biliary strictures

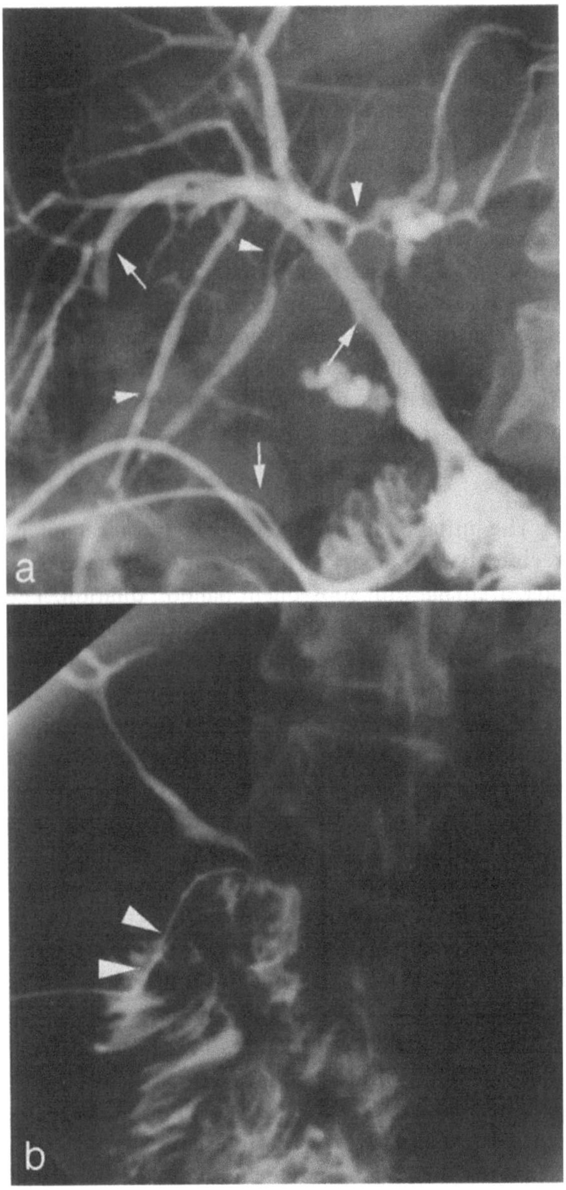

Fig. 21.2 a,b. Biliary strictures: balloon dilation using a retrograde transjejunal approach. **a** The afferent jejunal loop is punctured for the creation of a retrograde tract (*arrows*). Balloon dilation of biliary strictures (*arrowheads*) is then performed using this access. **b** After removal of the retrograde biliary catheter, the jejunostomy tract (*arrowheads*) allows retrograde catheterization of the biliary tree for performing balloon dilation of recurrent biliary strictures

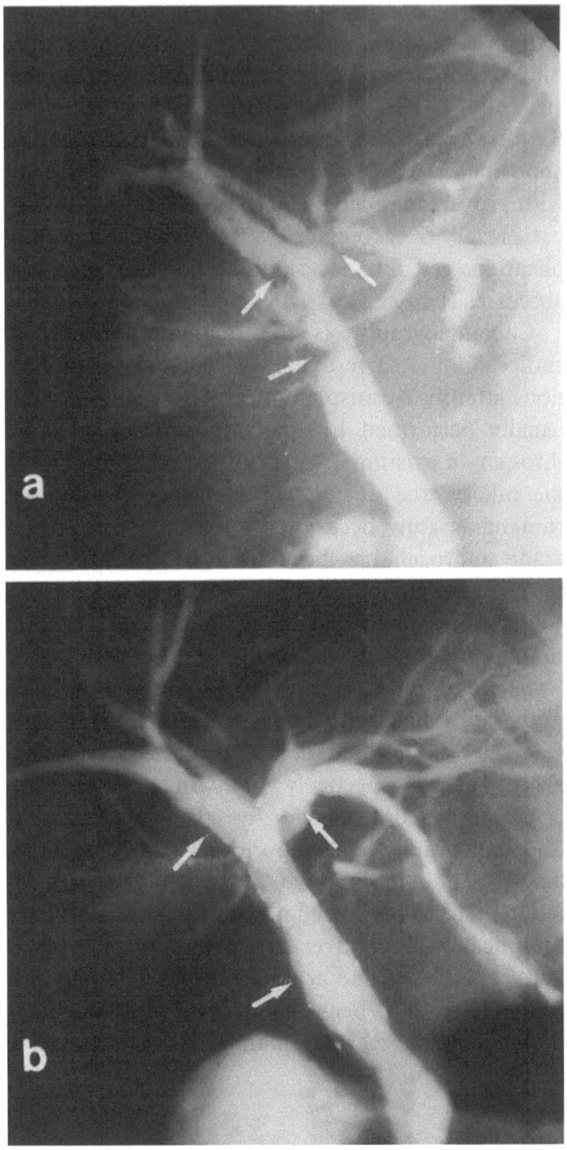

Fig. 21.3 a,b. Nonanastomotic biliary strictures: balloon dilation and stenting. **a** Biliary strictures are noted at the proximal portion of the common bile duct and at the distal portion of both common hepatic ducts (*arrows*). **b** Strictures recurred in spite of good immediate results after balloon dilation. Gianturco stents were successfully deployed in the areas of narrowing without complications

due to ischemia or sclerosing cholangitis were treated by retrograde catheterization of the biliary tree through a jejunostomy tract created by direct percutaneous puncture of the Roux-en-Y jejunal loop (LLERENA et al. to be published). The role of metallic stents in the biliary tree of hepatic transplants is still under investigation (Fig. 21.3).

21.2.2 Biliary Leaks

Anastomotic and nonanastomotic biliary leaks have been reported in 4%–23% of transplanted grafts (SHENG et al. 1994; OSORIO et al. 1993). Leaks at the anastomotic site have been associated with cold ischemia of the donor biliary tree, technical problems during surgical reconstruction, and, rarely, with ischemia due to hepatic artery complications. On the

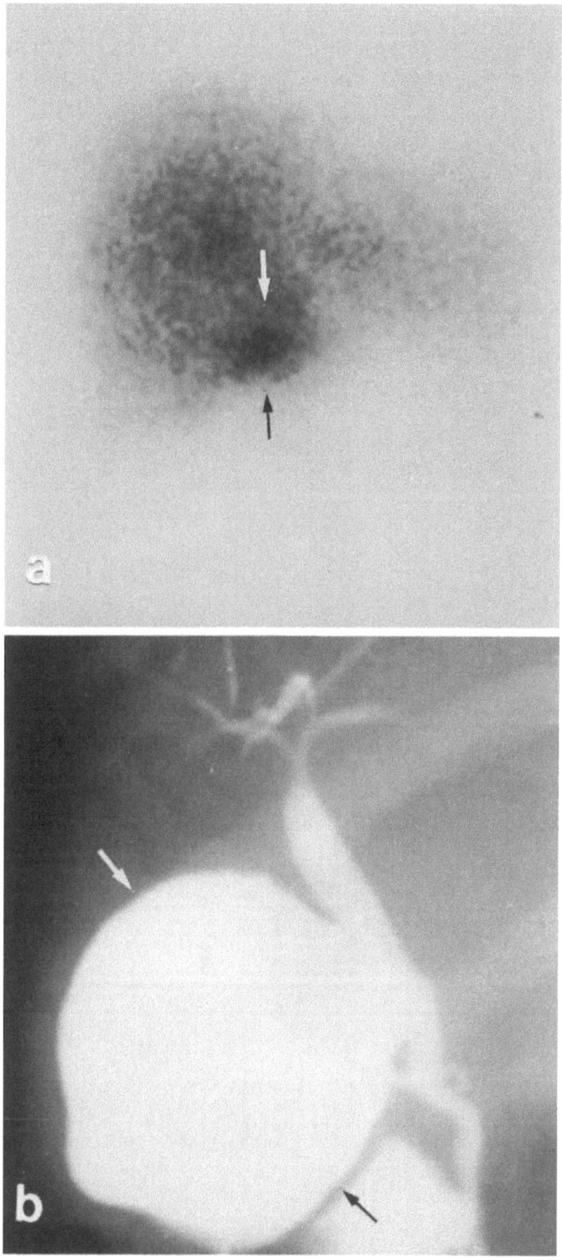

Fig. 21.4 a,b. Anastomotic leak: diagnosis. **a** Hepatic scintigram demonstrates an area of abnormal radionuclide activity projected over the lower portion of the liver parenchyma (*arrows*). **b** Cholangiogram confirms the diagnosis of large anastomotic biloma (*arrows*)

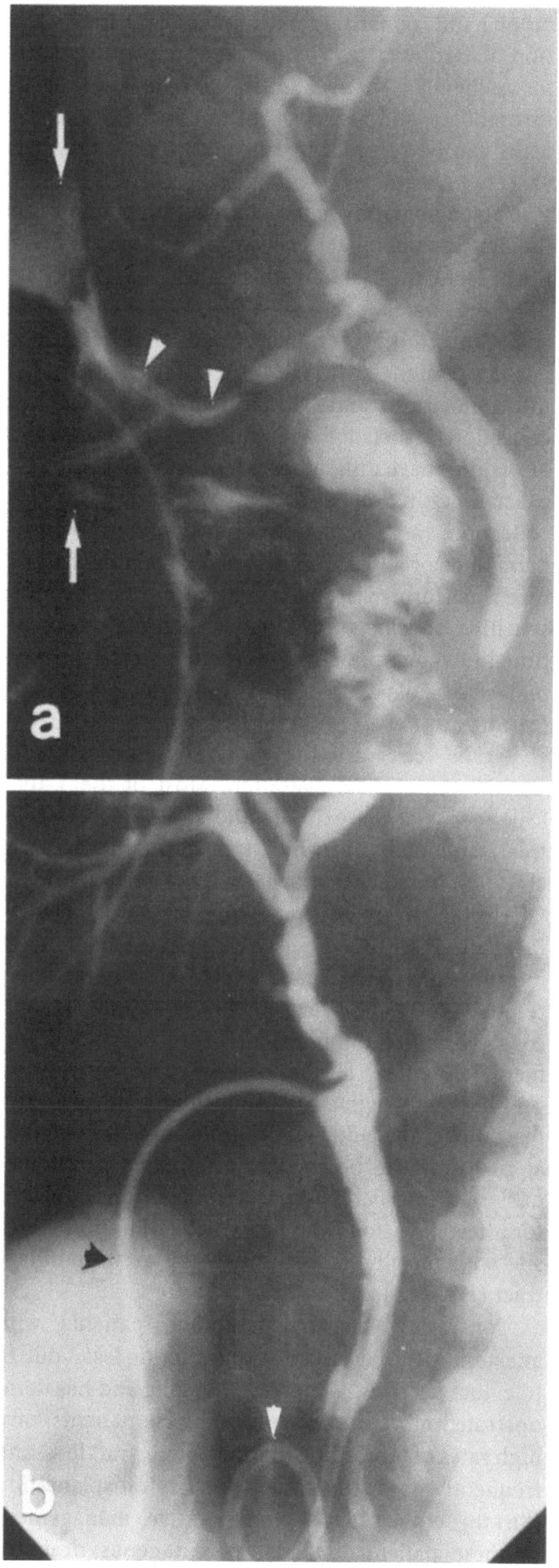

Fig. 21.5 a,b. T-tube dislodgement: percutaneous recanalization and drainage. **a** The T tube (*arrowheads*) was accidentally pulled back. Areas of leak were noted laterally and inferiorly to the dislodged T tube (*arrows*). **b** Good resolution of the T-tube leak after recanalization of the tract with a guidewire and placement of an internal-external biliary tube (*arrowheads*)

other hand, nonanastomotic leaks, with the exception of T-tube extravasations, are frequently related to ischemic changes due to hepatic artery thromboses and stenoses. Intra- and extrahepatic leaks can be secondarily infected, with subsequent abscess formation.

Cholangiography is the examination of choice for the diagnosis of biliary leak. When a T-tube access is not present, the percutaneous transhepatic approach may be technically more challenging, taking into account that the biliary tree is undilated because of decompression through the leak. Although ultrasonography and CT can demonstrate the fluid collection associated with the leak, both imaging modalities have difficulties not only in localizing the exact leak site, but also in characterizing the fluid collection; a biloma can be confused with a hematoma or abscess unless guided diagnostic puncture is performed. Hepatic scintigraphy can be more sensitive, specific, and accurate than the cross-sectional imaging modalities, by demonstrating abnormal areas of extravasated radionuclide activity (Mochizuki et al. 1991) (Fig. 21.4).

T-tube leaks are the most common cause of nonanastomotic extravasation. Most of the T-tube leaks occur at the exit site and can be successfully treated nonoperatively by T-tube drainage (Sheng et al. 1994). In patients with early leaks due to accidental dislodgement of the T tube, the tract can be recanalized with a guidewire, for placement of an internal-external biliary tube (Fig. 21.5). we routinely remove the T tube over a guidewire and under fluoroscopic control 3 months after transplantation. Immediately after T-tube removal, we perform a pull-back cholangiogram using an introducer sheath placed over the guidewire. Significant leaks detected at this time can be successfully treated percutaneously; an internal-external drainage tube is placed for a few days, followed by embolization of the leak site with metallic coils through the existing biliary tract (Fig. 21.6).

Conservative treatment among patients with anastomotic leaks, and nonanastomotic leaks due to bile duct necrosis, is more challenging and has demonstrated more limited results. These patients have high rates of morbidity, mortality, and graft loss, and frequently need surgical repair or retransplantation (Sheng et al. 1994). Nonoperative management of these patients includes percutaneous drainage of the biloma coupled with decompression of the biliary tree by percutaneous biliary drainage or sphincterotomy (Fig. 21.7). Conservative management has been used as definitive treatment or as a

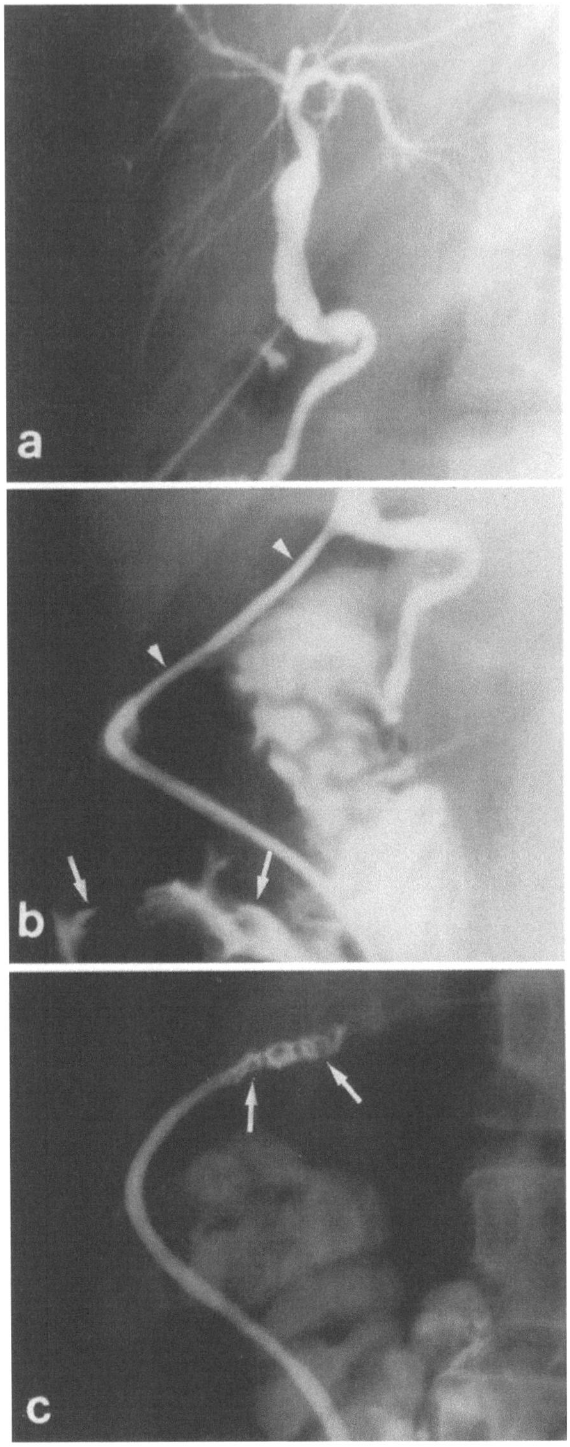

Fig. 21.6 a–c. Leak after T-tube removal: tract embolization. **a** Unremarkable cholangiogram 3 months after transplantation prior to T-tube removal. **b** Pull-back cholangiogram with a safety guidewire in place reveals intraperitoneal leak (*arrows*). The T-tube tract is well opacified (*arrowheads*). **c** The T-tube tract was successfully embolized with coils (*arrows*). No leak was noted after embolization

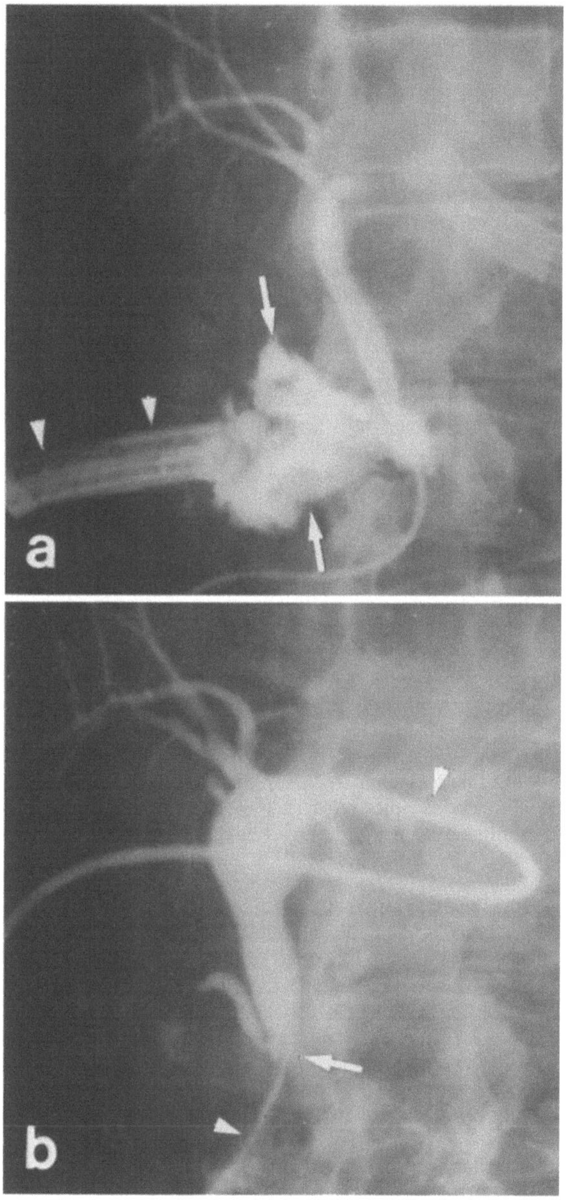

Fig. 21.7 a,b. Anastomotic leak: conservative treatment. a A large leak is noted at the anastomotic site (*arrows*). External drainage of the biloma was performed at this time (*arrowheads*). b Biliary drainage with an internal-external catheter was then established (*arrowheads*). Follow-up cholangiogram shows healing of the leak site (*arrow*). Moderate biliary dilation was finally treated with balloon angioplasty after healing of the narrowed anastomotic region

temporary measure prior to surgical reintervention (Osorio et al. 1993; Ward et al. 1991).

21.2.3 Miscellaneous Biliary Complications

Biliary sludge, caused by inspissate bile and/or necrotic duct tissue, has been found in 10%–29% of liver grafts between 6 days and 8 years after transplantation (Holland et al. 1991; Evans et al. 1990). Cold ischemia, infection, and rejection have been considered part of the multifactorial pathogenesis of biliary sludge and cast formation. Clinically, these patients may be asymptomatic, or they may present with life-threatening biliary obstruction and ascending cholangitis. A recent paper by Barton et al. (1995a) reviewed 51 patients with sludge-cast formation and confirmed that cholangiography was the imaging modality of choice in this group of patients. In this setting, ultrasonography and CT showed limited diagnostic value. Barton et al. (1995b) also evaluated the results of treatment in 49 patients with sludge-cast formation. Success rates of 40%, 63%, and 86% were found for oral chemolysis, interventional techniques (intraluminal chemolysis and basket extraction), and surgery respectively. The authors concluded that interventional techniques were effective and should be considered prior to surgical intervention.

Some T-tube complications, such as tube dislodgement and biliary leaks, have been already discussed in Sect. 21.2.2. Additional T-tube complications include migration, malposition, and occlusion by thickened bile. In most cases, T-tube malfunction manifests with biliary obstruction, which is commonly solved by tube removal. Contrarily, patients with choledocho-jejunostomies and obstructed internal stents may need percutaneous intervention. Orons and Zajko (1994) have treated such patients with transhepatic biliary drainage, followed by snaring and pushing of the stent into the jejunum.

21.3 Summary

Early diagnosis and treatment of biliary complications after orthotopic liver transplantation have contributed significantly to recent improvements in graft survival. Nonsurgical management of these complications using invasive radiologic techniques is adequate for the majority of patients with late bile leaks, leaks at the T-tube site, biliary strictures, and choledocholithiasis after liver transplantation. The role of metallic stents in the biliary tree of patients with hepatic transplants is still under investigation.

References

Anselmi M, Lancberg S, Deakin M et al. (1990) Assessment of the biliary tract after liver transplantation: T-tube cholangiography or IODIDA scanning. Br J Surg 77:1233–1237

Barton P, Maier A, Steininger R, Muhlbacher F, Lechner G (1995a) Biliary sludge after liver transplantation. I: Imaging findings and efficacy of various imaging procedures. AJR Am J Roentgenol 164:859–864

Barton P, Steininger R, Maier A, Muhlbacher F, Lechner G (1995b) Biliary sludge after liver transplantation. II. Treatment with interventional techniques versus surgery and/or oral chemolysis. AJR Am J Roentgenol 164:865–869

Belle SH, Beringer KC, Murphy JB, Detre KM (1993) The Pittsburgh-UNOS liver transplant registry. In: Terasaki PI, Cecka JM (eds) Clinical transplants 1992. UCLA Tissue Typing Laboratory, Los Angeles, p 1

Belle SH, Beringer KC, Detre KM (1994) Trends in liver transplantation in the United States. In: Terasaki PI, Cecka JM (eds) Clinical transplants 1993. UCLA Tissue Typing Laboratory, Los Angeles, p 19

Evans RA, Raby ND, O'Grady JG, Karani JB, Nunnerley HB, Calne RY, Williams R (1990) Biliary complications following orthotopic liver transplantation. Clin Radiol 41:190–194

Holland P, Morris E, Buckels J (1991) Cholangiography in liver transplantation: a comparison of two types of biliary reconstruction. Br J Radiol 64:983–989

Kuo PC, Lewis WD, Stokes K, Pleskow D, Simpson MA, Jenkins RL (1994) A comparison of operation, endoscopic retrograde cholangiopancreatography, and percutaneous transhepatic cholangiography in biliary complications after hepatic transplantation. J Am Coll Surg 179:177–181

Lerut J, Gordon RD, Iwatsuki S, et al. (1987) Biliary tract complications in human orthotopic liver transplantation. Transplantation 43:47–51

Lerut JP, Gordon RD, Iwatsuki S, Starzl TE (1988) Human orthotopic liver transplantation: surgical aspects in 393 consecutive grafts. Transplant Proc 20:603–607

Letourneau JB, Castañeda-Zuñiga WR (1990) The role of radiology in the diagnosis and treatment of biliary complications after liver transplantation. Cardiovasc Intervent Radiol 13:278–282

Letourneau JG, Day DL, Maile CW, Crass JR, Ascher NL, Frick MP (1987) Liver allograft transplantation: postoperative CT findings. AJR Am J Roentgenol 148:1099–1103

Llerena JM, Garcia-Medina V, Berna JD, Bjarnason H, Hunter DW, Castañeda-Zuñiga WR (1996) Balloon dilation of recurrent biliary strictures in liver transplant patients: value of the percutaneous transjejunal approach (in press)

Mochizuki T, Tauxe WN, Dobkin J, Shah AN, Shanker R, Todo S, Starzl TE (1991) Detection of complications after liver transplantation by technetium-99m mebrofenin hepatobiliary scintigraphy. Ann Nucl Med 5:103–107

O'Connor HJ, Vickers CR, Buckels JA, McMaster P, Neuberger JM, West RJ, Elias E (1991) Role of endoscopic retrograde cholangiopancreatography after orthotopic liver transplantation. Gut 32:419–423

Orons PD, Zajko AB (1994) Interventional procedures in the management of liver transplant complications. In: Cope C (ed) Current techniques in interventional radiology. Current Medicine, Philadelphia, p 1.2

Osorio RW, Freise CE, Stock PG, et al. (1993) Nonoperative management of biliary leaks after orthotopic liver transplantation. Transplantation 55:1074–1077

Pariente D, Bihet MH, Tammam S, et al. (1991) Biliary complications after transplantation in children: role of imaging modalities. Pediatr Radiol 21:175–178

Peclet MH, Ryckman FC, Pedersen SH, et al. (1994) The spectrum of bile duct complications in pediatric liver transplantation. J Pediatr Surg 29:214–219

Redvanly RD, Nelson RC, Stieber AC, Dodd GD (1995) Imaging in the preoperative evaluation of adult liver transplant candidates: goals, merits of various procedures, and recommendations. AJR Am J Roentgenol 164:611–617

Sanchez-Urdazpal L, Gores GJ, Ward EM, et al. (1992) Ischemic-type biliary complications after orthotopic liver transplantation. Hepatology 16:49–53

Sanchez-Urdazpal L, Gores GJ, Ward EM, et al. (1993) Diagnostic features and clinical outcome of ischemic-type biliary complications after liver transplantation. Hepatology 17:605–609

Sheng R, Sammon JK, Zajko AB, Campbell WL (1994) Bile leak after hepatic transplantation: cholangiographic features, prevalence, and clinical outcome. Radiology 192:413–416

Ward EM, Wiesner RH, Hughes RW, Krom RA (1991) Persistent bile leak after liver transplantation: biloma drainage and endoscopic retrograde cholangiopancreatographic sphincterotomy. Radiology 179:719–720

Westra SJ, Zaninovic AC, Hall TR, Busuttil RW, Kangarloo H, Boechat MI (1993) Imaging in pediatric liver transplantation. Radiographics 13:1081–1099

Zajko AB, Campbell WL, Logsdon GA, Bron KM, Tzakis A, Esquivel CO, Starzl TE (1987) Cholangiographic findings in hepatic artery occlusion after liver transplantation. AJR Am J Roentgenol 149:485–489

Zajko AB, Zemel G, Skolnick ML, Bron KM, Campbell WL (1988) Percutaneous transhepatic cholangiography rather than sonography as a screening test for postoperative biliary complications in liver transplant patients. Transplant Proc 20:678–681

Zajko AB, Sheng R, Zetti GM, Madariaga JR, Bron KM (1995) transhepatic balloon dilation of biliary strictures in liver transplant patients: a 10-year experience. J Vasc Interv Radiol 6:79–83

Zemel G, Zajko AB, Skolnick ML, Bron KM, Campbell WL (1988) The role of sonography and transhepatic cholangiography in the diagnosis of biliary complications after liver transplantation. AJR Am J Roentgenol 151:943–946

22 Recurrent Pyogenic Cholangitis: Pathology, Imaging, and Management by Interventional Radiology

M.C. HAN, J.K. HAN, B.I. CHOI, and J.H. PARK

CONTENTS

22.1 Introduction

Recurrent pyogenic cholangitis (RPC), also known as oriental cholangiohepatitis, oriental cholangitis, primary cholangitis, and intrahepatic pigment stone disease, is a distinct clinical entity characterized by recurrent attacks of fever, chills, abdominal pain, and jaundice caused by bacterial infection to the biliary tree (COOK et al. 1954; ONG 1962; SEEL and PARK 1983; CARMONA et al. 1984). The intra- and extrahepatic bile ducts show dilatation and stricture, and they may contain calculi, debris consisting mainly of bile pigments, epithelial cells, and mixed exudates, and sometimes frank pus. Intrahepatic stones are usually multiple, soft, muddy, and often tenaciously adherent to the duct wall. These stones cause progressive biliary obstruction and recurrent infection, resulting in the formation of multiple cholangitic abscesses, biliary strictures, and eventually severe destruction, cirrhosis, and portal hypertension (FAN et al. 1991).

As the name implies, the disease is prevalent in the oriental populations, but more cases have been reported in western countries because of increased immigration from Asian countries (LIM 1991). Patients usually experience several attacks of fever, chills, and jaundice before seeking medical attention. Most patients present with acute cholangitis and others complain of abdominal pain, pancreatitis, or jaundice, or have abnormal liver function tests.

M.C. HAN, MD, Professor, Department of Radiology, Seoul National University Hospital, 28, Yongon-dong, Chongno-gu, 110-744 Seoul, Korea
J.K. HAN, MD, Assistant Professor, Department of Radiology, Seoul National University Hospital, 28, Yongon-dong, Chongno-gu, 110-744 Seoul, Korea
B.I. CHOI, MD, Professor, Department of Radiology, Seoul National University Hospital, 28, Yongon-dong, Chongno-gu, 110-744 Seoul, Korea
J.H. PARK, MD, Professor, Department of Radiology, Seoul National University Hospital, 28, Yongon-dong, Chongno-gu, 110-744 Seoul, Korea

22.2 Etiology and Incidence

Although the exact cause of the disease is not known, it is postulated that portal bacteremia and metabolic derangement caused by poor hygiene, low socioeconomic status, and dietary insufficiency result in cholangitis and stone formation (ONG 1962; WENN and LEE 1972).

The primary cause is postulated to be portal bacteremia resulting from poor eating habits (high in carbohydrate, low in protein). Once the infection reaches the liver, it is excreted into the bile. Bacteria (mainly *Escherichia coli*) in bile have β-glucuronidase, which deconjugates bilirubin glucuronide and facilitates the formation of bilirubinate stone (CHOU and CHAN 1980). This happens especially when the intrahepatic environment is deficient in glucaric acid, a β-glucuronidase inhibitor which is lacking in a low protein diet (MATSUSHIRO et al. 1977). Low socioeconomic status plays a role in imposing poor eating habits and poor general hygiene.

a

b

c

Fig. 22.1. **a** Specimen photograph after a left lateral segmentectomy shows fibrotic thickening of bile duct wall (*arrows*). Multiple black pigment stones are removed from intrahepatic ducts. **b** The cut surface of the left hepatic lobe shows severe fibrotic thickening of the bile duct wall and a wide area of fibrosis in the hepatic parenchyma (*arrows*, white area). **c** Microscopic study (x 10) reveals fibrotic thickening of bile duct wall and infiltration of inflammatory cells. Epithelial lining cells are lost in multiple areas

In some areas, it is believed that infestation with certain parasites plays a role in the pathogenesis of RPC. Chronic infestation with parasites such as *Clonorchis sinensis* or *Ascaris lumbricoides* may induce ductal injury and stricture, leading to stone formation. Also, adult worm per se and clumps of eggs or desquamated epithelium block the bile flow and cause bile stasis. Bile stasis favors bacterial colonization, resulting in pyogenic cholangitis. When pyogenic cholangitis occurs, *C. sinensis* flukes are killed. Bile stasis and the presence of dead flukes and ova can form a nidus for the formation of biliary stones and this could initiate all the consequent pathologic processes, resulting in repeated bouts of cholangitis and stone formation. In areas where these parasites are endemic, many patients with RPC have evidence of their presence. However, many patients infested with the parasites do not have RPC. Moreover, ONG (1962) found that patients with RPC had an only 5% higher rate of infestation with *C. sinensis* than did patients in general outpatient clinics. Perhaps these parasites may play an initiating role in areas where they are endemic.

The incidence of RPC is reported to be as high as 20% in Taiwan and 15% in Korea. However, with recent advances in general economic status in both countries, the incidence of the disease is decreasing (SU et al. 1992; KIM et al. 1993).

22.3 Pathology

The primary pathologic changes in RPC are seen in the bile ducts. Microscopically, there is a proliferation of mucus-producing glandular elements of biliary epithelium. The surface lining of the dilated ducts in which stones are impacted is mostly denuded, and what epithelium still remains usually has a mildly hyperplastic appearance. Infiltration of inflammatory cells is noted in hepatic parenchyma as well as the periductal area. In severe cases, liver abscesses are also present. Fibrous tissue proliferates in the portal tracts, especially around the bile ducts (Fig. 22.1) (OHTA et al. 1991).

The intrahepatic ducts are dilated, but there are also multifocal strictures. Dilated ducts contain pigmented stones, pigmented mud, and debris consisting of desquamated cells, exudates, or pus. The liver is severely scarred and shrunken in the affected area; however, the unaffected area shows compensatory hypertrophy, resulting in a markedly distorted shape of the liver. The papilla is usually hypertrophied but patent, allowing easy passage of a large dilator (LIM 1991).

The bile ducts draining the left lateral and right posterior segments of the liver are the more frequently affected (ONG 1962; LIM et al. 1990; CHAN et al. 1989). This may be related to the more acute angulation of these ducts, which results in less efficient drainage. Some authors speculate that an anomalous intrahepatic ductal union may disturb the normal bile flow by sharp angulation, leading to bile stasis and stone formation (KOGA et al. 1987). However, a recent comparative study has shown that there are no significant differences in incidence of the type of union of intrahepatic bile ducts in patients with hepatolithiasis and cholecystolithiasis versus those with malignant biliary obstruction (YU et al. 1993). Thus, anatomic variations in the main intrahepatic ducts do not seem to be associated with hepatolithiasis, and it is unlikely that they are of etiologic significance in RPC.

The stones encountered in RPC are formed in the bile duct and are composed mostly of bile pigment with variable calcification. The stones are dark brown or black in color and have a soft consistency. They are multiple and are scattered throughout the liver. Occasionally the bile ducts are packed with stones.

Bile duct stones are present in 75%–80% of cases of RPC (LIM 1991). In the remaining cases, there is recurrent cholangitis in the absence of stone. Gallbladder stones are frequently found, and including figures for the incidence of previous cholecystectomy their incidence is about 50%–70% (LIM et al. 1990; CHAN et al. 1989; CHAN et al. 1987).

Recently, it has been suggested that the incidence of cholangiocarcinoma is higher (reportedly as high as 2.4%–12.5%) in patients with RPC. Chronic proliferative cholangitis, frequently seen in stone-bearing ducts, may transform into mucosal dysplasia, which is a precursor of the intraductal spreading type of peripheral cholangiocarcinoma. The prolonged exposure of hyperplastic epithelium to biochemically altered bile is thought to be the causative factor (KOGA et al. 1985; CHEN et al. 1989; OHTA et al. 1991).

22.4 Clinical Features

Men and women are affected almost equally by RPC, and the highest number of cases occur in persons 20–

40 years of age. Patients usually have histories of several attacks of fever, chills, and jaundice. The symptoms are abdominal pain, nausea, and vomiting accompanied by fever, chill, and jaundice. Physical examination reveals epigastric tenderness, rigidity, and hepatomegaly. In severe cases, patients present with manifestations of septic shock. Laboratory findings are leukocytosis and elevated levels of alkaline phosphatase and bilirubin.

Patients with acute cholangitis are managed conservatively with intravenous broad-spectrum antibiotics. Emergency procedures, either surgical or interventional, are reserved for those who do not respond to conservative management. Patients should undergo radiological examinations to evaluate the severity and extent of the disease (i.e., number and location of stones, strictures, lobar atrophy, secondary biliary cirrhosis, associated cholangiocarcinoma). Then, surgical treatment is planned according to the extent of the disease (FAN et al. 1991; LIM 1991).

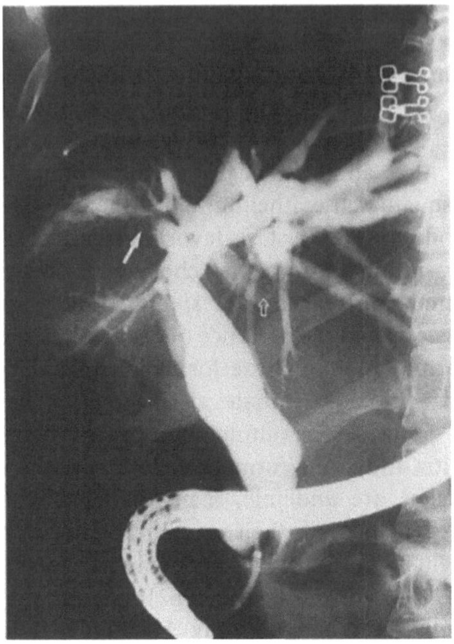

Fig. 22.2. Endoscopic retrograde cholangiopancreaticography shows a focal stricture of the right hepatic duct (*filled arrow*). There is a round filling defect in the left intrahepatic duct (*open arrow*). Because of a patulous sphincter of Oddi, an occlusion balloon catheter is used to fill the biliary tree. Note the small size of the right lobe of the liver, due to biliary cirrhosis

22.5 Radiologic Findings

Radiological assessment of the biliary tree is essential in planning the surgical treatment. Plain radiography frequently reveals air in the biliary tract. Signs of any intraperitoneal inflammatory process, such as paralytic ileus, are also noted. However, stones are only rarely demonstrated by plain radiography.

22.5.1 Cholangiographic Findings

Direct cholangiography, such as endoscopic retrograde cholangiopancreaticography (ERCP), percutaneous transhepatic cholangiography (PTC), operative cholangiography, T-tube cholangiography, or selective cholangiography, demonstrates in detail the shape of ductal changes and the location, number, and size of stones. Abnormal findings include dilated biliary trees, multifocal strictures, findings of cholangitis, and intraductal filling defects (Fig. 22.2).

Ductal dilatation is disproportionately severe in the extrahepatic bile duct and intrahepatic ducts show mild or no dilatation. Only stone-bearing ducts show severe dilatation. Intrahepatic ducts show multifocal strictures, acute tapering, rigidity, straightening, a right-angle branching pattern, and a decrease in arborization, suggesting cholangitis with fibrosis and scarring (WASTIE and CUNNINGHAM 1973; LAM et al. 1978; HONG et al. 1990; HAN et al. 1992). Stones produce filling defects. Sometimes, intrahepatic or extrahepatic stones are very large and stone-bearing ducts show cystic or fusiform dilatation (Fig. 22.3).

When the stones completely obstruct the orifice of the segmental duct or the duct is incompletely opacified because of a patulous sphincter of Oddi, the stones cannot be visualized and intrahepatic stones or stricture will fail to be diagnosed. This can be avoided by taking multiple oblique films and carefully analyzing the branching pattern to identify any missing (nonopacified) segment. For patients with a patulous sphincter of Oddi, use of a balloon catheter during ERCP helps to opacify intrahepatic biliary trees. In postoperative patients who have a T tube in place, selective cholangiography with a preshaped angulated catheter through the T-tube tract is very helpful in finding missing ducts. However, in about 10%–20% of cases cholangiography fails to detect missing ducts, and only computed tomography or ultrasound can demonstrate impacted intrahepatic stones (Fig. 22.4) (HAN et al. 1994; CHANGCHIEN et al. 1992).

A few patients (15%–25%) have no stones or pass stones spontaneously (ONG 1962). In these patients,

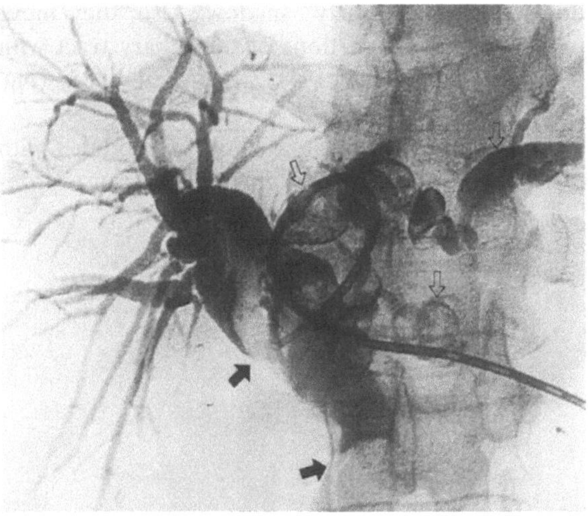

Fig. 22.3. Percutaneous transhepatic biliary drainage (PTBD) was performed in this patient because of acute cholangitis. Contrast injection through a PTBD tube shows markedly dilated left intrahepatic ducts filled with stones (*open arrows*). There are also large stones in the common hepatic and common bile duct (*filled arrows*). Compare to the left lobe, the right ducts do not have stones and show only mild dilatation

the characteristic biliary tree pattern will lead to the correct diagnosis (Lim 1991).

22.5.2 Sonographic Findings

Biliary trees are dilated in almost all patients. Extrahepatic biliary trees are dilated in about 85%–100% of patients (Lim et al. 1990; Chau et al. 1987). Central intrahepatic biliary trees are dilated in 66%–79% of patients. Usually, the peripheral ducts are only slightly dilated or not dilated at all unless they are packed with stones or have significant stenosis. Localized dilatation of intrahepatic ducts is closely related to the location of stricture and stone.

In most patients, a stone or stones can be detected with ultrasonography (Lim et al. 1990; Chau et al. 1987). The stones are either intrahepatic or extrahepatic or both. In some patients, stones are only in segmental duct(s). The most frequent sites are the left lateral segment (segments 2 and 3) and the right posterior inferior subsegment (segment 6) of the liver.

The stones are generally hyperechoic compared to surrounding hepatic parenchyma, but less commonly they may be isoechoic to the liver. Some stones have a posterior acoustic shadow and some do not, even in the same patient. The detectability of stones by ultrasonography depends on their size and location as well as their echogenicity and shadowing characteristics. Biliary mud or isoechoic stones filling the biliary tree as cast can be misdiagnosed because they can be misinterpreted as normal liver and overlooked or regarded as a soft tissue mass (Fig. 22.5). Echogenic stones with a strong posterior shadow impacted in an atrophic left lobe of the liver are sometimes misdiagnosed as overlying bowel gas. Biliary air resulting from a patulous sphincter of Oddi or previous choledochoenteric anastomosis either obscures visualization of stones or produces focal echogenic lesions which are sometimes difficult to differentiate from true intrahepatic stones. Biliary air collections usually are more echogenic, are located just posterior to the anterior wall of the bile duct because they are floating, and produce ring-

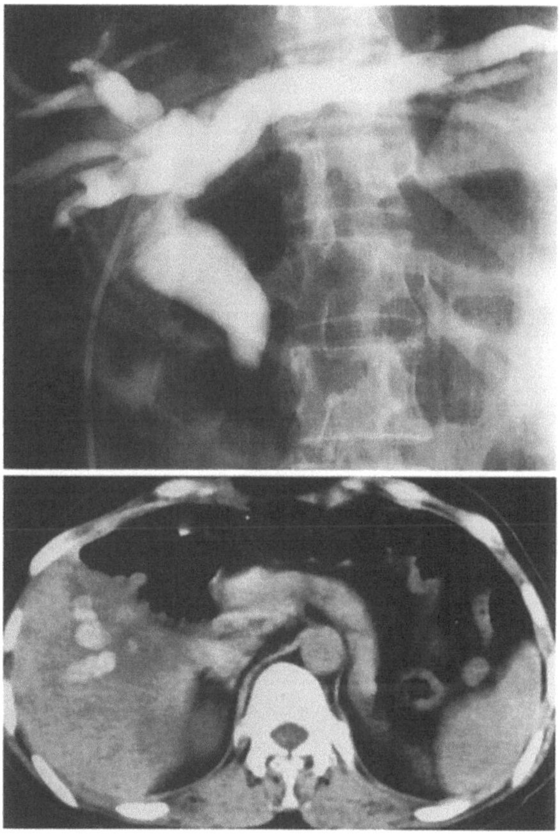

a

b

Fig. 22.4. **a** Selective cholangiography with a 9 French catheter shows a dilated biliary tree. Ducts are straight and rigid and peripheral branches are not filled well (decrease in arborization). The right lobe of the liver is small. However, we could not find any duct containing stone. **b** Non-enhanced CT shows multiple large stones in the right anterior segment of the liver. As this case demonstrates, it is sometimes difficult to recognize a missing duct with cholangiography alone. Nonenhanced CT or ultrasound can be of great help in finding missing ducts down artifact or "dirty" shadow. Also, they move to nondependent portions of the biliary tract with position change (Lim 1991; Choi et al. 1988) (Fig. 22.6).

Sometimes, hyperechoic spots or nodules with posterior acoustic shadow in the liver cause prob-

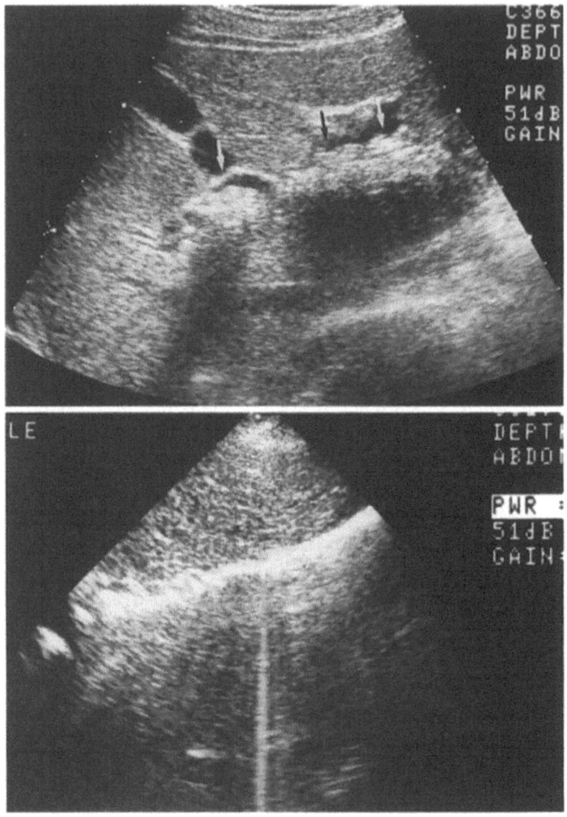

a

b

Fig. 22.6. a Transverse scan of the liver shows multiple intrahepatic stones (*arrows*). The stones have convex anterior borders. In most instances, there is a fluid meniscus between the anterior surface of the stone and the wall of the bile duct. The posterior shadow may be either strong or weak, but does not show any artifact. **b** Biliary air is linear, more echogenic, and floating toward the anterior wall of the bile duct, so that it does not have a fluid meniscus. The posterior acoustic shadow tends to be "dirty" because of ring-down or reverberation artifact

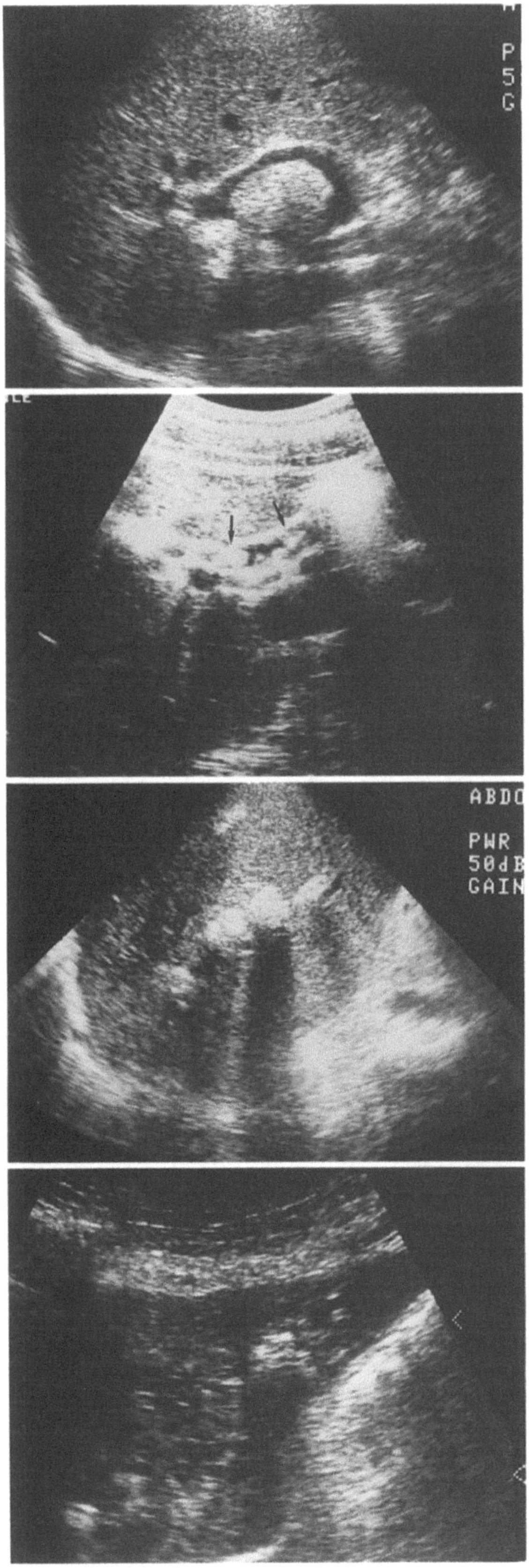

a

b

c

d

Fig. 22.5.a–d. Various sonographic findings in patients with recurrent pyogenic cholangitis. **a** A large echogenic lesion with a weak posterior shadow is noted in the common hepatic duct. The common hepatic duct shows fusiform dilatation. **b** Here, a transverse scan shows multiple stones (*arrows*) in slightly dilated ducts. The bile duct wall is thickened. **c** In the right lobe of this liver, there are multiple highly echogenic lesions with a strong posterior shadow in the liver. **d** Longitudinal scan of the left hepatic lobe shows weakly echogenic stone with a weak posterior shadow lems in differentiating hepatic granuloma from intrahepatic stones. Intrahepatic stones are frequently multiple and are accompanied by ductal dilatation or pneumobilia. In addition, the symptoms or liver function tests are of great help in differentiating stones from granuloma (LIN et al. 1989).

A higher proportion of nonshadowing stones are detected in the extrahepatic than in the intrahepatic biliary duct (LIM et al. 1990; CHAU et al. 1987). This is because the extrahepatic duct is dilated, providing a larger space in which it is easier to detect nonshadowing stones.

Hepatic abscess or biloma associated with cholangitis or ductal obstruction can be easily detected with ultrasonography.

22.5.3 Computed Tomographic Findings

Computed tomography (CT) can demonstrate the full extent of ductal dilatation and distribution of the stones. Because patients with RPC frequently have a patulous sphincter of Oddi' reflux of duodenal content into the biliary tree is frequent. The use of orally administered thin barium or Gastrografin is therefore not recommended, since, once refluxed into biliary trees, oral contrast media with high attenuation may obscure intrahepatic stones. Nonenhanced CT is more helpful in detecting stones, since some intrahepatic stones are faintly calcified and only detectable on nonenhanced CT, being frequently masked after intravenous injection of contrast medium (Fig. 22.7).

CT reveals intra- and extrahepatic biliary dilatation, strictures, calculi, pneumobilia, and segmental atrophy. Central intrahepatic ducts are more dilated as a result of progressive ductal destruction, loss of elastic fibers from previous inflammation, and fibrosis. This tendency of the central ducts to be more dilated than peripheral ones produces a characteristic arrowhead appearance on serial CT sections (Fig. 22.8). There may be additional uniform dilatation or segmental dilatation due to associated stricture or calculous obstruction (CHAN et al. 1989). Stones are detected in 70%–90% of cases (HAN et al. 1994; MENU et al. 1985; ITAI et al. 1980). Low- or isoattenuation stones are difficult to detect with CT. Sometimes, stones are masked by massive pneumobilia (Fig. 22.9).

CT can also demonstrate associated biliary cirrhosis by demonstrating splenomegaly, grossly distorted contour of the liver, and esophageal varices. During acute exacerbation of cholangitis, CT

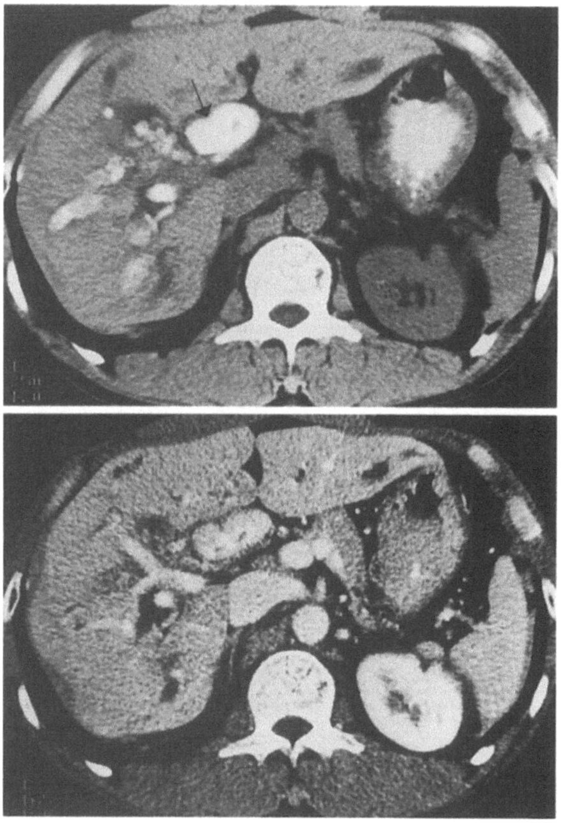

a

b

Fig. 22.7. a Nonenhanced CT clearly shows multiple right intrahepatic stones. The large, high-attenuation lesion in the porta hepatis is a left intrahepatic stone. With orally given contrast medium it could be mistaken for contrast-filled duodenum or gastric antrum. **b** On a CT scan at the same level after intravenous contrast injection, the hepatic parenchyma is enhanced so that intrahepatic stones are almost isodense to it and are difficult to recognize

can show additional features of localized ductal or parenchymal enhancement suggesting cholangitis or suppurative cholangiohepatitis as well as hepatic abscess, biloma, subphrenic abscess, or pleural effusion (Fig. 22.8).

22.5.4 Relative Roles of Imaging Modalities

Ultrasonography is noninvasive, easily available, and accurate. Thus, it is used as the screening procedure in patients who complain of right upper quadrant pain and fever. However, its usefulness is sometimes limited by the presence of biliary air, overlying bowel gas, surgical clips, drains, and surgical scar.

CT reveals the full extent of ductal abnormality and findings of cholangitis or suppurative cholangiohepatitis. However, biliary strictures and

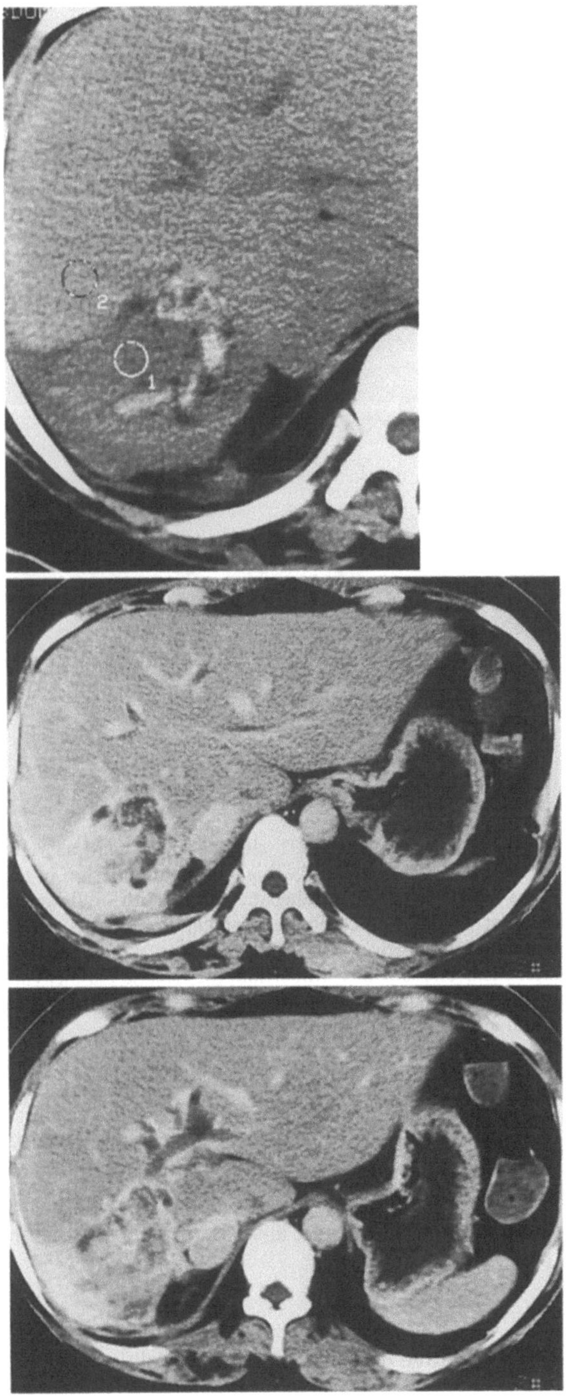

a

b

c

Fig. 22.8.a–c. CT scans show dilatation of the central intrahepatic bile ducts, but the ducts taper abruptly so that peripheral ducts are only minimally dilated. The right lobe of the liver is atrophic. Nonenhanced CT (**a**) shows multiple stones in crowded intrahepatic bile ducts. The hepatic parenchyma shows low attenuation. After intravenous contrast injection (**b,c**), the right lobe of the liver is strongly enhanced, suggesting suppurative cholangiohepatitis

ductal wall irregularities are difficult to evaluate with CT.

Direct cholangiography provides the most detailed anatomy of the biliary trees. It is mandatory as a road map in patients undergoing surgical or radiological intervention. However, it is invasive, and some ducts are not opacified due to severe obstruction or patulous sphincter. CT or ultrasound will provide valuable information for identifying these missing ducts. Thus, the three modalities are complementary to each other.

22.6 Interventional Management

After thorough radiological evaluation, surgical treatment is usually carried out. Cholecystectomy, choledochotomy or hepaticodochotomy and stone extraction or biliary drainage procedures such as choledochojejunostomy are usually performed. However, because of the multiplicity of stones and associated biliary strictures, removal of all intrahepatic stones is not always possible. Partial hepatectomy is the treatment of choice for the small proportion of patients who have severely damaged liver segments with localized intrahepatic stones, particularly if the left lobe is involved. Thus, the rate of residual stones after surgery is very high (range 42%–77%) despite the recent advances in surgical techniques and the increased use of the operative flexible choledochoscope (Fan et al. 1991; Kim et al. 1993; Chang and Passaro 1983).

There are also several factors which preclude further surgical treatment. The morbidity and the mortality associated with the second surgical procedure are higher, and due to dense adhesions surgery is more difficult than at the primary operation. In some patients, poor liver function due to parenchymal destruction and secondary biliary cirrhosis preclude general anesthesia and further surgical treatment. In addition, intrahepatic stones recur even after complete removal in 15% of patients (Fan et al. 1991) (Fig. 22.10). So, nonsurgical removal of the retained or recurrent stone and nonsurgical management of associated biliary stricture is appropriate in these patients.

22.6.1 Fluoroscopy-Guided Stone Removal

Roentgenologic techniques of nonoperative stone extraction in the ambulatory patient represent a significant improvement in the postoperative man-

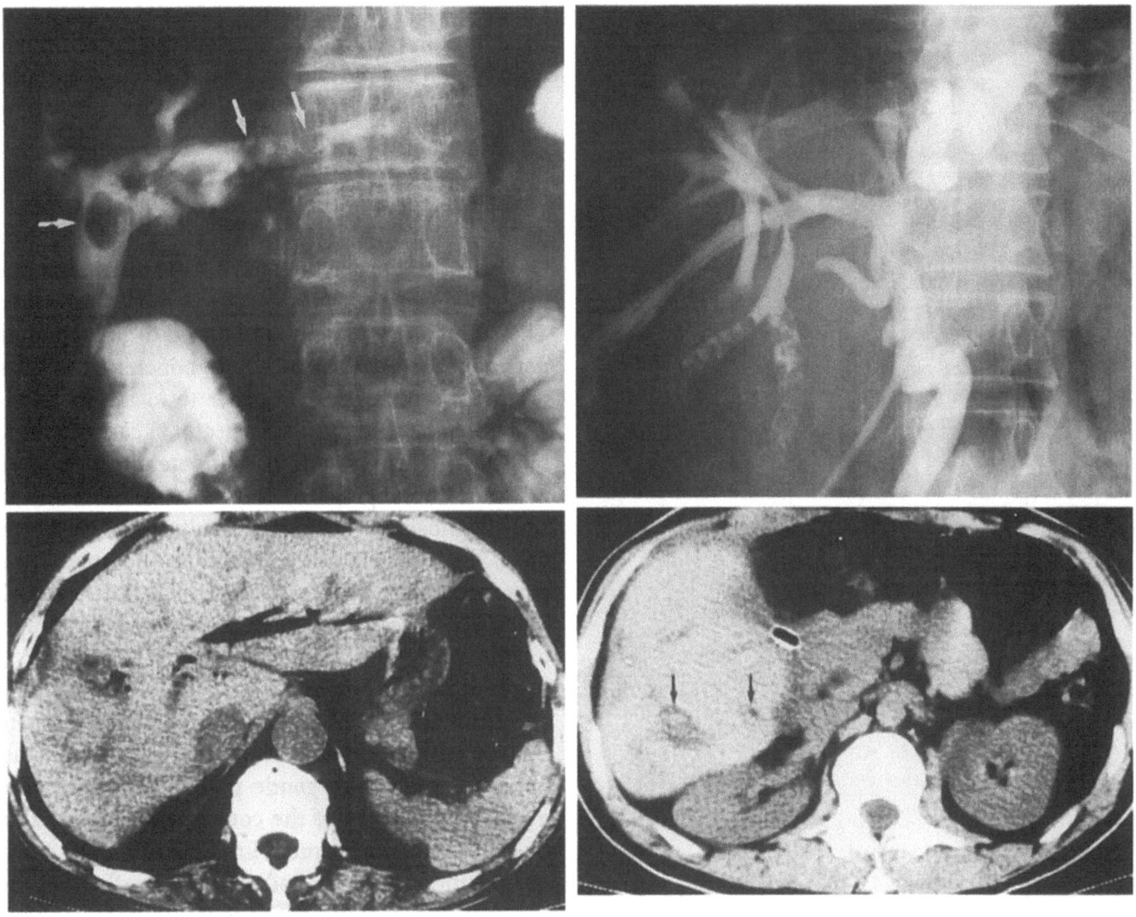

Fig. 22.9. **a** Selective cholangiogram shows multiple stones in the common hepatic and left hepatic ducts (in segment 2, *arrows*). **b** Nonenhanced CT of the same patient taken 1 h earlier at the level of the segment 2 bile duct shows only massive pneumobilia. **c** Selective cholangiogram of another patient shows multiple stones in the right lobe of the liver. Also note the multiple strictures, decreased arborization, and right-angle branching pattern. **d** Nonenhanced CT in the same patient shows isoattenuation lesions in the corresponding bile ducts (*arrows*). Without the cholangiogram, it would be hard to recognize these lesions as intrahepatic stones

agement of the patient with RPC. Trans-T-tube catheterization using standard vascular catheters and guide wire has been used to dislodge sediment and blood clots in order to reestablish drainage (Margulis et al. 1965; Short et al. 1971). Removal of the T-tube is indicated to extract the retained stones through the T-tube tract, the tube having been left in place for 4–5 weeks after the operation to establish a fibrous tract. Mondet extracted stones through mature T-tube tract as early as 1962 with a specially designed forceps. Margarey (1971) used the Dormia ureteral basket and small angiographic catheter. Burhenne (1973) developed a special steerable catheter and reported excellent results in the treatment of retained common duct stones.

While the use of the steerable catheter yielded excellent results in extrahepatic bile ducts, it was difficult to use in the relatively small, tortuous, and angulated intrahepatic bile ducts. So, Chen et al. (1980) used a fiberoptic choledochoscope and Park et al. (1987) an individually fitted preshaped angulated catheter to negotiate the tortuous intrahepatic ducts. Each procedure has its own advantages and drawbacks. In general, better results have been reported in choledochoscopy. However, the use of the choledochoscope requires expensive equipment and well-trained personnel, and carries a higher complication rate. Removal of the stone with the preshaped angulated catheter is relatively simple and economical. The preshaped catheter can negotiate acute angles and strictures of the bile duct so that

a

b

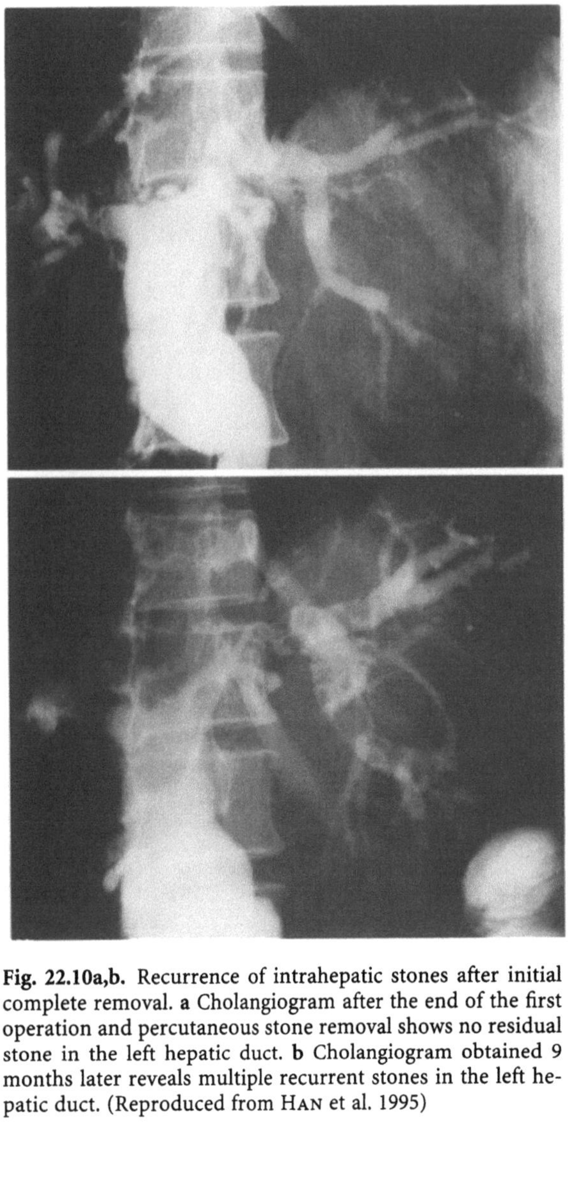

Fig. 22.10a,b. Recurrence of intrahepatic stones after initial complete removal. **a** Cholangiogram after the end of the first operation and percutaneous stone removal shows no residual stone in the left hepatic duct. **b** Cholangiogram obtained 9 months later reveals multiple recurrent stones in the left hepatic duct. (Reproduced from HAN et al. 1995)

it can approach even stones in small peripheral ducts. However, it has a lower success rate than the choledochoscope because of the lack of an effective means of crushing large stones, such as by intracorporeal shock wave lithotripsy, which can be used with the choledochoscope under direct visual control (HAN et al. 1992; CHOI et al. 1992). It also entails high radiation exposure to the patient and to the hand of the operator. In one report (BURHENNE 1973), the patient received about 350 mR per minute of fluoroscopy time to the entry skin at the table top, and the radiologist received 1.9–4.1 mR/min to the dorsum of his right hand and 0.6–1.2 mR/min at waist level outside his apron. To minimize the radiation exposure, one should reduce the size of the fluoroscopic field to as small as possible by tight collimation. The T tube must be inserted straight from the common bile duct to the right lateral abdominal wall, so that the radiologist's hand does not enter the radiation field.

22.6.1.1 Technique

Generally, the procedure starts 4–6 weeks after surgery, to allow maturation of the fibrous T-tube tract. The ambulatory patient is returned to the X-ray department. Before the procedure, nonenhanced CT of the liver is performed in order to locate the stone. The nonenhanced CT is very useful in cases of severe biliary stricture or patulous sphincter of Oddi, where a T-tube cholangiogram may result in a false-negative diagnosis because of difficulty in filling all the ducts, or a false positive diagnosis because of the presence of air bubbles in the ducts. After the CT scan is obtained, the patient is prepared and draped on the fluoroscopic table. Pethidine 50 mg is given intramuscularly prior to the procedure. Broad-spectrum antibiotics are given orally from just after the procedure for 3 days prophylactically. The T tube is then extracted by gentle pulling. This causes an average movement of the common duct by 1–2 cm, the duct returning almost immediately to its previous position. A preshaped angulated catheter – that is, a 9-F polyethylene tube about 30 cm long – and a guide wire with a flexible tip are inserted through the T-tube tract. Several types of angulated catheter with different tip angles can be made beforehand and the most appropriate one is selected according to the

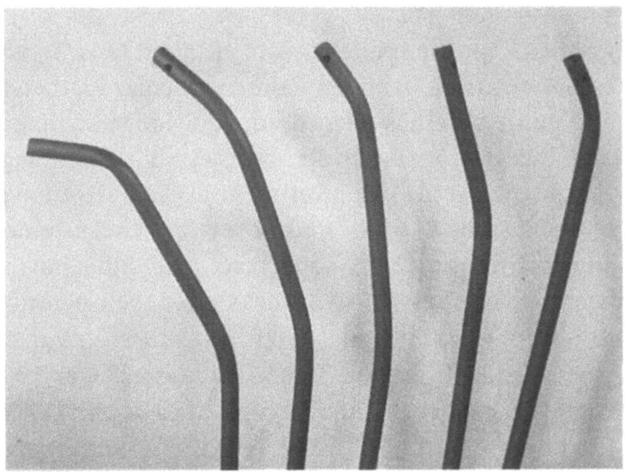

Fig. 22.11. Some frequently used preshaped catheters, made of 9 French polyethylene tubing. The small side hole is made near the tip to prevent adhesion of the catheter to the duct wall. (Reproduced from HAN et al. 1995)

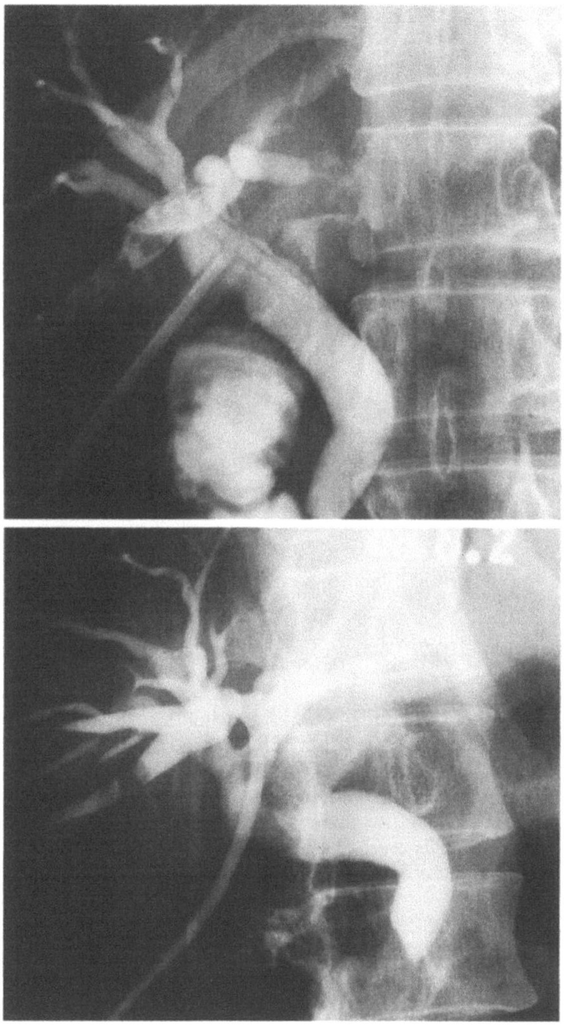
a

b

Fig. 22.12. a T-tube cholangiography shows multiple intrahepatic stones in the posterior inferior segment of the right hepatic lobe. **b** Deep right anterior oblique selective cholangiogram after removal of multiple stones shows that the stone-bearing duct is an aberrant right hepatic duct originating from the left hepatic duct at a steep angle. There is also a focal stenosis at the orifice. However, we were able to access the peripheral duct and remove all the residual stones with a 9 French catheter. (Reproduced from HAN et al. 1992)

anatomy and branching angle of the bile duct (Figs. 22.11, 22.12).

Selective cholangiograms are obtained by selection of individual ducts which may not be demonstrated by conventional T-tube cholangiography due to stricture or impacted stones. Nonenhanced CT can be a road map for the selection of individual ducts. After a thorough evaluation of the shapes of the bile ducts and the location of the stones, the selected preshaped catheter is inserted in the peripheral duct to beyond the stones. A Dormia stone basket is inserted through the catheter beyond the stone. The basket is then expanded by withdrawing the catheter. The stone is trapped by rotation of the basket and is extracted through the tube tract (Fig. 22.13). Usually Dormia baskets 9–15 mm in diameter are used; occasionally, larger sizes up to 25 mm in diameter are used. If the trapped stone is too large to be extracted, it can be crushed between the basket and the catheter by pulling, strongly on the basket while the catheter is held tightly. Usually, the intrahepatic stones are friable and easily crushed. After the crushing, fragments are extracted separately by basket or removed by repeated irrigation and suction. For the latter, a large Nelaton rubber tube (16–20 French outer diameter, 30 cm long) with many side holes is inserted into the biliary tree through the T-tube tract. Repeated infusion and aspiration with saline through the tube is carried out. Small particles are aspirated or passed into the duodenum, and larger fragments are impacted at the side hole of the tube and can be removed with the tube while continuous negative pressure is applied.

A large stone that cannot be trapped with a Dormia basket can be trapped and crushed using a mechanical lithotriptor basket or snare with a folded guide wire. In cases of severe stricture which impede the passage of the 9 French catheter, ductoplasty can be performed with a 5–7 French angioplasty balloon catheter. Usually a balloon catheter 6–10 mm in diameter and 2–4 cm long is applied at 5–10 atm for 1–3 min. For analgesia, 5–10 ml 2% lidocaine is instilled into the bile duct.

In cases of multiple small stones in peripheral ducts, the orifices of which are not stenotic, stones can be easily displaced in to the common duct by a 1-cm outer diameter Fogarty balloon catheter or an occlusion balloon. Simple follow-up observation usually reveals spontaneous passage of small fragments. Also, relocation can be easily induced by forceful injection or aspiration of normal saline through the catheter near the stone.

To remove multiple peripherally impacted stones, a percutaneous transhepatic biliary drainage (PTBD) procedure with an 8–9 French catheter can be done. The catheter is inserted into the duct containing the stones. The stones can then be trapped and crushed, or can be displaced to a more central duct where they can be easily trapped. Generally, extraction of large stones through a small PTBD tract is not recommended because damages the PTBD tract and one may lose the access route to the bile duct. Instead, we recommend crushing the stones into smaller fragments and removing the fragments with the irriga-

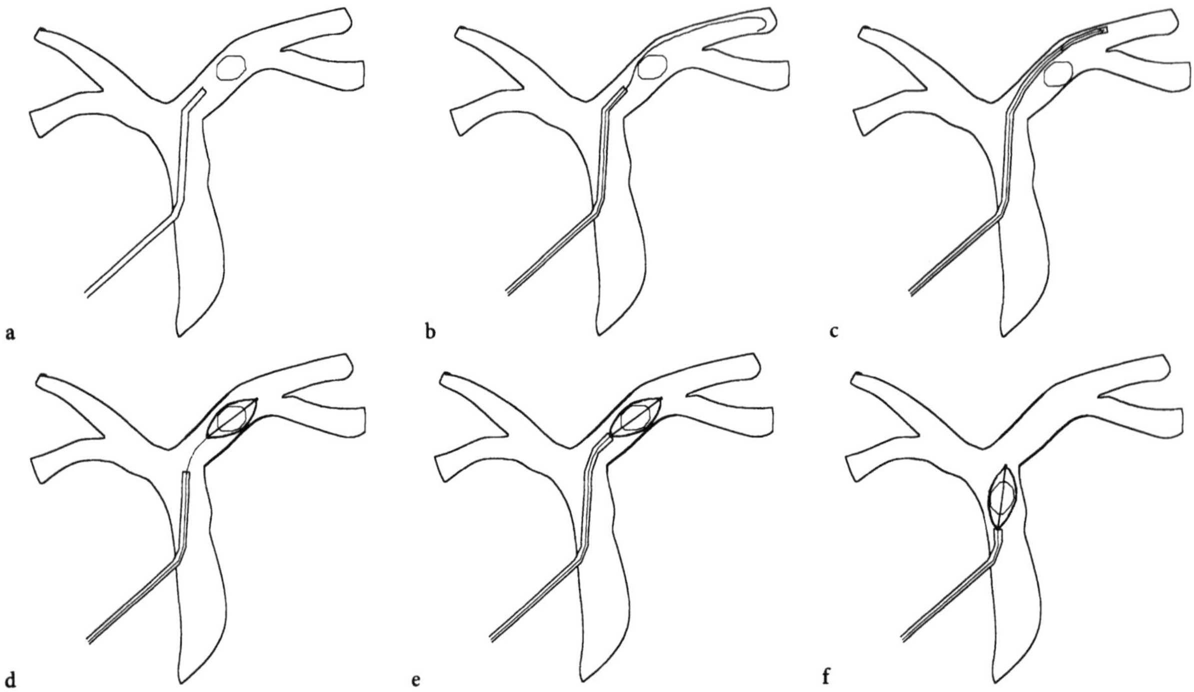

Fig. 22.13a–f. Technique used during percutaneous catheter removal of intrahepatic stones. **a** The stones are located by cholangiography after removal of the T-tube. **b** First, the guide wire is passed in to the distal intrahepatic duct beyond the stone. Then, the preshaped catheter is advanced following the guide wire. **c** After removal of the guide wire, a Dormia stone basket is inserted through the catheter to beyond the stone. **d** The basket is expanded by withdrawing the catheter, and the stone is trapped by rotation of the basket. **e** After the stone is trapped, it is securely held within the basket by readvancing the catheter and maintaining a firm pressure against the basket and stone. **f** The stone is either extracted through the T-tube tract or is crushed by pulling the basket strongly while keeping a light hold on the catheter

tion and suction technique (HAN et al. 1992; PARK et al. 1987; CHOI et al. 1992).

The size of the stone which can be extracted depends on the size of the tract. Thus, the use of a large-bore T-tube is recommended to surgeons to facilitate postoperative percutaneous removal. The tube should be placed at a right angle to the common duct, then through a lateral stab wound using a short, straight line to the skin (BURHENNE 1973).

22.6.1.2 *Extracorporeal Shock Wave Lithotripsy of Retained Stones*

Extracorporeal shock wave lithotripsy (ESWL) is a new procedure in the nonsurgical treatment of retained bile duct stones. With conventional methods, large stones that cannot be trapped or impacted stones which hinder the expansion of the basket are frequently encountered obstacles which prevent the successful removal of retained stones. Several nonoperative procedures such as laser lithotripsy and direct electrohydraulic lithotripsy have been developed (PICUS et al. 1989; FAN et al. 1989; DAWSON et al. 1992). However, laser or electrohydraulic lithotripsy requires close and effective physical contact of the probe with the stone to ensure efficacy and avoid perforation of the bile duct wall. Moreover, the need for additional devices, such as a choledochofiberscope, or a basket to hold the floating stone, may render these technically demanding procedures yet more difficult. With the introduction of ESWL, another technique has become available for the nonsurgical management of bile duct stones. The piezoelectric lithotriptor has low energy per pulse, a high pressure gain, and a small focal volume. Thus, it causes the least pain and does minimal tissue damage. Therefore, it may the most appropriate type of lithotriptor for intrahepatic stones (CHOI et al. 1991).

Intrahepatic stones are located with a real-time sonographic probe before the patient is placed on the lithotriptor. Once the patient is on the lithotriptor, the shock wave generator is coupled with the patient's skin via ultrasound gel. Either the prone position–subcostal route or the right lateral decubitus-intercostal route can be used. The patient position depends on the location of the stones and the sonic window which permits the best visualization of the stones. The stones are targeted by means of an in-line transducer and continuous ultrasound monitoring of the stones is done in real time throughout the course of ESWL. In the case of the piezoelectric lithotriptor (EDAP LT.01; EDAP, Croissy Beaubourg Marine La Val, France), the shock wave can be given at a rate of 5 shots per second at 30–50% power. Each session last 30–60 min (9000–18000 shock waves). After each procedure, a selective cholangiogram is obtained 1 day after ESWL. If the stone is fragmented and mechanical extraction is possible, then the ESWL is not repeated. However, if the fragmentation is not satisfactory, the ESWL procedure is repeated at weekly intervals until successful fragmentation is achieved. To facilitate early fragmentation and to locate and visualize the stones better, normal saline can be dripped into the bile duct

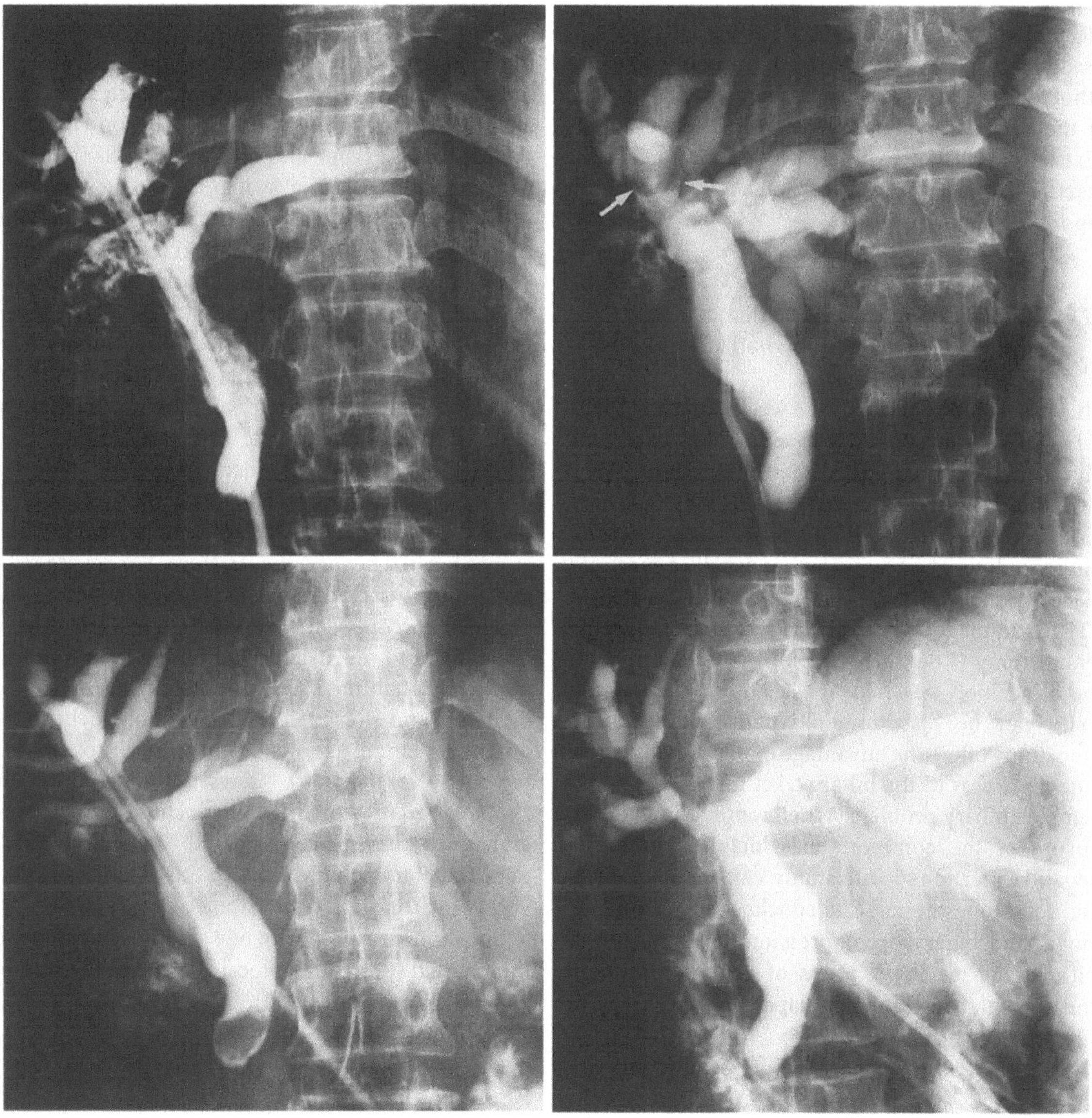

a b c d

Fig. 22.14a–d. Effect of extracorporeal shock wave lithotripsy (ESWL) on intrahepatic stones **a** Initial cholangiogram shows numerous intrahepatic stones in the right lobe of the liver. **b** After two percutaneous stone removal sessions, cholangiography reveals a stone (*arrows*) that could not be trapped with a Dormia stone basket because of its large size and because it was impacted. **c** After ESWL, the stone was displaced into the distal common duct. Although it was still large and was not fragmented, the relatively capacious diameter of the common duct made removal easy. **d** After seven ESWL sessions and eight percutaneous stone removal sessions, all residual stones were removed through the T-tube or PTBD tube during ESWL. Fragmentation of intrahepatic stones does not need to be as fine as for lithotripsy of gallbladder stones; breaking the stone into two or three fragments usually allows successful basket extraction. Sometimes, stones are not grossly fragmented, but may be rendered more fragile or be displaced to central ducts so that impaction or obstruction is relieved and mechanical extraction is possible (Fig. 22.14).

There is a certain amount of controversy regarding the upper limit of the number of shock waves that may be given safely without complication. Although the piezoelectric lithotriptor causes least pain and tissue damage compared to the other types of lithotriptor, it has caused some parenchymal damage in the liver and gallbladder in animal experiments (Ell et al. 1989). However, no significant adverse reaction caused by piezoelectric ESWL has yet been reported in humans.

22.6.2 Choledochoscopic Stone Removal and Intracorporeal Electrohydraulic Lithotripsy

This procedure is performed 4 weeks after the surgical exploration under no or mild parenteral diazepam sedation. General anesthesia is used only if a complex, traumatizing maneuver is planned. Antibiotics are given to all patients for 3–5 days following the procedure. The access route is usually a surgically placed T-tube tract. However, the tract created by PTBD drainage or a special hepaticocutaneous jejunostomy (Roux-en-Y hepaticojejunostomy with the blind loop attached to the anterior abdominal wall) can be also used. Hepaticocutaneous jejunostomy offers the advantage of permanent percutaneous access to the biliary tract in patients with a complex biliary problem. When reoperation and exposure of the common bile duct is required, a hepaticocutaneous jejunostomy can be opened under local anesthesia. Choledochoscopic stone removal and biliary decompression can be achieved without the metabolic stress of general anesthesia and the surgical trauma of laparotomy (Fan et al. 1993).

The choledochoscope has a steerable tip and channel for saline irrigation or for the passage of various instruments such as grasping forceps, stone basket, balloon catheter, or irrigation catheter. The flexibility of the catheter permits complete visualization of the common duct and up to second or tertiary radicals of intrahepatic ducts. Mud and small grains are removed with saline irrigation. If a stone is present, a stone basket is passed beyond the stone, opened, and manipulated until the stone is trapped. This is then tightened and retrieved, basket and endoscope together. The balloon catheter is used to dislodge stones impacted in the smaller ducts. Large stones are crushed with stone-crushing baskets or biopsy forceps into smaller bits, which are then removed with the basket or the grasping forceps. Intracorporeal electrohydraulic lithotripsy can also be done with a flexible electrode under direct visualization (Picus et al. 1989; Fan et al. 1989). A 5 French electrohydraulic probe is inserted through the working channel of the choledochoscope and directed until it is in contact with the stone. Care is taken to ensure that the probe is at least 2 cm beyond the tip of the choledochoscope and not touching the bile duct wall. At the press of a switch, it produces a high-voltage spark between two coaxial electrodes at the tip of the probe. The spark vaporizes the surrounding fluid, creating a high-amplitude hydraulic pressure wave which can fragment stones. Under direct vision, ductal injury is minimized. Preliminary trapping of stones is not absolutely necessary as the impacted stones are stationary and are not likely to be repelled by the shock wave.

When a stricture is encountered, dilatation is done with a Gruentzig balloon catheter to permit passage of the endoscope to remove the retained stones located beyond the stricture. For severe stricture which precludes the passage of a balloon or choledochoscope, PTBD of the strictured duct is done distal to the stricture and these stones are removed through the PTBD tract.

Before and after each session of choledochoscopy, a cholangiogram is taken. The whole procedure is repeated every 5–7 days until both the cholangiogram and the choledochoscope show all stones to have been cleared. When a hepaticocutaneous jejunostomy is used for choledochoscopy, once all the stones are removed the jejunostomy is closed with one seromuscular layer of interrupted suture under local anesthesia (Choi et al. 1986; Gandini et al. 1990).

22.6.3 Results

The rate of successful removal of retained stones depends on several factors: the number and location of the stones, the presence and degree of biliary strictures, and the anatomy of the bile ducts. Removal of extrahepatic bile duct stones is relatively easy and excellent results are reported – 95%–100% (Han et al. 1992; Park et al. 1987; Choi et al. 1992; Choi et al. 1986). However, for the removal of retained intrahepatic stones, success rates are lower. In fluoroscopy-guided removal, complete stone removal is possible in about 50% of patients; in 20% of the patients more than two-thirds of the stones can be removed and the patient is free of recurrent symptoms after tube clamping. Thus, the overall success rate is about 70%. The average number of sessions required is 3.7 ± 2.9, the range is from 1 to 22 (Choi et al. 1992). In choledochoscopy-guided removal, complete removal is possible in 82%–90%. The average number of sessions required is 4 and the range is from 1 to 22 (Fan et al. 1991; Choi et al. 1986; Gandini et al. 1990). Severe angulation deformities, strictures of the bile duct which impede the passage of the guide wire or devices, and impacted stones are the most frequently encountered problems (Table 22.1). With the recent advent of ESWL, these problems can be solved provided the stones can be aimed at with a large in-line transducer. In our experience at the Seoul National University Hospital, percutaneous fluoroscopy-guided removal combined with ESWL resulted in complete removal of retained stones in 25 and partially successful removal in 5 out of 35 patients who were treated in the last 2 years. Thus, the complete success rate was 71.4% and the overall success rate 85.7%. In combined treatment, the number of sessions for ESWL was 8.3 ± 5.7 and the number for stone removal was 5.9 ± 4.1.

Long-term follow-up results after successful stone removal have been reported for small groups. Stones recurred in 3% of patients with extrahepatic stones and 15.8% of those with intrahepatic stones (Fan et al. 1991; Gandini et al. 1990).

Table 22.1. Causes of failure of stone removal in 170 patients. (Reproduced from Choi et al. 1992)

Cause	No. of cases
Angulation	14
Stricture	24
Impacted stones	16
Large stones	2
Tortuous tract	2
Stone in short peripheral duct	1
Too many stones	1
Follow-up refused	5
Total[a]	65

[a]Some patients had more than one cause.

22.6.4 Complications and Their Prevention

As complications of catheter removal, right upper quadrant pain, nausea, and vomiting occur not infrequently during the procedure. Balloon dilatation is usually followed by minor hemobilia. However, these complications require only observation or may be treated with analgesics. Fever and chills caused by periprocedural infection are controlled with oral antibiotics and subside in 2 or 3 days. In our 10-year experience of catheter removal in 170 patients, two patients had an abscess in the liver and the T-tube tract, respectively, and one patient experienced hemobilia that required transfusion after traumatic catheterization. No deaths have yet been reported (Choi et al. 1992).

Choledochoscopic removal of stones carries a higher complication rate than fluoroscopy-guided catheter removal. Gandini et al. (1990) reported 3 fatalities and 13 complications among 97 patients. Four suffered pancreatitis, one acute tubular necrosis caused by contrast media, four patients had subphrenic fluid collections treated by percutaneous catheter drainage, and one suffered an intestinal perforation. Two patients had asymptomatic dissection of the main bile duct which resolved spontaneously by maintenance of internal drainage for 3 days. One patient developed persistent hemobilia that required a blood transfusion.

Gentle manipulation is required to avoid injury to the bile duct and the tube tract. The catheter should be advanced with caution and must follow the guide wire. Forceful injection of saline or contrast material into a bile duct, the orifice of which is severely strictured, results in an influx of bile or injected fluid into the liver parenchyma and blood vessels and leads to bacteremia or liver abscess. In some patients, the the tube tract is insufficiently mature after 4 weeks and one may lose a tract with even minor trauma, especially in patients being treated through the PTBD tract. In such cases, insertion of two guide wires, one as a safety guide wire and one as a working guide wire, is recommended.

22.6.5 Management of Biliary Strictures

Intrahepatic biliary strictures are a frequent associated finding in RPC. They are found in about 90% of patients and are a main cause of failure of conventional surgical treatment or postoperative stone removal as well as stone recurrence. Clinically, biliary stricture is characterized by repeated episodes of cholangitis with sepsis or progressive fibrosing inflammation of the involved bile ducts leading to biliary cirrhosis. To prevent progression of the inflammation and to minimize the recurrence of stones, it is mandatory to dilate the strictures and to extract the stones associated with the stricture (RYEON et al. 1993).

Recent data suggest that surgical reconstruction is the initial choice for most patients with primary postoperative benign stricture of extrahepatic bile ducts. Balloon dilatation is reserved for anastomotic stricture after biliary reconstruction or hilar stricture (MILLIS et al. 1992). However, the characteristics of the postoperative stricture (full-thickness fibrous stenosis) are different from those seen in RPC, which have more inflammatory mucosal lesions (JENG et al. 1990). Furthermore, the multiplicity of the lesions and the intrahepatic location of the stricture preclude surgical treatment. For multifocal intrahepatic strictures, dilatation with angioplasty balloon is the only possible option except for liver transplantation.

The balloon can be introduced through either a mature T-tube tract or a PTBD tract. Usually an angioplasty balloon 6–10 mm in diameter and 2–4 cm in length is used. Before the balloon is inflated, 50 mg pethidine is given intramuscularly and 2% lidocaine is instilled into the bile duct to reduce the pain. Broad-spectrum antibiotics (ampicillin 500 mg orally every 6 h) are given for 3 days afterward. The balloon is inflated and maintained for about 1 min. Dilatation of an intrahepatic biliary stricture greatly facilitates the removal of retained stones proximal to the stricture. After complete removal of stones, the stricture is stented by leaving a small-diameter Silastic catheter in place for 4–8 weeks. For patients who have a long stricture, we recommend keeping the stent in for longer.

Although long-term follow-up results after balloon dilatation of intrahepatic stricture are not available, current data suggest that the cumulative probability of recurrence of stricture is 4% at 2 years, 6% at 2.5 years, and 8% at 3 years (JENG et al. 1990). However, the progression of a benign stricture might be slow, and partial obstructions are sometimes completely asymptomatic for long periods, so longer follow-up is needed to appraise the efficacy of balloon dilatation in the treatment of biliary stricture. The most frequent complications of balloon dilatation are transient hemobilia just after the inflation of the balloon. Occasionally, cholangitis and massive bleeding may occur. Associated biliary cirrhosis increases the risk of massive hemobilia, owing to the hypervascularity around the proliferative bile ducts and the arterial hyperperfusion secondary to portal hypertension (MILLIS et al. 1992; JENG et al. 1990).

Self-expanding metallic stents have been used to treat intractable benign strictures. Initial results were favorable. However, as follow-up gets longer, more cases of delayed obstruction or stenosis of metallic stents are being reported. SUNG et al. (1993) reported an 18.7% incidence of stent obstruction after a median follow-up of 18 months (range 10–29 months) in 16 patients with hepatolithiasis. Although an obstructed metallic stent can be managed by the deployment of another stent within the lumen of occluded stent, the first stent may become an obstacle to further interventional procedures.

One thing we should bear in mind during the treatment of biliary stricture is the possibility that the stricture is malignant. Cholangiocarcinoma is found in 2%–10% of patients with RPC (KOGA et al. 1985; CHEN et al. 1989, 1993; OHTA et al. 1991; FAN and WONG 1992). The tumor can be either intra- or periductal, of the spreading type or the massive type. The massive type can be diagnosed easily with ultrasound or CT (Fig. 22.15). However, with the intra- or periductal spreading type it is quite difficult to differentiate a benign stricture from cholangiocarcinoma on the basis of radiological findings alone (Fig. 22.16).

For strictures that do not respond well to balloon dilatation, malignancy must be ruled out by a more aggressive work-up. Exfoliative cytology or brush

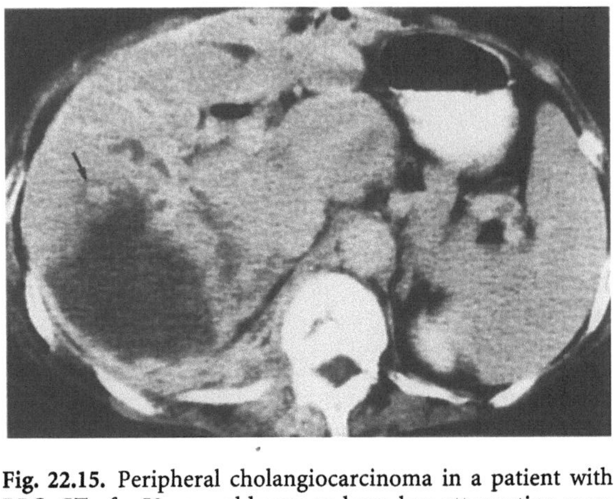

Fig. 22.15. Peripheral cholangiocarcinoma in a patient with RPC. CT of a 52-year-old woman large low-attenuation mass in the right lobe of the liver. Note also the presence of atrophy of the left hepatic lobe, hypertrophy of the caudate lobe, dilated right intrahepatic ducts, and intrahepatic stones (*arrow*)

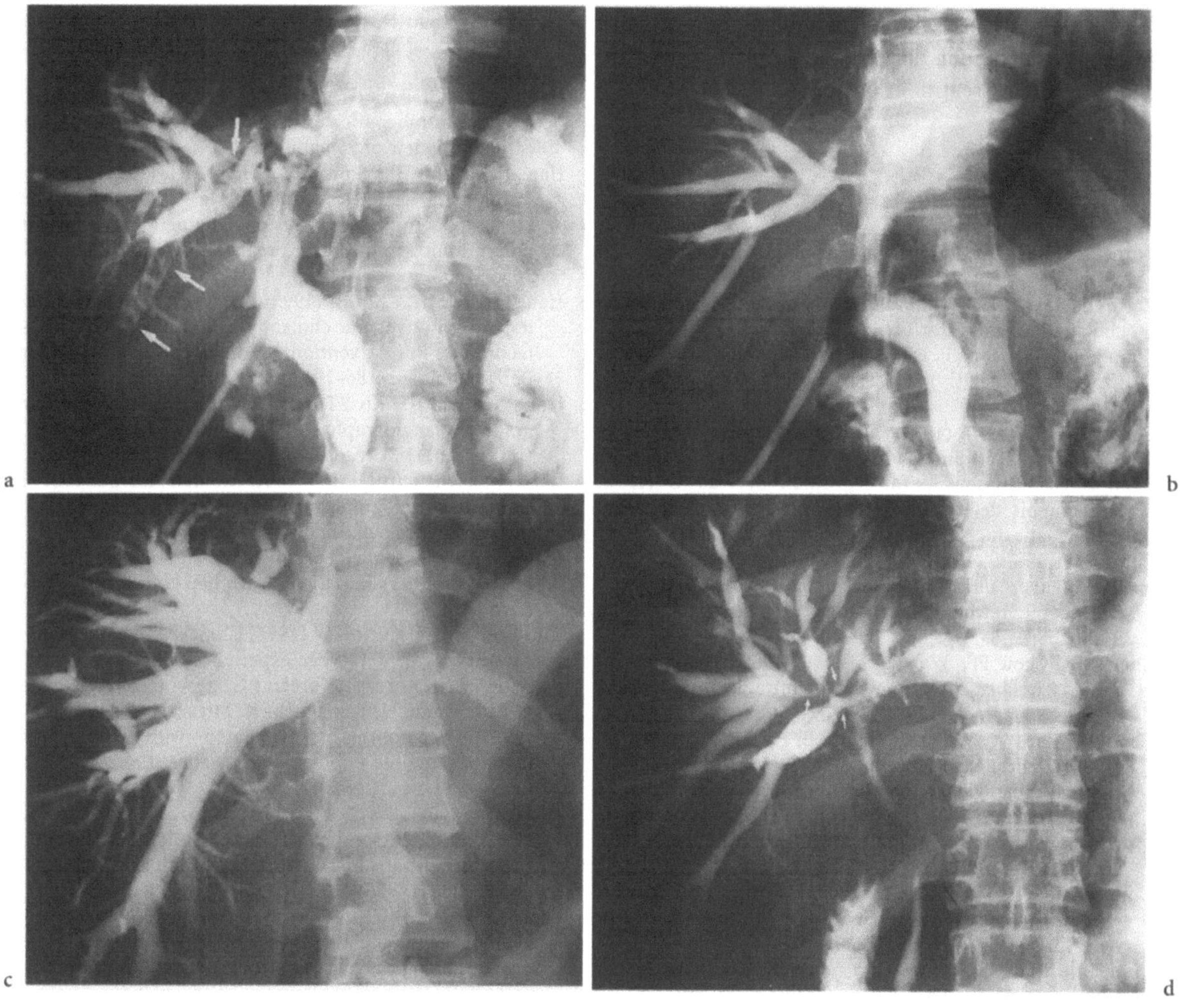

Fig. 22.16a–d. Hilar cholangiocarcinoma in a patient with RPC. **a** Cholangiogram of a 42-year-old woman shows multiple stones in the right hepatic duct (*arrows*). **b** After total removal of residual stones, there is a focal stenosis at the orifice of the right hepatic duct. A 14 French silastic tube was left as a stent and removed 2 months later. At this time, the stricture was believed to be benign and caused by recurrent cholangitis. **c** Six months after tube removal, the patient returned to the hospital because of jaundice. Percutaneous transhepatic biliary drainage was performed and the cholangiogram shows total obstruction of the right hepatic duct. **d** Four months later, the level of obstruction had progressed to the secondary branches (*arrows*). Abdominal CT (not shown here) revealed multiple lymphadenopathies in the periportal area

biopsy has a high spocificity but a low sensitivity (about 40%) for cholangiocarcinoma; however, three consecutive negative results greatly reduce the likelihood that cholangiocarcinoma is present (Rabinovitz et al. 1990). Forceps biopsy under fluoroscopic or choledochoscopic guidance can also be used to make a histological diagnosis.

22.7 Conclusions and Future Prospects

The combined use of all available modalities in each case of RPC is highly recommended to achieve better results. Despite multiple sessions and combination of many technical variables, results are not satisfactory at present. More efficient and safe methods of crushing stones are awaited.

Another problem remaining is the recurrence of intrahepatic stones in pathological liver and stenosed bile ducts even after complete removal of retained stones. Self-expanding metallic stents are used in some centers on an experimental basis to treat intractable biliary stricture. However, the interim results are not as satisfactory as expected and long-term follow-up is needed to for definitive conclusions to be drawn.

For the patients with recurrent cholangitis and severe parenchymal damage, liver transplantation could be a new treatment option.

References

Burhenne HJ (1973) Nonoperative retained biliary tract stone extraction, AJR 117:388-399

Carmona RH, Crass RA, Lim RC, Trunkey DD (1984) Oriental cholangitis. Am J Surg 148:117-124

Chan F, Man S, Leong L, Fan S (1989) Evaluation of recurrent pyogenic cholangitis with CT: analysis of 50 patients. Radiology 170:165-169

Chang TM, Passaro E (1983) Intrahepatic stones: the Taiwan experience. Am J Surg 146:241-244

Changchien C, Chen J, Tai D, Chiou S, Kuo C (1992) Sonographic detection of stones in poorly opacified left intrahepatic ducts. J Clin Ultrasound 20:121-125

Chau EMT, Leong LY, Chan Fl (1987) Recurrent pyogenic cholangitis: ultrasound evaluation compared with endoscopic retrograde cholangiopancreatography. Clin Radiol 38:79-85

Chen MF, Chou FF, Wang CS, et al (1980) Postoperative choledochofiberscopic removal of intrahepatic stones. J Formos a Med Assoc 79:700-705

Chen MF, Jan YY, Wang CS, Jeng LB, Hwang TL, Chen SC (1989) Clinical experience in 20 hepatic resections for peripheral cholangiocarcinoma. Cancer 64:2226-2232

Chen MF, Jan YY, Wang CS, et al (1993) Reappraisal of cholangiocarcinoma in patient with hepatolithiasis. Cancer 71:2461-2465

Choi TK, Lee M, Lui R, et al (1986) Postoperative flexible choledochoscopy for residual primary intrahepatic stones. Ann Surg 203:260-265

Choi BI, Han MH, Kim SH, Cho WH, Kim CW (1988) Intrahepatic stone vs gas: sonographic differentiation (in Korean). Korean J Med Ultrasound 7:7-10

Choi BI, Han JK, Park YH, et al (1991) Retained intrahepatic stones: treatment with piezoelectric lithotripsy combined with stone extraction. Radiology 178:105-108

Choi BI, Han JK, Han MC (1992) Percutaneous removal of retained intrahepatic stones utilizing combination of techniques with emphasis on a preshaped angulated catheter: review of 170 patients. Eur Radiol 2:199-203

Chou ST, Chan CW (1980) Recurrent pyogenic cholangitis: a necropsy study. Pathology 12:425-428

Cook J, Hou PC, Ho HC, McFadzean AJS (1954) Recurrent pyogenic cholangitis. Br J Surg 42:188-203

Dawson SL, Mueller PR, Lee MJ, et al (1992) Treatment of bile duct stones by laser lithotripsy: results in 12 patients. AJR 158:1007-1009

Ell CH, Kerzel W, Heyder N, et al (1989) Tissue reaction under piezoelectric shock wave application for the fragmentation of biliary calculi. Gut 30:680

Fan ST, Wong J (1992) Complications of hepatolithiasis. J Gastroenterol Hepatol 7:324-327

Fan ST, Choi TK, Wong J (1989) Electrohydraulic lithotripsy for biliary stones. Aust N Z J Surg 59:217-221

Fan ST, Choi TK, Lo CM, et al (1991) Treatment of hepatolithiasis: improvement of results by a systematic approach. Surgery 109:474-480

Fan ST, Mak F, Zheng SS, et al (1993) Appraisal of hepaticocutaneous jejunostomy in the management of hepatolithiasis. Am J Surg 165:332-335

Gandini G, Righi D, Regge D, et al (1990) Percutaneous removal of biliary stones. Cardiovasc Intervent Radiol 13:245-251

Han JK, Choi BI, Park JH, et al (1992) Percutaneous removal of residual intrahepatic stones with a pre-shaped angulated catheter: review of 96 patients. Br J Radiol 65:9-13

Han JK, Choi BI, Shin YM, Han MC (1994) Retained intrahepatic stones: comparative study of T-tube cholangiography, selective cholangiography and computed tomography (in Korean). J Korean Radiol Soc 30:493-498

Han JK, Park JH, Choi BI (1995) Interventional management of recurrent pyogenic cholangitis. In: Cope C (ed) Current techniques in interventional radiology. Current Medicine, Philadelphia, pp 186-196

Hong KS, Lim JH, Ko YT, Lee DH (1990) Endoscopic retrograde cholangiography in recurrent pyogenic cholangitis (in Korean). J Korean Radiol Soc 26:117-120

Itai Y, Araki T, Furui S, Tasaka A, Atomi Y, Kuroda A (1980) Computed tomography and ultrasound in the diagnosis of intrahepatic calculi. Radiology 136:399-405

Jeng KS, Yang FS, Ohta I, et al (1990) Dilatation of intrahepatic biliary strictures in patients with hepatolithiasis. World J Surg 14:587-593

Kim SW, Park YH, Choi JW (1993) Clinical and epidemiological analysis of a 10 year experience with 1719 gallstone patients (in Korean). Korean J Gastroenterol 25:159-167

Koga A, Ichyama H, Yamaguchi K, Miyazaki K, Nakayama F (1985) Hepatolithiasis associated with cholangiocarcinoma. Possible etiologic significance. Cancer 55:2826-2829

Koga A, Watanabe K, Takiguchi S, Miyazaki K, Nakayama F (1987) Etiologic significance of anatomic variations in the main intrahepatic bile ducts in hepatolithiasis. Acta Radiol 28:285-288

Lam SK, Wong KP, Chan PKW, Ngan H, Ong GB (1978) Recurrent pyogenic cholangitis: a study by endoscopic retrograde cholangiography. Gastroenterology 74:1196-1203

Lim JH (1991) Oriental cholangiohepatitis: pathologic, clinical and radiologic features. AJR 157:1-8

Lim JH, Ko YT, Lee DH, Hong KS (1990) Oriental cholangiohepatitis: sonographic findings in 48 cases. AJR 155:511-514

Lin H, Changchien C, Lin D (1989) Hepatic parenchymal calcifications: differentiation from intrahepatic stones. J Clin Ultrasound 17:411-415

Margarey CJ (1971) Non-surgical removal of retained biliary calculi. Lancet 1:1044-1046

Margulis AR, Newton TH, Najarian JS (1965) Removal of plugs from T-tube by fluoroscopically guided catheter: report of case. AJR 93:975-977

Matsushiro T, Suzuki N, Sato T, Mark T (1977) Effect of diet on glucaric acid concentration in bile and the formation of calcium bilirubinate stones. Gastroenterology 72:630-633

Menu Y, Lorphelin JM, Scherrer A, Grenier P, Nahum H (1985) Sonographic and computed tomographic evaluation of intrahepatic calculi. AJR 145:579-583

Millis JM, Tompkins RK, Zinner MJ, et al (1992) Management of bile duct strictures: an evolving strategy. Arch Surg 127:1077-1084

Mondet A (1962) Technica de la extraccion incruenta de los calculos la litiasis residual del coledoco. Bol Soc Cir B Aires 46:278-290

Ohta T, Nagakawa T, Ueda N, et al (1991) Mucosal dysplasia of the liver and the intraductal variant of peripheral cholangiocarcinoma in hepatolithiasis. Cancer 68:2217-2223

Ong GB (1962) A study of recurrent pyogenic cholangitis. Arch Surg 84:199–225
Park JH, Choi BI, Han MC, et al (1987) Percutaneous removal of residual intrahepatic stones. Radiology 163:619–623
Picus D, Weyman PJ, Marx MV (1989) Role of percutaneous intracorporeal electrohydraulic lithotripsy in the treatment of biliary tract calculi. Radiology 170:989–993
Rabinovitz M, Zajko A, Hassanein T, et al (1990) Diagnostic value of brush cytology in the diagnosis of bile duct carcinoma: a study in 65 patients with bile duct strictures. Hepatology 12:747–752
Ryeon HK, Sim JI, Park AW, et al (1993) Percutaneous transhepatic removal of biliary stones: clinical analysis of 16 cases (in Korean). J Korean Radiol Soc 29:1234–1239
Seel DJ, Park YK (1983) Oriental infestational cholangitis. Am J Surg 146:188–203
Short WF, Howard JM, Diven WF (1971) Trans T-tube catheterization. Arch Surg 102:136–138
Su CH, Lui WY, Peing FK (1992) Relative prevalence of gallstone disease in Taiwan: a nation wide cooperative study. Dig Dis Sci 37:764–768
Sung KB, Song HY, Kim MW, Lee SK (1993) Occlusion of metallic stents used for benign biliary strictures. Paper presented at the annual meeting of the Korean Radiological Society, 28–30 October 1993, Seoul
Wastie ML, Cunningham IGE (1973) Roentgenographic findings in recurrent pyogenic cholangitis. AJR 119:71–77
Wenn CC, Lee HC (1972) Intrahepatic stones: a clinical study. Ann Surg 175:166–177
Yu CS, Kim SW, Park YH, Han JK (1993) Cholangiographic features in hepatolithiasis (in Korean). Korean J Gastro enterol 25:140–150

Interventional Radiology
Malignant Diseases

23 Primary Cholangiocarcinoma of the Porta Hepatis: The Surgeon's Point of View

N.J. LYGIDAKIS

CONTENTS

23.1 Introduction

Despite the recent advances in diagnostic and surgical techniques, primary cholangiocarcinoma of the hepatic hilum remains a challenging clinical entity in regard to both diagnosis and treatment (LYGIDAKIS et al. 1988a,b; OTTOW et al. 1985).

Diagnostically, primary cholangiocarcinoma has to be differentiated from primary sclerosing cholangitis and from secondary fibroproliferative processes of the biliary tree as seen in patients with biliary lithiasis of long duration (Figs. 23.1, 23.2). The best diagnostic tool remains cholangiography, chiefly percutaneous transhepatic cholangiography, which is superior to endoscopic retrograde cholangiography in that it can outline the lesion and delineate the extent of the disease along the intrahepatic course of the main hepatic ducts more precisely (Fig. 23.3).

Cholangiocarcinoma of the hepatic hilum is classified as follows:

- Type I: Lesions confined below the bifurcation of the common hepatic duct
- Type II: Lesions confined to the bifurcation of the common hepatic duct
- Type III: Lesions confined above the bifurcation of the common hepatic duct (Fig. 23.4)

Preoperative assessment of resectability is difficult, at least in the majority of patients. Selective hepatic artery digital angiography and late-phase portography are of value in demonstrating anatomical abnormalities of the regional vessels, but are not good enough to allow accurate prediction of the resectability of the lesion. The same is true of endoscopic ultrasonography, which, although superior to conventional ultrasonography, cannot rule out resectability.

The prognosis of the disease is unpredictable. Although it is generally accepted that this is a slow-growing disease which rarely metastasizes, it has a high potential for invasion of nerves. This limits the possibilities for radical resection (LYGIDAKIS et al. 1988a,b). Additionally, it has a tendency to invade regional vessels, something that poses difficulties in any attempt at resection (LYGIDAKIS et al. 1988a). The frequent presence of jaundice is a factor that renders an operative procedure, particularly one including liver resection, risky and in many cases life-threatening.

Is preoperative biliary drainage necessary, and how long must it be carried out before surgery? Certainly preoperative internal biliary drainage is of great value, despite the arguments published in a number of studies on the value of external biliary drainage. Internal biliary drainage should be carried out at least 20–30 days before major surgery.

Despite the generally dismal figures, the prognosis of surgically treated cholangiocarcinoma is not so bad, and even after a nonradical resection the median survival rate is between 26 and 29 months (LYGIDAKIS et al. 1988b; OTTOW et al. 1985). These figures can be improved significantly by adjuvant locoregional immuno- and chemotherapy following resection of the tumor (LYGIDAKIS et al. 1995).

N.J. LYGIDAKIS, MD, FACS, Professor, Department of Hepatobiliary Pancreatic Surgery, St. Savas Hospital, P.O. Box 17160, 10024 Athens, Greece

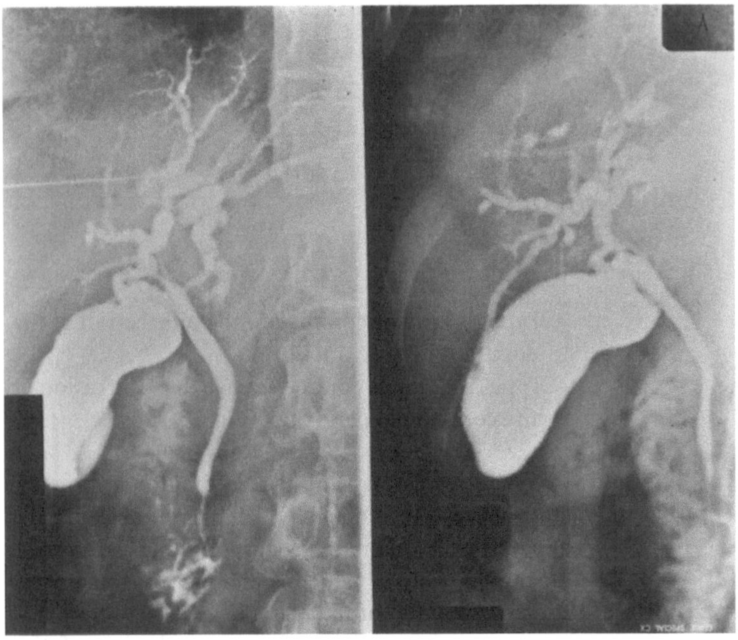

Fig. 23.1. Typical cholangiocarcinoma confined a location at and above the bifurcation of the common hepatic duct

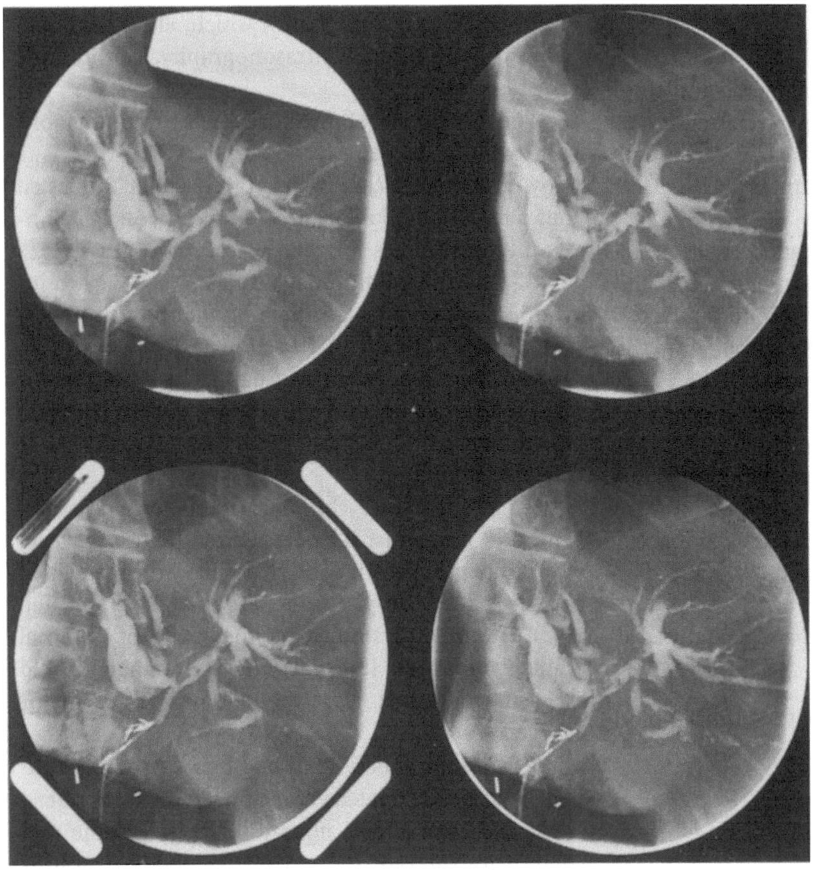

Fig. 23.2. Benign biliary structure in patients with biliary lithiasis confined to a site at and above the bifurcation of the common hepatic duct

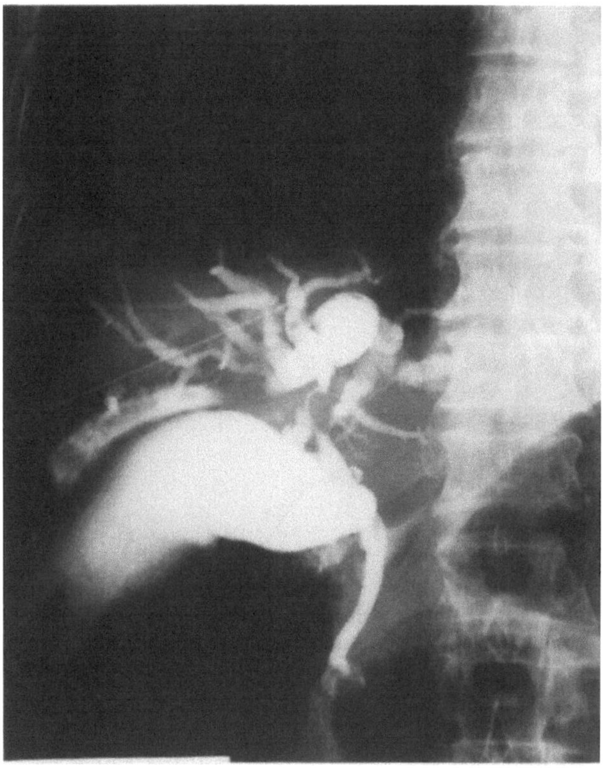

Fig. 23.3. Percutaneous transhepatic cholangiography reveals the site and extent of this lesion

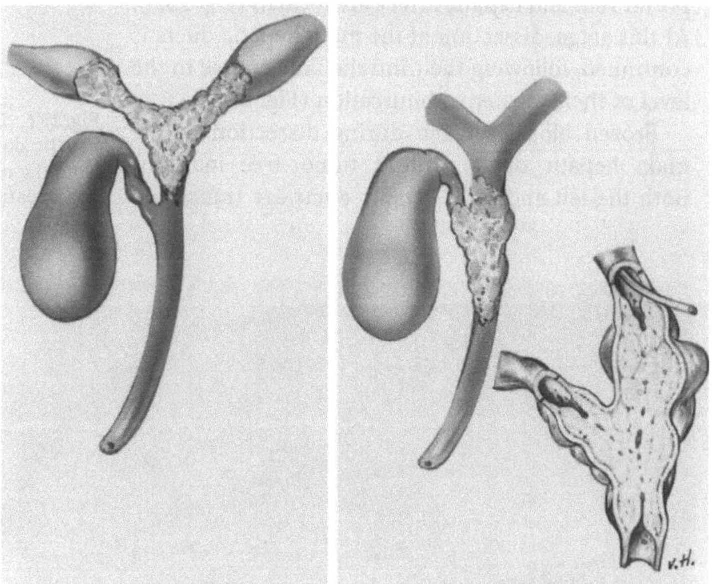

Fig. 23.4. *Left to right*: Stages I, II, and III of proximal cholangiocarcinoma of the porta hepatis

23.2 Surgical Techniques

23.2.1 Resection of the Tumor with the Bifurcation of the Common Hepatic Duct

With the patient under general anesthesia, a bilateral subcostal incision is made. The liver is inspected and palpated and the hepaticoduodenal ligament dissected. The hepatic artery, portal vein, and common bile duct are identified, dissected, and cleared of their lymphatics, then isolated with vessel loops.

The common bile duct is followed along its course to the distal end, where it is transected (Fig. 23.5). The distal end of the common bile duct is closed and

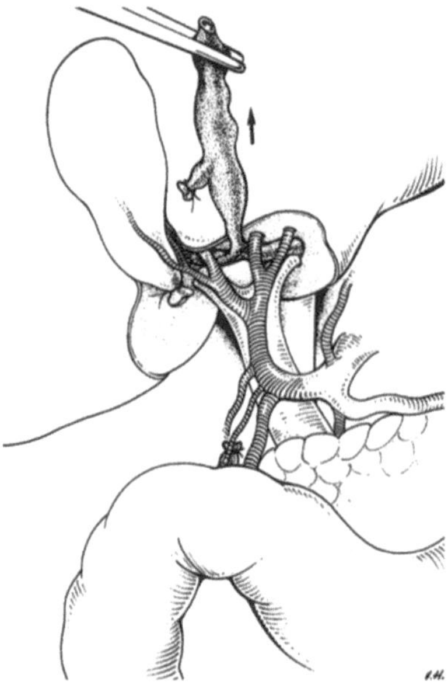

Fig. 23.5. The hepaticoduodenal ligament is dissected. The common bile duct is transected and reflected cephalad. Note the underlying confluence of the portal vein and hepatic artery

its proximal end reflected cephalad. Via a step-by-step cephalad dissection, the bifurcation of the common bile duct and the underlying bifurcation of the portal vein and hepatic artery are located (Fig. 23.5). At this stage, dissection of the main hepatic ducts is continued, following their intrahepatic course to the level of their segmental bifurcation (Fig. 23.6).

Frozen biopsies taken during dissection of the main hepatic ducts confirm tumor-free margins. Both the left and right hepatic ducts are transected proximal to the tumor and proximal to the bifurcation of the common hepatic duct, at the level of their segmental bifurcation – in other words, at the level of the segmental hepatic ducts.

Figure 23.7 shows the intrahepatic space left after resection of the tumor, the main hepatic ducts, the bifurcation of the common hepatic duct, the gallbladder, and the previously transected part of the common bile duct.

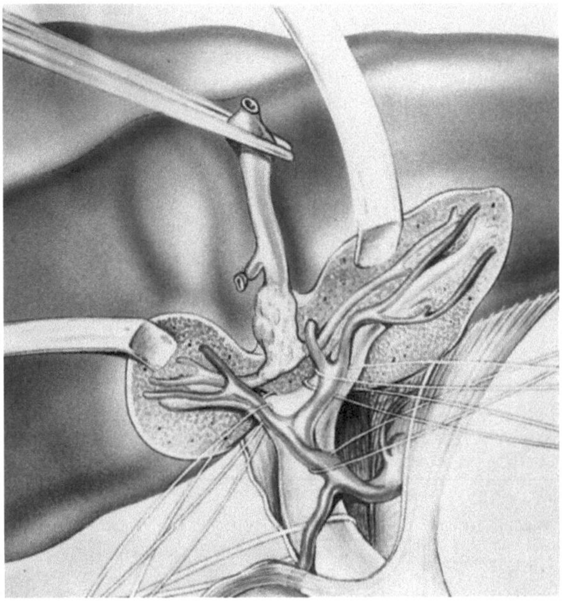

Fig. 23.6. Dissection along the intrahepatic course of the main hepatic duct. Continue to the level of segmental bifurcation. Note the underlying intrahepatic branches of the portal vein and hepatic artery

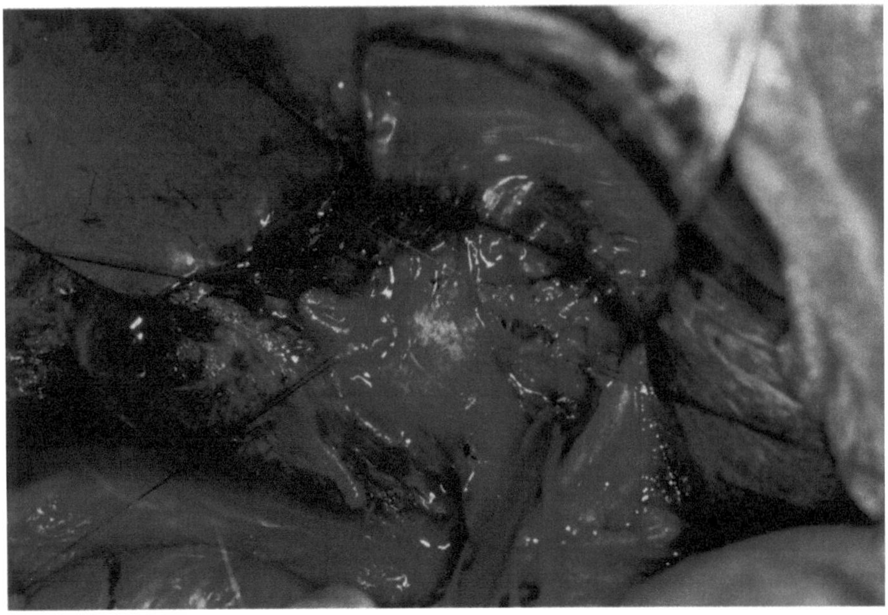

Fig. 23.7. Right and left segmental hepatic ducts after transection of the confluence of the common hepatic duct

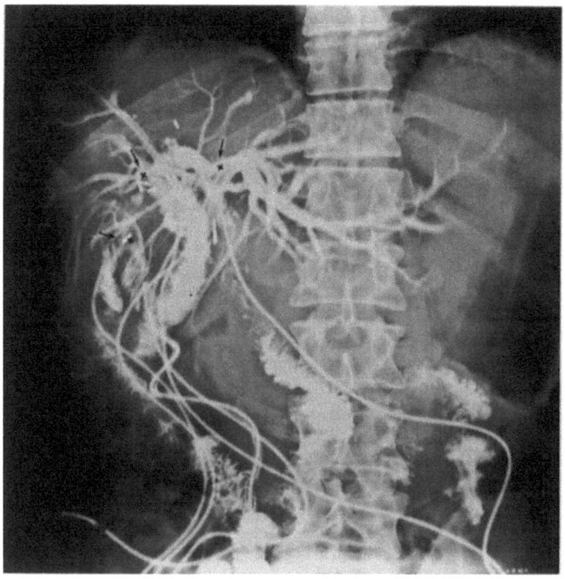

Fig. 23.8. Intrahepatic cholangiojejunostomies between segmental hepatic ducts and an isolated Roux-en-Y jejunal loop. In this patient three intrahepatic cholangiojejunostomies were carried out (*arrows*)

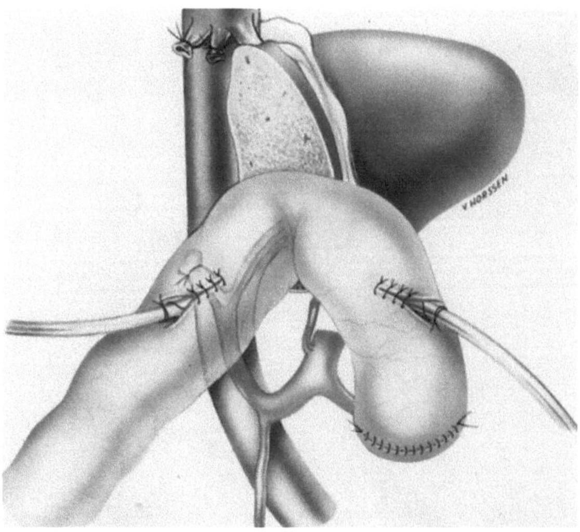

Fig. 23.9. Extended right hemihepatectomy. Reconstruction of the intrahepatic biliary tree by intrahepatic cholangiojejunostomies of segmental hepatic ducts: segments II and III and jejunal loop

The intrahepatic biliary tree is reconstructed with intrahepatic cholangiojejunostomies between segmental hepatic ducts and a Roux-en-Y jejunal loop (Fig. 23.8), using one-layer full-thickness interrupted absorbable sutures. Transanastomotic drains are not used and usually there are no problems with healing of the anastomoses. In general, the intrahepatic biliary tree can be drained with two or three intrahepatic cholangiojejunostomies (Fig. 23.8).

23.2.2 Resection of the Tumor with Liver Resection with or Without Regional Vascular Resection

For advanced cases of cholangiocarcinoma, tumor resection is combined with liver resection, using any type of anatomical hepatectomy. For a right or left hepatectomy or extended hepatectomy, reconstruct the biliary tree with intrahepatic cholangiojejunostomies between the segmental hepatic ducts of the residual liver and an isolated Roux-en-Y jejunal loop (Figs. 23.9, 23.10). In cases of bilateral involvement of the main hepatic ducts, resect the central liver (segments IV, V, and VIII) and drain the intrahepatic biliary tree via intrahepatic cholangiojejunostomies between the segmental hepatic ducts of the residual liver and an isolated Roux-en-Y jejunal loop.

23.2.3 Resection of the Tumor with Liver and Regional Vascular Resections

In cases where the tumor has invaded regional vascular structures, the portal vein, or the hepatic artery, combined tumor resection and vascular resection with or without liver resection is carried out. Vascular resection can include the hepatic artery, the portal vein, or both (Fig. 23.11). Vascular reconstruction is performed by end-to-end anastomosis or by interposing venous grafts between the transected margins of the portal vein or hepatic artery (Fig. 23.12). The internal iliac or the saphenous vein may be used for venous grafts (Fig. 23.13). The standard principles of vascular surgery are followed when fashioning the vascular anastomoses.

If liver resections performed, again the standard techniques of anatomical hepatoctomies are employed. The intrahepatic biliary tree is reconstructed by intrahepatic cholangiojejunostomies between the segmental hepatic ducts of the liver remnant and an isolated jejunal loop.

23.3 Results

Today, surgery in patients with proximal malignant biliary obstruction offers many options of treatment, depending on the existing anatomical and mechanical peculiarities of the individual patient. It is therefore reasonable to expect, in terms of survival, a mean survival ranging between 27 and 30 months, as reported by various centers (Ottow et al. 1985). It is certainly longer for those of our patients who have radical resections: we can expect a 5-year survival in

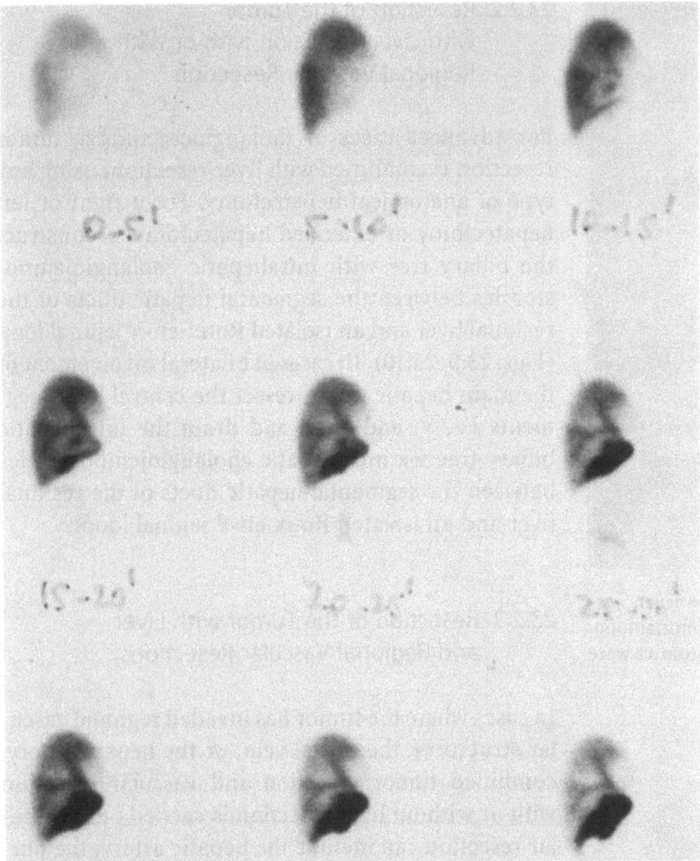

a

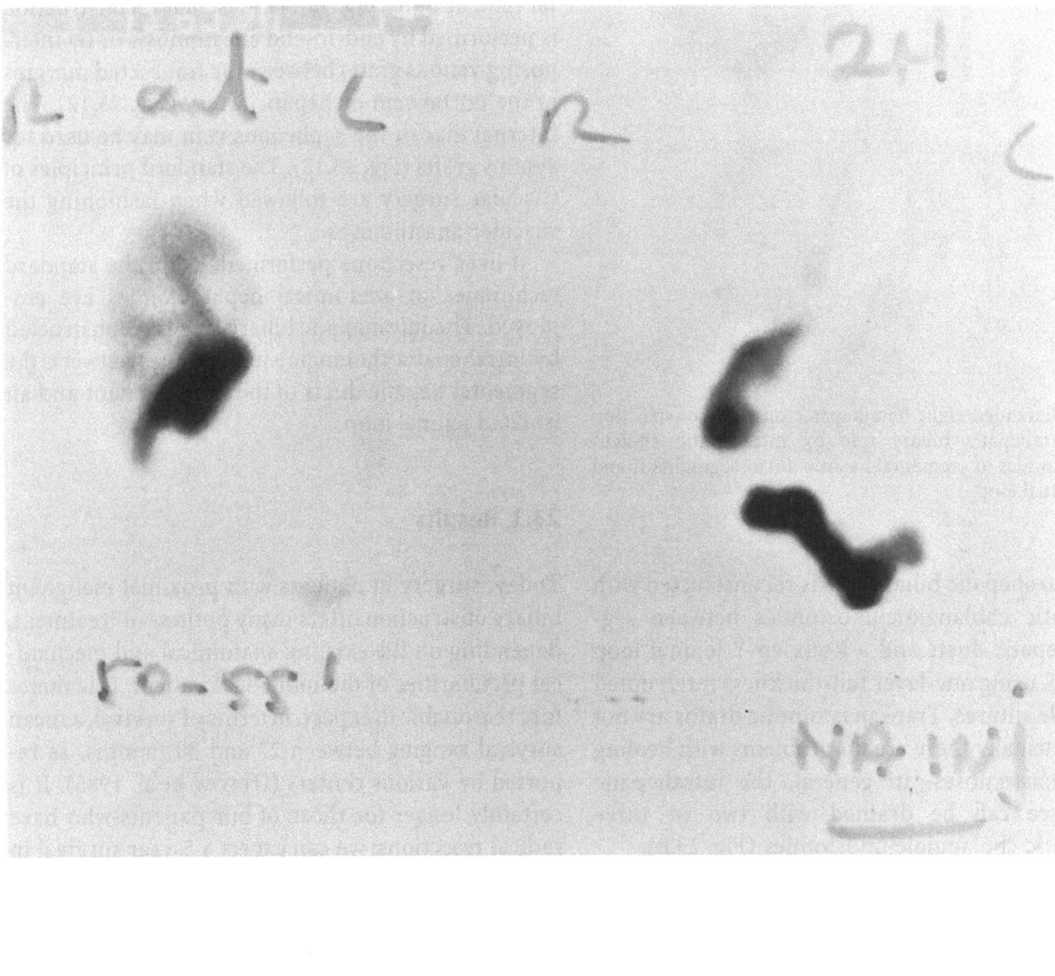

b

Fig. 23.10a,b. Extended left hemihepatectomy. Note visualizations in a hepatobiliary iminodiacetic acid scan study of intrahepatic cholangiojejunostomies between segmental hepatic ducts of the right liver and a jejunal loop. The further progression of bile along the isolated jejunal loop is shown

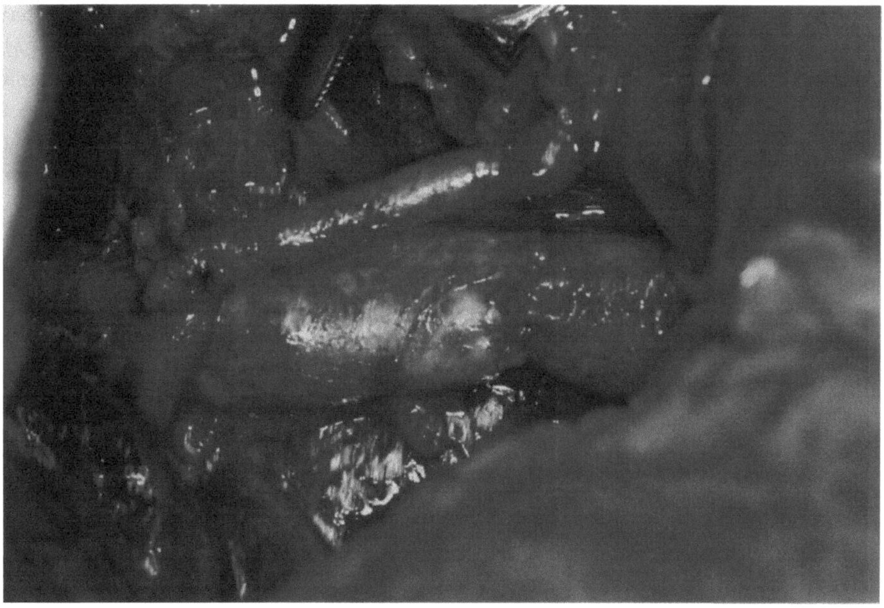

Fig. 23.11. Resection and reconstruction by use of venous grafts for both the portal vein and hepatic artery in a patient with cholangiocarcinoma of the porta hepatis

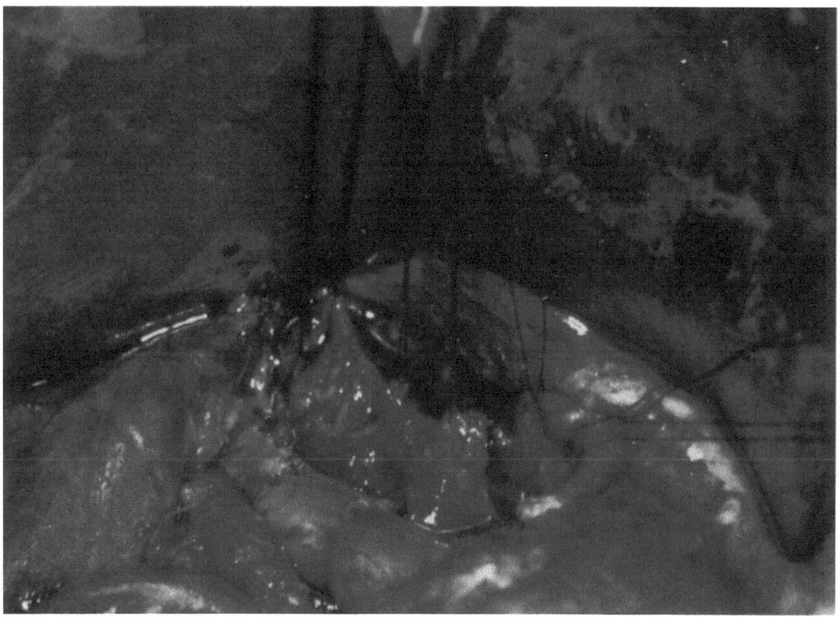

Fig. 23.12. Resection of both portal vein and hepatic artery reconstructed by venous graft, the hepatic artery also by an end-to-end anastomosis. Note the segmental hepatic ducts for both the left and right liver. In this case vascular resections were carried out without liver resection

them of around 40%. However, radical resection rates remain low despite the advances in surgical techniques, due to the fact that proximal cholangiocarcinoma has a tendency to invade nerves and blood vessels, rendering its eradication extremely difficult.

Recently, locoregional immuno- and chemotherapy have been seen to offer promising results even for patients with nonradical resections (LYGIDAKIS et al. 1995).

From my personal series, from 1983 onwards, I have operated upon 135 patients with a histologically proven diagnosis of cholangiocarcinoma of the porta hepatis. Of these, 93 patients underwent tumor resection with resection of the bifurcation of the common hepatic duct (group A), 22 patients received liver resection combined with tumor resection and resection of the bifurcation of the common hepatic duct (group B), and 20 patients underwent combined liver-tumor resection and resection of the regional vascular structures (group C), 15 having resection of both portal vein and hepatic artery and 5 having resection only of the portal vein.

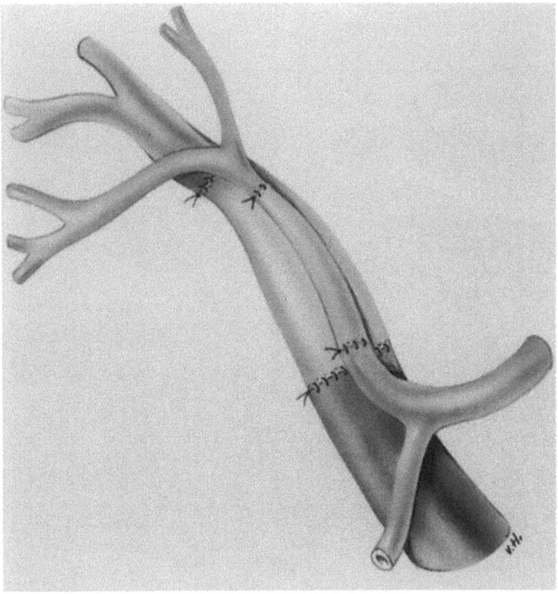

Fig. 23.13. Venous grafts from the internal iliac and saphenous vein for portal vein and hepatic artery reconstruction

Of group A patients, five died during the first 30 days after surgery; the surviving patients have a median survival of 28 months, within a range of 14-56 months. The quality of postoperative life was satisfactory. Among the total of 93 patients, resection was radical in only 7. All 7 patients survived for a long period, and 2 are alive 51 and 56 months respectively after surgery.

Of group B patients, five patients died during the first 30 days after surgery. Patients surviving longer than 30 days ($n = 17$) have a median survival of 33 months, with a range of 21-65 months. The quality of postoperative life was satisfactory, and out of the total of 22 patients, resection was radical in 11.

In group C patients, tumor resection was combined with liver and vascular resection. Thus, 15 patients had liver resection with concomitant resection and reconstruction of the portal vein and hepatic artery, and the remaining 5 patients had liver resection combined with resection and reconstruction of the portal vein only. Vascular reconstruction was carried out using venous grafts in 18 patients. In only two was vascular reconstruction carried out via end-to-end anastomosis of the transected margins of the hepatic artery. Of the 20 patients in group C, 4 died during the first 30 days after surgery. Surviving patients have a median survival of 30 months, with a range of 11-41 months. Out of the total of 20 patients radical resection was carried out in 6.

To appreciate these results, particularly for group B and C patients, it must be remembered that all these patients ($n = 42$) could be considered as ineligible for resection by the standard criteria of resectability (Fig. 23.14). In this case, the expected survival would not exceed about 4-6 months. On the other hand, internal biliary drainage, via either the endoscopic or the percutaneous transhepatic route, can offer a survival of around 10-12 months, but it is associated with a poor quality of life due to attacks of septic cholangitis caused by the presence of the endoprothesis in the biliary tree. Recently, we combined tumor resection with resection of the common hepatic duct confluence and postoperative adjuvant

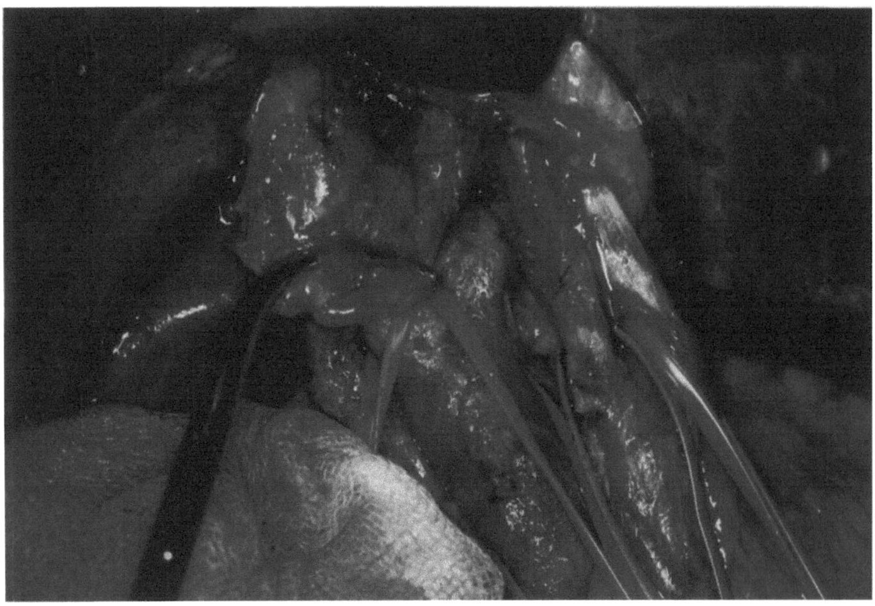

Fig. 23.14. Bilateral involvement by the tumor of both the portal vein confluence and the hepatic artery. This patient underwent tumor resection with resection of both the portal vein and the hepatic artery

therapy using combined targeting locoregional immuno- and chemotherapy, generating very promising results (Lygidakis et al. 1995). Indeed, 11 patients thus treated, in a separate study, have responded well to treatment, are free of disease with a median survival of 14 months, and are enjoying a satisfactory quality of life. In view of similar experiences with the clinical application of combined locoregional targeting immuno- and chemotherapy as an adjuvant to liver and pancreatic resection, I strongly recommend this treatment as an ancillary treatment to surgical resection.

23.4 Conclusion

Today we are entering a new era in medicine – the era of multimodality diagnosis and treatment. To meet current demands, we have to adjust our philosophy of the practice of surgery accordingly. In other words, we must work with determination and persistence, using all the new possibilities that current developments in oncology offer to surgery.

References

Lygidakis NJ, Van der Henden MN, Tytgat GN (1988a) Surgical approaches for the management of unresectable primary cholangiocarcinoma of the porta hepatis. Surg Gynecot Obstet 166:107–114

Lygidakis NJ, Van der Henden MN, Houthoff HJ (1988b) Surgical approaches to the management of primary biliary cholangiocarcinoma of the porta hepatis. The decision making dilemma. Hepatogastroenterology 35:261–267

Lygidakis NJ, Ziras N, Parissis J (1995) Resection versus resection combined with adjuvant pre- and post-operative chemotherapy-immunotherapy for metastatic colorectal liver cancer. A new look at an old problem. Hepatogastroenterology 42:155–161

Ottow R, August D, Supabaker P (1985) Treatment of proximal biliary carcinoma. An overview of techniques and results. Surgery 97:251–283

24 Critical Evaluation of the Different Imaging Modalities for Preoperative Staging of Bile Duct Carcinoma

J.W.A.J. REEDERS, O.M. VAN DELDEN, N.J. SMITS, and E.A.J. RAUWS

CONTENTS

24.1 Introduction

As the various radiological techniques improve, the radiologist is assuming a more active role in the diagnosis and management of patients with bile duct malignancies (FOLEY et al. 1980; MOORE and CLARK 1984; MUELLER and SIMEONE 1984; MAY et al. 1986). The rapid radiological developments have permitted more complete and detailed evaluation of the entire biliary tree, so that the accuracy with which tumor resectability can be predicted has increased considerably (BARON et al. 1982). The variety of techniques employed illustrates the fact that apparently no single imaging technique is the best or is adequate on its own. Techniques are often complementary to each other. Human, medical, and economic considerations should prompt the clinician to adopt a critical approach (VAN LEEUWEN and REEDERS 1989) towards the different imaging modalities for preoperative staging of bile duct carcinoma.

Bile duct malignancies have been diagnosed with increasing frequency, due not to an increase in incidence, but to the advent of these improved diagnostic methods of investigating the jaundiced patient (CARR et al. 1985). Only 5%–20% of patients with malignant biliary obstruction have surgically resectable tumors at the time of diagnosis (KALSER et al. 1985; LEA and STAHLGREN 1987; WARSHAW and SWANSON 1988; MICHELASSI et al. 1989; MERRICK and DOBELBOWER 1990; GULLIVER et al. 1992).

The spectrum of diseases that cause suprapancreatic biliary obstruction is considerably different from that in more distal (pancreatic) obstructions (NICHOLS et al. 1983; REIMAN et al. 1987). The choice between radical and palliative surgery depends on accurate assessment of the nature and extent of the lesion before treatment (REIMAN et al. 1987). The advent of ultrasound-guided percutaneous and endoscopic biliary stenting has created a need for accurate noninvasive staging, as most patients with unresectable tumors can now be treated nonoperatively with no adverse effect on survival rates (CLOUSE et al. 1977; FERRUCCI et al. 1980; FOLEY et al. 1980; FRANK et al. 1984). However, despite the advances in both preoperative staging of cholangiocarcinoma with improved investigation techniques and in surgical techniques, patient survival remains essentially unchanged (WARSHAW et al. 1986; TIO and TYTGAT 1986; CRIST et al. 1987; FREENY et al. 1988; VELLET et al. 1992), not only because of delayed symptomatic presentation but also because of ineffectual therapeutic options (KALSER and ELLENBERG 1985; VELLET et al. 1992).

24.2 General Considerations

The distal bile duct is the site of most obstructive processes (FREENY and LAWSON 1982; THORSEN et al. 1984). According to most authors (FOLEY et al.

J.W.A.J. REEDERS, MD PhD, Head, Department of Gastrointestinal Radiology and Hepato-pancreato-biliary Imaging, Academic Medical Center, Amsterdam, The Netherlands
O.M. VAN DELDEN, MD, Department of Gastrointestinal Radiology and Hepato-pancreato-biliary Imaging, Academic Medical Center, Amsterdam, The Netherlands
N.J. SMITS, MD, Department of Gastrointestinal Radiology and Hepato-pancreato-biliary Imaging, Academic Medical Center, Amsterdam, The Netherlands
E.A.J. RAUWS, MD PhD, Department of Gastroenterology, Academic Medical Center, Amsterdam, The Netherlands

1980; Tobin et al. 1987), primary pancreatic carcinoma is the most common cause of malignant common bile duct (CBD) obstruction, although some (Gibson et al. 1986) have found a higher incidence of cholangiocarcinoma. Isolated bile duct dilatation can be due to ampullary or periampullary carcinoma, chronic pancreatitis, pancreatic carcinoma, cholangiocarcinoma, or metastatic disease (breast, colon cancer) (Freeny et al. 1988). A *double duct* sign (dilatation of both distal pancreatic duct and CBD) is most frequently associated with pancreatic head malignancy. The different imaging modalities need to provide not only diagnostic and anatomic information about the level and cause of obstruction, but also an assessment of tumor resectability.

Conventional *intravenous cholangiography* has fallen out of favor, due to the better opacification of the biliary tree provided by endoscopic retrograde cholangiography (ERC) and ultrasound-guided percutaneous transhepatic cholangiography (PTC), and due to adverse reactions to the contrast material used (Ott and Gelfand 1981; Rholl et al. 1985; Stockberger et al. 1994). Intravenous cholangiography has been described as a sensitive method for outlining the biliary tract when used with helical computed tomography (*helical CT cholangiography*). It may be a clinically useful method for visualizing the bile ducts in patients with suspected biliary disease and in patients in whom attempts at ERC fail (Stockberger et al. 1994). In the clinical staging of bile duct tumors, *ultrasonography* (US) - perhaps the most operator-dependent imaging technique - will demonstrate dilated intra- and extrahepatic bile ducts and is superior to CT in the demonstration of masses in the extrahepatic bile ducts distal to the bifurcation (Itai et al. 1983; Thorsen et al. 1984; Nesbit et al. 1988). Both CT and US may show purely intrahepatic masses.

Even when no mass is depicted, both US and CT may provide important diagnostic information by accurately determining the level of obstruction, which they do in 100% of cases (Thorsen et al. 1984; Triller et al. 1994).

3D ultrasonography of the biliary tree and gallbladder has been developed, but has not been of proven clinical value so far (Gillams et al. 1991; Hamper et al. 1993; Patel and Less 1994). Further developments of this technique may facilitate teaching and allow ultrasonographers to communicate their findings in a form easily accessible to clinicians (Fine et al. 1991).

In the early 1980s, PTC, angiography, and conventional CT were regarded as the best imaging methods for assessing the resectability of proximal bile duct tumors (Voyles et al. 1983; Beazley et al. 1984; Looser et al. 1992), but they are of less clinical value in the late 1990s. Nowadays, duplex Doppler US, thin-section dynamic helical CT, ERCP, endoscopic US, and newer developments (intraductal US, intraductal cholangioscopy, laparoscopic US, 3D magnetic resonance cholangiography, etc.) seem more promising for accurate prediction of the resectability of tumors in patients with malignant biliary obstruction. However, tumor spread to normal-sized lymph nodes remains a well-known problem with the current imaging techniques. Detection of hepatic metastases smaller than 1 cm is a similar problem, and cannot be solved by CT and magnetic resonance imaging (MRI).

MRI at 1.5 T with respiratory motion compensation may be superior to CT in identification of intrahepatic bile duct tumors and assessment of vascular or perivascular invasion in the preoperative staging of pancreatic carcinoma (Vellet et al. 1992) although Megibow et al. in a recent study (1995) stress, that CT is recommended for initial imaging assessment.

Curative surgery seems to depend on the anatomical location and extension, morphologic type and pathological grade of the tumor, and vascular involvement. The various bile duct malignancies and the imaging techniques currently used to investigate them are discussed below.

24.3 Bile Duct Malignancies

Bile duct malignancies can be divided into various subgroups. The different imaging modalities for preoperative staging of each will be discussed in this section.

24.3.1 Primary Pancreatic Head Carcinoma

24.3.1.1 Ductal Adenocarcinoma

There has been a steady increase in the incidence of pancreatic ductal adenocarcinoma during the last six decades, and this tumor now accounts for 3% of all reported cases of cancer and 5% of cancer-related deaths (Silverberg and Lubera 1988; Vellet et al. 1992). It is estimated to be the eighth most common type of newly diagnosed cancer. At initial examination, regional or distant metastases are found in 85%

Table 24.1. Ultrasound (US) features of pancreatic head carcinoma

Mass lesion in pancreatic head:
<4 cm: homogeneous, hypoechogenic
>4 cm: heterogeneous with varying degrees of scattered internal echoes
Irregular pancreatic contour
Dilatation of common bile duct
Dilatation of pancreatic duct
Enlargement of gallbladder
Intrahepatic metastases
Regional lymphadenopathy
Peripancreatic vascular invasion; in presence of obstruction of the superior mesenteric and portal vein: abnormal Doppler signal (broadening of spectrum, high-velocity frequencies)

Modified after STRUNK et al. 1988.

of patients. Median survival from the time of diagnosis is 5-6 months and that after surgical resection not more than 10-20 months (EDIS et al. 1981; LONGMIRE and TRAVERSO 1981; FREENY et al. 1988).

Between 80% and 90% of the patients have unresectable lesions at diagnosis. Although FREENY et al. (1988) and MEGIBOW et al. (1991) consider CT as the best single imaging method for evaluation of patients with suspected pancreatic disease, it is generally agreed that *US* is preferable to CT as a preliminary screening procedure for pancreatic disease, since it is readily available, noninvasive, nonionizing, relatively inexpensive, and highly sensitive - if used in multiple planes - to parenchymal changes in echo texture, although highly dependent on the sonographer's technical skill (FEINBERG et al. 1977; LAWSON et al. 1978; MACKIE et al. 1978; SAUERBRUCH et al. 1979; POLLOCK and TAYLOR 1981; TAYLOR et al. 1981; BARON et al. 1982; HESSEL et al. 1982; SWOBODNIK et al. 1983; ORMSON et al. 1987; STRUNK et al. 1988). The disadvantage of US, however, is that clinicians are rarely competent to interpret US images, and even experienced ultrasonographers may experience difficulty in interpreting images produced by other operators. Ascites, obesity, or overlying bowel gas may preclude an adequate US study. Imaging of the pancreas is sometimes improved by oral administration of water and intravenous administration of glucagon, which may provide an acoustic window through the stomach. In emaciated patients US is favored because the pancreas can be visualized in multiple planes, while the efficacy of CT in diagnosing a pancreatic tumor may be limited, since it may be difficult to distinguish the pancreatic contour from adjacent organs due to the thinness of peripancreatic fat planes. US visualization of bile ducts may be improved by using a different approach, so that distal obstructing lesions can be optimally examined (LAING et al. 1986). For proximal duct visualization, a longitudinal scan must be performed with the patient in the supine left posterior oblique position. For investigation of the distal duct, a transverse scan with the patient in a semierect right posterior oblique position must be performed. Lesions in the head of the pancreas can be detected more accurately with US, while body/tail abnormalities are better recognized with (dynamic helical) CT. The ultrasonographic features of pancreatic head carcinoma are summarized in Table 24.1. In the literature US has a proven sensitivity of 50%-94% and a specificity of 74%-99% in the detection of pancreatic head carcinoma (Table 24.2).

The recent addition of pulsed Doppler to real-time US (*duplex Doppler US*) for initial diagnosis and staging has significantly shortened the delay and expense of diagnosis by eliminating the need for other diagnostic procedures such as CT, MRI, and angiography (FREENY et al. 1982; FREENY 1984; FREENY et al. 1988).

Duplex Doppler US has made a large contribution in assessing tumor involvement of the portal venous system, which is the main cause of unresectability in staging of pancreatic and large (peri)ampullary tumors and also Klatskin's tumors. Narrowing of the portal venous system, caused by compression or in-growth of tumor, results in changes in blood flow velocities and can be detected by pulsed Doppler (GARBER and LEES 1992; SMITS and REEDERS 1995) (Figs. 24.1-24.3). In contrast to pulsed Doppler in portal hypertension, where the mean velocity is studied (BOLONDI et al. 1990; KOSLIN et al. 1992), in pan-

Table 24.2. Diagnostic sensitivity and specificity of US in pancreatic head carcinoma as reported by different authors 1977-1990

	Sensitivity (%)	Specificity (%)
FEINBERG et al. 1977	90	-
LAWSON et al. 1978	90	96
MACKIE et al. 1978	87	74
SAUERBRUCH et al. 1979	89	-
POLLOCK and TAYLOR 1981	94	96
TAYLOR et al. 1981	94	99
HESSEL et al. 1982	56	-
SWOBODNIK et al. 1983	63	93
FRANK et al. 1984	50	97
ORMSON et al. 1987	85	-
STRUNK et al. 1988	88	98
CAMPBELL and WILSON 1988	90	-
J. ARIYAMA 1990, pers. comm.	93	-

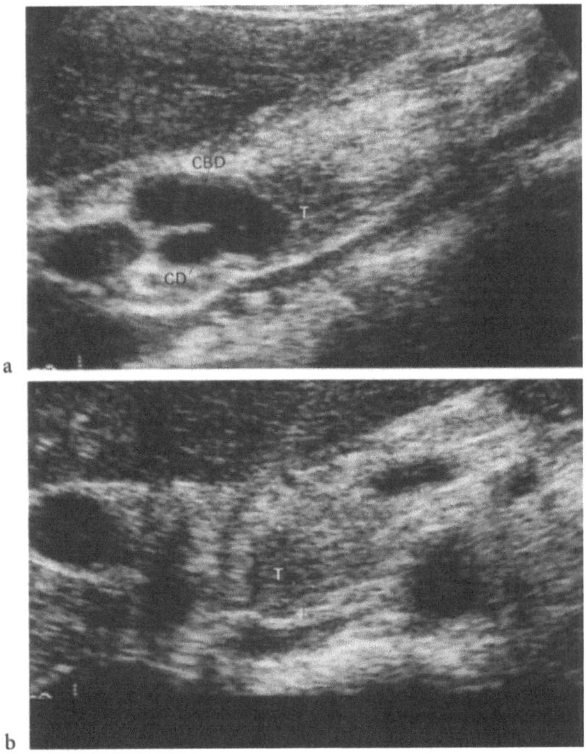

Fig. 24.1a,b. Resectable pancreatic head carcinoma. **a** Longitudinal, **b** transverse US scans: dilated common bile duct (*CBD*), cystic duct (*CD*), due to a well-delineated hypoechogenic mass (*T*) in the pancreatic head. The mass is free of the portal venous system

creatic and (peri)ampullary malignancies only the maximum velocity is of interest. Apart from portal venous involvement, other criteria of unresectability are: extrapancreatic infiltration of the tumor; regional lymph node metastases, liver metastases, and peritoneal metastases. An abnormal pulsed Doppler signal is highly suggestive of major involvement of the portal venous system and unresectability of the tumor. Because an abnormal pulsed Doppler signal can only be expected when a luminal narrowing of at least 30% exists, a normal pulsed Doppler signal does not rule out infiltration of the portal venous system. Loss of the hyperechoic interface between tumor and vessel is an indication of infiltration. A high Doppler shift due to narrowing of the portal venous system can be produced by chronic pancreatitis, lymphoma (non-Hodgkin lymphoma, AIDS-related disease, cat-scratch disease, etc.) and metastases (Figs. 24.4–24.7).

Since chronic pancreatitis can be associated with pancreatic carcinoma, and because lymphoma or metastasis can be mistaken for pancreatic head carcinoma, it is important to differentiate correctly between them. The respective sensitivity, specificity, and accuracy of pulsed Doppler, real-time US, and duplex US are summarized in Table 24.3.

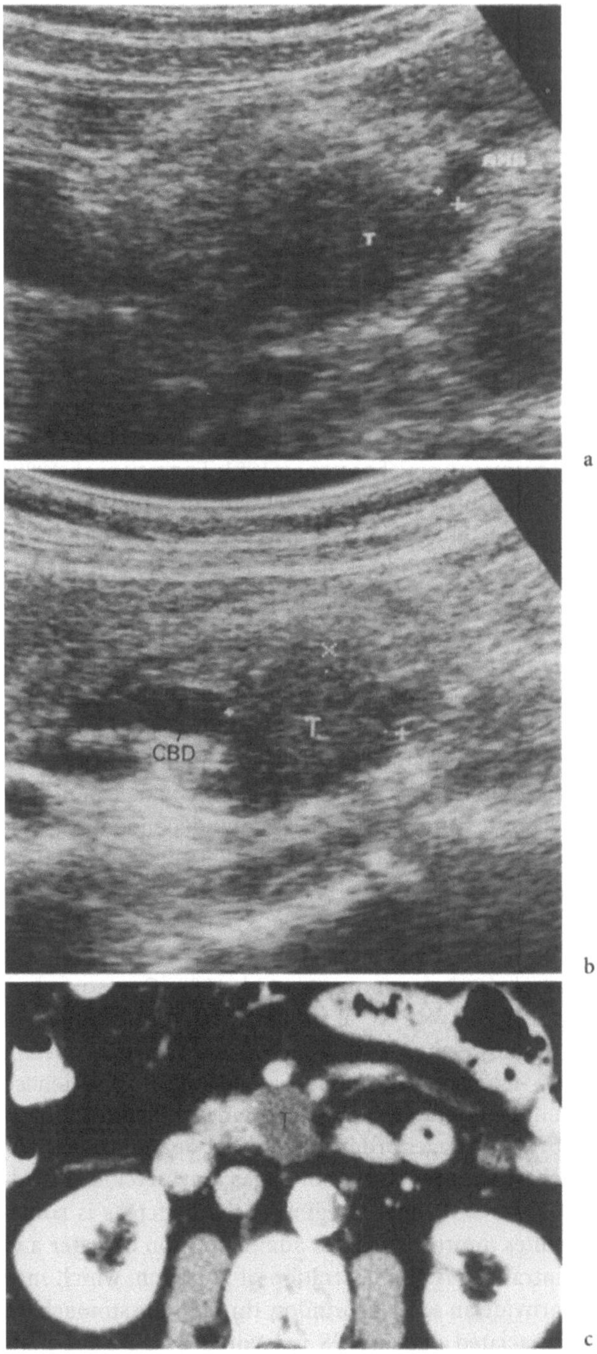

Fig. 24.2a,b. Unresectable pancreatic head carcinoma. **a** Transverse, **b** longitudinal US scans, **c** CT scan: a well-delineated hypoechogenic/hypodense low-attenuation mass (*T*) in the pancreatic head (uncinate process) with a broad area of contact with the superior mesenteric artery (*AMS*). Note dilation of the CBD

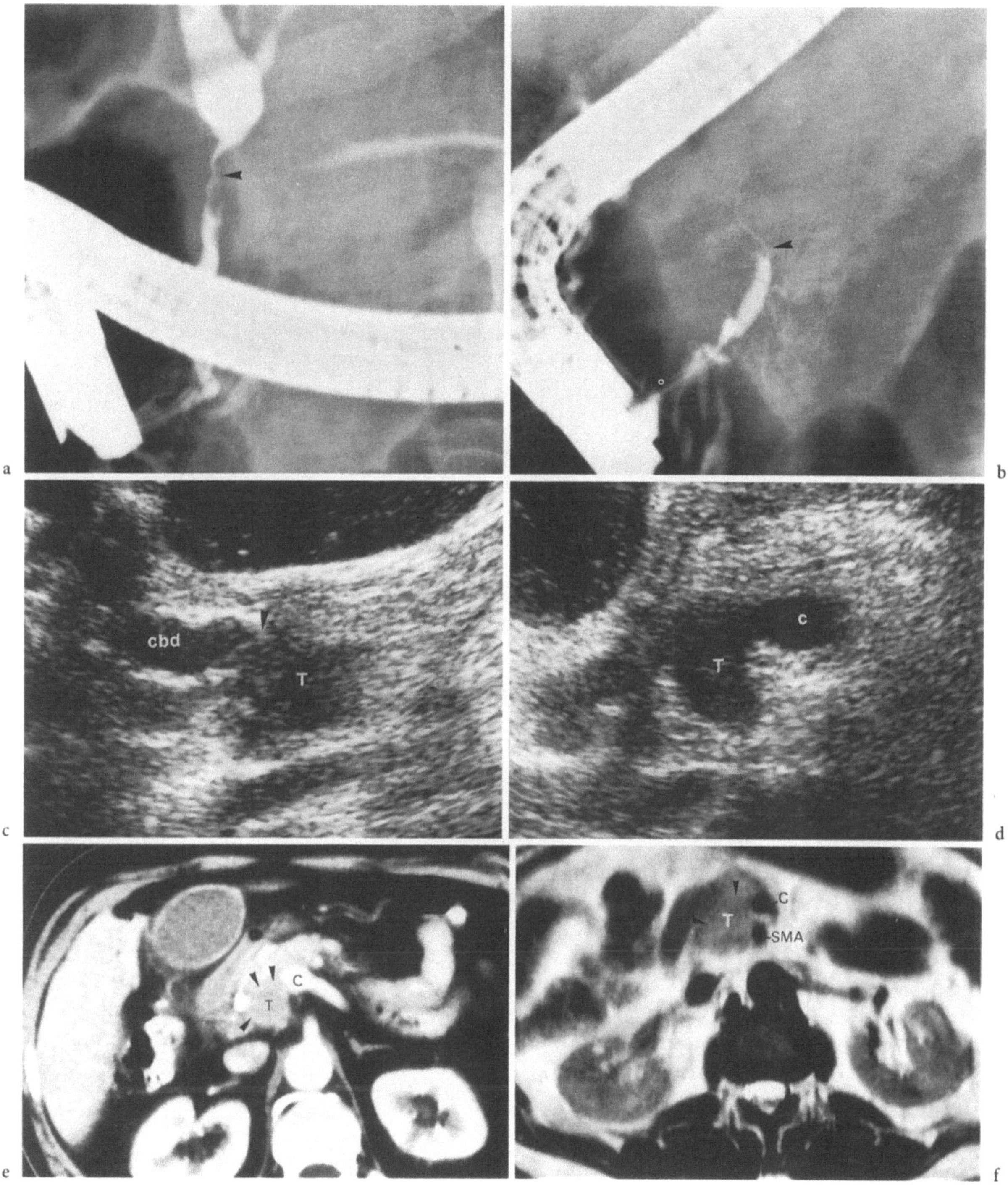

Fig. 24.3a–h. Unresectable pancreatic head carcinoma. **a, b** Endoscopic retrograde cholangiopancreatography (ERCP): *double duct* sign, compatible with pancreatic head carcinoma (*arrowheads*). **c,d** Oblique (**c**) and tranverse (**d**) US scans: dilated CBD with distal blunt obstruction (*arrowhead*) due to an ill-defined hypoechogenic mass (*T*) in the pancreatic head. The tumor clearly has a broad area of contact with the portal venous system (*c*). **e,f** Dynamic CT (**e**) and MRI (turbo T2, TR3200/TE90) (**f**): hypodense/low-attenuation mass (*T* and *arrowheads*) with inhomogeneous signal intensity, showing a broad area of contact with the confluence (*C*). **g** Doppler US: high Doppler shift (1.5 m/s) and spectral broadening beyond the stenosis. **h** Laparoscopic US, showing an obvious broad area of contact between tumor mass (*UP*) and superior mesenteric vein (*VMS*). The superior mesenteric artery is free (*AMS*)

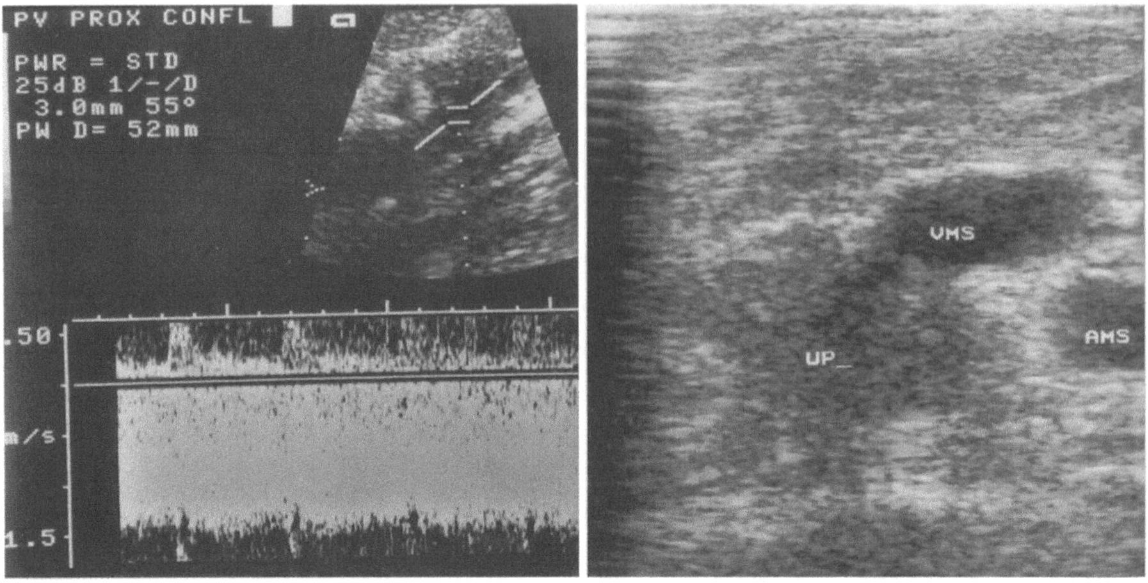

g h

Fig. 24.3g,h

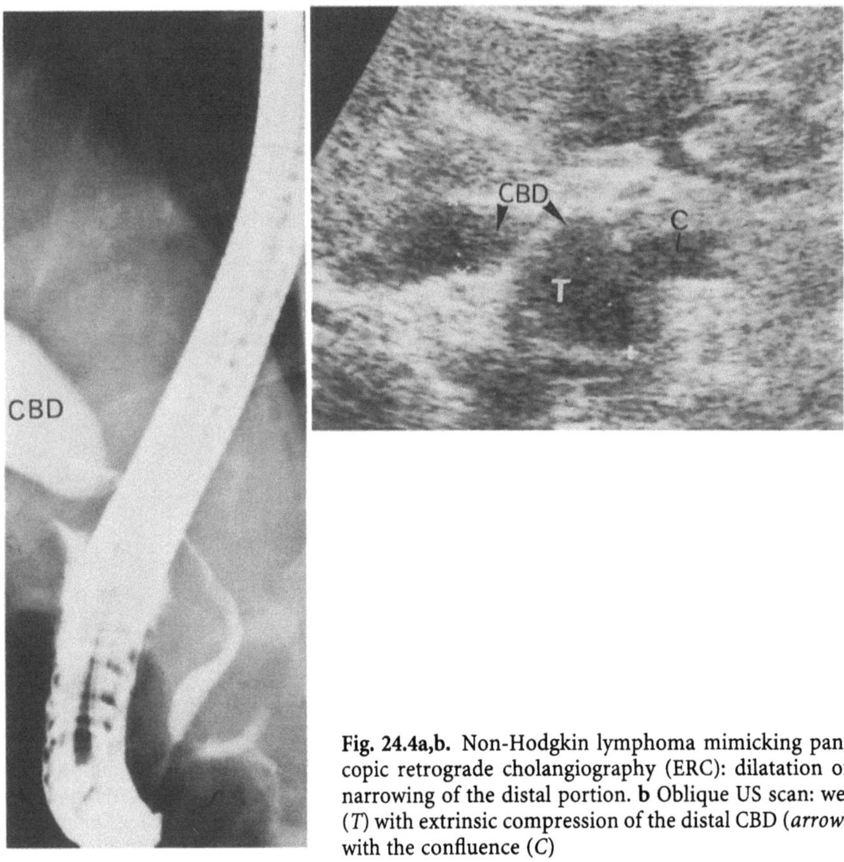

a b

Fig. 24.4a,b. Non-Hodgkin lymphoma mimicking pancreatic head carcinoma. **a** Endoscopic retrograde cholangiography (ERC): dilatation of the CBD with eccentric smooth narrowing of the distal portion. **b** Oblique US scan: well-delineated hypoechogenic mass (*T*) with extrinsic compression of the distal CBD (*arrowheads*) and a broad area of contact with the confluence (*C*)

The value of *CT* in the preoperative examination of patients with carcinoma of the head of the pancreas has been studied by several investigators, with varying results (Jafri et al. 1984; Freeny et al. 1988; Warshaw et al. 1990; Dooley et al. 1990; Roos et al. 1990; Gulliver et al. 1992). CT, however, is better than US in patients who are grossly obese or have had multiple previous operations. CT is more depen-

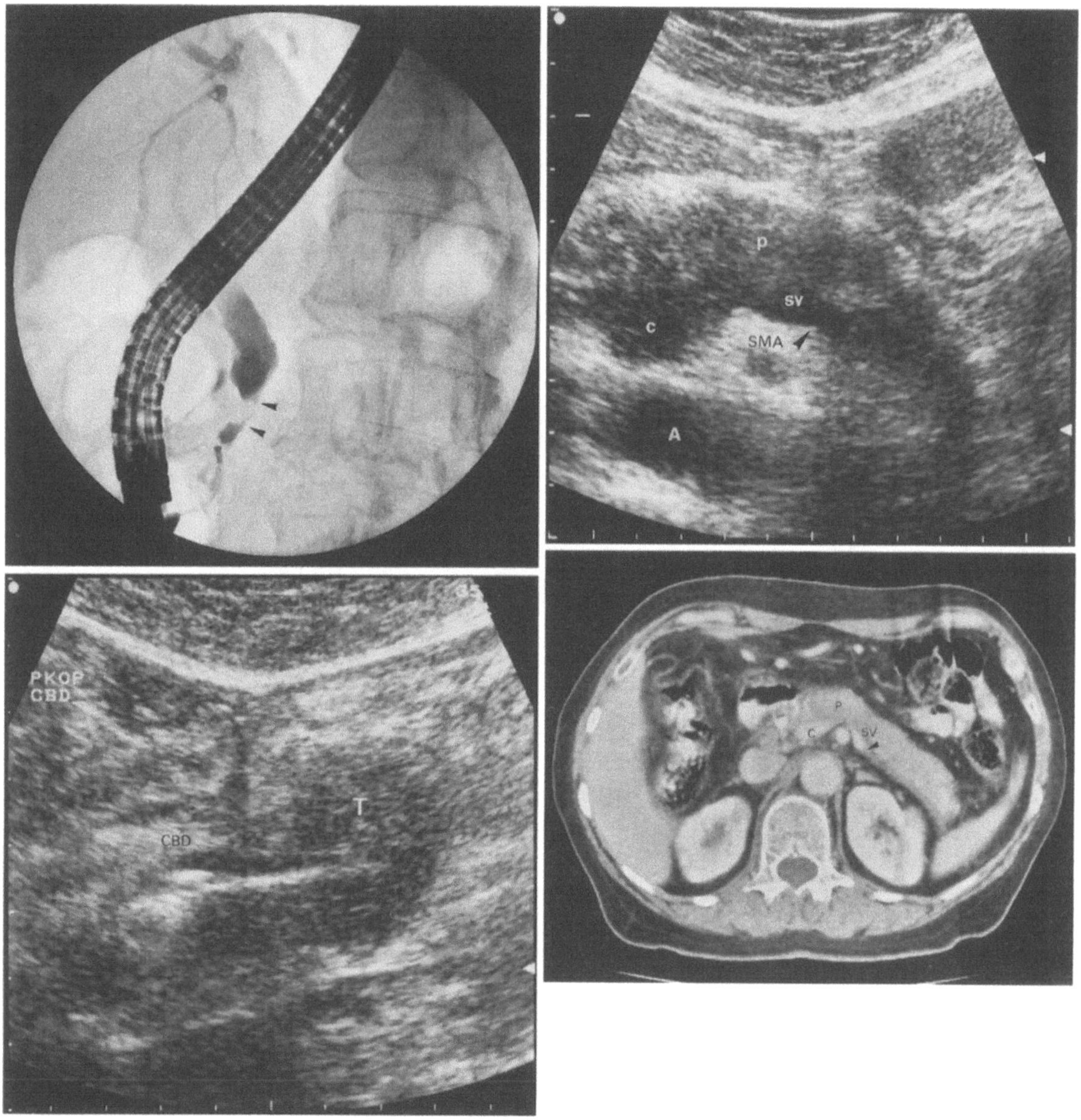

Fig. 24.5a–d. Chronic pancreatitis mimicking pancreatic head carcinoma. **a** ERC: eccentric irregular short stenosis (*arrowheads*) at the distal CBD, suggestive of malignant infiltration of the CBD. **b** Oblique US scan: enlarged hypoechogenic ill-defined mass (*T*) in the pancreatic head with subsequent tapering of the CBD, suggestive of pancreatitis. **c** Transverse US scan: the whole pancreas (*p*) is hypoechogenic and enlarged, suggestive of pancreatitis. Only the confluence (*c*) and distal splenic vein (*sv*) are visible. Stop at the midportion of the splenic vein (*arrowhead*). *SMA*, Superior mesenteric artery. **d** Dynamic CT: enlarged pancreas (*P*) with attenuation after intravenous injection of contrast medium. Evident obstruction of the midsplenic vein (*SV*) (*arrowhead*), which was verified with Doppler US. *C*, Confluence

dent on the quality of the equipment employed (Hosoki 1983) and less on the examiner. The literature indicates that these two cross-sectional imaging modalities are complementary rather than alternatives for the examination of the pancreas (Moossa 1982; Freeny 1988). The actual extent of intra- and extrapancreatic tumor is underestimated with CT; identification of the tumor within a complex pancreatic mass may not be possible. Many pitfalls encountered in evaluating CT scans of the pancreas can be avoided by a proper examination technique (Freeny 1991, 1993), using dynamic/helical CT scanning to enhance the pancreatic parenchyma and the major peripancreatic vascular structures (Fishman et al. 1992; Freeny 1995).

Dynamic/helical CT is more useful than conventional CT (Hosoki 1983; Freeny 1988; Freeny et al. 1988; Charnsangavej 1996) because it facilitates

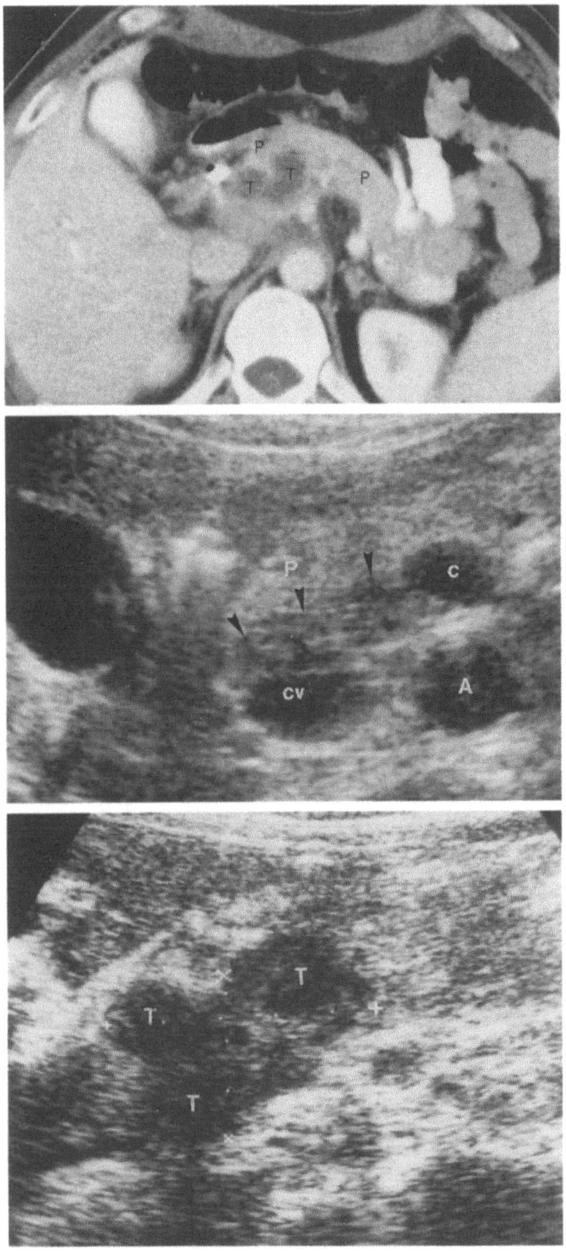

a

b

c

Fig. 24.6a–c. Cat-scratch disease mimicking pancreatic head malignancy. **a** Spiral CT after injection of contrast shows multiple low-attenuation lobulated lesions in the region of the pancreatic head, simulating an intrapancreatic tumor. *P*, Pancreas; *T*, tumor. **b** Transverse US scan: normal pancreatic head with a broad hypoechogenic linear band (*arrowheads*) at the ventral pancreas, which is a normal variation. *cv*, Inferior vena cava; *A*, aorta; *c*, confluence; *P*, pancreas. **c** Oblique US scan: multilobulated, well-delineated hypoechogenic mass (*T*), clearly of extrapancreatic origin

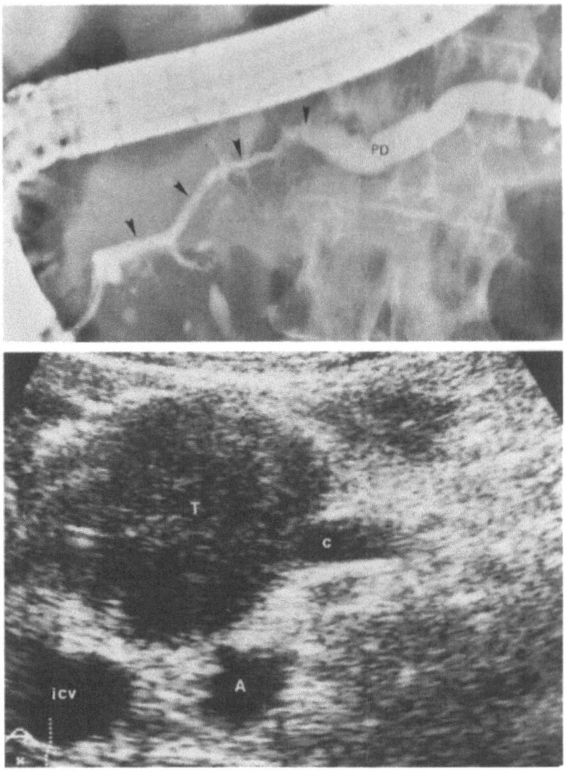

a

b

Fig. 24.7a,b. Apudoma of the pancreatic head mimicking pancreatic head adenocarcinoma. **a** ERP: eccentric irregular long stenosis of the pancreatic duct (*PD*) and side branches (*arrowheads*), compatible with malignancy. **b** Transverse US scan: large, well-delineated, hypoechogenic mass (*T*) in the pancreatic head with a broad area of contact with the portal venous system (*c*). *icv*, Inferior vena cava; *A*, aorta

Table 24.3. Results of pulsed Doppler ($n = 117$), real-time US imaging ($n = 105$), and duplex US ($n = 62$) in assessment of portal venous system involvement in carcinoma of the pancreatic head region, correlated with highly abnormal angiographic findings in nonoperated patients, surgical findings in nonresected tumors, and histopathologic findings in resected tumors

	Pulsed Doppler (%)	Real-time US Imaging (%)	Duplex US (%)
Sensitivity	44	90	82
Specificity	100	72	100
Accuracy	62	85	89
Positive predictive value	100	88	100
Negative predictive value	46	77	77

N.J. Smits and J.W.A.J. Reeders 1996, unpublished data.

reliable prediction of unresectability [local tumor extension, contiguous organ invasion, and vascular changes (portal venous invasion, infiltration of the celiac trunk and/or superior mesenteric arteries)] of tumors in patients with pancreatic head malignancy (Gulliver et al. 1992) (Figs. 24.2c, 24.3e).

The diagnostic accuracy of CT in pancreatic head carcinoma is illustrated in Table 24.4. Table 24.5 details the CT features of the disease.

Table 24.4. Diagnostic accuracy of CT in pancreatic head carcinoma as reported by different authors (%)

	Accuracy (%)
FOLEY et al. 1980	72
TOBIN et al. 1987	72
FREENY et al. 1988	91
MÜLLER et al. 1994	67

The most common contrast-enhanced CT finding (96%) in pancreatic carcinoma is a low-attenuation mass (FREENY et al. 1988). Occasionally normal anatomic variations in size and shape of the pancreas may simulate a mass on US and/or CT (ORMSON et al. 1987). In these cases, lack of a central zone of diminished attenuation, pancreatic duct and/or CBD dilatation, local tumor extension or distant metastases may suggest the correct diagnosis (FREENY et al. 1988). Therefore, an isolated pancreatic mass (with or without pancreatic duct or CBD dilatation) is indeterminate and may be caused by other primary - or metastatic - neoplasms of the pancreas (islet cell tumor, cystic neoplasms etc.), anatomic variants, or focal chronic pancreatitis (HOSOKI 1983).

One of the most difficult differential diagnoses with any investigative modality (WITTENBERG et al. 1982; PLUMLEY et al. 1982; NEFF et al. 1984; FREENY et al. 1988) is between focal or diffuse chronic pancreatitis and pancreatic carcinoma. Moreover, chronic pancreatitis and pancreatic carcinoma may coexist. Isolated pancreatic duct dilatation with no CBD dilatation and/or intraductal calculi is more frequently due to chronic pancreatitis (FREENY 1988; FREENY et al. 1988). False-positive diagnoses by US and/or CT may occur, since demonstration of CBD dilatation does not necessarily indicate obstruction, especially in patients who have had prior obstruction and/or biliary surgery (BARON et al. 1982, 1983). In our hospital, between 1983 and 1993, 220 patients underwent a subtotal pancreatoduodenectomy (Whipple's resection, n = 188) or a total pancreatectomy (n = 32) on the suspicion of pancreatic head cancer, based on clinical presentation and imaging studies. In 14 of these patients (6%) histopathological study of the specimen revealed a benign, inflammatory lesion diagnosed as chronic pancreatitis of greater or lesser severity or as chronic inflammation of the CBD (VAN GULIK et al. 1996).

A pancreatic head malignancy may locally invade the portal venous system, the medial wall of the duodenum, and may enlarge the periaortic and/or pericaval lymph nodes. Spread to lymph nodes and invasion of nonperivascular fat planes on CT are less reliable indications of unresectability. Obliteration of the retropancreatic fat planes and peripancreatic fascial planes between the pancreatic head and the vena is a reliable indicator of tumor extension. Retropancreatic extension of pancreatic carcinoma is more frequently found (60%) than extension into the anterior renal fascia in the lateral retroperitoneum (3%) - which carries a grave prognosis (FELDBERG et al. 1987).

The positive predictive value of the diagnosis of a resectable tumor was only 45% in a study by FREENY (1988), which reflects the poor accuracy of CT in detection of lymph node metastatses, microscopic local tumor extension, and small hepatic metastases (FREENY et al. 1986, 1988).

Endoscopic retrograde cholangio pancreatography (ERCP) - an invasive technique which is associated with a 2%–3% incidence of complications (pancreatitis, cholangitis) - has been described as 100% accurate in identifying a malignant pancreatic lesion (DIMAGNO et al. 1977; FEINBERG et al. 1977; FOLEY et al. 1980; TOBIN et al. 1987). The technical failure rate is still high (MOSS et al. 1980; MOOSSA 1982). The main ERCP feature is the *double duct* lesion: irregular, eccentric beading of a dilated pancreatic duct together with irregular, eccentric stenosis of the distal CBD with proximal dilatation, or abrupt distal termination of both pancreatic duct and CBD (Fig. 24.3a,b).

Table 24.5. CT features of pancreatic head carcinoma

Main features:
- Isolated (focal or diffuse) low-attenuation pancreatic mass (with areas of necrosis) (96%)
- Main pancreatic duct dilatation (88%)
- Intra-/extrahepatic biliary duct dilatation

Associated features:
- Local tumor extension (obliteration of retropancreatic fat planes between pancreatic head and inferior vena cava; involvement of peripancreatic fascial planes including the anterior limb of the fascia of Gerota)
- Contiguous organ invasion
- Metastases within liver/regional lymph nodes
- Ascites (peritonitis carcinomatosa)
- Involvement of peripancreatic arteries (celiac, splenic, hepatic, superior mesenteric, left renal, gastroduodenal artery) and/or veins (splenic, superior mesenteric, portal, left renal vein)
- Obstructive pseudocysts (due to pancreatic duct obstruction and pancreatitis and or rupture of side branches)

FREENY et al. 1988.

In pancreatic carcinoma, the pancreatic duct is not necessarily encased or obstructed. If the tumor originates in the accessory duct or the pancreatic duct distal to the junction, and the minor papilla of the accessory duct is open, it may not cause proximal pancreatic duct dilatation, since the duct can decompress via the unobstructed segment (FREENY 1988; FREENY et al. 1988).

The surgical resectability of pancreatic carcinoma depends upon local infiltrative tumor growth and the involvement of vessels; tumor size alone is not useful in predicting resectability or prognosis (FREENY et al. 1988), although in studies reported by STRUNK et al. (1988) 78% of tumors above 2–3 cm in size were stage T3 (42% resectable) and tumors larger than 5 cm were all stage T3 (0% resectable).

In 92% of cases of pancreatic carcinoma, associated features may be found on CT (FREENY 1988; FREENY et al. 1988) (Table 24.5). The presence of any of these features indicates tumor nonresectability. US- or CT-guided percutaneous cytodiagnosis of nonresectable malignancy is imperative to confirm the diagnosis of pancreatic carcinoma and to plan subsequent therapy (COHAN et al. 1986; HALL-CRAGGS and LEES 1986; FREENY et al. 1988). Preoperative fine-needle aspiration biopsy (FNAB) carries a potential risk of seeding metastasis (FROHLICH et al. 1986; FORNARI et al. 1989; VAN GULIK et al. 1996). In addition, the high percentage of false-negative results, as high as 60%, should be borne in mind when performing FNAB (PARSONS and PALMER 1989; RODRIQUEZ et al. 1992; VAN GULIK et al. 1996). For these reasons, in our institution we do not routinely undertake preoperative needle biopsy of the tumor when the diagnosis seems uncertain, but proceed with laparoscopy and, if that is negative, with resection. Molecular markers such as K-*ras* mutations and p53 may have the potential to increase the diagnostic yield of cytological aspiration biopsies, but their clinical application awaits further assessment (HRUBAN et al. 1994; VAN GULIK et al. 1996). Percutaneous needle placement has been guided by PTC (sensitivity 52%) (EVANDER et al. 1978; MCLOUGHLIN et al. 1978; NIMURA 1996), percutaneous cholangioscopy (sensitivity 66.7%) (NIMURA 1996), angiography (sensitivity range 61%–76%) (TYLEN et al. 1976; GOLDSTEIN et al. 1977; CLOUSE et al. 1977; EVANDER et al. 1978; MCLOUGHLIN et al. 1978; CONWAY et al. 1982), US (sensitivity range 70%–81%) (HANCKE et al. 1975; CLOUSE et al. 1977; GOLDSTEIN and ZORNOZA 1978; MCLOUGHLIN et al. 1978; PEREIRAS et al. 1978; ITOH et al. 1979; FERRUCCI et al. 1980; CONWAY et al. 1982), ERCP (sensitivity range 88%–93%) (HO et al. 1977; MCLOUGHLIN et al. 1978; FREENY et al. 1980), and CT (sensitivity range 89–95%) (FERRUCCI et al. 1980; HARTER et al. 1983; HUSBAND and GOLDBERG 1983). In our series (SMITS and REEDERS 1995), US-guided biopsy of the pancreatic head mass in nonresectable malignancies was positive in 93% of surgically confirmed cases of pancreatic carcinoma using the 21-gauge Rotex needle (Ursus Konsult AB, Stockholm, Sweden).

JAFRI et al. (1984) considered *angiography* and CT to be equivalent in the demonstration of vascular involvement, but found angiography to be superior to CT in a vessel-to-vessel comparison. In studies by FREENY et al. (1988) with high-resolution dynamic CT, the results of angiography and CT were similar in a vessel-to-vessel comparison. CT was more accurate than angiography in identifying tumor that surrounded or was contiguous with the vessels. For this reason, HOSOKI (1983) and FREENY et al. (1988) consider dynamic CT more accurate than angiography for staging pancreatic head carcinoma. It is difficult to accurately evaluate obliquely coursing arteries or veins on axial CT images (JAFRI et al. 1984).

WARSHAW et al. (1990), DOOLEY et al. (1990), ROOS et al. (1990), and ROSS et al. (1988) studied findings in patients with pancreatic cancer and periampullary tumors and found positive predictive values of CT findings that suggested resectability of 45%, 60%, 15%, and 38% respectively. Positive predictive values of CT findings that suggested unresectability were 92%, no result given, 85%, and 93% respectively (GULLIVER et al. 1992).

WARSHAW et al. (1990), DOOLEY et al. (1990), and MACKIE et al. (1979) investigated the value of angiography (celiac and superior mesenteric angiography using selective or superselective cannulation of pancreatic vessels) in predicting surgical resectability in patients with pancreatic and periampullary tumors. The positive predictive values in regard to tumors thought to be resectable on the basis of angiographic findings were 54%, 77%, and 59%, and positive predictive values in regard to those thought to be unresectable were 96%, 79%, and 80% respectively.

If there is encasement or stenosis of the superior mesenteric vein, splenic vein, or portal vein, the tumor is considered nonresectable (Figs. 24.8, 24.9). Involvement of the gastroduodenal artery is not a criterion of nonresectability. Currently, angiography is rarely used (WARSHAW et al. 1990; DOOLEY et al. 1990; ROOS et al. 1990; GULLIVER et al. 1992) in the

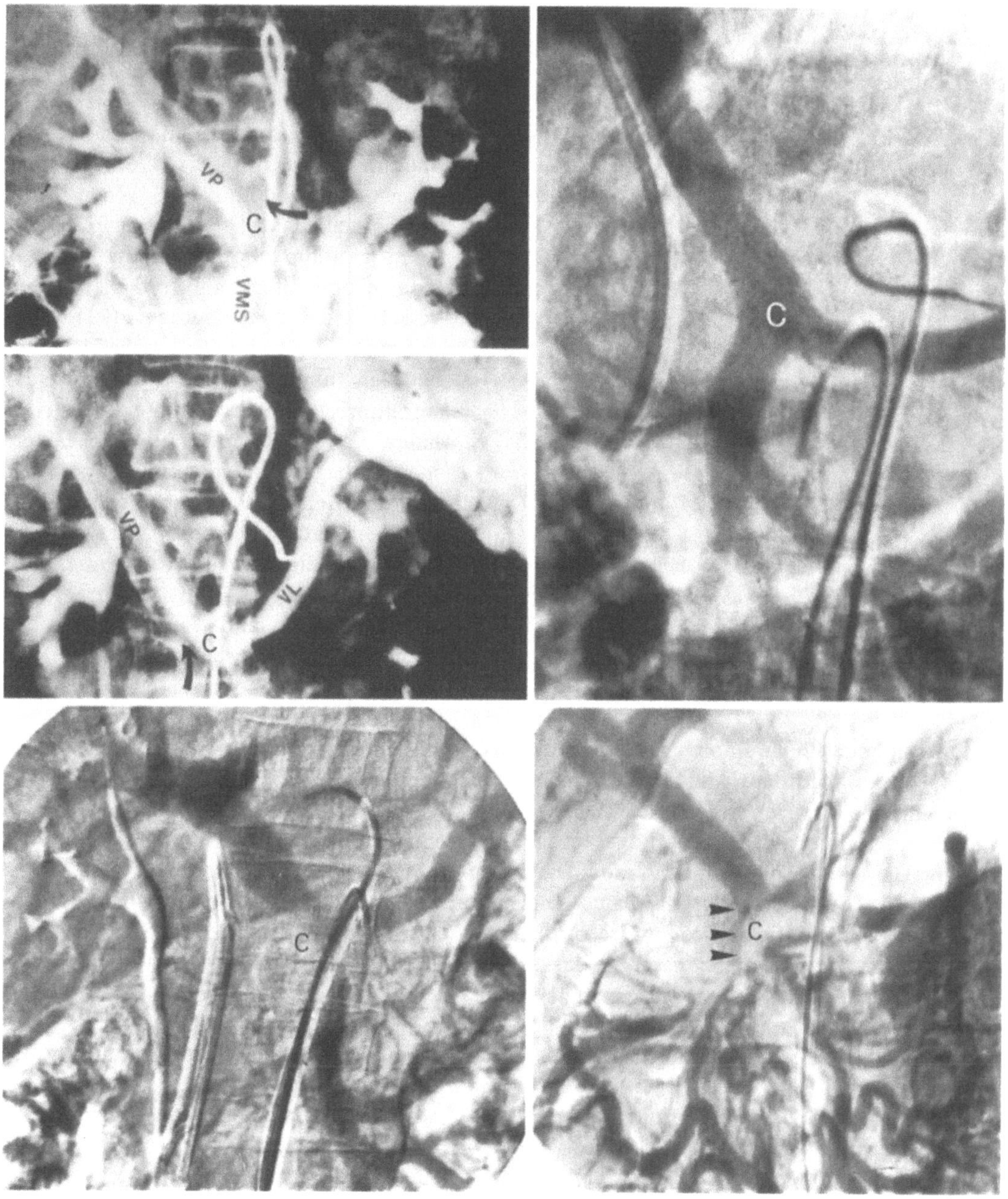

Fig. 24.8a–e. Angiography of the portal venous system. **a,b** Conventional angiography: single injection of contrast medium in the superior mesenteric artery (**a**) and celiac trunk (**b**), showing opacification of the superior mesenteric vein (*VMS*), portal vein (*VP*) and splenic vein (*VL*) respectively. Note the inflow of nonopacified blood via the splenic vein (**a**) and superior mesenteric vein (**b**) into the portal venous system. **c** Digital subtraction angiography (DSA): simultaneous injection of contrast medium via both catheters in the superior mesenteric artery and celiac trunk, eliminating the "inflow phenomena." **d** DSA: ingrowth of pancreatic head malignancy into the portal venous system (*C*). Simultaneous contrast injection using both catheters. **e** DSA: occlusion (*arrowheads*) of the superior mesenteric vein and splenic vein due to ingrowth of a large pancreatic head carcinoma. Note multiple collaterals

diagnosis and staging of pancreatic adenocarcinoma in our institution because the usual Doppler US and CT techniques provide accurate information. Angiography is used only for selected indications, such as in patients who have had recent exploration and/or are suspected of having anatomic variations of the hepatic artery, to provide a preoperative vascular road map (Mackie et al. 1979; Freeny 1989; Gulliver et al. 1992). At our institute, to exclude the inflow phenomena at the confluence of the splenic

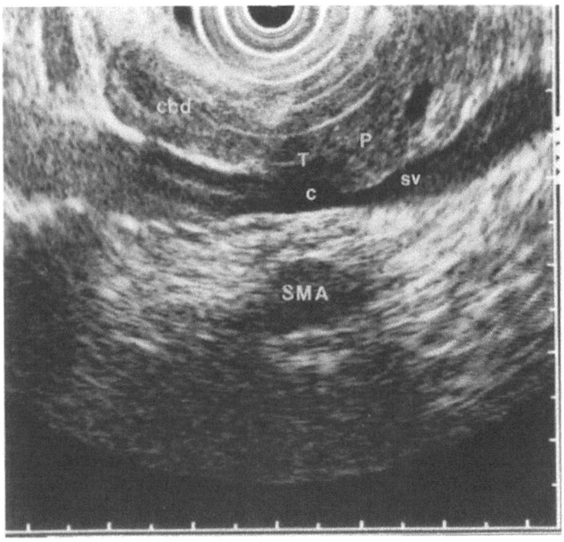

Fig. 24.9. Unresectable pancreatic head malignancy. EUS: small hypoechogenic tumor (*T*) in the head of the pancreas with contiguous invasion into the confluence (*c*) over a broad margin. *cbd*, Common bile duct, with a thickened wall and sludge; *SMA*, superior mesenteric artery; *sv*, splenic vein; *P*, normal pancreas. (Courtesy: Dr. P. Fockens, Amsterdam; The Netherlands)

and mesenteric vein we employ simultaneous injection of contrast medium via both catheters placed in the splenic and superior mesenteric arteries.

Helical CT offers an additional advantage to conventional CT because of its ability to provide imaging data that can be processed [multiplanar reformatting (MPR) and multiplane volume reconstruction (MPVR)] and can display 3D vascular images in a similar plane or format to that seen in angiography, without additional use of contrast material or exposure of the patients to radiation (Savader et al. 1990; Gulliver et al. 1992; Zeman et al. 1994; Charnsangavej 1996). The results of this *3D CT angiography* are promising, but more work needs to be done before this technique can be recommended routinely.

Endoscopic ultrasonography (EUS) is certainly the most accurate technique presently available for visualization of pancreatic malignant lesions. With EUS pancreatic carcinoma may present as mainly hypoechogenic mass lesions with irregular contours and inhomogeneous internal echoes, similar to what is seen using abdominal US. When scanning a pancreatic malignancy, special attention has to be paid to its relation to other organs (liver, duodenum, stomach, bile ducts) and to the large peripancreatic vessels (Tio et al. 1992; Rösch 1994). The diagnostic criteria of venous invasion are: an irregular interface between tumor and vessel, direct visualization of intraluminal growth, and total obstruction with collateral vessels (Rösch et al. 1994). Arterial encasement is diagnosed in the same way as with CT, when the tumor mass encircles the vessel. Due to its limited penetration depth, EUS staging of pancreatic carcinoma can only offer information about local regional tumor spread (EUS accuracy: T stage: 85%; N stage: 74%) and vascular invasion, but not about distant metastases (M stage) (Rösch et al. 1994). EUS is not accurate in the detection of distant metastases or carcinomatous involvement of the mesenteric axis and lymph nodes in the superior mesenteric axis. Small amounts of periduodenal or perigastric (lesser sac) ascitic fluid undetectable by conventional imaging methods can, however, often be seen with EUS in patients with large pancreatic tumors, and this may prompt the diagnosis of peritoneal seeding by subsequent laparoscopy (Rösch et al. 1994). Small peritoneal and omental metastases, typically measuring 1–2 mm, cannot be detected with EUS (Müller et al. 1994). EUS is more sensitive than dynamic CT and MRI, particularly in depicting tumors less than 3 cm in diameter. The accuracy of EUS versus other imaging modalities in the preoperative assessment of pancreatic carcinoma and involvement of the portal venous system is summarized in Table 24.6.

Table 24.6. Accuracy of endoscopic US and other imaging modalities in the preoperative assessment of pancreatic carcinoma and involvement of the portal venous system

Pathologic finding	MRI (%)	US (%)	CT (%)	Endoscopic US (%)	ERCP (%)	Angiography (%)
All pancreatic tumors	84[a]	69	67[a]–74	94–96[a]	91	–
Tumors <3 cm	67[a]	59	53	93[a]–100	85	–
Involvement of portal venous system	–	52	62	87	–	81

Updated from Rösch and Classen 1994 and [a]Müller et al. 1994.

The high rate of misdiagnoses of inflammatory mass as pancreatic carcinoma on the basis of EUS is due to the fact that distinguishing between focal pancreatitis and pancreatic carcinoma by imaging appearance alone is difficult (MÜLLER et al. 1994). However, EUS allows full delineation of the pancreatic duct and CBD, which may help differentiate focal pancreatitis (more irregularly dilated pancreatic duct with its side branches throughout the mass, small intraductal calcifications) from pancreatic carcinoma (MÜLLER et al. 1994). In the clinical setting, EUS should be used for the primary diagnosis of pancreatic carcinoma as a complementary method to ERCP in cases where US and CT have failed or yielded equivocal results (RÖSCH et al. 1994).

EUS-guided FNAB is possible using linear-type endoscopic transducers. In the future EUS may enable us to perform biopsy under endoscopic and US guidance in real time, which will probably allow more accurate sampling of the tumor (MÜLLER et al. 1994).

The wall of the portal venous system and the surrounding tissue can be clearly visualized by *intraportal US*, using specially designed catheters for scanning at 20 MHz and 30 MHz. With these catheters, high-resolution 360° real-time cross-sections of the vascular wall can be obtained. The future possibilities of this new investigation modality are promising. It may provide more detailed valuable information on changes in the portal vein than Doppler US, EUS, or arterial and/or direct portography (NOGUCHI 1993).

The same type of morphologic findings in pancreatic head carcinoma as are found on US and CT can be observed on *MRI* (STARK et al. 1984; STARK and MOSS 1984; ENGELHOLM et al. 1987, 1988). Recent evaluation of pancreatic MRI in several large series has shown that it is a valuable technique for the evaluation of pancreatic carcinoma (CHEZMAR et al. 1991; GEHL et al. 1991; MEGIBOW et al. 1991; MITCHELL et al. 1991; SEMELKA et al. 1991; VELLET et al. 1992; MITCHELL et al. 1992). However, bowel peristalsis, motion artifacts, and insufficient oral bowel opacification (oral magnetic particles; OMP, Nycomed, Oslo) are still major limitations (JENKINS et al. 1987; TSCHOLAKOV et al. 1987). Glucagon (administered i.m.) may reduce bowel peristalsis (TSCHOLAKOV et al. 1987). Pancreatic MRI is commonly performed using a high-field-strength magnet of about 1–1.5 T. Conventional T1- and T2-weighted spin-echo pulse sequences, fat-suppressed T1 images, and fast spin-echo or breathhold gradient-echo images with contrast enhancement using gadolinium-DTPA or manganese-DPDP should be acquired (FREENY 1995). These pulse sequences produce images with high contrast or signal-to-noise ratio, as well as depicting abnormalities owing to differential contrast enhancement (CHEZMAR et al. 1991; MITCHELL et al. 1991; GEHL et al. 1993; FREENY 1995).

Extrapancreatic infiltration of tumors into adjacent structures can be accurately identified with MRI (STARK and MOSS 1984; ANACKER et al. 1984; ENGELHOLM et al. 1987, 1988), but cannot be differentiated from inflammatory adhesions. The false-positive diagnosis of pancreatic carcinoma in patients with chronic pancreatitis remains a diagnostic dilemma with MRI. MRI is also unable to differentiate between pancreatic fibrosis in chronic pancreatitis and pancreatic desmoplasia in response to pancreatic carcinoma.

MRI demonstrates vascular and perivascular invasion, venous thrombosis, collaterals, extrinsic compression, and venous stasis (HAAGA 1984) particularly well.

VELLET et al. (1992) suggest that MRI (1.5 T) with respiratory motion compensation (pseudogating) is superior to CT in identification of tumor mass and assessment of vascular or perivascular invasion in the preoperative staging of pancreatic carcinoma. They also believe MRI to be superior to CT in assessment of intrapancreatic tumor size, extension of tumor, and portal venous extension (Table 24.7).

Surface coil MRI of the pancreas may improve the investigation of pancreatic carcinoma by MRI in the future (SIMEONE et al. 1985).

The development of a high-frequency *laparoscopic US* probe enables evaluation of solid organs and the retroperitoneal space for the presence of small intrahepatic metastases and local ingrowth of major vessels without the interposition of overlying bowel gas, and at a higher resolution (JAKIMOWICZ 1994; BEMELMAN et al. 1995) (Fig.

Table 24.7. Accuracy of MRI and CT in the preoperative staging of pancreatic ductal adenocarcinoma

Pathologic finding	MRI (%)	CT (%)
Tumor identification	84	71
Extrapancreatic extension	74	70
Invasion		
Portal vein	93	93
Duodenum	93	93

VELLET et al. 1992.

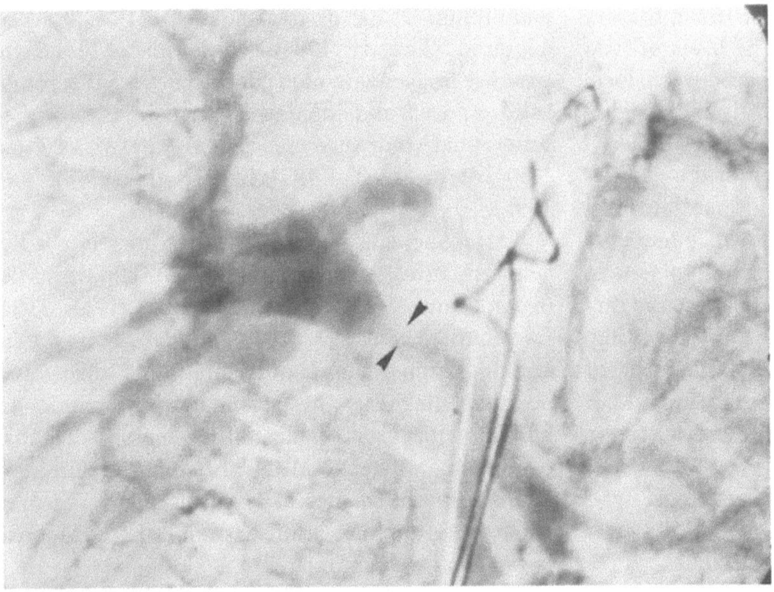

a

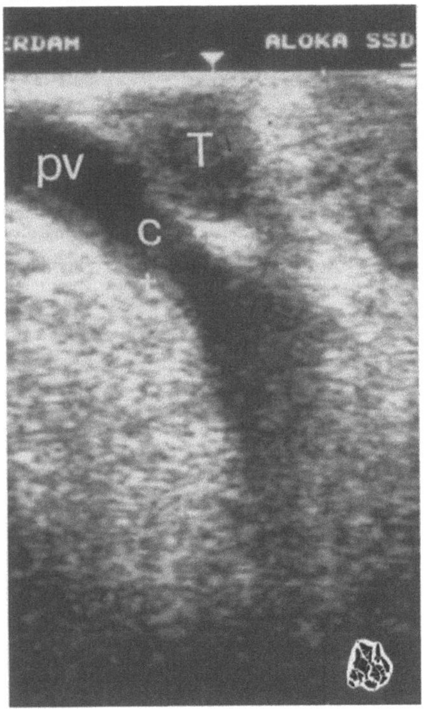

b

Fig. 24.10a,b. Laparoscopic US in pancreatic head carcinoma infiltrating the portal venous system. **a** DSA: narrowing of the portal venous system (*arrowheads*) due to tumor ingrowth from a mass in the pancreatic head. **b** Laparoscopic US: hypoechogenic mass (*T*) indenting the portal venous system (*pv*). *c*, confluence

24.10). The first results of staging of carcinoma of the pancreatic head region by laparoscopy with laparoscopic US have been encouraging (CUESTA et al. 1993; JOHN and GARDEN 1993; MURUGIAH et al. 1993; BEMELMAN et al. 1995).

Using a combination of CT, angiography, and laparoscopy, the positive predictive value for unresectability can be as high as 89% (WARSHAW et al. 1990; BEMELMAN et al. 1995). In the studies by BEMELMAN et al. (1995), laparoscopy combined with laparoscopic US as a single diagnostic modality showed a positive predictive value of 97% for unresectability.

Laparoscopic US is confirmed as an important tool in the assessment of intrahepatic liver metastases. (CUESTA et al. 1993; JOHN and GARDEN 1993; MURUGIAH et al. 1993; BEMELMAN et al. 1995.) In the studies by VAN DELDEN et al. (1996), laparoscopic US showed liver metastases in 14% and laparoscopic US combined with laparoscopy showed liver metastases in 23% of the patients with pancreatic head carcinoma preoperatively considered, on the basis of US

Table 24.8. ERCP features of ampullary carcinoma

Dilated CBD pancreatic duct
Eccentric or circumferential CBD pancreatic duct stenosis
Irregular "frayed" contour of the distal ducts, caused by tumor encroachment
Frank "apple core" stenosis with overhanging edges
Focal intraluminal mass in the CBD
Effacement of adjacent duodenum

ZEMAN and BURRELL 1987.

and/or CT, to have resectable disease. A pitfall in the detection of liver metastases was location of metastases on top of the dome of the upper border of the liver. The development of laparoscopic US probes equipped for accurate US-guided needle biopsy will probably improve histological confirmation of intrahepatic metastases in the near future. Unresectability as determined by laparoscopic US can be considered to be highly specific (BEMELMAN et al. 1995; VAN DELDEN et al. 1996).

24.3.1.2 Cystadenocarcinoma

Cystadenocarcinoma accounts for about 1% of pancreatic cancers (FRIEDMAN et al. 1983; CUBILLA and FITZGERALD 1979). The diagnosis and preoperative assessment of primary cystic malignant tumor is important, because it has a much better prognosis than the usually solid ductal pancreatic adenocarcinoma, and is curable by complete local resection so long as there are no distal metastases. Even if it is unresectable, the prognosis is better than that of pancreatic duct adenocarcinoma (FRIEDMAN et al. 1983).

Tiny central foci of calcification in the pancreatic region (due to subepithelial hemorrhage and necrosis) are seen on a plain X-ray film (FRIEDMAN et al. 1983). The cystic components of the lesion, the vascular involvement (vascular encasement and obstruction or displacement of the splenic vein), and secondary smooth compression of the distal CBD can be clearly demonstrated with the different imaging modalities. (WOLFMAN et al. 1982; ITAI et al. 1986). Percutaneous FNAB and/or endoscopic brush cytology may be helpful for confirming the radiologic diagnosis (WOLFMAN et al. 1982).

24.3.2 Ampullary Carcinoma

Ampullary or periampullary tumor – a polypoid adenocarcinoma protruding from the papilla – is less common than pancreatic carcinoma and may be quite difficult to distinguish preoperatively. The tumors are usually small (50% measuring less than 3 cm) when discovered, since the lumen of the CBD does not permit much encroachment before it becomes obstructed (LEVINE et al. 1979; MOORE and CLARK 1984). Early detection may allow curative resection (ZEMAN and BURRELL 1987; LYGIDAKIS and TYTGAT 1989).

An impacted stone with secondary edema may mimick a friable ampullary mass.

US and CT may demonstrate both a dilated CBD and a dilated pancreatic duct, and occasionally a mass lesion. The actual tumor may be small, but the dilated CBD may prolapse into the duodenum, thus mimicking a larger mass. The cholangiographic features of ampullary carcinoma are summarized in Table 24.8.

The advantage of *ERCP* over other diagnostic modalities is the possibility of obtaining brush biopsies for histology (ZEMAN and BURRELL 1987).

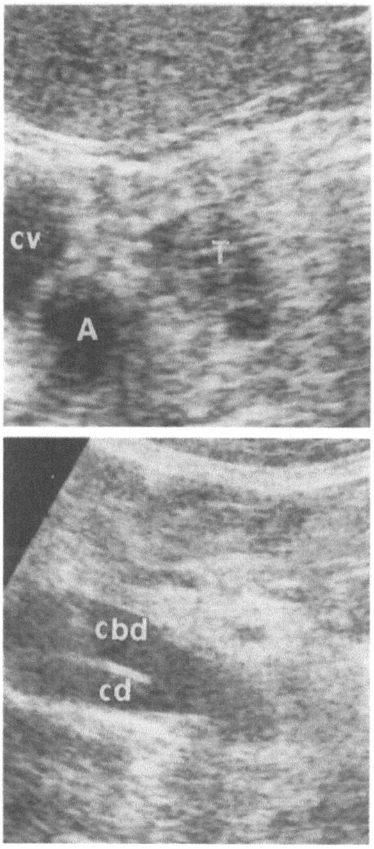

Fig. 24.11a,b. Ampullary malignancy. Oblique (**a**) and longitudinal US scans (**b**): small, round, well-delineated mass (*T*) at the ampulla with dilation of pancreatic duct, common bile duct (*cbd*), and cystic duct (*cd*). *cv*, Inferior vena cava

Table 24.9. Accuracy of US versus CT in determining the level and cause of obstruction of the CBD, as reported by various authors (1981–1994)

Authors	Level of obstruction		Cause of obstruction	
	US (%)	CT (%)	US (%)	CT (%)
PEDROSA et al. 1981b	–	97	–	94
HAUBEK et al. 1981	94	–	68	–
BARON et al. 1982	60	88	39	70
HONICKMAN et al. 1983	27	–	23	–
THORSEN et al. 1984	–	100	–	–
LAING et al. 1986	91.8	–	70.9	–
GIBSON et al. 1986	95	90	88	63
REIMAN et al. 1987	–	92	–	–
NESBIT et al. 1988	75	96	47	69
LOOSER et al. 1992	100	–	86	–
TRILLER et al. 1994	100	100	89	68

Tumors of the papilla of Vater are visualized with *EUS* as localized, mostly hypoechogenic and inhomogeneous lesions located in the duodenal wall and, with advancing malignancy, infiltrating through the wall into the pancreatic head (Fig. 24.11). EUS has been shown to be reliable in local staging, T-staging (accuracy 86%) being better than N-staging (70%) (RÖSCH 1994). The clinical role of EUS in assessing resectability, and its comparison with other procedures (e.g., ERCP and CT), is not yet entirely clear (RÖSCH 1994).

The differential diagnosis between benign ampullary stenosis and small tumors is hindered by the fact that inflammatory stenosis, especially after endoscopic sphincterotomy, can also lead to an hypoechogenic lesion mimicking a true neoplasm (RÖSCH et al. 1992).

Duplex Doppler US may be helpful in evaluating resectability preoperatively.

24.3.3 Primary Cholangiocarcinoma

Cholangiocarcinoma, a primary cancer of biliary ductal epithelium which accounts for 0.5%–1% of all cancers, is associated with an overall 5-year survival rate of 1%, and a 5-year survival rate of 20% among patients undergoing resection (FRAUMENI 1975;

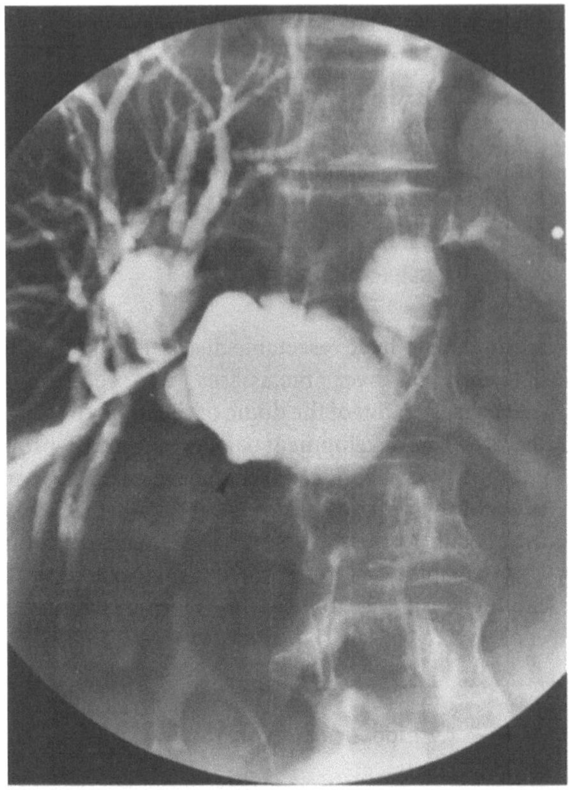

a

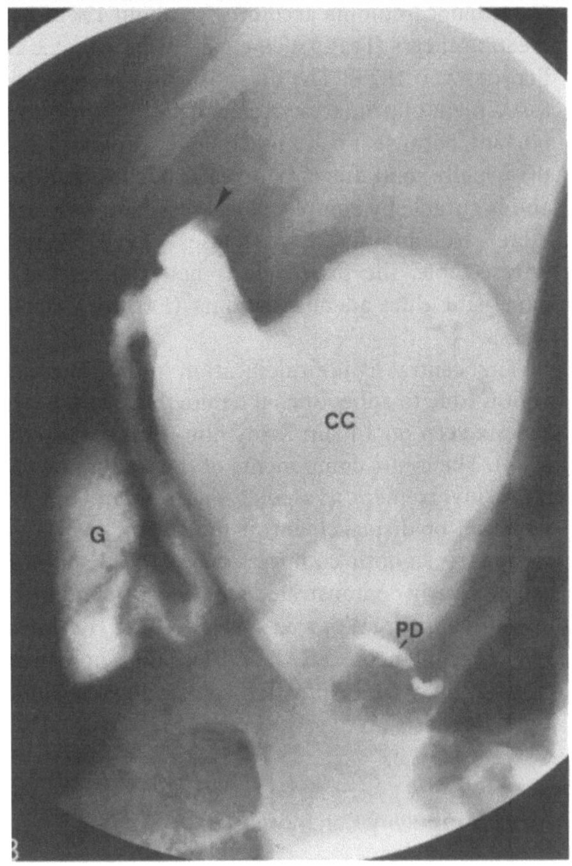

b

Fig. 24.12a,b. Choledochal cyst with malignant degeneration at the proximal CBD. **a** PTC: dilatation of the intrahepatic bile duct system of the right and left liver lobe with a stop at the proximal CBD (*arrowhead*). **b** ERCP: choledochal cyst with a proximal stop (*arrowhead*); *G*, Gallbladder; *PD*, pancreatic duct

GREENWALD and GREENWALD 1983; WHITE 1985; MACHAN et al. 1986; NESBIT et al. 1988).

Cholangiographically, three morphological types need to be considered (NICHOLS et al. 1983): focal stenotic carcinoma (pathologically high-grade tapered lesions, 1–3 cm long with secondary narrowing or obstruction of a bile duct segment), polypoid carcinoma, (pathologically low-grade papillary lesions protruding into the lumen of the bile duct), and high-grade diffuse sclerosing carcinoma (lesions obstructing multiple intrahepatic bile duct segments). Patients with an extrahepatic low-grade (polypoid) carcinoma have the best prognosis.

In differentiating obstructive from nonobstructive jaundice, *US*, which may confirm or exclude CBD dilatation, has an accuracy of over 90% (HAUBEK et al. 1981; GIBSON et al. 1986) and should be considered as the screening procedure of choice. In one report, information about the level of obstruction was obtained with US and CT in 100% of cases (TRILLER et al. 1994).

Table 24.9 shows that US is almost as accurate in determining the level and cause of obstruction as CT (HAVRILLA et al. 1977; LEVITT et al. 1977; GOLDBERG et al. 1978; PEDROSA et al. 1981a; DILLON et al. 1981; BARON et al. 1982, 1983; ITAI et al. 1983; MEYER and WEINSTEIN 1983; THORSEN et al. 1984; SUBRAMANYAM et al. 1984; CARR et al. 1985; MACHAN et al. 1986; NESBIT et al. 1988; TEEFEY et al. 1992). If US shows no abnormality, no further examination is usually necessary. If US shows dilatation, Doppler US, spiral CT, and ERCP should be performed for further evaluation (LEVINE et al. 1979; SHIMIZU et al. 1981; BARON et al. 1982; VAN LEEUWEN and REEDERS 1989) (Fig. 24.12).

A wall thickness greater than 5 mm has previously been shown to be characteristic of malignant disease, particularly cholangiocarcinoma (SCHULTE et al. 1990; SEMELKA et al. 1992). *CT* visualization of distal CBD tumors may be difficult because of volume averaging, since the tumors tend to be smaller and the extrahepatic duct system tends to run obliquely (GIBSON et al. 1986). Better imaging of tumor extent by CT may be possible if dynamic scanning is performed with 5-mm collimation at 5-mm intervals through the regions of interest.

Cholangiocarcinoma tends to cause abrupt termination of the CBD on US or CT, while pancreatitis or a benign stricture may cause gradual tapering over serial CT sections (PEDROSA et al. 1981a;

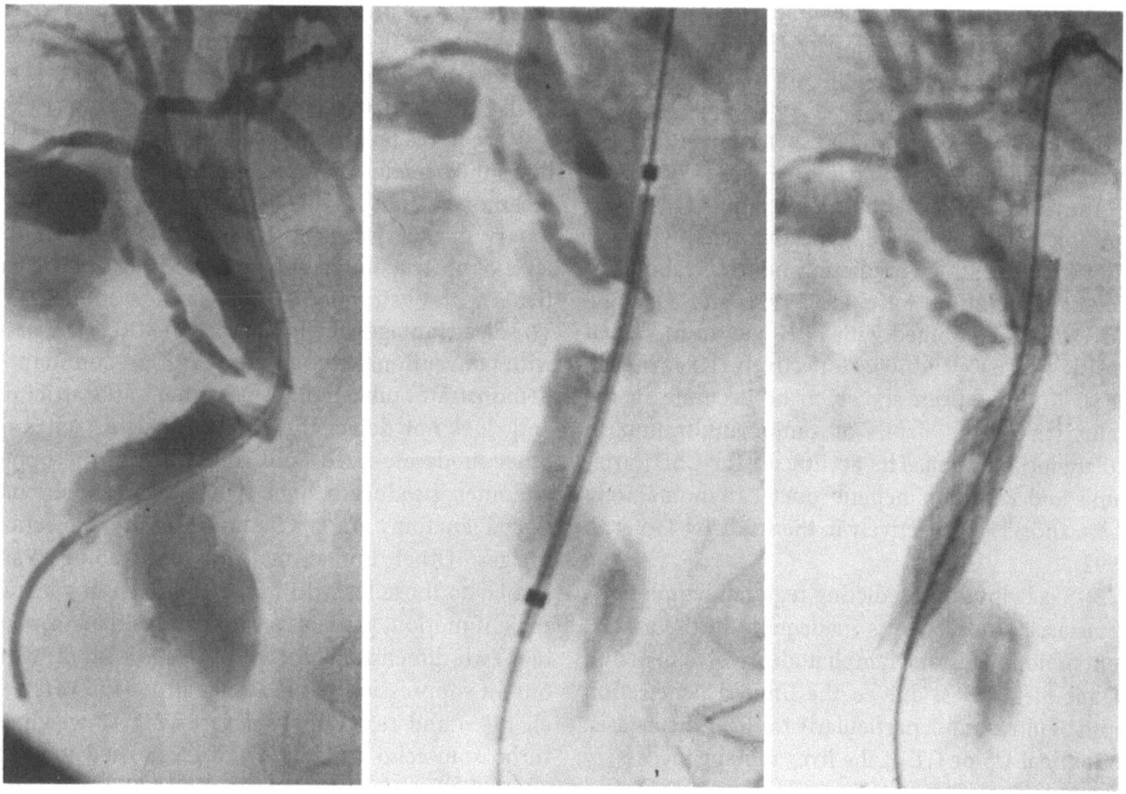

a,b c

Fig. 24.13a–c. Cholangiocarcinoma of the mid-CBD; palliative stenting. ERC: **a** eccentric irregular mid-CBD stenosis for which an metallic expandable stent has been inserted; **b** unexpanded stent; **c** expanded Memotherm stent (metallic expandable stent, Angiomed)

Table 24.10. US/CT features of primary cholangiocarcinoma

Dilated CBD, obstructive soft tissue mass (1.5–8 cm), hypoechoic/low-attenuation compared to normal liver
Abrupt ductal termination
Stenosis of CBD over 5–10 mm

Levine et al. 1979; Dillon et al. 1981; Thorsen et al. 1984.

Freeny 1982; Engelholm et al. 1988; Reiman et al. 1987).

Currently *ERC* is the definitive method for assessing intraductal tumor extent (Fig. 24.13), with a reported accuracy of 89%–96% (Nichols et al. 1983; Moore and Clark 1984; Gibson et al. 1986), but it is insensitive in detecting the size of exophytic tumor and extrabiliary tumor spread, which is an important determining factor of tumor resectability. Brush cytologic examinations may further enhance the diagnostic accuracy of ERC.

Percutaneous intraluminal US of the bile ducts is a safe and rapid procedure which may offer complementary or supplemental information about the extension of tumor and bile duct thickening beyond the fluoroscopically visualized lumen and mucosa in cholangiocarcinoma (van Sonnenberg et al. 1992). With future refinements and improvements this diagnostic technique may hold promise. US and CT complement ERC well. The US/CT features of primary cholangiocarcinoma are listed in Table 24.10.

In the series reported by Nesbit et al. (1988), CT correctly identified 54%, US 50%, and ERC 58% of the tumors as either resectable (40%) or nonresectable (60%). The respective sensitivities of CT, US, and ERC in predicting nonresectability of cholangiocarcinoma were 44%, 19%, and 43%. CT and US each combined with ERC had an increased sensitivity of 64% and 50% respectively (Nesbit et al. 1988). *EUS* has been shown to be accurate in the evaluation of the depth of tumor infiltration in cholangiocarcinoma. The accuracies for CBD carcinoma and common hepatic duct carcinoma were 82.8% and 85% respectively in the study by Tio et al. (1992).

EUS is helpful in predicting regional lymph node metastasis; however, it is inadequate in the assessment of nonmetastatic lymph node involvement and distant metastases, due to the limited penetration depth of ultrasound, particularly for liver metastasis. Abdominal US or CT of the liver remains necessary to complete staging.

MRI of bile duct malignancy has not yet been adequately assessed. Spin-echo sequences have been shown to be limited in their demonstration of cholangiocarcinoma, primarily because of phase artefacts (Dooms et al. 1986; Semelka et al. 1992).

MRI appears to provide greater information on the intrahepatic bile ducts, because the higher contrast resolution of MRI shows abnormal intrahepatic periportal tissue better than CT, whereas CT provides more information on the extrahepatic bile ducts, because it has higher spatial resolution and fewer artifacts than MRI (Semelka et al. 1992).

Improvement in techniques to compensate for respiratory movements in MRI studies may result in greater diagnostic utility of MRI. Based on these preliminary observations, further investigation into the role of new MRI sequences in the detection and characterization of biliary disease appears warranted (Semelka et al. 1992).

3D CT cholangiography is a new technique that provides noninvasive comprehensive demonstration of biliary anatomy and biliary tract obstruction. Attempts have been made to perform 3D US cholangiography, but the images are not yet diagnostic (Fine et al. 1991; Gillams et al. 1994).

MR cholangiography (MRC) is a novel method for depicting the intra- and extrahepatic biliary ducts (Wallner et al. 1991; Morimoto et al. 1992; Hall-Craggs et al. 1993; Ishizaki et al. 1993; Takehara et al. 1994; Guibaud et al. 1994; Barish et al. 1995). MRC requires no injection of contrast material, combines the benefits of cross-sectional and projectional techniques, and provides an overview of the entire biliary ductal system. MRC correctly identifies the level of obstruction in approximately 90% of the patients with obstruction (Hall-Craggs et al. 1993). The disadvantages and limitations of MRC compared with conventional studies are that it predominantly demonstrates dilated ducts proximal to the stricture and does not depict the stricture itself. Ascites or other moderate-sized fluid collections in the upper abdomen produce a high signal that can obscure ductal anatomy on maximum-intensity-projection images. Other limitations and disadvantages are similar to those for MRI of the abdomen in general: patient motion, peristalsis, and patient claustrophobia. Two-dimensional turbo spin-echo with or without fat suppression (Guibaud et al. 1994; Barish et al. 1995) and respiratory-triggered 3D T2-weighted turbo spin-echo sequences have been used to overcome the limitations (Barish et al. 1995). For these reasons, MRC is better for showing ductal anatomy proximal to occlusive lesions, especially in cases of high-grade obstruction in which retrograde filling is not possible or not advisable because of the risk of sepsis or pancreatitis.

Diagnostic MRC was successful in 97% of the patients in the study of BARISH et al. (1995). MRC may be important especially in patients in whom previous biliary surgery or drainage procedures make endoscopic access impossible. PTC may be reserved for patients in whom US, CT, and MRI fail or are equivocal, for those in whom ERC is not possible, and for preliminary percutaneous biliary decompression (BARON et al. 1982).

24.3.4 Primary Gallbladder Carcinoma

Primary gallbladder carcinoma, which is responsible for less than 1.4% of all biliary tract operations and represents 1%–3% of all malignancies (KLEIN and FINCK 1972; KEILL and DEWEESE 1973), is a rapidly progressive malignancy carrying a poor prognosis, with a 5-year survival rate of 1%–5%. In autopsy studies gallstones have been found in 65%–95% of all gallbladder carcinomas (ROBBINS 1967; HSU CHONG YEH 1979). If confined to the gallbladder, the tumor can be easily resected, but it usually is not detected until it reaches an advanced stage and has extended beyond the gallbladder (ADOSON 1973; PIEHLER and CRICHLOW 1978; MORROW et al. 1983; MUIR and MORRIS 1986; SAGOH et al. 1990).

For successful treatment of gallbladder carcinoma, it is important to detect this tumor at an early stage (ROBERTS and DAUGHERTY 1986; SAGOH et al. 1990), that is, to detect small protrusions or irregular gallbladder wall thickening. Gallbladder carcinoma readily extends into the hepatoduodenal ligament and para-aortic region by means of various routes: lymph node metastasis and direct, vascular, and perineural spread. Most radiological examinations (*oral cholecystography, intravenous cholangiography, PTC*) have failed to permit a definitive diagnosis of gallbladder cancer, owing to the absence of specific findings and the inability to delineate the gallbladder (HSU CHONG YEH 1979; ITAI et al. 1980). *US* is the most useful modality, with its ease of performance, low costs, and the availability of the equipment (DALLA PALMA et al. 1980; YUM and FINK 1980; ITAI et al. 1980; WEINER et al. 1984; THORSEN et al. 1984; KOGA et al. 1985; COOPERBERG and GIBNEY 1987; SAGOH et al. 1990). US has also been reported to have limited potential for the evaluation of tumor infiltration in this region (BAKER et al. 1987; SAGOH et al. 1990).

Despite the prevalence of US screening, the majority of cases of gallbladder carcinoma are still not diagnosed until the tumor has extended beyond the gallbladder (MUIR and MORRIS 1986; SAGOH et al. 1990).

As previously reported (BAKER et al. 1987; SAGOH et al. 1990), tumor contour and the relationship of the tumor mass (particularly the infiltrative form) to the surrounding structures, such as blood vessels and the pancreas, are often unclear on CT due to motion artifacts, poor contrast between the tumor and the surrounding structures, and inadequate contrast enhancement techniques (SAGOH et al. 1990).

Gallbladder carcinoma has been described on conventional MR sequences (SAGOH et al. 1990; SEMELKA et al. 1992). *MRI* may perform at least as well as CT or US in the detection of direct liver invasion and distant metastases (ROSSMANN et al. 1987; SAGOH et al. 1990). These are well displayed on MRI, although it is sometimes difficult to differentiate infiltrative tumor from inflammatory disease (ITAI et al. 1980; YUM and FINK 1980; WEINER et al. 1984; THORSEN et al. 1984; KOGA et al. 1985; COOPERBERG and GIBNEY 1987; SAGOH et al. 1990). MRI is inaccurate in demonstrating duodenal invasion due to respiratory motion artifacts, partial volume effects, and paucity of fat, as with CT.

Unlike CT, MRI depicts tumor spread well and clearly demonstrates its contour on T1-weighted images, with high contrast between the tumor and the surrounding fat tissue (SAGOH et al. 1990).

MRI may be useful for the differential diagnosis in selected cases when US shows marked gallbladder wall thickening in the absence of clinical signs of acute cholecystitis or other causes of gallbladder wall thickening.

Table 24.11. US/CT features of gallbladder carcinoma

Main features:
Mass filling the gallbladder
Fungating mass
Infiltrating mass
Discontinuous thickened gallbladder wall with irregularity and liver invasion
Associated features:
Dilated biliary ducts
Stones?
Liver mass
Direct invasion
Metastases
Retroperitoneal/peripancreatic mass

HSU CHONG YEH 1979; ITAI et al. 1980; SOLOMON and RUBINSTEIN 1981.

The US/CT features of gallbladder carcinoma are shown in Table 24.11.

Angiography is nowadays rarely used for investigation of gallbladder carcinoma. It may show neovascularization from right hepatic arterial branches, early hepatic venous filling, tumor blush in the arterial phase, and encasement of the celiac artery (Kido et al. 1974).

The use of US- or CT-guided percutaneous FNAB should improve the qualitative and quantitative accuracy of the diagnosis (Itai et al. 1980).

24.3.5 Primary Intrahepatic Cholangiocarcinoma

Intrahepatic cholangiocarcinoma is frequently classified into two types according to site of origin:

- Primary hilar cholangiocarcinoma, also called Klatskin's tumor, originating from the main or first-order branches of the bile ducts.
- Primary peripheral cholangiocarcinoma, originating proximal to second-order branches of the bile ducts.

The initial symptoms of the peripheral type are abdominal pain, malaise, anorexia, and fever. The incidence of jaundice is relatively low. The hilar type, however, presents with obstructive jaundice and generalized malaise. The clinical symptoms accompanying the hilar type are similar to those of extrahepatic bile duct carcinoma (Thorbjarnarson 1959; Cruickshank 1961; Honda et al. 1993). The two types will be discussed separately.

24.3.5.1 Primary Hilar Cholangiocarcinoma (Klatskin's Tumor)

Most authors consider carcinoma arising at the bifurcation of the hepatic duct (Klatskin's tumor) (Klatskin 1965) and tumors of the extrahepatic bile ducts as separate entities. Some consider them together (Nichols et al. 1983). Klatskin's tumors (at the hepatic hilum) are reported to represent 10.5%–26.3% of all bile duct carcinomas. Their reported incidence in autopsy studies is 0.01%–0.46% (Looser et al. 1992; Alexander et al. 1994). There is an increased incidence in chronic ulcerative colitis (sclerosing cholangitis), reported at 29%, in Crohn's disease, cystic biliary disease (Caroli's syndrome), and in patients with liver fluke (*Opisthorchis sinensis*) (Juttijudata et al. 1982; Altaee et al. 1991).

Klatskin's tumors are usually small (diameter 1–3 cm), slow-growing tumors of the bile duct bifurcation, with distant metastasis occurring at a late stage of the disease (21.4%). Local invasion of the proximal and distal bile ducts and in the liver may occur early, with involvement of adjacent portal structures of the liver. Liver metastases are rare.

Death is frequently the result of hepatocellular failure and cholangitis due to obstruction rather than to metastatic disease.

Hilar cholangiocarcinomas are usually divided into three macroscopic types: sclerotic, nodular, and papillary (Blumgart 1988; Nesbit et al. 1988; Choi et al. 1989; Yamashita et al. 1992). Tumor type correlates moderately well with the level of bile duct involved: the papillary type is located predominantly in the distal part of the common duct and the nodular type in the mid part of the CBD (Weinbren and Mutum 1983; Adam and Benjamin 1992). The most common appearance of hilar cholangiocarcinoma is as a sclerosing mass extending along the bile ducts. This infiltrating type of neoplasm, which invades adjacent liver tissue, is pathologically a sclerotic lesion with abundant fibrous tissue and exhibits a desmoplastic and inflammatory stroma response and perineural growth (Weinbren and Mutum 1983; Bosma 1990; Yamashita et al. 1992).

Klatskin's tumor is resectable in only up to 20% of cases (Wagner et al. 1987). In resections which are intended to be radical, complete tumor exision can be achieved in only 8% of the cases, with diseased dissection planes in up to 85% (Bosma 1990). Although the surgical results are poor, with a median survival rate of 6–22 months (Bengmark et al. 1988; Blumgart and Benjamin 1989), complete excision is still considered to be the treatment of choice for this tumor.

Criteria of unresectability are (Adam and Benjamin 1992; Looser et al. 1992):

- Cholangiographic evidence of bilateral extension to the segmental branches of the intrahepatic bile ducts)
- Involvement of the main portal vein
- Involvement of both branches of the portal vein
- Involvement of the portal vein to one side of the liver combined with involvement of the hepatic artery to the contralateral side of the liver
- Combination of vascular involvement to one side of the liver with extensive cholangiographic involvement of the contralateral side
- Hepatic or nodal metastases

For determining unresectability, extensive biliary and vascular involvement are considered the most important factors. The aim of radiological

a

b

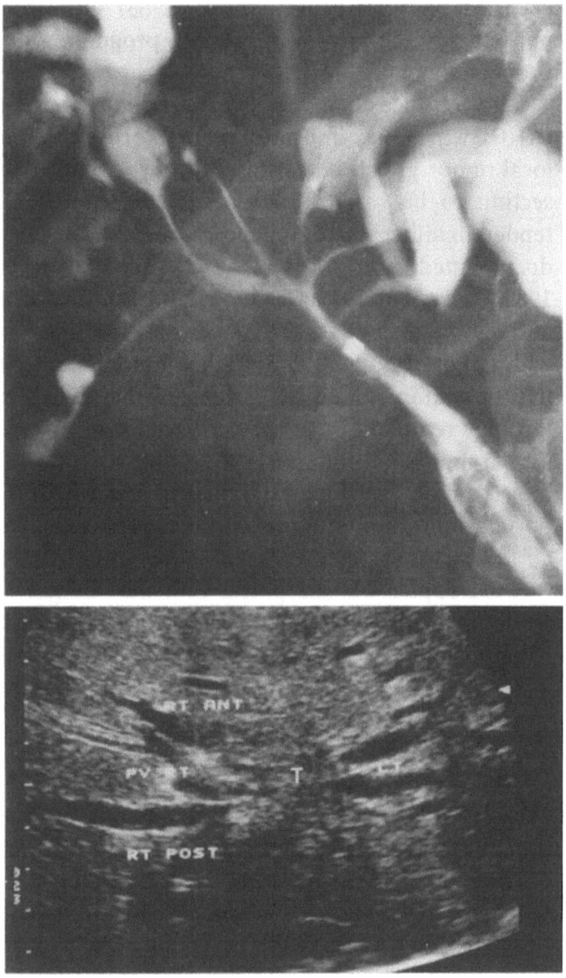

Fig. 24.14a,b. Hilar cholangiocarcinoma (Klatskin's tumor, type IV). **a** ERC: multiple strictures of the proximal CBD, right and left liver main and segmental branches. **b** Oblique US scan: abutting of the intrahepatic segmental branches of both systems due to an (invisible) isoechogenic tumor mass at the liver hilum

investigation is early detection and accurate preoperative assessment of the resectability of these tumors in terms of curative or palliative surgery (TRILLER et al. 1994).

ERC is the definitive investigation for detecting these lesions (accuracy 93%) and determining their location (MÜLLER et al. 1994) (Fig. 24.14). Beaded or long segmental stricture(s) may be found. However, cholangiography defines only the intraluminal tumor extent and has inherent limitations on showing the size of the extraluminal portion of the tumor and allowing proper assessment of suprahilar tumor extension. In the series of TRILLER et al. (1994), assessment of suprahilar tumor extension with ERC was correct in 43% of the cases, while in 57% no evaluation of proximal tumor extension was possible because of insufficient contrast filling of the proximal bile duct system, due to severe stenosis at the confluence. US-guided PTC, however, can correctly show diagnostic evidence of tumor (accuracy 94%) and intraluminal tumor extension into the right and left hepatic ducts or a liver segment (accuracy 80%) (TRILLER et al. 1994). Insufficient filling of the intrahepatic bile ducts with contrast medium and chronic cholangitis with secondary changes of the bile ducts can lead to an incorrect diagnosis (TRILLER et al. 1994).

Percutaneous cholangioscopy (PTCS) may improve the preoperative staging of biliary tumors, and cholangioscopic biopsy enables us to make an exact mapping of the intraductal extent of superficially spreading carcinoma of the bile duct. PTCS can be carried out 2 weeks after percutaneous biliary drainage through the sinus tract of a 12- to 15-F catheter, using a cholangiofiberscope. PTCS has been performed by Nimura in 665 patients from 1977–1994 from which 335 (50.4%) patients showed a malignancy (NIMURA 1996).

Cholangioscopic observation reveals irregularly dilated and tortuous vessels ("tumor vessels") on the biliary stenosis. Tumor development can be clearly observed in the lumen of the obstructive duct (KUBOTA et al. 1992; NEUHAUS 1992; NIMURA et al. 1993; NIMURA 1996).

Cholangioscopic biopsy was performed in 286 out of 335 malignancies, with a positive rate of 94.6% in bile duct carcinoma.

Tumor site, invasion of the liver parenchyma (visible on US as loss of the hyperechogenic interface between bile ducts and liver) and surrounding vasculature (particularly the portal venous system), and the presence of metastases can all be accurately assessed by (duplex Doppler) US and are important in determining tumor resectability (TRILLER et al. 1994; LOOSER et al. 1992; SMITS and REEDERS 1996, unpublished).

The *US/CT* features of Klatskin's tumor (MEYER and WEINSTEIN 1983; CARR et al. 1985) are shown in Table 24.12. US can correctly visualize the tumor size in 83%–89% of patients, the extent of ductal tumor invasion in 64%, and lobar segmental tumor extension in 80% (TRILLER et al. 1994). Duplex US can correctly diagnose vascular infiltration in 85%–91% of cases; angiography does the same in 82%–86% (TRILLER et al. 1994; SMITS and REEDERS 1995). This means that there is no need to carry out the invasive procedure of angiography. Duplex Doppler US may assist in the selection of patients for at-

Table 24.12. US/CT features of Klatskin's tumor

Focal/segmental intrahepatic duct dilatation
Normal CBD
Nonunion of the right and left hepatic ducts at the hilum
Small solid intraductal hyperechogenic to isoechogenic mass (with low-attenuation and minimal contrast enhancement on CT) at the confluence of the right and intrahepatic ducts
Enlarged perihilar lymph nodes
Local spread into the liver
Liver atrophy

MEYER and WEINSTEIN 1983; CARR et al. 1985; GARBER et al. 1993.

tempted resection and, more importantly, may noninvasively identify patients whose tumors are clearly unresectable but who might be considered as candidates for nonoperative palliation (CARRASCO et al. 1985; IRVING et al. 1989; NEUHAUS 1992; LOOSER et al. 1992; TRILLER et al. 1994) (Fig. 15). Doppler US can give false-negative results, firstly for technical reasons, e.g., if the portal vein (especially the left one) courses deeply or the Doppler angle is not shallow enough, and secondly if tumor infiltration of the vessel wall does not result in narrowing of the vessel. Doppler US can give a false-positive result if narrowing of the portal vein is caused by compression without tumor infiltration.

CT and US can depict intrahepatic masses better than previous reports suggested, showing them in 63%–69% and 47%–89% of cases respectively (THORSEN et al. 1984; CARR et al. 1985; GIBSON et al. 1986; NESBIT et al. 1988; TRILLER et al. 1994). The discrepancy in results may be due to the inclusion of CT examinations without intravenous contrast enhancement in earlier studies and to improvements in sonographic technology and image resolution (DILLON et al. 1981; MACHAN et al. 1986; NESBIT et al. 1988). Retrospective analysis of US scans is more difficult than that of CT studies, and more of these tumors are isoechogenic than are isodense relative to the liver parenchyma. Because most tumors (42%–56%) are more or less isoechogenic to the adjacent liver parenchyma, they are difficult to delineate on US (Fig. 24.14b). The sonographic character of these tumors can be hyperechogenic (16%–42% of cases), hypoechogenic (16%–27% of cases), or mixed (1% of cases) (TRILLER et al. 1994; SMITS and REEDERS 1995).

Lobar atrophy, due to longstanding severe biliary obstruction and/or severe portal venous stenosis or occlusion (BRAASCH et al. 1972; NESBIT et al. 1988), is an important predictive factor arguing unresectability (NESBIT et al. 1988) but does not affect the overall patient prognosis disproportionately.

The presence of an atrophy/hypertrophy complex (AHC), which may indicate a marked progression of local tumor growth, determines the type of liver resection to be carried out (hemihepatectomy, extended hemihepatectomy) and can be appropriately documented with CT (BEAZLEY and BLUMGART 1985; CARR et al. 1985; GIBSON et al. 1986; HADJIS et al. 1989; LOOSER et al. 1992; TRILLER et al. 1994), or with (color) Doppler US if CT is indeterminate. No liver resection should be performed that leaves an atrophic remnant! In the presence of lobar atrophy, axial rotation or compression by the contralateral hypertrophic liver lobe can result in stenosis of the portal vein without tumorous infiltration (LOOSER et al. 1992; TRILLER et al. 1994).

On Doppler US, the portal vein may be gracile without involvement and may show low-velocity hepatopetal blood flow or even a hepatofugal blood flow (N.J. SMITS et al. unpublished). Tumor infiltration of the hepatoduodenal ligament is inadequately shown by US and/or CT.

Intraportal US may provide more valuable information on changes of the portal venous wall than duplex Doppler, arterial portography, and/or direct portography (NOGUCHI 1993).

The accuracy of a judgment of nonresectability in Klatskin's tumor (GIBSON et al. 1986; NESBIT et al. 1988) is higher when different methods (US, CT, ERC, angiography) are combined (70%–90%) than with any single modality (42%–71%) (Table 24.13).

Angiographic findings of the porta hepatis in cases of Klatskin's tumor include tiny, thin neoplastic vessels with irregular encased or obstructed arteries and stenosis or occlusion of the portal venous system (Fig. 24.15c). There are no unusual hypervascular tumor vessels, arteriovenous shunts, or thrombi in veins, as in hepatocellular carcinoma. If the tumor is unresectable, pathologic confirmation of a suspected

Table 24.13. Accuracy of a judgment of nonresectability of cholangiocarcinoma/Klatskin's tumor on the basis of different diagnostic modalities

Reference	CT (%)	US (%)	ERC (%)	Angiography (%)	Combined (%)
GIBSON et al. 1986	42	71	58	25	90
NESBIT et al. 1988	54	50	58	-	68–70

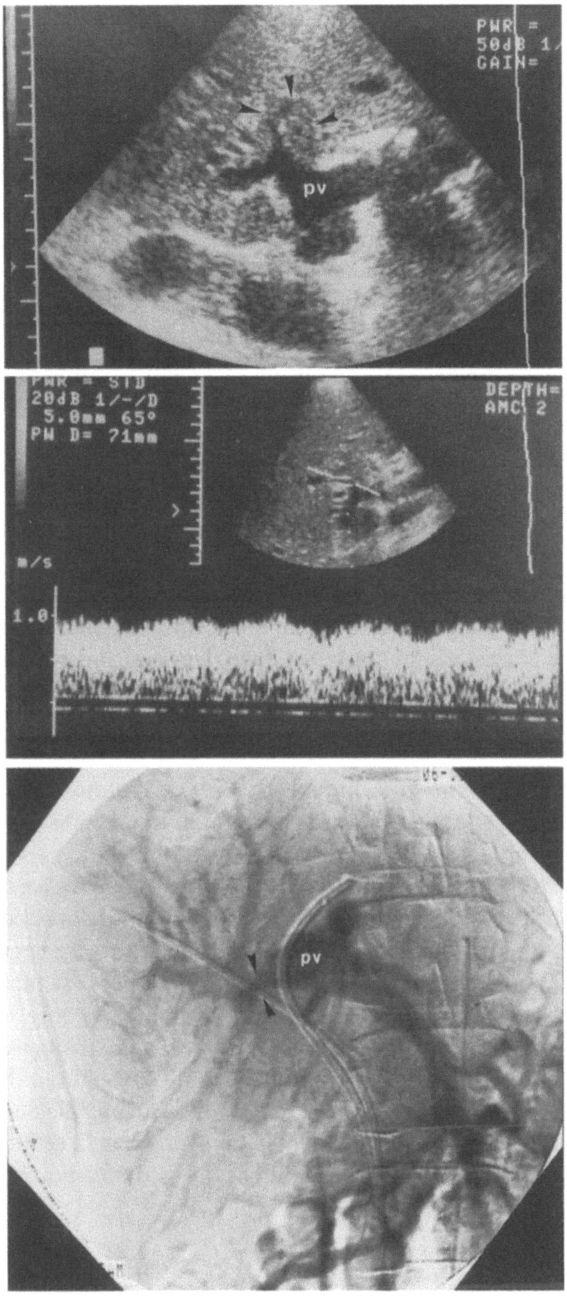

a

b

c

Fig. 24.15a–c. Unresectable hilar cholangiocarcinoma (Klatskin's tumor) with infiltration of the portal vein. **a** Oblique US scan: well-delineated hypoechogenic mass circumferentially infiltrating the anterior branch of the right portal vein (*pv*) (*arrowheads*). **b** Doppler US: high Doppler shift (1 m/s) at the site of ingrowth. **c** DSA: narrowing of the anterior branch (*arrowheads*) of the right portal vein (see **a**) PV = portal vein

hilar tumor mass by percutaneous FNAB is necessary, although difficult if there is extensive fibrosis surrounding the tumor mass.

In hilar cholangiocarcinoma, regional lymph node metastases are reported in approximately 30% of cases. The nodes most commonly involved are the nodes of the foramen of Winslow, the superior pancreatoduodenal nodes, the posterior pancreatoduodenal chain, and the nodes on the celiac trunk (Engels et al. 1989). The nodes of the foramen of Winslow are situated on the posterolateral aspect of the portal vein, anterior to the inferior vena cava, the superior pancreatoduodenal nodes immediately posterior to the apex of the duodenal bulb, accompanied by the gastroduodenal artery, and the posterior pancreatoduodenal chain posterolateral and posterior to the pancreatic head (Engels et al. 1989). US and CT are not reliable for the indication of lymph node metastases, since lymph nodes that are metastatically invaded but not enlarged cannot be detected with these methods. At our institute we investigate regional lymph node status preoperatively using laparoscopy combined with *laparocopic US* (see Sect. 24.3.1.1).

The various imaging methods (CT, US, ERC) are less accurate in detecting a mass in a patient with primary sclerosing cholangitis (PSC) (Carr et al. 1985; Gibson et al. 1986; Majoie et al. 1991, 1995). Segmental diffuse fibrosis in the periportal region with periductal extension and irregularity, a multiplicity of strictures, and small filling defects may masquerade as a malignancy. However, there are no cholangiographic features that specifically indicate malignant degeneration.

Cholangiographic findings in patients with PSC alone and in those with PSC within a carcinoma are summarized in Table 24.14.

A study of serial cholangiograms (ERC) by MacCarty et al. (1985) indicates that for 1.5–6.5 years some 80% of patients with PSC will show no

Table 24.14. ERC findings in primary sclerosing cholangitis (PSC) versus PSC within a carcinoma

Finding	PSC (%)	PSC + carcinoma (%)
Polypoid mass	7	46
Marked ductal dilatation	24	100
Progressive stricture in follow-up studies	18	100
Progressive dilatation in follow-up studies	2	100

MacCarty et al. 1985.

a
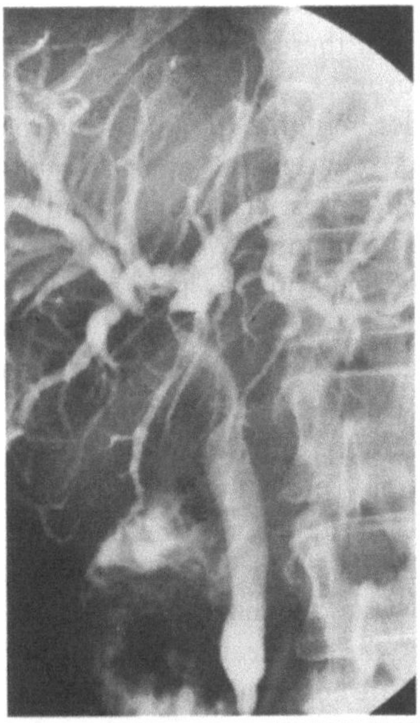

b
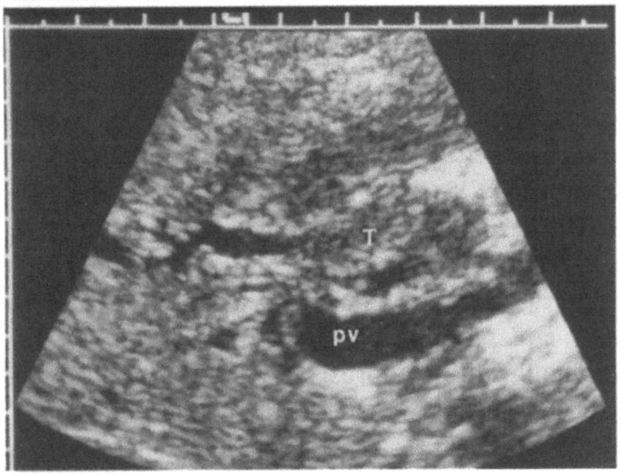

Fig. 24.16a,b. Benign postcholecystectomy fibrosis of the liver hilum, mimicking a Klatskin's tumor. **a** ERC: irregular eccentric stenosis of the proximal CBD and right and left liver lobe main branches, suggestive of malignancy. **b** Oblique US scan: inhomogeneous partly hypo-, partly hyperechogenic mass (*T*) at the liver hilum. pv, Portal vein

progression of disease. The appearance of progressive stricture with associated progressive dilatation of ductal segments should suggest the possibility of complicating malignancy. Cholangiocarcinoma is usually highly obstructive and more likely to result in marked duct dilatation above the obstruction than PSC alone (MacCarty et al. 1985).

Benign fibrosing or localized sclerosing tumors at the hepatic confluence, although notorious for mimicking proximal bile duct cancer, have rarely been described in the literature (Hadjis et al. 1985; Wetter et al. 1991). Of the 82 patients who underwent resective surgery under the presumptive diagnosis of hilar cholangiocarcinoma (Klatskin's tumor) at our institute in the period from 1989 to 1990, 11 (13.4%) proved to have benign fibrosing or localized sclerosing lesions (Verbeek et al. 1992) (Fig. 24.16).

Percutaneous intraluminal US of the bile ducts (van Sonnenberg et al. 1992) and percutaneous cholangioscopy (Nimura 1996) can discriminate reliably between benign and malignant hilar lesions and may help in differentiating between PSC and carcinoma in the future.

Dynamic MRI findings of Klatskin's tumors after intravenous injection of a gadolinium chelate have been reported (Hamrick-Turner et al. 1992; Fan et al. 1993; Soyer et al. 1995). The typical appearance of a Klatskin's tumor on MRI is as a well-delineated, nonencapsulated tumor with vascular encasement, of heterogeneous signal intensity in 53% of cases and homogeneous in 47%. In 88% it is hypointense to normal liver parenchyma on T1-weighted SE images, and in 100% it is hyperintense on T2-weighted SE images (Soyer et al. 1995). Central scarring, bile duct dilatation within the tumor, and associated retraction of the liver capsule adjacent to the tumor are

Table 24.15. Differential diagnosis of obstruction at the liver hilum

Cause of obstruction	Clues to diagnosis
Primary cholangiocarcinoma of proximal CBD/hepatic duct confluence (Klatskin's tumor)	US, ERC, or PTC, incremental/helical CT, MRI, brush cytology (3D-MRC)
Hilar invasion of gallbladder carcinoma	ERC/PTC, US (gallbladder); incremental/helical CT, MRI, brush cytology
PSC with cholangiocarcinoma	Sudden worsening of PSC, sudden changes of ERC findings in ulcerative colitis patients
Metastatic carcinoma	History, ERC, tumor markers
Malignant lymphoma	History, lymphoma elsewhere, ERC (extrinsic lesions) US, incremental/helical CT

PSC, Primary sclerosing cholangitis; 3D MRC, three-dimensional magnetic resonance cholangiography.

Table 24.16. US/CT features of primary peripheral cholangiocarcinoma

Well-defined round cystic lesions with internal papillary projections
Localized intrahepatic biliary dilatation without definite mass lesion
Ill-defined, hypoechogenic (low density) mass on early images, hyperdense on delayed images (80%)

ITAI et al. 1983.

suggestive features. MRI with FSE T2-weighted sequences and enhanced breathhold fast multiplanar spoiled gradient (FMPSPGR) sequences may provide additional clinically useful information (LOW et al. 1994).

3D MR cholangiography may be able to delineate the dilated biliary system irrespective of the serum bilirubin level (MORIMOTO et al. 1992). Nonopacified areas on the direct cholangiogram can be depicted clearly. Thus, as a noninvasive method for constructing the projection image of intrahepatic bile duct radicles, 3D MR cholangiography may have great potential for the future in the preoperative assessment of Klatskin's tumor (WALLNER et al. 1991; MORIMOTO et al. 1992; MCDERMOTT and NELSON 1995).

The clues to the diagnosis and differential diagnosis of obstruction at the liver hilum are summarized in Table 24.15.

24.3.5.2 Primary Peripheral Cholangiocarcinoma

Peripheral intrahepatic cholangiocarcinoma arises predominantly in noncirrhotic livers. It is a primary adenocarcinoma of the liver arising from the intrahepatic bile ducts and is a relatively rare tumor

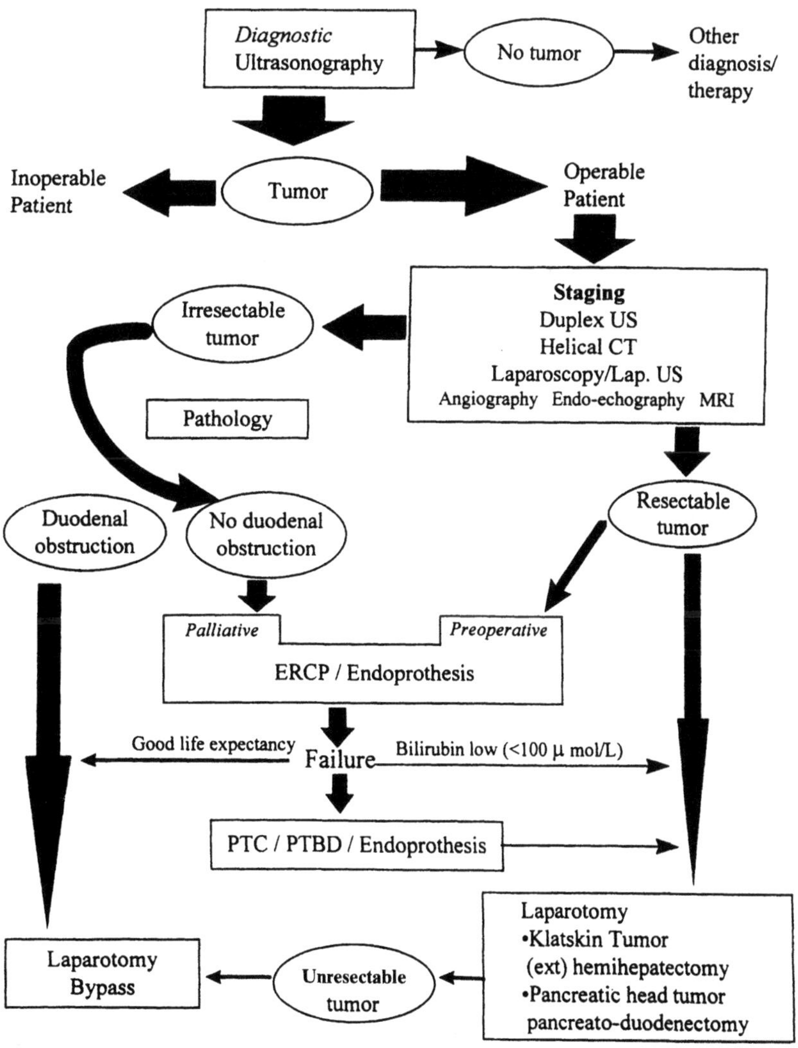

Fig. 24.17. Diagnostic work-up algorithm (1-week program) for assessing patients with presumptive CBD obstruction

(Choi et al. 1988; Ros et al. 1988; Choi et al. 1995; Lee et al. 1995). There are no descriptions of comparative studies between US, CT, and MRI in the evaluation of peripheral cholangiocarcinoma. It has the same gross massive nodular or diffuse configuration as hepatocellular carcinoma. The US/CT features of primary peripheral cholangiocarcinoma (Itai et al. 1983) are shown in Table 24.16.

CT may provide important diagnostic information for the assessment of peripheral cholangiocarcinoma by depicting the tumor and focal duct dilatation around the tumor and delineating the extent of the tumor. The tumor may exhibit various enhancement patterns, being hypodense on early images, isodense and hyperdense (80% the latter) on delayed images (Honda et al. 1993). This characteristic enhancement pattern may be due to the large amount of fibrous tissue and neovascularization in the tumor (Hamrick-Turner et al. 1992; Fan et al. 1993; Choi et al. 1995). The mechanism of the prolonged enhancement of fibrous tumors has not been clarified, but the slow washout of extravascular contrast agent in fibrous stroma may play a major part (Yoshikawa et al. 1992; Choi et al. 1995).

MRI may be superior to CT in demonstrating a small tumor and evaluating vascular invasion.

If the existence of another primary tumor has already been established, bile duct obstruction and lymph node enlargement are more likely to be the results of metastasis than to be a second primary tumor. Lymphoma and tumors of the gastrointestinal tract (rectosigmoid), breast, pancreas, and gallbladder all show a propensity for spread to the porta hepatis.

24.4 Conclusion

The referral pattern of patients with obstructive jaundice at our institution includes a diagnostic work-up (1-week program) starting with an assessment of the level and cause of obstruction, with US as the initial examination. In some cases the US findings may obviate more invasive radiologic examination. If the bile duct is obstructed, our approach is as shown in Fig. 24.17, which shows the role of the different imaging modalities. When high-resolution US indicates that surgical resection is feasible, Doppler US, incremental spiral CT, and laparoscopic US are performed as adequate procedures for preoperative TNM staging of the tumor process. When surgical resection does not seem feasible (i.e., if there is ingrowth into the mesenteric, splenic or portal venous system, or liver or peritoneal/lymph node metastases are present), the insertion of an endoprosthesis [polyethylene stent, metallic expandable stent (Wallstent, Angiomed stent, etc.) during ERC or US-guided PTC], with or without irradiation of the tumor mass, is the only possible palliative treatment. Diagnostic PTC is rarely necessary, and is reserved for those cases in which the Doppler US and/or CT findings are indeterminate and ERC is impossible.

Whereas in the 1970s the diagnosis of obstructive jaundice by US was an indication either to abandon further treatment or to refer the patient to the surgeon for confirmation of unresectability by an explorative laparotomy, followed by a bypass, in the late 1990s it is a reason for the hepatico-pancreatico-biliary team (radiologist, gastroenterologist, hepatologist, and surgeon) to discuss who should proceed with the next diagnostic and therapeutic steps (May et al. 1986; van Leeuwen and Reeders 1989).

References

Adam A, Benjamin IS (1992) Review. The staging of cholangiocarcinoma. Clin Radiol 46:299–303

Adoson MA (1973) Carcinoma of the gallbladder. Surg Clin North Am 53:1203–1216

Alexander F, Rossi RL, O'Bryan M, et al (1994) Biliary carcinoma. A review of 109 cases. Am J Surg 147:503–509

Altaee MY, Johnson PJ, Farrant JM, Williams R (1991) Etiologic and clinical characteristics of peripheral and hilar cholangiocarcinoma. Cancer 68:2051–2055

Anacker H, Rupp N, Reiser M (1984) Magnetic resonance (MR) in the diagnosis of pancreatic disease. Eur J Radiol 4:265–269

Baker ME, Silverman PM, Halvorson RA Jr, et al (1987) Computed tomography of masses in peritoneal/hepatoduodenal ligament. J Comput Assist Tomogr 11:258–263

Barish MA, Ycel EK, Soto JA, Chuttani R, Ferruci JT (1995) MR cholangiopancreatography: efficacy of three-dimensional turbo spin-echo technique. AJR Am J Roentgenol 165:295–300

Baron RL, Stanley RJ, Lee JKT, Koehler RE, Melson GL, Balfe DM, Weyman PJ (1982) A prospective comparison of the evaluation of biliary obstruction using computed tomography and ultrasonography. Radiology 145:91–98

Baron RL, Stanley RJ, Lee JKT, Koehler RE, Levitt RG (1983) Computed tomographic features of biliary obstruction. AJR Am J Roentgenol 140:1173–1178

Beazley RM, Blumgart LH (1985) Malignant stricture at the confluence of the biliary tree: diagnosis and management. Ann Surg 17:125–150

Beazley RM, Hadjis N, Benjamin IS, Blumgart LH (1984) Idiopathological aspects of high bile duct cancer. Experience with resection and bypass surgical treatments. Ann Surg 207:120–125

Bemelman WA, de With LTH, van Delden OM, Smits NJ, Obertop H, Rauws EJA, Gouma DJ (1995) Diagnostic laparoscopy combined with laparoscopic ultrasonography in staging of cancer of the pancreatic head region. Br J Surg 82:820–824

Bengmark S, Ekberg H, Evander A, et al (1988) Major liver resection for hilar cholangiocarcinoma. An Surg 207:120–125

Blumgart LH (1988) Cancer of the bile duct. In: Blumgart LH (ed) Surgery of the liver and biliary tract. Churchill Livingstone, Edinburgh, p 829

Blumgart LH, Benjamin IS (1989) Liver resection for bile duct cancer. Surg Clin North Am 69:323–337

Bolondi L, Li Bassi S, Gaiani S, Barbara L (1990) Doppler flowmetry in portal hypertension. J Gastroenterol Hepatol 5:459–467

Bosma A (1990) Surgical pathology of cholangiocarcinoma of the liver hilus (Klatskin tumor). Semin Liver Dis 10:85–90

Braasch JW, Whitcomb FF Jr, Watkins E Jr, Maguire RR, Khazei AM (1972) Segmental obstruction of the bile duct. Surg Gynecol Obstet 134:915–920

Campbell JP, Wilson SR (1988) Pancreatic neoplasms: how useful is evaluation with US? Radiology 267:341–344

Carr DH, Hadjis NS, Banks LM, Hemingway AP, Blumgart LH (1985) Computed tomography of hilar cholangiocarcinoma: a new sign. AJR Am J Roentgenol 145:53–56

Carrasco CH, Wallace S, Charnsangavej C, Richli W, Wright KC, Fanning T, Gianturco C (1985) Expandable biliary endoprosthesis: an experimental study. AJR Am J Roentgenol 145: 1279–1281

Charnsangavej C (1996) Computed tomography in hepatobiliary and pancreatic disease. In: Van Leeuwen DJ, Reeders JWAJ, Ariyama J (eds) Imaging in hepatobiliary and pancreatic disease: a practical clinical guide. Saunders, London, (in press)

Chezmar JL, Nelson RC, Small WC, Bernardino ME (1991) Magnetic resonance imaging of the pancreas with gadolinium-DPTA. Gastrointest Radiol 16:139–142

Choi BI, Park JH, Kim YI, et al (1988) Peripheral cholangiocarcinoma and clonorchiasis: CT findings. Radiology 169:149–153

Choi BI, Lee JH, Han MC, Kim SH, Yi JS, Kim CW (1989) Hilar cholangiocarcinoma. Comparative study with sonography and CT. Radiology 172:689

Choi BI, Han JK, Shin YM, Baek SY, Han MC (1995) Peripheral cholangiocarcinoma: comparison of MRI with CT. Abdom Imaging 20:357–360

Clouse ME, Gregg JA, McDonald DG, Legg MA (1977) Percutaneous fine needle aspiration biopsy of pancreatic carcinoma. Gastrointest Radiol 2:67–69

Cohan RH, Illescas FF, Braun SD, Newman GE, Dunnick NR (1986) Fine needle aspiration biopsy in malignant obstructive jaundice. Gastrointest Radiol 11:145–150

Conway S, Roman T, Tuft RJ (1982) Fine needle aspiration biopsy of malignant tumors of the abdomen. S Afr Med J 62:637–641

Cooperberg PL, Gibney RG (1987) Imaging of the gallbladder, Radiology 163:605–613

Crist DW, Sitzmann JV, Cameron JL (1987) Improved hospital morbidity, mortality, and survival after the Whipple procedure. Ann Surg 206:358–365

Cruickshank AH (1961) The pathology of 111 cases of primary hepatic malignancy collected in the Liverpool region. J Clin Pathol 14: 120–131

Cubilla AL, Fitzgerald PJ (1979) Classification of pancreatic cancer (non-endocrine). Mayo Clin Proc 54:449–458

Cuesta MA, Meijer S, Borgstein PJ, Sibinga Mulder L, Sikkenk AC (1993) Laparoscopic ultrasonography for hepatobiliary and pancreatic malignancy. Br J Surg 80:1571–1574

Dalla Palma L, Rizzato G, Pozzi-Mucelli RS, et al (1980) Grey-scale ultrasonography in the evaluation of carcinoma of the gallbladder. Br J Radiol 53:662–667

Dillon E, Peel ALG, Parkin GJS (1981) The diagnosis of primary bile duct carcinoma (cholangiocarcinoma) in the jaundiced patient. Clin Radiol 32:311–317

DiMagno EP, Malagelada JR, Taylor WF (1977) A prospective comparison of current diagnostic tests for pancreatic cancer. N Engl J Med 297:737–741

Dooley WC, Cameron JL, Pitt HA, Lillemoe KD, Yue NC, Venbrux AC (1990) Is preoperative angiography useful in patients with periampullary tumors. Ann Surg 211:649–655

Dooms GC, Kerlan RK Jr, Hricak H, Wall SD, Margulis AR (1986) Cholangiocarcinoma: imaging by MR. Radiology 159:89–94

Edis AJ, Kiernan PD, Taylor WF (1981) Attempted curative resection of ductal carcinoma of the pancreas: review of Mayo clinic experience, 1951–1975. Mayo Clin Proc 55:531–536

Engelholm L, De Toeuf J, Zalcman M, Jeanmart J, Nasr A, Matos C, Cremer M, Segebarth C (1987) Tomographie computée et resonance magnetique dans le cancer du pancréas. Comparaison avec le cholangiowirsungiography. Acta Gastroenterol Belg 50:195–210

Engelholm L, Segelbart C, De Toeuf J, Zalcman M (1988) IRM du pancréas. In: Vasile N (ed) IRM corps entier. Vigot, Paris, pp 219–235

Engels JT, Balfe DM, Lee JK (1989) Biliary carcinoma: CT evaluation of extrahepatic spread. Radiology 172:35–40

Evander A, Inso I, Lundergquist A, Tylen U, Akkerman M (1978) Percutaneous cystodiagnosis of carcinoma of the pancreas and bile duct. Ann Surg 188:90–92

Fan ZM, Yamashita Y, Harada M, et al (1993) Intrahepatic cholangiocarcinoma: spin-echo and contrast-enhanced dynamic MR imaging. AJR Am J Roentgenol 161:313–317

Feinberg S, Schreiber D, Goodale R (1977) Comparison of ultrasound pancreatic scanning and endoscopic retrograde cholangiopancreaticograms. J Clin ultrasound 5:96–99

Feldberg MAM, Hendriks MJ, van Waes PFGM, Sung KJ (1987) Pancreatic lesions and transfascial perirenal spread: computed tomographic demonstration. Gastrointest Radiol 12:121–127

Ferrucci JT, Wittenberg J, Mueller PR, Simeone JF, Harbin WP, Kirkpatrick RH, Taft PD (1980) Diagnosis of abdominal malignancy by radiological fine needle aspiration biopsy. AJR Am J Roentgenol 134:323–330

Fine D, Perring S, Herbetko J, Hacking CN, Fleming JS, Dewbury KC (1991) Technical note. Three-dimensional (3D) ultrasound imaging of the gallbladder and dilated biliary tree: reconstruction from real-time B-scans. Br J Radiol 64:1056–1057

Fishman EK, Wyatt SH, Ney DR, Kuhlman JE, Siegelman SS (1992) Spiral CT of the pancreas with multiplanar display. AJR Am J Roentgenol 159:1209–1215

Foley WD, Stewart ET, Lawson TL, Geenan J, Loguidice J, Maher L, Unger GF (1980) Computed tomography, ultrasonography and endoscopic retrograde cholangiopancreatography in the diagnosis of pancreatic disease: a comparative study. Gastrointest Radiol 5:29–35

Fornari F, Civardi G, Cavanna L, Di Strasi M, Rossi S, Sbolii G, Buscarni L (1989) The Cooperative Italian Study Group. Complications of ultrasonically guided fine-needle abdominal biopsy. Results of a multicenter Italian study

and review of the literature. Scand J Gastroenterol 24: 949-955
Frank K, Linhart P, Schmidt H (1984) Rationalisiering der Diagnostik abdomineller Erkrankungen durch die Sonographie. Forschungsbericht technologische Forschung und Entwicklung, Gesellschaft zur Förderung der Forschung an der DKD
Fraumeni JF (1975) Cancers of the pancreas and biliary tract: epidemiological considerations. Cancer Res 35:3437-3446
Freeny PC, Kidd R, Ball TJ (1980) ERCP guided percutaneous fine needle pancreatic biopsy. West J Med 132:283-287
Freeny PC, Lawson TL (1982) Radiology of the pancreas. Springer, New York Berlin Heidelberg
Freeny PC (1984) Computed tomography of the pancreas. Clin Gastroenterol 13:791-818
Freeny PC (1988) Radiology of the pancreas: two decades of progress in imaging and intervention. AJR Am J Roentgenol 150:975-981
Freeny PC (1989) Radiologic diagnosis and staging of pancreatic ductal adenocarcinoma. Radiol Clin North Am 27:121-128
Freeny PC (1991) Angio-CT: diagnosis and detection of complications of acute pancreatitis. Hepatogastroenterology 38:109-115
Freeny PC (1993) Incremental dynamic bolus computed tomography of acute pancreatitis - state of the art. Int J Pancreatol 13:147-158
Freeny PC (1995) Cross-sectional imaging of the pancreas. In: Tytgat GNJ, Reeders JWAJ (eds) Baillière's clinical gastroenterology, vol. 9, I. Baillière Tindall, London, pp 125-151
Freeny PC, Marks WM, Ball TJ (1982) Impact of high-resolution computed tomography of the pancreas on utilization of ERCP and angiography. Radiology 142:35-39
Freeny PC, Marks WM, Ryan JA, Bolen JW (1986) Colorectal carcinoma evaluation by CT: preoperative staging and detection of postoperative recurrence. Radiology 158:347-353
Freeny PC, Marks WM, Ryan JA, Traverso LW (1988) Pancreatic ductal adenocarcinoma: diagnosis and staging with dynamic CT. Radiology 166:125-133
Friedman AC, Lichtenstein RE, Dachman AH (1983) Cystic neoplasms of the pancreas; radiological/pathological correlation. Radiology 149:45-50
Frohlich E, Fruhmorgen P, Seeliger H (1986) Hautimpfmetastase nach Feinnadelpunktion eines Pankreaskarzinoms. Ultraschall Med 7:141-144
Garber SJ, Lees WR (1992) The characterization of pancreatic bile duct tumours by duplex Doppler. Clin Radiol 45:181-184
Garber SJ, Donald JJ, Lees WR (1993) Cholangiocarcinoma: ultrasound features and correlation of tumor position with survival. Abdom Imaging 18:66-69
Gehl HB, Urhahn R, Bohnhof K, et al (1991) Mn-DPDP in MR imaging of pancreatic carcinoma: initial clinical experience. Radiology 186:795-798
Gibson RN, Yeung E, Thompson JN, Carr DH, Hemingway AP, Bradpiece HA, Benjamin IS, Blumgart LH, Allison DJ (1986) Bile duct obstruction: radiologic evaluation of level, cause and tumor resectability. Radiology 160:43-47
Goldstein HM, Zornoza J, Wallace S, Anderson JH, Bree RL, Samuels BJ, Lukeman J (1977) Percutaneous fine needle aspiration biopsy of paraaortic and other abdominal masses. Radiology 123:319-322
Gillams A, Lees WR, Richards R (1991) Three-dimensional demonstration of biliary anatomy in obstructive jaundice. Br J Radiol 64[Suppl]:37
Gillams A, Gardener J, Richards R, Tan AC, Linney A, Lees WR (1994) Three-dimensional computed tomography cholangiography: a new technique for biliary tract imaging. Br J Radiol 67:445-448
Goldberg HI (1994) Helical cholangiography: complementary or substitute study for endoscopic retrograde cholangiography? Radiology 192:615-616
Goldberg HI, Filly RA, Korobkin M, Moss AA, Kressel HY, Callen PW (1978) Capability of CT body scanning and ultrasonography to demonstrate the status of the biliary ductal system in patients with jaundice. Radiology 129:731-737
Goldstein HM, Zornoza J (1978) Percutaneous transperitoneal aspiration biopsy of pancreatic masses. Dig Dis 23:840-843
Greenwald ED, Greewald ES (1983) Cancer epidemiology. Medical Examination Publishing, New York, p 87
Guibaud L, Bret PM, Reinhold C, Atri M, Barkum AN (1994) Diagnosis of choledocholithiasis: value of MR cholangiography. AJR Am J Roentgenol 163:847-850
Gulliver DJ, Baker ME, Cheng CA, Meyers WC, Pappas WN (1992) Malignant biliary obstruction: efficacy of thin-section dynamic CT in determining resectability. AJR Am J Roentgenol 159:503-507
Haaga JR (1984) Magnetic resonance imaging of the pancreas. Radiol Clin North Am 22:869-877
Hadjis NS, Collier NA, Blumgart LH (1985) Malignant masquerade at hilum of the liver. Br J Surg 72:659-661
Hadjis NS, Adam AA, Gibson R, Blenkharn JI, Benjamin IS, Blumgart LH (1989) Nonoperative approach to hilar cancer determined by the atrophy-hypertrophy complex. Am J Surg 157:395-399
Hall Craggs MA, Lees WR (1986) Fine needle aspiration biopsy in pancreatic and biliary tumors. AJR Am J Roentgenol 147:399-403
Hall-Craggs MA, Allen CM, Owens CM, et al (1993) MR cholangiography: clinical evaluation in 40 cases. Radiology 189:423-427
Hamper UM, Trapanotto V, Sheth S (1993) Three-dimensional ultrasound: evaluation of abdominal and superficial organs. Radiology 189(P):342
Hamrick-Turner J, Abbitt PL, Ros PR (1992) Intrahepatic cholangiocarcinoma: MR appearance. AJR Am J Roentgenl 158:77-79
Hancke S, Holm HH, Kock F (1975) Ultrasonically guided percutaneous fine needle biopsy of the pancreas. Surg Gynecol Obstet 140:361-364
Harter LP, Moss AA, Goldberg HI, Gross BH (1983) CT guided fine needle aspirations for diagnosis of benign and malignant disease. AJR Am J Roentgenol 140:363-367
Haubek A, Pedersen JH, Burcharth F, Gammelgaard J, Hancke S, Willumsen L (1981) Dynamic sonography in the evaluation of jaundice. AJR Am J Roentgenol 136:1071-1074
Havrilla TR, Haaga JR, Alfidi RJ, Reich NE (1977) Computed tomography and obstructive biliary disease. AJR Am J Roentgenol 128:765-768
Hessel S, Siegelman S, McNeil B, Sanders R, Adams D, Alderson P, Finberg H, Abrams H (1982) A prospective evaluation of the CT and US of the pancreas. Radiology 143:129
Ho CS, McLoughlin MJ, McHattie JD, Tao LC (1977) Percutaneous fine needle aspiration biopsy of the pancreas following endoscopic retrograde cholangiopancreatography. Radiology 125:351-353
Honda H, Onitsuka H, Yasumori K, Hayashi T, Ochiai K, Gibo M, Adachi E, Matsumata T, Masuda K (1993) Intrahepatic peripheral cholangiocarcinoma: two-phased dynamic in-

cremental CT and pathologic correlation. J Comput Assist Tomogr 17(3):397-402
Honickman SP, Mueller PR, Wittenberg J, Simeone JF, Ferrucci JT Jr, Cronan JJ, van Sonnenberg E (1983) Ultrasound in obstructive jaundice: prospective evaluation of site and cause. Radiology 147:511-515
Hosoki T (1983) Dynamic CT of pancreatic tumors. AJR Am J Roentgenol 140:959-965
Hruban RH, DiGiuseppe JA, Offerhaus GJA (1994) K-ras mutations in pancreatic ductal proliferative lesions. Am J Pathol 145:1547-1550
Hsu Chong Yeh (1979) Ultrasonography and computed tomography of carcinoma of the gallbladder. Radiology 133:167-173
Husband JE, Goldberg SJ (1983) The role of computed tomography guided needle biopsy in the oncology service. Clin Radiol 34:255-260
Irving JD, Adam AA, Dick R, Dondelinger RF, Lunderquist A, Roche A (1989) Gianturco expandable metallic biliary stents: results of a European clinical trial. Radiology 172:321-326
Ishizaki Y, Wakayama T, Okada Y, Kobayashi T (1993) Magnetic resonance cholangiography for evaluation of obstructive jaundice. Am J Gastroenterol 88:2072-2077
Itai Y, Araki T, Yoshikawa K, Furui S, Yashiro N, Tasaka A (1980) Computed tomography of gallbladder carcinoma. Radiology 137:713-718
Itai Y, Araki T, Furui S, Yashiro N, Ohtomo K, Iio M (1983) Computed tomography of primary intrahepatic biliary malignancy. Radiology 147:485-490
Itai Y, Ohhashi K, Nagai H, Murakami Y, Kokubo T, Makita K, Ohtomo K (1986) "Ductectatic" mucinous cystadenoma and cystadenocarcinoma of the pancreas. Radiology 161:697-700
Itoh K, Yomanaka T, Kasa-hora K, Koike M, Nakamuva A, Itayaski A, Kimura K, Morioka Y, Kowai T (1979) Definitive diagnosis of pancreatic carcinoma with percutaneous fine needle aspiration biopsy under ultrasonographic guidance. Am J Gastroenterol 71:469-472
Jafri SZH, Aisen AM, Glazer GM, Weiss CA (1984) Comparison of CT and angiography in assessing resectability of pancreatic carcinoma. AJR Am J Roentgenol 142:525-529
Jakimowicz JJ (1994) Technical and clinical aspects of intraoperative ultrasound applicable to laparoscopic ultrasound. Endosc Surg Allied Technol 2:119-126
Jenkins JPR, Braganza JM, Hickey DS, Isherwood I, Machin M (1987) Quantitive tissue characterisation in pancreatic disease using magnetic resonance imaging. Br J Radiol 60:333-341
John TG, Garden OJ (1993) Pancreatic cancer. Part I. Assessment of pancreatic cancer. In: Cuesta MA, Nagy AG (eds) Minimally invasive surgery in gastrointestinal cancer. Churchill Livingstone, London, pp 95-111
Juttijudata P, Chiemchaisri C, Palavatana C, Churnraranakul S (1982) A clinical study of cholangiocarcinoma caused cholestasis in Thailand 1982. Surg Gynecol Obstet 155:373-376
Kalser MH, Ellenberg SS (1985) Pancreatic cancer: adjuvant combined radiation and chemotherapy following curative resection. Arch Surg 120:889-903
Kalser MH, Barkin J, MacIntyre JM (1985) Pancreatic cancer: assessment of prognosis by clinical presentation. Cancer 56:397-402
Keill RH, DeWeese MS (1973) Primary carcinoma of the gallbladder. Am J Surg 125:726-729
Kido C, Hibino K, Kaneko M (1974) Angiography of gallbladder carcinoma. Nippon Acta Radiol 34:1-11
Klatskin G (1965) Adenocarcinoma of the hepatic duct at its bifurcation within the porta hepatis. An unusual tumor with distinctive clinic and pathological features. Am J Med 38:241-256
Klein JB, Finck FM (1972) Primary carcinoma of the gallbladder. Arch Surg 104:769-772
Koga A, Yamaguchi S, Izumi Y, et al (1985) Ultrasonographic detection of early and curable carcinoma of the gallbladder. Br J Surg 72:728-730
Koslin DB, Mulligan SA, Berland LL (1992) Duplex assessment of the portal venous system. Semin Ultrasound CT MR 13:22-33
Kubota Y, Seki T, Yamaguchi K, et al (1992) Bilateral internal drainage of biliary hilar malignancy via a single percutaneous track. Role of percutaneous transhepatic cholangioscopy. Endoscopy 24:194-198
Laing FC, Jeffrey RB, Wing VW, Nyberg DA (1986) Biliary dilatation: defining the level and cause by real time US. Radiology 160:39-42
Lawson TH, Berland L, Foley W, Stewart E, Geenan J, Hogan W (1978) Ultrasonic visualization of the pancreatic duct. Radiology 144:865
Lea MS, Stahlgren LH (1987) Is resection appropriate for adenocarcinoma of the pancreas? A cost-benefit analysis. Am J Surg 154:651-654
Lee SK, Choi BI, Cho JM, Han JK, Kim YI, Han MC (1995) Cystic peripheral cholangiocarcinoma: sonography and CT. Abdom Imaging 20:131-132
Levine E, Maklad NF, Wright CH, Lee KR (1979) Computer tomography and ultrasonic appearances of primary carcinoma of the common bile duct. Gastrointest Radiol 4:147-151
Levitt RG, Sagel SS, Stanley RJ, Jost RG (1977) Accuracy of computed tomography of the liver and biliary tract. Radiology 124:123-128
Longmire WP Jr, Traverso LW (1981) The Whipple procedure and other standard operative approaches to pancreatic cancer. Cancer 47[Suppl]:1706-1711
Looser C, Stain SC, Baer HU, Triller J, Blumgart LH (1992) Staging of hilar cholangiocarcinoma by ultrasound and duplex sonography: a comparison with angiography and operative findings. Br J Radiol 65:871-877
Low RN, Sigeti JS, Francis IR, Weinman D, Bower B, Shimakawa A, Foo TKF (1994) Evaluation of malignant biliary obstruction: efficacy of fast multiplanar spoiled gradient-recalled MR imaging vs spin-echo MR imaging, CT and cholangiography. AJR Am J Roentgenol 162:315-323
Lygidakis NJ, Tytgat GNJ (eds) (1989) Hepatobiliary and pancreatic malignancies. Diagnosis, medical and surgical treatment. Thieme, Stuttgart
MacCarty RL, La Russo NF, May GR, Bender CE, Wiesner RH, King JE, Coffey RJ (1985) Cholangiocarcinoma complicating primary sclerosing cholangitis: cholangiographic appearances. Radiology 156:43-46
Machan L, Müller NL, Cooperberg PL (1986) Sonographic diagnosis of Klatskin tumors. AJR Am J Roentgenol 147:509-512
Mackie C, Cooper M, Lewis M, Moossa A (1978) Prospective evaluation of grey-scale ultrasonography in the diagnosis of pancreas cancer, Am J Surg 136:575
Mackie CR, Lu CT, Noble HG, et al (1979) Prospective evaluation of angiography in the diagnosis and management of patients suspected of having pancreatic cancer. Ann Surg 189:11-17

Majoie CBLM, Reeders JWAJ, Sanders JB, Huibregtse K, Jansen PLM (1991) Primary sclerosing cholangitis: a modified classification of cholangiographic findings. AJR Am J Roentgenol 157:495–497

Majoie CBLM, Smits NJ, Phoa SSKS, Reeders JWAJ, Jansen PLM (1995) Primary sclerosing cholangitis: sonographic findings. Abdom Imaging 20:109–112

May GR, James EM, Bender CE, Williams HJ, Adson MA (1986) Diagnosis and treatment of jaundice. Radiographics 65:847–890

McDermott VG, Nelson RC (1995) Re: MR cholangiopancreatography: efficacy of three-dimensional turbo spin-echo technique. AJR Am J Roentgenol 165:301–302

McLoughlin MJ, Ho CS, Langer B, McHattie J, Tao LC (1978) Fine needle aspiration biopsy of malignant lesions in and around the pancreas. Cancer 41:2413–2419

Megibow AJ, Walsh SJ, Francis IF, et al (1991) Comparison of CT and MR imaging in evaluation of patients with pancreatic adenocarcinoma: report of the diagnostic radiology oncology group. Radiology 181:259

Megibow AJ, Zhou XH, Rotterdam H, Francis IR, Zerhouni EA, Balfe DM, Weinreb JC, Aisen A, Kuhlman J (1995) Pancreatic adenocarcinoma: CT versus MR imaging in the evaluation of resectability: report of the radiology diognostic oncology group. Radiology 195(2):327–332

Merrick HW, Dobelbower RR (1990) Aggressive therapy for cancer of the pancreas: does it help? Gastroenterol Clin North Am 19:935–962

Meyer DG, Weinstein BJ (1983) Klatskin tumors of the bile ducts: sonographic appearance. Radiology 148:803–804

Michelassi F, Erroi F, Dawson PJ, et al (1989) Experience with 647 consecutive tumors of the duodenum, ampulla, head of the pancreas, and distal common bile duct. Ann Surg 210:544–556

Mitchell DG, Vinitski S, Saponaro S, et al (1991) Liver and pancreas: improved spin-echo T1 contrast by shorter echo time and fat suppression at 1.5 T. Radiology 178:67–71

Mitchell DG, Shapiro M, Schuricht A (1992) Pancreatic disease: findings on state-of-the-art MR images. AJR Am J Roentgenol 159:533–538

Moody F, Thorbjarnarson B (1964) Carcinoma of the ampulla of Vater. Am J Surg 107:572–579

Moore PT, Clark RA (1984) An update of interventional biliary radiology, Sem Ultrasound CT MR 5:349–368

Moossa AR (1982) Pancreatic cancer: Approach to diagnosis, selection for surgery and choice of operation. Cancer 50:2689–2698

Morimoto K, Shimoi M, Shirakawa T, Aoki Y, Choi S, Miyata Y, Hara K (1992) Biliary obstruction: evaluation with three-dimensional MR cholangiography. Radiology 183:578–580

Morrow CE, Sutherland DER, Florack G, et al (1983) Primary gallbladder carcinoma: significance of subserosal lesions and results of aggressive surgical treatment and adjuvant chemotherapy. Surgery 94:709–714

Moss AA, Federle M, Shapiro HA (1980) The combined use of computed tomography and ERCP in the assessment of suspected pancreatic neoplasm: a blind clinical evaluation. Radiology 134:159–163

Muir IM, Morris DL (1986) Carcinoma of the gallbladder. Br J Hosp Med 36:278–280

Müller MF, Meyenberger C, Bertschinger P, Schaer R, Marincek B (1994) Pancreatic tumors: evaluation with endoscopic US, CT, and MR imaging. Radiology 190:745–751

Mueller PR, Simeone JF (1984) New concepts in biliary ultrasound. Semin ultrasound CT MR 5:338–348

Murugiah M, Paterson-Brown S, Windsor JA, Miles WF, Garden OJ (1993) Early experience of laparoscopic ultrasonography in the management of pancreatic carcinoma. Surg Endosc 7:177–181

Neff CC, Simeone JF, Wittenberg J, Muller PR, Ferrucci JT Jr (1984) Inflammatory pancreatic masses: problems in differentiating focal pancreatitis from carcinoma. Radiology 150:35–38

Nesbit GM, Johnson CD, James EM, MacCarty RL, Nagorney DM, Bender CE (1988) Cholangiocarcinoma: diagnosis and evaluation of resectability by CT and sonography as procedures complementary to cholangiography. AJR Am J Roentgenol 151:933–938

Neuhaus H (1992) Cholangioscopy. Endoscopy 24:125–132

Nichols DA, MacCarty RL, Gaffey TA (1983) Cholangiographic evaluation of bile duct carcinoma. AJR Am J Roentgenol 141:1291–1294

Nimura Y (1993) Staging of biliary carcinoma. Cholangiography and cholangioscopy. Endoscopy 25:76–80

Nimura Y (1996) Cholangioscopy. In: van Leeuwen DJ, Reeders JWAJ, Ariyama J (eds) Imaging in hepatobiliary and pancreatic disease: a practical clinical guide Saunders, London (in press)

Nimura Y, Hayakawa N, Kamiya J, Kondo S, Shionoya S (1990) Hepatic segmentectomy with caudate lobe resection for bile duct carcinoma of the hepatic hilus. World J Surg 14:535–544

Noguchi T (1993) Intraportal US with 20-MHz and 30-MHz scanning catheters. Radiology 186:203–205

Ormson MJ, Charboneau JW, Stephens DH (1987) Sonography in patients with a possible pancreatic mass shown on CT. AJR Am J Roentgenol 148:551–555

Ott D, Gelfand D (1981) Complications of gastrointestinal radiologic procedures. II. Complications related to biliary tract studies. Gastrointest Radiol 6:47–56

Parsons I, Palmer CH (1989) How accurate is fine-needle biopsy in malignant neoplasms of the pancreas? Arch Surg 124:681–683

Patel U, Lees WR (1994) Current clinical potential of abdominal ultrasound. In: Tytgat GNJ, Reeders JWAJ (eds) Baillière's clinical gastroenterology, vol 8, IV. Baillère Tindall, London, pp 595–602

Pedrosa CS, Casanova R, Lezana AH, Fernadez MC (1981a) Computed tomography in obstructive jaundice. II: the cause of obstruction. Radiology 139:635–645

Pedrosa CS, Casanova R, Rodriguez R (1981b) Computed tomography in obstructive jaundice. I: the level of obstruction. Radiology 139:627–634

Pereiras RV, Meiers W, Kunhardt B, Troner M, Hutson D, Barkin JS, Viamonte M (1978) Fluoroscopically guided thin needle aspiration biopsy of the abdomen and retroperitoneum. AJR Am J Roentgenol 131:197–202

Piehler JM, Crichlow RW (1978) Primary carcinoma of the gallbladder. Surg Gynecol Obstet 147:929–942

Plumley TF, Rohrmann CA, Freeny PC, Silverstein FE, Ball TJ (1982) Double duct sign: reassessed significance in ERCP. AJR Am J Roentgenol 138:31–35

Pollock D, Taylor K (1981) Ultrasound scanning in patients with clinical suspicion of pancreatic cancer: a retrospective study. Cancer 47:1662

Reiman TH, Balfe DM, Weyman PJ (1987) Suprapancreatic biliary obstruction: CT evaluation. Radiology. 163: 49–56

Rholl K, Smathers R, McClennan B, Lee J (1985) Intravenous cholangiography in the CT era. Gastrointest Radiol 10:69–74

Robbins SL (1967) Pathology. 3rd edn. Saunders, Philadelphia, pp 957–959

Roberts JW, Daugherty SF (1986) Primary carcinoma of the gallbladder. Surg Clin North Am 66:743–749

Rodriquez J, Kasberg C, Nipper M, Schoolar J, Riggs MW, Dyck WP (1992) CT-guided needle biopsy of the pancreas: retrospective analysis of diagnostic accuracy. Am J Gastroenterol 87:1610–1613

Roos WK, Welvaart K, Bloem JL, Hermans J (1990) Assessment of resectability of carcinoma of the pancreatic head by ultrasonography and computed tomography: a retrospective analysis. Eur J Surg Oncol 16:411–416

Ros PR, Buck JL, Goodman ID, Ros AMV, Olmsted WW (1988) Intrahepatic cholangiocarcinoma: radiologic-pathologic correlation. Radiology 167:689–693

Rösch T, Classen M (1994) Endosonographic possibilities in the pancreatobiliary area. In: Tytgat GNJ, Reeders JWAJ (eds) Ballière's clinical gastroenterology, vol 8 IV. Baillière Tindall, London, pp 621–633

Rösch T, Dittler HJ, Lorenz R (1992) The role of endoscopic ultrasonography in the diagnosis and staging of tumors of the papilla of Vater. Gastrointest Endosc 38:259–260 (abstract)

Rösch T, Dittler HJ, Kunte M (1994) Comparison of a sector-type with a linear type echoendoscope in the staging of GI cancer (abstract). Gastrointest Endosc 40:53

Ross CB, Kaufman AJ, Sharp KW, Andrews T, Willimans LF (1988) Efficacy of computerized tomography in the preoperative staging of pancreatic carcinoma. Am Surg 54:221–226

Rossmann MD, Friedman AC, Radecki PD, Caroline DF (1987) MR Imaging of gallbladder carcinoma. AJR Am J Roentgenol 148:143–144

Sagoh T, Itoh K, Togashi K, et al (1990) Gallbladder carcinoma: evaluation with MRI imaging. Radiology 174:131–136

Sauerbruch T, Wotzka R, Rehkamp D, Rummel T (1979) Upper abdominal ultrasonography and ERCP. Dtsch Med Wochenschr 104:169

Savader BL, Fishman EK, Savader SJ (1990) Comparison of CT angiography and routine angiography in assessing resectability of pancreatic tumors. Radiology 177(P):192 (abstract)

Schulte SJ, Baron RL, Tefey SA, Rohrmann CA Jr, Freeny PC, Shuman WP, Foster MA (1990) CT of the extrahepatic bile ducts: wall thickness and contrast enhancement in normal and abnormal ducts. AJR Am J Roentgenol 154:79–85

Semelka RC, Shoenut JP, Kroeker MA, Kricak H, Minuk GY, Yaffe CS, Micflikier AB (1991) Bile duct disease: prospective comparison of ERCP, CT, and fat suppression MRI. Gastrointest Radiol 17:347–352

Shimizu H, Ida M, Tahayama S (1981) The diagnostic accuracy of CT in obstructive biliary disease: a comparative evaluation with direct cholangiography. Radiology 138:411–416

Silverberg E, Lubera JA (1988) Cancer statistics. CA Cancer J Clin 38:5–22

Simeone JF, Edelman RR, Stark DD, Wittenberg J, White EM, Butch RJ, Mueller PR, Brady TJ, Ferrucci JT (1985) Surface coil MR imaging of abdominal viscera. III: the pancreas. Radiology 157:437–441

Smits NJ, Reeders JWAJ (1995) Current applicability of duplex Doppler ultrasonography in pancreatic head and biliary malignancies. In: Tytgat GNJ, Reeders JWAJ (eds) Baillière's clinical gastroenterology, vol 9 I. Ballière Tindall, London, pp 153–172

Solomon A, Rubinstein ZJ (1981) Carcinoma of the gallbladder: a CT diagnosis. Fortschr Geb Rontgenstrahlen Necien Bildgeb Verfahr 135(5):622–623

Soyer P, Bluemke DA, Sibert A, Laissy JP (1995) MR imaging of intrahepatic cholangiocarcinoma. Abdom Imaging 20:126–130

Stark DD, Moss AA (1984) Magnetic resonance imaging of the pancreas. Semin Ultrasound CT MR 5(4):428–436

Stark DD, Moss AA, Goldberg HI, Davis PL, Federle MP (1984) Magnetic resonance and CT of the normal and diseased pancreas: a comparative study. Radiology 150:153–162

Stockberger SM, Wass JL, Sherman S, Lehman GA, Kopecky KK (1994) Intravenous cholangiography with helical CT: comparison with endoscopic retrograde cholangiography. Radiology 192:675–680

Strunk H, Kuhn FP, Weibler U, Frank K, Heinz A (1988) Sonographie beim Pancreaskarzinom; Sonomorphologie, Treffsicherheit und Tumorstaging. Radiologe 28:277–283

Subramanyam BR, Raghavendra BN, Balthazar EJ, Horri SC, LeFleur RS, Rosen RJ (1984) Ultrasonic features of cholangiocarcinoma. J Ultrasound Med 3:405–408

Swobodnik W, Meyer W, Brecht-Kraus D, Wechsler J, Geiger S (1983) Ultrasound, computed tomography and ERCP in the morphologic diagnosis of pancreatic disease. Klin Wochschr 61:291

Takehara Y, Ichijo K, Tooyama N, et al (1994) Breath-hold MR cholangiopancreatography with a long-echo-train fast spin-echo sequence and a surface coil in chronic pancreatitis. Radiology 192:73–78

Taylor K, Buchin P, Viscont G, Rosenfield A (1981) Ultrasonographic scanning of the pancreas. Radiology 138:211

Teefey SA, Baron RL, Schulte SJ, Patten RM, Molloy MH (1992) Patterns of intrahepatic bile duct dilatation at CT: correlation with obstructive disease processes. Radiology 182:139–142

Thorbjarnason B (1959) Carcinoma of the bile ducts. Cancer 12:708–713

Thorsen MK, Quiroz F, Lawson TL, Smith DF, Foley WD, Stuart ET (1984) Primary biliary carcinoma: CT evaluation. Radiology 152:479–483

Tio TL (1988) Endosonography in gastroenterology, Springer, Berlin Heidelberg New York pp 33–50

Tio TL, Tytgat GN (1986) Endoscopic ultrasonography in staging local resectability of pancreatic and periampullary malignancy. Scand J Gastroenterol 21:135–142

Tio TL, Mulder CJJ, Eggink WF (1992) Endosonography in staging early carcinoma of the ampulla of Vater. Gastroenterology 102:1392–1395

Tobin RS, Vogelzang RL, Gore RM, Keigley B (1987) A comparative study of computed tomography and ERCP in pancreaticobiliary disease. J Comput Assist Tomogr 11:261–266

Triller J, Losser C, Baer HU, Blumgart LH (1994) Hilar cholangiocarcinoma: radiological assessment of resectability. Eur J Radiol 4:9–17

Tscholakoff D, Hricak H, Thoeni R, Winkler ML, Margulis AR (1987) MR imaging i the diagnosis of pancreatic disease. AJR Am J Roentgenol 148:703–709

Tylen U, Arnesjo B, Lindberg LG, Lunderquist A, Akerman M (1976) Percutaneous biopsy of carcinoma of the pancreas guided by angiography. Surg Gynecol Obstet 142:737–739

Van Delden OM, Smits NJ, Bemelman WA, de Wit Lth, Gouma DJ, Reeders JWAJ (1996) Comparison of laparoscopic ultrasound and transdominal ultrasound in staging of cancer of the pancreatic head region. J Ultrasound Med 16:207–212

van Leeuwen D, Reeders JWAJ (1989) Part I: Diagnostics in hepatobiliary-pancreatic malignancies In: Lygidakis NJ, Tytgat GNT (eds) Hepatobiliary and pancreatic malignan-

cies: diagnosis, medical and surgical treatment. Thieme, Stuttgart
Van Gulik TM, Reeders JWAJ, Bosma A, Moojen TM, Smits NJ, Allema JH, Rauws EAJ, Offerhaus GJA, Obertop H, Gouma DJ (1996) Pancreatoduodoenectomy performed for an inflammatory lesion in the pancreatic head masquerading as pancreatic cancer: error or reality? Gut (in press)
Van Sonnenberg E, D'Agostino HB, Sanchez RL, Goodacre BB, Esch OG, Easter DE, Gosink BB (1992) Percutaneous intraluminal US in the gallbladder and bile ducts. Radiology 182:693–696
Vellet AD, Romano W, Bach DB, Passi RB, Taves DH, Munk PL (1992) Adenocarcinoma of the pancreatic ducts: comparative evaluation with CT and MR imaging at 1.5T. Radiology 183:87–95
Verbeek PCM, van Leeuwen DJ, de Wit Th, Reeders JWAJ, Smits NJ, Bosma A, Huibregtse K, van der Heyde MN (1992) Benign fibrosing disease at the hepatic confluence mimicking Klatskin tumors. Surgery 112:866–871
Voyles CR, Bowley NJ, Allison DJ, Benjamin IS, Blumgart LH (1983) Carcinoma of the extrahepatic biliary tree. Radiologic assessment and therapeutic alternatives. Ann Surg 197:188–194
Wagner HE, Ballmer FT, Blumgart LH (1987) Hepatic portal tumor - today. Schweiz Med Wochenschr 117:2104–2112
Wallner BK, Schumacher KA, Weidenmaier W, Freidrich JM (1991) Dilated biliary tract: evaluation with MRI cholangiography with a T2-weighted contrast-enhanced fast-sequence. Radiology 181:805–808
Warshaw AL, Tepper JE, Shipley WU (1986) Laparoscopy in the staging and planning of therapy for pancreatic cancer. Am J Surg 151:76–80
Warshaw AL, Swanson RS (1988) Pancreatic cancer in 1988. Ann Surg 208:541–551
Warshaw AL, Gu ZY, Wittenberg J, Waltman AC (1990) Preoperative staging and assessment of resectability of pancreatic cancer. Arch Surg 125:230–233
Weinbren K, Mutum SS (1983) Pathological aspects of cholangiocarcinoma. J Pathol 139:217–238
Weiner SN, Koenigsberg M, Morehous H, et al (1984) Sonography and computed tomography in the diagnosis of carcinoma of the gallbladder. AJR Am J Roentgenol 142:735–739
Wetter LA, Ring EJ, Pelligrini CA, Way LW (1991) Differential diagnosis of sclerosing cholangiocarcinomas of the common hepatic duct (Klatskin tumors). Am J Surg 161:57–62
White TT (1985) Management of bile duct tumors. In: Najarian JS, Delaney JP (eds) Advances in hepatic, biliary and pancreatic surgery. Year Book Medical Publishers, Chicago, pp 335–344
Wittenberg J, Simeone JF, Ferucci JT Jr, Muller PR, van Sonnenberg E, Neff CC (1982) Non-focal enlargement in pancreatic carcinoma. Radiology 144:131–135
Wolfman NT, Ramquist NA, Karstaedt N, Hopkins MB (1982) Cystic neoplasms of the pancreas: CT and sonography. AJR Am J Roentgenol 138:37–41
Yamashita Y, Takahashi M, Kanazawa S, Charnsangavej C, Wallace S (1992) Hilar cholangiocarcinoma: an evaluation of subtypes with CT and angiography. Acta Radiol 33:351–355
Yoshikawa J, Matsui O, Kadoya Y, Gabata T, Arai K, Takashima T (1992) Delayed enhancement of fibrotic areas in hepatic masses: CT-pathologic correlation. J Comput Assist Tomogr 16:206–211
Yum HY, Fink AH (1980) Sonographic findings in primary carcinoma of the gallbladder. Radiology 134:693–696
Zeman RK, Burrell MI (1987) Gallbladder and bile duct imaging: a clinical radiological approach Churchill Livingstone, New York
Zeman RK, Davros WJ, Berman P, Weltman DI, Silverman PM, Cooper C, Evans SRT, Buras RR, Stahl TJ, Nauta RJ, Al-Kawas F (1994) Three-dimensional models of the abdominal vasculature based on helical CT: usefulness in patients with pancreatic neoplasms. AJR Am J Roentgenol 162:1425–1429

25 Percutaneous Management of Malignant Biliary Tract Obstruction

A.F. Watkinson and A. Adam

CONTENTS

25.1 Introduction

In patients with malignant biliary obstruction it is important to assess the patient clinically, biochemically, and radiologically using ultrasonography (US) and/or computed tomography (CT). The most common causes of malignant obstruction of the bile duct are carcinoma of the pancreas, cholangiocarinoma, and secondary hilar deposits. Histological confirmation of malignancy and identification of the cell type are important, as the prognosis is slightly different in each case, and are also helpful in planning further management. Cross-sectional imaging may provide information concerning the level and probable cause of bile duct obstruction but usually cannot establish the definitive diagnosis. In such instances, in order to assess tumour resectability and plan the most appropriate approach to surgical or non-surgical biliary decompression, direct cholangiography is often required, either percutaneous transhepatic cholangiography (PTC) or endoscopic retrograde cholangiography (ERC).

In many centres ERC has substantially replaced PTC for diagnostic purposes. However, in some cases ERC may not be technically possible, and it does not always demonstrate the intrahepatic radicles completely. PTC thus retains an important and integral role in the radiological evaluation of the obstructed biliary tree, especially in hilar obstruction, where complete visualization of the biliary tree by ERC is frequently impossible. PTC provides excellent delineation of the intra- and extrahepatic bile ducts. It is also the initiating step in percutaneous biliary drainage (PBD), which still plays an important role in patient management. Although recent advances in the endoscopic placement of stents (Boender et al. 1990) have led to widespread abandonment of PBD for low biliary strictures, except in cases where endoscopy has failed, hilar obstruction due to bile duct tumours or enlarged lymph nodes in non-surgical candidates is still best treated by PBD. The exception to this is in patients with obstructive jaundice who have multiple peripheral intrahepatic metastases; in these patients drainage is generally ineffective in relieving symptoms and should therefore be avoided.

25.2 Indications and Contraindications for PTC and PBD

The main indication for direct cholangiography and drainage is bile duct obstruction demonstrated by US or CT but where the information provided by these studies is insufficient for diagnostic purposes or for planning treatment. The choice between PTC and ERC is often determined by local expertise, but

A.F. Watkinson, MD, Consultant Radiologist and honorary Senior Lecturer, X-Ray Department, Royal Free Hospital, Pond Street, Hampstead, London NW3, UK
A. Adam, MD, Guy's and St. Thomas' Medical and Dental School, Division of Radiological Sciences, Interventional Radiology, Guy's Hospital, London Bridge, London SE1 9RT, UK

in general ERC is preferable when there is no intrahepatic duct dilatation or when low obstruction of the common bile duct is suspected. PTC is indicated in cases where ERC is unsuccessful or when high obstruction is suspected, when ERC may provide incomplete demonstration of the intrahepatic ducts. PTC is the preferred method of investigation if segmental duct obstruction is suspected or if the second part of the duodenum is inaccessible to endoscopy.

In general PTC is said to be successful in approximately 98% of patients with dilated ducts and in more than 70% of patients with non-dilated ducts (Harbin et al. 1980). In specialist centres it is virtually always possible to obtain a percutaneous cholangiogram, even in the absence of duct dilatation. It may be that, as imaging modalities improve, spiral CT and magnetic resonance imaging will provide all the information required for diagnostic and treatment planning purposes; however direct percutaneous cholangiography is still required in many patients at present.

Contraindications to PTC are the following:

1. Uncorrectable coagulopathy. In such cases the procedure can be carried out under fresh frozen plasma cover.
2. Marked ascites. In such cases an endoscopic approach is preferable in the first instance. If this is not possible, or is unsuccessful, the ascites can be drained prior to percutaneous drainage. It is best to use a left approach, as most of the remaining fluid will be found in the patient's flanks.
3. Unsafe access route, due to either interposed bowel in the right flank or intrahepatic lesions making passage of a catheter unsafe.
4. Presence of numerous obstructed and noncommunicating intrahepatic biliary segments.

25.3 Patient Preparation

1. If the patient's blood coagulation is abnormal, vitamin K should be administered. If the prothrombin time remains more than 2 s above control level, fresh frozen plasma should be given during the procedure.
2. If the number of platelets is less than 80×10^9/l, a platelet infusion should be given before the procedure.
3. Intravenous access should be established. During PBD the patient should be given an infusion of physiological saline as a prophylactic measure against hepatorenal failure.
4. Prophylactic antibiotics are administered 1 h prior to the procedure (80 mg gentamicin i.v., 2 mg piperacillin i.v.). This is because of the high incidence of bacterial colonization of obstructed biliary systems (Keithley 1977). Antibiotic therapy should be modified according to sensitivities assertained from positive bile or blood cultures.
5. The patient should be connected to a pulse oximeter. Oxygen should be administered nasally with close monitoring of the patient's respiratory and circulatory status.
6. Sedation and analgesia are used. In most patients a combination of midazolam and fentanyl is employed.
7. The patient's pre-procedure imaging should be studied to define as precisely as possible:
   (i) The proximal extent of the bile duct stricture
   (ii) The presence of intrahepatic tumour. This may affect the choice between a right or a left lobe approach.
   (iii) The presence of lobar liver atrophy. CT is more accurate in assessing this condition, which may be present with hilar obstruction, particularly with cholangiocarcinoma. Lobar atrophy more commonly affects the left lobe. It is recognized on cholangiography by crowding of dilated bile ducts and is usually associated with a reduction in overall size of the lobe.
   (iv) The presence of intrahepatic abscesses. CT is preferable to US for assessing this.
   (v) The presence of ascites, which is a relative contraindication to PBD (see above).

25.4 Technique of PTC

The procedure is performed with sterile technique under fluoroscopic guidance. Lateral screening is not essential but is very helpful in selecting the duct to be entered during the definitive procedure. Ultrasound guidance is helpful in patients with segmental duct obstruction or those with unusual anatomy (e.g., after liver resection).

PTC can be performed from a right- or a left-sided approach, although in most institutions the initial diagnostic study is usually from the right. Lignocaine 2% is used for local anaesthesia. The initial puncture is with a 15-cm-long, 22-gauge (0.7 mm) flexible Chiba needle. Puncturing is easier to perform if the intrahepatic ducts are dilated, but it is virtually always possible to insert a needle into the biliary tree even if they are not, although this may require more

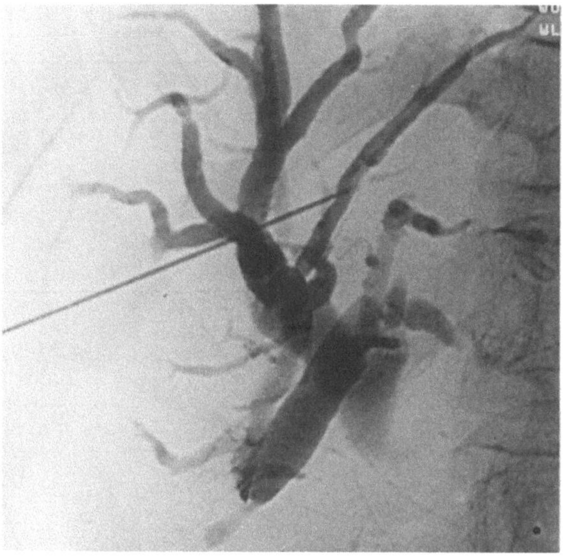

Fig. 25.1. Percutaneous transhepatic cholangiography (PTC). A 22-gauge Chiba needle has been inserted in a cephalad direction parallel to the table top in the mid-axillary line. The tip is seen to lie lateral to the thoracic spine. Non-ionic contrast medium has been injected. This early cholangiogram shows a dilated biliary system. Reproduced with kind permission from Watkinson A, Adam A (1996) Interventional radiology: A practical guide, Radcliffe Medical Press Ltd, Oxon, UK

needle passes. Under fluoroscopic guidance the range of liver movement is monitored during respiration, and a puncture point is chosen at a level below the lateral costophrenic angle on full inspiration and superior to the hepatic flexure of the colon on full expiration. The needle is inserted one rib space anterior to the mid-axillary line and advanced in mid-inspiration parallel to the table top in a 20°–30° cephalad direction to a point lateral to the thoracic spine (Fig. 25.1). Specific vertebral bodies cannot be cited for guidance as their relationship to the liver is variable. Subsequent needle movements should be performed during suspension of respiration.

When the bile ducts are dilated, aspiration is preferable to contrast medium injection, as diffuse parenchymal contrast obscures the field of view and may be painful, particularly around the hilum and within the peritoneum. If the biliary tree is not dilated, injection of contrast medium during withdrawal is preferable, as aspiration is frequently unsuccessful. If the aspiration technique is being used, the needle stylet is removed and, using a syringe and connecting tube, suction is applied as the needle is withdrawn. Once a duct has been punctured, bile flows freely into the connecting tube.

Injection of contrast medium into portal or hepatic veins is frequent and is easily recognizable by a rapid flow of contrast away from the needle tip. Occasionally injection into lymphatic vessels occurs, especially around the hilum; these have a characteristic "beaded" appearance (Fig. 25.2). Injection into the bile ducts is immediately recognizable by a slow, "oil-like" flow of contrast medium away from the needle tip. With multiple needle passes it is common to produce haemobilia. When this occurs, blood-stained bile is aspirated, which is more viscous than frank venous blood and drips like oil rather than water.

The number of needle passes needed for opacification of the ducts is mainly determined by the degree of intrahepatic dilatation: usually only one to three are required in very dilated systems, but an undilated biliary tree will often require more. No correlation has been demonstrated between the number of passes needed and the frequency of complications. It is important to ensure adequate analgesia and sedation if a large number of passes is anticipated.

In most cases fluoroscopic guidance is adequate, but ultrasound guidance is useful in cases of segmental duct obstruction or where the anatomy of the liver is unusual (following surgery or in the presence of lobar atrophy).

Once a duct has been punctured, bile is aspirated for bacteriological and cytological studies before more contrast medium is introduced to outline the

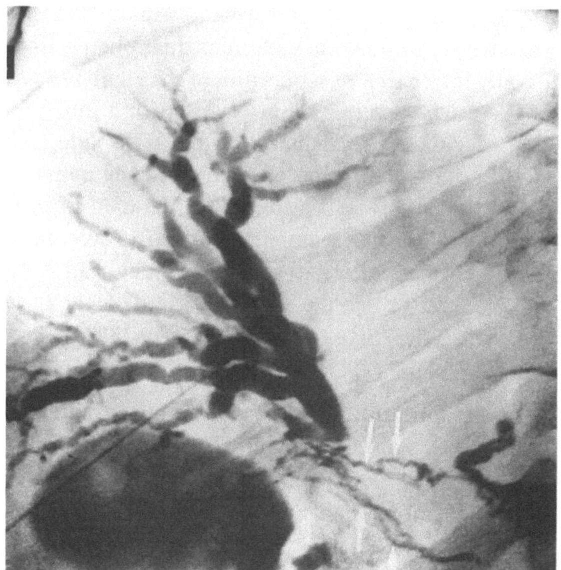

Fig. 25.2. PTC. Injection of non-ionic contrast medium in a patient with obstructive jaundice demonstrates filling of right lobe dilated intrahepatic ducts. There is no cross-filling into the left hepatic ducts; however, the classical "beaded" appearance of lymphatic vessels (*arrows*) is seen in the region of the porta hepatis

biliary tree. Water-soluble contrast medium (200–300 mg iodine per millilitre) is injected in sufficient quantities to fill the intra- and extra-hepatic biliary tree as much as possible without using undue pressure. If possible, it is desirable to aspirate bile as contrast medium is introduced, in approximately equal amounts, so as not to overdistend the biliary tree and increase the risk bacteraemia. Concentrated contrast medium is used for PTC. However, this may later need to be diluted by injecting saline in order not to obscure stones or other filling defects and to visualize catheters and guidewires.

If a right duct is punctured, it may be necessary to rotate the patient to a left-lateral or even semi-prone position to demonstrate the left ducts. It is very important to obtain lateral and oblique as well as anterior-posterior views for full demonstration and correct identification of ductal anatomy. It is not unusual to see temporary cessation of flow of contrast medium at the hilum of the liver in patients with low common bile duct obstruction. This "pseudo-obstruction" can mislead the inexperienced and may be recognized by the presence of a "hazy" margin at the level of apparent obstruction. A tilting table can be useful in this situation, as tipping the patient 30°–40° feet down encourages distal flow of contrast medium and allows identification of the level of obstruction.

If only the right-sided ducts are opacified despite patient rotation and table tilt, then a separate left duct puncture is performed, usually via the epigastrium. In patients with hilar duct obstruction, not only may there be separation of the right and left hepatic ducts, but extension of stricturing into second-order ducts, particularly on the right side, can lead to non-opacification of obstructed segments. It is vital that the interventionist has detailed knowledge of the normal segmental pattern (Table 25.1) and the normal pattern of segmental duct branching (Fig. 25.3) as well as the normal variants (Fig. 25.4) (Couinaud 1957; Healey and Schroy 1953).

A left-sided PTC is indicated if:

(i) A left lobe cholangiogram is required and this lobe has not been demonstrated by the right-sided PTC.
(ii) Drainage of the right lobe is inappropriate because of previous lobar resection, lobar atrophy, or extensive tumour infiltration (Fig. 25.5).

Table 25.1. Segmental nomenclature of the liver. (After Couinaud 1957 and Healey and Schroy 1953)

I	Caudate lobe
II	Left lateral superior segment
III	Left lateral inferior segment
IV	Left medial segment or quadrate lobe
V	Right anterior inferior segment
VI	Right posterior inferior segment
VII	Right posterior superior segment
VIII	Right anterior superior segment

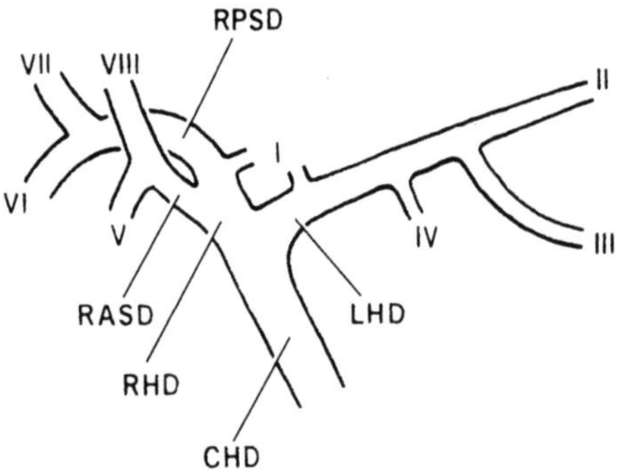

Fig. 25.3. Standard intrahepatic ductal anatomy. Segments numbered according to Couinaud's system (see Table 25.1). *RPSD*, Right posterior sectoral duct; *RHD*, right hepatic duct; *CHD*, common hepatic duct; *RASD*, right anterior sectoral duct; *LHD*, left hepatic duct. (Reproduced from Adam and Gibson 1994 by permission of Edward Arnold)

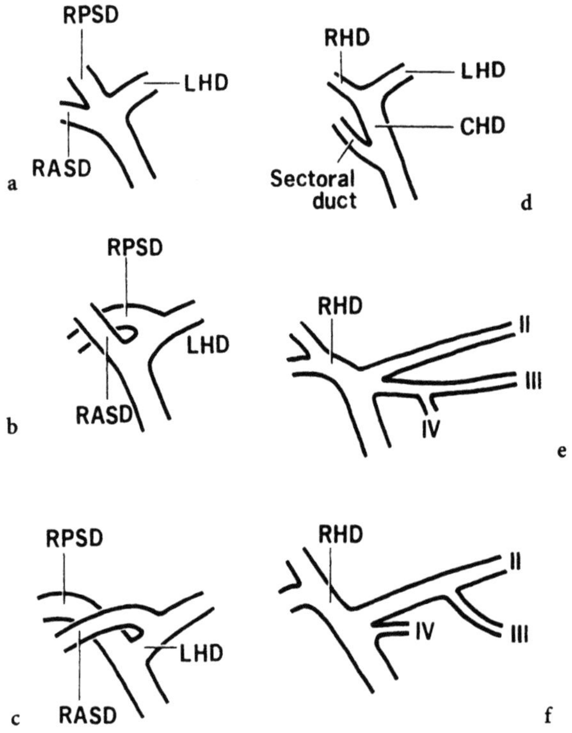

Fig. 25.4a–f. Variations of perihilar ductal anatomy. Segments are numbered according to Couinaud's system (see Table 25.1). (Reproduced from Adam and Gibson 1994 by permission of Edward Arnold)

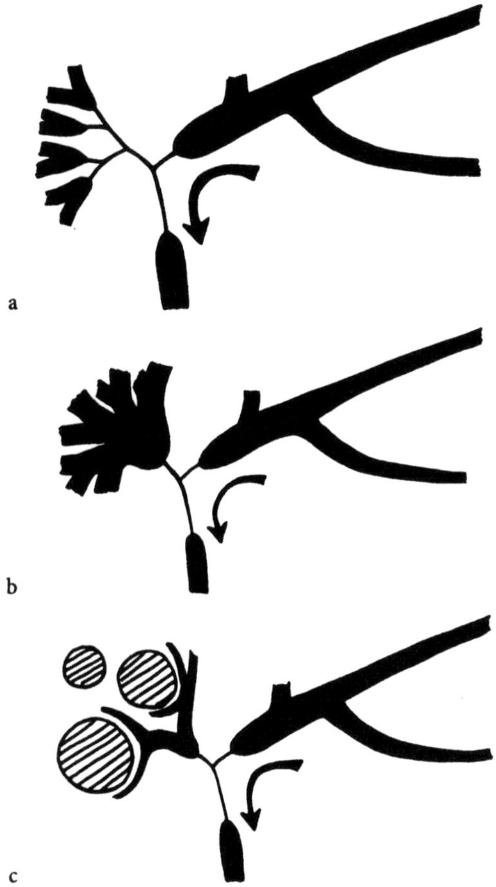

Fig. 25.5a–c. Factors affecting the choice of a right versus left lobe approach to biliary drainage. The presence in one lobe of **a** stricturing extending into third- or fourth-order hepatic ducts, producing multiple isolated segements, **b** significant atrophy or **c** extensive tumour dictates that drainage should be of the contralateral lobe. (Reproduced from ADAM and GIBSON 1994 by permission of Edward Arnold)

Some radiologists use the left lobe approach routinely, both for the initial PTC and for subsequent drainage, but the present authors do not. This is because the left hepatic duct tends to form a more acute angle with the common duct than the right, making catheter guidewire manipulation more difficult. In addition, it is harder for the operator to keep his or her hands out of the primary X-ray beam.

An epigastric approach is used. The position of the left lobe is variable and difficult to locate under fluoroscopy, and the initial puncture is best performed using ultrasound guidance. Following cleansing and draping of the skin and infiltration with local anaesthetic, the chosen duct is punctured and direct cholangiography performed (see Fig. 25.7a) using non-ionic contrast medium and taking care not to overdistend the intrahepatic biliary tree. Since the left lobe is quite close to the anterior abdominal wall, it is best to use a 22-gauge spinal needle for the initial puncture as this allows the image intensifier to be closer to the patient.

25.5 Percutaneous Biliary Drainage

In patients with unresectable biliary tract malignancy, percutaneous biliary catheterization is performed to provide access for transhepatic placement of biliary endoprostheses or, depending on local practice, to provide transhepatic assistance for endoscopic retrograde stent placement. Long-term external tube drainage is not advocated, and routine pre-operative PBD has been abandoned in most centres as it does not lower the overall perioperative morbidity and mortality. Early retrospective studies suggested that a period of preoperative PBD reduced postoperative mortality in jaundiced patients (NAKAYAMA et al. 1978), but subsequent prospective randomized studies failed to show any advantage (MCPHERSON et al. 1984).

25.5.1 Patient Selection and Preparation

In patients with tumour which is unresectable, because of either tumour extent or medical risk factors, a number of questions need to be addressed prior to further management:

1. *Is biliary decompression indicated?* Most patients require biliary drainage to alleviate the symptoms of jaundice, i.e. pruritus or cholangitis, rather than the jaundice itself. Consequently, in patients with jaundice who are relatively asymptomatic and have a very limited prognosis, intervention may be inappropriate. In addition, patients with multiple obstructed and isolated biliary segments, e.g. due to liver metastases or extensive hilar cholangiocarcinoma, should not in general undergo PBD.
2. *Is surgical or non-surgical decompression more appropriate?* The decision on this question has often been made prior to referral to the radiology department. Surgical decompression is more appropriate if gastric outlet obstruction is present.
3. *Is percutaneous or endoscopic intervention more appropriate?* Providing there is endoscopic access to the second part of the duodenum, endoscopy is the method of choice for mid and low common bile duct strictures. For hilar lesions the choice is less clear and may be governed by local expertise.

For hilar lesions extending into first- or second-order right or left hepatic ducts, selection of the most appropriate lobe and segment for drainage is more reliably achieved by a percutaneous approach.

Cytological or histological confirmation of suspected malignant obstruction should be obtained. Percutaneous needle aspiration or biopsy can be performed using cholangiographic guidance under fluoroscopy or using US or CT if there is a demonstrable mass (HALL-CRAGGS and LEES 1985). Obtaining a positive biopsy is difficult in some instances, particularly in cases of cholangiocarcinoma, and brushings or biopsy via a percutaneous catheter track can be used (DAVIDSON et al. 1992) (see Sect. 25.5.2).

25.5.2 Technique

PBD is usually performed as an extension of PTC. In most cases it can be performed with intravenous analgesia and sedation, with close monitoring of the respiratory and circulatory status; however, in some cases anaesthetic assistance is required.

In patients with hilar duct obstruction where there is partial or complete separation of the right and left ducts, it is important to study the CT with the diagnostic PTC carefully in order to assess the proximal extent of the stricture and the presence or absence of lobar atrophy. The choice of approach may be decided by the presence of atrophy, extensive tumour or multiple isolated segments in one lobe; in any of these situations the contralateral lobe should be drained (Fig. 25.5). If "white bile" is drained from one lobe and "yellow bile" from the opposite lobe, the latter should be drained as the obstruction on that side is probably not as longstanding. In patients with clinical cholangitis the side yielding purulent bile needs to be drained to resolve the sepsis. In these circumstances it may be advantageous to drain both lobes.

The aim of biliary drainage is to drain as much tumour-free and non-atrophic liver as possible. Fewest errors will be encountered if a complete cholangiogram is obtained before selection of the duct for puncture. Once the needle cholangiogram has been obtained, the duct for access is chosen. Lateral fluoroscopy is very valuable in identifying the most appropriate duct. Ideally, the duct selected for drainage should form a fairly obtuse angle with the common hepatic duct, as this facilitates subsequent manoeuvres. PBD can be performed by approaching the right or the left lobe; in many patients an appropriate puncture point is the confluence of the ducts of segments VI and VII on the right (Fig. 25.6a) and the duct of segment III on the left (Fig. 25.7). The right-sided approach is preferred by the present authors in most instances, particularly for obstructions of the common duct, as it provides a straighter approach and enables the operator to keep his or her hands out of the primary X-ray beam.

25.5.2.1 *Right Lobe Approach*

The skin puncture site is similar or slightly posterior to that used for the PTC and modified to give the most direct approach to the selected duct. It is important to study the cholangiogram to identify variations in the intrahepatic pattern, in particular whether the right posterior or right anterior sectoral duct drains into the left hepatic duct; this occurs in 22% and 6% of patients respectively (Fig. 25.4b,c) (COUINAUD 1957). If either of these ducts is catheterized it creates an "S-bend" approach to the common duct, making manipulation through the strictures difficult. A 21-gauge Chiba needle is used for the puncture, and duct entry is characterized by distortion of the wall and a perceptible "give' as the needle enters the duct. The anterior-posterior relationship of the needle and the duct can be assessed using lateral fluoroscopy. Intraductal location is indicated by aspiration of bile, which may be blood-stained from the initial PTC, and confirmed by injection of contrast medium into the duct.

Fig. 25.6. a PTC. A second puncture is usually required to access an appropriate duct for percutaneous biliary drainage. In this case, a 21-gauge Chiba needle has been inserted on the right side to enter the intrahepatic biliary system at the confluence of the ducts from segments VI (*straight arrow*) and VII (*curved arrow*). **b** A coaxial catheter system (Accustick, Neff set) is shown, comprising a 22-gauge Chiba needle, an 0.018-inch (0.45-mm) platinum-tipped guidewire and a coaxial system of inner metal stiffener with 3F and 5F sheaths. There is an additional 0.018-inch guidewire in case the platinum tip fractures. **c** The magnified tip of a biliary manipulation catheter, demonstrating the 45° angulation. **d** PTC. Here, a malignant hilar stricture could not be traversed at the initial attempt and an external drainage catheter has been left for 2–3 days to allow oedema to settle with decompression. **e** A Simp-loc internal-external biliary drainage catheter. Note the side holes along the length of the sheath. **f** PTC. A malignant stricture in the lower common bile duct has been traversed and an 8F internal-external Simp-loc biliary catheter inserted. It is important to ensure that the distal loop lies in the duodenum and that sufficient sideholes lie above the stricture for effective drainage. Reproduced with kind permission from Watkinson A, Adam A (1996) Interventional radiology: A practical guide, Radcliffe Medical Press Ltd, Oxon, UK

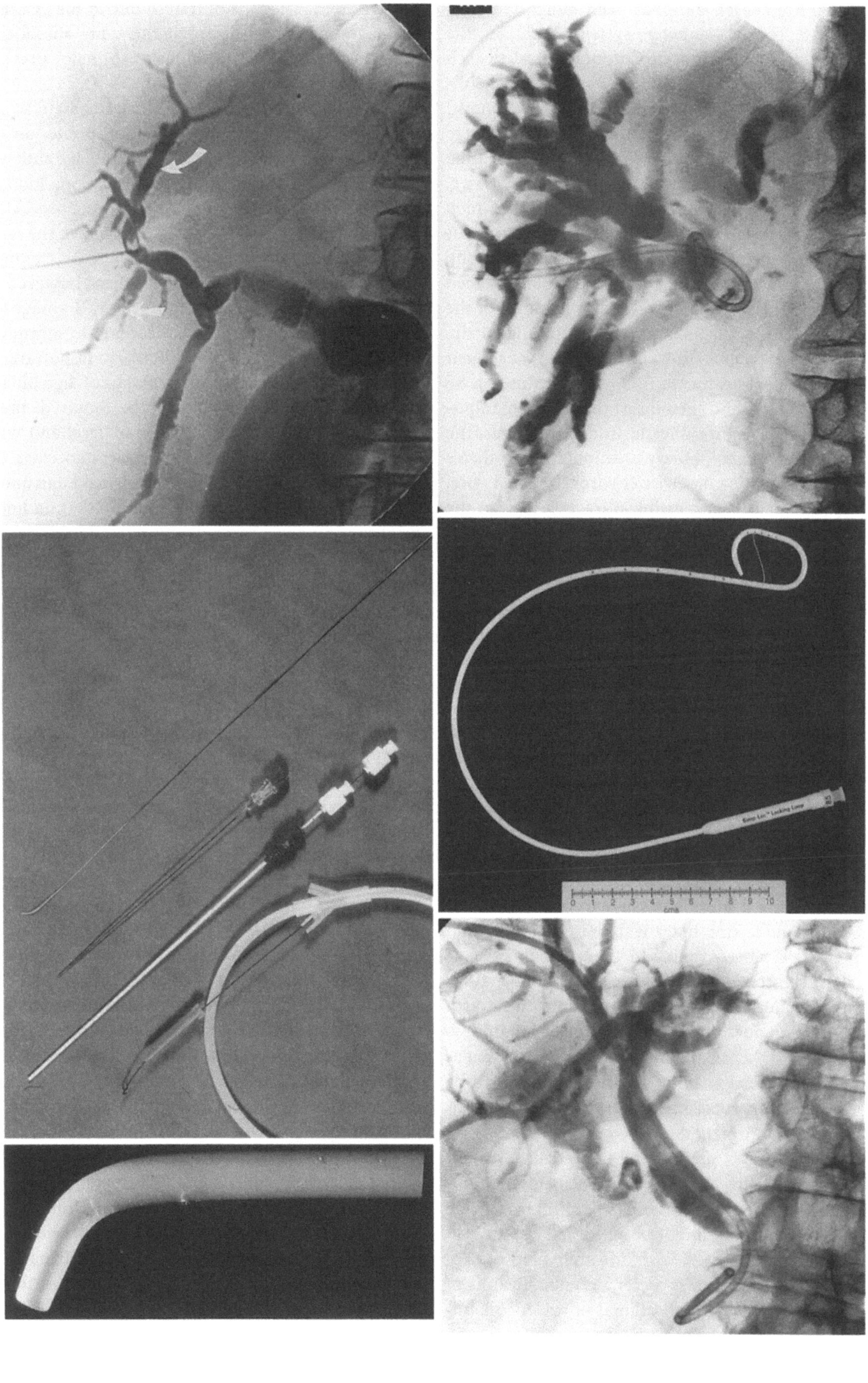

a b c d e f

Once the chosen duct has been punctured (Fig. 25.6a), a coaxial catheter system such as the Accustick (Boston Scientific Corporation, Medi-tech Inc., Watertown, MA 02172, USA) or Neff set [William Cook (Europe) A/S, Bjaeverskov, Denmark] (Fig. 25.6b) is used to access the biliary tree. These systems enable a puncture with a 21-gauge Chiba needle, using an 0.018-inch (0.45-mm) mandril wire, to be converted into an indwelling 5F sheath which will accept an 0.035-inch (0.89-mm) J guidewire. A 6.5F biliary manipulation catheter [William Cook (Europe) A/S] (Fig. 25.6c) can then be introduced over this guidewire and advanced to the level of the stricture. This catheter is short, has an angled tip, and allows excellent torque control. If advancement of the catheter is difficult, it is helpful to use an 8F peelaway sheath and then insert the biliary manipulation catheter via the sheath. In difficult cases, the combination of the biliary manipulation catheter and a hydrophilic guidewire (hydrogel-coated wire, Terumo, Japan) is successful in traversing even the tightest strictures. The catheter is positioned just proximal to the stricture and the wire used to probe gently to find the lumen. Even if no contrast medium passes through the stricture, the wire will frequently advance with gentle probing. If the stricture cannot be traversed at the first session, an 8F Simp-Loc nephrostomy catheter is placed above the stricture and left on external drainage for 2–3 days to decompress the biliary tree (Fig. 25.6d). This allows oedema to settle and a complete obstruction will convert to an incomplete obstruction, allowing the stricture to be crossed at a second attempt. If a hydrophilic wire is used to traverse the stricture, it must be replaced with a conventional guidewire for subsequent dilatation and catheter placement. In patients with very tight strictures it may be possible to introduce a hydrophilic guidewire but the biliary manipulation catheter will not advance. In these instances a 4F straight catheter will usually pass, enabling a stiffer Amplatz guidewire to be advanced into the fourth part of the duodenum. Subsequent insertion of low-profile hydrophilic balloon catheters (Boston Scientific Corporation) enables stricture dilatation to be performed with subsequent passage of larger catheters and endoprostheses.

If cholangitis is present and the stricture cannot be traversed easily, it is best to keep catheter manipulations to a minimum, establish external drainage with intravenous antibiotic cover for a few days, and then attempt to cross the stricture after a few days of drainage. If haemobilia is present, regular flushing of the catheter with normal saline is advised. It is important to monitor the output of bile in the external drainage bag and ensure that the same amount of fluid is replaced intravenously to help prevent hepatorenal failure.

If the stricture is traversed but immediate insertion of an endoprosthesis is inappropriate, an 8F internal-external Simp-Loc biliary drainage catheter (Fig. 25.6e,f) is placed with the distal Cope loop in the duodenum. This has a number of sideholes along the distal shaft which should extend above the level of obstruction, allowing the catheter to be clamped and the bile to drain internally. It is best, however, to always leave the catheter on free gravity drainage for the first 1–2 days after the procedure, to minimize the risk of sepsis. This is particularly important in patients with cholangitis. For drainage of high biliary obstruction extra sideholes may be required; these can be cut using a hole-puncher or freehand with scissors or a scalpel. Too many holes too close together can weaken the catheter and make introduction difficult. It is important to ensure that no holes lie within the liver parenchyma to avoid the risk of haemobilia from flow of blood into a catheter sidehole lying within a blood vessel. Subsequent stent insertion can then be performed after 2–3 days.

25.5.2.2 Left Lobe Approach

The left lobe approach is used in situations described in Fig. 25.5. Duct puncture and catheterization are carried out as described previously. The initial approach is via the epigastrium. A second puncture is usually required to optimize the angle of approach to the hilum for biliary drainage. Usually the segment III duct is punctured (Fig. 25.7), as the segment II duct passes posteriorly and superiorly, creating unfavourable angles for catheter-guidewire manipulation through a hilar stricture. However, the choice of duct in each case depends on the individual circumstances. Generally, the point of puncture for introduction of the drainage catheter is inferior and lateral to the puncture used for the initial diagnostic cholangiogram. This allows a more peripheral entry into the segment III duct and enables pushing forces to be directed medially.

Once the obstruction has been traversed, the principles of catheter and stent placement are as described for the right-sided approach. Drainage of both lobes is sometimes indicated for hilar duct obstruction, if the amount of liver parenchyma in a single lobe alone is not thought to be enough to relieve symptoms (see below).

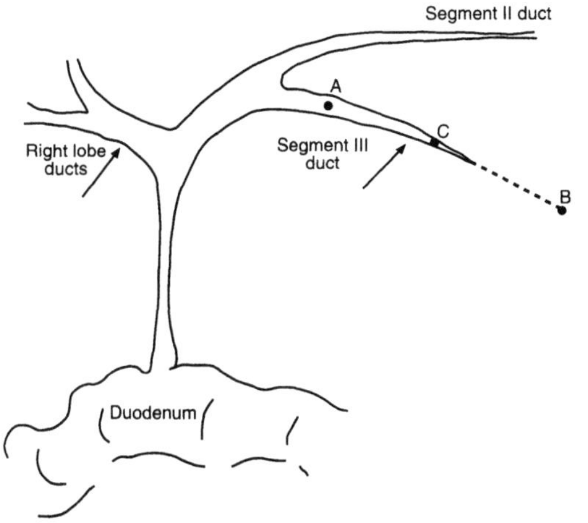

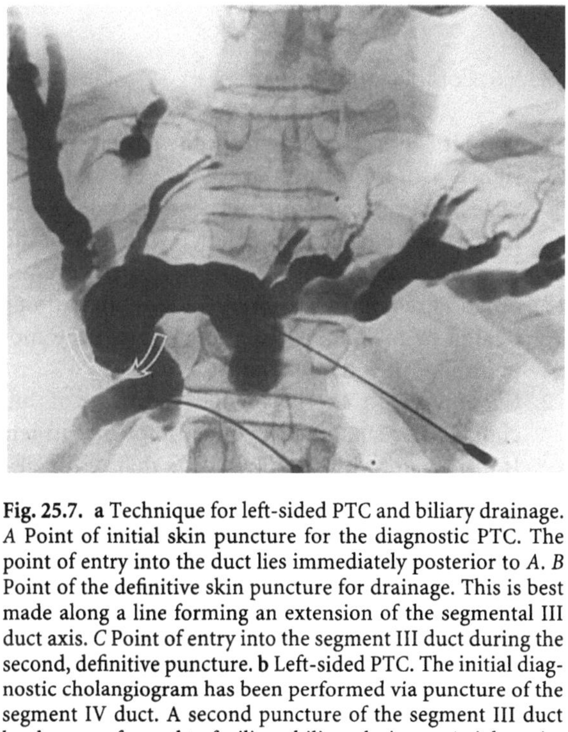

Fig. 25.7. **a** Technique for left-sided PTC and biliary drainage. *A* Point of initial skin puncture for the diagnostic PTC. The point of entry into the duct lies immediately posterior to *A*. *B* Point of the definitive skin puncture for drainage. This is best made along a line forming an extension of the segmental III duct axis. *C* Point of entry into the segment III duct during the second, definitive puncture. **b** Left-sided PTC. The initial diagnostic cholangiogram has been performed via puncture of the segment IV duct. A second puncture of the segment III duct has been performed to facilitate biliary drainage. A tight stricture at the porta hepatis is seen obstructing the left hepatic duct (*curved arrow*). Reproduced with kind permission from Watkinson A, Adam A (1996) Interventional radiology: A practical guide, Radcliffe Medical Press Ltd, Oxon, UK

25.5.2.3 Multiple Drains or a Single Drain in Hilar Biliary Obstruction?

In patients with hilar biliary obstruction there is frequently separation of the right and left lobe bile ducts and often isolation of segmental ducts. The aim in these cases is to drain the largest segment of tumour-free and non-atrophic liver. It is usually possible to provide effective palliation by draining one side only. Drainage of more than one lobe or segment is necessary:

1. If both lobes contain infected bile causing clinical sepsis.
2. If contrast medium has been introduced into both lobes during a failed attempt at endoscopic stent insertion. In this case drainage of both sides is mandatory, due to the high risk of cholangitis in an "invaded" but undrained closed system.
3. If the initial catheter or endoprosthesis fails to resolve symptoms effectively and a considerable amount of tumour-free liver remains undrained.

25.5.3 Problems on Catheter Insertion

Guidewire Looping Outside the Liver. This usually occurs during advancement of a catheter over a guidewire when the catheter meets resistance. The risk of this occurring can be minimized by:

1. Selecting the correct duct for initial puncture, thus avoiding angles and tight curves
2. Preliminary stricture dilatation
3. Using the Amplatz extra-stiff guidewire
4. Keeping the guidewire straight with some applied tension and advancing the catheter off it, rather than advancing both catheter and guidewire together
5. Using a stiffening cannula such as with the Simp-Loc external and internal-external catheters

Catheter will not Advance Through the Stricture. This can be usually corrected by:

1. Dilating the stricture further
2. Using a stiffer guidewire or stiffening cannula
3. Changing to a smaller-diameter catheter (e.g., 4F straight)
4. Manipulating the catheter through a peelaway sheath.

25.5.4 Catheter Fixation

At the end of the procedure the catheter needs to be fixed securely to the skin. A number of fixation devices are available but none of these are reliable; we prefer to use sutures to ensure that the catheter is not displaced. Sterile Elastoplast is wound around the catheter at the skin puncture site. A 2/0 silk suture is then used to close the skin wound adjacent to the catheter and firmly tied or sutured along

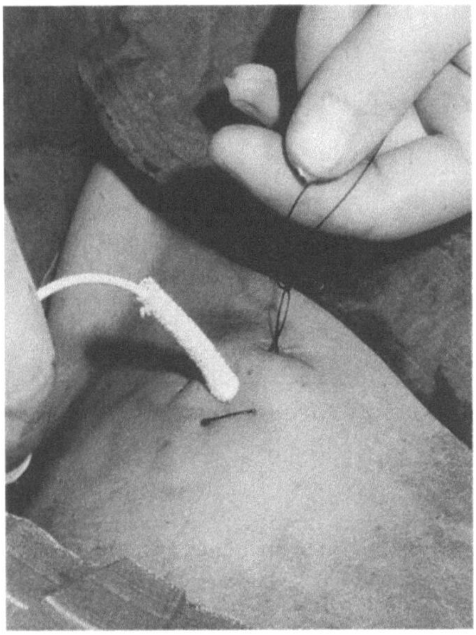

Fig. 25.8. The technique of internal-external catheter fixation using Elastoplast and 0/0 silk sutures. The Elastoplast is firmly adhered to the catheter and the sutures tied around the Elastoplast to firmly attach the catheter to the skin. Reproduced with kind permission from Watkinson A, Adam A (1996) Interventional radiology: A practical guide, Radcliffe Medical Press Ltd, Oxon, UK

the length of the Elastoplast (Fig. 25.8). Caution is taken not to puncture the catheter during this part of the procedure.

The use of subcutaneous bupivacaine hydrochloride as a longlasting (approximately 6–8 h) local anaesthetic ensures minimal discomfort to the patient overnight. In most institutions the goal is internalization of the biliary drainage by insertion of an endoprosthesis. Long-term external or internal-external drainage should only be used in the small percentage of patients (<5%) in whom the stricture cannot be traversed even after a period of decompression.

25.5.5 Postprocedural Care

25.5.5.1 Immediate

The drainage catheter is left on external drainage for at least 24 h. During this time the patient should be kept under close observation with adequate analgesia and the catheter should be flushed with 10 ml saline every 6 h. The volume of bile drained is charted and added into calculations of daily fluid replacement to counter dehydration and electrolyte loss. Antibiotic therapy is continued if sepsis is present, and is modified according to bile culture and sensitivity.

Plastic endoprostheses are usually inserted after 2–5 days of catheter drainage, whereas metallic stents are inserted during the initial session in the majority of cases. If the patient is unfortunate enough to be discharged with a catheter in place, it is important that he or she is instructed about catheter care and domiciliary nursing care arranged.

25.5.5.2 Long-Term

If the catheter becomes occluded, bile leakage may occur along the tract, the patient may develop cholangitis, and jaundice usually recurs with symptoms of pruritus. This is ultimately inevitable, and catheter replacement is advised every 2–3 months.

25.5.6 Complications

1. *Pain.* It is not uncommon for the patient to experience some pain, especially if the local anaesthesia is inadequate. If concentrated or infected bile comes into contact with the peritoneum the pain can be severe, and intravenous opiates are usually required. Balloon dilatation of benign or malignant strictures is often the most painful part of the procedure.
2. *Cholangitis.* If there is good drainage of bile, cholangitis usually settles on antibiotic treatment. If it does not, this usually means that drainage is inadequate or there is an isolated lobe or segment. Diagnostic PTC should be performed in this latter case and further drainage performed if infected bile is aspirated.
3. *Bacteraemia and endotoxic shock.* Bacteraemia and endotoxic shock are unusual during diagnostic PTC but may occur during PBD. Their incidence is minimized by using prophylactic intravenous antibiotics.
4. *Haemobilia.* A relatively common occurrence is haemobilia, particularly if the patient's coagulation is abnormal. It is recognized initially by the presence of bloodstained bile and later by the presence of intraluminal filling defects on the cholangiogram (Fig. 25.9). However it is usually slight and self-limiting. More significant bleeding, which is usually arterial, may, rarely, require embolization.
5. *Pneumothorax.* If a high right puncture is carried out, pneumothorax may occur. In thin patients with low-lying livers, a left-sided approach may be preferable.
6. *Biliary peritonitis.* Biliary peritonitis is most likely to occur if catheter sideholes are inadvertently left

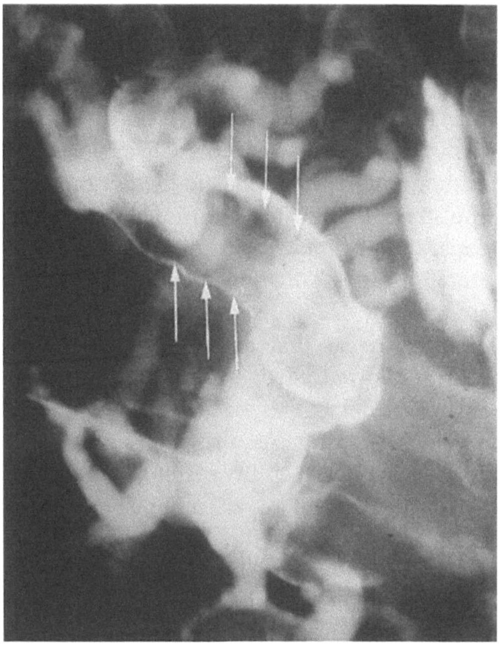

Fig. 25.9. PTC. Non-ionic contrast medium has been introduced into a dilated biliary system. Intraluminal filling defects (*arrows*) are seen, consistent with haemobilia. Reproduced with kind permission from Watkinson A, Adam A (1996) Interventional radiology: A practical guide, Radcliffe Medical Press Ltd, Oxon, UK

outside the liver, allowing free flow of bile into the peritoneum. Catheter repositioning and drainage of the fluid collection are mandatory.

7. *Bile effusion or subphrenic collection.* If there are signs of sepsis and ultrasound scanning shows the presence of fluid collections, these should be drained.
8. *Intrahepatic abscess.* Intrahepatic abscesses may develop secondary to cholangitis. If small and communicating with a decompressed biliary tree, they usually settle on antibiotics. Larger abscesses not draining into the biliary tree require percutaneous catheter drainage.

Other complications include bowel puncture, gall bladder puncture and duodenal perforation.

25.5.7 Endoprostheses

If palliative treatment of malignant biliary tract obstruction is decided upon, then the goal in most institutions is internal drainage via an indwelling endoprosthesis. If a 12F plastic stent is used, insertion is performed as a two-stage procedure with an 8F internal-external drainage catheter across the stricture for 2–5 days to predilate the tract. If a self-expanding metallic endoprosthesis is used, this is mounted on a 7F introducing catheter and can be inserted as a one-stage procedure without a period of internal-external catheter drainage. Internal-external catheter drainage has ceased to be current practice since the advent of indwelling endoprostheses, except in patients who have pre-operative biliary drainage. This is because long-term catheter drainage is associated with bile leakage, infection and pain around the catheter entry site. In addition, the psychological discomfort inflicted on the patient is considerable, with the catheter as a permanent reminder of the malignant condition and the short life expectancy. The main objection to the use of indwelling endoprostheses has been that they are likely to become occluded; however, techniques for overcoming this problem are freely available. Plastic stent replacement is now a routine procedure either endoscopically or percutaneously. Metallic stents tend to remain patent longer – in most instances beyond the life expectancy of this patient group. However, if they do become occluded, they cannot be removed. If the lower end projects into the duodenum, a plastic stent can be introduced endoscopically through the occluded metallic stent lumen. This justifies insertion of the longest metallic stent available with the lower end in the duodenum, to provide subsequent access if stent occlusion occurs. If the lower end of the metallic stent is higher than this and is not accessible from below, percutaneous techniques are available to re-establish patency and these are described below.

25.5.7.1 Choice of Endoprosthesis

This has been a controversial subject in recent years. In our view, metallic endoprostheses are superior to plastic stents in most situations. However, if a plastic stent is to be used, the 12F Miller double mushroom endoprosthesis [William Cook (Europe) A/S] is a suitable device. The most commonly used metallic stent is the 10-mm Wallstent [Schneider (Europe) AG, Bulach, Switzerland], which is self-expanding. Balloon expandable stents have not been successful in the biliary system. The main advantage of metallic stents is that they can be introduced in their contracted state through a small-calibre tract and achieve a large internal lumen following expansion (Fig. 25.10). This reduces the incidence of bile encrustation and also increases stability such that the migration rate is negligible.

The Wallstent biliary endoprosthesis is identical in design to endovascular Wallstents: it consists of surgical grade stainless steel alloy filaments woven in

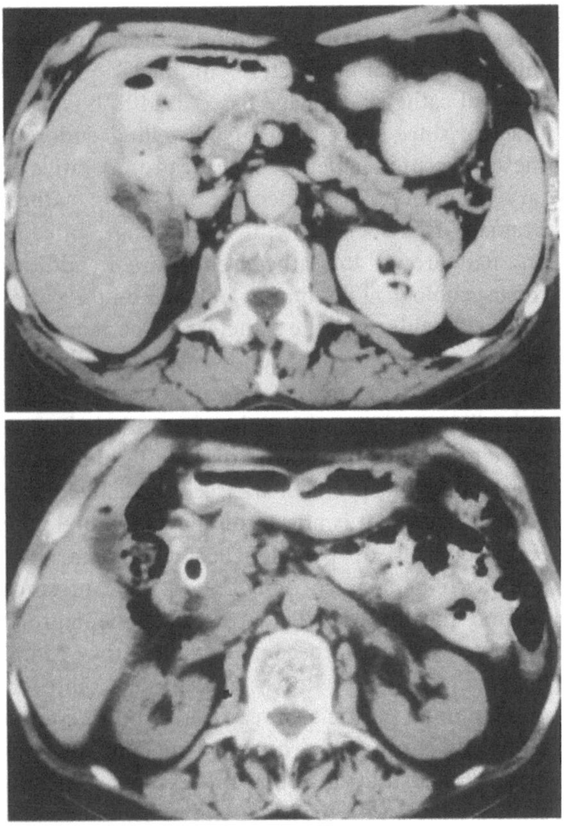

a

b

Fig. 25.10a,b. CT scans of two patients with carcinoma of the pancreas with (**a**) a plastic endoprosthesis and (**b**) a metallic self-expanding Wallstent across the malignant stricture. Note the significant difference in lumen diameter

Fig. 25.11. Wallstent endoprosthesis (fully expanded)

a tubular fashion (Fig. 25.11). Since the crosspoints of the filaments are not soldered together, the stent is pliable, self-expanding and flexible in the longitudinal axis (Zollikofer et al. 1988). It is so elastic and pliable that its diameter can be substantially reduced by moderate elongation, which allows it to be mounted on a special catheter for introduction into the biliary system. Until recently the Wallstent endoprosthesis was mounted on a 7F delivery system with a doubled-over plastic membrane to compress the stent. This required an inflation device to allow separation of the two layers and release of the stent. Re-application of the vacuum by deflating the inflation device allowed proximal repositioning of a partially released stent. The use of an 8F peelaway sheath allowed advancement and distal repositioning. This device has been superseded by a system which does not require the above preparation manoeuvres [Placehit Wallstent, Schneider (Europe) AG]. Withdrawal of an outer sheath releases the stent. Recovery of the stent is possible by re-advancing this outer sheath, enabling both proximal and distal repositioning. The stent is available in a range of lengths up to 90 mm with a maximum diameter of 10 mm.

There are several other metal stents manufactured for biliary use [Gianturco stent, William Cook (Europe) A/S; Strecker stent, Boston Scientific Corporation; Angiomed Memotherm stent, Angiomed UK, Repton, Derby, UK]. Early experience with the Gianturco device suggested it was inappropriate for malignant biliary strictures (Irving et al. 1989) because ingrowth of tumour results in rapid occlusion in a high proportion of cases. There have been satisfactory results with the Strecker device (Wagner and Knyrim 1993), and early data on the Angiomed stent are promising. The greatest amount of information and data available is on the Wallstent, and at present that is the most widely used endoprosthesis.

25.5.7.2 Metallic Stents

Placement Technique

Intravenous sedation is usually required; we generally use fentanyl and midazolam. Access to the biliary tree is obtained and the stricture traversed using the coaxial technique previously described (Fig. 25.12a,b). Bile aspiration and brushings can be taken at this point if previous percutaneous biopsy has been unsuccessful. The combination of analysis of exfoliative bile and brush cytology provides confirmation of the presence of malignancy in more than 60% of cases (Davidson et al. 1992).

The Wallstent is introduced over a 0.035-inch (0.89-mm) Amplatz stiff guidewire across the predilated stricture. Eight- to ten-millimetre high-pressure (12–17 atm) balloon catheters are used (Fig. 25.12c). This tends to be the most uncomfortable part of the procedure, and an intravenous injection of fentanyl immediately before the dilatation is advisable. Radio-opaque metal markers indicate the proximal and distal ends of the Wallstent in its compressed form (Fig. 25.12d) and a third marker indicates the shortened length of stent in its fully expanded state. The stent is positioned to traverse the stricture with the distal end in the duodenum. Although the precise choice of stent size depends on the length and site of the stricture, in general long stents are preferable in order to reduce the problem of tumour overgrowth. The upper end of the stent is placed in the intrahepatic radicles and the distal end in the duodenum.

A double-contrast duodenogram helps to locate the ampulla of Vater accurately and the stent is then partially deployed and withdrawn until the expanded distal end projects 1–2 cm into the duodenum. The stent is then fully deployed (Fig. 25.12e). If the stent does not expand fully and there is still a tight stricture after stent placement, balloon dilatation can be performed. A small straight catheter (4F) is left within the stent overnight (Fig. 25.12f). A tube cholangiogram is performed the following day. If the stent is fully patent, the biliary tree decompressed, the patient afebrile, and there is free flow into the duodenum (Fig. 25.12g), the catheter is removed.

The small size of the introducing catheter enables the Wallstent to be placed, in most instances, during the initial procedure. If there is severe sepsis, or significant haemobilia during the procedure, it is best to defer stent placement and flush the biliary tree with normal saline before proceeding to stent placement at a subsequent session.

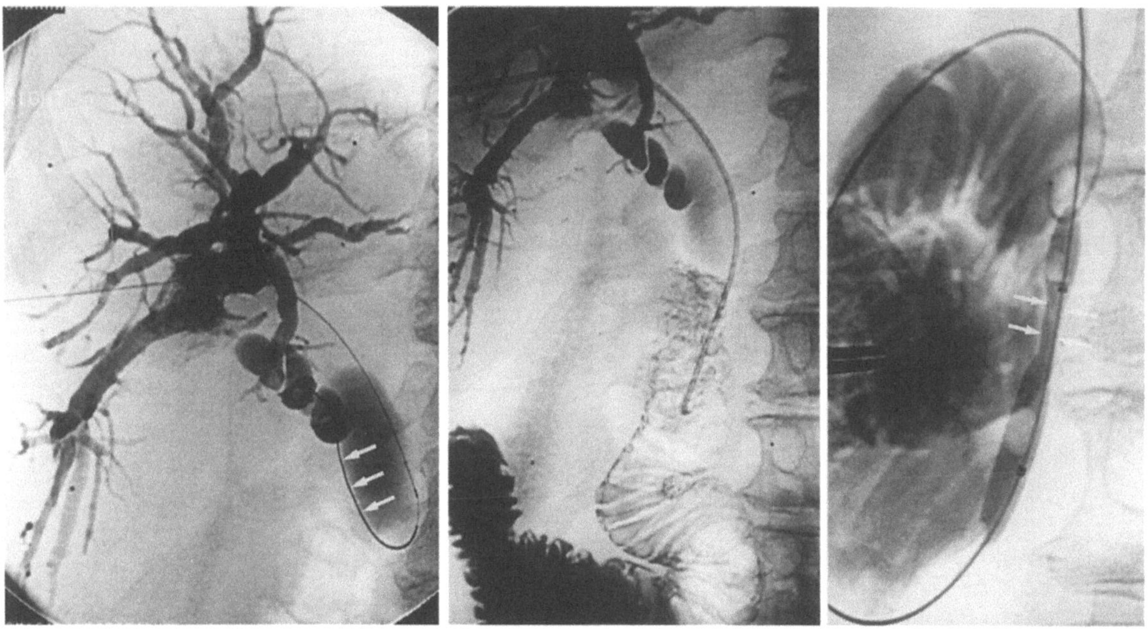

a,b c

Fig. 25.12a–g. Treatment of malignant biliary tract obstruction by insertion of a self-expanding metallic endoprosthesis: technique. **a** PTC demonstrating malignant obstruction of the common bile duct. An 0.018-inch (0.45-mm) mandril wire has been advanced to the level of obstruction. The floppy radio-opaque platinum tip (*arrows*) has looped back up the cystic duct. **b** The stricture has been traversed using a catheter-guidewire technique and a biliary manipulation catheter has been advanced into the duodenum. Air and non-ionic contrast medium have been injected to produce a double contrast duodenogram. **c** Predilatation of the malignant stricture using an 8-mm balloon. Waisting (*arrows*) is seen in the centre of the balloon at the site of the stricture. **d** The metallic endoprosthesis (Wallstent) has been introduced transhepatically through a peelaway sheath and positioned across the stricture. There are three radio-opaque markers, the most proximal and distal markers representing the length of the unreleased stent (*straight arrows*). On stent deployment the distal end of the stent shortens proximally to a third marker (*curved arrow*) with full expansion. **e** The fully released stent is seen with a residual waist (*curved hollow arrows*) at the site of the stricture. The stent markers are again seen, labelled as in **d**. The distal end of the stent has not shortened completely back to the middle marker due the residual waist. **f** A safety catheter is left in the transhepatic tract overnight to provide further access if required and enable a check cholangiogram to be performed to confirm stent patency. This should be left on free gravity drainage overnight. **g** A check cholangiogram demonstrating free flow into the duodenum after 24 h. The catheter can now be withdrawn. Reproduced with kind permission from Watkinson A, Adam A (1996) Interventional radiology: A practical guide, Radcliffe Medical Press Ltd, Oxon, UK

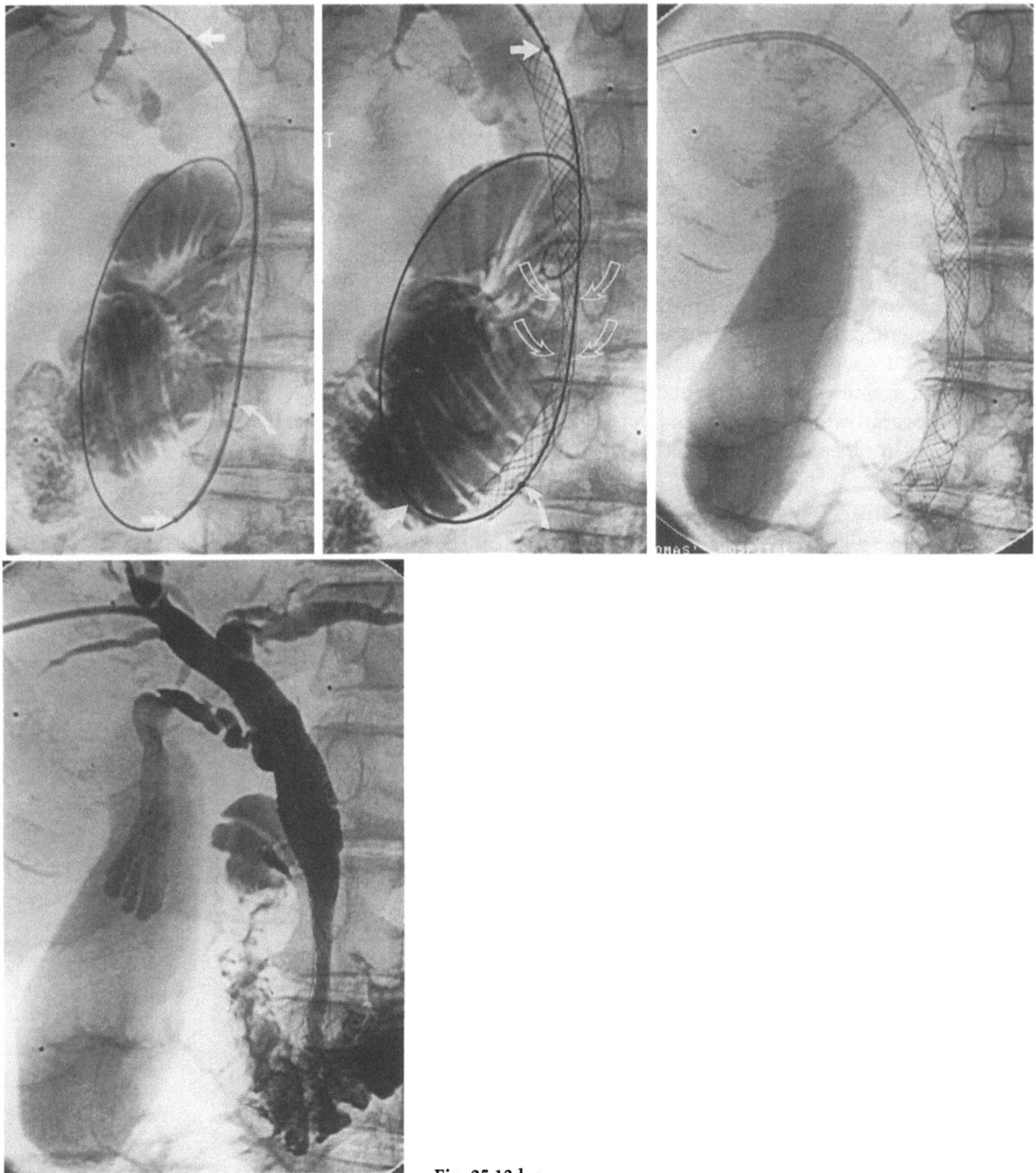

d,e f

g

Fig. 25.12d–g

The main complication of metallic stents is tumour overgrowth. Migration, and occlusion secondary to bile encrustation, are uncommon problems.

Tumor Overgrowth

The most effective way of overcoming tumour overgrowth is to avoid it in the first instance. If a long endoprosthesis is used initially with the distal end in the duodenum and the proximal end in the intrahepatic radicles, then tumour overgrowth can be minimized if not completely avoided. If it does occur, a PTC is performed to demonstrate dilated ducts. The needle is guided under fluoroscopy to the upper end of the metallic stent and contrast medium injected (Fig. 25.13a–c) to demonstrate the dilated ducts. Then a catheter can be inserted into the biliary tree through as peripheral a puncture as possible and advanced to the level of obstruction. Injection of contrast medium through this catheter may show a track through the tumour. If no track is seen, the combination of a biliary manipulation catheter and a hydrophilic angled guidewire (hydrogel-coated wire, Terumo, Japan) will usually cross the obstruction and advance into the duodenum. The catheter can then be advanced and the guidewire exchanged for an Amplatz superstiff wire [William Cook (Europe) A/S]. A second overlapping Wallstent can then be introduced (Fig. 25.13d) such that the proximal end of the stent lies well above the tumour overgrowth.

In cases of advanced hilar malignancy extending into the liver, initial drainage is often of the dominant lobe. Bilateral drainage is only indicated at the initial procedure if:

1. There is no obvious dominant lobe and it is thought that the whole liver is required to recruit enough hepatocytes to relieve symptoms.
2. Contrast medium has entered the contralateral lobe. This is particularly relevant in patients who have had ERC with a failed attempt at endoscopic stent insertion. In such cases bila-teral drainage reduces the risk of cholangitis, and percutaneous drainage is indicated within 48 h.

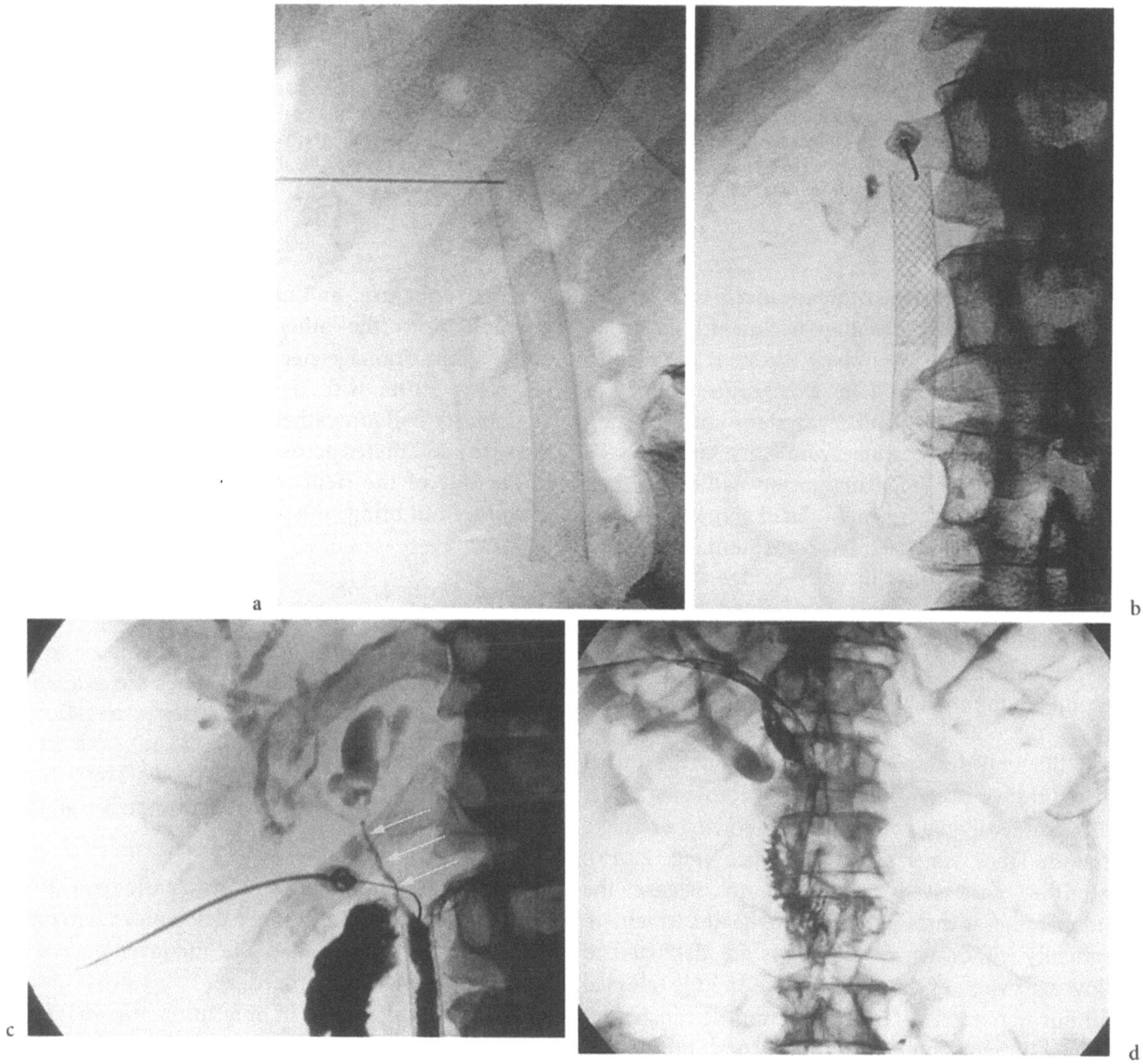

Fig. 25.13a–d. Management of an occluded metallic self-expanding Wallstent. **a,b** A Chiba needle has been advanced to the proximal end of the Wallstent (anteroposterior and lateral views). **c** Contrast medium has been injected, demonstrating occlusion of the upper end of the stent from tumour overgrowth (*arrows*). **d** A second, overlapping stent has been inserted to re-establish drainage

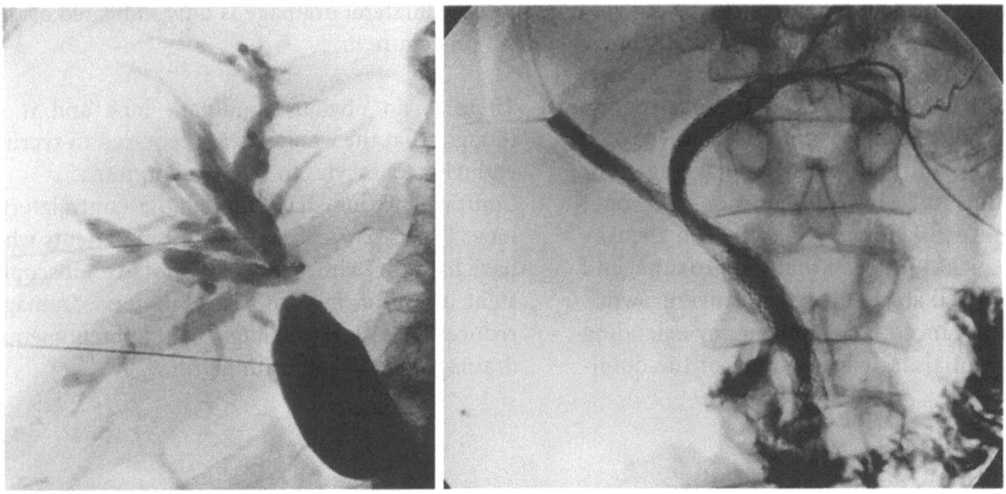

a b

Fig. 25.14a,b. Bilateral punctures with insertion of overlapping Wallstents for malignant hilar obstruction. **a** A right sided PTC has initially punctured the gallbladder (the needle should not be withdrawn until drainage has been established). A second puncture has demonstrated a dilated right intrahepatic biliary system to level of the right main hepatic duct, with no communication with either the left main or the common hepatic duct. The underlying obstruction was due to a cholangiocarcinoma in the region of the porta hepatis. **b** A subsequent third puncture has been made to access the left lobe of the liver and bilateral overlapping Wallstents have been inserted. Cholangiograms at 24 h via 4F safety catheters demonstrate free flow with bilateral decompressed intrahepatic biliary ducts

If a single stent inserted for advanced hilar malignancy subsequently becomes occluded, it may be necessary not only to unblock the stent but also to drain the contralateral lobe. This requires puncture of the contralateral biliary system and passing through the stent struts from this side with a guidewire. A low-profile angioplasty catheter is then advanced over the wire and inflated across the struts to create a hole in the side of the stent. A metallic stent can then be introduced to bridge the gap and decompress the previously undrained lobe. Occasionally it is possible to advance a wire by the side of the initial stent into the duodenum and deploy the second endoprosthesis side-by-side with the first one.

If bilateral drainage is performed at the initial procedure we prefer side-by-side deployment (Fig. 25.14) rather than a configuration allowing drainage of both lobes via a single puncture (Fig. 25.15). In either case it is important to release the endoprostheses through peelaway sheaths which are gradually withdrawn as the stents are deployed to allow expansion of the stents. If one stent is released without the second stent in position, it can be extremely difficult to advance the second, unreleased stent alongside. Although side-by-side deployment requires bilateral punctures, the management of occlusion of the stents is much more straightforward as the endoprostheses can be recannulated from both sides as described above. If two stents are inserted via a single puncture, and the stent projecting from one lobe into the other becomes occluded, reestablishing drainage necessitates passing through the stent struts with a guidewire. A low-profile angioplasty balloon catheter is then advanced over the wire and inflated across the struts to create a hole in the side of the stent. A third stent can then be introduced to bridge the gap.

25.5.7.3 Plastic Stents

Plastic stents were the first endoprostheses to be used in the biliary tree. Several types are available, the most commonly used being the Carey-Coons and Miller endoprostheses (Fig. 25.16), and most have sideholes along their length. They have several disadvantages compared to metallic stents, including:

1. The necessity of a larger transhepatic tract (usually 12F), resulting in more discomfort, a greater risk of haemobilia, and the inconvenience and cost of a two-stage procedure.
2. Increased incidence of migration. In an effort to combat this problem various designs have been introduced, including the incorporation of mushroom tips at either end of the stent and subcutaneous plastic buttons attached to the proximal end of the endoprosthesis (Fig. 25.16). These have reduced the rate of migration; how-

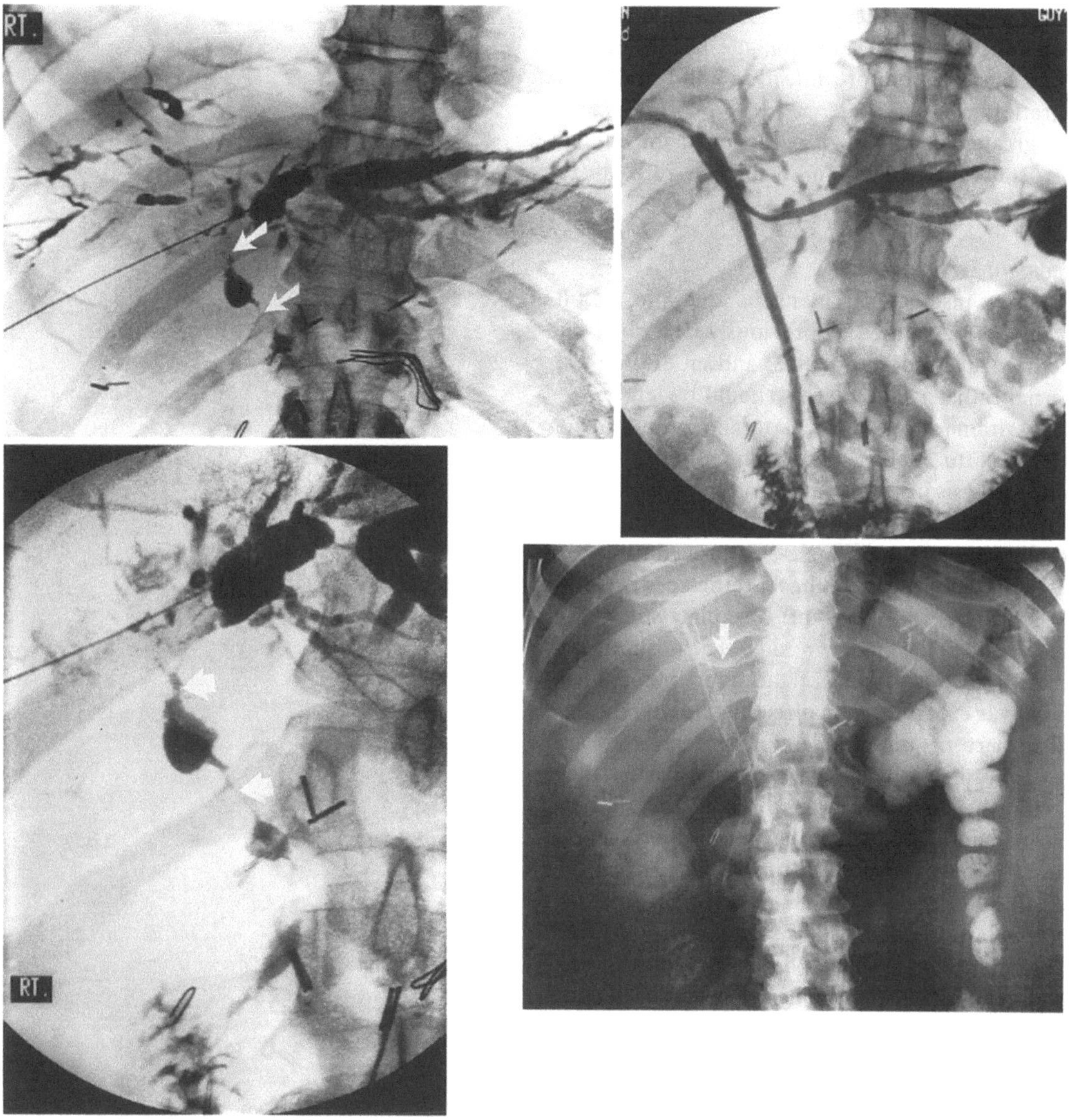

a
b
c
d

Fig. 25.15a,d. A "Y" arrangement of overlapping Wallstents for malignant hilar obstruction. **a,b** A PTC has been performed in a 57-year-old male patient with multiple intra- and extrahepatic metastases from colonic carcinoma. Numerous intrahepatic strictures can be seen in both lobes. In addition, strictures can be seen in the common hepatic and common bile ducts (*arrows*). **c** Bilateral Wallstents have been inserted in a "T" arrangement from a single right-sided percutaneous puncture. A cholangiogram at 24 h via a 4F safety catheter demonstrates good drainage of both lobes. **d** This radiograph of the right upper quadrant has been taken 24 h after stent insertion, demonstrating that the Wallstent into the left lobe has not yet fully expanded (*arrow*) due the tight hilar stricture

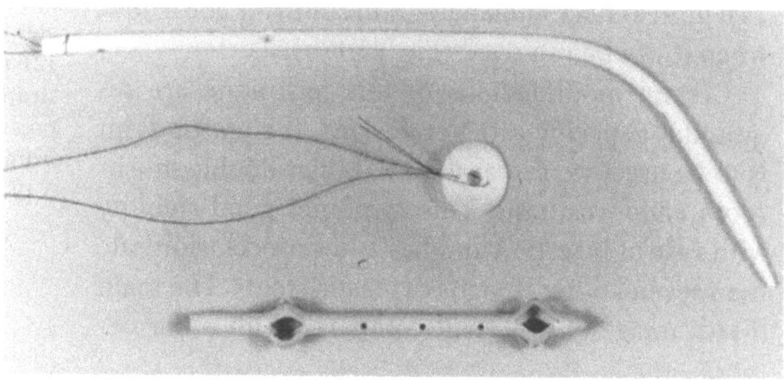

Fig. 25.16. The Carey-Coons and Miller biliary endoprostheses. The former is attached with silk sutures to a plastic button which can be buried subcutaneously to prevent migration of the stent. Reproduced with kind permission from Watkinson A, Adam A (1996) Interventional radiology: A practical guide, Radcliffe Medical Press Ltd, Oxon, UK

ever, the rate is still higher than that reported with metallic stents. In addition, the plastic buttons can cause discomfort, infection and tumour migration to the skin.
3. A greater likelihood of occlusion from bile encrustation, resulting in a higher rate of re-intervention.

Placement Technique

In the first instance percutaneous access to the biliary tree is performed as previously described, the stricture traversed and dilated, and an 8F internal-external catheter placed. This should be left on free gravity drainage for 2–3 days to allow the transhepatic tract to mature. This time interval makes endoprosthesis insertion much easier, faster and less painful for the patient, and is associated with a lower complication rate than single-step procedures.

At a second procedure the 8F catheter is removed over an 0.035-inch (0.89-mm) J Amplatz stiff guidewire. The tip of the guidewire should be positioned in the third or fourth part of the duodenum. If any difficulty is experienced in directing the wire into the fourth part of the duodenum, a femorovisceral (cobra) catheter will facilitate this manoeuvre. The transhepatic tract is then dilated to 12F using biliary dilators; the patient may require intravenous analgesia and sedation during this part of the procedure.

The plastic endoprosthesis is then introduced. Each stent design is slightly different, but most consist of a central plastic stiffener, the stent itself and a pusher which has a flat end and is of the same diameter as the stent. A loop of suture material passed through the proximal end of the endoprosthesis enables the stent to be withdrawn if it is introduced too far. The endoprosthesis is then introduced coaxially over an 0.035-inch Amplatz guidewire until the stent is in the desired position (Fig. 25.17). Flow through the stent is assessed fluoroscopicaly by injection of contrast medium through a small (4F) catheter placed above the stent. This catheter is then advanced until it lies just within the stent and the guidewire is removed. It remains in position for 24 h until a check cholangiogram confirms good flow, when it can be removed (Fig. 25.17b).

Certain modifications of this technique are required for specific catheter designs. If a plastic stent is to be used, we prefer the 12F Miller double mushroom endoprosthesis. This combines good stability with ease of insertion and has a lower occlusion rate than several other designs of plastic stents. The main disadvantage of this stent compared with other de-

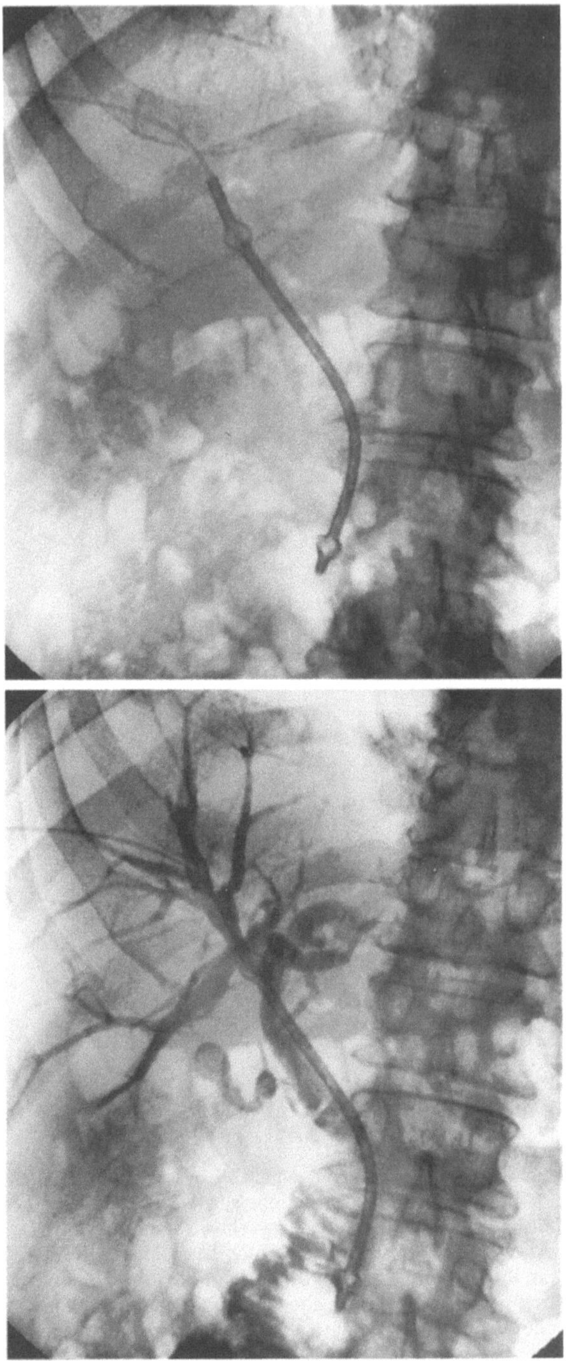

a

b

Fig. 25.17. a Abdominal radiograph showing a Miller endoprosthesis in the right upper quadrant. A 4F safety catheter can be seen transhepatically with the tip lying within the 12F plastic endoprosthesis. **b** A check cholangiogram demonstrating satisfactory positioning of the stent with a decompressed biliary system. Reproduced with kind permission from Watkinson A, Adam A (1996) Interventional radiology: A practical guide, Radcliffe Medical Press Ltd, Oxon, UK

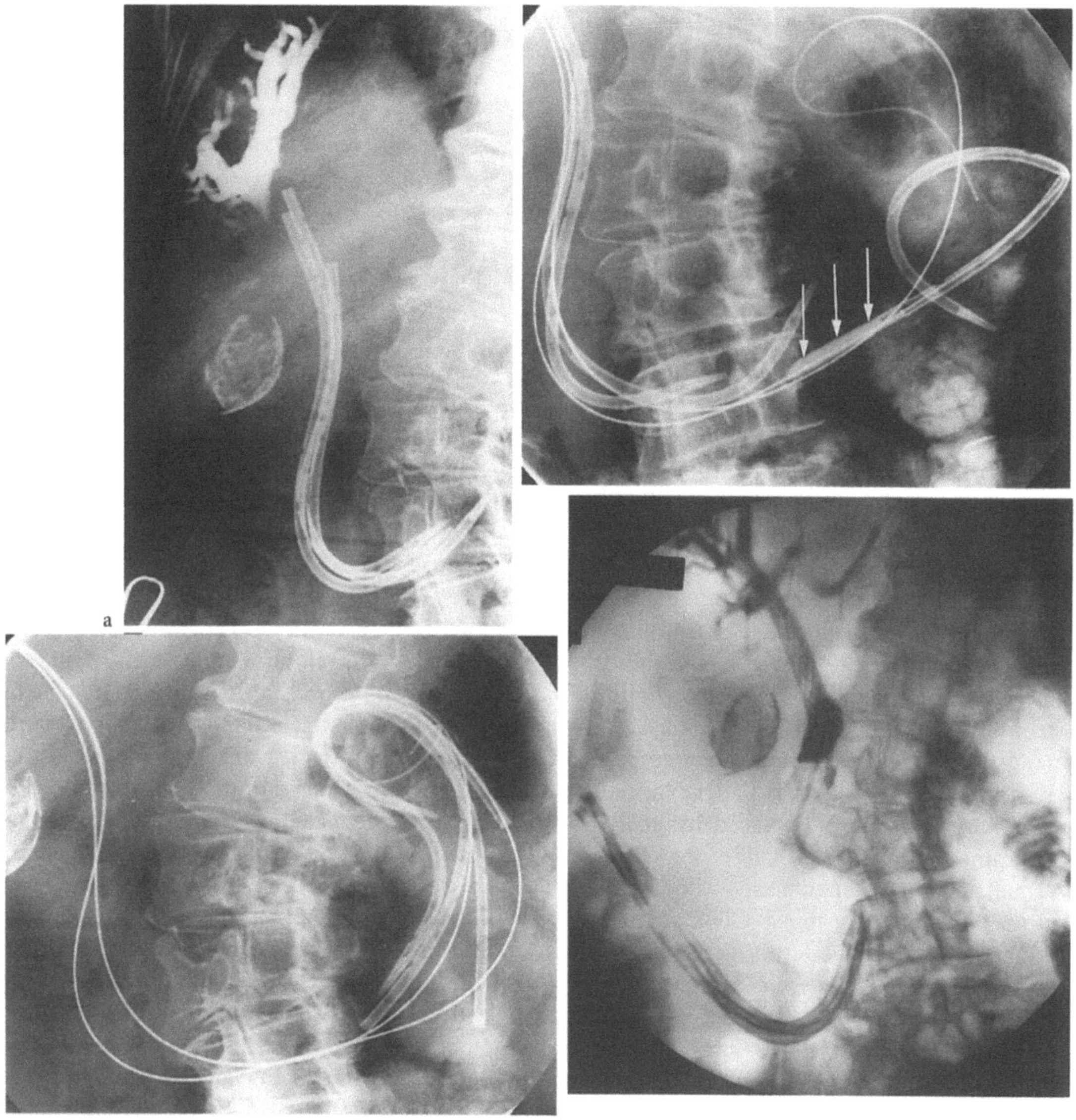

a b c d

Fig. 25.18. **a** PTC in a patient with gallbladder carcinoma causing obstructive jaundice. Note the calcification in the gallbladder wall. Three plastic endoprostheses, on three separate occasions, have been inserted percutaneously over a period of 12 months to relieve symptoms. Endoscopic assess was unsuccessful due to a previous gastrojejunostomy with a long afferent loop. The cholangiogram demonstrates filling of an obstructed right biliary system with obstruction of all three plastic endoprostheses. **b** One of the plastic endoprostheses has been cannulated with an extrastiff guidewire and an angioplasty balloon catheter introduced coaxially. The balloon has been inflated (*arrows*) and the endoprosthesis advanced into the duodenum. A second guidewire can be seen as a safety precaution to maintain access whilst cannulating the upper lumen of the occluded endoprosthesis. **c** The three plastic endoprostheses have been advanced using a similar technique and are seen to lie in the region of the gastrojejunostomy. These should transit the small and large bowel within 2–5 days with no symptoms. **d** A metallic endoprosthesis has been inserted and an 8F safety catheter left for 24 h. A cholangiogram demonstrates a decompressed biliary system. The plastic endoprostheses can be seen in the transverse colon

signs is that it is slightly more difficult to replace percutaneously.

The main problems of plastic endoprostheses are migration and occlusion. Tumour ingrowth does not occur and overgrowth is uncommon. Duodenal perforation is rare.

Migration

If migration occurs early, whilst the safety catheter is still in position, the stent can be cannulated with a guidewire and subsequently with a small, low-profile angioplasty balloon [Olbert 4.8F catheter carrying a 4-mm balloon (Meadox Inc.)]. After inflation of the balloon the endoprosthesis can be repositioned (Harries-Jones et al. 1982). If late migration occurs into the small bowel, the stent is left to pass through the bowel and a repeat procedure is performed.

Occlusion

Obstruction of plastic biliary endoprostheses occurs in 20%–30% of patients who live longer than 6–18 months (Teplick 1991). It is usually due to bacterial contamination causing decomposition of bile and subsequent encrustation. This results in stent malfunction and jaundice, often associated with cholangitis. The easiest technique for removing an occluded stent is via retrograde endoscopic access; a replacement endoprosthesis can be inserted at the same time. If endoscopic replacement is unsuccessful, percutaneous management of an occluded stent is possible. This involves coaxial recannulation prior to withdrawal, or advancement into the duodenum. The proximal end of the stent is cannulated with a 21-gauge Chiba needle under biplane fluoroscopic guidance. The stent is then washed out with contrast medium and saline and an 0.018-inch (0.45-mm) mandril wire advanced through the stent lumen. The 5F sheath, with its dilator, is then advanced and the inner wire and dilator replaced by an 0.035-inch (0.89-mm) Amplatz stiff guidewire (Fig. 25.18a). A small, low-profile angioplasty balloon [Olbert 4.8F catheter carrying a 4-mm balloon (Meadox Inc.)] is then introduced into the stent lumen and the balloon inflated. The stent can then be withdrawn through the transhepatic tract or pushed through into the duodenum to pass into the large bowel (Fig. 25.18b–d).

25.5.7.4 Results

A very large number of factors influence the outcome of patients who have had biliary endoprostheses inserted. Comparison of results from different series is difficult and complex due to the use of different stents, incomplete follow-up of patients in some series and differing tumour types and locations. Consequently, our practice is influenced by our own experience, and the techniques and preferences described in this chapter reflect this.

Plastic endoprostheses have been in use since the 1980s (Coons and Carey 1983; Dooley et al. 1984; Lammer and Neumayer 1986; Mueller et al. 1985). Introduction of these stents requires a percutaneous tract of 12F or greater, with a high incidence of patient discomfort and associated morbidity. Early complications (migration, haemobilia, intrahepatic abscess and portobiliary fistula) occur in 13%–51% of patients and the late complications of migration and stent occlusion in 3%–10% and 6%–25% respectively (Coons and Carey 1983; Dooley et al. 1984; Lammer and Neumayer 1986; Mueller et al. 1985).

The recent introduction of metallic endoprostheses (Adam et al. 1991; Gillams et al. 1990; Lameris et al. 1991; Lammer et al. 1990; Lee et al. 1992), which require a smaller percutaneous tract for introduction (7F), has resulted in less patient discomfort, has enabled the stent to be inserted at the time of the initial procedure, and has reduced the early and late complication rates to 5%–16% and 7%–14% respectively. The disadvantage of the high cost of the device itself is counterbalanced by a reduced reintervention rate, more “single-step” procedures and a shorter hospital stay (A. Adam, unpublished data, 1994; Davids et al. 1992; Knyrim et al. 1993). The reintervention rate is governed by stent failure due to occlusion or migration. Late occlusion of metallic stents is, in most instances, not due to bile encrustation (as it is in plastic endoprostheses) but to tumour progression. This can cause ingrowth through the metallic mesh or overgrowth, occluding the proximal or distal end of the stent. Tumour ingrowth has not been a major problem in recent series (0%–1.5%) (Adam et al. 1991; Lameris et al. 1991; Jackson et al. 1991) and may be eliminated completely by the development of plastic-covered metallic stents which are currently under trial. It remains to be seen whether these devices will have an increased migration rate and a higher frequency of occlusion by encrusted bile than uncovered stents. Tumour overgrowth is the main cause of reintervention in patients with metallic stents and occurs in 2.5%–12% of patients (Adam et al. 1991; Gillams et al. 1990; Jackson et al. 1991; Harries-Jones et al. 1982; Lee et al. 1992). This can be minimized by the use of long stents (A. Adam, unpublished data, 1995).

References

Adam A, Chetty N, Roddie M, et al (1991) Self-expanding stainless steel endoprostheses for treatment of malignant bile duct obstruction. AJR 156:321–325
Adam A, Gibson RN (1994) Interventional radiology of the hepatobiliary system and gastrointestinal tract. London, Edward Arnold
Boender J, Nix GA, Schulte HE, et al (1990) Malignant common bile duct obstruction: Factors influencing the success rate of endoscopic drainage. Endoscopy 23:259–262

Coons HG, Garey PH (1983) Large-bore, long biliary endoprostheses (stents) for improved drainage. Radiology 148:89–94
Couinaud C, Le Foie (1957) Etudes Anatomiques et Chirurgicales. Masson, Paris
Davids PHP, Groen AK, Rauws EAJ, Tygat GNJ, Huibregtse K (1992) Randomized trial of self-expanding metal stents versus polyethylene stents for distal malignant biliary obstruction. Lancet 340:1488–1492
Davidson BR, Varsamidakis Dooley JS T, Deery A, Dick RN (1992) Kurzawinski Hobbs KEF Value of exfoliative cytology for investigating bile duct strictures. Gut 33:10;1408–1411
Dooley JS, Dick R, George P, Kirk RM, Hobbs K, Sherlock S (1984) Percutaneous transhepatic endoprosthesis for bile duct obstruction. Gastroenterology 86:905–909
Gillams A, Dick R, Dooley JS, et al (1990) Self-expandable stainless steel braided endoprosthesis for biliary strictures. Radiology 174:137–140
Hall-Craggs MA, Lees WR (1985) Fine needle aspiration biopsy: pancreatic and biliary tumours. AJR 147:399–403
Harbin WP, Mueller PR, Ferrucci JT (1980) Transhepatic cholangiography - complications and use patterns of the fine needle technique. A multi-institution survey. Radiology 135:15–22
Harries-Jones EP, Fataar S, Tuft RJ (1982) Repositioning of biliary endoprosthesis with Gruntzig balloon catheters. AJR 138:771–772
Healey JE, Schroy PC (1953) Anatomy of the biliary ducts within the human liver. Am Med Assoc Arch Surg 66:599 616
Irving JD, Adam A, Dick R, et al (1989) Gianturco expandable metallic biliary stents; Results of a European clinical trial. Radiology 172:321–326
Jackson JE, Roddie ME, Chetty N, Benjamin IS, Adam A (1991) The management of occluded metallic self-expandable biliary endoprostheses. AJR 157:291–292
Keithley RB (1977) Micro-organisms in the bile. A preventable cause of sepsis after biliary surgery. Ann R Coll Surg Engl 59:328–334
Knyrim K, Wagner HJ, Pausch J, Vakil N (1993) A prospective randomised, controlled trial of metal stents for malignant obstruction of the common bile duct. Endoscopy 25(3):207–212
Lameris JS, Stoker J, Nijs HGT, et al (1991) Malignant biliary obstruction: percutaneous use of self-espandable stents. Radiology 179:703–707
Lammer J, Neumayer K (1986) Biliary drainage endoprostheses: Experience with 201 placements. Radiology 159:625–629
Lammer J, Klein GE, Kleinert R, et al (1990) Obstructive jaundice: use of expandable metal endoprosthesis for biliary drainage. Radiology 177:789–792
Lee MJ, Dawson SL, Mueller PR, et al (1992) Palliation of malignant bile duct obstruction with metallic biliary endoprostheses: technique, results, and complications. J Vasc Interv Radiol 3:665–671
McPherson GAD, Benjamin IS, Hodgson HJF, Bowley NB, Allison DJ (1984) Blumgart LH Pre-operative percutaneous transhepatic biliary drainage: results of a controlled trial. Br J Surg 71:371–375
Mueller P, Ferrucci JT Jr, Teplick S, vanSonnenberg E, Haskin PH, Butch SJ (1985) Papanicolaou N. Biliary stent endoprosthesis: analysis of complications in 113 patients. Radiology 156:637–639
Nakayama T, Ikeda A (1978) Okuda K Percutaneous transhepatic drainage of the biliary tract. Gastroenterology 74:544–559
Teplick SK (1991) Percutaneous transhepatic insertion of biliary endoprostheses. In: Current practice of interventional radiology. 557–562. Edited by Kadir S.B.C. Decker Inc., Philadelphia
Wagner HJ, Knyrim K (1993) Relief of malignant obstructive jaundice by endoscopic or percutaneous insertion of metal stents. Imaging 60(2):76–82
Zollikofer CL, Largiader I, Bruhlmann WF, Uhlschmid GK, Marty AH (1988) Endovascular stenting of veins and grafts: Preliminary clinical experience. Radiology 167:707–712

26 Percutaneous Hepatogastric Biliary Drainage

L. TIPALDI

CONTENTS

26.1 Introduction

The improvements in stent material and design, together with the progress achieved in implantation techniques (OTT et al. 1992; LAI et al. 1992; FERRUCCI et al. 1980), have made it possible over the past 5 years to achieve better results with percutaneous and endoscopic biliary drainage. Nevertheless, a large number of complications related to drainage system dysfunctions have prompted further investigation into the palliative nonsurgical treatment of malignant obstructive jaundice. Despite the low morbidity rates related to the procedure, clogging of plastic stents with sludge is a major complication in 21%–54% of cases (FORTNER et al. 1989; MUELLER et al. 1985; DAVIDS et al. 1992), while the use of an expandable metallic stent across the tumor site is limited in 10%–53% of cases by reocclusion due to tumor ingrowth or overgrowth and build-up of sludge (ADAM et al. 1991; COONS 1989, 1983; LAMMER et al. 1990; YOSHIORA et al. 1990).

Moreover, in some patients, it may prove impossible to pass the obstruction or to dilate the stricture, which often necessitates multiple attempts and/or combined percutaneous and endoscopic approaches (LAMMER et al. 1990; YOSHIORA et al. 1990; TIPALDI et al. 1993). The drainage procedure, therefore, may prove very painful for the patient, and it is not always possible to establish an internal biliary drainage which may ultimately result in long-term drainage (LAMMER et al. 1990; YOSHIORA et al. 1990; TIPALDI et al. 1993).

L. TIPALDI, MD, Servizio di Radiologia e Diagnostica per Immagini, Istituto Regina Elena, Viale Regina Elena 291, 00161 Rome, Italy

Percutaneous hepatogastric biliary drainage is a palliative treatment for malignant biliary obstruction consisting of a peripheral biliary diversion in which a bile duct of the left biliary tree is anastomosed to the lesser curvature of the stomach. This technique represents an alternative to long-term external drainage and could have applications in patients with a long life expectancy in whom, because it is placed extratumorally, it is less liable to reocclusion of the stent by tumor ingrowth.

26.2 Technique

Two different techniques are described in the literature for carrying out this procedure. With the first, it is carried out in the CT room and under fluoroscopy in the radiology intervention room, while with the second it is performed under combined fluoroscopic, endoscopic, and laparoscopic guidance in the radiology intervention room with the patient under general anesthesia.

The first technique (Fig. 26.1) was developed at our institution in 1992 and presented at the Giornate di Radiologia Diagnostica Oncologica (GRADO) meeting in Rome and at the 18th Meeting of the Cardiovascular and Interventional Radiology Society of Europe (CIRSE) in Budapest in 1993 (TIPALDI et al. 1993; TIPALDI 1995). In this experience the decision to perform the hepatogastric drainage was taken after several unsuccessful attempts to cross the obstruction: all patients treated underwent at least two attempts to negotiate the malignant biliary stricture using wires with hydrophilic coating via the right transhepatic approach. Before performing the hepatogastric drainage, a right-to-left external drainage is established with a 7-F catheter. The hepatogastric drainage procedure is then performed as a two-stage procedure. First, the air-filled stomach is punctured percutaneously under CT guidance through the left lobe of the liver with a 18-gauge needle (Figs. 26.1a, 26.2); the gastric wall is then brought into contact with the liver by inserting a

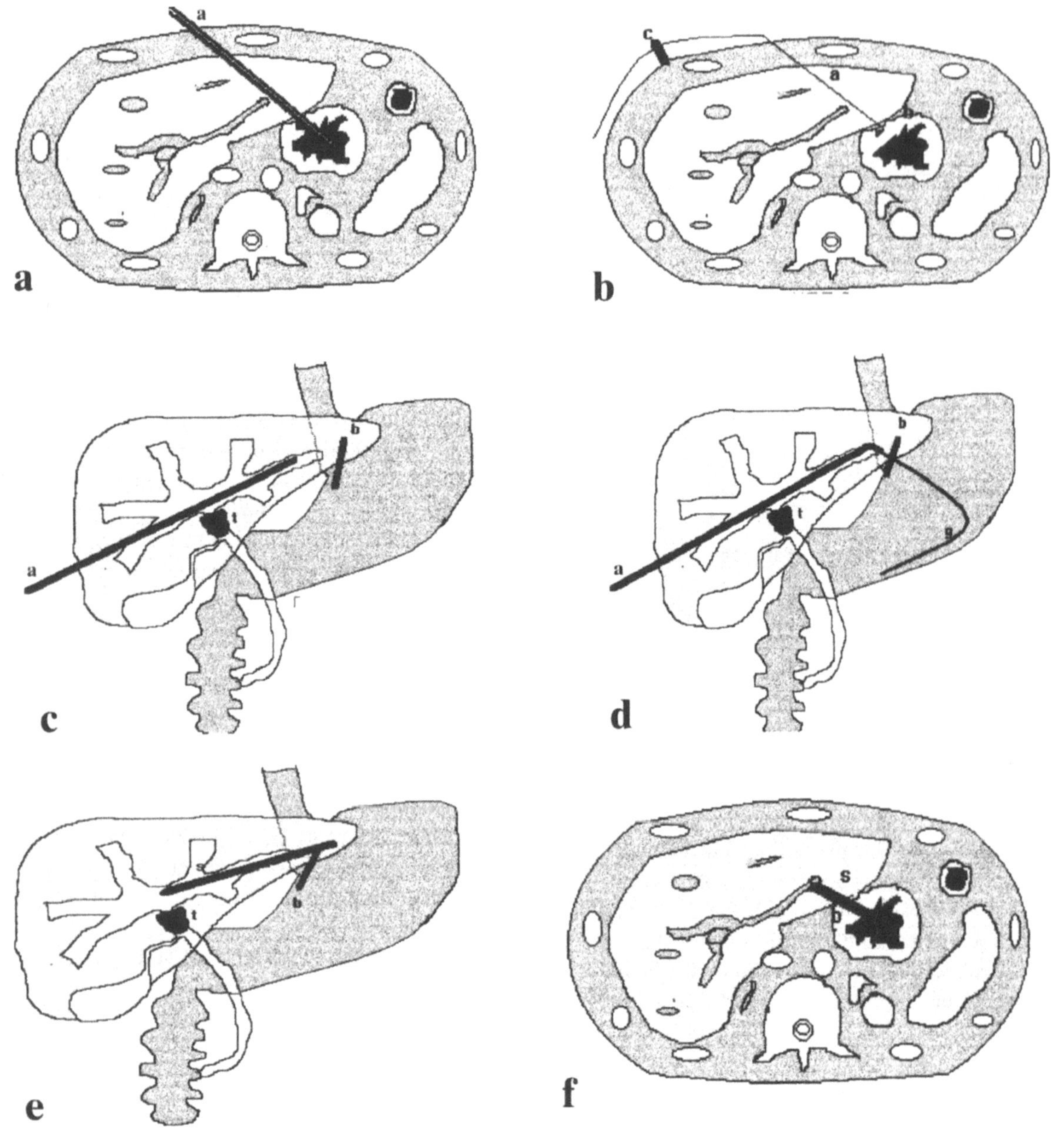

Fig. 26.1a–f. Technique of hepatogastric drainage. **a** An 18-gauge needle (*a*) is inserted into the stomach through the left lobe of the liver under CT guidance. **b** The gastric wall is brought in contact with the liver by inserting a Gias-Cope anchor system. *a*, Thread; *b* , anchor; *c*, suture. **c** A 9-F rigid sheath (*a*) is positioned from the right to the left biliary tree with the tip directed towards the anchor (*b*). *t*, tumor. **d** A needle is advanced into the stomach under fluoroscopic guidance, and then a guidewire (*g*) is advanced deep into the stomach. *a*, Sheath; *b*, anchor; *c*, wire; *t*, tumor. **e** Frontal view of the hepatogastrostomy. *s*, Metallic stent; *b*, anchor; *t*, tumor. **f** Axial view of the hepatogastrostomy. *s*, Metallic stent; *b*, anchor

Gias-Cope anchor system (Cook Inc., Bloomington, IN, USA). The anchor system is pulled towards the liver and the thread sutured 5–6 cm laterally to the puncture site (Figs. 26.1b, 26.3). After several days, the external drainage catheter is replaced by a 9-F rigid sheath. A Colapinto needle is then introduced into the sheath and oriented towards the metallic anchor in the stomach under fluoroscopic guidance. The stomach is filled with air and the relationship between the needle tip and anchor checked on frontal and lateral projections (Figs. 26.1c, 26.4). Once the position is correct, the needle is advanced through

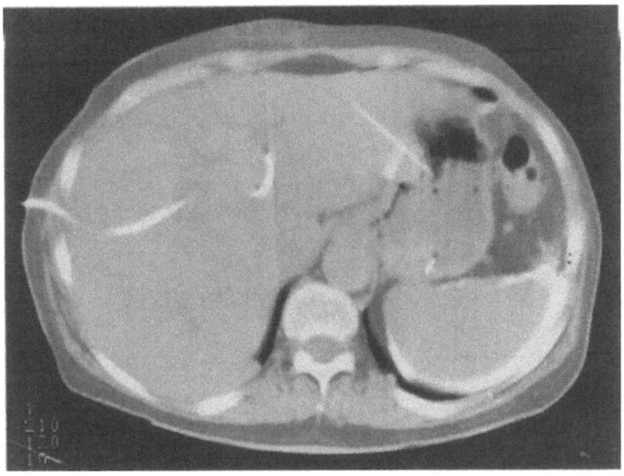

Fig. 26.2. CT scan shows the path of the needle and part of the wire inserted into the stomach. The biliary drainage catheter is also seen going from the right to the left biliary tree

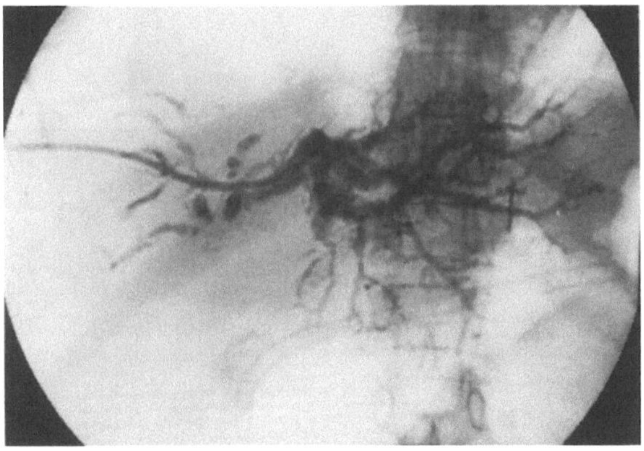

Fig. 26.3. Cholangiogram showing a high biliary obstruction and the anchor located on the gastric wall

the liver parenchyma into the stomach. An extrastiff guidewire (Figs. 26.1d, 26.5) is then inserted through the biliary tree into the stomach; over it an 8.3-F Ring catheter for internal external drainage is positioned and maintained for 3–4 days (Fig. 26.6). The Ring catheter is then replaced by a metallic stent (Wallstent or Strecker stent) in order to establish continuous internal biliary drainage into the stomach (Fig. 26.7). A 5-F catheter is maintained in the right biliary tree for check cholangiograms during the following 24 h.

The second technique was described by Soulez et al. in 1994 and has been mainly carried out in the operating room. Once a 6.5-F external biliary drainage catheter is inserted into a distal bile duct branch of liver segment II, the patient undergoes laparoscopic assessment and, if a palliative procedure is indicated, the radiologist perforates the left

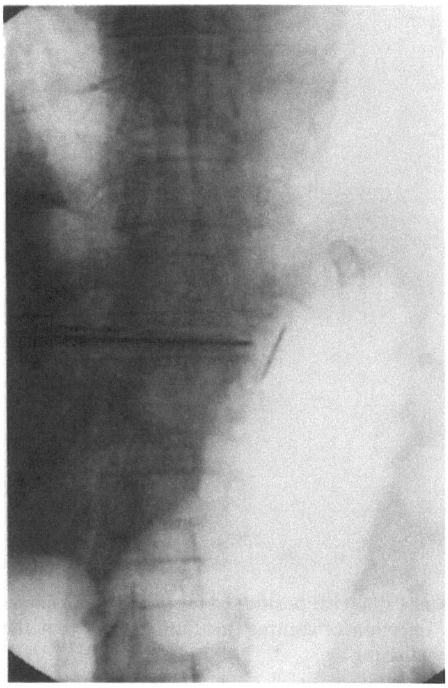

Fig. 26.4. Plain film showing the relation between the 9-F sheath with the needle inside and the anchor. Note the air-filled stomach

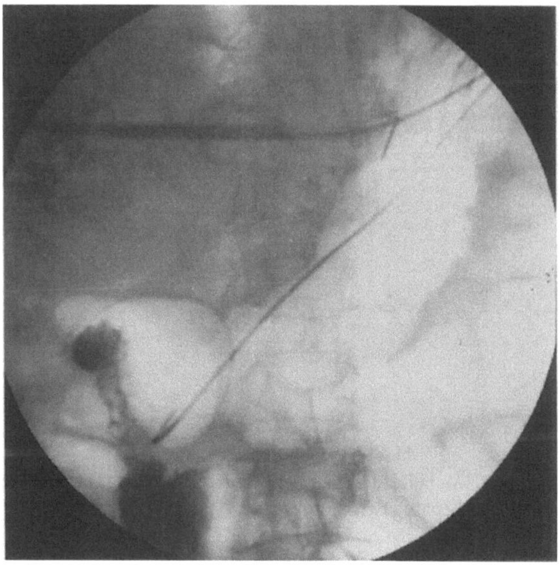

Fig. 26.5. After the puncture of the stomach a small angiographic catheter is advanced deep into the stomach

lobe of the liver with a needle under fluoroscopic guidance using the external drainage catheter pathway. The position of the needle tip is assessed under laparoscopic guidance: the surgeon holds the gastric pouch with a forceps while the radiologist perforates

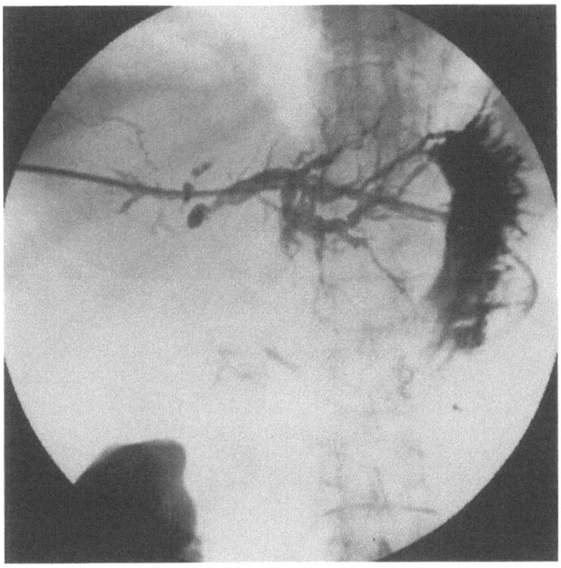

Fig. 26.6. A Ring catheter is positioned for internal and external drainage. Injection of contrast medium reveals that the drainage is functioning

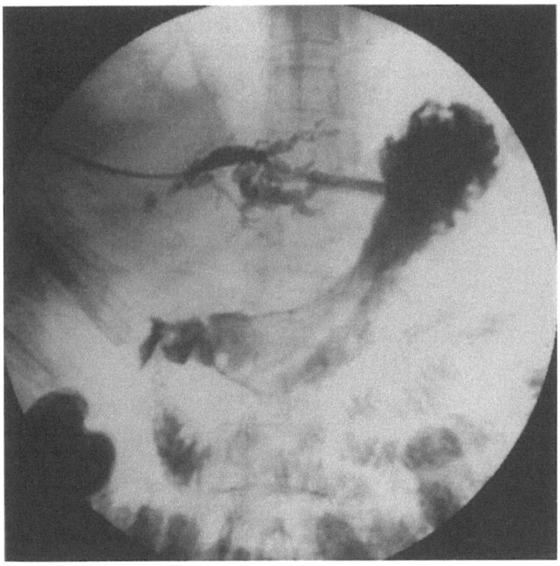

Fig. 26.7. Metallic Strecker stent replaces the Ring catheter for definitive hepatogastrostomy

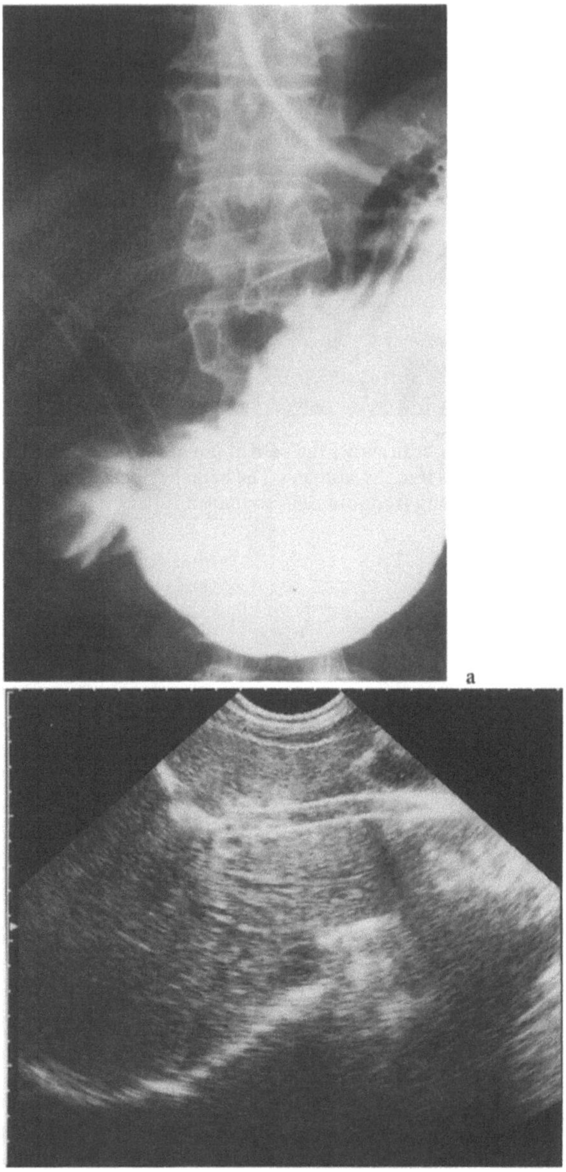

a

b

Fig. 26.8a,b. Follow-up studies. **a** Upper gastrointestinal imaging series of a patient 11 months after hepatogastric drainage placement. Two Wallstents are visible: one on the (patient's) right is draning the right biliary tree into the duodenum, while the left one drains the left biliary tree into the stomach. No reflux is seen. **b** Ultrasonogram. The transverse subxiphoid scan shows the metallic stent in place. No bile duct dilation is seen

the stomach with a 0.38-gauge needle. A snare is then introduced into the stomach and grasped by an endoscopist, who pulls it out through the patient's mouth and binds it to a 20-F percutaneous endoscopic gastrostomy tube. The radiologist then pulls the wire and the gastrostomy tube is drawn into the stomach across the transhepatic tract. The surfaces of both the stomach and the inferior face of the left hepatic lobe are coated with fibrin glue by the surgeon and adherence of the surfaces is ensured by maintaining slight traction on the tube, which is affixed to the skin. The catheter is left in place for internal-external drainage for 15 days and then retrieved by the endoscopist. Patency of the hepatogastric anastomosis is maintained by a Gianturco stent positioned by the radiologist through the transhepatic approach.

The same authors also describe a technique by which the hepatogastrostomy was performed using only fluoroscopic and endoscopic guidance, without laparoscopy, in one patient (SOULEZ et al. 1994).

26.3 Results and Discussion

Hepatogastric drainage was first described by the Italian surgeon DOGLIOTTI in 1952 in a textbook of surgery. It was a technically a difficult surgical procedure to carry out, especially in regard to locating the left hepatic duct, and for this reason the procedure was abandoned. Recently, RAMESH (1992) reported the case of patient treated surgically with intrahepatic cholangiogastrostomy; the drainage was excellent and no serious complications occurred.

Up to December 1994, 34 hepatogastric biliary drainages were reported without significant complications. The quality of drainage was always high and neither obstruction nor dislocation of stents occurred. Bile gastritis - a potential complication due to the presence of bile acid in the stomach - was not observed in any case.

The main advantage of hepatogastric drainage over the conventional percutaneous drainage procedure is the possibility of passing through normal hepatic parenchyma instead of tumor tissue. For this reason, the likelihood of neoplastic growth through the stent is small. From a technical point of view, too, the procedure seems simpler to perform, since it is not necessary to dilate the stenosis with a scalar or balloon catheter before positioning either the drainage catheter or the endoprosthesis.

Problems may arise from the difficulty of passing from the right to the left biliary tree in the presence of a neoplasm involving the hepatic hilum. This accounts for the fact that in some cases the procedure would not allow drainage of both hemisystems, thus limiting its range of possible applications. However, hepatogastrostomy can be utilized only in the left system in association with conventional transtumoral biliary drainage of the right system.

A further drawback of this technique is represented by the passage of bile into the stomach instead of the duodenum. Up to present time no clinical problems have occurred, but should patients with a long life expectancy be treated, appropriate treatment to prevent biliary gastritis is suggested.

Follow-up of a patient with a hepatogastric drainage is by ultrasonography to detect residual or recurrent bile duct dilation (Fig. 26.8). Barium studies usually fail to show any reflux of contrast medium from the stomach into the biliary tree. If dysfunction occurs, it is usually simpler to check a hepatogastric drainage than to check a transtumoral drainage protruding in the duodenum, since gastroscopy easily reaches the extremity of the stent in order to cannulate it for check cholangiography.

References

Adam A, Ghetty N, Roddie M, Yeung E, Benjamin IS (1991) Self expandable stainless steel endoprotheses for treatment of malignant bile duct obstruction. AJR 156:321–325

Coons HG (1989) Self expanding stainless steel biliary stents. Radiology 170:979–983

Davids PH, Groen AK, Rauws EA, Tytgat NJ, Muibregtse K (1992) Randomised trial of self expanding metal stents versus polyethylene stents for distant malignant biliary obstruction. Lancet 340:1488–1492

Ferrucct JT Jr, Mueller PR, Harbin WA (1980) Percutaneous transhepatic biliary drainage. Radiology 135:1–13

Fortner JG, Vitelli CE, Maclean BJ (1989) Proximal extrahepatic bile duct tumors. Arch Surg 124:1275–1279

Lai EC, Chu RM, Lo CY, Fan ST, Lo CM, Wong S (1992) Choice of palliation for malignant hilar biliary obstruction. Am J Surg 163:208–212

Lammer J, Klein GE, Kleinert R, Hausegger K, Einspieler R (1990) Obstructive jaundice: use of expandable metal endoprosthesis for biliary: preliminary clinical evaluation. Radiology 177:253–257

Mueller PR, Ferrucci JT Jr, Teplick SK, et al (1985) Biliary stent endoprothesis: analysis of complications in 113 patients. Radiology 156:637–639

Ott DJ, Gillian JH, Zagoria RS, Young GP (1992) Interventional endoscopy of the biliary and pancreatic ducts: current indications and methods. AJR 158:243–250

Ramesh H (1992) Intrahepatic cholangiogastrostomy for malignant biliary obstruction at the hilum. Br J Surg 79:1349–1350

Soulez G, Gagner M, Therasse E, et al (1994) Malignant biliary obstruction: preliminary results of palliative treatment with hepaticogastrostomy under fluoroscopic, endoscopic and laparoscopic guidance. Radiology 192:241–246

Tipaldi L (1995) A simplified percutaneous hepatogastric drainage. Technique for malignant biliary obstruction. Cardiovasc Intervent Radiol 18:333–336

Tipaldi L, Santoro E, Squillact S (1993) The hepatogastric drainage: a new percutaneous bilioenteric anastomosis (AB). Cardiovasc Intervent Radiol 16 (Suppl):950

Yoshiora K, Klein GE, Kleinert R, Hausegger K, Einspieler R (1990) Obstructive jaundice: use of expandable metal endoprosthesis for biliary drainage. Radiology 177:789–792

27 Management of Malignant Biliary Strictures

M. CREMER, J. DEVIERE, and J.-L. VAN LAETHEM

CONTENTS

27.1 Papillary Carcinoma

Diagnosis of obstructive jaundice has improved recently with magnetic resonance cholangiopancreatography (MRCP) to show the precise site of the obstruction (SOTO et al. 1996). In papillary carcinoma, both the common bile duct (CBD) and the pancreatic duct are enlarged down to their distal part (Fig. 27.1), except in patients presenting with a dominant dorsal duct and a functional minor papilla. In such patients, the differential diagnosis in those presenting with single impacted stone at the papilla remains difficult. Using secretin stimulation to provide further hypotonic duodenography, it would be possible to detect an enlarged papilla in those presenting with a tumor protruding into the duodenum. Endoscopy, however, remains the most reliable method to obtain the final diagnosis, via macrobiopsies performed before or after endoscopic sphincterotomy, especially for tumors located within the papilla. Because of the higher risk of bleeding with these tumors, significantly more coagulation current is necessary to avoid hemorrhage, while benign tumors and papillary stenosis do not bleed after sphincterotomy.

M. CREMER, MD, Professor, Medical Surgical Department of Gastroenterology and Hepatopancreatology, Hôpital Erasme, Université Libre de Bruxelles, 1070 Brussels Belgium
J. DEVIERE, MD PhD, Medical Surgical Department of Gastroenterology and Hepatopancreatology, Hôpital Erasme, Université Libre de Bruxelles, 1070 Brussels Belgium
J.-L. VAN LAETHEM, MD PhD, Medical Surgical Department of Gastroenterology and Hepatopancreatology, Hôpital Erasme, Université Libre de Bruxelles, 1070 Brussels Belgium

The "tumorotomy" has to be performed with an incision as large as the tumor in order to reach the noninfiltrated choledochoduodenal wall.

Stenting is only indicated for tumors with deep infiltration of the lower CBD, which are more likely to be pancreatic cancer or a lower cholangiocarcinoma.

After biliary sphincterotomy, a pancreatic septotomy has to be performed in order to obtain adequate drainage of the obstructed pancreatic duct.

Photodynamic treatment has been recently proposed as an alternative to tumor coagulation (ABULAFI et al. 1995). After sphincterotomy, resection of the tumor can be attempted using both a polypectomy snare and argon beam coagulation.

Both nasobiliary and nasopancreatic drainage catheters are inserted at the end of the procedure to avoid further obstruction due to blood clots. Once this "preoperative" drainage is in place, assessment of local and metastatic tumor spread is carried out during the next few days, while the jaundice is regressing. Curative duodenopancreatectomy, which gives the best prognosis for all biliopancreatic tumors, must be performed within the next 3 weeks. Candidates for surgery may be of any age but must be in good clinical condition. If surgery is contraindicated, the palliative management has to be followed by further endoscopic treatment when cholestasis recurs.

Prophylactic antibiotic treatment has (NIEDERA et al. 1994; BYL et al. 1995) been shown to be mandatory before endoscopic retrograde cholangiopancreatography (ERCP) in patients with obstructive jaundice.

27.2 Pancreatic Head Carcinoma

Pancreatic cancer is the most frequent cause of obstructive jaundice and is also detected easily by MRCP, showing the classical "double duct sign" (Fig. 27.2) while both CT and MRI demonstrate the tumor

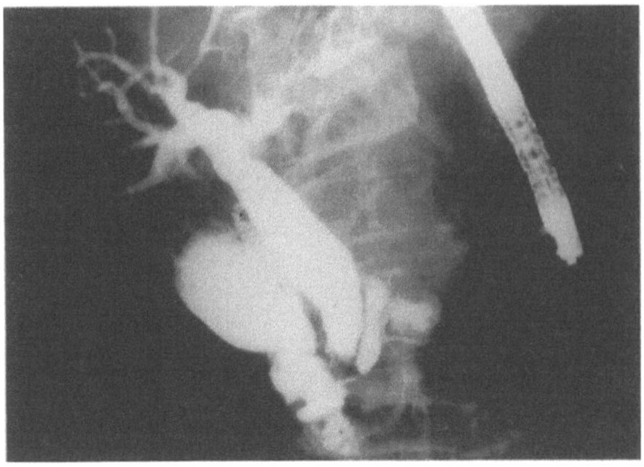

Fig. 27.1. Endoscopic retrograde cholangiopancreatography (ERCP) showing both the common bile duct (CBD) and the pancreatic duct to be enlarged down to their distal part in a patient with papillary carcinoma

in the head of the pancreas. Furthermore, MR angiography may in the same session provide good information about its resectability.

Unfortunately, duodenopancreatectomy is considered as curative in only 10% of patients with this tumor (Sorr and Cameron 1984). Bypass surgery has been shown to result in higher morbidity and mortality than endoscopic stenting (Sheperd et al. 1988), which is presently the technique of choice for unresectable tumors and for patients with metastases or who are inoperable (Sheperd et al. 1988; Smith et al. 1994; Huibregtse and Tytgat 1984).

Stenting the CBD was first described in 1980 using 7 French stents (Soehendra and Reynders-Frederix 1980). In our department, where we presently use only videoduodenoscopes with a 4.2-mm operative channel, plastic stents of 10 F, with an internal diameter of 2.4 to 2.6 mm, are the only plastic stents used for biliary drainage, either preoperatively or for definitive palliation. Stents sets are produced by several manufacturers (Oasis, Wilson-Cook; Salopass, Microvasive), accelerating placement by using a three-layer technique: the stent and the pusher tube are slid over a 6-F inner guiding catheter, which itself lies over a 0.035-inch Teflon-coated metal guidewire. Stents, whether straight or bent (the better to fit the angulated bile duct), have flaps with or without ("tannenbaum") side holes.

Sphincterotomy before stenting is usually performed after deep cannulation (using a cannulotome from Wilson-Cook, or a fluorotome from Microvasive) with a guidewire. Then brush cytology is performed for the tissue diagnosis. While conventional brush cytology has a limited sensitivity (around 60%) and good specificity (90 to 100%), evaluation of brushing specimens for ploidy - the presence of Ki-*ras* gene mutation on codon 12 - appears to increase the sensitivity (Lee et al. 1995; van Laethem et al. 1995). Dilatation of the stricture is then performed using Soehendra bougies (up to 11.5 F) or Gruntzig balloons. Finally, the stent is inserted and remains in a transpapillary position, allowing easy replacement (Fig. 27.3). There has been some recent debate about the necessity of sphincterotomy before stenting, the idea being to avoid its complications. However, it should at least be done in all cases where placement of multiple stents (biliary and pancreatic) or expandable metal

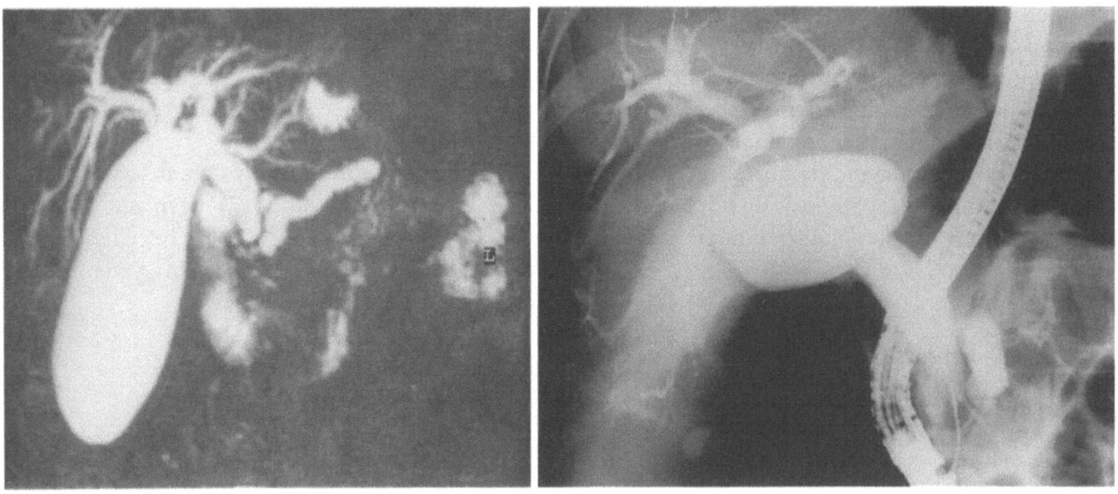

a b

Fig. 27.2a,b. Classical "double duct sign" as demonstrated by magnetic resonance cholangiopancreatography (MRCP; **a**) and ERCP (**b**) in a patient with an obstructive tumor of the head of the pancreas. (MRCP kindly provided by Dr. Celso Matos) a,b 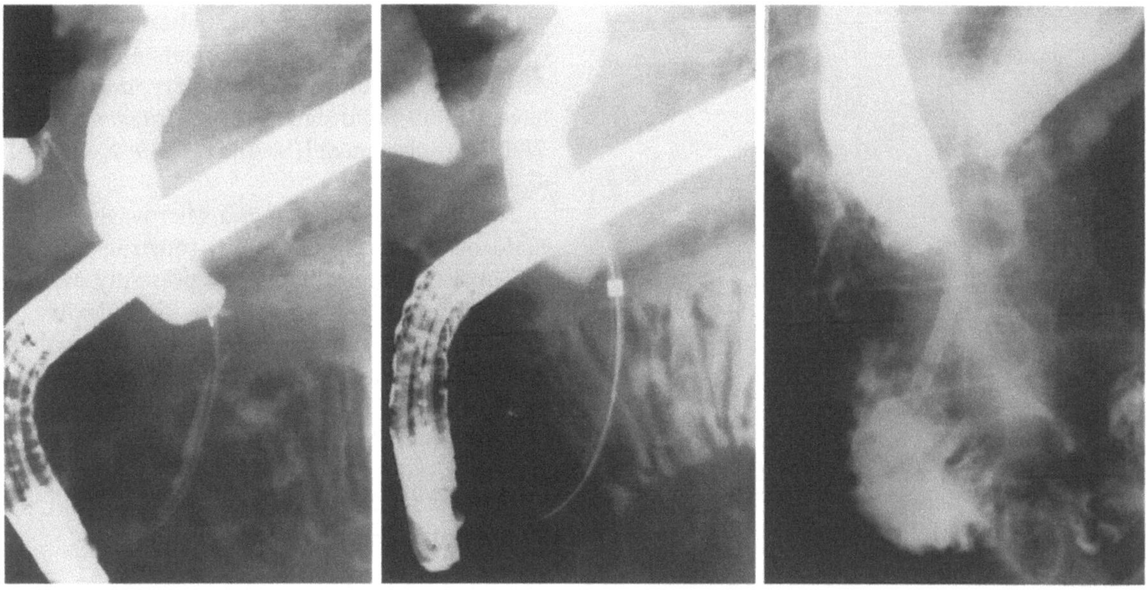 c

Fig. 27.3a–c. Endoscopic drainage of a malignant stricture of the CBD. **a** Deep cannulation of the CBD demonstrating a long stricture. **b** After sphincterotomy, dilatation of the stricture is achieved using biliary bougies. **c** The last step is the insertion of an anatomic bent stent which remains in a transpapillary position. Contrast flow rapidly reaches the second duodenum

Table 27.1. Results of prospective randomized studies comparing expandable metal stents and polyethylene stents for unresectable distal biliary malignancies

Authors	Type of stent	No. of patients	Early complications (%)	Late complications (%)	Stent patency (days)	Survival (days)
KNYRIM et al. 1993	W	31	3	22	189[a]	
	PE	31	3	43	140	
DAVIDS et al. 1992	W	49	12	33	273[b]	175[b]
	PE	56	11	54	126	147

W, Wallstent; PE, Polyethylene stent.
[a] Mean value.
[b] Median values.

stents is intended, to facilitate access to the bile duct or ensure that access to the pancreas remains possible (Fig. 27.4).

The success rate of implantation reaches as high as 95%. If passage of the stricture is unsuccessful, the “rendezvous” technique is used.

The usefulness of plastic stents is still limited by their clogging after 3–6 months; no improvement has been achieved by using 11.5-F instead of 10-F stents (SHERMAN et al. 1995).

Metal expandable biliary stents (Wallstents, Schneider) have been used since 1989, have been shown to occlude less frequently and less rapidly than standard plastic stents, and have been advocated for patients with a long life expectancy (O'BRIEN et al. 1995; GORDON et al. 1992). The Wallstent is made of braided steel mesh fixed to a catheter by a double membrane. It is easier to implant than a plastic stent because dilatation has to be achieved up to only 10 F. The 9-F instrument is passed directly over a guidewire. The insertion catheter is then pressurized with contrast medium to 75 mmHg to separate its two membranes, and the outer layer is progressively withdrawn to allow stent expansion. Full expansion to 10 mm (30 F) occurs spontaneously over the next few days without the need for further dilatation (Fig. 27.5).

Recanalization of Wallstents by electrocoagulation with a 10-F cystoenterostome has been suggested (Endoflex, Germany) (CREMER et al. 1992) (Fig. 27.6), but most authors place a plastic stent into the Wallstent for further biliary drainage (Fig. 27.7).

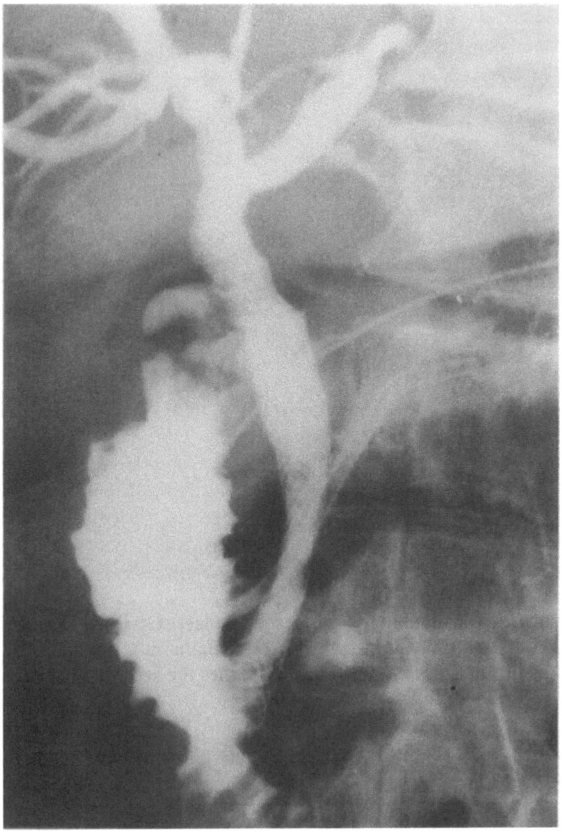

Fig. 27.4. Pancreatic head carcinoma: double stenting was performed after a double sphincterotomy performed to facilitate access to the bile duct (Wallstent metal expandable stent) and to the pancreatic duct (stenting the minor papilla using a Costamagna sigmoid plastic stent). In this case, pancreatic stenting was indicated in order to relieve pain syndrome due to overpressure above the head stricture

Two prospective randomized studies comparing the expandable metal stents (Wallstents and polyethylene stents for unresectable biliary malignancies) have demonstrated significant advantages related to Wallstents plaçement (DAVIDS et al. 1992; KNYRIM et al. 1993) (Table 27.1).

New metal stents (Diamond, Microvasive, Boston; Endocoil Instent, Minneapolis) constructed of nitinol (nickel-titanium alloy) are presently available. The Diamond stent, which is the more flexible, has less radial force and may delay tumor ingrowth and distal mesh impaction (Fig. 27.8). The Endocoil stent, the more difficult to insert, is exclusively indicated for strictures of the CBD (Fig. 27.9).

27.3 Cholangiocarcinoma

Cholangiocarcinoma of the lower two-thirds of the CBD is usually an indication for surgical resection, but preoperative endoscopic drainage using plastic stents may improve the outcome after surgery and is mandatory if diagnostic ERCP is performed (Fig. 27.10). On the other hand, cholangiocarcinomas arising at the confluence (Klatskin's tumors) are much more difficult to treat surgically, and the best results are obtained in patients in whom an enlarged right or a left hepatectomy can be performed (Bismuth types II and III). For Bismuth type III or IV, some authors propose liver transplantation for patients without any sign of tumor spread or lymph node involvement.

a,b

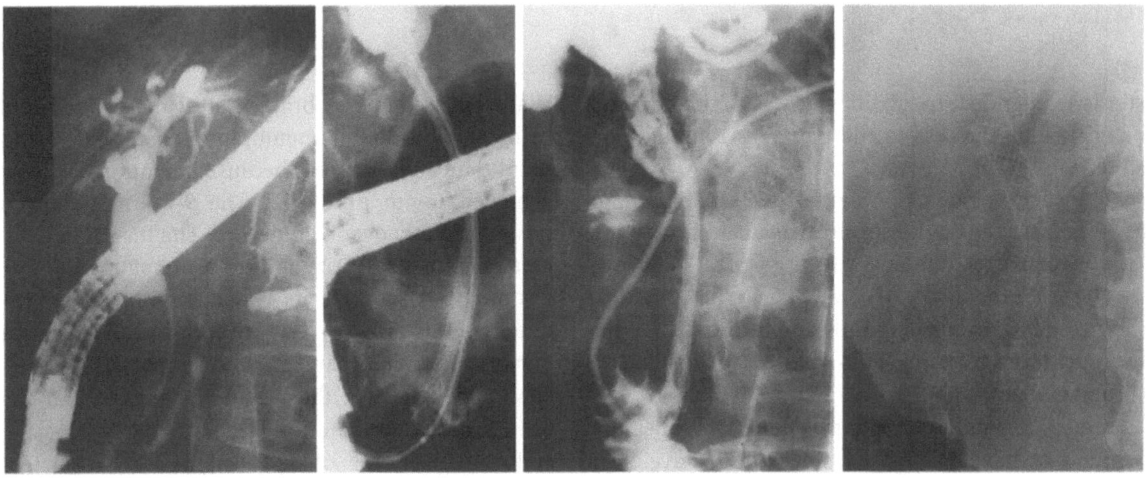

c,d

Fig. 27.5a–d. Pancreatic head carcinoma: endoscopic drainage using an expandable metal stent (Wallstent). a A guidewire is passed through the biliary stricture. b The 9-F instrument is passed over the guidewire and the stent is released. c By the end of the procedure, the stent is progressively expanding; d 24 h later, it has completely expanded to nearly 10 mm (plain film)

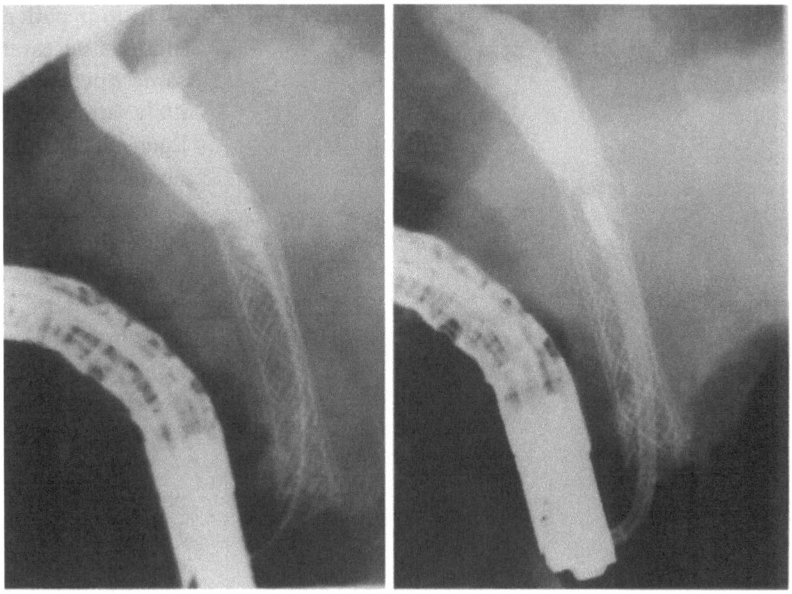

Fig. 27.6a,b. Pancreatic head carcinoma: recurrent tumor ingrowth into two Wallstents (a) treated by recanalization using electrocoagulation with a 10-F cystoenterostome (b)

a,b

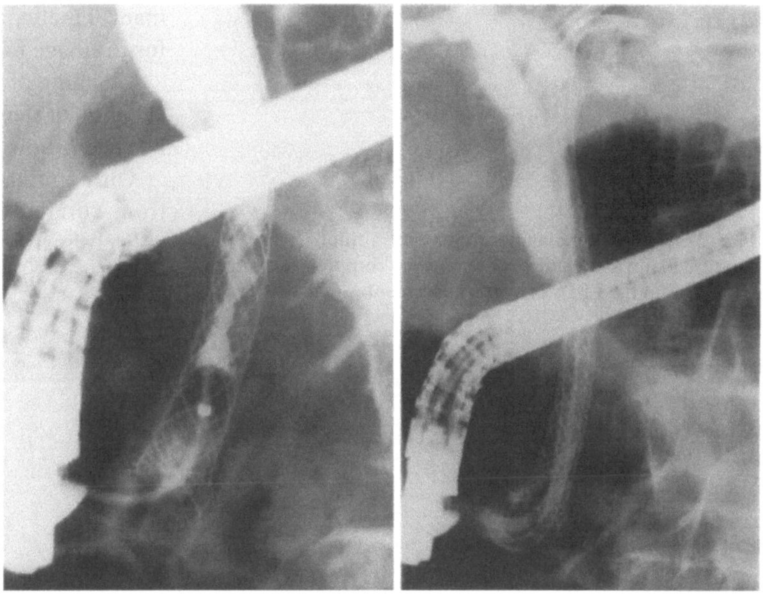

Fig. 27.7a,b. Pancreatic head carcinoma. a Tumor ingrowth into the metal stent as demonstrated by contrast injection through an inflated balloon catheter. b Biliary drainage using two plastic 10-F hilar stents inserted into the left hepatic duct

a,b

Preoperative drainage can be achieved endoscopically in 60% of the patients using two or three plastic hilar stents (Fig. 27.11), but the rendezvous technique using the percutaneous approach is necessary in the remaining 40% (MARTIN 1994). Although some authors maintain that one stent can be sufficient in Bismuth types II–IV to relieve jaundice, our experience shows that cholangitis of the nondrained segments most often has dramatic consequences (DEVIERE et al. 1988).

With good technique, it is possible to place two stents endoscopically, the major problem remaining the selective cannulation of the obstructed ducts. The rules to follow are: always cannulate first the duct that is most difficult to reach (most often the left hepatic duct), using soft hydrophilic guidewires (Terumo, Japan) and torque guidewires (Zebra, Microvasive). Preformed catheters, like those used for angiography, would also be useful. If both ducts are difficult, it is better to place one guidewire in each stricture before balloon dilatation of each side, and then to insert an 8.5-F stent on the left side followed by a 10-F stent on the right side.

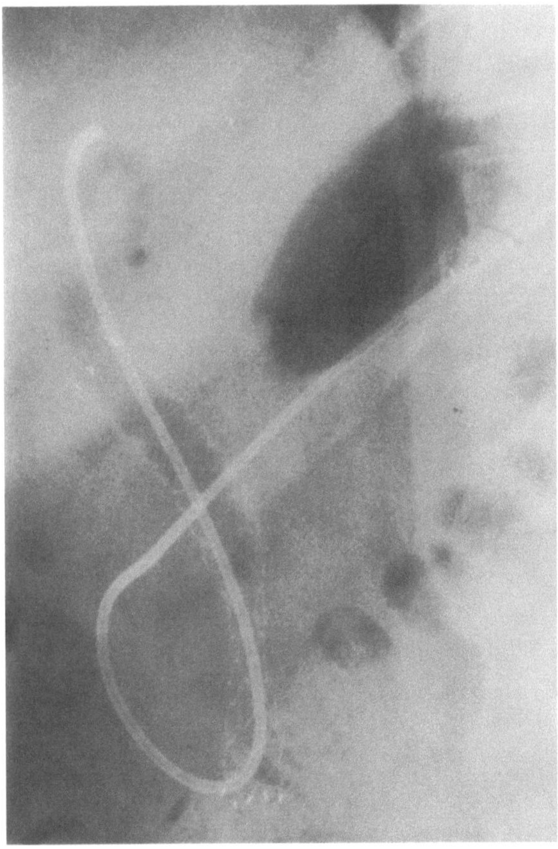

Fig. 27.8. Distal cholangiocarcinoma: nitinol stent (Diamond). Plain X-ray film shows nearly complete expansion with air into the CBD, 24 h after placement

In four patients out of ten, the rendezvous technique is necessary, and has to be performed by the same operators, on the same X-ray table, *immediately* after failure of endoscopic cannulation. A longwell coaxial needle catheter is advanced percutaneously to reach the ducts of segment VI, already opacified endoscopically. Then, a Terumo 145-cm guidewire is usually used, inserted through the stricture of the relevant segment, followed by the catheter. The wire is then exchanged for an Amplatz 180-cm (Cook) wire, which is pushed to the duodenum. An 8-F Ring internal-external catheter is then placed over this guidewire and kept in for drainage for the next 48 h (Fig. 27.12).

Stenting of the corresponding stricture is then performed, after inserting a 400-cm straight metal Teflon-coated 0.035-inch guidwire which is taken outside the scope using a snare loop. The stent is then threaded endoscopically, as in any classical stenting, and adjusted precisely in the relevant segment. Finally, a 7-F external catheter is maintained for drainage for the next 24 h.

Clogging of plastic stents occurs even faster than clogging of stents placed for more distal strictures, as has been show (Cheung and Lai 1995).

Only a minority of patients (7%–10%) can benefit from curative resection; self-expandable stents are entirely justified for patients in whom surgery will not be curative. In our own series, comparing pa-

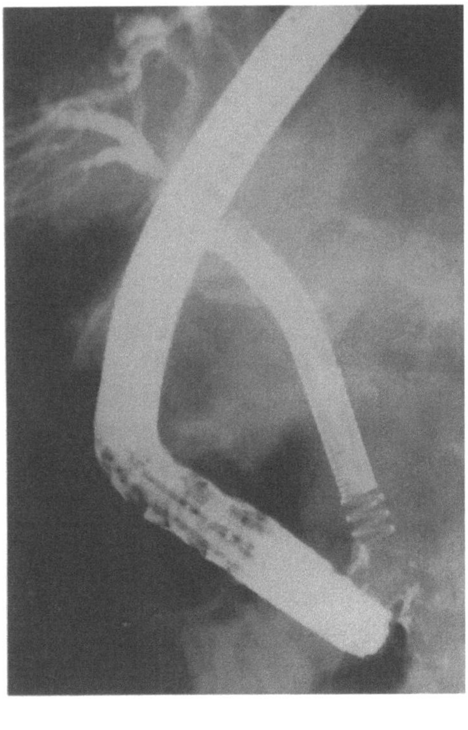

a

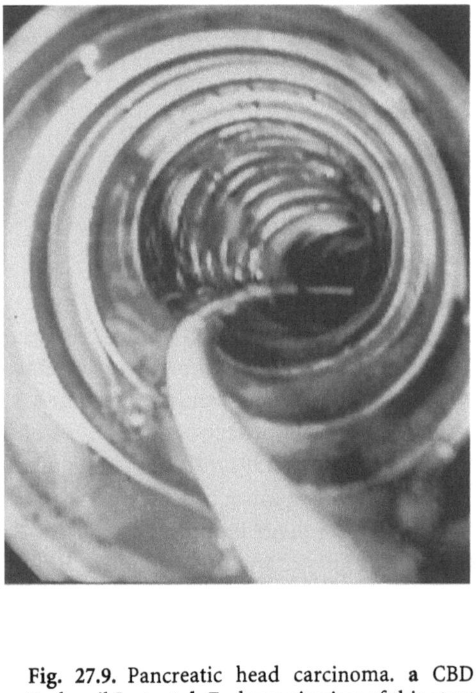

b

Fig. 27.9. Pancreatic head carcinoma. **a** CBD stricture treated an Endocoil Instent. **b** Endoscopic view of this stent just after insertion

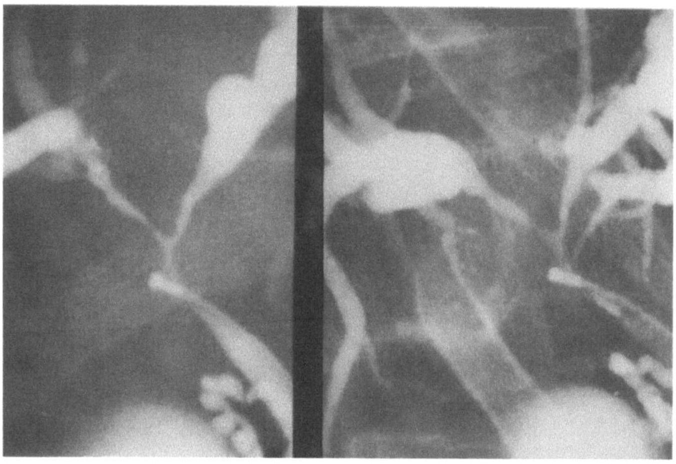

Fig. 27.10. Klatskin's tumor: radiographic definition of the Bismuth type III hilar cholangiocarcinoma using retrograde catheterization and successive radiologic angles of view, the patient being supine

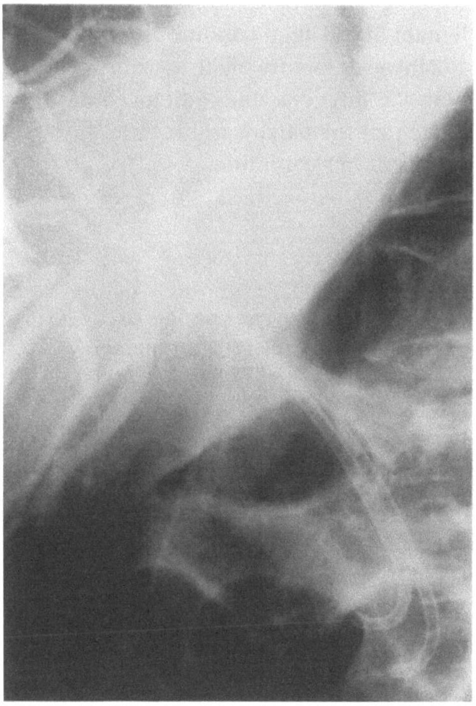

Fig. 27.11. Hilar cholangiocarcinoma treated by placing two plastic "anatomic" 10-F hilar stents

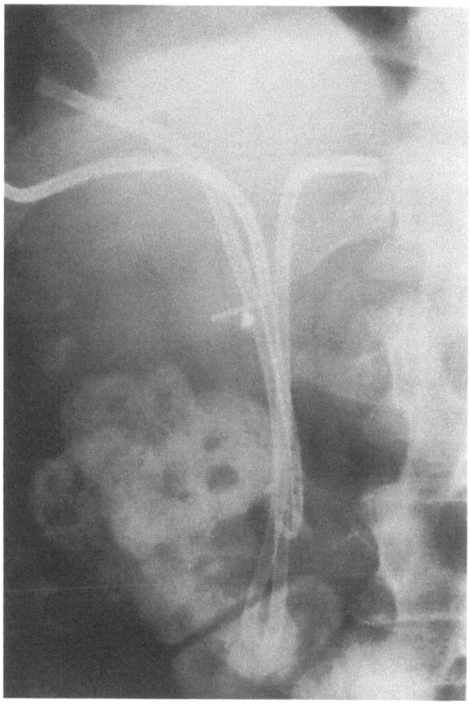

Fig. 27.12. Bismuth III type hilar stricture treated by placement of two plastic hilar stents and a combined percutaneous approach: an 8-F Ring internal-external catheter is placed into the ducts of segment VI, allowing the further insertion of a third stent using the rendezvous technique

tients presenting with Bismuth type II–IV Klatskin's tumors treated with either multiple plastic stents or Wallstents, we observed not only much lower morbidity and mortality but also longer mean patient survival (13 months against 6 months) (Cremer et al. 1995). Furthermore, the insertion of Wallstents being much easier than 10-F plastic stents, more than half of the patients were treated endoscopically using two Wallstents (Fig. 27.13). For three Wallstents, except in one patient, the rendezvous technique was been used, the percutaneous Wallstent being delivered at the same time as the other two Wallstents.

As an adjunct to palliative treatment, intraductal radiotherapy combined with external irradiation has been proposed in association with biliary stenting. In patients with unresectable cholangiocarcinoma,

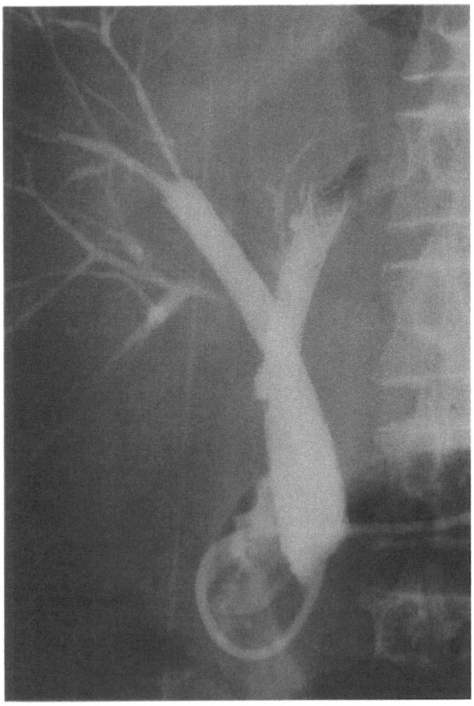

Fig. 27.13. Hilar cholangiocarcinoma drained by two expandable metal stents (Wallstents)

intraluminal irradiation with iridium-192 wires and external radiotherapy (30–45 Gy) were administered after jaundice had been relieved by stenting (plastic or metal (EDE et al. 1989; LEVIT et al. 1988). Retrospective studies have shown a possible benefit from this procedure, with increased median survival and symptom-free intervals. However, the possible role of this treatment awaits definitive results of randomized, controlled trials.

27.4 Gallbladder Carcinoma

Gallbladder cancer invading the CBD almost never allows curative surgery. Surgical bypass procedures are often difficult in these situations (CUBERTAFOND and GAINAND 1988). In our experience (median survival 5.3 months), the results of patients who receive stents are at least as good as those of patients who undergo surgical palliation (median survival 4.3 months), although surgical candidates were younger and in better clinical condition prior to treatment. Two stents are immediately used here also, even in patients with a Bismuth type I stricture, because most of these tumors rapidly enter the confluence.

27.5 Conclusions

The advent of MRCP has improved the quality of the diagnosis of biliopancreatic diseases and provides an anatomy of the biliopancreatic ducts before any invasive management is decided on. The development of new videoduodenoscopes and other new devices, and improved knowledge of the indications for drainage and its potential complications, have dramatically improved the possibilities of endoscopic therapy in the pancrease and biliary tract. The combined endoscopic and percutaneous transhepatic approach to the biliary tree is now well delineated and has improved the success rate of stenting in case of difficult and multiple strictures.

Further developments of stents is continuing, with the aim of better long-term success in both malignant and benign conditions.

However, before all else, teamwork (gastroenterologists, surgeons, and radiologists) is the best guarantee that the patient will receive the best treatment for his or her condition.

References

Abulafi AM, Alladice JT, Williams NS, et al (1995) Photodynamic therapy for malignant tumors of the ampulla of Vater. Gut 36:853–856

Byl B, Deviere J, Struelens M, Roucloux I, de Coninck A, Thys JP, Cremer M (1995) Antibiotic prophylaxis for infectious complications after therapeutic endoscopic retrograde cholangiopancreatography: a randomized, double-blind, placebo-controlled study. Clin Infect Dis 20:1236–1240

Cheung KL, Laie C (1995) Endoscopic stenting for malignant biliary obstruction. Arch Surg 130:204–207

Cremer M, Deviere J, Sugai B, Baize M (1992) Expandable biliary metal stents for malignancies: endoscopic insertion and diathermic cleaning for tumor ingrowth. Gastrointest Endosc 36:451–457

Cremer M, Hendlisz A, Sugai B, Deviere J (1995) Prospective nonrandomized stenting for cholangiocarcinoma Bismuth type II to IV. Personal communication, IVth United European Gastroenterology Week, Berlin, September 1995

Cubertafond P, Gainand A (1988) Les cancers des voies biliaires extrahépatiques. Monograph of the Association Française de Chirurgie, Paris, pp 44–57

Davids PHP, Green AK, Ramos EAJ, Tytgat GNJ, Huibregtse K (1992) Randomized trial of self-expanding metal stents versus polyethylene stents for distal malignant biliary obstruction. Lancet 340:1488–1492

Deviere J, Baize M, de Toeuf J, Cremer M (1988) Long-term follow-up of patients with hilar malignant stricture treated by endoscopic internal biliary drainage. Gastrointest Endosc 34:95–101

Ede RJ, Williams SJ, Hatfield ARW, McIntyre S, Mair G (1989) Endoscopic manaement of inoperable cholangiocarcinoma using iridium-192. Br J Surg 76:867–869

Gordon RL, Ring EJ, Laberge JM, Doherty MM (1992) Malignant biliary obstruction: treatment with expandable metallic stents: follow-up of 50 consecutive patients. Radiology 182:697–701
Huibregtse K, Tytgat GNJ (1984) Endocopic placement of biliary prosthesis. In: Salmon PR (ed) Gastrointestinal endoscopy: advances in diagnosis and therapy. Chapman and Hall, London, pp 219–223
Knyrim K, Wagner HJ, Pausch J, Vakil N (1993) A prospective, randomized, controlled trial of metal stents for malignant obstruction of the common bile duct. Endoscopy 25:207–212
Lee J, Leung JW, Cotton PB, et al (1995) Diagnostic utility of K-ras mutational analysis on bile obtained by endoscopic retrograde cholangiopancreatography. Gastrointest Endosc 42:317–320
Levit MD, Laurence BH, Cameron F, Kleung PFB (1988) Transpapillary iridium-192 wire in the treatment of malignant bile duct obstruction. Gut 29:149–152
Martin DF (1994) Combined percutaneous and endoscopic procedures for bile duct obstruction. Review. Gut 35:1011–1012
Niederau C, Pohlmann U, Lubke H, Thomas L (1994) Prophylactic antibiotic treatment in therapeutic or complicated diagnostic ERCP: results of a randomized controlled clinical study. Gastrointest Endosc 40:533–537
O'Brien S, Hatfield ARW, Craig PI, Williams SP (1995) A three year follow-up of self-expanding metal stents in the endoscopic palliation of long term survival with malignant biliary obstruction. Gut 36:618–621
Sheperd HA, Royle G, Ross AP, Diba A, Arthur M, Colin-Jones D (1988) Endoscopic biliary endoprosthesis in the palliation of malignant obstruction of the distal common bile duct: a randomized trial. Br J Surg 75:1166
Sherman S, Lehman G, Earle D, et al (1995) Multicenter randomized trial of 10 French versus 11.5 French plastic stents for malignant bile duct obstruction (abstract). Gastrointest Endosc 41:415A
Smith AC, Dowsett JF, Russel RC (1994) Randomized tiral of endoscopic stenting versus surgical bypass in malignant low bile duct obstruction. Lancet 344:1655–1660
Soehendra N, Reynders-Frederix V (1980) Palliative bile duct drainage: a new endoscopic method of introducing a transpapillary drain. Endoscopy 12:8–11
Sorr MG, Cameron JL (1984) Surgical palliation of unresectable carcinoma of the pancreas. World J Surg 8:906–918
Soto JA, Barish MA, Yucel EK, Siegenberg D, Ferrucci JT, Chuttani R (1996) Magnetic resonance cholangiography: comparison with endoscopic retrograde cholangiopancreatography. Gastroenterology 110:589–597
van Laethem JL, Vertongen P, Deviere J, van Rampelbergh J, Rickaert F, Cremer M, Robberecht P (1995) Detection of c-Ki-ras gene codon 12 mutations from pancreatic duct brushings in the diagnosis of pancreatic tumors. Gut 36:781–787

28 Comparison of Various Types of Metallic Biliary Stents

J. Lammer

CONTENTS

28.1 History of Metal Stents

Intraluminal metal stents were initially designed to prevent relapse of balloon-dilated arterious or venous obstructions. Dotter first introduced the concept of an endovascular "splint" in 1964 and subsequently, in 1969, he published a new technique of transluminally placed coilspring tube grafts in canine arteries. However, due to technical limitations this idea was not further developed until 1982, when Maass et al. started to experiment with a self-expanding "double helix" spiral prosthesis. Between 1983 and 1987, Dotter et al. (1983), Cragg et al. (1983), Gianturco (1985), Palmaz et al. (1985), Wallsten (1987), Strecker et al. (1987), and Rabkin et al. (1991) independently developed metal stents for endovascular use. However, Carrasco et al. were the first to use intraluminal metal stents as biliary endoprostheses, in an animal model in 1985.

28.2 Technical Characteristics of Metal Stents

Endoluminal metal stents can be divided into four types: balloon-expanded stents, self-expandable spring-loaded stents, shape-memory alloy stents, and covered stents.

28.2.1 Balloon-Expanded Stents

Palmaz Stent. This stent is a seamless tube of electropolished, medical-grade stainless steel with staggered parallel slots etched through the wall (Johnson & Johnson Interventional Systems, Warren, NJ, USA). The diameter of the tube wall is 0.15 mm. The stent is available in lengths of 10–40 mm. For placement, the stent has to be crimped on a Gruentzig or Olbert balloon (Fig. 28.1). It is then passively expanded to the desired diameter by inflation of the balloon. It can be expanded up to a diameter of 9–12 mm. At a pressure of 4.5 bar within the balloon, the stent is dilated up to 88% under in vitro conditions. Full expansion reduces the length of the stent by 17%. When fully expanded, the diamond-shaped open area between the metal struts increases to a maximum of 88% of the tube surface. This promotes rapid incorporation into the bile duct wall, which may be advantageous in benign strictures. However, in malignant obstructions, it has the disadvantage that rapid ingrowth of tumor tissue is likely to occur. The stent is rigid in the longitudinal axis and is highly resistant to circular load: for a 1-mm decrease in circumference, a resistance force of 4.0 N was measured. The stent is fairly radiopaque. It can be inserted through a 7-French introducer sheath when mounted on a 5-French balloon.

Tantalum Strecker Stent. This stent is made of electropolished tantalum wire (T205) measuring 0.13 mm in diameter (Boston Scientific International, Watertown, MA, USA). The wire mesh tube is knitted with six loops per circumference. The stent is mounted by the manufacturer on a 5-French Gruentzig-type balloon catheter (Fig. 28.2). The prosthesis is held by silicone sleeves at either end of the balloon to cover the ends of the tubular stent. During balloon expansion, the stent shortens by 19%

J. Lammer, MD, Universitätsklinik-Allgemeines Krankenhaus Wien, Währinger Gürtel 18-20, 1090 Vienna, Austria

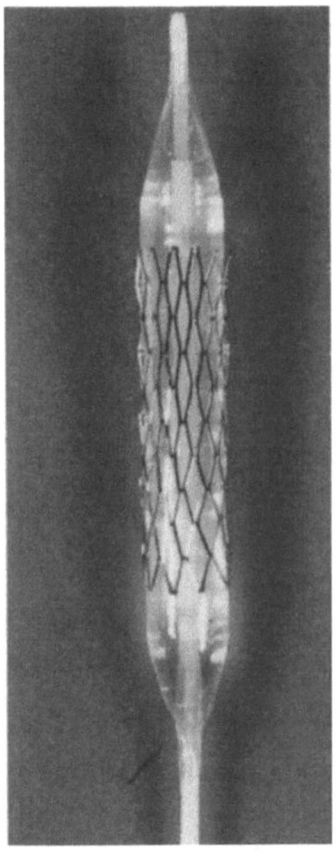

Fig. 28.1. Balloon-expandable Palmaz stent

and thus becomes released from the silicone sleeves. The shortening and relaxation after dilatation mean that a stent with a nominal length of 80 mm and a diameter of 8 mm measures 65 mm in length and 7.1 mm in diameter after expansion, in vitro. The open area between the wire loops is 81% of the stent surface. The stent is flexible, both longitudinally and radially. However, its resistance to circular load is low (0.25 N for 1 mm decrease in circumference). The biliary stent is highly radiopaque and can be inserted through an 8-French introducer sheath.

28.2.2 Self-Expandable Stents

Wallstent. This stent consists of monofilaments made of surgical steel (cold-formed cobalt-chromium-nickel-molybdenum-iron alloy) measuring 0.12–0.14 mm in diameter (Schneider/Europe AG, Bülach, Switzerland). The wire mesh is braided out of 24 wires in a criss-cross tubular pattern without fixed crossing points. The stent is mounted on a catheter and fixed by a rolling, retractable membrane (Fig. 28.3). The diameter of the instrument is 7 French. Biliary stents are available at restrained lengths of 100, 125 or 150 mm. After release, the stent expands to 8 or 10 mm diameter and shortens to 68, 85, or 102 mm, respectively, in length (32% shortening). The stent itself has low radiopacity. However, the addition of tungsten to some filaments improves its fluoroscopic visibility. The stent has longitudinal and radial elasticity. Once it is fixed in the common bile duct, the stent has good resistance to circular load (0.64 N for 1 mm decrease in circumference). The premounted stent can be inserted through a 7-French introducer sheath.

Gianturco-Rösch Z Stent. This stent is constructed from zigzag stainless steel wire measuring 0.1 mm in diameter (Cook Inc. Bloomington, IN, USA). The wires are soldered together with a lead-free solder to form a closed ring. A retained suture (4-0 nylon

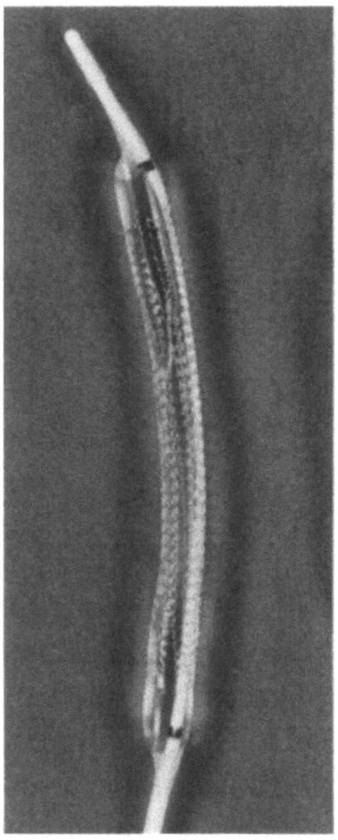

Fig. 28.2. Balloon-expandable Strecker stent

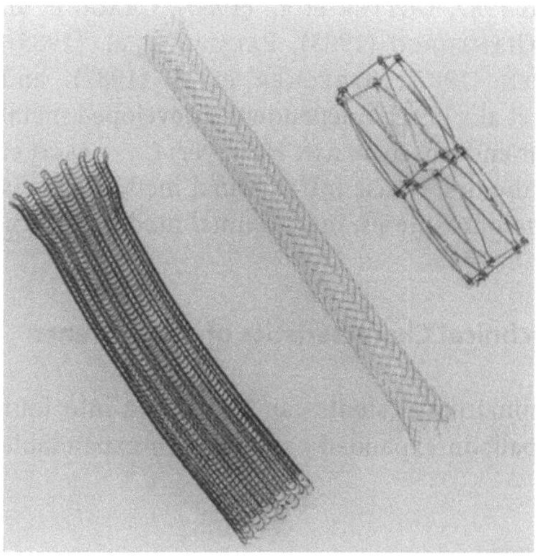

Fig. 28.3. Self-expandable stents. *From left to right:* Nitinol Strecker stent, Wallstent, Gianturco-Rösch Z stent

monofilament) through eyelets at each end controls the diameter of the stent. These stents are available as single-body to six-body stents with a retaining suture woven through the eyelets of each stent (Fig. 28.3). The stents expand to a diameter of 6–12 mm. Each body has a length of 1.5 cm, which shortens by less than 5% during expansion. The stent is rigid in the longitudinal axis, but elastic in the radial axis, with low resistance against circumferential stress; the in vitro resistance of the stent against circumferential load was only 0.07 N (1 mm constriction). The stent can be inserted through an 8.5-French sheath, 32 cm in length, and is delivered into the correct position by retracting the sheath while steadying the stent with a coaxial pusher measuring 34 cm in length.

28.2.3 Shape Memory Alloy Stents

Cragg Stent. The Cragg stent is made of a nitinol wire measuring 0.27 mm in diameter (Min Tec Inc., Grand Bermuda). Nitinol is an alloy of nickel and titanium, with thermal recovery properties. If nitinol wire is constrained to a desired shape and annealed at 500°C, it will memorize this configuration. When cooled in icewater, the wire becomes soft and can be formed without destroying its memory. If the stent is warmed to its transition temperature (30°–60°, depending on the alloy), it rapidly resumes its original configuration. For the Cragg stent, the nitinol wire is initially bent in a zigzag configuration. The zigzag wire is bent and tied together in a spiral to form a tube. In a compressed state, this tube has an outer diameter of 2.5 mm and a length of 62 mm. During expansion at body temperature, the stent dilates to 8 or 10 mm and shortens to 58 mm (7% shortening). The stent has longitudinal and radial elasticity. The resistance forces against 1 mm concentric compression were 0.3 N. The tantalum component gives the stent moderate radiopacity. The stent is preloaded in a plastic capsule. Directly from this capsule it can be inserted through an 8-French introducer sheath.

Nitinol Ultraflex Strecker Stent. This stent is also made of a nickel-titanium (nitinol) alloy wire, measuring 0.13 mm in diameter (Medi-Tech/Boston Scientific International, Watertown, MA, USA). The wire is knitted into a cylindrical, flexible mesh with looped ends (Fig. 28.3). When expanded, the stent is 40–80 mm long and 10 mm in diameter. The moderate radiopacity is mainly due to the titanium. The stent is flexible in its longitudinal axis. When it opens it shortens by 43%, which makes positioning difficult. However, it is fixed at the catheter tip by six prongs, such that it shortens from the proximal end to the distal end at the tip. The introducing system is 10 French in diameter and made of a shaft (on which the stent is mounted) and an outer sheath that holds the constrained stent in place.

28.2.4 Covered Stents

Covered stents are composed of a metal stent which serves as a frame and provides the stent's expansile force and elasticity. It is covered by a plastic membrane which is intended to prevent ingrowth of hyperplastic or tumor tissue.

Covered Wallstent (Prototype). The framework of this stent is a biliary Wallstent measuring 10 mm in diameter and 80 mm in length. The metallic mesh is covered by an elastic polyurethane membrane (Fig. 28.4). The length of the covering is 60 mm, leaving the proximal and distal ends (10 mm each) uncovered to provide stability. Because of the cover, the introducing instrument is 9 French in size.

Covered Gianturco-Rösch Stent. The covered Gianturco-Rösch stent is made of a quadruple body Rösch stent measuring 10 mm in diameter and 50 mm in length. The stent is fully covered by a sili-

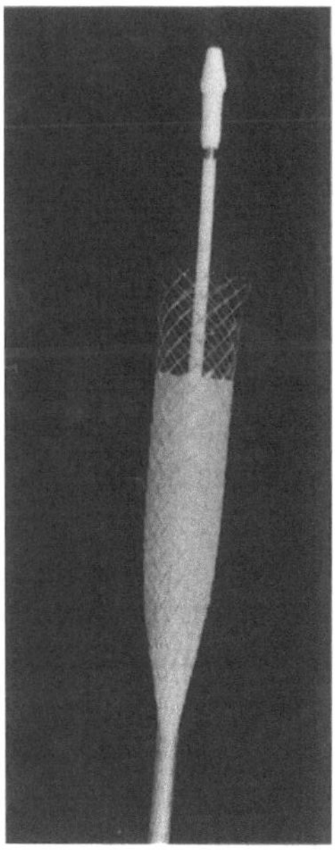

Fig. 28.4. Prototype covered Wallstent. The stent is covered by a polyurethane membrane to prevent tissue ingrowth

cone membrane. It can be inserted through a 10-French introducer sheath.

Cragg Endo Pro System 1. This stent is made of a Cragg nitinol zigzag stent measuring 10 mm in diameter and 60 mm in length. The outer surface of the metal framework is completely covered by a polyester tube. It can be inserted through a 10-French introducer catheter.

28.3 Biocompatibility of Metallic Biliary Stents

All metal stents used in the biliary system were designed and tested primarily for endovascular use. However, stents in the common bile duct are exposed to various problems. Initially, after placement of the stent the bile duct epithelium is partially destroyed by pressure-induced necrosis. Edematous swelling of the epithelium and subepithelial stroma causes luminal narrowing, which can be seen on control cholangiograms 48 h after stent placement. Bacterial contamination is common in bile duct foreign bodies such as stents. Bacteria may be already present from an infected bile duct system, may invade via reflux through a transpapillary stent, or may come through the enterohepatic circulation. Bacteria cause deposition of glycoproteins on the stent struts and deconjugation of bilirubin. This may lead to sludge formation. Edematous narrowing of the bile duct lumen, mucosal detritus, and sludge can obstruct the stent even a few days or weeks after implantation. Early reobstruction of Wallstents due to sludge or food impaction was observed in 3.5%–8% of cases by Boguth et al. (1994) and Rossi et al. (1994). Later, the stents are partially or completely incorporated into the submucosal tissue. Directly adjacent to the struts of the stent, Hausegger et al. (1992) observed fibrosis as well as a few multinucleated giant cells. These signs of foreign body reactions develop within the first year. Reactive mucosal hyperplasia has been found in animals but only rarely in humans.

28.4 Clinical Applications of the Various Stent Designs

28.4.1 Malignant Biliary Obstruction

Carcinoma of the pancreas or lymph node metastases in the hepatoduodenal ligament cause obstruction of the common bile duct by external compression in the majority of patients. In these patients, direct invasion of tumor tissue into the lumen of the metal stent is less probable. In patients with primary biliary tract neoplasms or hepatocellular carcinoma, tumor ingrowth is more likely. Evidently, in all stents, the area between the stent struts is large enough for tumor ingrowth. However, clinical results have demonstrated that stents with a dense wire mesh, such as the Wallstent and the nitinol Ultraflex Strecker stent, have lower reobstruction rates than do stents with large open areas such as the Gianturco-Rösch stent. The frequency of reobstruction with the Wallstent has been reported to range from 13% to 24% (Lammer 1990; Gordon et al. 1992; Boguth et al. 1994; Rossi et al. 1994). The mean duration of patency until reobstruction ranged from 171–196 days. For the nitinol Ultraflex Strecker stent, a reobstruction rate of 17% was reported after a mean period of 259 days (Bezzi et al. 1994). For the Gianturco-Rösch stent, Irving et al. (1989), Mathieson et al. (1994), and Rossi et al. (1994) reported obstruction rates ranging from 35% to 67% after an average period of 70–279 days. For the tantalum Strecker stent, reobstruction rates ranging from 14% to 100% have been reported after a mean follow-up period of 101–125 days (Jaschke et al. 1992; Rossi et al. 1994). A European multicenter study on metallic biliary stents showed significantly higher patency rates for the Wallstent and nitinol Ultraflex Strecker stents than for the Gianturco-Rösch Z stent or the tantalum Strecker stent ($P < 0.1$ and $P < 0.001$, respectively) (Rossi et al. 1994). An advantage of the Wallstent is, of course, the fact that it requires a smaller introducing system (7 French) than the nitinol Ultraflex Strecker stent (10 French). This makes primary implantation of the Wallstent feasible, a fact which reduces the 30-day mortality rate as well as the length of hospital stay.

28.4.2 Benign Biliary Stenosis

Inflammatory stenoses or postsurgical stenoses of bile duct or bilioenteric anastomoses may not respond to repeat surgery or balloon dilatation. Recurrence rates after surgery or balloon dilatation are reported to be as high as 15%–25%. In these cases, stenting of the stricture was shown to be the alternative treatment of choice.

Various stent designs were used for the purpose. Basically the ideal stent should be short, because benign strictures are usually very short. Good visibility is mandatory for accurate placement. The stent should resist high external pressure and should be incorporated into the bile duct wall within a short period of time without hyperplastic tissue reaction. Animal studies have been performed with the Gianturco stent, Palmaz stent, and Wallstent (Carrasco et al. 1985; Alvarado et al. 1989; Vorwerk et al. 1993). However, hyperplastic epithelial growth was observed with all designs of stent. Vorwerk et al. (1993) reported complete covering of stents after 9–12 months and a decrease of mucosal hyperplasia once the stent had been incorporated in the submucosal tissue layer. Howerer, clinical experience on this subject is limited. Maccioni et al. (1992) reported a 68.7% patency rate after three years (11/16 patients). Wallstents, Gianturco-Rösch Z stents, and tantalum Strecker stents were used, and hyperplastic mucosal reaction was proven by biopsy in all three types of stents.

28.5 Conclusion

A large variety of stent designs are currently available for biliary stenting. Depending on the technical characteristics of the stents and clinical experience, specific stent designs can be used for palliation of malignant obstructions and treatment of benign strictures.

References

Adam A, Chetty N, Roddie M. et al (1991) Self-expandable stainless steel endoprostheses for treatment of malignant bile duct obstruction. AJR Am J Roentgenol 156:321–325

Alvarado R, Palmaz J, Garzia O, Tio F, Rees C (1989) Evaluation of polymer-coated balloon-expandable stents in bile ducts. Radiology 170:975–978

Bezzi M, Salvatori FM, Maccinoni F, Ricci P, Rossi P (1991) Biliary metallic stents in benign strictures. Semin Intervent Radiol 8:321–330

Bezzi M, Orsi F, Salvatori FM, Maccioni F, Rossi P (1994) Self-expanding nitinol stent for the management of biliary obstruction: long-term clincal results. J Vasc Interv Radiol 5:287–293

Blumgart LH, Kelley CJ, Benjamin IS (1984) Benign bile duct stricture following cholecystectomy: critical factors in management. Br J Surg 71:836–843

Boguth L, Tatolovic S, Antonucci F, Heer M, Susler H, Zollikofer CL (1994) Malignant biliary obstruction: clinical and histopathologic correlation after treatment with self-expanding metal prostheses. Radiology 192:669–674

Carrasco H, Wallace S, Charnsangavej C, et al (1985) Expandable biliary endoprostheses: an experimental study. AJR Am J Roentgenol 145:1279–1281

Coene P, Groen A, Cheng J, et al (1990) Clogging of biliary endoprostheses: a new perspective. Gut 31:913–917

Coons H (1992) Metallic stents for the treatment of biliary obstruction: a report of 100 cases. Cardiovasc Intervent Radiol 15:367–374

Cragg A, Lund G, Rysavy J, et al (1983) Nonsurgical placement of arterial endoprostheses: a new technique using nitinol wire. Radiology 147:261–263

Cragg AH, Dake MD (1993) Percutaneous femoropopliteal graft placement. J Vasc Interv Radiol 4:455–464

Dawson SL, Lee MJ, Mueller PR (1991) Metal endoprostheses in malignant biliary obstruction. Semin Intervent Radiol 8:242–251

Dotter CT (1969) Transluminally placed coilspring endarterial tube grafts: long-term patency in canine popliteal artery. Invest Radiol 4:327–332

Dotter CT, Buschmann RW, McKinney MK, Rösch J (1983) Transluminally expandable nitinol coil stent grafting: preliminary report. Radiology 147:259–260

Flueckiger F, Sternthal H, Klein GE, et al (1994) Strength, elasticity and plasticity of expandable metal stents: in vitro studies with three types of stress. J Vasc Interv Radiol 5:745–750

Gordon RL, Ring EJ, La Berge JM, Doberty MM (1992) Malignant biliary obstruction: treatment with expandable metallic stents: follow-up of 50 consecutive patients. Radiology 182:697–701

Hausegger KA, Kleinert R, Lammer J, Klein GE, Flückiger F (1992) Malignant biliary obstruction: histologic findings after treatment with self-expandable stents. Radiology 185:461–464

Irving JD, Adam A, Dick R, Dondelinger RF, Lunderquist A, Roche A (1989) Gianturco expandable metallic biliary stents: results of a European clinical trial. Radiology 172:321–326

Jaschke W, Klose KJ, Strecker EP (1992) A new balloon-expandable tantalum stent (Strecker stent) for the biliary system: preliminary experience. Cardiovasc Intervent Radiol 15:356–359

Lammer J (1990) Biliary endoprostheses. Plastic versus metal stents. Radiol Clin North Am 28:1211–1222

Lammer J, Stoeffler G, Petek G, Hoefler H (1986) In vitro long-term perfusion of different materials for biliary endoprostheses. Invest Radiol 21:329–331

Lammer J, Klein GE, Kleinert R, Hausegger K, Einspieler R (1990) Obstructive jaundice: use of expandable metal endoprosthesis for biliary drainage. Radiology 177:789–792

Lammer J, Flueckiger F, Hausegger KA, Klein GE, Aschauer M (1991) Biliary expandable metal stents. Semin Intervent Radiol 8:233–241

Lossef SV, Lutz RJ, Mundorf J, Barth KH (1994) Comparison of mechanical deformation properties of metallic stents with use of stress-strain analysis. J Vasc Interv Radiol 5:341–350

Maass D, Kropf L, Egloff L, et al (1982) Transluminal implantation of intravascular "double helix" spiral prosthesis: technical and biological considerations. ESAO Proc 9:252–256

Maccioni F, Rossi M, Salvatori FM, Ricci P, Bezzi M, Rossi P (1992) Metallic stents in benign biliary strictures: 3 year follow up. Cardiovasc Intervent Radiol 15:360-366

Martin EC, Laffey KJ, Bixon R, Getrajohnan GI (1990) Gianturco-Rösch biliary stents: preliminary experience. J Vasc Interv Radiol 1:101–105

Mathieson JR, McLoughlin RF, Cooperberg PL, et al (1994) Malignant obstruction of the common bile duct: long-term results of Gianturco-Rösch metal stents used as initial treatment. Radiology 192:663–667

Mueller PR, Van Sonnenberg E, Ferrucci JT Jr, et al (1986) Biliary stricture dilatation: multicenter review of clinical management in 73 patients. Radiology 160:17–22

Palmaz JC, Sibbitt RR, Reuter STR, Tio FO, Rice WJ (1985) Expandable intraluminal graft: a preliminary study. Radiology 156:73–77

Rabkin JK, Natzvlishvili ZG, Kavteladze ZA (1991) Seven years' experience with Rabkin technology nitinol prostheses for vessels after balloon, laster and rotor recanalization. National Research Center of Surgery, Moscow

Rossi P, Bezzi M, Rossi M, et al (1994) Metallic stents in malignant biliary obstruction: results of a multicenter European study of 240 patients. J Vasc Interv Radiol 5:279–285

Rousseau H, Pucl J, Joffre F, et al (1987) Self-expanding endovascular prosthesis: an experimental study. Radiology 164:709–714

Strecker EP, Berg C, Weber H, et al (1987) Experimentelle Untersuchungen mit einer neuen perkutan einführbaren und aufdehnbaren Gefäßendoprothese. Fortschr Geb Rontgenstrahlen Neuen Bildgeb Verfahr 147:669–672

Vorwerk D, Kissinger G, Handt S, Günther RW (1993) Long-term patency of Wallstent endoprostheses in benign biliary obstructions: experimental results. J Vasc Interv Radiol 4:625–634

Wright KC, Wallace S, Charnsangavej C, Carrasco CH, Gianturco C (1985) Percutaneous endovascular stents: an experimental evaluation. Radiology 156:69–72

Zollikofer CL, Antonucci F, Stuckmann G, Mattias P, Salomonwitz EK (1992) Historical overview on the development and characteristics of stents and future outlooks. Cardiovasc Intervent Radiol 15:272–278

29 Metallic Stents in Malignant Biliary Obstruction: Autopsy Findings on Causes of Obstruction

C.L. ZOLLIKOFER

CONTENTS

29.1 Introduction

Endoscopic and/or transhepatic insertion of expandable metal endoprostheses is increasingly becoming the preferred method of palliative treatment for malignant biliary obstruction. Several stent designs are currently available for treatment of malignant jaundice. The most commonly used for transhepatic placement are the Wallstent (Schneider Europe AG, Bülach, Switzerland), the modified Gianturco-Rösch stent (Cook Inc., Bloomington, Indiana, USA) and the Strecker balloon-expandable tantalum or self-expanding nitinol stent (Meditec, Boston Scientific, Watertown, Massachusetts, USA) (LAMMER et al. 1990, 1991; IRVING et al. 1989; YOSHIOKA et al. 1990; GILLAMS et al. 1990; ADAM et al. 1991; LAMERIS et al. 1991; MARTIN et al. 1990; BEZZI et al. 1994; ROSSI et al. 1994; JASCHKE et al. 1992). The Wallstent may also be implanted via the retrograde endoscopic approach (BOGUTH et al. 1994).

Although metal endoprostheses have a considerably larger inner diameter than do conventional plastic stents, occlusions still occur in a significant percentage of treated patients. Rates of occlusion range from 7% to 28% for Wallstents (LAMMER et al. 1990; ADAM et al. 1991; ROSSI et al. 1994; BOGUTH et al. 1994; LEE et al. 1993; MATHIESON et al. 1994), and from 24% to 67% for Gianturco stents (IRVING et al. 1989; YOSHIOKA et al. 1990; MATHIESON et al. 1994). Strecker tantalum stents show occlusion rates of 19%–100% (ROSSI et al. 1994; JASCHKE et al. 1992) and Strecker nitinol stents of about 16% (ROSSI et al. 1994).

Potential causes of stent occlusion are sludge formation and bile incrustation, impaction of food particles in transpapillary stents, tumor ingrowth, tumor overgrowth, or a combination of these (BEZZI et al. 1994; ROSSI et al. 1994; JASCHKE et al. 1992; BOGUTH et al. 1994; MATHIESON et al. 1994; HAUSEGGER et al. 1992). Rarer causes are kinking of the bile duct due to stiffness of the metallic stent, bleeding after stent placement, formation of biliary duct stones (BOGUTH et al. 1994), and stent migration (MATHIESON et al. 1994). Mucosal hyperplasia has mainly been found in benign disease (VOWERK et al. 1993; MACCIONI et al. 1992).

Few data are available about the biological reaction of the bile duct wall adjacent to the stent and its influence on tumor progression (BOGUTH et al. 1994; HAUSEGGER et al. 1992). Evaluation of tissue reaction to stent placement in animal experiments is also limited to a few reports (VOWERK et al. 1993; ALVARADO et al. 1989; CARRASCO et al. 1985; unpublished data by the author). Since 1988, therefore, we have been collecting autopsy data on the macroscopic and histologic examination of individuals who had been treated with endoscopically or transhepatically placed Wallstents for malignant biliary obstruction at the Kantonsspital Winterthur.

29.2 Autopsy Data

29.2.1 Subjects

Of a total of 110 patients undergoing biliary drainages with Wallstents, 31 who died were available for autopsy and closer macroscopic and microscopic inspection. All patients had been treated for obstructive jaundice and 90% had intractable pruritus. There were 16 female and 15 male subjects, age 55–93

C.L. ZOLLIKOFER, MD, Department of Radiology, Kantonsspital, Brauerstrasse 15, 8401 Winterthur, Switzerland

Table 29.1. Macroscopic autopsy findings in metallic stents implanted in 31 patients for malignant biliary obsbruction (survival 5–575 days, mean 199 days)

Patent, no jaundice:	27 (4 after second intervention: for inspissated bile in 2, food particles in 1, and duodenal obstruction in 1)
Tumor overgrowth (jaundice):	2 (proximal obstruction, stent itself patent)
Acute occlusion:	1 (blood clots)
Obstruction by food:	1 (food particles proximal to stent, stent itself patent)
Tumor ingrowth, no jaundice:	2 (nonobstructing)
Significant sludge:	7

years (mean 73 years). The etiology of the obstruction was cholangiocarcinoma including carcinoma of the gallbladder in 16 cases, pancreatic carcinoma in 11, and metastatic lymph node compression in 4. In 23 patients the stents were inserted percutaneously and in 8 patients a retrograde endoscopic approach was used. Twelve stents were placed with the distal end extending 5–10 mm through the papilla into the duodenum. Four patients had hilar lesions, 2 patients stenoses of the common hepatic duct, and 25 patients hepatic duct and common bile duct stenoses.

29.2.2 Autopsies and Macroscopic Findings

The 31 autopsied patients survived from 4 days to 575 days (average 199 days) after stent placement. At the time of autopsy 27 of the 31 patients were found to have patent stented bile ducts (87%), including 4 patients which previously had undergone secondary interventions for intermittent recurrence of jaundice (Table 29.1).

Bile duct occlusion was found in four patients and was due to proximal tumor overgrowth over the stents in two, acute occlusion of the stent secondary to bleeding in one, and blockage of the bile ducts and proximal part of the stent by food particles in the last one. The secondary interventions in the four patients who had intermittent bouts of jaundice were performed by retrograde endoscopy in three cases and a percutaneous transhepatic approach in one. The retrograde endoscopic maneuvers consisted of clearing of the stent and proximal bile ducts of inspissated bile and debris in one case and clearing out food particles in another case. In a third patient two attacks of jaundice were treated with a second Wallstent and endoscopic placement of plastic stents. In all three cases retrograde cholangiography was suggestive of tumor invasion as the cause of occlusion, but on autopsy no macroscopic tumor invasion was found (Fig. 29.1). In a fourth patient the transhepatic cholangiogram had shown a patent stent; however, marked duodenal obstruction was found just distal to the papilla, which was successfully treated with a transhepatic Ringcatheter placed beyond the duodenal obstruction.

On macroscopic inspection partial stent obstruction due to tumor overgrowth at the proximal end of the stent was found in two patients. One patient had died because of acute bleeding (unintentional heparinization) 4 days after stent placement, with complete obstruction of the bile ducts due to blood clot which had caused cholangitis and sepsis. Finally, another patient who had died from cholangitis was found to have obstruction at the proximal end of the stent caused by impaction of food particles and debris (Fig. 29.2). Again, the retrograde cholangiogram had shown stent obstruction of a nonspecific appearance. The stent itself had remained patent. Substantial amounts of sludge (without obstruction) were seen in seven cases (Table 29.1).

Macroscopically the stents were found to be totally covered by tissue lining in ten patients (32%), the shortest period since stent insertion being 118 days and the longest 575 days (mean 332 days). Macroscopic ingrowth of the tumor without having caused significant obstruction or jaundice was found in two of these cases. Partial covering was seen in 11 patients (36%) in whom the time since stenting ranged from 53 to 346 days (mean 221 days). Ten patients (32%) in whom the time since stenting ranged from 4 days to 129 days (mean 43 days) did not show any tissue covering of the stent filaments. The relation between survival time after stent placement and macroscopic tissue reaction is shown in Table 29.2.

29.2.3 Microscopic Findings

In 24 of the 31 autopsy cases specimens of the bile ducts and stented areas were available for histologic evaluation (Table 29.3).

The main findings in the early period (less than 3 months) after stent implantation were destruction of

Fig. 29.1A–D. A 63-year-old patient with carcinoma of the gallbladder treated with bilateral stents. There were two episodes of stent obstruction, 5 and 10 months after stent placement. The patient was treated with one additional Wallstent in the right bile duct and further plastic stents bilaterally. She died without jaundice 13.5 months after stent placement. **A** Cholangiogram 5 months after stent placement shows complete obstruction of both stents, compatible with the presence of either tumor obstruction or sludge. The patient was treated with a Wallstent on the right and a double pigtail endoscopically on the left. **B** Macroscopic specimen shows patent stents with some debris and only partial tissue covering. No tumor ingrowth is seen. Note large tumor (*T*) infiltrating the liver and multiple liver metastases. **C,D** Microscopic section shows the holes of the removed stent filaments to be surrounded by circularly arranged connective tissue. Note clusters of tumor cells peripheral to the connective tissue (*blue*). A clear boundary between tumor and connective tissue is seen on higher magnification (**D**)

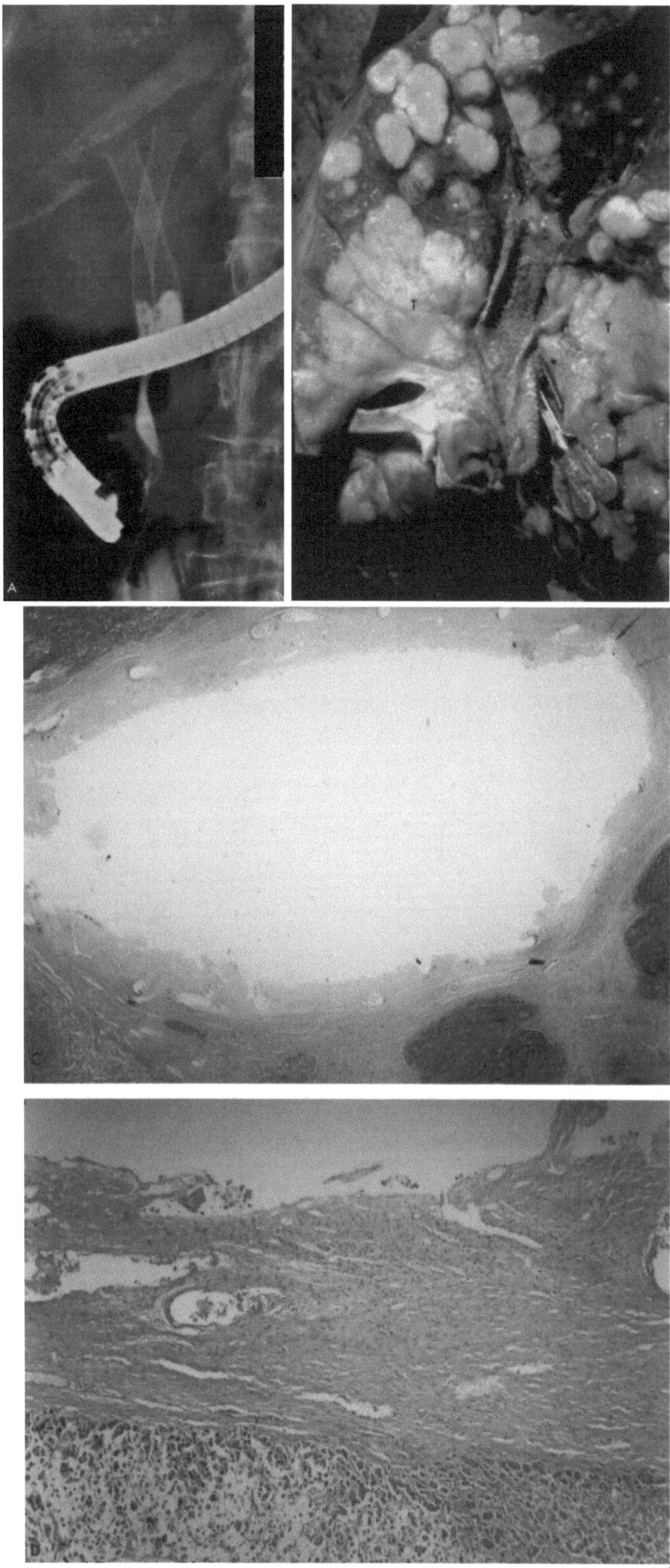
B
C
D

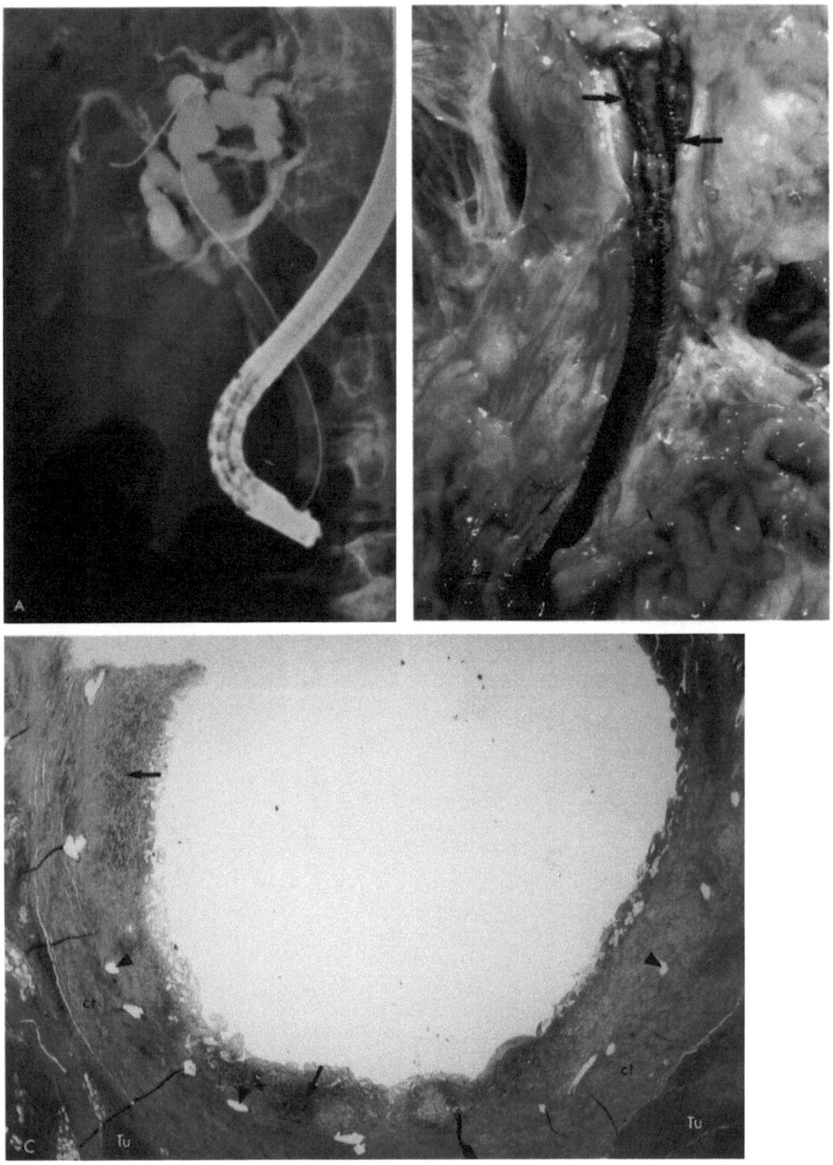

B

Fig. 29.2A–C. A 76-year-old patient with recurrent jaundice 18 months after implantation of a Wallstent for a well-differentiated cholangiocarcinoma of the common hepatic duct. **A** Endoscopic retrograde cholangiography shows nonspecific obstruction of the Wallstent placed 18 months previously. The cholangiogram does not permit a specific diagnosis of the nature of the obstruction. **B** Macroscopic specimen shows the transected Wallstent to be patent and completely covered by a thin layer of tissue. Note particles of food (*arrows*) reaching into the stent from the proximal, nonstented part of the hepatic duct, causing the obstruction. **C** Microscopic section of the common bile duct after removal of the stent filaments. Towards the lumen the stent is covered by a layer of tissue with areas of inflammation (*arrows*). There is no normal mucosa. Note that tumor cells (*Tu*) are only seen outside, peripheral to the circumferential ring of connective tissue (*ct*) outside the stent filaments (*arrowheads*)

the normal epithelium and a nonspecific inflammatory reaction with signs of necrosis. This reaction, particularly superficial necrosis, was seen as much as 1 year after stent placement. Within 3 months, five of the six stents did not show any covering of the stent filaments at all; in only one patient was partial covering noted, as there was tumor invasion through the stent filaments from an undifferentiated cholangiocarcinoma of the common bile duct (Fig. 29.3).

By 6 months, four of the six patients showed partial (2) or total (2) stent covering, in two with tumor invasion of the stent filaments (Table 29.3, Figs. 29.4, 29.5).

Table 29.2. Macroscopic autopsy findings of tissue covering of metallic stents implanted in 31 patients

Patient survival after implantation	No covering	Partial covering	Complete covering	Total
Up to 90 days	8 (73%)	3	0	11
Up to 180 days	2 (29%)	3	2	7
Up to 360 days	0	4	4	8
>360 days	0	1	4	5
Total	10 (32%)	11 (36%)	10 (32%)	31

Table 29.3. Histologic autopsy findings of tissue covering (connective tissue with or without tumor cells)[a] in metallic stents implanted in 24 patients

Patient survival after implantation	No covering	Partial covering	Complete covering	Total
Up to 90 days	5	0/1	–	6
Up to 180 days	2	2/0	0/2	6
Up to 360 days	–	1/2	3/1	7
>360 days	–	0/1	4/0	5
Total	7 (29%)	3/4	7/3	24

[a] 7/17 (41%) stents covered with tissue lining showed tumor invasion.

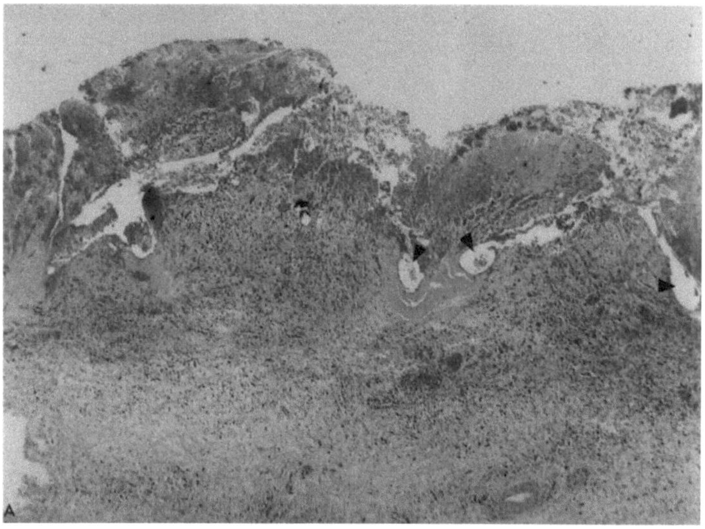

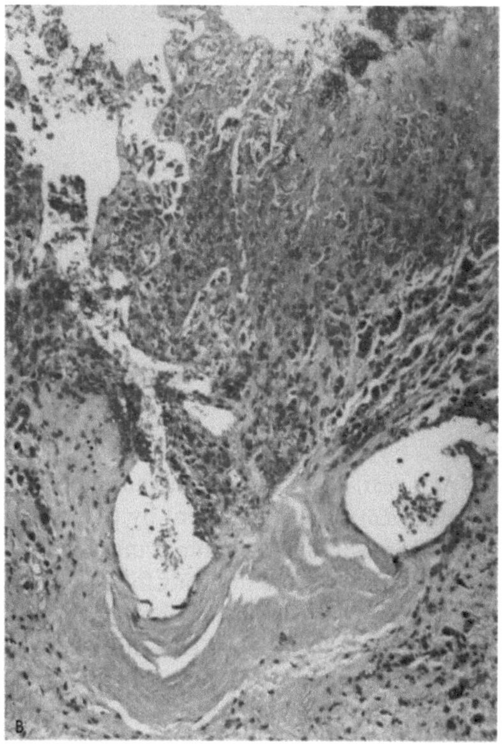

Fig. 29.3A,B. An 80-year-old patient with undifferentiated cholangiocarcinoma of the middle part of the common bile duct, 57 days after stent implantation. Microscopic sections (H&E). **A** There is tumor invasion inside the holes of the removed stent filaments. Note irregular and necrotic luminal surface. Connective tissue reaction is limited to around the stent holes (*arrowheads*). **B** At higher magnification the tumor invasion is well shown with sloughing of superficial cellular elements

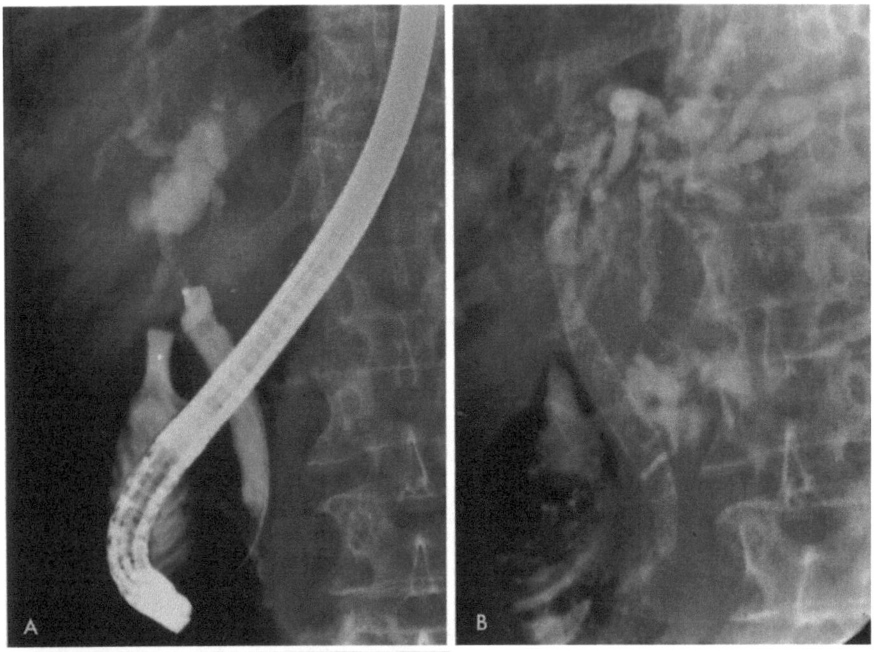

Fig. 29.4A–E. An 82-year-old patient with carcinoma of the gallbladder involving the proximal bile ducts. The patient died 193 days after endoscopic implantation of a 10-mm Wallstent. **A,B** Retrograde cholangiogram (**A**) shows tight obstruction of the common hepatic duct close to the bifurcation, treated with a 10-mm Wallstent (**B**). **C** Macroscopic specimen shows the proximal part of the stent surrounded by tumor to be covered by tissue, while the distal part of the stent in the common bile duct (*CBD*) is free. **D,E** Microscopic sections of the covered (**D**) and bare (**E**) parts of the stent. Note circumferential connective tissue formation, also involving the luminal aspect. No mucosa and no tumor cell invasion are seen in **D**. The location of the stent filaments (*arrows*) is clearly seen. Superficial necrosis is the main feature in the distal common bile duct, where the stent filaments were not covered by tissue (**E**). *Arrows* mark the places where the stent filaments were removed

At 1 year, all seven patients showed partial (3) or total (4) coverage of the stent by tissue (Figs. 29.1, 29.2). Three of these had tumor invasion.

Of the five patients surviving for more than 1 year, four had total covering with no tumor infiltration, and only one had partial covering with histologic tumor infiltration inside the stent filaments.

The main tissue reaction in most patients consisted of circular connective tissue formation, sharply demarcated against the tumor cells in all cases where there was no tumor invasion (Figs. 29.1, 29.2). In patients in whom histologic tumor invasion inside the stent filaments was seen, the circular connective tissue reaction was invaded by tumor cells to various degrees (Figs. 29.3, 29.5). Sometimes, especially with marked tumor cell infiltration, the connective tissue reaction seemed to be less obvious (Figs. 29.3, 29.5). In addition, some nonspecific in-

Fig. 29.4D–E

flammatory reaction (mainly lymphocytic) was seen in the majority of cases (Fig. 29.2). In only two cases did the surface of the mucosa covering the stents show a pattern resembling a neomucosa; in no other patients could any normal mucosa be found, even in areas where the stent extended several centimeters beyond the tumor. The layer covering the stents exhibited clear necrotic areas in about half the cases (Figs. 29.2, 29.4), but no frank necrosis was seen towards the periphery (deeper layers outside the stent). Tumor invasion was seen only in 29% of the 24 patients studied by histologic analysis, but in 41% of the stents that showed covering by tissue lining, tumor invasion could be detected microscopically. All these patients with tumor invasion had undifferentiated tumors (four carcinomas of the pancreas, two cholangiocarcinomas, and one undifferentiated carcinoma of the stomach). However, not every undifferentiated tumor (11 of 17) caused malignant stent invasion. Furthermore, only one of five patients surviving for more than 1 year showed tumor invasion within the stent filaments,

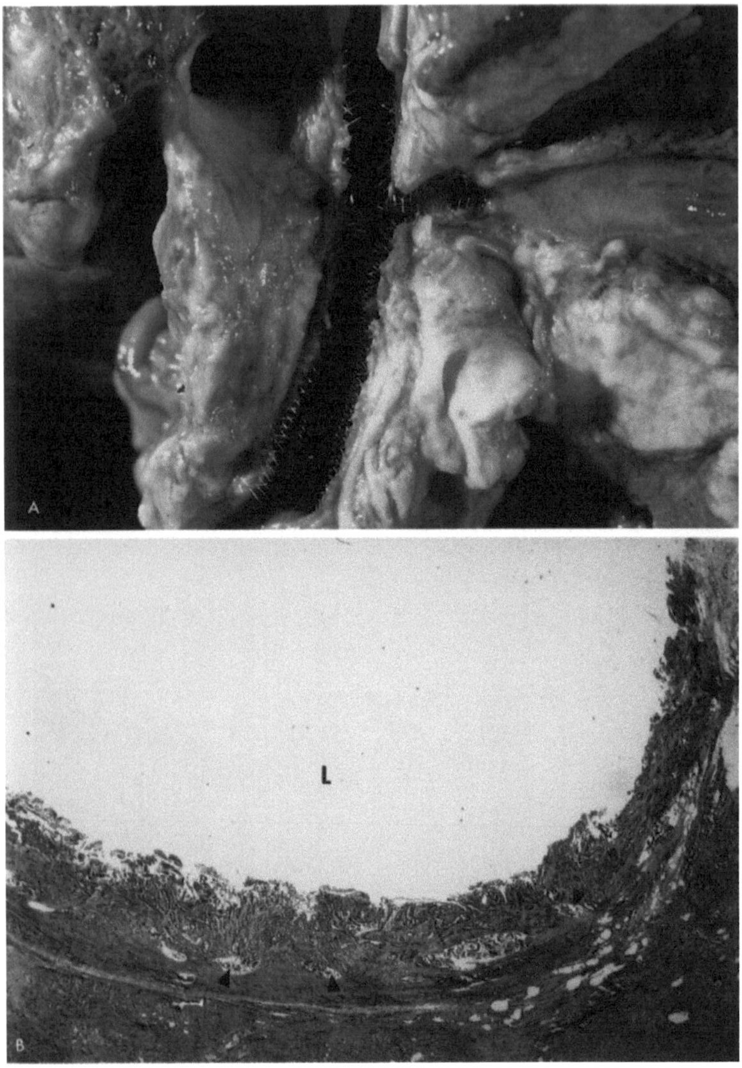

Fig. 29.5A,B. A 61-year-old patient 4 months after bilateral stent placement for Klatskin's tumor. **A** Macroscopic specimen shows partial covering of the stent, mainly at the bifurcation. The distal stent filaments in the common bile duct are not covered. **B** Microscopic section after removal of the stent filaments demonstrates moderate circular connective tissue reaction peripheral to the stent holes (*arrowheads*). The tumor cells have infiltrated between the stent filaments with proliferation of tumorous mucosa. However, there is no obstruction of the lumen (*L*)

although three of these five patients had undifferentiated tumors.

Covering of the stents by tissue occurred first at the areas where tumor had caused the original obstruction. Parts of bare stent struts were seen as long as 16 months after implantation in areas of the bile ducts not involved by tumor. Stent lining seemed to increase with time, as shown in Table 29.2.

29.3 Interpretation of the Data

Tissue reaction to stents in malignant disease obviously varies quite substantially from animal experiments, where mucosal hyperplasia is the main feature (Vowerk et al. 1993; Carrasco et al. 1985). In our study of human specimens, circumferential connective tissue formation was the main reaction. Similar findings were also described by Hausegger et al. (1992) and by ourselves in our previous study (Boguth et al. 1994). Whether this reaction is influenced by the continuous expanding force of the stent causing some ischemic fibrous response cannot be proven. However, it may be speculated that a combination of continuous ischemia and reactive fibrous connective tissue formation may form some sort of barrier against tumor cell invasion, at least for differentiated and less aggressive tumors. The close wire mesh used in the Wallstents may further contribute to this barrier against tumor invasion, particularly in the case of well-differentiated tumors, as suggested by Hausegger et al. (1992). With marked tumor proliferation (Figs. 29.3, 29.5) the connective tissue reaction seemed to be somewhat less pronounced. However, even then the tumor invasion never led to luminal obstruction in any of the seven cases with undifferentiated tumors found inside the stent structs. Interestingly enough, four of these seven patients survived for at least 11 months. Looking at the survival of all 31 patients, no correlation could be identified between survival time and degree of tumor differentiation. This means that patients with well-differentiated tumors did not survive for a significantly longer period after stenting. Furthermore, no significant correlation was found between the time and/or extent of tissue covering of the stent and the histologic differentiation of the obstructing tumor.

The surface of the tissue covering the stents had a clearly abnormal pattern in all cases, with a sparsity of glandular tissue in practically all cases. In only two patients was some degree of neomucosa seen. Frequently, areas of some necrosis at the surface with sloughing were seen, or tumor cell proliferation in patients who had tumor invasion. This may explain why, as we believe, occlusion by sludge and debris is more common than is usually realized: the absence of a normal mucosa probably increases the rate of sludge formation by exfoliation of superficial tumor cells. This absence of normal mucosa in areas of stent placement may also trigger, or at least contribute to, bacterial colonization of the biliary system, which itself probably contributes to sludge formation (Hausegger et al. 1992; Lammer et al. 1986). At macroscopic inspection also, substantial amounts of sludge and debris were found within the stents at autopsy.

Our experience further confirms that it is difficult to distinguish tumor ingrowth from debris and sludge on cholangiography or other imaging modalities (Becker et al. 1993; Gerard et al. 1992; Kelly et al. 1993). We therefore disagree with Stoker and Lameris' statement that occlusion by sludge has become a rare cause of obstruction with the use of Wallstents (Stoker and Lameris 1993). Probably many reported cases of stent occlusion thought to be caused by tumor ingrowth were in fact secondary to sludge and bile incrustation. Why stents become occluded by sludge in some patients and not in others remains unclear. We found no correlation between sludge formation and partial or total of overgrowth of the wire mesh by tissue. In transpapillary stent placement, however, food particles caused obstruction in 2 of 12 cases.

Whether covered stents will result in a higher patency rate seems rather questionable at this time, since stent obstruction seems not significantly lower (Bezzi et al. 1996). In our own, though limited, personal experience of four cases, two patients had tumor invasion of the covered stent proved by percutaneous biopsy. Furthermore, covered stents are of limited application since they are probably not suitable for lesions close to or involving the hilum, because of the potential for obstructing major biliary radicles.

The three conclusions which may be drawn from autopsy studies of metallic stents in cases of malignant biliary obstruction are:

1. The main tissue reaction consists of fibrous tissue formation, and tumor cell invasion is correlated with the degree of dedifferentiation of the tumor. However, tumor invasion (tumor ingrowth) is most likely a rare cause of mechanical stent occlusion: rather, debris and sludge are the main reasons for stent occlusion not related to tumor overgrowth. It therefore seems doubtful whether covered stents will significantly improve stent patency.
2. Stent coverage by tissue reaction starts in the areas of original tumor obstruction; extension to complete stent coverage may take more than 1 year. The continous expanding force of the Wallstent together with its close wire mesh provoke a fibrous tissue reaction which probably forms some sort of barrier against tumor invasion, at least in the case of less aggressive tumors. This confirms the findings published recently by Hausegger et al. (1992).
3. Cholangiography does not allow reliable differentiation between tumor ingrowth and stent occlusion by debris, sludge, or food particles.

References

Adam A, Chetty N, Roddie M, Yeung E, Benjamin IS (1991) Self-expandable stainless steel endoprostheses for treatment of malignant bile duct obstruction. AJR 156:321–325

Alvarado R, Palmaz J, Garcia O, Tio F, Rees C (1989) Evaluation of polymercoated balloon-expandable stents in bile ducts. Radiology 170:975–978

Becker CD, Glättli A, Maibach R, Baer HU (1993) Percutaneous palliation of malignant obstructive jaundice with the Wallstent endoprosthesis: follow-up and reintervention in patients with hilar and non-hilar obstruction. J Vasc Interv Radiol 4:597–604

Bezzi M, Orsi F, Salvatori FM, Maccioni F, Rossi P (1994) Self-expandable nitinol stent for the management of biliary obstruction: long-term clinical results. J Vasc Interv Radiol 5:287–293

Bezzi M, Panzetti C, Bonomo G, Pedicini V, Salvatori FM, Rossi P (1996) Plastic covered Wallstents in malignant biliary obstructions: evaluation in 17 patients (abstract). J Vasc Interv Radiol 7:115

Boguth L, Tatalovic S, Antonucci F, Heer M, Sulser H, Zollikofer CHL (1994) Malignant biliary obstruction: clinical and histopathologic correlation after treatment with self-expanding metal prostheses. Radiology 192:669–674

Carrasco H, Wallace S, Charnsangavej C, et al. (1985) Expandable biliary endoprothesis: an experimental study. AJR 145:1279–1281

Gerard PS, Siegmann R, Albert J, Wetter E (1992) Inspissated bile within the common bile duct simulating cholangiocarcinoma. Case report. Clin Imaging 16:190–193

Gillams A, Dick R, Dooley JS, Wallsten H, El-Din A (1990) Self-expandable stainless steel braided endoprosthesis for biliary strictures. Radiology 174:137–140

Hausegger KA, Kleinert R, Lammer J, Klein GE, Flückiger F (1992) Malignant biliary obstruction: histologic findings after treatment with self-expandable stents. Radiology 185:461–464

Irving JD, Adam A, Dick R, Dondelinger RF, Lunderquist A, Roche A (1989) Gianturco expandable metallic biliary stents: results of an European clinical trial. Radiology 172:321–326

Jaschke W, Klose KJ, Strecker EP (1992) A new balloon-expandable tantalum stent (Strecker stent) for the biliary system: preliminary experience. Cardiovasc Intervent Radiol 15:356–359

Kelly IMG, Lees WR, Russell RCG (1993) Tumefactive biliary sludge: a sonographic pseudotumour appearance in the common bile duct. Clin Radiol 47:251–254

Lameris JS, Stoker J, Niis HGT, et al (1991) Malignant biliary obstruction: percutaneous use of self-expandable stents. Radiology 179:703–707

Lammer J, Stöffler G, Petek WW, Höfler J (1986) In vitro long-term perfusion of different materials for biliary endoprostheses. Invest Radiol 21:329–331

Lammer J, Klein GE, Kleinert R, Hausegger K, Einspieler R (1990) Obstructive jaundice: use of expandable metal endoprosthesis for biliary drainage. Radiology 177:789–792

Lammer J, Flückiger F, Klein CE, Hausegger KA, Waltner F, Aschauer M (1991) Plastic versus expandable biliary metal endoprostheses: randomized trial, phase 1 (abstract) Radiology 181(P):215

Lee MJ, Dawson SL, Mueller PR, et al (1993) Percutaneous management of hilar biliary malignancies with metallic endprostheses: results, technical problems and causes of failure. Radio graphics 13:1249–1263

Maccioni F, Rossi M, Salvatori F, Ricci P, Bezzi M, Rossi P (1992) Metallic stents in benign biliary strictures: three-year follow-up. Cardiovasc Intervent Radiol 15:360–366

Martin EC, Laffey KJ, Bixon R, Getrajdman GI (1990) Gianturco-Rösch biliary stents: preliminary experience. J Vasc Interv Radiol 1:101–105

Mathieson JR, McLoughlin RF, Cooperberg PL, Prystai CC, Stordy SN, MacFarlaine JK, Schmidt N (1994) Malignant obstruction of the common bile duct: long-term results of Gianturco-Rösch metal stents used as initial treatment. Radiology 192:663–667

Rossi P, Bezzi M, Rossi M, Adam A, Chetty N, Roddie M, Iacari V, Cwikiel W, Zollikofer CHL, Antonucci F, Boguth L (1994) Metallic stents in malignant biliary obstruction: results of a multicenter European study of 240 patients. J Vasc Interv Radiol 5:279–285

Stoker J, Lameris JS (1993) Complications of percutaneously inserted biliary Wallstents. J Vasc Interv Radiol 4:767–772

Vowerk D, Kissinger G, Handt S, Günther RW (1993) Long-term patency of Wallstent endoprostheses in benign biliary obstructions: experimental results. J Vasc Interv Radiol 4:625–634

Yoshioka T, Sakaguchi H, Yoshimura H, et al (1990) Expandable metallic biliary endoprostheses: preliminary clinical evaluation. Radiology 177:253–257

30 Endoluminal Radiation Therapy and Infusion Chemotherapy for Malignant Biliary Strictures

H. Yoshimura, H. Sakaguchi, and H. Uchida

CONTENTS

30.1 Introduction

Relief of obstructive jaundice by insertion of a biliary endoprosthesis consisting of a tube or metallic stent is an extremely useful palliative therapy for inoperable advanced malignant biliary strictures and has come to be widely employed (Uchida et al. 1975; Yoshioka et al. 1990). Since 1987 we have used metallic stents as biliary endoprostheses in approximately 200 patients with inoperable advanced malignant biliary strictures. However, to prevent tumor ingrowth into the metallic stent and to enhance the therapeutic results, antitumor treatment is needed, and recently a therapeutic effect has been achieved with the combination of radiation therapy and infusion chemotherapy. Since all the causes of malignant biliary stricture, such as bile duct cancer, gallbladder cancer, pancreatic cancer, and lymph node metastases, have in common: (1) the presence in most cases of radioresistant well-differentiated adenocarcinomas, (2) the presence of adjacent organs such as liver and bowel that have a low tolerance to radiation, and (3) the disadvantage of being subject to physiological movements such as respiration, the role of routine external radiation therapy (ERT) is limited. Furthermore, since sensitivity to anticancer agents is low, little effect can be anticipated from the systemic administration of such agents (Falkson et al. 1984; Harvey et al. 1984; Oberfield and Rossi 1988; Taal et al. 1993; Kajanti and Pyrhonen 1994; Okada et al. 1994).

Endoluminal radiation therapy (ELRT), in which a small radiation source is inserted within the tumor, overcomes these problems when bile duct external-internal drainage is also used to treat the obstructive jaundice which frequently complicates bile duct cancer (Ikeda et al. 1979). This approach has been employed in our institution since 1987 in the belief that it would expand the indications for radiation therapy and enhance the therapeutic results (Yoshimura et al. 1989; Tamada et al. 1991). Also, using ELRT for difficult-to-treat hepatic hilar bile duct cancer, it has been possible to perform concurrent intermittent arterial infusion chemotherapy (IAIC) using a vascular access port without interfering with the patient's activities of daily living as an easily repeatable outpatient procedure. Recently, with the introduction of combined interventional radiology consisting of the use of a biliary endoprosthesis using metallic stent, ELRT, and IAIC for inoperable advanced bile duct cancer, we have experienced a number of cases in

H. Yoshimura, MD, Associate Professor Department of Oncoradiology, Nara Medical University, 840 Shijo-cho, Kashihara, Nara, 634 Japan
H. Sakaguchi MD, Department of Radiology Nara Medical University, 840 Shijo-cho, Kashihara, Nara, 634 Japan
H. Uchida, MD, Professor, Department of Radiology, Nara Medical University, 840 Shijo-cho, Kashihara, Nara, 634 Japan

which survival was prolonged and quality of life enhanced. In this chapter we describe our experience with combined interventional radiology consisting of ELRT, IAIC, and metallic stent for the treatment of bile duct cancer.

30.2 Basic Aspects

30.2.1 Endoluminal Radiation Therapy

ELRT has a characteristic dose distribution, namely a concentric gradient; its center is the radiation source, which is inserted into the central portion of the tumor, and the dose decreases abruptly as the distance from the source increases (Figs. 30.2c, 30.8d). Accordingly, normal tissues adjacent to the tumor are exposed to only low radiation doses, and a strong local effect can be achieved by selective irradiation of the tumor with a high dose. ELRT is effective in inhibiting growth when the tumor is located in or near the bile duct wall, and when the tumor is limited to this area ELRT alone may constitute sufficient treatment. In addition, ELRT is also used to treat localized residual foci near the bile duct wall after resection and localized recurrences, as well as prophylactically to prevent recurrence (Cameron et al. 1982; Fritz et al. 1994). On the other hand, when the tumor has spread beyond the bile duct wall, e.g., to adjacent lymph nodes and/or around nerves, the effect of ELRT alone is very slight, and the combined use of ERT becomes necessary.

30.2.2 Intermittent Arterial Infusion Chemotherapy via Port

IAIC increases antitumor effectiveness owing to the high local concentration of the anticancer drug achieved in the tumor. Moreover, the amount of drug entering the systemic circulation after arterial infusion decreases to less than the administered dose because of drug elimination in the tumor and adjacent liver, thereby reducing the severity of systemic side effects. Since the upper common hepatic duct and hilar hepatic duct receive blood flow via the peribiliary arterial plexus from the proper hepatic artery and its distal intrahepatic arteries, it can be anticipated that a drug infused into the proper hepatic artery will be distributed at a high concentration to tumors in the hepatic hilum. On the other hand, since the middle and lower common bile ducts receive blood flow from complicated small arteries branching from the posterior pancreaticoduodenal and proper hepatic arteries, side effects from distribution of the drug to the pancreas and duodenum are unavoidable, and thus IAIC is generally not indicated for tumors in these locations.

30.2.3 Biliary Endoprosthesis Using Metallic Stent

Biliary endoprosthesis placement using a metallic stent is indispensable for combined interventional radiology for bile duct cancer. Metallic stents and cancer treatments such as ELRT, ERT, or IAIC exert a cooperative effect. The freedom from jaundice achieved by the metallic stent improves the general condition and liver functions of the patient, allowing adequate cancer treatment, while on the other hand effective cancer treatment will maintain long-term patency of the stent by preventing tumor growth into or around the stent. In addition, the stent will prevent the inflammatory stenoses of bile duct caused by radiotherapy and/or IAIC.

30.3 Practical Aspects

30.3.1 Selection of Treatment Method

The treatment method is selected according to the site of the tumor and its extent. The selection criteria and indications used at our institution for radiotherapy for bile duct cancer, including ELRT and IAIC, are outlined below (Fig. 30.1).

30.3.1.1 Middle–Lower Common Bile Duct Cancer

Although theoretically stages 0–I (UICC staging system, 1987) are curable with ELRT alone (30 Gy), the degree of wall infiltration is difficult to determine strictly by imaging studies, and so ERT (30 Gy) is also applied. For stages II–IV, combined therapy with ELRT (30 Gy) and ERT (30 Gy) is administered. IAIC is not generally indicated.

30.3.1.2 Hepatic Hilar Bile Duct Cancer

Stages 0–I are treated in the same way as middle-lower common bile duct cancer. For stages II–III, ELRT (30 Gy) and ERT (30 Gy) are combined, and for stage IV (hepatic parenchyma infiltration), ERT (50 Gy) is used in conjunction with IAIC.

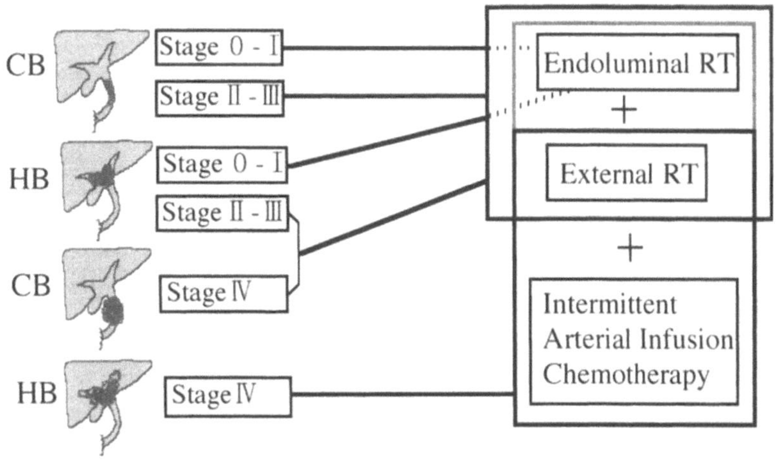

Fig. 30.1. Selection of therapy according to tumor site and stage of bile duct cancer. *CB*, Middle–lower common bile duct; *HB*, hepatic hilar bile duct (including upper common bile duct); *RT*, radiotherapy;

Since January 1995, we have started using the combination of ELRT (30 Gy), ERT (30 Gy), and IAIC for stage IV.

30.3.2 Endoluminal Radiation Therapy

30.3.2.1 Radiation Sources

The radiation sources used for ELRT are high-dose-rate sources like ^{60}Co (Yoshimura et al. 1989; Tamada et al. 1991) and ^{192}Ir (Fritz et al. 1994), low-dose-rate sources like ^{192}Ir wire and seed (Ikeda et al. 1979; Fletcher et al. 1983; Buskirk et al. 1984; Fields and Emami 1987; Hendrickx et al. 1990), and ^{198}Au grains. The high-dose-rate radiation sources require only a short irradiation time, and with the use of a remote afterloading system exposure of medical personnel to radiation can be avoided. A further advantage is that a radioisotope room to isolate the patient is not required. On the other hand, low-dose-rate radiation sources have the radiobiological advantages of a large therapeutic ratio and small oxygen enhancement ratio, but require a long irradiation time during which the patient cannot leave the radioisotope room, and they are associated with infection and displacement of the radiation source as well as exposing medical personnel to radiation. To alleviate the physical burden on the patient and side effects associated with the insertion of the radiation source, when administering ELRT the thinnest possible applicator must be used. We have developed a 14F applicator for ^{60}Co (3 mm diameter) for use with a remote afterloading system, but its diameter is thick and so its insertion into multiple intrahepatic bile ducts is difficult. Since the external-internal drainage tube used to insert ^{192}Ir wire (0.3 mm diameter) and ^{198}Au grains (0.8 mm diameter) is thin (8.3F), it can be used for insertion into two or more intrahepatic bile ducts and so is suitable for use in the intrahepatic bile ducts (Figs. 30.3b, 30.8c). Furthermore, recently, a ^{192}Ir-high-dose-rate source with remote afterloading system has been developed in which the radiation source has a diameter of 1.0 mm and is insertable with a 6F applicator, which has further lessened the burden on the patient.

30.3.2.2 Techniques

After percutaneous bile duct drainage, ELRT is performed by placing the radiation source in the tube used to create a passage for internal drainage through the occluded portion of the bile duct.

High-Dose-Rate ELRT with ^{60}Co Remote Afterloading System

After creating a passage for internal drainage through the occluded portion of the bile duct, the drainage tract is gradually expanded using a dilator to permit the insertion of a 14F applicator. Next, using a guidewire, the drainage tube is replaced by an applicator for ELRT use. The applicator has a double lumen, consisting of a thick lumen with a blind end and an internal diameter of 12F into which a ^{60}Co radiation source can be inserted, and a thin lumen into which a 0.035-inch (0.89 mm) guidewire can be inserted and easily exchanged for a drainage tube without contaminating the ^{60}Co radiation source.

The irradiation field is determined by inserting a dummy radiation source, and the irradiation is performed with remote afterloading a system with by moving the ^{60}Co radiation source by 1 cm (Fig.

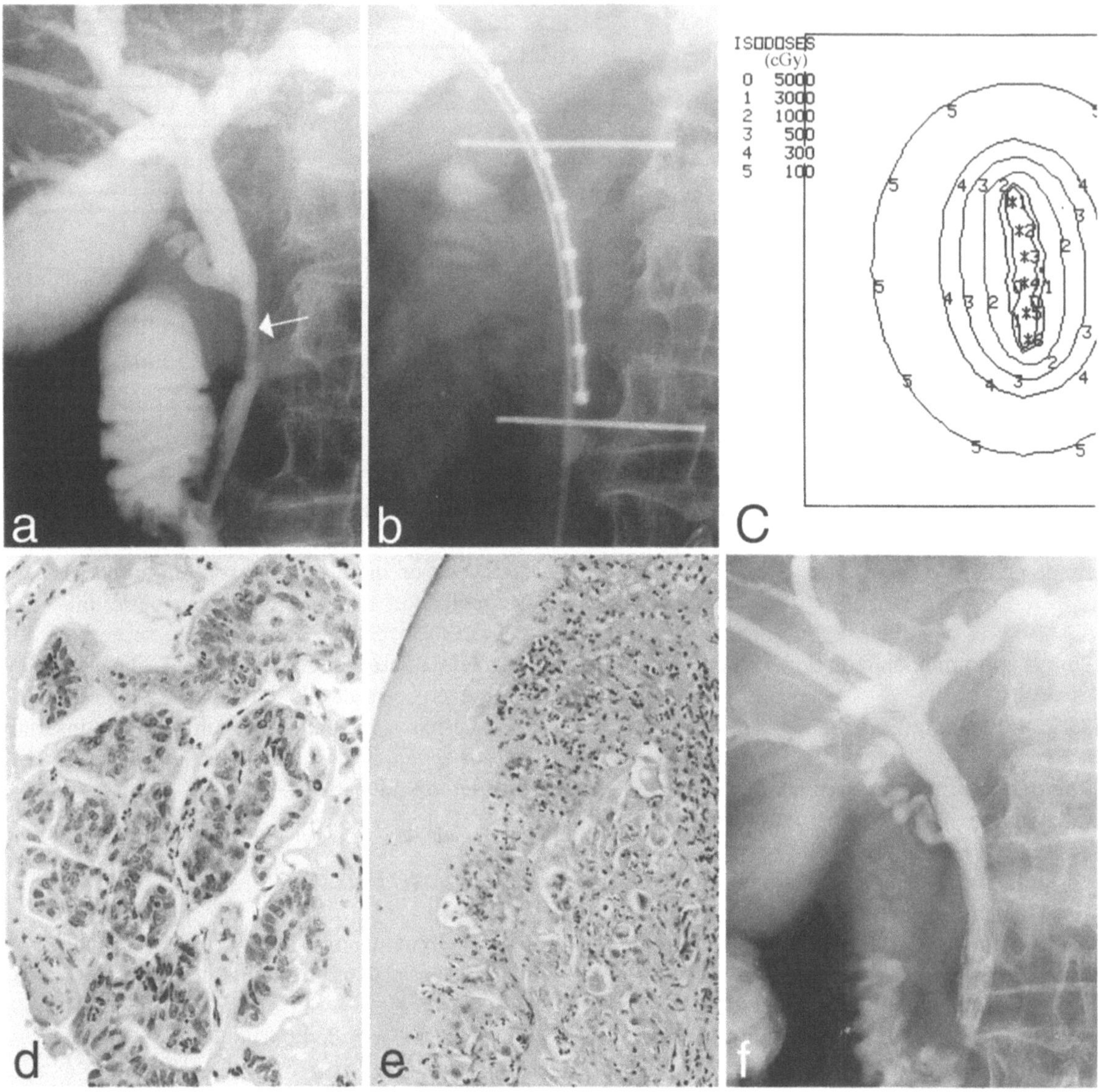

Fig. 30.2a–f. High-dose-rate endoluminal radiotherapy (ELRT) with ^{60}Co remote afterloading system. **a** Cholangiogram showed severe stenosis of the middle common bile duct (*arrow*). **b** Plain *x*-ray showed the inserted ^{60}Co dummy radiation source to determine the irradiation length of the bile duct. **c** Isodose distribution of ELRT. The total radiation dose of 30 Gy (1 cm from the center of the radiation source) was delivered as 7.5 Gy per session twice a week for a total of four times. **d** Microscopy of biopsied specimen before radiotherapy showed adenocarcinoma of the bile duct. **e** Microscopy of biopsied specimen after completion of ERT and ELRT showed disappearance of viable cancer cells with degeneration and necrosis, indicating a good therapeutic effect. **f** A biliary endoprosthesis (spiral z-stent) was placed after ELRT. Cholangiogram shows good patency of the biliary tract

30.2b). The total radiation dose amounts to 30 Gy (1 cm from the center of the radiation source), delivered as 7.5 Gy per session twice a week for 2 weeks for a total of four times (Fig. 30.2c). Each session of high-dose-rate radiation takes several minutes. Prior to ELRT, ERT is performed with anterior-posterior parallel opposing portals delivering five divided fractions of 2 Gy each per week for 3 weeks for a total radiation dose of 30 Gy.

Low-Dose-Rate ELRT with ^{198}Au Grains or ^{192}Ir Wire

After inserting an 8.3F internal drainage tube, ^{198}Au grains or ^{192}Ir wire is sealed into a 3.2F plastic tube, which is then inserted into the internal drainage tube (Figs. 30.3b, 30.8c). The number of ^{198}Au grains (10–27 grains) or length of ^{192}Ir wire (2, 3, or 5 cm) is selected according to the length of the area to be irradiated. A total radiation dose of 30 Gy (1 cm from the center of the radiation source) is delivered by the

ELRT (Fig. 30.8d). A period of 2–7 days is needed for the ELRT, during which the patient is isolated in the radioisotope room. As for ELRT with ^{60}Co, ERT delivering a total radiation dose of 30 Gy is administered prior to ELRT.

Interstitial Radiation Therapy with ^{198}Au Grains

For cancer infiltrating the hepatic parenchyma for which an adequate radiation dose cannot be given with ELRT, interstitial radiotherapy is performed by placing ^{198}Au grains as a permanent insertion radiation source within the tumor under ultrasound guidance. If the indications are carefully considered, this modality may be a powerful component of multidisciplinary therapy.

30.3.3 Intermittent Arterial Infusion Chemotherapy via Port

30.3.3.1 Anticancer Drugs

It is important to select drugs to which the tumor is sensitive and which have a high elimination rate in the liver and tumor. FE(A)M, combining 5-fluorouracil (5-FU), epirubicin or adriamycin, and mitomycin C, considered to be effective against adenocarcinoma, is the regimen of first choice.

30.3.3.2 Techniques and Planning of Treatment

Catheter Placement and Implantation of Infusion Port

To make possible long-term (1–2.5 years) intravascular placement, a heparin-coated catheter with antithrombotic properties (Anthron, Toray, Tokyo) is used. The route of insertion is via either the femoral artery or the subclavian artery, using either a percutaneous approach with Seldinger's method or a surgical cut-down approach. Our first choice is a percutaneous approach via the femoral artery. The catheter tip is placed in the common hepatic artery (CHA) or gastroduodenal artery (GDA) with a side hole opened at a site corresponding to the CHA (Figs. 30.3b, 30.4a). To prevent the influx of anticancer drugs to surrounding organs such as the pancreas and stomach, the GDA and right gastric artery are occluded with metallic coils (Figs. 30.3b, 30.4a). Then the proximal end of the catheter is attached to the port. In the approach from the femoral artery the port is embedded in the lower abdominal wall (Fig. 30.4b), while in that from the subclavian artery it is embedded in the subcutaneous tissue of the chest wall.

Flow Check of Anticancer Drugs

To confirm that the infused anticancer drugs are being distributed to the tumor-bearing area and not to

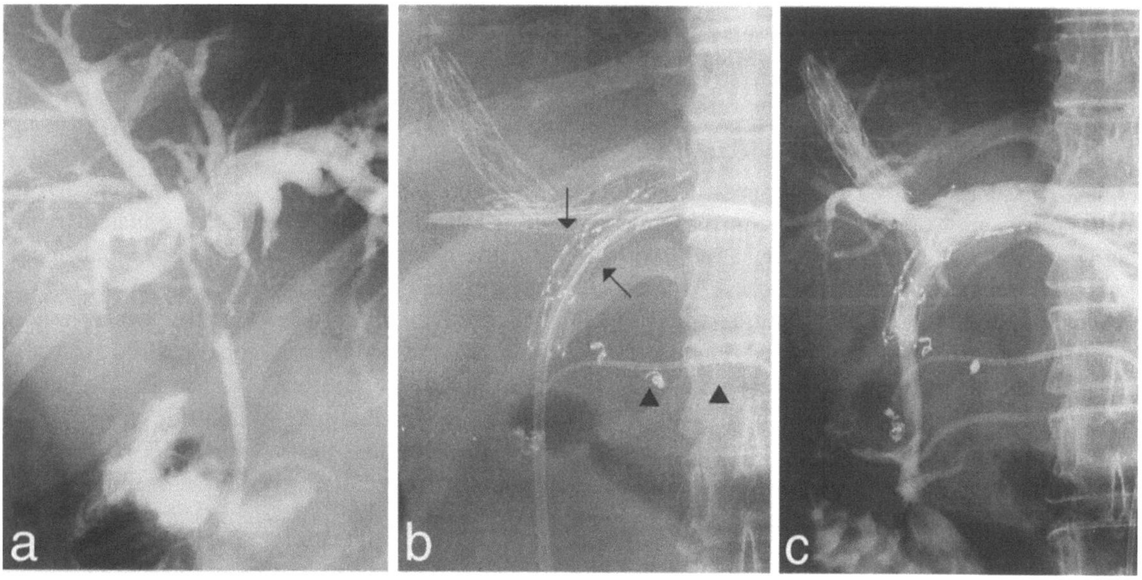

Fig. 30.3a–c. Low-dose-rate ELRT with ^{192}Ir wire. **a** Cholangiogram showed no opacification of the hepatic hilar bile duct, indicating severe stenosis of the right and left hepatic ducts as far as the upper common bile duct. **b** ^{192}Ir wires sealed into 3.2F plastic tubes were inserted in 8.3F internal drainage tubes (*arrow*) after biliary endoprosthesis placement (spiral and modified z-stents) and catheter placement for IAIC (*arrowhead*), and a total radiation dose of 30 Gy was delivered to the site 1 cm from the radiation source. **c** Cholangiogram after ELRT showed good patency of the biliary tract

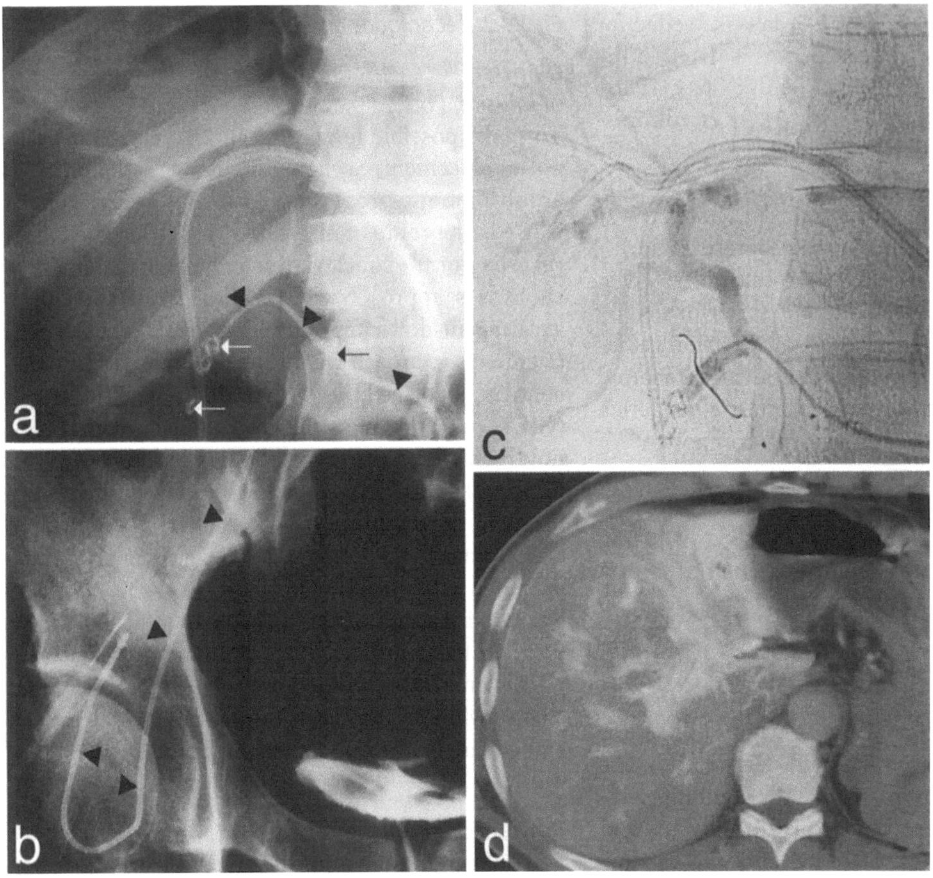

Fig. 30.4a–d. Catheter placement and flow-check study for IAIC. **a,b** Plain X-ray showed a heparin-coated catheter placed via the right femoral artery to the gastroduodenal artery (GDA) (*arrowhead*) with a side hole open at a site corresponding to the common hepatic artery. The proximal end of the catheter is attached to the port, which is embedded in the lower abdominal wall. The GDA and right gastric artery were occluded with metallic coils (*arrow*) to prevent influx of anticancer drugs into the pancreas and stomach. **c** Angiography via the catheter placed into the GDA showed the entire intrahepatic arterial branches and occlusion of the GDA. **d** CT arteriography performed during infusion of contrast medium from the port showed distribution to the entire liver and not to the surrounding organs

the surrounding organs, it is imperative that a flow check is performed. For this purpose either scintigraphy is performed by infusing a radioisotope (^{99m}Tc-macroaggregated albumin, $^{99m}Tc\text{-}O_4^-$) from the port, or digital subtraction angiography (Figs. 30.4c, 30.9d–g) or CT arteriography (Fig. 30.4d) is performed during infusion of contrast medium from the port. The former is a good method to check the relation between the injection rate and drug distribution, while the latter is convenient to estimate drug distribution from the enhanced region.

Infusion of Anticancer Drugs

In the case of IAIC, anticancer drugs are infused as a rule once every 2 weeks in the out-patient clinic. Epirubicin or adriamycin (20 mg/body) is used together with mitomycin C (4 mg/body) and 5-FU (500 mg/body). Hematological examinations are performed prior to each intraarterial infusion to confirm the absence of myelosuppression and hepatic dysfunction. The administered dose of each of the drugs is reduced to one-half when any of the following laboratory abnormalities is noted: white blood cell count ≤3000/mm^3, platelets ≤30 000/mm^3, glutamine-oxaloacetic transaminase or glutamic-pyruvic transaminase ≥100 IU/l, total bilirubin ≥2.0 mg/dl. If any further worsening is noted on hematological tests 2 weeks later, drug administration is terminated and restarted only after the laboratory values improve. When tumor regrowth and/or reelevation of tumor markers is evident, a flow check is performed and the state of drug distribution to the tumor area should be reconfirmed. If the drug distribution is found to be appropriate, drug resistance to FE(A)M is judged to have developed and the drug regimen is changed to cisplatinum (CDDP) and 5-FU infusion or Lipiodol-AI which is the infusion of Lipiodol ultrafluid mixed with epirubicin.

30.3.4 Biliary Endoprosthesis Using Metallic Stent

30.3.4.1 Choice of Metallic Stent

Several kinds of metallic stent, such as the Gianturco z-stent, modified z-stent (Fig. 30.3b), spiral z-stent (Figs. 30.2f, 30.3b, 30.8e, 30.9c) (MAEDA et al. 1992), Wallstent, and Nitinol Strecker stent, are generally used for biliary strictures. It is important to select stents appropriate to the site in the bile duct where they are to be placed. We have had experience with all these kinds of metallic stent since 1987. On the basis of our experience, we can say that any kind of metallic stent may be useful in the hilar or common bile duct. In the intrahepatic bile ducts, however, we prefer to use the spiral z-stent, because it has the characteristics of good flexibility and less shortening (Figs. 30.3b, 30.8e) (MAEDA et al. 1992).

30.3.4.2 Timing of Metallic Stent Placement in Relation to Radiotherapy

We used to place metallic stents after the completion of radiotherapy (Figs. 30.2f, 30.8e, 30.9c). Recently, we have started to place the stent before completion of radiotherapy, to prevent bile stasis during ELRT and to reduce in-hospital stay (Fig. 30.3b). On that occasion, it was confirmed by our basic study that the influence of secondary electrons and scattering caused by radiotherapy after metallic stent placement are negligible (TAMADA et al. 1995).

30.3.5 Results

From June 1987 to December 1994, 30 patients with bile duct cancer were treated by combined interventional radiology, and the results of treatment were investigated.

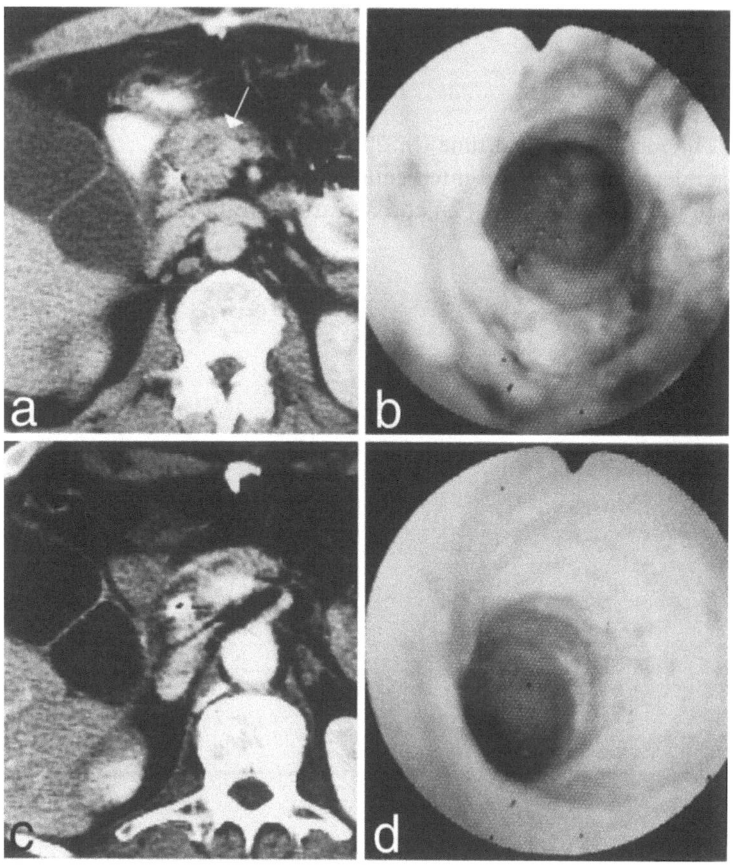

Fig. 30.5a–d. Evaluation of the therapeutic effect after the combination of ELRT and ERT. **a** CT before radiotherapy showed the tumor invading the portal vein in the middle portion of the common bile duct (*arrow*). **b** Endoscopically, an irregular surface and stenosis of the bile duct wall with easy bleeding were seen before radiotherapy. **c** CT after radiotherapy showed disappearance of the tumor, indicating a good therapeutic effect. **d** Endoscopically, a smooth surface of the bile duct wall without bleeding was seen after radiotherapy

30.3.5.1 Evaluation of Therapeutic Effect

Therapy-induced tumor shrinkage could be evaluated in 22 patients in whom the extent of the tumor could be determined by diagnostic imaging techniques such as cholangiography, CT (Fig. 30.5a,c), MRI, ultrasonography, and angiography (Fig. 30.9d–g). However, in 8 patients with wall infiltration, in whom the tumor could not be clearly visualized, determination of tumor shrinkage could not be easily done.

Changes in tumor marker (CA19-9) levels were useful in evaluating therapeutic effect and development of recurrence. Sixteen of the 21 patients with initially elevated values showed decreases following therapy.

Biliary endoscopy is useful in assessing treatment results (Fig. 30.5b,d), but since dilatation to 13F is required, it was possible only in a minority of patients. Less invasive diagnosis of mucosal surface extension, depth of mural infiltration, and extramural extension of the tumor is possible using a thin-diameter bile duct endoscope (8F) and/or endoluminal ultrasound probe (7.2F), and it is expected that use of these instruments will contribute to a very precise evaluation of the therapeutic effect.

30.3.5.2 Survival

The median survival time (MST) of the 30 patients undergoing combined interventional radiology was 10 months, with the 1-, 2-, and 3-year survival rates amounting to 43%, 13%, and 3% respectively (Fig. 30.6). Of these patients, the MST of the 22 (12 with middle–lower common bile duct cancer and 10 with hepatic hilar bile duct cancer) subjected to a combination of ELRT and ERT was 11.4 months, with 1-, 2-, and 3-year survival rates of 39%, 18%, and 6% respectively (Table 30.1, Fig. 30.6).

In the study of survival rates according to tumor site and stage, 12 patients with middle–lower common bile duct cancer (stage I: 2 patients; stage III: 3 patients; stage IV: 7 patients) who received the combination of ELRT and ERT achieved 1- and 2-year survival rates of 50% and 25% respectively (Fig. 30.7). Among them, five patients with stage I or III tumors achieved the favorable 1- and 2-year survival rates of 80% and 40% respectively, including two who are still alive and tube-free at present.

On the other hand, ten patients with hepatic hilar bile duct cancer (stage III: one patient; stage IV: nine patients) who received both ELRT and ERT achieved the 1- and 2-year survival rates of 20% and 10% respectively, with the one stage III patient surviving for 3 years (Fig. 30.7). In contrast, six patients (all with stage IV tumors) with hepatic hilar bile duct cancer who received IAIC and ERT achieved the 1- and 2-year survival rates of 83% and 17% respectively, with an MST of 16.3 months (Table 30.1, Fig. 30.7). These results suggested the efficacy of the combined use of IAIC for stage IV hepatic hilar bile duct cancer associated with hepatic parenchyma infiltration.

Few reports are available on ELRT and IAIC for bile duct cancer, and objectively assessable therapeutic results have not been described. Table 30.1 sum-

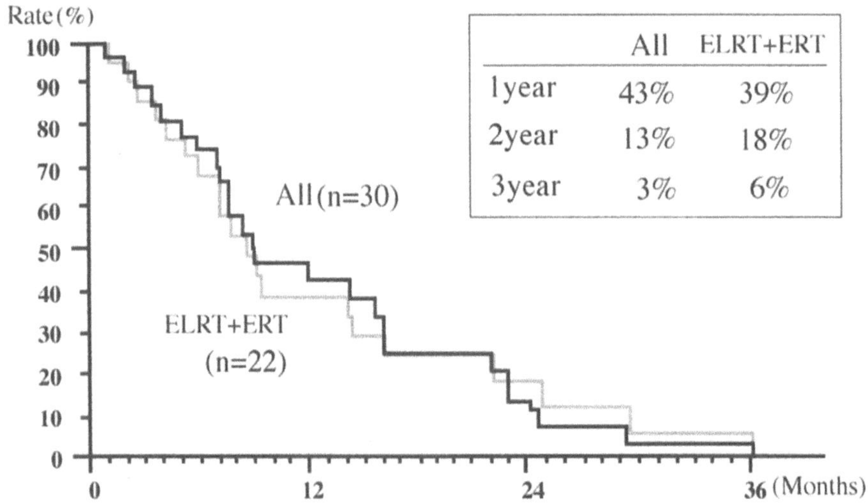

Fig. 30.6. Cumulative survival of all patients with bile duct cancer and those treated with ELRT plus ERT (Kaplan-Meier method)

Table 30.1. Treatment results in bile duct cancer: a review of the literature

Treatment	Reference	*n*	ELRT			ERT dose (Gy)	IAIC Drug	Survival time (months)	
			Radiation source	Dose (Gy)	Ref. point (mm)			Median	Range
ELRT	FLETCHER et al. (1983)	18	^{192}Ir-LDR	40–48	5	–	–	11	4–38
	HENDRICKX et al. (1990)	22	^{192}Ir-LDR	60–80	10	–	–	8	1–20
ELRT + ERT	IKEDA et al. (1979)	2	^{192}Ir-LDR	15–33	5	33–34	–	7.5	7–8
	BUSKIRK et al. (1984)	5	^{192}Ir-LDR	50	5–10	50	–	15	7–20
	FIELDS and EMAMI (1987)	8	^{192}Ir-LDR	22–50	10	22-50	–	15	1.5–34
	Our cases (1995)	22	^{60}Co-HDR ^{198}Au-LDR	30	10	30	–	11.4	1–36
ELRT + ERT ±surgery	CAMERON et al. (1982)	26(26)[a]	^{192}Ir-LDR	25	–	30–50	–	18.8	1–16
	FRITZ et al. (1994)	30(9)[a]	^{192}Ir-HDR	20–45	10	40–45	–	10	1–69
IAIC ± ERT	SMITH et al. (1984)	4		–		–	FM	6.2	4–9
	Our cases(1995)	6		–		50	FE(A)M	16.3	7–24

ELRT, Endoluminal radiotherapy; ERT, external radiotherapy; IAIC, intermittent arterial infusion chemotherapy; LDR, low dose rate; HDR, high dose rate; F, 5-fluorouracil; M, mitomyain C; E, epirubicin; A, adriamycin.
[a]Numbers in parentheses are surgical cases.

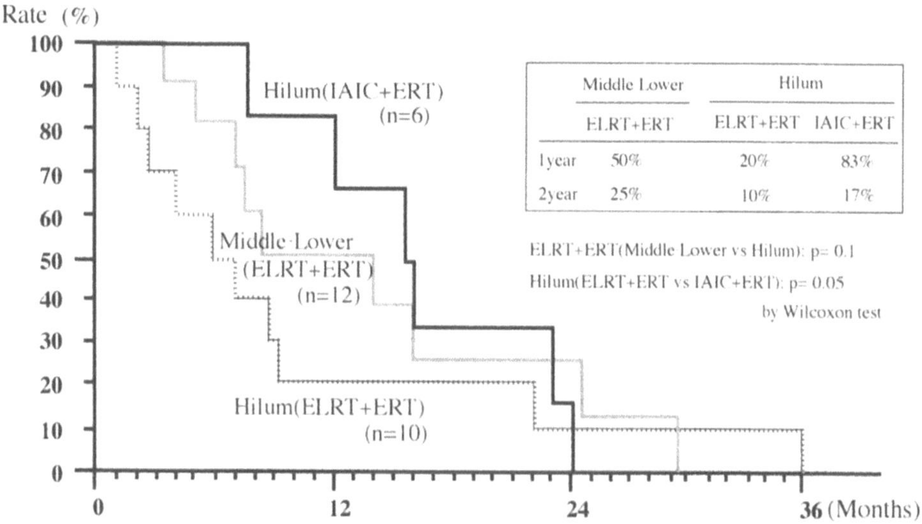

Fig. 30.7. Cumulative survival in bile duct cancer according to tumor site (hilum vs middle-lower common bile duct) and therapeutic method (ELRT + ERT vs IAIC + ERT; Kaplan-Meier method)

marizes representative therapeutic results in bile duct cancer as hitherto noted in the literature. With the exception of CAMERON et al. (1982), who by combining ELRT and ERT as adjuvant therapy after resection of bile duct cancer achieved an MST of approximately 1.5 years, most series achieved an MST of only about 1 year. Thus, improvements in the prognosis will require major breakthroughs in treatment methods including radiotherapy. On the other hand, SMITH et al. (1984) achieved a disappointing MST of approximately 6 months with IAIC used alone without radiotherapy. Our present results obtained with the combination of IAIC and radiotherapy resulted in an MST of approximately 16 months, suggesting that some improvement in patient prognosis can be anticipated with this therapeutic approach.

30.3.5.3 Complications

During the administration of ELRT, cholangitis and/or liver abscess may develop due to cholestasis. In five of our patients, cholangitis developed during

ELRT. In four of these cases antibiotics were effective and the ELRT could be continued, while in one case the ELRT was interrupted and restarted only after alleviation of the cholangitis following the placement of a metallic stent. In two patients in whom ERT and IAIC were used in conjunction, liver abscess developed in association with tumor necrosis, but it healed in both cases after percutaneous abscess drainage.

Duodenal ulcer related to ELRT represents another serious complication. Fritz et al. (1994) reported duodenal ulcer in 7 of 30 patients (23%), 5 of whom had received radiotherapy concurrently with ERT and ELRT. Duodenal ulcer was noted in 1 of our 22 patients (4.5%) in whom ELRT was performed. This patient received ELRT and ERT (both 30 Gy), and it was considered unlikely that this represented a side effect of the radiotherapy alone, but rather that the stress of being confined to the radioisotope room also contributed to its development. To prevent such duodenal ulcers it is important not to deliver a high dose of radiation to the duodenum within a short period and to avoid stress.

Mild leukopenia due to myelosuppression was noted in 50% of the patients receiving the combination of radiotherapy and IAIC, and resolved in every case when the drug dose was reduced to one-half.

Other possible complications, although not experienced in our series, include late bile duct scarring stricture induced by radiotherapy and/or intra-arterial anticancer drug infusion and arterial injury and subcutaneous abscess caused by the catheter or port placed to facilitate the IAIC.

30.4 Representative Cases

30.4.1 Case 1

The patient was a 74-year-old man with hepatic hilar bile duct cancer (T3N0M0, stage IV) treated with ELRT and ERT (Fig. 30.8). Obstructive jaundice developed in November 1991, at which time percutaneous transhepatic cholangiography revealed occlusion of the right hepatic duct and stenosis of the upper common bile duct from the left hepatic duct, leading to a diagnosis of inoperable hepatic hilar bile duct cancer (Fig. 30.8a). After performing external and internal drainage from the right and left hepatic ducts with a total of three tubes, ELRT and ERT were given from January 4 to February 6, 1992 to treat the hepatic hilar tumor. ERT was administered to a $5 \times 4\,cm^2$ field using a 10 MVX linear accelerator with anterior-posterior parallel opposing portals with which a total dose of 30 Gy was delivered in daily single doses of 2 Gy (Fig. 30.8b). ELRT was performed by sealing ^{198}Au grains (185 MBq × 23) into three drainage tubes and then isolating the patient in the radioisotope room for 3 days so that a total radiation dose of 30 Gy was delivered to the site 1 cm from the radiation source (Fig. 30.8c,d). After completion of the radiotherapy the CA19-9 value dropped from 1791 U/ml to a normal value of 16 U/ml. In addition, biliary endoprosthesis placement using metallic stent was performed using four spiral z-stents in the right and left hepatic duct–common bile duct (Fig. 30.8e,f), and the patient could be discharged from hospital on February 28, 1992. The patient was subsequently followed as an outpatient. In November 1993 local recurrence of the tumor was noted, and abscess and disseminated intravascular coagulation supervened, leading to the patient's death approximately 2 years after the initiation of therapy.

30.4.2 Case 2

The patient in this case was a 61-year-old man with hepatic hilar bile duct cancer (T3N0M0, stage IV) treated with ERT and IAIC (Fig. 30.9).

In October 1992, obstructive jaundice developed, and cholangiography revealed occlusion from the left and right hepatic ducts to the middle common bile duct, leading to a diagnosis of inoperable hepatic hilar bile duct cancer (Fig. 30.9a). After ERT (total dose 50 Gy) was administered to a $5 \times 5\,cm^2$ irradiation field in the hepatic hilum using a 10 MVX linear accelerator with anterior-posterior parallel opposing portals (Fig. 30.9b), the CA19-9 value dropped from 1225 U/ml to 983 U/ml. Spiral z-stents were deployed

Fig. 30.8a–f. Representative case 1: a 74-year-old man with hepatic hilar bile duct cancer (T3N0M0, Stage IV) treated with ELRT and ERT. **a** Cholangiogram showed occlusion of the right hepatic duct and stenosis of the upper common bile duct from the left hepatic duct. **b** External-internal drainage from the right and left hepatic ducts was performed with a total of three tubes. ERT was administered to a $5 \times 4\,cm^2$ field using a 10 MVX linear accelerator to a total dose of 30 Gy. **c** ELRT was performed by sealing ^{198}Au grains (185 MBq × 23) into three drainage tubes. **d** Isodose distribution of ELRT. A total radiation dose of 30 Gy was delivered to the site 1 cm from the radiation source. **e** Biliary endoprosthesis placement using spiral z-stent was performed, using four spiral z-stents in the right and left hepatic duct–common bile duct. **f** Cholangiography showed good patency of the biliary tract after radiotherapy and biliary endoprosthesis using metallic stent

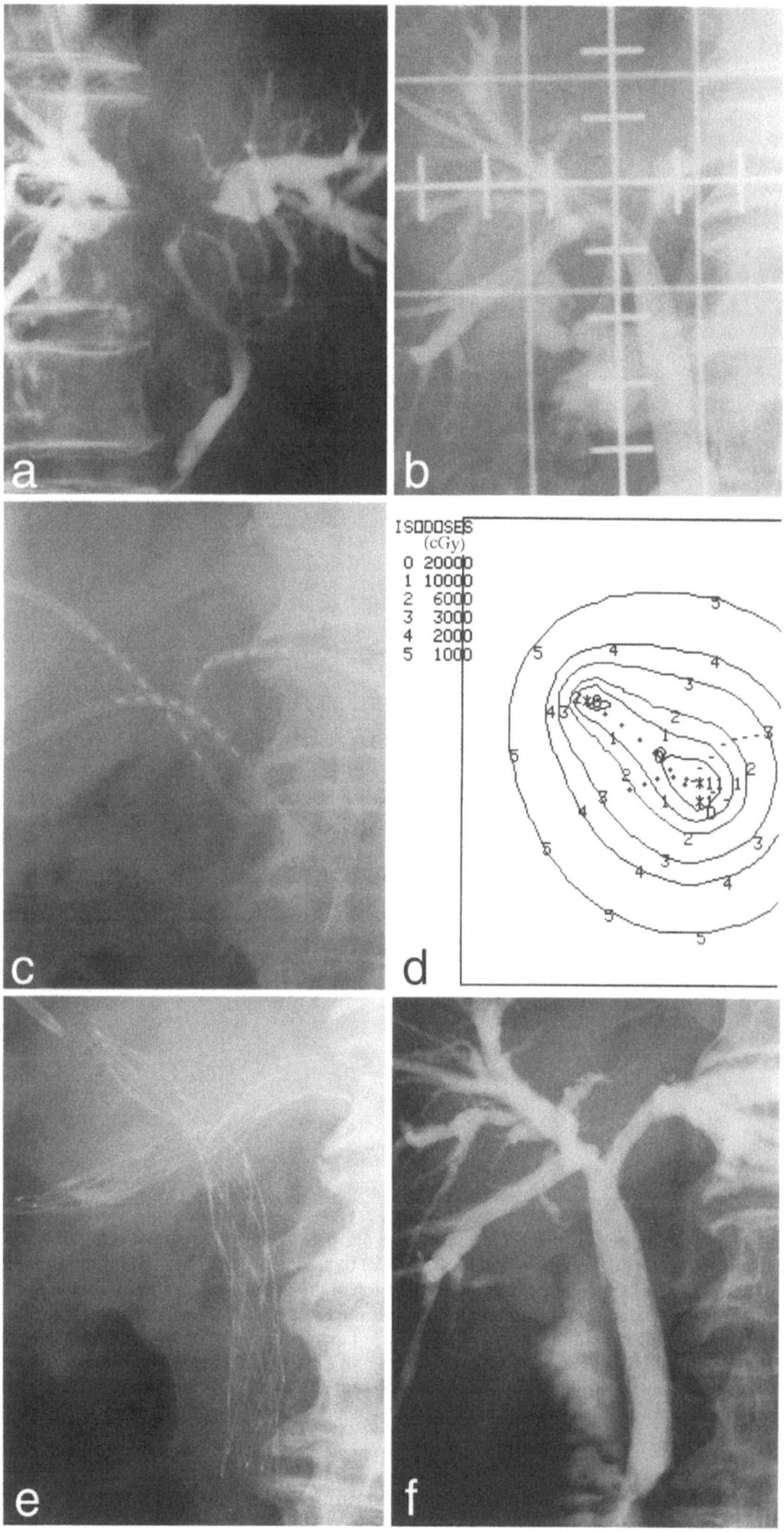
a
b
(cGy)
0 20000
1 10000
2 6000
3 3000
4 2000
5 1000
c
d
e
f

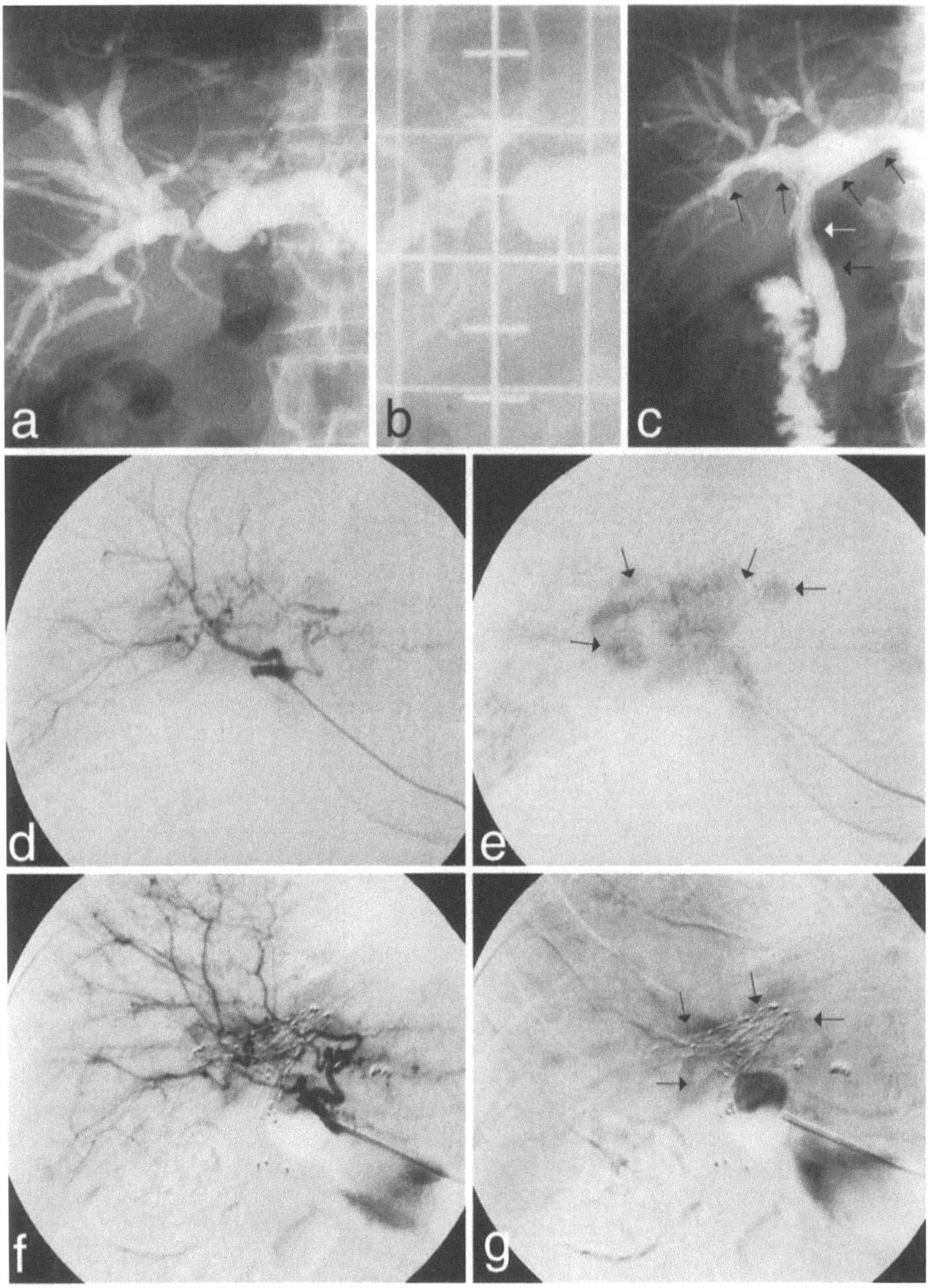

Fig. 30.9a–g. Representative case 2: a 61-year-old man with hepatic hilar bile duct cancer (T3N0M0, stage IV) treated with ERT and IAIC. **a** Cholangiogram showed occlusion of the right and left hepatic ducts, and the common hepatic duct was not opacified. **b** ERT was administered to a 5 × 5 cm² field using a 10 MVX linear accelerator with a total dose of 50 Gy. **c** Cholangiogram after biliary endoprosthesis placement using spiral z-stents (*arrows*) showed good patency of the biliary tract. **d,e** A catheter was placed in the common hepatic artery for IAIC 3 months after ERT because of regrowth of the tumor. Angiography via the placed catheter showed faint tumor stain at the hepatic hilar region (*arrows*). **f,g** Angiography 3 months after the initiation of IAIC (the sixth IAIC session) showed marked shrinkage of the tumor stain (*arrows*). During the patient's 24-month survival period, IAIC was performed a total of 27 times

into the right and left hepatic duct–common bile duct (Fig. 30.9c), and on December 25 the patient was discharged. However, the CA19-9 value rose again, and so regrowth of the tumor was diagnosed (Fig. 30.9d,e), and on March 5, 1993, a catheter was placed in the common hepatic artery so that IAIC could be performed. FE(A)M arterial infusion was commenced via the port once every 2 weeks and this IAIC was continued subsequently in the outpatient clinic. Three months after the initiation of IAIC (the sixth IAIC session) marked shrinkage of the tumor stain was noted on angiography (Fig. 30.9f,g), and the CA19-9 value was also found to have dropped to 71 U/ml. During the 24-month period from the initiation of IAIC until the patient's death from cancer IAIC was performed a total of 27 times. During most of this period (19 months) the patient was able to live at home.

30.5 Conclusion

No definitive treatment strategy for inoperable bile duct cancer has yet been established. Combined interventional radiology by appropriate use of various combinations of metallic stent, ELRT, ERT, and IAIC according to the features of the individual case is a new treatment approach that shows considerable promise in improving both the quality of life and the prognosis of the patient. To further enhance the therapeutic results and quality of life, more accurate diagnosis of the degree of tumor spread based on diagnostic imaging modalities and the establishment of optimal irradiation and arterial infusion chemotherapy regimens will be required.

References

Buskirk SJ, Gunderson LL, Adson MA, et al (1984) Analysis of failure following curative irradiation of gallbladder and extrahepatic bile duct carcinoma. Int J Radiat Oncol Biol Phys 10:2013–2023

Cameron JL, Broe P, Zuidema GD (1982) Proximal bile duct tumors. Ann Surg 196:412–419

Falkson G, MacIntyre JM, Moertel C (1984) Eastern cooperative oncology group experience with chemotherapy for inoperable gallbladder and bile duct cancer. Cancer 54:965–969

Fields JN, Emami B (1987) Carcinoma of the extrahepatic biliary system – results of primary and adjuvant radiotherapy. Int J Radiat Oncol Biol Phys 13:331–338

Fletcher MS, Brinkley D, Dawson JL, et al (1983) Treatment of hilar carcinoma by bile drainage combined with internal radiotherapy using ^{192}iridium wire. Br J Surg 70:733–735

Fritz P, Brambs H, Schraube P, et al (1994) Combined external beam radiotherapy and intraluminal high dose rate brachytherapy on bile duct carcinomas. Int J Radiat Oncol Biol Phys 29:855–861

Harvey JH, Smith FP, Schein PS (1984) 5-Fluorouracil, mitomycin, and doxorubicin (FAM) in carcinoma of the biliary tract. J Clin Oncol 2:1245–1248

Hendrickx P, Luska G, Junker D et al (1990) Ergebnisse der perkutan transluminalen Bestrahlung von Gallenwegskarzinomen mit 192-iridium. Strahlenther Onkol 166:392–396

Ikeda H, Kuroda C, Uchida H, et al (1979) Intraluminal irradiation with iridium-192 wires for extrahepatic bile duct carcinoma – a preliminary report. Nippon Acta Radiol 39:1356–1358

Kajanti M, Pyrhonen S (1994) Epirubicin-sequential methotrexate-5-fluorouracil-leucovorin treatment in advanced cancer of the extrahepatic biliary system. Am J Clin Oncol 17:223–226

Maeda M, Timmermans HA, Uchida BT, et al (1992) In vitro comparison of the spiral z stent and the Gianturco z stent. J Vasc Interv Radiol 3:565–569

Oberfield RA, Rossi RL (1988) The role of chemotherapy in the treatment of bile duct cancer. World J Surg 12:105–108

Okada S, Ishii H, Nose H, et al (1994) A phase II study of cisplatin in patients with biliary tract carcinoma. Oncology 51:515–517

Smith JW, Bukowski RM, Hewlett JS, et al (1984) Hepatic artery infusion of 5-fluorouracil and mitomycin C in cholangiocarcinoma and gallbladder carcinoma. Cancer 54: 1513–1516

Taal BG, Audisio RA, Bleiberg H, et al for the Gastrointestinal Tract Cancer Cooperative Group (1993) Phase II trial of mitomycin C in advanced gallbladder and biliary tree carcinoma. An EORTC gastrointestinal tract cancer cooperative group study. Ann Oncol 4:607–609

Tamada T, Yoshimura H, Yoshioka T, et al (1991) High dose rate ^{60}Co-RALS intraluminal radiation therapy for advanced biliary tract cancer with obstructive jaundice. J Jpn Soc Ther Radiol Oncol 3:251–263

Tamada T, Yoshimura H, Iwata K, et al (1995) The influence of the metallic stent for irradiation – experimental study by phantom and normal bile duct of dogs. J Jpn Soc Ther Radiol Oncol 7:39–46

Uchida H, Kuroda C, Nakamura H, et al (1975) Percutaneous external and internal drainage of biliary tract with special reference to technique and diagnostic evaluation of follow-up cholangiography. Nippon Acta Radiol 35:53–67

Yoshimura H, Sakaguchi H, Tamada T, et al (1989) Afterloading intracavitary irradiation and expandable stent for malignant biliary obstruction. Radiat Med 7:36–41

Yoshioka T, Sakaguchi H, Yoshimura H, et al (1990) Expandable metallic biliary endoprostheses: preliminary clinical evaluation. Radiology 177:253–257

31 Controversies in Biliary Intervention

H. Coons

CONTENTS

31.1 Introduction: Reviewing the Literature

Biliary interventions are some of the most difficult and challenging procedures the interventional radiologist is asked to perform. Not only are they often technically difficult, but deciding which procedure or device is best for your patient can be a challenge. In this chapter an attempt will be made to clarify some of the issues and, by using a common sense approach, give the reader a way of evaluating the literature and making an appropriate decision for the particular patient to be treated.

Reviewing the literature is often confusing rather than enlightening. In part, this is due to the common practice of reporting an inhomogeneous group of patients in the same paper and drawing conclusions which do not apply to all the patients equally (Adam et al. 1991; Gordon et al. 1992; Lameris et al. 1991; Lammer et al. 1990; Lee et al. 1992; Salomonowitz et al. 1992). Stated another way, only by comparing results in patients with similar disease conditions and similar life expectancies can the efficacy of a particular procedure or device be realistically determined. The "lumping" of patient groups together in one paper may be in part to increase the total number of patients included, or may simply reflect the author's feeling that separation is not needed or critical. A single paper reporting the results of a particular technique, such as using expandable metallic stents for treating biliary obstruction, can lump together patients with as diverse a set of disease processes as benign postoperative stricture, sclerosing cholangitis, carcinoma of the pancreas, hilar obstruction secondary to metastatic disease, and cholangiocarcinoma. Is it really possible to draw reliable conclusions from such a widely diverse group with widely different life expectancies, such that readers can use them as a help in deciding what is best for a given patient? The answer to this question clearly is no. Ideally, authors should recognize this problem and separate their patients into appropriate categories, so that the conclusions which are drawn can be used in making decisions for individual patients. It is unlikely that this will happen, however, so it is important to read a paper critically and really try to see how the results apply to your particular patient.

Why is this an issue that needs to be discussed? When trying to decide which procedure or device is best for your patient, you need to see results in patients with similar disease or problems. This is particularly important when there are conflicting reports or new technologies. In addition, you must always individualize therapy to the unique circumstances of your particular patient. For instance, "metallic stents should not be used in treating benign biliary strictures" (Coons 1992) is an appropriate generalization, but there are cases in which the use of a metallic stent is the best available option. This will be discussed later in this chapter. As with any procedure, be it surgical or interventional, you need to assess the risks of the procedure, including long-term complications, against its potential benefits. The "risk-benefit ratio" must also be applied to alternative therapies. Only then can an appropriate therapy be offered to the patient and his or her family.

31.2 Treatment of Cholangiocarcinoma

The problem that this indiscriminate grouping of patients creates becomes visible if we look at the

H. Coons, MD, Department of Radiology, Sharp Memorial Hospital, 7901 Frost Street, San Diego, CA 92123-2788, USA treatment of cholangiocarcinoma. There is controversy about the best treatment for cholangiocarcinoma, partly due to the grouping of these patients with those who have hilar obstruction from metastatic disease or, worse, with patients who have malignant common duct obstruction from carcinoma of the pancreas (MARTIN 1992). The trouble is that the life expectancies and disease processes are so very different in these groups of patients that comparison is meaningless. Patients with carcinoma of the pancreas have an average life expectancy of 4 months after the initial diagnosis. Patients with metastatic hilar obstruction likewise have a very short life expectancy. Patients with cholangiocarcinoma, on the other hand, have a much longer life expectancy - but even in patients with cholangiocarcinoma there is a wide range of variation, depending on how advanced the disease process is. In most instances, common sense tells us that a patient with very advanced disease involving intrahepatic radicles in both the right and left systems (Bismuth class IV) will do less well than a patient with an isolated lesion in the common duct (Bismuth class I). How does all this affect our decision making?

First, we must realize that the occlusion rates (or, conversely, the patency rates) of catheters or stents are the most common way of evaluating their efficacy. If patients die while their stent is patent, the stent can be counted as "patent forever" in statistics. Obviously, the longer a patient lives, the greater chance that eventually the catheter or stent will become obstructed.

Next, one must review the cause of obstruction of the stents. The most common cause of obstruction of plastic stents is infection, with clogging of the stent with bacterial slime (LEUNG et al. 1988). So how can we keep infection rates down? Antibiotic therapy seems to be a natural choice; however, there are pitfalls in its long-term use. Very resistant strains of bacteria such as *Pseudomonas* or *Klebsiella* can become established. In addition, fungal infections, which are extremely difficult to treat, often occur in cases where long-term antibiotic therapy has been used. It is better to attack the source of the infection rather than to try to treat the infection once it has occurred. So what is the source of the recurrent infection? Inadequately drained systems nearby which are bacterially contaminated. One can be sure that the system is contaminated if contrast is seen to go into the system. Even if no contrast refluxes into the system, however, bacteria can make their way from the contaminated system in which there is a biliary drainage catheter through a very tiny communication with the nearby obstructed system. Anyone who has a biliary drainage catheter in place for any length of time will have a contaminated biliary tree, as it is simple for the bacteria to enter the biliary tree along the catheter.

For good long-term patency rates, the need to adequately drain the entire biliary tree cannot be overemphasized. Again, this is a common sense issue. It is always preferable to drain all of the biliary tree rather than only a portion. Past convention has been that left-sided drainage is adequate for patients with hilar obstruction, but GORDON et al. (1981) clearly defined the problem that an undrained, infected system causes as early as 1980. What is "adequate" for the patient? Certainly, the bilirubin levels will decrease with left-sided drainage, and if the patient is suffering from pruritus, the symptoms may well go away. However, such patients will have many more episodes of cholangitis and in most cases will be chronically infected and not feel well, and ultimately survival is shortened. Draining the left side only is often performed because there are multiple obstructions on the right side and having multiple catheters seems less desirable. There is a technique, however, which allows one to drain all of the system through a single or, at most, two access points. If one initially drains the left system, it is often possible to enter the right system in a retrograde fashion and, by placing metallic stents, get adequate drainage of all of the major biliary systems. Using an angled catheter, such as the Kumpe catheter (Cook, Inc.), and an angled guidewire, such as the Terumo Glidewire, the right system can be accessed in the vast majority of cases. If this fails, one can perform a right-sided cholangiogram for roadmapping and either advance a catheter from the right side into the common duct or access the right system in a retrograde fashion (Fig. 31.1). Once a guidewire has been successfully advanced into the obstructed system, a small angioplasty balloon, such as a 4-mm balloon, can be used to create more room. Any angioplasty performed, however, will have short-term success in terms of dilating the obstructed area.

Now that we have a guidewire into the obstructed system, what do we do? Although Wallstents provide excellent drainage, they cannot be placed at right angles to one another, and usually only two stents can be placed. This, obviously, will provide inadequate drainage if both the right anterior and the posterior systems are obstructed as well as the left system. Again, leaving an obstructed system leaves you with a recurrent cholangitis problem and a re-

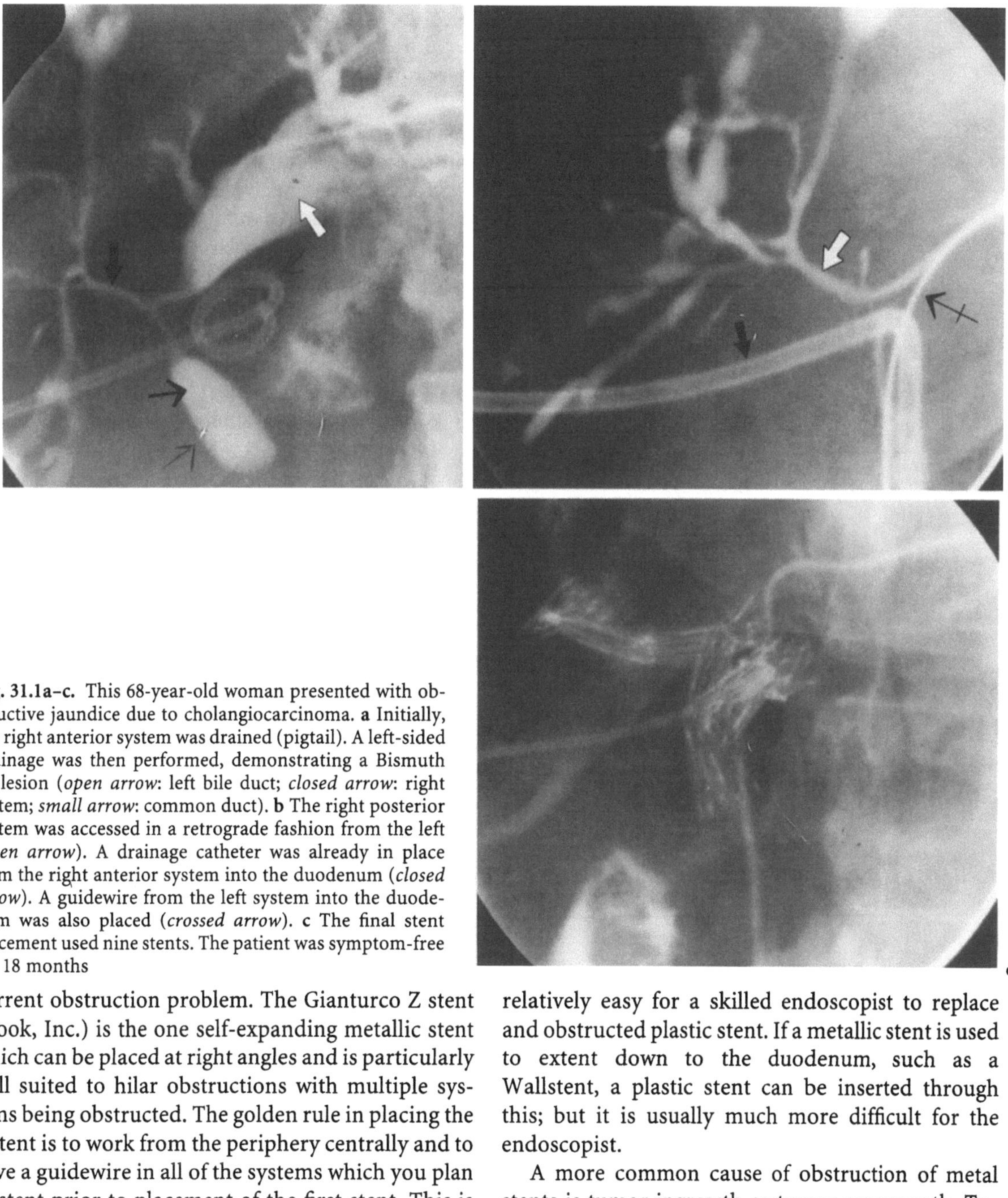

Fig. 31.1a–c. This 68-year-old woman presented with obstructive jaundice due to cholangiocarcinoma. **a** Initially, the right anterior system was drained (pigtail). A left-sided drainage was then performed, demonstrating a Bismuth IV lesion (*open arrow*: left bile duct; *closed arrow*: right system; *small arrow*: common duct). **b** The right posterior system was accessed in a retrograde fashion from the left (*open arrow*). A drainage catheter was already in place from the right anterior system into the duodenum (*closed arrow*). A guidewire from the left system into the duodenum was also placed (*crossed arrow*). **c** The final stent placement used nine stents. The patient was symptom-free for 18 months

current obstruction problem. The Gianturco Z stent (Cook, Inc.) is the one self-expanding metallic stent which can be placed at right angles and is particularly well suited to hilar obstructions with multiple systems being obstructed. The golden rule in placing the Z stent is to work from the periphery centrally and to have a guidewire in all of the systems which you plan to stent prior to placement of the first stent. This is critical, as placement of the first stent will often obstruct the neighboring system to the extent that one will not be able to reaccess it once the stent has been released (Fig. 31.2). A combination of the metallic Z stents in the right systems and a plastic stent extending from the left system into the duodenum is often used. The advantage of this combination is that if there is an obstruction, the plastic stent can be removed endoscopically and replaced without the need to have a second intervention through the liver (Fig. 31.3). This is certainly easier on the patient, and it is relatively easy for a skilled endoscopist to replace and obstructed plastic stent. If a metallic stent is used to extent down to the duodenum, such as a Wallstent, a plastic stent can be inserted through this; but it is usually much more difficult for the endoscopist.

A more common cause of obstruction of metal stents is tumor ingrowth or tumor overgrowth. Tumor overgrowth – the extension of tumor beyond the stents – is best avoided by providing wide margins proximal and distal to the known tumor. In patients with cholangiocarcinoma, intracavitary iridium treatment to slow the ingrowth has been very effective. Life is prolonged by decreasing the infection and occlusion rates, and the quality of life is also much better. Martin et al. (1991) concluded that metallic stents were not effective in treating cholangiocarcinoma, while Coons (1992) suggested that they are. In the first study rather short stents were used and the tumor was not irradiated with iridium seeds. In the second study there was a large margin of stent beyond the tumor and iridium was used. The appropriate conclusion is that in cholangiocarcinoma the tumor should be pretreated with iridium seeds and wide stent margins are necessary for good long-term results. This point is emphasized when the causes of failure in the first study are analyzed: the stents occluded from tumor overgrowth – the result of stents that are too short – and tumor ingrowth – due to lack of pretreatment. Seen this way, controversy over the best treatment for cholangiocarcinoma seems less controversial.

To summarize: the most effective treatment for cholangiocarcinoma includes the following: (1) Adequate drainage of the entire obstructed biliary system, facilitated by approaching from the left side and entering the right system in a retrograde fashion. (2) Treating the tumor with iridium seeds to cut down the rate of ingrowth through the stent. (3) Using Gianturco Z stents and deploying them from the periphery toward the center after placing a guidewire in each system that is to be stented. (4) Allowing wide margins at both ends of the stent in expectation of continued tumor extension.

31.3 Metallic Versus Plastic Stents

Since the introduction of the metallic stents, there has been a debate about the superiority of metallic stents over plastic stents in treating biliary obstruction. Are metallic stents better than plastic stents? The answer is yes, no, and maybe. Let us explore the matter and see how this can be.

Clearly, metallic stents are preferable in treating hilar obstruction if they are used in the way described in the previous section. Their use in carcinoma of the pancreas with distal common duct obstruction is less clear. Common sense tells us that a 30-French metallic stent is going to have a better patency rate than a 12- or 14-French thick-walled tube, and therefore metallic stents should be better. A paper by Adam (1992) tends to bear out this intuitive feeling; however, earlier papers suggested that

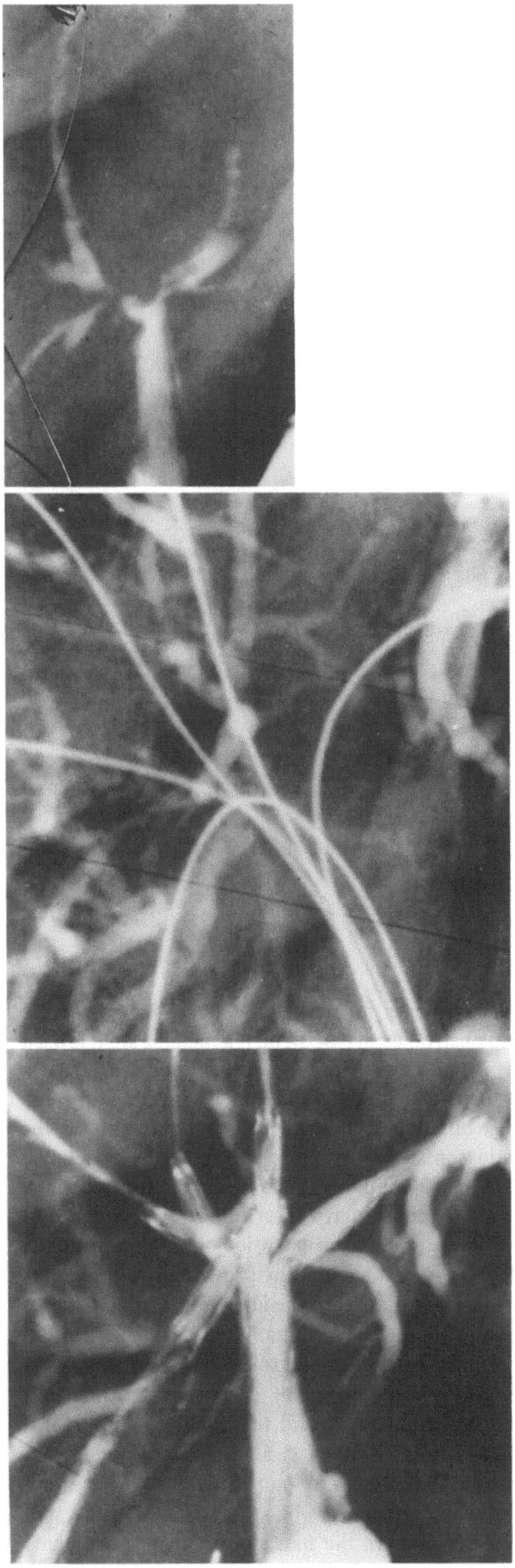

a b c

Fig. 31.2a–c. This 72-year-old woman remained jaundiced after a cholecystectomy. **a** T-tube cholangiogram demonstrated cholangiocarcinoma at the confluence. **b** Multiple obstructed systems were accessed in a retrograde manner via the T-tube tract and guidewires placed. **c** After stent placement a cholangiogram demonstrated free flow

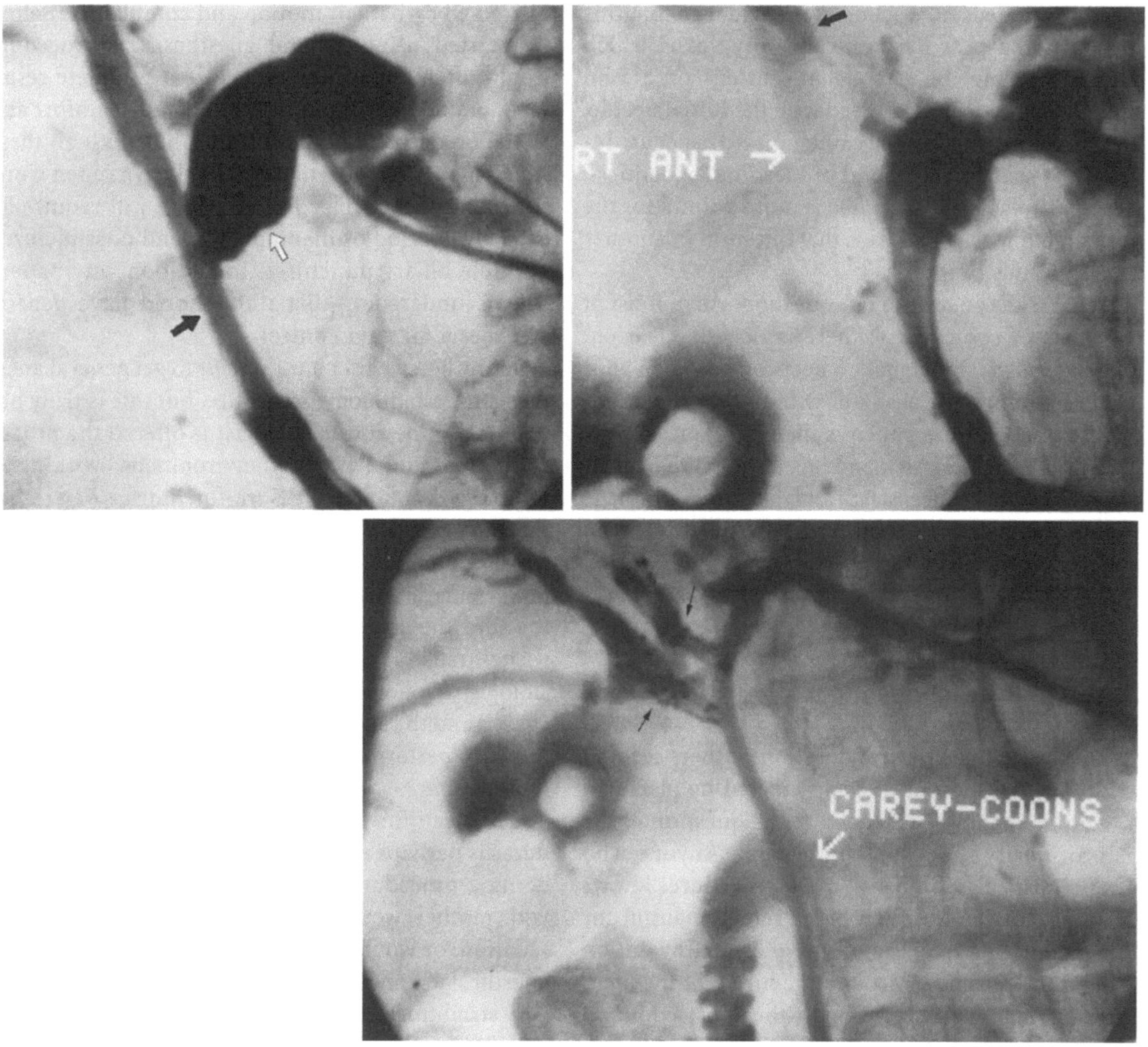

Fig. 31.3. **a** This 69-year-old man had an endoscopically placed Teflon stent (*closed arrow*). His jaundice returned and a left-sided cholangiogram was performed (*open arrow*). **b** The endoscopic stent was removed and cholangiography showed obstruction of the right anterior and posterior systems (*closed arrow*). **c** The obstructed right systems were entered in a retrograde fashion and Gianturco-Rösch Z stents were deployed (*arrow*) and a plastic Carey-Coons stent placed from the left system to the duodenum. He remains without obstruction at 16 months

there was not a significant difference in patency rates between the two.

A number of factors need to be considered. First, the life expectancy of the patients is very short and most plastic tubes will remain patent for up to 4 or 5 months. Another factor is cost. The metallic stents are significantly more expensive. But one must look at the true cost of the care of the patient rather than the isolated cost of one stent. Reintervention for an obstructed plastic stent nullifies any savings one might have made in using it initially. However, as stated previously, if a plastic stent becomes obstructed, it can be removed and replaced endoscopically at a reduced cost in stress to the patient; it is certainly less traumatic and painful than a second biliary intervention. It is often difficult for the endoscopist to place a plastic stent within a previously placed metallic stent, particularly when it extends well into the duodenum. If a metallic stent becomes obstructed and requires a second biliary intervention, the cost savings then shift back to the plastic stent.

Another factor is the size of the hole necessary to place the stent. With the metallic stents a 7- or 8-French hole is adequate; with the plastic stents one must go to the size of the stent being used, 12- or 14-French. Although it has been implied that a larger tube increases the frequency of complication, such as bleeding, there is no documentary evidence that this in fact is the case. Theoretically, any vessel which causes bleeding, be it a hepatic artery or a vein branch, has been transgressed with the initial needle, and therefore whether the catheter is 8-French or 14-French is size the likelihood of bleeding is identical. Certainly, placing a large tube is more painful for the patient, but we shall address that question in another section of this chapter.

What is the answer to our question? As in most of the areas of controversy, there is no one clear-cut answer, and you must make a decision based on your local economics, the availability of the various products to you, and the patient's life expectancy. If the patient is extremely sick and has a short life expectancy, the quickest, easiest, and cheapest drainage should be performed. In most cases this is an endoscopically placed stent.

31.4 General Anesthesion for Biliary Interventions

As we answer one group of questions, others seem to arise. The question of the pain created by placing a large tube should be extended to the question of the pain of biliary procedures in general. Anyone who has performed a number of these procedures knows that on occasion they can be extremely painful for the patient, particularly if there is extravasation around the capsule or around the hilum of the liver. Although conscious sedation is the most common pain control used, it is inadequate in such cases. After a number of cases of this kind, we have chosen to use general anesthesia for all initial biliary drainage procedures where a cholangiogram is to be performed and a catheter to be placed.

There are many advantages to this. The most significant advantage is that the patient does not suffer during the procedure. Although patients often have amnesia regarding the procedure after conscious sedation has been used, during the procedure they tend to be uncooperative and agitated. This makes a difficult procedure even more difficult and is an unnecessary hindrance. Although there is a risk related to the general anesthesia, and a cost, they are certainly very small relative to the advantages that general anesthesia provides for the interventionalist and the patient. Mainly, it allows one to focus completely on the task at hand rather than continually having to break concentration to evaluate the patient's vital signs and administer additional medication, or simply being distracted by patient motion and complaints. Before we started using general anesthesia, a procedure would often be terminated before a complete result was obtained, owing to the patient's discomfort and to the operator's fatigue from dealing with all these side issues. This is particularly likely to happen if one is dealing with a hilar lesion, which will require an extensive period of manipulation and possibly even multiple biliary punctures. In addition, any patient who is undergoing dilatation should have general anesthesia for pain control.

Many believe that it is difficult to get general anesthesia for a radiology procedure, but this is truly not the case. If the anesthesiologist is offered the proper equipment and a friendly environment, working in radiology is often a pleasure for him or her rather than an ordeal. We can all relate to the problem: if we are asked to perform a difficult procedure in an unfamiliar surrounding with inadequate equipment, we too find the stress levels very high. If a radiologist wishes to use anesthesia, it is best to ask the anesthesiologist for advice as to the equipment to be purchased and the environment that is necessary for him or her to perform the job in an adequate and competent manner.

The need for general anesthesia for these procedures is perhaps most easily grasped if one considers doing a procedure on a family member: if this was your mother, would you want her to have conscious sedation, or would you want her to be under general anesthesia? For me, the answer to that question is very strightforward. My mother would be under general anesthesia if she were to undergo a biliary intervention.

31.5 Large- Versus Small-Bore Catheters

Under general anesthesia, obviously, the passage of a large-bore catheter is no more painful than the passage of a small-bore catheter. The feeling that smaller is better when it comes to tube size is understandable, but it may perhaps be a prejudice which is not borne out by fact. If a choledochoscope is to be used for stone removal, a relatively large-bore sheath must be placed; even with the newer small-sized choledochoscopes, it is necessary to place a generous-sized sheath. Complex stone cases, such as in patients with oriental cholangiohepatitis, require the use of a choledochoscope to adequately fragment and remove all of the many stones present (Bower et al. 1990). If one is concerned with a bleeding complication from the placement of a large catheter, this can be alleviated by checking the track after the initial puncture of the biliary tree. If one is using a coaxial entry system, such as the Accustick system (Medi-tech), checking for transgression of blood vessels is relatively easy: all you need to do is to place a Touhy-Borst-Y connector on the end of the catheter with a guidewire in place, tighten the diaphragm, and inject through the side arm as the catheter is gradually withdrawn through the liver substance. Obviously, the guidewire is going to move as well, so one must have enough guidewire in the biliary tree so that one does not lose position within the biliary tree when withdrawing the catheter and guidewire. If a major vein or an arterial branch has been crossed, the biliary tree can be repunctured at a different location and significant bleeding eliminated by not placing a large catheter across this transgressed vessel. A repuncture is often desirable once the biliary tree has been initially opacified. Ideally, the biliary tree is entered very peripherally and through a small radicle. Unless you are just lucky on your original pass, however, this is usually not where you have entered the tree during the cholangiogram, and most commonly we repuncture the biliary tree at a better site after partial opacification. A word of caution: overdistention of the biliary tree with contrast increases the likelihood of septicemia.

31.6 Use of Metallic Stents in Benign Postoperative Strictures

The use of metallic stents for treating benign postoperative strictures was mentioned earlier in this chapter. The early results were very promising and soon after their initial appearance it became widespread practice to use metallic stents for benign strictures. As long-term results began to be reported, however, it became clear that there were problems with metallic stents in benign postoperative strictures. Patency rates at 1 year were excellent, but at 5 years the picture had changed completely: the foreign body reaction, with subsequent mucosal hypertrophy, had caused reocclusion in the vast majority of patients. This was usually signaled by an episode of ascending cholangitis, and restudy revealed obstruction which was not scar tissue but rather hypertrophied mucosa. Not uncommonly the patients also formed intrabiliary stones as a result of the infection.

The conclusion that metallic stents were not a good choice in treating benign biliary strictures was an appropriate one. This does not mean, however, that no patient with a benign postoperative stricture should have a metallic stent. A case in point is the first patient in our practice to receive a metallic stent as treatment for benign postoperative stricture. This woman was 65 years of age when she first presented to us with obstructive jaundice. She had undergone a Whipple procedure some 3 years previously and had two subsequent revisions of the choledochojejunostomy because of recurrent strictures. Initially, a biliary drainage was performed and a balloon dilatation gave her good relief. Over the next 3 years, however, she returned approximately every 9 months with recurrence of symptoms and recurrence of her stricture. Multiple dilatations were performed and, in addition, electrocuting of the stricture, none of which permanently relieved her symptoms. Nine years ago we placed a metallic stent for the first time, and the lady was symptom-free for 4 years and 8 months. She then reappeared with ascending cholangitis, and study revealed obstruction at the level of the stent. A second stent was placed at that time, after which she was symptom-free for an additional 30 months. A reintervention was performed and she was found to have formed stones above the biliary stents; there was also mucosal hypertrophy. The stents were cleared, and after a period of 2 months in which she had an external catheter, the biliary tree was again clear and the catheter was removed. She is now 16 months after the latest intervention and is symptom-free. At this time she is 77 years of age and doing well.

Most likely this lady will return with obstructive jaundice and require an additional intervention. Although this is certainly not ideal, she has been symptom-free for the vast majority of the 12 years that she has been treated. A worse-case scenario would have her with an external drainage catheter for the remainder of her life. Is this good treatment or not? From the patient's standpoint, it certainly is. The three surgeries she had had prior to our first intervention were very difficult, with very long recovery times. Her life has been modified very little by the need to have additional procedures over the subsequent years. Certainly a more ideal solution would be a device which would keep the biliary tree open and require no further interventions, but to date such a device is not available. This is a demonstration that decisions must be made on an individual patient basis, and if the best option, under the circumstances, is to place a metallic stent, then certainly it should be used.

31.7 Conclusion

Controversy arises when there is no clear-cut best procedure or device for treating a specific problem. In most cases, the best one can do is to research the literature to see the experience others have had with a similar problem, to evaluate all of the options available for the patient, and to make the best decision that one can. Each alternative must be scrutinized and honestly evaluated in terms of its complications, its potential for success, and its long-term efficacy. The critical question to ask oneself is "What procedure would I recommend if the patient were a family member?" Answering this question honestly often resolves a difficult dilemma in which there appears to be no good choice or certainly no optimal choice in treating the patient.

This chapter could have taken the opposite view on each of these issues and presented a very strong case for that viewpoint. I am sure other chapters in this volume voice convincingly opinions that conflict with those expressed here. The intention in this chapter was to try not merely to identify controversial issues and to give an opinion on them, but to offer an approach to these difficult questions which will allow you to choose the right procedure or device for your patients. The controversial issues enumerated here may soon have clear-cut resolutions, but others will certainly arise as progress is made in both techniques and devices. For two patients with essentially identical clinical stories there may be two very different solutions, each correct for the patient concerned. It all depends on the talent of the interventional radiologist, the surgeon, endoscopist, and others caring for the patient, as well as the operator's own personal experience in dealing with the problem at hand. Critical review of the literature with the application of common sense and an appropriate evaluation of the patient should give you the tools with which to make good decisions in each specific case.

References

Adam A, Chetty N, Roddie M, Young E, Benjamin IS (1991) Self-expandable stainless steel endoprosthesis for treatment of malignant bile duct obstruction. AJR Am J Roentgenol 156:321–325

Bower BL, Picus D, Hicks ME, Darcy MD, Rollins ES, Kleinhoffer MA, Weyman PJ (1990) Choledochoscopic stone removal through a T-tube tract: experience in 75 consecutive patients. J Vasc Interv Radiol 1:107–111

Castaneda WR, Tadavarthy SM, Laerum F, Amplatz K (1981) Anterior approach for biliary duct drainage. Radiology 139:746–748

Coons H (1992) Metallic stents for treatment of biliary obstruction: report of 100 cases. Cardiovasc Intervent Radiol 15:360–367

Gordon RL, Ring EJ, La Berge JM (1992) Malignant biliary obstruction: treatment with expandable metallic stents – follow-up of 50 consecutive patients. Radiology 182:697–701

Lameris JS, Stoker J, Nijs HGT, Zonderland HM, Terpstra OT, von Blankenstein M, Schulte HE (1991) Malignant biliary obstruction: percutaneous use of self-expandable stents. Radiology 179:703–709

Lammer J, Neumayer K (1986) Biliary drainage endoprostheses: experience with 201 placements. Radiology 159:625–629

Lammer J, Klein GE, Kleinert R, Hausegger K, Einspieler R (1990) Obstructive jaundice: use of expandable metal prostheses for biliary drainage. Radiology 177:789–792

Lee MJ, Dawson SL, Mueller PR, Krebs TL, Saini S, Hahn PF (1992) Palliation of malignant bile duct obstruction with metallic biliary endoprostheses: technique, results, and complications. J Vasc Interv Radiol 3:665–671

Lee MJ, Dawson SL, Mueller PR, Saini S, Hahn PF (1993) Percutaneous management of hilar biliary malignancies with metallic endoprostheses: results, technical problems, causes of failure. Radiographics 13:1249–1253

Leung JWC, Ling TKW, Kung JLS, Vallance-Owen J (1988) Clogging of stents by bacterial infection. Gastrointest Endosc 34:19–22

Martin EC, Laffey KJ, Bixon R, Getrajdman GI (1990) Gianturco-Rösch biliary stents: preliminary experience. J Vasc Interv Radiol 1:102–105

Mendez GJ, Russell E, Le Page JR, Guerra JJ, Posniak RA, Trefler M (1984) Abandonment of endoprosthetic drainage technique in malignant biliary obstruction. AJR Am J Roentgenol 143:617–620

Salomonowitz EK, Antonucci F, Heer M, Stuckmann G, Eglof B, Zollikofer CL (1992) Biliary obstruction: treatment with self-expanding metal prostheses. J Vasc Interv Radiol 3:365–370

Subject Index

List of Contributors

ANDY ADAM, MD
Guy's and St. Thomas's Medical and Dental School
Division of Radiological Sciences
Interventional Radiology
Guy's Hospital
London Bridge
London SE1 9RT
UK

FLAMINIA M. ARATA, MD
Department of Radiology
University of Rome "La Sapienza"
Policlinico Umberto I
00161 Rome
Italy

ALBERT L. BAERT, MD
Department of Radiology
University Hospitals K.U. Leuven
Herestraat 49
3000 Leuven
Belgium

HELENA BALON, MD
Department of Nuclear Medicine
William Beaumont Hospital
3601 West 13 Mile Road
Royal Oak, MI 48073-6769
USA

MASSIMO BAZZOCCHI, MD
Director
Cattedra di Radiologia
Policlinico Universitario
Piazzale S. Maria della Misericordia
33100 Udine
Italy

MARIO BEZZI, MD
Assistant Professor
Department of Radiology
University of Rome "La Sapienza"
Policlinico Umberto I
00161 Rome
Italy

GUIDO BONOMO, MD
Resident in Diagnostic Radiology
Department of Radiology
University of Rome "La Sapienza"
Policlinico Umberto I
00161 Rome
Italy

HANS-JÜRGEN BRAMBS, MD
Department of Radiology
University Hospital
Steinhövelstrasse 9
89075 Ulm
Germany

PATRICE M. BRET, MD
Department of Diagnostic Radiology
Montreal General Hospital
1650 Cedar Avenue
Montréal
Quebec H3G 1A4
Canada

LAURA BROGLIA, MD
Department of Radiology
University of Rome "La Sapienza"
00161 Rome
Italy

MARIA CARLA CASSINIS, MD
Istituto di Radiodiagnostica
Ospedale Maggiore
28100 Novara
Italy

WILFRIDO R. CASTAÑEDA-ZUÑIGA, MD
Professor and Chairman
Department of Radiology
Louisiana State University Medical Center
1542 Tulane Avenue
New Orleans, LA 70112
USA

Byung Ihn Choi, MD
Professor
Department of Radiology
Seoul National University Hospital
28 Yongon-dong, Chongno-gu
Seoul 110-744
Korea

Harold Coons, MD
Department of Radiology
Sharp Memorial Hospital
7901 Frost Street
San Diego, CA 92123-2788
USA

Guido Costamagna
Professor
Department of Surgery
Catholic University School of Medicine
Largo Gemelli 8
00168 Rome
Italy

Michel Cremer, MD
Professor
Medical Surgical Department of Gastroenterology
and Hepatopancreatology
Université Libre de Bruxelles
Hôpital Erasme
1070 Brussels
Belgium

Steven L. Dawson, MD
Department of Radiology
Massachusetts General Hospital
Harvard Medical School
32 Fruit Street
Boston, MA 02114
USA

Jacques Deviere, MD
Medical Surgical Department of Gastroenterology
and Hepatopancreatology
Hôpital Erasme
1070 Brussels
Belgium

Raffaela Di Nardo, MD
Senior Staff Radiologist
Department of Radiology
University of Rome "La Sapienza"
00161 Rome
Italy

Kenneth E. Fellows, MD
Department of Radiology, 3rd Floor
Children's Hospital of Philadelphia
34th and Civic Center Blvd.
Philadelphia, PA 19104
USA

Darlene Fink-Bennett, MD
Department of Nuclear Medicine
William Beaumont Hospital
3601 West 13 Mile Road
Royal Oak, MI 48073-6769
USA

Fausto Fiocca, MD
Department of Surgery
University of Rome "La Sapienza"
Policlinico Umberto I
00161 Rome
Italy

Paolo Fonio, MD
Istituto di Radiodiagnostica
Ospedale Maggiore
28100 Novara
Italy

Roberto Galeotti, MD
Istituto di Radiologia
Università di Ferrara
Corso Giovecca 203
44100 Ferrara
Italy

Giovanni Gandini, MD
Professor
Istituto di Radiodiagnostica
Ospedale Maggiore
28100 Novara
Italy

L. Grenacher, MD
Abteilung Radiodiagnostik
Radiologistche Universitatsklinik Heidelberg
Im Neuenheimer Feld 110
69120 Heidelberg
Germany

Joon Koo Han, MD
Assistant Professor
Department of Radiology
Seoul National University Hospital
28 Yongon-dong, Chongno-gu
Seoul 110-744
Korea

Man Chung Han, MD
Professor
Department of Radiology
Seoul National University Hospital
28 Yongon-dong, Chongno-gu
Seoul 110-744
Korea

Johannes Lammer, MD
Universitätsklinik
Allgemeines Krankenhaus Wien
Währinger Gürtel 18-20
1090 Vienna
Austria

Jose M. Llerena, MD
Assistant Professor
Louisiana State University
University Medical Center
Department of Radiology
2390 West Congress Street
Lafayette, LA 70506
USA

Davide Lomanto, MD
Assistant Professor
Clinica Chirurgical II
University of Rome "La Sapienza"
Policlinico Umberto I
Viale del Policlinico, 155
00161 Rome
Italy

Nicolas J. Lygidakis, MD
Professor
Department of Hepatobiliary Pancreatic Surgery
St. Savas Hospital
P.O. Box 17160
10024 Athens
Greece

Jochen Maass, MD
Istituto di Radiodiagnostica
Ospedale Maggiore
28100 Novara
Italy

Francesca Maccioni, MD
Department of Radiology
University of Rome "La Sapienza"
Policlinico Umberto I
00161 Rome
Italy

Paolo Mannella, MD
Professor
Istituto di Radiologia
Università di Ferrara
Corso Giovecca 203
44100 Ferrara
Italy

Roberto Merlino, MD
Department of Radiology
University of Rome "La Sapienza"
Policlinico Umberto I
00161 Rome
Italy

Peter R. Mueller, MD
Department of Radiology
Massachusetts General Hospital
Harvard Medical School
32 Fruit Street
Boston, MA 02114
USA

Massimiliano Natrella, MD
Istituto di Radiodiagnostica
Ospedale Maggiore
28100 Novara
Italy

Yuji Nimura, MD
Professor and Chairman
The First Department of Surgery
Nagoya University School of Medicine
65 Tsurumai-cho Showa-ku
Nagoya 466
Japan

Jae Hyung Park, MD
Professor
Department of Radiology
Seoul National University Hospital
28 Yongon-dong, Chongno-gu
Seoul 110-744
Korea

Alexandre Pissas, MD
Laboratoire d'Anatomie
Faculté de Médicine
34059 Montpellier
France

E. Quaia, MD
Cattedra di Radiologia
Policlinico Universitario
Piazzale S. Maria della Misericordia
33100 Udine
Italy

Pierre Rabischong, MD
Professor
Laboratoire d'Anatomie
Faculté de Médicine
34059 Montpellier
France

E.A.J. Rauws, MD, PhD
Department of Gastroenterology
Academic Medical Center
Meibergdreef 9
1105 AZ Amsterdam
The Netherlands

Douglas C.B. Redd, MD
Department of Radiology, 3rd Floor
Children's Hospital of Philadelphia
34th and Civic Center Blvd.
Philadelphia, PA 19104
USA

Jacques W.A.J. Reeders, MD, PhD
Head, Department of Gastrointestinal Radiology
and Hepato-Pancreato-Biliary Imaging
Academic Medical Center
Meibergdreef 9
1105 AZ Amsterdam
The Netherlands

Caroline Reinhold, MD
Department of Diagnostic Radiology
Montreal General Hospital
1650 Cedar Avenue
Montréal, Quebec H3G 1A4
Canada

Götz M. Richter, MD
Abteilung Radiodiagnostik
Radiologische Universitätsklinik Heidelberg
Im Neuenheimer Feld 110
69120 Heidelberg
Germany

A. Rieber, MD
Department of Radiology
University Hospital
Steinhövelstrasse 9
89075 Ulm
Germany

A. Rigamonti, MD
Cattedra di Radiologia
Policlinico Universitario
Piazzale S. Maria della Misericordia
33100 Udine
Italy

Dorico Righi, MD
Istituto di Radiodiagnostica
Ospedale Maggiore
28100 Novara
Italy

Michele Rossi, MD
Department of Radiology
University of Rome "La Sapienza"
Policlinico Umberto I
00161 Rome
Italy

Plinio Rossi, MD
Professor and Chairman
Department of Radiology
University of Rome "La Sapienza"
Policlinico Umberto I
00161 Rome
Italy

Hiroshi Sakaguchi, MD
Department of Radiology
Nara Medical University
840 Shijo-cho
Kashihara, Nara 634
Japan

Filippo M. Salvatori, MD
Department of Radiology
University of Rome "La Sapienza"
Policlinico Umberto I
00161 Rome
Italy

N.J. Smits, MD
Department of Gastrointestinal Radiology
Academic Medical Center
Meibergdreef 9
1105 AZ Amsterdam
The Netherlands

Vincenzo Speranza, MD
Professor
Clinica Chirurgica II
University of Rome "La Sapienza"
Policlinico Umberto I
Viale del Policlinico, 155
00161 Rome
Italy

Luigi Tipaldi, MD
Servizio di Radiologia e Diagnostica
per Immagini
Istituto Regina Elena
Viale Regina Elena 291
00161 Rome
Italy

Alessandra Tortora, MD
Department of Radiology
University of Rome "La Sapienza"
00161 Rome
Italy

Hideo Uchida, MD
Professor
Department of Radiology
Nara Medical University
840 Shijo-cho
Kashihara, Nara 634
Japan

Dirk Vanbeckevoort, MD
Department of Radiology
University Hospitals
Herestraat 49
3000 Leuven
Belgium

Otto J.M. van Delden, MD
Department of Gastrointestinal Radiology
and Hepato-Pancreato-Biliary Imaging
Academic Medical Center
Meibergdreef 9
1105 AZ Amsterdam
The Netherlands

Lieven van Hoe, MD
Department of Radiology
University Hospitals K.U. Leuven
Herestraat 49
3000 Leuven
Belgium

Jean-Luc van Laethem
Medical Surgical Department of Gastroenterology
and Hepatogastroenterology
Hopital Erasme
1070 Brussels
Belgium

Robert L. Vogelzang, MD
Professor of Radiology
Northwestern University Medical School
and Chief
Vascular and Interventional Radiology
Northwestern Memorial Hospital
710 North Fairbanks Court
Chicago, IL 60611
USA

Anthony F. Watkinson, MD
Consultant Radiologist and Senior Lecturer
X-Ray Department
Royal Free Hospital
Pond Street
Hampstead
London NW3
UK

Hitoshi Yoshimura, MD
Associate Professor
Department of Oncoradiology
Nara Medical University
840 Shijo-cho
Kashihara, Nara 634
Japan

Arthur L. Zerbey, MD
Department of Radiology
Massachusetts General Hospital
Harvard Medical School
32 Fruit Street
Boston, MA 02114
USA

Christoph L. Zollikofer, MD
Department of Radiology
Kantonsspital
Brauerstrasse 15
8401 Winterthur
Switzerland

C. Zuiani, MD
Cattedra di Radiologia
Policlinico Universitario
Piazzale S. Maria della Misericordia
33100 Udine
Italy

MEDICAL RADIOLOGY – Diagnostic Imaging and Radiation Oncology

Titles in the series already published

Diagnostic Imaging

Innovations in Diagnostic Imaging Edited by J.H. Anderson

Radiology of the Upper Urinary Tract Edited by E.K. Lang

The Thymus – Diagnostic Imaging, Functions, and Pathologic Anatomy Edited by E. Walter, E. Willich and W.R. Webb

Interventional Neuroradiology Edited by A. Valavanis

Radiology of the Pancreas Edited by A.L. Baert, co-edited by G. Delorme

Radiology of the Lower Urinary Tract Edited by E.K. Lang

Magnetic Resonance Angiography Edited by I.P. Arlart, G.M. Bongartz and G. Marchal

Contrast-Enhanced MRI of the Breast S.H. Heywang-Köbrunner and R. Beck

Spiral CT of the Chest Edited by M. Rémy-Jardin and J. Rémy

Radiological Diagnosis of Breast Diseases Edited by M. Friedrich and E.A. Sickles

Radiology of the Trauma Edited by M. Heller and A. Fink

Biliary Tract Radiology Edited by P. Rossi. Co-edited by M. Bezzi

Radiation Oncology

Lung Cancer Edited by C.W. Scarantino

Innovations in Radiation Oncology Edited by H.R. Withers and L.J. Peters

Radiation Therapy of Head and Neck Cancer Edited by G.E. Laramore

Gastrointestinal Cancer – Radiation Therapy Edited by R.R. Dobelbower, Jr.

Radiation Exposure and Occupational Risks Edited by E. Scherer, C. Streffer and K.-R. Trott

Radiation Therapy of Benign Diseases – A Clinical Guide S.E. Order and S.S. Donaldson

Interventional Radiation Therapy Techniques – Brachytherapy Edited by R. Sauer

Radiopathology of Organs and Tissues Edited by E. Scherer, C. Streffer and K.-R. Trott

Concomitant Continuous Infusion Chemotherapy and Radiation Edited by M. Rotman and C.J. Rosenthal

Intraoperative Radiotherapy – Clinical Experiences and Results Edited by F.A. Calvo, M. Santos and L.W. Brady

Radiotherapy of Intraocular and Orbital Tumors Edited by W.E. Alberti and R.H. Sagerman

Interstitial and Intracavitary Thermoradiotherapy Edited by M.H. Seegenschmiedt and R. Sauer

Non-Disseminated Breast Cancer Controversial Issues in Management Edited by G.H. FLETCHER and S.H. LRVITT

Current Topics in Clinical Radiobiology of Tumors Edited by H.-P. BECK-BORNHOLDT

Practical Approaches to Cancer Invasion and Metastases
A Compendium of Radiation Oncologists Responses to 40 Histories Edited by A.R. KAGAN with the Assistance of R.J. STECKEL

Radiation Therapy in Pediatric Oncology Edited by J.R. CASSADY

Radiation Therapy Physics Edited by A.R. SMITH

Late Sequelae in Oncology Edited by J. DUNST and R. SAUER

Mediastinal Tumors. Update 1995 Edited by D.E. WOOD and C.R. THOMAS, Jr.

Thermoradiotherapy and Thermochemotherapy
Volume 1: Biology, Physiology, and Physics
Volume 2: Clinical Applications
Edited by M.H. SEEGENSCHMIEDT, P. FESSENDEN and C.C. VERNON

Carcinoma of the Prostate. Innovations in Management Edited by Z. PETROVICH, L. BAERT, and L.W. BRADY

MIX
Papier aus verantwortungsvollen Quellen
Paper from responsible sources
FSC® C105338

If you have any concerns about our products,
you can contact us on
ProductSafety@springernature.com

In case Publisher is established outside the EU,
the EU authorized representative is:
Springer Nature Customer Service Center GmbH
Europaplatz 3, 69115 Heidelberg, Germany

Printed by Libri Plureos GmbH
in Hamburg, Germany